International Handbook of the Learning Sciences

The *International Handbook of the Learning Sciences* is a comprehensive collection of international perspectives on this interdisciplinary field. In more than 50 chapters, leading experts synthesize past, current, and emerging theoretical and empirical directions for learning sciences research. The three sections of the handbook capture, respectively: foundational contributions from multiple disciplines and the ways in which the learning sciences has fashioned these into its own brand of use-oriented theory, design, and evidence; learning sciences approaches to designing, researching, and evaluating learning broadly construed; and the methodological diversity of learning sciences research, assessment, and analytic approaches. This pioneering collection is the definitive volume of international learning sciences scholarship and an essential text for scholars in this area.

Frank Fischer is Professor of Educational Psychology and Educational Sciences in the Department of Psychology and Director of the Munich Center of the Learning Sciences (MCLS) at Ludwig Maximilian University, Munich, Germany.

Cindy E. Hmelo-Silver is Barbara B. Jacobs Chair of Education and Technology and Professor of Learning Sciences at Indiana University Bloomington, USA.

Susan R. Goldman is Distinguished Professor of Psychology and Education and Co-Director of the Learning Sciences Research Institute at the University of Illinois at Chicago, USA.

Peter Reimann is Professor of Education at the CoCo Research Centre and Co-Director of the Centre for Research on Learning & Innovation (CRLI) at the University of Sydney, Australia.

International Handbook of the Learning Sciences

Edited by Frank Fischer, Cindy E. Hmelo-Silver, Susan R. Goldman, and Peter Reimann

NEW YORK AND LONDON

First published 2018
by Routledge
711 Third Avenue, New York, NY 10017

and by Routledge
2 Park Square, Milton Park, Abingdon, Oxon, OX14 4RN

Routledge is an imprint of the Taylor & Francis Group, an informa business

© 2018 Taylor & Francis

The right of Frank Fischer, Cindy E. Hmelo-Silver, Susan R. Goldman, and Peter Reimann to be identified as the authors of the editorial material, and of the authors for their individual chapters, has been asserted in accordance with sections 77 and 78 of the Copyright, Designs and Patents Act 1988.

All rights reserved. No part of this book may be reprinted or reproduced or utilized in any form or by any electronic, mechanical, or other means, now known or hereafter invented, including photocopying and recording, or in any information storage or retrieval system, without permission in writing from the publishers.

Trademark notice: Product or corporate names may be trademarks or registered trademarks, and are used only for identification and explanation without intent to infringe.

Library of Congress Cataloging-in-Publication Data
Names: Fischer, Frank, 1965-editor.
Title: International handbook of the learning sciences / edited by
    Frank Fischer, Cindy E. Hmelo-Silver, Susan R. Goldman
    & Peter Reimann.
Description: New York, NY : Routledge, 2018.
Identifiers: LCCN 2017055182 (print) | LCCN 2018005558
    (ebook) | ISBN 9781138670594 (hardback) | ISBN
    9781138670563 (pbk.) | ISBN 9781315617572 (ebk)
Subjects: LCSH: Learning, Psychology of—Handbooks, manuals, etc.
Classification: LCC LB1060 (ebook) | LCC LB1060 .I575 2018
    (print) | DDC 370.15/23—dc23
LC record available at https://lccn.loc.gov/2017055182

ISBN: 978-1-138-67059-4 (hbk)
ISBN: 978-1-138-67056-3 (pbk)
ISBN: 978-1-315-61757-2 (ebk)

Typeset in Bembo
by Swales & Willis Ltd, Exeter, Devon, UK

Dedication

To the past, present, and future of the international community of the learning sciences.

Contents

List of Contributors	xii
Foreword	xx
Janet L. Kolodner	

1 Introduction: Evolution of Research in the Learning Sciences   1
    Frank Fischer, Susan R. Goldman, Cindy E. Hmelo-Silver, and Peter Reimann

SECTION 1
Historical Foundations and Theoretical Orientations of the Learning Sciences   9

2 A Short History of the Learning Sciences   11
    Christopher Hoadley

3 Epistemic Cognition and Epistemic Development   24
    Clark Chinn and William Sandoval

4 Cognitive and Sociocultural Perspectives on Learning: Tensions and Synergy in the Learning Sciences   34
    Joshua A. Danish and Melissa Gresalfi

5 Apprenticeship Learning   44
    Julia Eberle

6 Expertise   54
    Peter Reimann and Lina Markauskaite

7 Cognitive Neuroscience Foundations for the Learning Sciences   64
    Sashank Varma, Soo-hyun Im, Astrid Schmied, Kasey Michel, and Keisha Varma

Contents

8 Embodied Cognition in Learning and Teaching: Action, Observation, and Imagination — 75
Martha W. Alibali and Mitchell J. Nathan

9 Learning From Multiple Sources in a Digital Society — 86
Susan R. Goldman and Saskia Brand-Gruwel

10 Multiple Representations and Multimedia Learning — 96
Shaaron Ainsworth

11 Learning Within and Beyond the Disciplines — 106
Leslie R. Herrenkohl and Joseph L. Polman

12 Motivation, Engagement, and Interest: "In the End, It Came Down to You and How You Think of the Problem" — 116
K. Ann Renninger, Yanyan Ren, and Heidi M. Kern

13 Contemporary Perspectives of Regulated Learning in Collaboration — 127
Sanna Järvelä, Allyson Hadwin, Jonna Malmberg, and Mariel Miller

14 Collective Knowledge Construction — 137
Ulrike Cress and Joachim Kimmerle

15 Learning at Work: Social Practices and Units of Analysis — 147
Sten Ludvigsen and Monika Nerland

16 Complex Systems and the Learning Sciences: Implications for Learning, Theory, and Methodologies — 157
Susan A. Yoon

SECTION 2
Learning Environments: Designing, Researching, Evaluating — 167

17 4C/ID in the Context of Instructional Design and the Learning Sciences — 169
Jeroen J. G. van Merriënboer and Paul A. Kirschner

18 Classroom Orchestration — 180
Pierre Dillenbourg, Luis P. Prieto, and Jennifer K. Olsen

19 Research on Scaffolding in the Learning Sciences: A Methodological Perspective — 191
Iris Tabak and Eleni A. Kyza

20 Example-Based Learning — 201
Tamara van Gog and Nikol Rummel

21	Learning Through Problem Solving *Cindy E. Hmelo-Silver, Manu Kapur, and Miki Hamstra*	210
22	Inquiry Learning and Opportunities for Technology *Marcia C. Linn, Kevin W. McElhaney, Libby Gerard, and Camillia Matuk*	221
23	Supporting Informal STEM Learning with Technological Exhibits: An Ecosystemic Approach *Leilah Lyons*	234
24	Intelligent Tutoring Systems *Arthur C. Graesser, Xiangen Hu, and Robert Sottilare*	246
25	Simulations, Games, and Modeling Tools for Learning *Ton de Jong, Ard Lazonder, Margus Pedaste, and Zacharias Zacharia*	256
26	Supporting Teacher Learning Through Design, Technology, and Open Educational Resources *Mimi Recker and Tamara Sumner*	267
27	Games in the Learning Sciences: Reviewing Evidence From Playing and Making Games for Learning *Deborah A. Fields and Yasmin B. Kafai*	276
28	The Maker Movement and Learning *Erica Halverson and Kylie Peppler*	285
29	Knowledge Building: Theory, Design, and Analysis *Carol K. K. Chan and Jan van Aalst*	295
30	Collective Inquiry in Communities of Learners *James D. Slotta, Rebecca M. Quintana, and Tom Moher*	308
31	Computer-Supported Argumentation and Learning *Baruch B. Schwarz*	318
32	Theoretical and Methodological Frameworks for Computer-Supported Collaborative Learning *Heisawn Jeong and Kylie Hartley*	330
33	Scaffolding and Scripting (Computer-Supported) Collaborative Learning *Ingo Kollar, Christof Wecker, and Frank Fischer*	340
34	Group Awareness Tools for Computer-Supported Collaborative Learning *Daniel Bodemer, Jeroen Janssen, and Lenka Schnaubert*	351

Contents

35	Mobile Computer-Supported Collaborative Learning *Chee-Kit Looi and Lung-Hsiang Wong*	359
36	Massive Open Online Courses (MOOCs) and Rich Landscapes of Learning: A Learning Sciences Perspective *Gerhard Fischer*	368

SECTION 3
Research, Assessment, and Analytic Methods   381

37	Design-Based Research (DBR) *Sadhana Puntambekar*	383
38	Design-Based Implementation Research *Barry Fishman and William Penuel*	393
39	Participatory Design and the Learning Sciences *Kimberley Gomez, Eleni A. Kyza, and Nicole Mancevice*	401
40	Assessment of and for Learning *James W. Pellegrino*	410
41	Learning Progressions *Ravit Golan Duncan and Ann E. Rivet*	422
42	Measuring Competencies *Stefan Ufer and Knut Neumann*	433
43	Mixed Methods Research as a Pragmatic Toolkit: Understanding Versus Fixing Complexity in the Learning Sciences *Filitsa Dingyloudi and Jan-Willem Strijbos*	444
44	Multivocal Analysis: Multiple Perspectives in Analyzing Interaction *Kristine Lund and Daniel Suthers*	455
45	Ethnomethodology: Studying the Practical Achievement of Intersubjectivity *Timothy Koschmann*	465
46	Interactional Ethnography *Judith L. Green and Susan M. Bridges*	475
47	Video Research Methods for Learning Scientists: State of the Art and Future Directions *Sharon J. Derry, Lana M. Minshew, Kelly J. Barber-Lester, and Rebekah Duke*	489

48 Quantifying Qualities of Collaborative Learning Processes   500
    Freydis Vogel and Armin Weinberger

49 Learning Analytics in the Learning Sciences   511
    Carolyn P. Rosé

50 Epistemic Network Analysis: Understanding Learning
    by Using Big Data for Thick Description   520
    David Williamson Shaffer

51 Selecting Statistical Methods for the Learning Sciences and
    Reporting Their Results   532
    Bram De Wever and Hilde Van Keer

Index   *542*

Contributors

Ainsworth, Shaaron, Professor, Learning Sciences, Director of the Learning Sciences Research Institute at the University of Nottingham, United Kingdom.

Alibali, Martha W., Vilas Distinguished Achievement Professor of Psychology and Educational Psychology, University of Wisconsin–Madison, Wisconsin, United States.

Barber-Lester, Kelly J., Ph.D. Candidate, Learning Sciences and Psychological Studies at the School of Education, University of North Carolina at Chapel Hill, NC, United States.

Bodemer, Daniel, Professor, Research Methods in Psychology – Media-Based Knowledge Construction. University of Duisburg-Essen, Germany.

Brand-Gruwel, Saskia, Professor, Dean of the Faculty Psychology and Educational Sciences at the Open University, the Netherlands.

Bridges, Susan M., Associate Professor, Assistant Dean (Curriculum Innovation) Faculty of Education/Centre for the Enhancement of Teaching and Learning at the University of Hong Kong, HKSAR, China.

Chan, Carol K. K., Professor, Division of Learning, Development and Diversity, Faculty of Education at the University of Hong Kong, HKSAR, China.

Chinn, Clark, Professor, Department of Educational Psychology at the Graduate School of Education, Rutgers University, NJ, United States.

Cress, Ulrike, Professor, Department of Psychology at the University of Tübingen, Germany. Director of the Leibniz-Institut für Wissensmedien (Knowledge Media Research Center) in Tübingen, Germany, and Head of the Knowledge Construction Lab at the Leibniz-Institut für Wissensmedien.

Danish, Joshua A., Associate Professor, Learning Sciences and Cognitive Science, and Program Coordinator of the Learning Sciences Program, School of Education, Indiana University, IN, United States.

Contributors

de Jong, Ton, Professor, Departments of Instructional Technology Educational Sciences, Faculty of Behavioral, Management and Social Sciences, University of Twente, the Netherlands.

De Wever, Bram, Professor, Department of Educational Studies, Ghent University, Ghent, Belgium.

Derry, Sharon J., Professor Emeritus, Department of Educational Psychology at the University of Wisconsin-Madison; Thomas James Distinguished Professor of Experiential Learning (retired), School of Education, University of North Carolina – Chapel Hill, NC, United States.

Dillenbourg, Pierre, Professor, Learning Technologies, School of Computer and Communication Sciences, École Polytechnique Fédérale de Lausanne (EPFL), Switzerland.

Dingyloudi, Filitsa, Assistant Professor, Department of Educational Sciences, Faculty of Behavioural and Social Sciences, University of Groningen, the Netherlands.

Duke, Rebekah, Doctoral Student, Learning Sciences and Psychological Studies, at the School of Education, University of North Carolina at Chapel Hill, NC, United States.

Duncan, Ravit Golan., Associate Professor, Department of Learning and Teaching at the Graduate School of Education, and the Department of Ecology, Evolution, and Natural Resources, School of Environmental and Biological Science, Rutgers University, New Brunswick, NJ, United States.

Eberle, Julia, Senior Research Scientist, Institute of Educational Research at the Ruhr-Universität Bochum, Germany.

Fields, Deborah A., Temporary Assistant Professor, Instructional Technology and Learning Sciences, College of Education and Human Services at the Utah State University, UT, United States.

Fischer, Frank, Professor, Educational Psychology and Educational Sciences, Department of Psychology, Director of the Munich Center of the Learning Sciences (MCLS), Ludwig Maximilian University, Munich, Germany.

Fischer, Gerhard, Professor Emeritus, Institute of Cognitive Science, Department of Computer Science, Director of the Center for Lifelong Learning and Design (L3D) at the University of Colorado, Boulder, CO, United States.

Fishman, Barry, Arthur F. Thurnau Professor, Learning Technologies, School of Information and School of Education, University of Michigan, Ann Arbor, MI, United States.

Gerard, Libby, Research Scientist, Graduate School of Education, University of California at Berkeley, CA United States.

Goldman, Susan R., Distinguished Professor of Liberal Arts and Sciences, Psychology and Education, Co-Director of the Learning Sciences Research Institute at the University of Illinois – Chicago (UIC), IL, United States.

Gomez, Kimberley, Professor, Graduate School of Education & Information Studies at the University of California, Los Angeles (UCLA), CA, United States.

Contributors

Graesser, Arthur C., Professor, Department of Psychology, Institute for Intelligent Systems at the University of Memphis, TN, United States.

Green, Judith L., Professor Emeritus, Department of Education, Gevirtz Graduate School of Education (GGSE) at the University of California, Santa Barbara, CA, United States.

Gresalfi, Melissa, Associate Professor, Mathematics Education and Learning Sciences & Learning Environment Design, Department of Teaching & Learning at Vanderbilt University, Nashville, TN, United States.

Hadwin, Allyson, Professor, Educational Psychology, Co-Director of the Technology Integration and Evaluation (TIE) research lab at the University of Victoria, BC, Canada.

Halverson, Erica, Professor, Department of Curriculum & Instruction, School of Education, University of Wisconsin-Madison, WI, United States.

Hamstra, Miki P., Director of Graduate Programs, Robert H. McKinney School of Law, Indiana University, Indianapolis, IN, United States.

Hartley, Kylie, Instructor, School of Education, Indiana University, IN, United States.

Herrenkohl, Leslie R., Professor, Learning Sciences & Human Development at the University of Washington, Seattle, WA, United States.

Hmelo-Silver, Cindy E., Barbara B. Jacobs Chair in Education and Technology, Professor, Learning Sciences, Director of the Center for Research on Learning and Technology at the School of Education, Indiana University, IN, United States.

Hoadley, Christopher, Associate Professor, Learning Sciences & Educational Technology, Steinhardt School of Culture, Education, and Human Development; Director of dolcelab, the Laboratory for Design of Learning, Collaboration & Experience, New York University (NYU), New York, NY, United States.

Hu, Xiangen, Professor, Department of Psychology, Computer Science, Electrical and Computer Engineering, and Senior Researcher at the Institute for Intelligent Systems (IIS) at the University of Memphis (UofM); Professor and Dean of the School of Psychology at Central China Normal University, Wuhan, China.

Im, Soo-hyun, Ph.D. Candidate, Department of Educational Psychology, University of Minnesota–Twin Cities, Minneapolis, MN, United States.

Janssen, Jeroen, Associate Professor, Faculty of Social and Behavioural Sciences, Department of Education at Utrecht University, the Netherlands.

Järvelä, Sanna, Professor, Learning and Educational Technology, Head of the Learning and Educational Technology Research Unit (LET), Department of Educational Sciences and Teacher Education at the University of Oulu, Finland.

Jeong, Heisawn, Professor, Department of Psychology, Hallym University, South Korea.

Kafai, Yasmin B., Professor, Teaching, Learning, and Leadership Division at the Graduate School of Education, University of Pennsylvania, PA, United States.

Kapur, Manu, Professor, Learning Sciences and Higher Education, Department of Humanities, Social and Political Sciences, ETH Zurich, Switzerland.

Kern, Heidi M., Student Research Assistant, Educational Psychology, Swarthmore College, PA, United States.

Kimmerle, Joachim, Distinguished University Professor, Deputy Head of the Knowledge Construction Lab at the Leibniz-Institut für Wissensmedien (Knowledge Media Research Center), Tübingen, Germany; Adjunct Professor in the Department of Psychology at the Eberhard Karls University Tübingen, Germany.

Kirschner, Paul, A. Professor, Educational Psychology at the Open University of the Netherlands; Visiting Professor and Honorary Doctor, Educational Sciences at Oulu University, Finland.

Kollar, Ingo, Professor, Educational Psychology at the University of Augsburg, Germany.

Kolodner, Janet L., Professor (Visiting), Lynch College of Education, Boston College; Regents' Professor Emerita, Computing and Cognitive Science, Georgia Institute of Technology; Editor-in-Chief Emerita, *Journal of the Learning Sciences*, Chestnut Hill, MA, United States

Koschmann, Timothy, Professor Emeritus, Department of Medical Education at Southern Illinois University, IL, United States.

Kyza, Eleni A., Associate Professor in Information Society, Faculty of Communication and Media Studies, Department of Communication and Internet Studies, Cyprus University of Technology, Limassol, Cyprus.

Lazonder, Ard W., Professor, Behavioural Science Institute, Radboud University, Nijmegen, the Netherlands.

Linn, Marcia C., Professor, Graduate School of Education, University of California at Berkeley, CA, United States.

Looi, Chee-Kit, Professor, Learning Sciences and Technologies, and Head of the Learning Sciences Lab at the National Institute of Education, Nanyang Technological University, Singapore.

Ludvigsen, Sten, Professor, Learning and Technology, Faculty of Educational Sciences, Dean of the Faculty of Educational Sciences at the University of Oslo.

Lund, Kristine, Senior Research Engineer, Interactions, Corpora, Learning, Representations (ICAR) Research Lab, Centre National de la Recherche Scientifique (CNRS), Ecole Normale Supérieure Lyon, University of Lyon 2, France; Coordinator of the Academic College "Education, Cognition, Language" for Lyon-St. Etienne.

Lyons, Leilah, Professor, Learning Sciences at the Learning Sciences Research Institute and Computer Science, University of Illinois at Chicago (UIC), IL; Director of Digital Learning Research at the New York Hall of Science (NYSCI), NY, United States.

Malmberg, Jonna, Postdoctoral Researcher, Learning and Educational Technology Research Unit (LET) at the University of Oulu, Finland.

Contributors

Mancevice, Nicole, Ph.D. Candidate, Graduate School of Education & Information Studies at the University of California, Los Angeles (UCLA), CA, United States.

Markauskaite, Lina, Professor, eResearch (Educational and Social Research Methods) in the Centre for Research on Computer Supported Learning and Cognition (CoCo) at Sydney School of Education and Social Work, University of Sydney, Australia.

Matuk, Camillia, Assistant Professor, Educational Communication and Technology, Department of Administration, Leadership and Technology, Steinhardt School of Culture, Education and Human Development, New York University (NYU) Steinhardt, New York, NY, United States.

McElhaney, Kevin, W., Associate Director of Science and Engineering Education Research, Education Division, SRI International, Menlo Park, CA, United States.

Michel, Kasey, Ph.D. Candidate, Department of Educational Psychology, University of Minnesota–Twin Cities, Minneapolis, MN, United States.

Miller, Mariel, Technology Integrated Learning Manager. Division of Learning and Teaching Support and Innovation, University of Victoria, Victoria, BC, Canada.

Minshew, Lana M., Ph.D. Candidate, Learning Sciences and Psychological Studies at the University of North Carolina at Chapel Hill.

Moher, Tom, Professor Emeritus, Computer Science, Learning Sciences, and Education, Department of Computer Science at the College of Engineering, University of Illinois at Chicago (UIC), IL, United States.

Nathan, Mitchell J., Professor, Learning Sciences, Department of Educational Psychology, School of Education, University of Wisconsin-Madison, WI, United States.

Nerland, Monika B., Professor, Faculty of Educational Sciences, Department of Education, University of Oslo, Norway.

Neumann, Knut, Professor, Physics Education, Department of Physics Education at the Leibniz Institute for Science and Mathematics Education (IPN), Kiel, Germany.

Olsen, Jennifer K., Researcher, Human-Computer Interaction Institute, at Carnegie Mellon University, Pittsburgh, PA, United States.

Pedaste, Margus, Professor, Educational Technology, Institute of Education, Faculty of Social Sciences at the University of Tartu, Estonia.

Pellegrino, James, Distinguished Professor of Liberal Arts and Sciences, Psychology and Education, Co-Director of the Learning Sciences Research Institute at the University of Illinois at Chicago (UIC), IL, United States.

Penuel, William, Professor, Learning Sciences and Human Development at the School of Education, University of Colorado Boulder, CO, United States.

Peppler, Kylie, Associate Professor, Learning Sciences, Department of Counselling and Educational Psychology at the School of Education, Indiana University, Bloomington, IN, United States.

Polman, Joseph L., Professor, Learning Sciences and Science Education, Associate Dean for Research at the School of Education, University of Colorado Boulder, CO, United States.

Prieto, Luis P., Senior Research Fellow, Center of Excellence for Educational Innovation (HUT) at the School of Educational Sciences, Tallinn University (TLU), Estonia.

Puntambekar, Sadhana, Professor, Educational Psychology, Department of Educational Psychology, Wisconsin Center for Education Research at the School of Education, University of Wisconsin-Madison, WI, United States.

Quintana, Rebecca, Learning Experience Designer, Office of Academic Innovation at the University of Michigan, MI, United States.

Recker, Mimi, Professor, Instructional Technology and Learning Sciences, Department of Instructional Technology & Learning Sciences at the Emma Eccles Jones College of Education and Human Services, Utah State University, Logan, UT, United States.

Reimann, Peter, Professor, Education, Co-Director of the Centre for Research on Learning & Innovation (CRLI) at the Sydney School of Education and Social Work, University of Sydney, Australia.

Ren, Yanyan, Student Research Assistant, Department of Educational Studies at Swarthmore College, PA, United States.

Renninger, K. Ann, Dorwin P. Cartwright Professor of Social Theory and Social Action, and Chair of the Department Educational Studies at Swarthmore College, PA, United States.

Rivet, Ann E., Professor, Science Education at Teachers College Columbia University, New York, NY; Program Director, Division of Research on Learning in Formal and Informal Settings, Directorate for Education and Human Resources, National Science Foundation, Alexandria, VA, United States.

Rosé, Carolyn P., Professor, Carnegie Mellon University, Language Technologies Institute and Human-Computer Interaction Institute, Carnegie Mellon University, Pittsburgh, PA, United States.

Rummel, Nikol, Professor, Educational Psychology at the Institute of Educational Research, Ruhr-Universität Bochum, Germany; Adjunct Professor, Carnegie Mellon University, Human-Computer Interaction Institute, Carnegie Mellon University, Pittsburgh, PA, United States.

Sandoval, William A., Professor, Graduate School of Education & Information Studies, University of California, Los Angeles, CA, United States.

Schmied, Astrid, Ph.D. Candidate, Department of Educational Psychology, University of Minnesota–Twin Cities, Minneapolis, MN, United States.

Schnaubert, Lenka, Researcher, Research Methods in Psychology – Media-Based Knowledge Construction, University of Duisburg-Essen, Germany.

Contributors

Schwarz, Baruch B., Professor, School of Education, Hebrew-University of Jerusalem, Israel.

Shaffer, David W., Vilas Distinguished Achievement Professor of Learning Science, Department of Educational Psychology, University of Wisconsin-Madison, Madison, WI, United States; Obel Foundation Professor of Learning Analytics, Department of Education, Learning and Philosophy, Aalborg University, Copenhagen, Denmark.

Slotta, James D., Professor and Associate Dean for Research, Lynch School of Education, Boston College, MA, United States.

Sottilare, Robert A., Head of Adaptive Training Research, US Army Research Laboratory, Human Research and Engineering Directorate (HRED), Adelphi, MD, United States.

Strijbos, Jan-Willem, Professor, Department of Educational Sciences, Faculty of Behavioural and Social Sciences, University of Groningen, the Netherlands.

Sumner, Tamara, Professor, Cognitive and Computer Science, Director of the Institute of Cognitive Science at the University of Colorado, Boulder, CO, United States.

Suthers, Daniel, Professor, Department of Information and Computer Sciences at the University of Hawai'i at Manoa.

Tabak, Iris, Senior Lecturer, Department of Education at the Ben-Gurion University of the Negev, Israel.

Ufer, Stefan, Professor, Mathematics Education, Department of Mathematics at Ludwig Maximilian University and Munich Center of the Learning Sciences (MCLS), Munich, Germany.

van Aalst, Jan, Associate Professor, Associate Dean for Research, Faculty of Education at the University of Hong Kong, China.

van Gog, Tamara, Professor, Educational Sciences, Department of Education at Utrecht University, the Netherlands.

Van Keer, Hilde, Professor, Department of Educational Studies, Ghent University, Ghent, Belgium.

van Merriënboer, Jeroen J. G., Professor, Learning and Instruction, Research Director, School of Health Professions Education (SHE), Maastricht University, Maastricht, the Netherlands.

Varma, Keisha, Associate Professor, Department of Educational Psychology, University of Minnesota–Twin Cities, Minneapolis, MN, United States.

Varma, Sashank, Associate Professor, Department of Educational Psychology, University of Minnesota–Twin Cities, Minneapolis, MN, United States.

Vogel, Freydis, Research Scientist, Department of Teacher Education, TUM School of Education, Technical University of Munich and Munich Center of the Learning Sciences, Germany.

Wecker, Christof, Professor, Psychology, University of Passau, Germany.

Weinberger, Armin, Professor, Educational Technology and Knowledge Management at Saarland University, Germany.

Wong, Lung-Hsiang, Senior Research Scientist, Learning Sciences Lab at the National Institute of Education, Nanyang Technological University, Singapore.

Yoon, Susan, Professor, Teaching, Learning, and Leadership Division at the Graduate School of Education, University of Pennsylvania, PA, United States.

Zacharia, Zacharias, Professor, Science Education, Research in Science and Technology Education Group, Department of Educational Sciences, University of Cyprus, Cyprus.

Foreword

When I began working on creating the *Journal of the Learning Sciences* (*JLS*) at the end of 1989, there was a small group of American researchers in the cognitive sciences (cutting across science and math education, educational and cognitive psychology, educational and instructional technology, computer science, and anthropology) who were working on issues related to learning that went way beyond what their fields and what the core cognitive science and education communities were working on.[1] They were asking questions their fields had not asked before and requiring new methodologies for answering their questions. They were investigating learning in real-world situations to understand how education might be reimagined, designing new ways of educating, and putting those new approaches into practice and investigating what happens developmentally when new approaches are used. They needed methods for investigating learning *in situ* that embraced the complexity of the real world. They needed a venue for sharing their crazy ideas and what was being learned from their investigations. My editorial board and I envisioned *JLS* as a venue for proposing new pedagogies and reporting on how to make them work, proposing new ways of using technology and reporting on how to make them work, explicating complex combinations of processes involved in really deeply coming to understand something and becoming masterful at complex skills, showing how skills, practices, and understanding evolve over time and what scaffolding best fosters that development, proposing new methodologies for studying learning and reporting on how to make them work, and more. A major goal of the new journal was to "foster new ways of thinking about learning and teaching that will allow cognitive science disciplines to have an impact on the practice of education."[2]

From the beginning, *JLS*'s articles focused on learners as active agents – how can we empower learners and give them agency? How can technology help with that? What mental processes are involved in asking questions, making predictions, and making explanations? How can we help learners do those things more productively, and how can we help them become masterful at such practices? How can we impact the goals learners have so that they will actively take on learning? What kind of classroom culture, teacher practices, and learning materials are needed for that, and so forth? Articles focused on learning kinds of content and skills we learn in school (e.g., how cells work, the water cycle, how to argue a point) and those learned more informally (e.g., how to play basketball or dominoes). They focused on learning itself and how to draw learners in and engage them enthusiastically in learning. Researchers learned that learners' beliefs about themselves and others and their identities, the ways they use their bodies and interact with the world, the cultural context of the venue where they are learning, the form of aid they receive while learning and who provides that aid, and more, all play roles in what they will pay attention to, give time to, talk about, and ultimately learn.

This community's focus was on the intricacies and complexities of learning processes and mechanisms and how to affect them, not merely the trends. We wanted to learn how to take advantage of the processes and mechanisms people use when they are interested in something to figure out how to make learning in school, in communities, in museums, and so forth more engaging and productive. We wanted to learn how to do that for every participant. And we assumed there would not be a one-size-fits-all way of doing that – that we would have to learn about, if not individual differences, then the varieties of differences that might be encountered among learners so that all could be addressed.

A hallmark of the learning sciences, then, is a research tradition that embraces complexity. From the field's beginning, we have wanted to understand as full a range of influences as possible on how learning happens and the practices that enable learning. In the 27 years since the first issue of *JLS*, the community has developed much theory and methodology that allow us to address these complexities. The learning sciences community in 2018 is international: we have a wide variety of conferences we attend; we attract many hundreds of people to learning sciences conferences; we write for a wide variety of journals; and we have influenced research practices and the questions asked in other research communities. We have discovered much about the mental and social processes involved in learning along with the cultural, systemic, and other factors that affect those processes. Informed by that knowledge, and working with others, learning sciences community members have designed curriculum approaches, pedagogies, technologies, means of facilitating and scaffolding, and a full variety of educational resources. The results of studying the use of those products in practice had led to their refinement, to new understanding about how to adapt products to particular populations and educational ecosystems, and to new insights about mental and social processes. Learning scientists have adapted existing methodologies and developed new ones for studying learning *in situ* and for identifying what is responsible for that learning (and what discourages it).

Along with that growth and rich activity has come a need: an easy-to-access resource that allows newcomers to the learning sciences community as well as researchers and practitioners in other disciplinary communities to learn about the origins of our field, its theoretical underpinnings, and the wide variety of constructs, methodologies, analytic frameworks, and findings related to learning and to fostering learning that we have come to understand (and often take for granted) over the past two and a half decades.

I am therefore delighted by the contents and organization of this volume. The editors have done a yeoman's job of deconstructing the complexities of both the theoretical foundations we draw on and what we have come to understand. The first section's 15 chapters detail the origins of the field, the intellectual traditions we draw from, the ways we've learned to integrate those traditions, and what we've learned about human learning processes and mechanisms through that integration and through research that keeps in mind the complex interactions between processes, mechanisms, and other influences on human learning as we interact in the world. What roles do our interactions with others play? What about the culture around us? How do we use our bodies and the ways they interact with the world as we learn? How do our beliefs (about ourselves, about the world, about others) affect our learning? How do our goals and personal interests and passions affect what we will give time to and pay attention to? How do we form representations, and what affects that? How do human learners make sense of the complexity of the world? How does all of this (interactions, goals, the ways we use our bodies, beliefs, representations we form, and so on) affect our cognitive processing?

The second section's 20 chapters detail what we've learned about how to foster learning – in particular, how to design learning environments so that participants will both enthusiastically engage and productively come to new understandings and capabilities. This section focuses on both pedagogical approaches and the design and use of technology to engage learners and help them learn. And it focuses not only on those we might call "students" but also on teachers and how to help them learn, on participants young and old in informal learning venues, and on learning among a collective of learners and the ways learning across a collective affects the capabilities of individuals within the collective. What roles might technology play? How might the possibilities of new technologies be harnessed to make learning more engaging and productive?

The third section (15 chapters) focuses on our methods: what we've come to know about how to do research in contexts of immense complexity and how to extract from those complex environments understandings of how people learn and guidelines for designing ways of fostering learning. A big contribution of the learning sciences to research methodology has been our development of methods that make iterative design a first-class research activity: the design of the environment in which data is collected affects what can be collected, what can be learned and how the data can be interpreted.

I invite the community to dig in and enjoy; I take deep pleasure in seeing a compendium that brings it all together – both because of my pride in how far the community has progressed and my thoughts about how I will use these chapters as I mentor new members of the community.

Janet L. Kolodner
Professor (Visiting), Lynch College of Education, Boston College
Regents' Professor Emerita, Computing and Cognitive Science, Georgia Institute of Technology
Editor-in-Chief Emerita, *Journal of the Learning Sciences*
Chestnut Hill, MA

Notes

1 See, e.g., Kolodner, J. (1991). Editorial: "The Journal of the Learning Sciences": Effecting Changes in Education. *Journal of the Learning Sciences, 1*(1), 1–6. See also Hoadley (this volume).
2 Ibid.

1

Introduction

Evolution of Research in the Learning Sciences

Frank Fischer, Susan R. Goldman, Cindy E. Hmelo-Silver, and Peter Reimann

---

Over the past 25 years, the interdisciplinary field of learning sciences has emerged as an important nexus of research on how people learn, what might be important for them to learn and why, how we might create contexts in which such learning can occur, and how we can determine what learning has occurred and for whom. At the same time this emergence has prompted repeated attempts to probe and elucidate how learning sciences is similar to, as well as differentiated from, long-established disciplinary research areas, such as anthropology, cognitive psychology, cognitive sciences, curriculum and instruction, educational psychology, and sociology. This is a difficult question to answer, in part because the learning sciences builds on the knowledge base of many of these disciplinary research areas while at the same time taking a "use oriented" perspective on the knowledge base. That is, much as Stokes (1997) distinguished between basic research and research oriented toward solving practical problems, (i.e., research in Pasteur's quadrant), research in the learning sciences is often situated in problems of practice that occur in a range of "learning" contexts, including formal or informal settings dedicated to schooling, workplace, or leisure/entertainment goals.

Because the learning sciences are "use oriented," they are also holistic; practically useful knowledge needs to be coherent (Bereiter, 2014). When we speak of the learning sciences as aiming for a holistic understanding of human learning, we take both epistemic and systems views. The epistemic perspective is that learning can be studied from multiple perspectives. By claiming that human learning is a systems phenomenon, we assume that learning is brought about by the coordination of biological learning with socio-cultural knowledge and tool production. Just imagine—and this means asking for the impossible—how different human learning would be if we would not have language to communicate, would not have writing systems, including those for mathematics and music, would not have invented technologies, from tables to tablets. None of these essential elements of (and for) human learning depend on a particular brain function; instead, each extends the brain—the biological system—into a bio-socio-cultural hybrid system that is the locus of human learning, and generally for human cognition (Clark, 2011).

In a similar vein, concerning methodology, the learning sciences have resisted crude reductionism. Instead, what is often practiced is a kind of dialectical reductionism, for lack of a better word. To produce good explanations for learning, the learning process(es) under study needs to be decomposed into parts, and the explanation runs 'upwards' from the components, and their configurations and coordinations, to the process that gets explained. At the same time, the lower level processes get meaning only when seen from the higher level: moving an arm up or down is part of a dance move or directing traffic cannot be determined from analyzing the motor control processes in the brain, as little as from analyzing the muscle contractions in the arm. Any human action, other than reflexes, can serve multiple—indeed, infinitely many—purposes. Furthermore, most of our actions are tool-mediated, which makes them mediated by the culture that provides the tool and the community of practice in which a specific way of using the tool makes sense (Wertsch, 1998). Although learners may neither be aware of the body and brain processes, nor of the cultural pedigree of their actions, to understand human learning, and to shape it, all of these need to be taken into account.

Purposes

The overarching purposes of this handbook are to bring together international perspectives on theoretical and empirical work (1) that has informed the research agenda of the learning sciences, with its emphasis on design and how learning technologies (computer and internet based and otherwise) can support learning and its assessment; (2) that comprise signature and unique contributions of learning sciences design research cycles to understanding how, what, why, and for whom learning is happening; and (3) that comprises the multiple and complementary methods of examining learning and how it happens within learning sciences research traditions. In so doing, we hoped to create an internationally oriented up-to-date resource for research and teaching in the learning sciences.

We intend the handbook to serve as a resource to the burgeoning number of post-baccalaureate programs in the learning sciences. In the past decade, the numbers of programs describing themselves as providing advanced degrees in learning sciences has gone from just a handful to more than 50 worldwide (see isls.naples.com). Many more programs in education, psychology, and related fields include specializations or subprograms in learning sciences. The programs are geographically distributed across North America, Europe, Asia, and Australia, with emerging interest from South America and Africa.

We also intend the handbook to provide a compendium of past, current, and future research trends. The contributors to this handbook are actively participating in learning sciences research and graduate preparation programs; have served as editors or editorial board members of the premier journals of the learning sciences, the *Journal of the Learning Sciences* and the *International Journal of Computer Supported Collaborative Learning*; or have played key roles in the activities of the International Society of the Learning Sciences, including the annual meetings. They are thus well positioned to both introduce newcomers to the learning sciences to its major theories, methods, and empirical findings, as well as to provide for more seasoned members of the learning sciences communities well-informed and reflective perspectives on major trends and future directions in their specific areas of expertise. In soliciting authors, we provided content guidelines that we hoped would provide some consistency across a diverse set of topical areas. We asked that the authors provide a brief historical introduction to the topic and discuss the relevance or intersection of their topic with the learning sciences. They were asked to refer to empirical studies and their findings to support claims and/or provide examples of the application of particular research methods or analytic strategies. We encouraged the authors to avoid the "not invented here" syndrome and seek out international scholars working on the specific topic.

Given the limited length of each chapter, we asked the authors to include four or five further readings with a brief annotation, along with the citations included in the body of the chapter. In addition, most of the chapters include a section with links (URLs) to specific video resources, most of them part of the NAPLeS (Network of Academic Programs in the Learning Sciences) collection of webinars, interviews, and short videos. We encourage you to take advantage of these additional resources.

As editors of the volume our purpose in the remainder of this introduction is to provide an overview of the three sections from the perspective of what we hoped to capture and reflect in each. We focus on overall trends across sets of chapters rather than providing summaries of each. We conclude with several emergent trends and directions for learning sciences, including greater attention to social responsibility and research that speaks to issues of equity.

Organization of the Handbook

Learning sciences is an interdisciplinary field that works to further scientific understanding of learning as well as engages in the design and implementation of learning innovations in methodologies and learning environments intended to improve learning processes and outcomes. Conceptions of learners, learning spaces and places, the time span over which learning occurs, what manner of processes and outcomes are defined as evidence of learning all reflect the interdisciplinarity of the learning sciences. The first section of the handbook, 'Historical Foundations and Theoretical Orientations of the Learning Sciences,' endeavors to reflect foundational contributions to this interdisciplinarity as well as the particular way in which the learning sciences has taken up these contributions and then used them to create its own brand of use-oriented theory, design, and evidence. The second section, 'Learning Environments: Designing, Researching, Evaluating,' turns to various configurations of places, spaces, time frames, tasks, processes, and outcomes that constitute the what of learning sciences research, design, and evaluation. The third section, 'Research, Assessment, and Analytic Methods,' reflects the methodological diversity of the learning sciences. We discuss each section in turn and conclude with the themes and trends that emerge for the future.

Section 1: Historical Foundations and Theoretical Orientations of the Learning Sciences

The history of science is replete with paradigm shifts stimulated by the accumulation of evidence that simply did not "fit" extant theoretical paradigms. Such was the case with the "cognitive revolution" in psychology during the 1960s (see Miller, 2003). Similarly, the learning sciences emerged in part as a response to evidence and phenomena of learning emanating from different disciplines. However, rather than a paradigm shift within a single discipline (e.g., psychology), the seemingly inconsistent evidence and phenomena were emanating from different disciplines, leading to a shift to a more *inter*disciplinary conception of learning. For example, the juxtaposition of sophisticated quantitative reasoning in everyday situations seemed at odds with data indicating that people were far less successful in such reasoning in formal school mathematics (Lave, 1998; Saxe, 1991). As Hoadley indicates in his "short history," four themes emerged as characteristic of the learning sciences and form the foundations of the epistemology as depicted in Chinn and Sandoval. The next set of three chapters (Danish & Gresalfi; Eberle; Reimann & Markauskaite) detail productive tensions of efforts to look at learning, development, and expertise from individual "in the head" as well as socio-cultural perspectives. Two chapters draw attention to the importance of looking at multiple systems in which learning occurs, in particular the neural system (S. Varma, Im, Schmied, Michel, & K. Varma) and the motor/kinesthetic system reflected visibly in action and gesture (Alibali & Nathan). This work points to productive future directions for work in the learning sciences in attempting to understand learning as a multi-level phenomenon.

A theme of the next four chapters reflects the increasing consideration in the learning sciences of the purposes and goals for which people interact with and try to make sense of the various forms of information that are ubiquitous in the 21st century. Why do people turn to certain information resources whether in everyday life, academic, or professional endeavors? This theme runs through the next four chapters—by Goldman and Brand-Gruwel; Ainsworth; Herrenkohl and Polman; and Renninger, Ren, and Kern—from different perspectives, ranging from general interest to disciplinary inquiry in formal and informal settings. They discuss the influence and interconnections between learners' perspectives on the purposes and functions of their efforts, how they define and how deeply they engage with information they decide is relevant to their purposes, and what they learn. Furthermore, epistemic goals and values emerge in interaction with others, as a collaborative and collective activity, whether in educational or workplace settings (Järvelä, Hadwin, Malmberg, & Miller; Cress & Kimmerle; Ludvigsen & Nerland). Learners do not operate in isolation from the people and objects in their worlds. They build shared understandings via processes that require regulation and modulation in interaction with others and as shaped by and shaping the contributions, knowledge, and beliefs of self and others. Section 1 concludes with a very apropos chapter on complex systems (Yoon), addressing the issue that many of the properties of complex systems (e.g., emergence, structure/function relationships, causality, scale, and self-organization) pose substantial teaching and learning challenges in K-12 education.

Section 2: Learning Environments: Designing, Researching, Evaluating

Much of the work in the learning sciences is concerned with designing learning situations that are challenging for learners, that ask them to grapple with situations, tasks, and problems for which they do not have rote solutions or for which they cannot simply call upon a memorized factoid. They are asked to work just beyond their comfort zones, in what Vygotsky (1978) referred to as the Zone of Proximal Development (ZPD). To be successful working in the ZPD, learners require supports. The second section contains chapters describing approaches to designing learning environments that support learners' engagement in ways that lead to knowledge and dispositional outcomes that prepare them to be able subsequently to use what they have learned in conditions different from those of the original learning. Learning activities typically have an inquiry or problem-solving orientation and more often than not involve both independent and collaborative work.

Learning environment designs reflect a variety of learning contexts, pedagogical approaches, and supports for learning. Contexts range across formal and informal educational institutions; informal, opt-in spaces and places (e.g., sports clubs, after school, affinity groups); home, work, and other institutional settings. Pedagogically, designs run the gamut from prescriptive to co-designed to learner-centered. Supports for learners, broadly referred to as scaffolds, may be built into tasks and task sequences provided to do the task, guidance or other forms of coaching and feedback. Scaffolding requires the presence of a "more knowledgeable other", a role that may be played by humans (e.g., peers, tutors, teachers, parents), computers, or a mix of the two. The chapters in this section attempt to reflect the diversity of designs that result from various combinations of contexts, pedagogies, and forms of support.

The first four chapters present relatively broad, big picture-perspectives on pedagogical designs (van Merriënboer & Kirschner; Dillenbourg, Prieto, & Olsen), scaffolding (Tabak & Kyza), and one specific genre of scaffolding, examples (van Gog & Rummel). These four chapters provide some general considerations for design across a variety of tasks and disciplinary contexts. The focus then shifts to particular forms of inquiry learning, raising considerations of the timing and specificity of guidance and feedback (Hmelo-Silver, Kapur, & Hamstra; Linn, Gerard, McElhaney, & Matuk). Both chapters imply important roles for technology. Indeed, the subsequent chapters in this section involve technologies for supporting learning in a variety of different contexts, disciplines, and learner configurations (individual, small group, whole class). Specifically, Lyons examines issues that arise in introducing technologies in different types of informal learning institutions and the importance of considering the institution as an ecosystem. Computers as intelligent tutoring systems (Graesser, Hu, & Sottilare) and vehicles for providing learners with experiential learning through simulations, games, and modeling (de Jong, Lazonder, Pedaste, & Zacharia) have been used to support learning in a variety of content areas, most frequently in the sciences and mathematics, sometimes focusing on individual learners and sometimes supporting multiple learners working together. The contexts and situations reflect design characteristics of inquiry and problem solving.

Design activities as a vehicle *for* learning are the focus of the next three chapters in this section. Recker and Sumner discuss how teachers' learning through their instructional design efforts is enabled and supported by resources available on the internet. Fields and Kafai review major findings from research on game-based learning, showing that designing games can be highly effective for learning, especially if learners engage in scaffolded design activities. Halverson and Peppler analyze the maker movement and identify two characteristics as core features: authenticity and purpose in making, and self-selection and agency in choosing a particular maker activity. These features of Makerspaces may make them particularly interesting sites for attending to equity and diversity in learning.

The next eight chapters discuss various ways in which collaboration and knowledge building have been major goals of design efforts in the learning sciences from its inception. Indeed, a seminal computer-based system to support collaboration around the development of ideas, CSILE (Computer Supported Intentional Learning Environment), arose out of Scardamalia and Bereiter's efforts to foster knowledge-transforming rather than knowledge-telling learning opportunities (Bereiter & Scardamalia, 1987a, 1987b; Chan & van Aalst). Since this seminal work and the profusion of design and research projects that it has spawned internationally, there have been a number of parallel design and research efforts that emphasize creating and generating knowledge in scaffolded communities of inquiry (Slotta, Quintana, & Moher) or through dialogic and dialectic argumentation (Schwarz). After a theoretical and methodological overview of these efforts in CSCL research (Jeong & Hartley), the following two chapters discuss approaches that have attended to more specific issues. The designs are typically realized in computer- or internet-based systems and provide various types of support for collaborative knowledge construction, including scripting and scaffolding for groups (Kollar, Wecker, & Fischer) or group awareness information supposed to help groups regulate their activity themselves (Bodemer, Janssen, & Schnaubert). The final two chapters of that section emphasize collaborative learning opportunities that are already at scale in K–12 as well as higher education: mobile learning (Looi and Wong) and massive open online courses (G. Fischer). In these two chapters, the authors discuss the potential for linking formal and informal learning but also emphasize the value and importance of bringing a learning sciences perspective to design issues in these spaces.

Section 3: Research, Assessment, and Analytic Methods

The learning sciences first distinguished itself from other approaches by combining participant observation with systematic design and refinement efforts. Research focused on designing based on extant theory but researching the design in action for purposes of determining how to improve the design in situ was a signature characteristic of design-based research (DBR). The reflective redesign was the main vehicle for developing and generating theoretical positions beyond those from which the design had originated. In the first chapter in this section, Puntambekar details this history as well as directions that DBR has moved. Early DBR research reflected the "heavy hand" of the researcher on the design: researchers created for teachers and students, and watched what happened as teachers and students implemented the designs, consulted them about their experiences and suggestions, and then did the redesign. Similar enactments of DBR have occurred in informal institutional contexts and in game design. Lessons learned from these efforts have increasingly led to design approaches that involve the implementers in the process from the beginning, including design-based implementation research (DBIR) (Fishman & Penuel), and participatory co-design (Gomez, Kyza, & Mancevice). The process is one of working and designing *with*, rather than designing *for*, those who are intended users/implementers of what is designed.

It seems obvious that designers need to know what the designs they are creating are intended to produce in terms of processes and outcomes. Two aspects of this statement may not be so obvious. First, the level of specificity in process and outcome that seems sufficient to begin with, quickly prove to be too global and vaguely specified to really help with critical design decisions (Ko et al., 2016). Second, there is a paucity of longitudinal and even cross-sectional research for much of the disciplinary knowledge that is the target of designs intended for use in formal schooling contexts. Thus, design-based research efforts go hand in hand with assessment approaches that ask what we want students to know and be able to do and how we will know if they are making adequate progress toward what we want them to know and be able to do at specific points in their development and schooling. Pellegrino's chapter speaks to the centrality of these issues in both the design of instruction and the assessment of what students are learning, and the next two chapters (Duncan & Rivet; Ufer & Neumann) indicate approaches to addressing them with a longitudinal, cumulative perspective on learning.

The chapters on mixed methods (Dingyloudi & Strijbos) and on multivocality (Lund & Suthers) indicate the value of exploring data sets (in whole or in part) from multiple perspectives using different methods in combination to rigorously and systematically develop sound evidentiary arguments for empirical and theoretical claims. The next three chapters concern analytic approaches to various forms of qualitative data often captured in video and audio traces of interactions (Koschmann; Green & Bridges; Derry, Minshew, Barber-Lester, & Duke). The final four chapters discuss quantitative methods for the analysis of interaction (Vogel & Weinberger) that can be harnessed in service of constructing descriptive as well as predictive and causal patterns in the data of learning and instruction (Rosé; Shaffer). The final chapter in the section is on statistics for the learning sciences and suggests analytic strategies responding to the multi-level nature of the phenomenon of learning (De Wever & Van Keer).

Themes and Trends for the Future

We conclude this introduction by highlighting six themes and trends for future emphases in the learning sciences.

1. Increasing recognition of learning as a complex systems phenomenon. Learning and learning mechanisms operate at different levels and as semi-independent self-organized systems. For example, behavioral evidence of learning in an individual may be emergent from mechanisms operating in self-organized cognitive, affective, and kinesthetic/motoric systems, each of which has a neural signature. Furthermore, the individual is part of a larger socio-historical cultural system, and as such influences and is influenced by that system. Even as learning is attributed at the individual level, it is an accomplishment and attributable to the community collective(s) of which the individual is a member. While at the present time we have neither the theoretical, empirical, or analytic tools to investigate connections across more than a few of these multi-leveled systems, statistical methods like multi-level analyses and latent growth models are moving in directions that are much more attuned to the complex and dynamic multi-level phenomena of learning in real-world contexts that are of interest to learning scientists. The learning sciences is thus increasingly well positioned to elucidate at least some of the connections and emergent properties across levels.

2. Increasing emphasis on more precise and longitudinal explication of what learners are expected to know and be able to do, and what indicators would provide evidence pertinent to the targeted competencies. Research on learning progressions in a variety of domains will figure prominently in identifying targeted competencies. The results will be a tighter connection between the design of learning environments and the assessment of the learning that the designs are intended to support and promote. Thus, rather than assessment being external to the learning environment, it is part and parcel of its design from the beginning. Assessment positioned in this way can foster critical reflective practices in individuals and in groups of learners and potentially contribute to greater agency and self-direction. DBR and DBIR are excellent vehicles for incorporating this perspective on assessment. As well, new approaches to analyzing process and product data (e.g., talk, gesture, actions, problem solutions, written work) to identify conceptually meaningful patterns will make important contributions to these efforts. In many cases these

approaches will increasingly rely on automated or semi-automated analyses that capitalize on computer-based technologies.

3. More adaptive technologies in support of learning. From its inception, the learning sciences has incorporated various technologies in support of individual and collaborative learning. Automated analyses of response patterns reflecting behavioral, cognitive, and affective processes during learning and problem-solving activities are becoming increasingly sophisticated; they can provide a basis for more adaptive feedback to individuals and groups of learners but also support the teachers in their monitoring and intervention of students' learning. The nature of such automated analyses may enable the strategic selection of scaffolds based on detected patterns, a "just in time" provision of critical but apparently absent information, or guidance (e.g., hints or prompts) for learners' reflection and strategic decision making about productive next steps. However, to realize these potentials we need to combine these algorithmic approaches with our knowledge of learning and teaching. Learning analytics may become a success story to the extent that these combinations will be achieved. We are convinced that the interdisciplinary collaborations in the learning sciences are optimal pre-conditions to master this challenge.

4. In regard to methodology, the learning sciences have been developing a distinct blend of evidence forms, combining ethnomethodological and ethnographical research methods with more quantitatively oriented approaches to dialog analysis and experimental research. Increasingly, the different approaches are used in conjunction, as part of "mixed methods" strategies in interdisciplinary research projects. Although the chapter authors differ widely in the approaches they suggest for generating scientific knowledge, there is a noticeable convergence towards a balanced combination of qualitative and quantitative methods. This blend includes case studies and the detailed analyses of dialog and artifacts, as well as experimental (and quasi-experimental) variation of instructional conditions and contexts of learning. As such, the learning sciences are well placed to overcome the often-claimed incompatibility of qualitative and quantitative methods that has dominated methodological debates in education research.

5. Researchers engaged in design-based research increasingly emphasize designing *with* rather than *for*, of acting *with* rather than acting *on*. Historically, design and improvement efforts in the learning sciences—and more broadly in educational research—in large measure have been developed externally to the context of implementation. Generally, there has been minimal consultation with—and input from—those charged with implementing the resulting designs (typically teachers) and those whose learning is supposed to be impacted by those designs (typically students). Although the DBR cycle is intended to address such concerns, initial designs (and in some cases subsequent iterations) only partially addressed the issue of designing *for* rather than *with*. Designing *for* rather than *with* more often than not results in surface-level enactments and lack of ownership or investment in the success or sustainability of the effort. The learning sciences has begun to address these issues through greater use of participatory design and consultation, student-initiated and student-directed designs, and research focused on understanding the learning processes of researchers/designers as well as of teachers and students. We see this trend toward participatory design increasing as evidence accrues of its benefits to sustainable change.

6. Increasing attention to issues of social justice and equity. Moving forward the learning sciences needs to thoughtfully consider issues of equity and power as they shape and are shaped by our designs. We need to consider whether and how particular disciplinary content, epistemic practices, and outcome measures reinforce existing inequities and power structures, benefiting some learners but not others (Booker, Vossoughi, & Hooper, 2014; Politics of Learning Writing Collective, 2017). We have often failed to adequately examine context in ways that call attention to how power is circulating in learning spaces, where learners are coming from, and where learners are going. To fully realize its potential impact on teachers, policy makers, communities, and learners, learning sciences scholarship needs to more directly consider "issues of power and privilege, because power is always already there, in our research contexts, in our

articles and books, in our conferences, and in our classrooms" (Esmonde & Booker, 2016, p. 168). By thoughtfully considering issues of equity and power, we open up space for the learning sciences to more productively contribute to conversations about whether and how our research perpetuates existing power structures and reifies educational inequities. Doing so entails understanding and designing democratically for multiple levels of context: "the immediate setting in which individuals participate; relationships across the multiple settings that people navigate; the broader cultural, political, economic, and, indeed, ideological belief systems and institutional configurations in which these micro-level settings exist" (Lee, 2012, p. 348). We hope that future editions of this handbook will include more voices of those who explicitly focus on issues of social justice, equity, and power as well as engaging authors of other chapters in conversations about who is and is not being served by learning sciences research.

Acknowledgments

We thank Suraj Uttamchandani for extensive help in thinking about issues of social justice and equity. We thank Elena Janßen for her excellent support in editing this handbook.

References

Bereiter, C. (2014). Principled practical knowledge: Not a bridge but a ladder. *Journal of the Learning Sciences*, 23, 4–17.
Bereiter, C., & Scardamalia, M. (1987a). *The psychology of written composition*. Hillsdale, NJ: Lawrence Erlbaum Associates.
Bereiter, C., & Scardamalia, M. (1987b). An attainable version of high literacy: Approaches to teaching higher-order skills in reading and writing. *Curriculum Inquiry*, 17(1), 9–30.
Booker, A., Vossoughi, S., & Hooper, P. (2014). Tensions and possibilities for political work in the learning sciences. In J. Polman, E. Kyza, D. K. O'Neill, I. Tabak, W. R. Penuel, A. S. Jurow, et al. (Eds.), Proceedings of the International Conference of the Learning Sciences, (Vol. 2, pp. 919–926). ISLS.
Clark, A. (2011). *Supersizing the mind. Embodiment, action, and cognitive extension*. Oxford, UK: Oxford University Press.
Ko, M., Goldman, S. R., Radinsky, J. R., James, K., Hall, A., Popp, J., et al. (2016). Looking under the hood: Productive messiness in design for argumentation in science, literature and history. In V. Svhila & R. Reeve (Eds.), *Design as scholarship: Case studies from the Learning Sciences* (pp. 71–85). New York: Routledge.
Lave, J. (1988). *Cognition in practice: Mind, mathematics, and culture in everyday live*. New York: Cambridge University Press.
Lee, C. D. (2012). Conceptualizing cultural and racialized process in learning. *Human Development*, 55(5–6), 348–355.
Miller, G. A. (2003). The cognitive revolution: a historical perspective. *Trends in Cognitive Sciences*, 7(3), 141–144.
Politics of Learning Writing Collective. (2017). The learning sciences in a new era of U.S. nationalism. *Cognition and Instruction*, 35(2), 91–102.
Saxe, G. B. (1991). *Culture and cognitive development: Studies in mathematical understanding*. Hillsdale, NJ: Lawrence Erlbaum.
Stokes, D. E. (1997). *Pasteur's quadrant: Basic science and technological innovation*. Washington, DC: The Brookings Institution.
Vygotsky, L. S. (1978). *Mind in society: The development of higher mental process*. Cambridge, MA: Harvard University Press.
Wertsch, J. V. (1998). *Mind as action*. New York: Oxford University Press.

Section 1
Historical Foundations and Theoretical Orientations of the Learning Sciences

2

A Short History of the Learning Sciences

Christopher Hoadley

The learning sciences is a field that studies how people learn and how to support learning. It is a relatively young scholarly community whose history reflects the influence of the many other disciplines that are concerned with learning and how to support it (e.g., anthropology, education, psychology, philosophy). Disciplinary communities reflect not just epistemological, intellectual, and methodological commitments in the abstract. Rather, as is well documented in the sociology of science, research fields reflect the people in them and both their interconnections and disconnections from other communities. Understanding these as well as their origins is enlightening with respect to what aspects of a field are core commitments, what aspects are hidden assumptions, and what aspects might merely be accidents of history. For these reasons, this introduction to the history of the learning sciences will be primarily about a community of people who dub themselves "learning scientists." And, like most historical accounts, this history reflects the perspective of the author. As a U.S.-based academician in the field for approximately 30 years, my familiarity is greatest with the North American parts of this story, and is almost entirely limited to the portions that were accessible through English-language research literature. As such, this chapter is best understood as "a" history, not "the" history.

My perspective is that the learning sciences are empirical, interdisciplinary, contextualized, and action-oriented. Throughout this narrative, I hope to illustrate the forms and functions through which the field of learning sciences manifests these four characteristics. Like most historical unfoldings, the path is twisted not straight. I will try to highlight how and when elements of these four characteristics start to emerge.

Seeds of Learning Sciences

Explorations of how best to teach are centuries old, but the scientific study of the nature of the mind and how it learns has its origins in philosophy and medicine. Around the beginning of the 20th century, there were several developments that marked what one might call the emergence of modern-day empirical approaches to the study of learning. On one hand, drawing on medical models, psychology began to emerge as independent of philosophy with different motivations and methods. For example, the physician Wilhelm Wundt used the methods of experimental natural science to understand phenomena such as human perception of color and sound. Sigmund Freud began to address so-called "nervous disorders" by trying to understand the nature of the mind—his empirical investigations involved introspection, leading to the invention of Freudian psychoanalysis.

Ivan Pavlov, the Russian physiologist, investigated the nature of conditioning in shaping learning after discovering physiological responses that preceded physical stimuli (such as dogs salivating before food was actually present in anticipation of a meal). In the early 20th century the biologist Jean Piaget studied learning as a manifestation of development, likening the maturing of children to the ways a flower might bloom, with biologically constrained possibilities emerging from the intersection of nature and nurture. Maria Montessori, trained as a physician, investigated children with disabilities near the turn of the 20th century and children's responses to various stimuli led her to begin creating the techniques used to the present day in Montessori schools. On the other hand, one countervailing force to these physiologically based approaches was a more contextualized way of conducting empirical research that emerged from philosophy, exemplified by the philosopher John Dewey. Dewey founded a laboratory school at the University of Chicago to study education within an authentic social context, one in which teachers and researchers were the same people. It is important to note that this was a time when many disciplines were working in parallel to formulate their core epistemologies and methods of approaching learning and education, including problems of explanation, prediction, and application to practical problems. The early to mid-20th century saw the empirical disciplines become more solidified and differentiated in academic institutions. For instance, not only did psychology become its own discipline, distinct from medicine, but psychology began to distinguish between experimental and clinical psychology. This posed an interesting question for how education would be institutionalized.

Generally, the shift from education as an applied profession to a legitimate area for empirical research was a contested one. In the United States, education in the form of teacher preparation was taught in 'normal schools' through the end of the 19th century, but this gradually was displaced by the notion of a school of education as a center for, not only practical training of teachers, but academic research relevant to the problems of education. By the mid- to late 20th century, most universities in the US had a school or college of education with a dual mission of preparing teachers and conducting educational research. However, Lagemann (2000) chronicles the history of educational research as contested terrain—at every level, especially methodology and epistemology. These tensions linked strongly to the characteristics of discipline, empiricism, contextualization, and action-orientation. From a disciplinary perspective, disputes focused on whether education was an intellectual (scientific) discipline unto itself, an application area to be colonized by 'real' disciplines, or a crossroads in which interdisciplinary inquiry could flourish. From an epistemological perspective, Dewey saw "an intimate and necessary relation between the processes of actual experience and education" (Dewey, 1938/1997, p. 20), and advocated for a holistic, pragmatic approach to the science of learning, while behaviorists like Tolman and Skinner saw human experiences as epiphenomenal and an invitation to pseudo-science. Skinner argued for the importance of the human environment as a source of conditioning for the individual, but saw the processes underlying learning as entirely universal, while Dewey saw learning as an inherently social, cultural, and societally embedded phenomenon. Methodological rifts were consistent with the epistemological: In the name of objectivity, the behaviorists advocated an arms-length, objective science of learning while the pragmatists and progressive education researchers took the positions of participant-observers in research. According to Lagemann (2000), behaviorism claimed the highest status among these epistemologies, but behaviorist theories increasingly ran up against phenomena that required hypothesizing hidden internal stimuli and other workarounds to keep the mind's "black box" closed while explaining human behavior. Openings to cognitivism also came from developmental psychologists' (most notably Piaget [1970] and Bruner [1966]), proposals that thoughts in the head mattered and that the stages of human development were less fixed than previously thought (see Olson, 2007).

By the 1970s and 1980s, and relevant to the challenge of "opening the black box" of the mind, two trends dramatically changed the landscape for people studying thinking (and by extension, learning): the advent of computing and the emergence of cognitive science. Beginning in the 1930s and 1940s, technology advances had led to both the field of cybernetics (studying the nature of dynamic systems that had self-regulating properties) and the use of metaphors like the telephone switchboard for thought and thinking. The emergence of digital computing birthed computer science, a field concerned with not only calculation of numbers but symbol manipulation more generally. Early on, the subfield of artificial intelligence emerged to design and study ways in which artificial symbol manipulation systems (electronics, digital computers, and software) could mimic intelligent behaviors exhibited by natural organisms like people. This approach included not only emulation of intelligent behavior, but using the computer as a model of the mind (Block, 1990).

In parallel, by the 1950s, debates about whether mental events and representations were empirically measurable had begun to chip away at Skinner's conception of thoughts as epiphenomenal, most notably by linguists like Noam Chomsky, who argued that language development was demonstrably not explainable with behaviorist theories (Gardner, 1985). Chomsky argued that mental machinery innately constrained language development beyond mere conditioning. Between the 1950s and 1970s, interdisciplinary examinations of thought started to reveal that not only the contents, but the mechanisms or machinery of thinking could be studied. Researchers began to overcome the limitations of introspection as the sole method of studying internal mental processes by drawing on techniques from a range of fields. For instance, Chomsky's argument was bolstered by much earlier medical research showing that damage to specific areas of the brain yielded very particular disabilities in producing or comprehending language. Reaction time studies and the methods of experimental psychology were used to attempt to infer the internal processes of thinking, from perception to attention to memory. The combination of computational perspectives focused on how to simulate or model thinking with artificial computational systems, and cognitive perspectives that viewed the contents and processes of thought as inspectable (breaking open the 'black box' with which behaviorists viewed thinking) created the conditions under which an interdisciplinary and empirical field calling itself cognitive science emerged (Gardner, 1985).

The degree to which cognitive science viewed thought as linked to the context "outside the head" increased over time; a special issue published in the journal *Cognitive Science* (Cognitive Science Society, 1993) posed a debate on how (much) cognition was "situated," i.e., inextricable from both physical and sociocultural context. On the one hand, you had the information processing psychology view which opened up cognition to inspection compared to behaviorism, but still treated the outside world as 'inputs.' On the other hand, you had the situated view, which helped establish a contextualized science for learning, in which learning at the minimum required investigating the social and cultural contexts of learning, and at the maximum treated learning as inherently a phenomenon not in the head but in the relationships between person and their context. Thus, prior to the beginning of the learning sciences, the cognitive science revolution helped establish more interdisciplinary approaches to thinking (and learning), with two effects: It laid the groundwork for empirical studies of learning to grow beyond black-box models, and it paved the way for examining learning as a product of context.

Early Learning Sciences (1980s–1990s)

The dilemma of how to leverage the interdisciplinary, empirical methods of the cognitive sciences for designing learning environments (action-orientation) while dealing with the messiness of learning-in-context, arguably led to the birth of what we now call learning sciences. The early history of the learning sciences was a time when the action-orientation and contextualization characteristics of educational research in cognitive science were being worked out. In 1989, I was at MIT pursuing what now might be called a learning sciences agenda while obtaining a cognitive science degree. I was working for Seymour Papert's Learning and Epistemology group at the MIT Media Lab, and simultaneously with developmental psychologist Susan Carey's research group studying conceptual change and scientific reasoning. I vividly recall a week in which colleagues in both quarters questioned why I was bothering with the other. The mantra at the Media Lab, "demo or die," contrasted with the traditional "publish or perish" in the psychology program. This question of which was more important—innovation and creative design versus scientific explanation and prediction—paralleled the difference between engineering and science. The tension I felt was more about how separating these two endeavors impoverished each.

Near the time of the arrival of more situated theories of thinking and learning, education researchers working in cognitive science grew somewhat frustrated with the degree to which cognitive science was distancing itself from cognition 'in the wild' (to borrow a term from Hutchins, 1995). The late 1980s and early 1990s can be marked as the birth of the term 'learning sciences,' and the field as such. Janet Kolodner, the computer scientist who founded the *Journal of the Learning Sciences* in 1991 clearly displayed an action-oriented stand in describing some of the motivations for the journal and the field. These included "need[ing] concrete guidelines about what kinds of educational environments are effective in what kinds of situations" and the need to make use of such guidelines "to develop more innovative ways to use computers" (Kolodner, 1991). In a retrospective history, she described how the cognitive scientists working at the Institute for the Learning Sciences founded in 1990 at Northwestern University were fundamentally as interdisciplinary as the cognitive sciences, but with additional linkages to educational psychology and curriculum and instruction (Kolodner, 2004). She also noted a frustration in the community with the lack of connection between what theories of cognition could predict (for example, with AI production systems) and what might be educationally relevant in real contexts. Kolodner highlighted that the action-oriented design mandate of learning sciences might contrast it with much of the cognitive science community in the 1990s for whom the design of artificial intelligence systems was primarily in service of generating theories and models of thinking. In that same issue of *Educational Technology* in which the Kolodner piece appeared, Smith (2004), a graduate of the first cohort of the Northwestern Learning Sciences Ph.D. program, drew a distinction between 'neat' as in lab-based and 'scruffy' as in field-based studies of learning. His characterization of learning sciences as 'scruffy' highlights the contextualized nature of the learning sciences' action orientation, and distinguishes the design research conducted by learning scientists from that done in the instructional systems design field. As well, each article described some of the milestones of the era leading to the creation of a community.

It was in 1991 as well that the first International Conference of the Learning Sciences was spearheaded by Northwestern's Director of the Institute for Learning Sciences, Roger Schank. Essentially, Schank renamed and refocused what was supposed to have been an Artificial Intelligence in Education conference. This renaming sparked interest in learning sciences particularly in the US, but had long-term consequences that made it more difficult to establish an international society of the learning sciences. During this same period of time, a community was coalescing around interests in computer support for collaborative learning (CSCL) with commitments to interdisciplinarity, and an action-oriented, empirical, and contextualized view of learning (Stahl, Koschmann, & Suthers, 2014). Following a workshop in 1989 in Maratea, Italy, in 1991, a workshop on CSCL was held in Carbondale, Illinois, hosted by Tim Koschmann, underwritten by John Seely Brown and sponsored by Xerox Parc. The workshop yielded a 1992 special issue of the newsletter of the Association for Computing Machinery (ACM) Special Interest Group on Computer Uses in Education. In 1995, the first biennial conference on Computer-Supported Collaborative Learning was held in Bloomington, Indiana, under the auspices of the ACM and the AACE (Association for the Advancement of Computers in Education) with an explicit attempt to alternate years with the ACM Computer-Supported Cooperative Work (CSCW) conference. Victor Kaptelinin from Umeå in Sweden gave a keynote on cultural-historical activity theory, and Marlene Scardamalia discussed knowledge-building communities, cementing the connection between the CSCL conference and sociocultural theories of learning and technology use.

Coincident with this five-year period of emerging conferences in CSCL and ICLS, there was an explosion of technologies that invited not only interdisciplinarity between technologists and educators, but also an action-orientation towards creating technology-mediated learning environments.

In the early 1990s, the web emerged (with the popularity of the Mosaic browser), as did the capacity to include video in consumer-grade computers (with the creation of Apple's QuickTime). Teleconferencing technologies were just barely getting out of the lab (for example, the CU-SeeMe software from Cornell). The emergence of commercial internet service providers at this time ensured that networked technologies, critical to collaboration, were widespread, and interest in educational applications grew beyond high-end training, government, and higher education settings to include learners at home, in grade school, and the general citizenry. Many members of the ICLS and CSCL program committees had appointments in computer science or informatics departments, a sharp distinction between these two conferences and most education conferences (even those with a focus on educational technology).

Several institutions played a role in bringing technology, design, and a contextualized view of learning together. The Institute for Research on Learning (IRL), began with initial funding from the Xerox Palo Alto Research Center (PARC). The IRL, directed by Jim Greeno, took culture and anthropology as seriously as technology and design. Similarly, the Lab for Comparative Human Cognition led by Michael Cole advocated a socially construed perspective on both learning research and learning design, and was an early adopter of technologies as a means to bring an action-orientation to social context (Cole, 1999). Many of the institutions which became known for learning sciences in the 1990s were places where interdisciplinary groups of faculty examined new methods for studying and designing learning settings, including notably Stanford and Berkeley on the West Coast, the Cognition and Technology Group at Vanderbilt, and so on. Each, of course, was different but in many cases these groups were supported by funding from the U.S. National Science Foundation or the McDonnell Foundation in projects that shared the four characteristics I've described of an interdisciplinary, empirical, contextualized, and action-oriented approach to understanding learning. The McDonnell Foundation alone, through its Cognitive Studies in Education Program (CSEP) funded approximately 50 such projects located in the US, Canada, or Europe over a 10-year period (1987–1997). CSEP was also foundational in building a learning sciences community through its annual meetings of grantees. In those early days of the learning sciences in the US, most of the theoretical stances were either cognitive or somewhat situative (rather than socio-political, cultural-historical, etc.). But interesting interventions implemented in the field were twinned with interesting learning theories that were design-relevant, including Brown and Campione's fostering communities of learning, Bransford's anchored instruction (Cognition and Technology Group at Vanderbilt, 1990), Brown, Collins, and Duguid's (1989) cognitive apprenticeship, Papert's constructionist environments for learning (Harel & Papert, 1991), Scardamalia and Bereiter's (1994) knowledge-building communities, Anderson's cognitive tutors (e.g., Andreson, Conrad, & Corbett, 1989), and Lave and Wenger's (1991) communities of practice. In each case, important claims about learning were asserted and tested by creating new genres of (mostly technology-mediated) learning environments. The particular mix of disciplines, theories, and approaches to action and context were different in other regions; for instance, Scandinavian researchers often drew on cultural-historical activity theory and participatory design approaches in this era. But one can argue that, in Europe as well as North America, there was a confluence of researchers representing these four characteristics, and that this challenged to varying degrees the particular entrenchments of "mainstream" educational research (for example, attempts to make education research generally more like the discipline of educational psychology).

Institutionalization of Learning Sciences (1990s–2000s)

By the late 1990s, both the ICLS conference and CSCL conference had established themselves. CSCL cemented itself as a field in two volumes edited by Tim Koschmann (Koschmann, 1996; Koschmann, Hall, & Miyake, 2002) and the *Journal of the Learning Sciences* was achieving outsized impact given its youth. Key contributions came in: cognition and learning (including elaboration of how conceptual change could be supported with technology scaffolding, both cognitively and interpersonally, the role of self-explanation, mental causal models, convergent conceptual change, and new theories of transfer); new methodologies such as interaction analysis, microgenetic analysis, and design experiments; and new approaches to technology including microworlds, tools for fostering communities of learners, tools for scaffolded inquiry, new models of intelligent tutoring systems and goal-based scenarios. Theories of situated activity, co-construction of knowledge, and distributed intelligence helped connect learning to its contexts. In general, all of this research fit the profile of interdisciplinary, empirical, contextualized, and action-oriented. For example, the LeTUS project at Northwestern and the University of Michigan attempted to scale up ideas about using technology to support inquiry science in the large urban school districts of Detroit and Chicago.

It was around this time that the learning sciences as a moniker for an interdisciplinary field began to take hold, as evidenced by data from the Google Books Ngram viewer shown in Figure 2.1. Figure 2.1 shows the prevalence within the Google Books corpus of the literal capitalized phrase "Learning Sciences" for the period 1980–2008, the last year for which data are available. Within North America, many scholars began to attend ICLS and CSCL in alternating years. At the time, ICLS had been held solely in the US, while CSCL had been held in the US and Canada. Both conferences were to some extent international, with attendees from Europe and, to a lesser extent, Asia and Australia (Hoadley, 2005). Although the conferences were organized informally, with the hosting university taking on financial management, there were real questions about the sustainability of this approach, which led Janet Kolodner (then editor-in-chief of *JLS*), Tim Koschmann (still considered a founding father of CSCL) and me (a newly minted Ph.D. with the job title 'research cognitive and computer scientist') to begin organizing a formal professional society that could house these three activities, support continuity, increase visibility and legitimacy, and provide financial stability. We began discussing the idea at the business meetings of each of the conferences in 1999–2000, and elicited experts from North America, Europe, and Asia who could serve on an interim advisory board to guide the founding of a society in 2000–2001.

Early attempts by the advisory board to define and name the organization revealed important differences in how different groups defined "the field" and felt about the two conferences and the journal. While CSCL had a track record of attracting an international audience, and the first European CSCL conference (dubbed "ESCSCL") was held in Maastricht in the Netherlands in 2001, the ICLS had had less success at attracting an international audience. Within Europe, strong networks of researchers were institutionalizing through formal networks such as the Intermedia project in Norway, the DFG Priority Program Net-based knowledge communication in groups, and

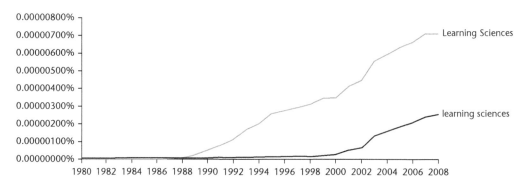

Figure 2.1 Prevalence of the Literal Phrases "Learning Sciences" and "learning sciences" in Works Indexed by Google Books, 1980–2008

Source: Google Ngram viewer. Retrieved July 27, 2017 from https://books.google.com/ngrams/interactive_chart?content=learning+sciences&case_insensitive=on&year_start=1980&year_end=2008&corpus=15&smoothing=3

the EU Kaleidoscope Network of Excellence on technology-mediated learning, which were formed in the early 2000s. In the US, a network called CILT (Center for Innovative Learning Technology) was funded by the U.S. National Science Foundation. These networks capitalized on many disciplinary networks in education research, including educational psychology and instructional design, but also helped incorporate technologists from computer sciences as well as human–computer interaction and informatics/information sciences. At the CSCL 2002 meeting in Boulder, Colorado, the interim board discussed the negative connotation that the name 'learning sciences' carried for some, given its connection to Roger Schank's unilateral renaming and co-opting of what was supposed to have been an AI in Ed conference. These connotations concerned many both in North America and Europe, although perhaps for different reasons. Indeed, many in Europe who were frequent CSCL conference participants had no affinity to either *JLS* or the ICLS conference. However, the interim board failed to identify a better alternative name for the society and the field it intended to support, and voted not to formally define the field, instead allowing the *JLS* and conferences to speak for themselves.

We continued our work and incorporated the organization in mid-2002 as the International Society of the Learning Sciences. Nine months later, during the ICLS 2002 meeting in Seattle, many participants were concerned that the Society should hold elections as soon as possible to allow wider participation in governance. We attempted to do so, and this backfired spectacularly. At the CSCL 2003 conference in Bergen, Norway, many Europeans saw the ISLS as an American takeover of a quintessentially European conference and scholarly community, a view exacerbated by the attempt to hold elections quickly. This led to a contentious business meeting and a negotiated agreement that CSCL would have a leadership committee within ISLS that was elected by the CSCL community, with some budget autonomy and a formal role in CSCL conference organization. Part of what had happened was that members of the community had been more insular than they realized. North Americans tended to go to both ICLS and CSCL but Europeans tended not to go to ICLS. They saw much of the work in CSCL emerging from European research, whether the traditions of participatory design in Scandinavia, with its strong tradition of research in cultural historical activity theory informing collaborative technologies, the experimental psychology research on collaborative learning processes in Germany, Belgium and the Netherlands, or some of the groundbreaking technology work coming from Europe in CSCW. Through an empirical analysis of the CSCL organizing committees and presenters, I was able to document that CSCL up to that point was truly an international, interdisciplinary conference, but that international collaboration was less strong than you might expect: the majority of CSCL papers were coauthored, but less than 10% of co-authorships were international collaborations (Hoadley, 2005).

In part to foster further internationalization and to avoid worsening any tensions between U.S. and European scholars, the next CSCL was held in Taipei, an important step towards truly internationalizing the society. Since then, both the ICLS and CSCL conferences have rotated among North America, Europe, and Asia or Australia. This has had a number of important outcomes over the years, including solidifying international exchange of scholarship (Kienle & Wessner, 2006). It appears from my perspective that the interdisciplinarity of the field has had a different texture in different parts of the world. For instance, in the United States, instructional design and learning sciences were different, whereas in the Netherlands, educational design and educational sciences were more connected. In the United States, it's quite common for schools of education to have departments of educational psychology, whereas in other areas of the world the psychology researchers might have less connection to schools of education and more to traditional psychology departments. Although the origins of the terminology 'learning sciences' may have been contentious, and there may still be debate about whether the field of CSCL is a subfield or a sibling of the field of learning sciences, the institutionalization of the professional society has been echoed in a shift in terminology in the published literature, in the naming of degree programs and institutes, and became a label for a stable and growing worldwide community of scholars.

Another important institutionalization of the field was reflected in the increase of visibility of design-based research methods as a core methodology for the learning sciences. Following on initial description of 'design experiments' by Collins (1992) and Brown (1992), in the mid-1990s a Design Experiment Consortium was founded, with many partners recognizable as members of the nascent learning sciences community. In the late 1990s, the Spencer Foundation funded the Design-Based Research Collective (Design-Based Research Collective, 2003) and a variety of other researchers began elaborating the method with special issues of the journals *Educational Researcher* (Jan/Feb issue, 2003), *Educational Psychologist* (2004, issue 4), *Educational Technology* (Jan/Feb issue, 2005) and the *Journal of the Learning Sciences* (2004, issue 1). This blending of design and research, while not universal in the learning sciences, nonetheless became identified with the community (Hoadley, 2004; Hung, Looi, & Hin, 2005) and helped entrench both the action-oriented and contextualized ways that the field conducts empirical research. As the principal investigator for the Design-Based Research Collective, I experienced firsthand how, while these methods often produced useful findings, they challenged core beliefs in the education community that followed from the tensions Lagemann had identified between the Deweyan and Thorndikean approach to studying learning. While now it is far less controversial to suggest that a designer of curriculum might be able to use their involvement in the creation and adjustment of interventions as they unfold in context to more effectively guide research, at that time it was seen as a gross violation of a notion of rigor that depended on a rigid separation between the 'objective' scientist and the educator or designer.

Finally, several books had an important impact on cementing the learning sciences. Initially published in 1999, the book *How People Learn* was written by a committee convened by the U.S. National Academy of Sciences, including a number of scholars active in the learning sciences community. This book helped consolidate in an authoritative way both the known findings about education and learning, and helped legitimize linking scientific research and practice (design) in education, advocating as one of its five core principles that we should "conduct research in teams that combine the expertise of researchers and the wisdom of practitioners" (National Research Council Committee on Learning Research and Educational Practice, Bransford, Pellegrino, & Donovan, 1999, p. 3). It also provided a framework that legitimized the role of context in both fostering and studying learning. Around the same time, Pierre Dillenbourg founded a CSCL book series at the publisher Kluwer (later absorbed into Springer). Two of the first volumes in the series were *What We Know About CSCL and Implementing It in Higher Education* (Strijbos, Kirschner, & Martens, 2004) and *Arguing to Learn* (Andriessen, Baker, & Suthers, 2003). And, in 2006, the first edition of the *Cambridge Handbook of the Learning Sciences* was published (Sawyer, 2006) (although, sadly, this edition contained almost exclusively U.S.-based authors). Thus, we see in this period a consolidation of the learning sciences as a field supporting interdisciplinary, empirical research that was both action-oriented and sensitive to the contextualized nature of learning. Although still in its infancy, methodologies, representative interventions, and core perspectives were emerging. To a large extent this period can be characterized as learning scientists finding each other and the common label for what they do.

Flowering of the Learning Sciences (2000s–present)

By the 2000s, the learning sciences, including CSCL, were flowering globally, with increasing institutionalization through the ISLS. Key achievements included the launch of the *International Journal of Computer Supported Collaborative Learning* at Springer and formal arrangements with the ACM's digital library to support archiving and indexing of society conference proceedings. The success of doctoral consortia associated with ICLS and CSCL conferences spawned the creation of conference-related workshops for early career faculty. As well, with the support of the ISLS, a Network of Academic Programs in the Learning Sciences (NAPLeS) was initiated. Both JLS and IJCSCL achieved impact factors that put them among the top five journals in educational research worldwide. Exchange programs started to crop up; for example, the US NSF (National Science Foundation) and the

German DFG (Deutsche Forschungsgemeinschaft) created a series of international workshops. The ISLS leadership began outreach efforts to articulate better with related societies such as the AI in Ed, Learning Analytics & Knowledge, and Educational Data Mining. After some drifting apart between learning sciences and computer science, funding agencies in several parts of the world were prioritizing work at the intersection of learning and computer science. These programs helped encourage new partnerships between computer scientists and education researchers at the forefront of both fields; in the US, this intersection was termed 'cyberlearning.'

Some of the key conceptual achievements of the field during this time included a deepening of the insights linking context and learning and further interdisciplinarity. For example, Gerry Stahl led efforts at Drexel University to examine cognition at the small group level through close study and design of software environments to support mathematics learning, leading to his theory of group cognition (Stahl, 2006). As well, an interdisciplinary, international team of psychologists, designers, and computer scientists released an important book on productive multivocality (Suthers, Lund, Rosé, Teplovs, & Law, 2013). Importantly, an edited book on critical and socio-cultural theories of learning brought new disciplinary perspectives on power and privilege to the Learning Sciences community (Esmonde & Booker, 2017).

The commitments in the Learning Sciences to action-orientation in conjunction with empirical research in context led to new developments in methodologies. Design-based research was augmented by design-based implementation research drawing on literature from improvement sciences (Bryk, Gomez, Grunow, & LeMahieu, 2015) and new models of research practice partnerships (Penuel, Allen, Farrell, & Coburn, 2015). The field also began examining new video-based technologies for studying learning (Goldman, Pea, Barron, & Derry, 2007; Derry, Minshew, Barber-Lester, & Duke, this volume), interfaces for detecting emotional states learners (Calvo & D'Mello, 2011), and big data approaches (Larusson & White, 2014; Rosé, this volume). Each of these techniques will undoubtedly be important in learning research generally, but each has come from individuals and groups with ties to the global learning sciences community and helps demonstrate the eclecticism, both in disciplines and in epistemology, that supports an empirical, contextualized, action-oriented interdisciplinary research community in the learning sciences. As I edit this, a new article has come out in *JLS* surveying the breadth of what self-described learning scientists do (Yoon & Hmelo-Silver, 2017). It demonstrates that the field is interdisciplinary with strong ties to both empirical research and design, using a broad variety of methodologies and mixed-method approaches, suggesting a sensitivity to contexts of learning.

Summing up: What Are the Learning Sciences Today and What Will They Be in the Future?

In an earlier paper, I described research communities as defined by scope and goals, theoretical commitments, epistemology and methods, and history (Hoadley, 2004). To these four, I would add a fifth today, coming directly from the word 'community': communion, i.e., being not only in communication, but also recognition and acceptance of each other's stances. When I was a junior scholar, *JLS* was one of the only places where eclectic methodologies were welcomed. At present, the Learning Sciences remain a community or field but not a discipline: People in the community retain allegiances to disciplines they call home, whether it is computer science, psychology, design, or any number of other disciplines. The Learning Sciences does not claim to have a monopoly on interdisciplinary approaches to studying education. Nevertheless, my claim is that, globally, learning scientists form a cohesive, yet diverse, community of scholars with enduring characteristics of interdisciplinarity, empiricism, attention to researching learning in context rather than in the lab, and action-orientation—the desire not only to study, but also to invent, environments for learning. Table 2.1 summarizes this evolutionary history to date.

Table 2.1 Evolution of Four Characteristics of the Learning Sciences Over Time (Empirical, Interdisciplinary, Contextualized, and Action-Oriented)

	Empirical	Interdisciplinary	Contextualized	Action- or design-oriented
Early 20th century	Empirical study of learning is emerging from medicine, biology, physics	Education moves from pre-disciplinary to becoming a discipline	Contested: Deweyan vs behaviorist approaches	Educational interventions just beginning to be connected to research
1950s–1980s	Experimental paradigm entrenched in educational psychology.	Psychology is established as a discipline. Education becomes a quasi-discipline, with major branches in curriculum and instruction, and educational psychology. Cognitive science begins bringing disciplines together	"Methods wars" show tension between quantitative and qualitative (contextualized) approaches in education schools. Most research attempts to explain culture within cognitive framing	Instructional design and curricular design well entrenched in U.S. schools of education but separated from development of learning theory
1990–2000s	Education moves towards randomized, controlled clinical trials as "gold standard." Other forms of empiricism are contested. In contrast, learning sciences embraces eclectic empiricism, including new methods	Education is entrenched as a discipline. Learning sciences explicitly draws on cognitive sciences and computer science	Situated cognition becomes a mainstay in learning sciences as well as in mainstream education Learning sciences links to older theories in cultural-historical activity theory and ecological psychology.	Learning sciences differentiates itself from education research writ large by linking design and research through novel methodologies (design-based research). Learning sciences considers applied research in schools even as cognitive science becomes less applied
2000s–present	Learning sciences continues to link to new forms of empiricism, including new ways of modeling through learning analytics and educational data mining	As a community, learning scientists become more established while residing in many disciplinary departments (computer science, education, communication, psychology, information science, etc.)	Learning sciences moves from primarily investigating individual cognition to a much greater emphasis on practices, groups, culture and language, and identity	Learning sciences' design orientation continues to embrace school settings and technologies, but also moves towards designing learning environments across contexts and through the lifespan. Design-based research and variants are taken up by other disciplines

Bibliometric analyses by Lund and colleagues indicate that education is one of the most crosscutting intellectual areas within social sciences generally, outstripping fields such as psychology and anthropology (Lund, 2015; Lund, Jeong, Grauwin, & Jensen, 2015). Importantly, the analyses of Lund et al. (2015) indicate that many of the seminal publications related to the foundations and flowering of the learning sciences themselves are more likely to be cited across disciplines.

As I reflect back on the community, I am grateful. Intellectually, I came of age at a time when Learning Sciences was able to create an exciting space for action and reflection, science and design, innovation and insight. Some of the battles in creating this space were hard-won, including legitimizing the role of design knowledge as a valid product of scholarship (Edelson, 2002; Hoadley & Cox, 2009), insisting that we attempt to internationalize the community of researchers, and successfully navigating the tension between being a discipline versus an interdisciplinary field. The creation of a vibrant professional organization and maintaining the exceptional quality of two Society-affiliated journals has taxed a phalanx of the best scholars in the field. They have set aside their own work to edit, review, run conferences, and so on, usually without the built-in respect that would come from doing that work in the more discipline-based venues that align with the names of their academic departments. And, I keep coming back to this idea of communion—of being willing to recognize and embrace the epistemologies, methods, and theories of disciplines that are not one's own. As new students interested in learning encounter the community for the first time, they are often as excited as I was at the possibilities when we try to both understand and engineer learning with all the tools at our disposal. However, they also are frequently nervous about transgressing the norms of their home discipline. After 30 years of participating in this community, it is easy for me to tell them the results are worth it.

Further Readings

Gardner, H. (1985). *The mind's new science: A history of the cognitive revolution.* New York: Basic Books.
Cognitive science is a major influence in the learning sciences. This is a good introduction to its history and hence to some important intellectual roots of learning scientists.

Hoadley, C. (2004). Learning and design: Why the learning sciences and instructional systems need each other. *Educational Technology, 44*(3), 6–12.
Another discussion of how learning sciences overlaps with other fields, and some of the characteristics that define it.

Kolodner, J. L. (2004). The learning sciences: Past, present, and future. *Educational Technology, 44*(3), 37–42.
An analysis and a vision for the young field of the learning sciences, written by one of its foundational scholars.

Lagemann, E. C. (2000). *An elusive science: The troubling history of education research.* Chicago, IL: University of Chicago Press.
An excellent source for those who want to understand the philosophical, theoretical, and methodological tensions in educational research.

Stahl, G., Koschmann, T., & Suthers, D. (2014). Computer-supported collaborative learning: An historical perspective. In R. K. Sawyer (Ed.), *Cambridge handbook of the learning sciences* (2nd ed., pp. 479–500). Cambridge, UK: Cambridge University Press.
This chapter focuses on the development of CSCL as a research field and community, both as part of the learning sciences and beyond.

NAPLeS Resources

Dillenbourg, P., *Evolution of research on CSCL* [Webinar]. In *NAPLeS video series.* Retrieved October 19, 2017, from http://isls-naples.psy.lmu.de/intro/all-webinars/dillenbourg_video/index.html

Hoadley, C., *A short history of the learning sciences* [Webinar]. In *NAPLeS video series.* Retrieved October 19, 2017, from http://isls-naples.psy.lmu.de/intro/all-webinars/hoadley_video/index.html

References

Anderson, J. R., Conrad, F. G., & Corbett, A. T. (1989). Skill acquisition and the LISP tutor. *Cognitive Science*, 13(4), 467–505.

Andriessen, J., Baker, M., & Suthers, D. (Eds.). (2003). *Arguing to learn: Confronting cognitions in computer-supported collaborative learning environments*. Dordrecht: Springer Science & Business Media.

Block, N. (1990). The computer model of the mind. In D. N. Osherson & E. E. Smith (Eds.), *Thinking: An invitation to cognitive science* (Vol. 3., pp. 247–289). Cambridge, MA: MIT Press.

Brown, A. L. (1992). Design experiments: Theoretical and methodological challenges in creating complex interventions in classroom settings. *Journal of the Learning Sciences*, 2(2), 141–178.

Brown, A. L., & Campione, J. C. (1994). *Guided discovery in a community of learners*. Cambridge, MA: MIT Press.

Brown, J. S., Collins, A., & Duguid, P. (1989). Situated cognition and the culture of learning. *Educational Researcher*, 18(1), 32–41.

Bruner, J. S. (1966). *Toward a theory of instruction*. Cambridge, MA: Harvard University Press.

Bryk, A. S., Gomez, L. M., Grunow, A., & LeMahieu, P. G. (2015). *Learning to improve: How America's schools can get better at getting better*. Cambridge, MA: Harvard Education Press.

Calvo, R. A., & D'Mello, S. (Eds.). (2011). *New perspectives on affect and learning technologies*. New York: Springer.

Cognition and Technology Group at Vanderbilt (1990). Anchored instruction and its relationship to situated cognition. *Educational Researcher*, 19(6), 2–10.

Cognitive Science Society. (1993). Special issue on situated action. *Cognitive science*, 17(1).

Cole, M. (1999). Cultural psychology: Some general principles and a concrete example. In Y. Engeström, R. Miettinen, & P. Raija-Leena (Eds.), *Perspectives on Activity Theory* (pp. 87–106). New York: Cambridge University Press.

Collins, A. (1992). Toward a design science of education. In E. Scanlon & T. O'Shea (Eds.), *New directions in educational technology* (pp. 15–22). New York: Springer.

Derry, S., Minshew, L. M., Barber-Lester, K., & Duke, R. (2018). Video research methods for Learning Scientists: state-of-the-art and future directions. In F. Fischer, C. E. Hmelo-Silver, S. R. Goldman, & P. Reimann (Eds.), *International handbook of the learning sciences* (pp. 489–499). New York: Routledge.

Design-Based Research Collective. (2003). Design-based research: An emerging paradigm for educational inquiry. *Educational Researcher*, 32(1), 5–8, 35–37.

Dewey, J. (1938/1997). *Experience and Education* (First Touchstone Edition ed.). New York: Simon & Schuster.

Edelson, D. C. (2002). Design research: What we learn when we engage in design. *Journal of the Learning Sciences*, 11(1), 105–121.

Esmonde, I., & Booker, A. N. (Eds.). (2017). *Power and privilege in the Learning Sciences: Critical and cociocultural theories of learning*. New York: Routledge.

Gardner, H. (1985). *The mind's new science: A history of the cognitive revolution*. New York: Basic Books.

Goldman, R., Pea, R., Barron, B., & Derry, S. J. (Eds.). (2007). *Video research in the learning sciences*. Mahwah, NJ: Lawrence Erlbaum.

Harel, I. E., & Papert, S. E. (1991). *Constructionism*. New York: Ablex Publishing.

Hoadley, C. (2004). Learning and design: Why the learning sciences and instructional systems need each other. *Educational Technology*, 44(3), 6–12.

Hoadley, C. (2005). The shape of the elephant: Scope and membership of the CSCL community. In T. Koschmann, D. D. Suthers, & T.-W. Chan (Eds.), *Computer-Supported Collaborative Learning (CSCL) 2005* (pp. 205–210). Taipei, Taiwan: International Society of the Learning Sciences.

Hoadley, C., & Cox, C. D. (2009). What is design knowledge and how do we teach it? In C. diGiano, S. Goldman, & M. Chorost (Eds.), *Educating learning technology designers: Guiding and inspiring creators of innovative educational tools* (pp. 19–35). New York: Routledge.

Hung, D., Looi, C.-K., & Hin, L. T. W. (2005). Facilitating inter-collaborations in the learning sciences. *Educational Technology*, 45(4), 41–44.

Hutchins, E. (1995). *Cognition in the wild*. Cambridge, MA: MIT Press.

Kienle, A., & Wessner, M. (2006). The CSCL community in its first decade: development, continuity, connectivity. *International Journal of Computer Supported Collaborative Learning*, 1(1), 9–33.

Kolodner, J. L. (1991). The *Journal of the Learning Sciences*: Effecting changes in education. *Journal of the Learning Sciences*, 1(1), 1–6.

Kolodner, J. L. (2004). The Learning Sciences: Past, present, and future. *Educational Technology*, 44(3), 37–42.

Koschmann, T. D. (Ed.). (1996). *CSCL, theory and practice of an emerging paradigm*. Mahwah, NJ: Lawrence Erlbaum Associates.

Koschmann, T. D., Hall, R., & Miyake, N. (2002). *CSCL 2, carrying forward the conversation*. Mahwah, NJ: Lawrence Erlbaum Associates.

Lagemann, E. C. (2000). *An elusive science: The troubling history of education research.* Chicago, IL: University of Chicago Press.

Larusson, J. A., & White, B. (Eds.). (2014). *Learning analytics: From research to practice.* New York: Springer.

Lave, J., & Wenger, E. (1991). *Situated learning: Legitimate peripheral participation.* New York: Cambridge University Press.

Lund, K. (2015, September 25). *Revealing knowledge bases of educational research,* cyberlearning webinar series [video file], Arlington, VA. Retrieved from www.nsf.gov/events/event_summ.jsp?cntn_id=136349&org=CISE

Lund, K., Jeong, H., Grauwin, S., & Jensen, C. (2015). *A scientometric map of global educational research,* research report. Retrieved from HAL archives-ouvertes.fr

National Research Council Committee on Learning Research and Educational Practice, Bransford, J., Pellegrino, J. W., & Donovan, S. (Eds.). (1999). *How people learn: Bridging research and practice.* Washington, DC: National Academy Press.

Olson, D. (2007). Bruner's psychology and the cognitive revolution. In D. Olson (Ed.), *Jerome Bruner: The cognitive revolution in educational theory* (pp. 13–30). London: Continuum.

Penuel, W. R., Allen, A.-R., Farrell, C., & Coburn, C. E. (2015). Conceptualizing research–practice partnerships as joint work at boundaries. *Journal for Education of Students at Risk, 20*(1–2), 182–197.

Piaget, J. (1970). Piaget's theory (G. L. Gellerier & J. Langer, Trans.). In P. Mussen (Ed.), *Carmichael's manual of child psychology* (3rd ed., Vol. 1, pp. 703–732). New York: Wiley.

Rosé, C. (2018) Learning analytics in the Learning Sciences. In F. Fischer, C. E. Hmelo-Silver, S. R. Goldman, & P. Reimann (Eds.), *International handbook of the learning sciences* (pp. 511–519). New York: Routledge

Sawyer, R. K. (2006). *The Cambridge handbook of the learning sciences.* In Cambridge: Cambridge University Press.

Scardamalia, M., & Bereiter, C. (1994). Computer support for knowledge-building communities. *Journal of the Learning Sciences, 3*(3), 265–283.

Smith, B. K. (2004). Instructional systems and learning sciences: When universes collide. *Educational Technology, 44*(3), 20–25.

Stahl, G. (2006). *Group cognition: Computer support for building collaborative knowledge.* Cambridge, MA: MIT Press.

Stahl, G., Koschmann, T., & Suthers, D. (2014). Computer-supported collaborative learning: An historical perspective. In R. K. Sawyer (Ed.), *Cambridge handbook of the learning sciences, revised version* (pp. 479–500). Cambridge, UK: Cambridge University Press.

Strijbos, J.-W., Kirschner, P. A., & Martens, R. (Eds.). (2004). *What we know about CSCL and implementing it in higher education.* Boston, MA: Kluwer Academic Publishers.

Suthers, D. D., Lund, K., Rosé, C. P., Teplovs, C., & Law, N. (Eds.). (2013). *Productive multivocality in the analysis of group interactions.* New York: Springer.

Yoon, S. A., & Hmelo-Silver, C. E. (2017). What do learning scientists do? A survey of the ISLS membership. *Journal of the Learning Sciences.* Retrieved from http://dx.doi.org/10.1080/10508406.2017.1279546

3
Epistemic Cognition and Epistemic Development

Clark Chinn and William Sandoval

Epistemic Cognition and the Learning Sciences

Scholarship on topics such as epistemic cognition, epistemic beliefs, epistemic development, and epistemic practices has flourished in the past five decades (Chinn, Buckland, & Samarapungavan, 2011; Greene, Sandoval, & Bråten, 2016; Hofer & Pintrich, 1997; Kuhn, Cheney, & Weinstock, 2000; Sandoval, Greene, & Bråten, 2016). Much of this research has defined its scope as people's beliefs, stances, or theories related to knowledge and knowing.

In this chapter, we begin by discussing distinctive features of learning sciences (LS) approaches to epistemic cognition (EC). Second, we illustrate these approaches to epistemic cognition in the domain of science. Third, to illustrate the LS emphasis on the situativity of cognition within disciplines, we contrast EC in science with EC in history. Finally, we point to what we see as productive areas for new research.

Distinctive Features of Learning Sciences Approaches to Epistemic Cognition

In this section, we discuss distinctive features of learning scientists' approaches to epistemic cognition. Where appropriate, we contrast these features with the features of psychological work on EC, given that much EC research has been conducted by psychologists. We will not review psychological work in detail; current reviews can be found in Greene, Sandoval, and Bråten (2016). Taken collectively, the distinctive features of LS research on EC include: (1) emphasizing multidisciplinary research, (2) broadening the range of questions, (3) challenging normative assumptions, (4) a focus on practices, (5) the thoroughly social nature of EC, and (6) its situativity.

Multidisciplinary and Interdisciplinary Research

It is a definitional feature of the learning sciences to embrace multiple disciplinary approaches to investigating learning and thinking, including anthropology, psychology, sociology, computer science, epistemology, and the history and philosophy of specific disciplines (e.g., the sciences, mathematics, history). EC work may accordingly involve interdisciplinary teams (e.g., Goldman et al., 2016) or otherwise draw on scholarship from multiple disciplines. A number of the features discussed below derive in part from the multi- and interdisciplinarity of LS.

Broader Range of Questions

Psychological work has often analyzed epistemic cognition as people's beliefs related to the nature of knowledge and the processes of knowing (Hofer & Pintrich, 1997); the latter has typically been operationalized in terms of whether people believe that knowledge is sourced or justified by authority, experience, or some other source. Drawing on ideas from a broader range of disciplines, including philosophy, learning scientists have expanded the scope of epistemic cognition beyond this to focus on the many different practices (e.g., observation, scientific methods, community processes such as peer review) used by individuals and communities to create knowledge and a variety of other epistemic products (e.g., models, arguments, evidence) (Chinn et al., 2011).

Challenging Normative Assumptions

Learning scientists have challenged the explicit and implicit normative assumptions made in some mainstream psychological work on EC. For example, Chinn et al. (2011) and Bromme, Kienhues, and Porsch (2009) have questioned the frequent assumption that relying on "authorities" is a poor epistemic approach. Both point out that most human knowledge is derived from testimony from others; further, one cannot be an expert in all areas and so must rely on the expertise of others in their domains of expertise. Chinn and Rinehart (2016) noted that the common developmental assumption that realism (the view that one's theories or ideas correspond to what is really in the world) is unsophisticated is contradicted by the fact that many scientists and most philosophers adopt realist stances. Barzilai & Chinn (in press) have developed a normative analysis of the goals of epistemic education.

A Focus on Practices

Psychological work has often measured people's epistemic beliefs or stances using Likert-scale questionnaires or interviews with short vignettes (e.g., asking why experts differ on the safety of food additives; King & Kitchener, 1994). Some learning scientists (Samarapungavan, Westby, & Bodner, 2006; Sandoval, 2005) have argued that such general beliefs are too general and abstract to have much impact on people's practical thinking. Accordingly, much learning sciences research emphasizes epistemic practices as the appropriate focus of investigation. If one wants to understand, for example, why people make a jury decision, one must know how they evaluate, discuss, and integrate particular kinds of evidence that arise in the particular situation of a given trial.

By epistemic *practices*, we refer to socially normed activities that people carry out to accomplish epistemic aims such as developing evidence, arguments, theories, and so on. Practices are social, in the sense that the norms used to evaluate the practices and the products they produce are socially developed, shared, and applied by communities. Practices are also tightly intertwined with the material (e.g., the laboratories, equipment, chemicals, carefully bred laboratory animals, etc., in science) (Kelly, 2016).

Although some have interpreted epistemic practices in ways that minimize the role of metacognitive reflection (e.g., Berland et al., 2016), others have pointed to a role for metacognition, especially metacognition at the practical level of epistemic activity (e.g., Barzilai & Chinn, in press; Barzilai & Zohar, 2014). For example, people's specific metacognitive beliefs about whether and how biases and error can enter into scientific observations can affect critical choices about how to conduct observations (e.g., double blind studies, etc.).

Thoroughly Social Nature of Epistemic Cognition

Processes of creating and evaluating knowledge are thoroughly social (A. Goldman, 1999). This is obviously so in the case of scientists creating and publishing knowledge in teams, and evaluating each other's work in communities of critique. But it is also true of seemingly individual reasoning; for example, the individual evaluating information about medical treatments is relying on information provided by others.

Greene et al. (2016) distinguished three relevant levels of research in epistemic cognition: the individual, the individual in interaction, and the community/system level. At the individual level, researchers studying history classes might investigate how individual students draw conclusions about historical events using primary and secondary sources. At the individual-in-interaction level, researchers could study how students argue with each other, how their arguments influence later arguments and positions, how particular forms of argumentation spread in classes, and how collective norms for argumentation emerge and take shape. Some analyses would treat groups as the unit of analysis. At the level of community/system, investigators might examine the emergence of community norms that govern what is counted as a strong argument or a good mathematical solution and look at how these community norms are sustained or revised over time (Cobb, Stephan, & McClain, 2001).

Situativity

Learning scientists emphasize the situativity of EC, by which we mean that EC can vary (within the same person or group of people) from one situation to another. There are many dimensions of situations along which EC is situated; we note two exemplars. First, EC is not only discipline but even topic specific (Chinn et al., 2011). Engaging in epistemic practices on a topic requires deep, specific knowledge of that topic—not just theoretical knowledge (e.g., knowing cell processes and structures) but also methodological knowledge without which evidence cannot be evaluated or produced (such as accepted processes for preparing slides for electronic microscopy). Accordingly, people's epistemic judgments can be expected to differ from topic to topic (Elby & Hammer, 2001). Gottlieb and Wineburg (2012) demonstrated that religious historians use sharply different ways of thinking about historical texts on the biblical Exodus versus texts on the first American Thanksgiving.

Second, even within the same topic, epistemic practices of individuals and groups can vary sharply according to how the task is framed or introduced (Kienhues, Ferguson, & Stahl, 2016). Rosenberg et al. (2006) investigated eighth-grade science students working on answering the question "How are rocks formed?" They dramatically changed their epistemic approach to this task following a simple suggestion from the teacher to focus on what they know about rocks. This prompted the students to shift from making vocabulary lists to developing a causal story of how different kinds of rocks could form.

Hammer, Elby, and their colleagues have developed a resources-based model of EC to account for its situativity (e.g., Hammer & Elby, 2002). EC is composed of epistemological resources, "fine-grained pieces of epistemology that are sensitive to context in their activation" (Rosenberg et al., 2006). Examples of resources that can be activated flexibly in different situations are: "knowledge is stuff transferred from one person to another," "knowledge is fabricated stuff," "knowledge is accumulated as lists of facts," "knowledge involves causal stories," "knowledge can be created by imagining," and so on. Different clusters of resources are activated in different contexts.

Learning Sciences Methodologies

The focus on practices in LS means that EC is typically investigated not through questions about beliefs about knowledge in general or even beliefs about knowledge in disciplines but instead through providing people with practical reasoning tasks and analyzing their reasoning as they engage with these tasks—often in collaboration with peers. Methods involve detailed analyses of discourse and interactions, such as analyzing categories of epistemic discourse (Herrenkohl & Cornelius, 2013), examining the emergence of mathematical norms (Cobb et al., 2001), or using network analysis to understand how students share epistemic responsibility (Zhang, Scardamalia, Reeve, & Messina, 2009). Students require some inferencing to move from observed talk to conclusions about epistemic commitments and practices, but the tasks involve what learning scientists would regard as authentic epistemic activity.

Learning scientists do not completely eschew methodological approaches of other fields, such as interviews, but are likely to use rich tasks with multiple pieces of evidence or multiple documentary sources to afford the opportunity to engage in thinking that connects more deeply with disciplinary knowledge and that affords opportunities to look at variation in reasoning across different situations (e.g., Gottlieb & Wineburg, 2012).

Learning Sciences Approaches to Studying Epistemic Practice

Learning scientists recognize that epistemic practices and their development function across the levels of the community or system, the level of the individual in interaction, and the level of the individual as a cognitive agent (cf. Rogoff, 1995), as well as between microgenetic, sociogenetic, and ontogenetic scales of activity (Saxe, 2004). Levels vary in both the number of people who might constitute the unit of analysis and the temporal scale over which activity might be analyzed. As an example, we consider practices of explanation and argumentation in science.

Community/System Level

Scientific communities share and enforce criteria that govern acceptance of proffered explanations, and methods are expected to follow reliable processes established by the community. A wide range of processes and criteria operate at the level of the community, such as peer review and standards of critique and uptake of ideas. While there is broad consensus on some of these processes across the sciences, they are also differentially specified within fields (Knorr-Cetina, 1999). For example, while there is a broad view that controlled experiments are ideal for establishing causal relationships, there are many fields in which this form of experimentation is unavailable, leading those fields to develop alternative standards for justifying causal claims.

Promoting communities within school classrooms that pursue similar aims and develop versions of these reliable processes requires aims focused on both construction and critique of explanations (Ford & Forman, 2006). This includes opening up all aspects of practice, from the questions investigated to the means for investigating them, to the same contestation and stabilization seen in professional science (Manz, 2015). Manz pointed out that, in classrooms where aspects of practice naturally become contested, argumentation emerges as a functional practice for stabilizing resolutions.

Learning sciences research is rich with examples of the development of such classroom communities. The Cheche Konnen project explicitly drew connections between children's everyday life and home language (Haitian creole) and more formally scientific ways of talking and thinking (Rosebery, Warren, & Conant, 1992). Another example is Lehrer and Schauble's long-running work in which students encounter problems of measurement and modeling they must work to resolve, and in doing so develop shared classroom practices and standards for building and evaluating models (Lehrer & Schauble, 2000, 2004).

A key feature of such projects in the learning sciences is the emergence of accountability to collective norms of practice. In professional science communities, arguments function to resolve real disagreements. In classrooms, for authentic forms of argument to emerge, disagreements must be legitimate. Such legitimacy is a consequence of students being supported to become active authors of the epistemic aims and practices pursued in the classroom.

Interactional Level

A great deal of LS work on scientific argumentation and explanation focuses on individuals in interaction, at least in part because this level of analysis is where practice is most easily seen. Community norms and aims are manifested through particular interactions, and versions of collective practice are understood through analysis of how they play out in specific interactions among community members. A key indicator, in fact, of how students learn practices of argumentation is to analyze how interactions change over time, as participants appropriate versions of practice (Lehrer & Schauble, 2004; Rosebery et al., 1992; Ryu & Sandoval, 2012).

An analysis of scientists in interaction reveals the broad array of social, technical, and semiotic resources with which they interact, and through which scientific knowledge is constructed, typically with difficulty and uncertainty (Pickering, 1995). Argumentative interactions in classrooms similarly rely on the social, material, and semiotic resources available. These resources are used in relation to community level norms and practices. Learning sciences analyses of interaction show both how children are often attuned to the affordances of material and symbolic resources in making arguments, and the importance of support for making meaning from them (e.g., Engle & Conant, 2002). Interaction analyses show that students' practices are sensitive to how teachers frame the purpose and nature of instructional activity (Rosenberg et al., 2006). Efforts to build tools specifically to support argumentation also show how those tools structure both the practice and the products of argumentation (Bell & Linn, 2000; Clark & Sampson, 2007; Iordanou & Constantinou, 2014; Sandoval & Reiser, 2004). More broadly, structured opportunities for students to engage in science practices that problematize activities such as measurement promote a "grasp of practice" (Ford, 2005).

Individual Level

The cognitive practices taken up by individual scientists are situated in those practices used within particular groups and communities of scientists. Individuals learn argumentative practices through apprenticeship into the work of their specific field, rather than through some direct instruction in the nature of disciplinary arguments (Goodwin, 1994; Longino, 1990). Learning sciences research similarly shows that the cognitive practices learned by students are tied to the instructional activities in which they take part. First graders learn to identify sources of error and uncertainty through designing their own experiments (Metz, 2011). Students who develop models over the course of a school year develop epistemic conceptions of tentativeness and measurement uncertainty directly traceable to their own efforts (Lehrer et al., 2008). Middle school students identify a range of criteria for evaluating explanatory models that seem tied to their schooling experiences (Pluta, Chinn, & Duncan, 2011). Elementary children's improvement in justification practices is directly tied to persistent focus on justification in science lessons (Ryu & Sandoval, 2012).

LS research on practices of explanation and argumentation in science shows that children from a very early age display cognitive capabilities consonant with professional science practices, and that such early competencies can be extended and refined through appropriately structured instruction. A good deal is now known of the features of such instruction for a range of science practices. A number of questions remain open concerning how understanding of practices generalize, how generalization is tied to contexts of learning, and how students perceive relations between the science they do in school, professional science, and science as encountered in everyday activity.

Comparisons with Other Disciplines: History in Contrast to Science

LS approaches are of course applicable to EC within any discipline. As discussed earlier, learning scientists generally view EC as varying across disciplines and topics. For example, Goldman et al. (2016) presented a detailed analysis of reading practices across literary reading, science, and history, including differences along epistemic dimensions. In the following sections, we analyze some specific differences in epistemic practices between science and history.

To illustrate an additional LS approach to EC, we use a model developed by Chinn et al. (2014), the AIR model, as a lens for this analysis. This model specifies three principal components of EC: (1) aims and values—the goals that people set in particular situations (e.g., to know, understand, develop a model, etc.) and what they value (e.g., valuing knowledge that solves societal problems); (2) ideals—the criteria or standards applied to evaluate epistemic products such as models and evidence; and (3) reliable epistemic processes—the processes that are used to create knowledge and other epistemic products (e.g., processes of testimony, observation, statistical analysis, argumentation, peer review, etc.). We discuss differences in aims, ideals, and reliable processes across science and history.

Epistemic Aims

A diversity of epistemic aims prevails in both disciplines. LS approaches to EC assume a diversity of aims. Scientists aim to create general models, establish laws, test the reliability of methods, or estimate parameters such as Planck's constant. Historians seek to establish the authenticity of a particular document, develop a historical narrative, understand the perspective of historical actors, or provide a broad explanation of events (Tucker, 2011).

Some aims appear to be unique to one discipline. Developing models is central to the practice of science, but models do not figure in discussions of historiography (see Tucker, 2011). Conversely, some movements in historiography uniquely emphasize the aim of constructing historical narratives with plots, settings, characters, and literary devices (Green & Troup, 1999).

When aims look similar at a first glance, deeper differences may appear. Scientists aim to develop general explanations (the general causes of a disease) as well as particular ones (how a particular person developed a disease). Although some historians have sought general explanations in history (e.g., how economic systems change), others have objected to such explanations, given the particularity of historical events, so that explanations of historical events must always be particular (Tucker, 2011).

Epistemic Ideals

Some epistemic ideals (or criteria) for evaluating epistemic products seem relatively specific to one field or another. In science, philosophers have argued that scientists evaluate explanations using ideals such as fit with evidence, simplicity, and making new predictions of future events (see Pluta et al., 2011). Historians would likely agree that fit with evidence is an ideal in their field (although the nature of evidence across fields differs vastly), but the complex, rich, contextual descriptions prized by many historians are not simple, and most contemporary historians do not hold their histories accountable to predicting future events (Tucker, 2011). Similarly, contextualization (weaving the rich chronological, social, and personal contexts surrounding events) is an ideal that appears to be distinctive to history (Wineburg, 2001).

Other ideals that appear to be the same on the surface may differ very substantively, such as corroboration. In science, marks of corroboration might include statistical meta-analyses, which would not figure in evaluation of whether historical accounts are well corroborated. In contrast, historical corroboration would involve careful textual comparison of primary source documents (Goldman et al., 2016).

Reliable Processes for Achieving Epistemic Aims

Scientists and historians use a variety of processes to reliably achieve their epistemic aims. Some processes are shared: some scientists and some historians use statistical analyses, though the problems faced by historians with missing data mean that historians need to use approaches not needed by scientists with more complete data (Howell & Prevenier, 2001). Other processes differ sharply. Scientists conduct controlled experiments, make live observations of ongoing behaviors and activities (such as animal behavior or chemical reactions), and use techniques for combining evidence such as meta-analysis; all of these appear to be absent in historical inquiry. In contrast, historians engage in processes such as taking historical actors' perspectives through empathy (Breisach, 2007) and developing extensive counterfactual scenarios to support claims (Weinryb, 2011).

Sourcing is central to both history and science. There is a difference, however. Both historians and scientists evaluate the trustworthiness of their peers as "secondary sources"—i.e., historians evaluate other historians, and scientists evaluate other scientists (Collins, 2014). Historians also evaluate the trustworthiness of primary sources who produce the diaries and other documents that are the primary data for their research. However, scientists do not typically evaluate the trustworthiness of their "primary sources"—the lab assistants who report results of research—except in rare cases of suspected fraud. Thus, the processes used by historians to evaluate primary sources (e.g., evaluating human motivations, biases, contextual positioning, and so on) are typically not salient when scientists evaluate their primary sources (lab assistants, etc.), who are assumed to use procedures that render these personal factors irrelevant.

Disciplinary Differences in Inquiry by Learners

To this point, we have noted differences between inquiry by *experts* in science versus history. These differences also appear in research with *learners* engaged in the practices of scientists and historians. As noted earlier, Goldman et al. (2016) developed a detailed analysis of goals for instruction based on analyses of disciplinary differences in epistemic practices and have developed efficacious curricula based on their analysis. Herrenkohl and Cornelius (2013) described class interactions in curricula developed for history and science that indicated that fifth and sixth graders can learn to develop distinct aims, justificatory practices, and processes for constructing knowledge across the two disciplines.

Conclusion and Implications

The distinctive features of LS research on EC suggest directions for productive new research to advance the field (see also Sandoval et al., 2016; Barzilai & Chinn, in press). In accord with the value LS places on interdisciplinary and multidisciplinary research, EC research would benefit by more extensive collaborations, e.g., by philosophers working with psychologists and educators. (2) LS researchers should fully explore the broader range of questions that have been opened up by recent LS scholarship, such as a broader range of epistemic aims, deeper explorations of practices used productively by experts and laypeople, and so on. (3) The LS work on normative assumptions should be expanded into detailed normative accounts that can be used to establish productive goals of epistemic education. (4) Although LS researchers are leaders in investigating practical and social aspects of EC, more research is needed particularly at the individual-in-interaction level and the community/systems level, both to understand effective modes of knowledge production by expert communities and to understand how to promote learning in schools and other settings. LS researchers could also investigate further the proper roles of metacognition in sophisticated EC. (5) Finally, research on EC would benefit from more systematic analyses of the ways in which context affects EC, as well as how people can learn to be effective thinkers across multiple contexts.

Learning scientists should also be leaders in designing effective learning environments to promote epistemic growth. Such work would systematically examine implications of theories of EC for setting goals for education and then for achieving these goals, using the field's understanding of effective scaffolding, methods of collaborative learning, and other features of design to promote achievement of these goals.

Further Readings

Chinn, C. A., Buckland, L. A., & Samarapungavan, A. (2011). Expanding the dimensions of epistemic cognition: Arguments from philosophy and psychology. *Educational Psychologist, 46*, 141–167.
A model of EC grounded in a broad review of philosophical work. It is the precursor to the AIR model (Chinn et al., 2014) discussed in this chapter, and the article provides readers with a broad range of

philosophical references that can be consulted. It also argues for a strong contextual-sensitivity of EC and for social components of EC.

Goldman, S. R., Britt, M. A., Brown, W., Cribb, G., George, M., Greenleaf, C., et al. (2016). Disciplinary literacies and learning to read for understanding: A conceptual framework for disciplinary literacy. *Educational Psychologist, 51*, 219–246.

An explanatory account of differences in the critical literacy practices across three disciplines—literary reading, science, and history—along dimensions including epistemic dimensions of epistemology; inquiry practices and strategies of reasoning; and forms of information. The article exemplifies interdisciplinary scholarship and points to important aspects of disciplinary situativity in EC.

Manz, E. (2015). Representing student argumentation as functionally emergent from scientific activity. *Review of Educational Research, 85*(4), 553–590.

A review of research on the epistemic practice of argumentation, emphasizing the embedding of argumentation in the activity systems of communities and the central role of community norms for argumentation practice. Emphasizing the material and representational aspects of science, Manz shows that it is necessary for students to find critical features of investigations to be genuinely problematic to engage in argumentation to stabilize their scientific work.

Rosenberg, S., Hammer, D., & Phelan, J. (2006). Multiple epistemological coherences in an eighth-grade discussion of the rock cycle. *Journal of the Learning Sciences, 15*(2), 261–292.

A detailed account of how the resources theory of Hammer and Elby (2002) can be applied to explain the epistemic practices of eighth graders discussing the rock cycle. This paper illustrates both contrastive analyses of cases and analyses of discourse, and it provides a helpful elaboration of what resources are and how they figure in two kinds of epistemic practices.

Sandoval, W. A., Greene, J. A., & Bråten, I. (2016). Understanding and promoting thinking about knowledge: Origins, issues, and future directions of research on epistemic cognition. *Review of Research in Education, 40*, 457–496.

A comprehensive review of the origins of research on epistemic cognition and the conflicts and convergences among different traditions of scholarship. They point to the need to comparatively test competing models of epistemic cognition, pursue methodological nuance, and connect analyses of EC across settings and time.

NAPLeS Resources

Chinn, C., *Epistemic cognition* [Webinar]. In *NAPLeS Video series*. Retrieved October 19, 2017, from http://isls-naples.psy.lmu.de/intro/all-webinars/chinn_all/index.html

Chinn, C., *Interview about epistemic cognition* [Video file]. In *NAPLeS video series*. Retrieved October 19, 2017, from http://isls-naples.psy.lmu.de/video-resources/interviews-ls/chinn/index.html

Sandoval, W. 15 minutes about situating epistemological development [Video file]. In NAPLeS video series. Retrieved October 19, 2017, from http://isls-naples.psy.lmu.de/video-resources/guided-tour/15-minutes-sandoval/index.html

Sandoval, W., Situating *epistemic development* [Webinar]. In *NAPLeS Video series*. Retrieved October 19, 2017, from http://isls-naples.psy.lmu.de/intro/all-webinars/sandoval_all/index.html

References

Barzilai, S., & Chinn, C. A. (in press). On the goals of epistemic education: Promoting apt epistemic performance. *Journal of the Learning Sciences*.

Barzilai, S., & Zohar, A. (2014). Reconsidering personal epistemology as metacognition: A multifaceted approach to the analysis of epistemic thinking. *Educational Psychologist, 49*(1), 13–35.

Bell, P., & Linn, M. C. (2000). Scientific arguments as learning artifacts: Designing for learning from the web with KIE. *International Journal of Science Education, 22*(8), 797–817.

Berland, L. K., Schwarz, C. V., Krist, C., Kenyon, L., Lo, A. S., & Reiser, B. J. (2016). Epistemologies in practice: Making scientific practices meaningful for students. *Journal of Research in Science Teaching, 53*, 1082–1112.

Breisach, E. (2007). *Historiography: Ancient, medieval, and modern*. Chicago, IL: University of Chicago Press.

Bromme, R., Kienhues, D., & Porsch, T. (2009). Who knows what and who can we believe? Epistemological beliefs are beliefs about knowledge (mostly) to be attained from others. In L. A. Bendixen & F. C. Feucht (Eds.), *Personal epistemology in the classroom: Theory, research, and implications for practice* (pp. 163–193). Cambridge, UK: Cambridge University Press.

Chinn, C. A., & Rinehart, R. W. (2016). Epistemic cognition and philosophy: Developing a new framework for epistemic cognition. In J. A. Greene, W. A. Sandoval, & I. Bråten (Eds.), *Handbook of epistemic cognition* (pp. 460–478). New York: Routledge.

Chinn, C. A., Buckland, L. A., & Samarapungavan, A. (2011). Expanding the dimensions of epistemic cognition: Arguments from philosophy and psychology. *Educational Psychologist, 46*, 141–167.

Chinn, C. A., Rinehart, R. W., & Buckland, L. A. (2014). Epistemic cognition and evaluating information: Applying the AIR model of epistemic cognition. In D. Rapp & J. Braasch (Eds.), *Processing inaccurate information: Theoretical and applied perspectives from cognitive science and the educational sciences* (pp. 425–453). Cambridge, MA: MIT Press.

Clark, D. B., & Sampson, V. (2007). Personally-seeded discussions to scaffold online argumentation. *International Journal of Science Education, 29*(3), 253–277.

Cobb, P., Stephan, M., & McClain, K. (2001). Participating in classroom mathematical practices. *Journal of the Learning Sciences, 10*, 113–163.

Collins, H. (2014). *Are we all scientific experts now?* Cambridge, UK: Polity.

Elby, A., & Hammer, D. (2001). On the substance of a sophisticated epistemology. *Science Education, 85*, 554–567.

Engle, R. A., & Conant, F. R. (2002). Guiding principles for fostering productive disciplinary engagement: Explaining an emergent argument in a community of learners classroom. *Cognition and Instruction, 20*(4), 399–483.

Ford, M. (2005). The game, the pieces, and the players: Generative resources from two instructional portrayals of experimentation. *Journal of the Learning Sciences, 14*(4), 449–487.

Ford, M., & Forman, E. A. (2006). Redefining disciplinary learning in classroom contexts. *Review of Research in Education, 30*, 1–32.

Goldman, A. I. (1999). *Knowledge in a social world*. Oxford, England: Oxford University Press.

Goldman, S. R., Britt, M. A., Brown, W., Cribb, G., George, M., Greenleaf, C., et al. (2016). Disciplinary literacies and learning to read for understanding: A conceptual framework for disciplinary literacy. *Educational Psychologist, 51*, 219–246.

Goodwin, C. (1994). Professional vision. *American Anthropologist 96*(3), 606–633.

Gottlieb, E., & Wineburg, S. (2012). Between *veritas* and *communitas*: Epistemic switching in the reading of academic and sacred history. *Journal of the Learning Sciences, 21*, 84–129.

Green, A., & Troup, K. (Eds.). (1999). *The houses of history: A critical reading in twentieth century history and theory*. New York: New York University Press.

Greene, J. A., Sandoval, W. A., & Bråten, I. (Eds.). (2016). *Handbook of epistemic cognition*. New York: Routledge.

Hammer, D., & Elby, A. (2002). On the form of a personal epistemology. In B. K. Hofer & P. R. Pintrich (Eds.), *Personal epistemology: The psychology of beliefs about knowledge and knowing* (pp. 169–190). Mahwah, NJ: Erlbaum.

Herrenkohl, L. R., & Cornelius, L. (2013). Investigating Elementary Students' Scientific and Historical Argumentation. *Journal of the Learning Sciences, 22*(3), 413–461.

Hofer, B. K., & Pintrich, P. R. (1997). The development of epistemological theories: Beliefs about knowledge and knowing and their relation to learning. *Review of Educational Research, 67*, 88–140.

Howell, M., & Prevenier, W. (2001). *From reliable sources: An introduction to historical methods*. Ithaca, NY: Cornell University Press.

Iordanou, K., & Constantinou, C. P. (2014). Developing pre-service teachers' evidence-based argumentation skills on socio-scientific issues. *Learning and Instruction, 34*, 42–57.

Kelly, G. J. (2016). Methodological considerations for the study of epistemic cognition in practice. In J. A. Greene, W. A. Sandoval & I. Bråten (Eds.), *Handbook of epistemic cognition* (pp. 393–408). New York: Routledge.

Kienhues, D., Ferguson, L., & Stahl, E. (2016). Diverging information and epistemic change. In J. A. Greene, W. A. Sandoval & I. Bråten (Eds.), *Handbook of epistemic cognition* (pp. 318–330). New York: Routledge.

King, P. M., & Kitchener, K. S. (1994). *Developing reflective judgment: Understanding and promoting intellectual growth and critical thinking in adolescents and adults*. San Francisco, CA: Jossey-Bass.

Knorr-Cetina, K. (1999). *Epistemic cultures: How the sciences make knowledge*. Cambridge, MA: Harvard University Press.

Kuhn, D., Cheney, R., & Weinstock, M. (2000). The development of epistemological understanding. *Cognitive Development, 15*, 309–328.

Lehrer, R., & Schauble, L. (2000). Developing model-based reasoning in mathematics and science. *Journal of Applied Developmental Psychology*, *21*(1), 39–48.

Lehrer, R., & Schauble, L. (2004). Modeling natural variation through distribution. *American Educational Research Journal*, *41*(3), 635–679.

Lehrer, R., Schauble, L., & Lucas, D. (2008). Supporting development of the epistemology of inquiry. *Cognitive Development*, *23*(4), 512–529.

Longino, H. (1990). *Science as social knowledge*. Princeton, NJ: Princeton University Press.

Manz, E. (2015). Representing student argumentation as functionally emergent from scientific activity. *Review of Educational Research*, *85*(4), 553–590.

Metz, K. E. (2011). Disentangling robust developmental constraints from the instructionally mutable: Young children's epistemic reasoning about a study of their own design. *Journal of the Learning Sciences*, *20*(1), 50–110.

Pickering, A. (1995). *The mangle of practice: Time, agency, and science*. Chicago: University of Chicago Press.

Pluta, W. J., Chinn, C. A., & Duncan, R. G. (2011). Learners' epistemic criteria for good scientific models. *Journal of Research in Science Teaching*, *48*(5), 486–511.

Rogoff, B. (1995). Observing sociocultural activity on three planes: participatory appropriation, guided participation, and apprenticeship. In J. V. Wertsch, P. d. Rio, & A. Alvarez (Eds.), *Sociocultural studies of mind* (pp. 139–164). Cambridge: Cambridge University Press.

Rosebery, A. S., Warren, B., & Conant, F. R. (1992). Appropriating scientific discourse: Findings from language minority classrooms. *Journal of the Learning Sciences*, *2*(1), 61–94.

Rosenberg, S., Hammer, D., & Phelan, J. (2006). Multiple epistemological coherences in an eighth-grade discussion of the rock cycle. *Journal of the Learning Sciences*, *15*(2), 261–292.

Ryu, S., & Sandoval, W. A. (2012). Improvements to elementary children's epistemic understanding from sustained argumentation. *Science Education*, *96*(3), 488–526.

Samarapungavan, A., Westby, E. L., & Bodner, G. M. (2006). Contextual epistemic development in science: A comparison of chemistry students and research chemists. *Science Education*, *90*, 468–495.

Sandoval, W. A. (2005). Understanding students' practical epistemologies and their influence on learning through inquiry. *Science Education*, *89*, 634–656.

Sandoval, W. A. (2016). Disciplinary insights into the study of epistemic cognition. In J. A. Greene, W. A. Sandoval, & I. Bråten (Eds.), *Handbook of Epistemic Cognition* (pp. 184–194). New York: Routledge.

Sandoval, W. A., & Reiser, B. J. (2004). Explanation-driven inquiry: Integrating conceptual and epistemic supports for science inquiry. *Science Education*, *88*, 345–372.

Sandoval, W. A., Greene, J. A., & Bråten, I. (2016). Understanding and promoting thinking about knowledge: Origins, issues, and future directions of research on epistemic cognition. *Review of Research in Education*, *40*, 457–496.

Saxe, G. B. (2004). Practices of quantification from a sociocultural perspective. In A. Demetriou & A. Raftopoulos (Eds.), *Cognitive developmental change: Theories, models, and measurement* (pp. 241–263). Cambridge, UK: Cambridge University Press.

Strømsø, H. I., Bråten, I., & Samuelstuen, M. S. (2008). Dimensions of topic-specific epistemological beliefs as predictors of multiple text understanding. *Learning and Instruction*, *18*(6), 513–527.

Tucker, A. (Ed.). (2011). *A companion to the philosophy of history and historiography*. Malden, MA: Wiley-Blackwell.

Weinryb, E. (2011). Historical counterfactuals. In A. Tucker (Ed.), *A companion to the philosophy of history and historiography* (pp. 109–119). Malden, MA: Wiley-Blackwell.

Wineburg, S. (2001). *Historical thinking and other unnatural acts: Charting the future of teaching the past*. Philadelphia: Temple University Press.

Zhang, J., Scardamalia, M., Reeve, R., & Messina, R. (2009). Designs for collective cognitive responsibility in knowledge-building communities. *Journal of the Learning Sciences*, *18*(1), 7–44.

4
Cognitive and Sociocultural Perspectives on Learning
Tensions and Synergy in the Learning Sciences

Joshua A. Danish and Melissa Gresalfi

Introduction

Since its inception as a field, the interdisciplinary nature of the Learning Sciences has led researchers to leverage, develop, and refine a wide variety of theories to better understand how to predict and support learning across diverse contexts. At the heart of this process has been a debate—sometimes implicit, and often quite explicit—between those who subscribe to so-called cognitive versus sociocultural theories of learning. Broadly speaking, cognitive theories focus on the *mental processes* of the individual learner, while sociocultural theories focus on the *participation* of learners in the social practices within a particular context. A number of well-known articles and chapters (Anderson, Reder, & Simon, 1996; Greeno, 1997; Greeno & Engeström, 2014; Sfard, 1998) have addressed the differences between these two approaches, often highlighting the perceived strengths or weaknesses of one approach over the other. Our goal is not to reproduce those debates. Rather, we believe that a defining characteristic of the Learning Sciences as a field lies in how scholars have used these tensions to advance theories of learning, and to demonstrate their utility in understanding and designing for learning. Through this process, scholars have not only advanced the respective fields of cognitive and sociocultural theory, but have also demonstrated the overlaps and synergies that exist between the perspectives. Our goal in this chapter is to briefly summarize the unique contributions of each theoretical perspective and how they have shaped our perception as a field, and then to describe what we view as promising synergies that have arisen. In doing so, we are influenced by work that has highlighted that experience within a discipline involves refining one's perception, necessarily shifting what one notices or disregards in the world (Goodwin, 1994; Sherin & van Es, 2009; Stevens & Hall, 1998). We want to explore similarly how adopting cognitive, sociocultural, or mixed theoretical frameworks may lead scholars to look at or ignore key aspects of learning in context. We begin with a brief summary of the core theoretical differences before focusing on how we see these theoretical assumptions have been taken up in research and design.

Core Theoretical Assumptions

Below we present some generalizations regarding the core assumptions of each theory and its application to practice. We recognize that a great deal of work in the Learning Sciences moves beyond these generalizations in productive ways, and also blurs these lines. Nevertheless, we see the noted patterns as driving a great deal of debate over the last few decades, and thus present them here.

For each theory we present its approach to knowing and learning, transfer, and motivation (Greeno, Collins, & Resnick, 1996).

Cognitive Theories

We use the term "cognitive" to refer to theories that aim to model mental processes—the perception, encoding, storage, transformation, and retrieval of information—within individual minds. Scholarship in this space has been referred to generally as cognitive science and includes schema theory, information processing, and constructivism,[1] as well as more recent work within cognitive neurosciences. These approaches share a focus on developing empirically testable models, often reminiscent of computer architecture, which can explain and predict cognitive processes. As a result, the focus is commonly on how an individual mind works. In fact, many early studies within this space focused on the individual to the exclusion of all else, typically treating the environment solely as a "variable" to be controlled. However, work over the last few decades has increasingly addressed how cognition occurs within rich environments, recognizing that knowledge impacts our perception as well as our actions, thus shaping our engagement with the environment, which is also continually changing and thus triggering new responses. Furthermore, scholarship within the field of embodied cognition has been particularly focused on exploring the role of the body as a source of knowledge within the environment (Alibali & Nathan, this volume).

Knowing. Broadly speaking, cognitive approaches view knowledge as the representation of information within an individual mind. Cognition, from this perspective, is the manipulation, transformation, and retrieval of these representations. The distinctions between specific cognitive theories lie in how knowledge is represented and transformed. Models of cognition also predict how the processes of representing and transforming knowledge are visible within experimental conditions. One of the strengths of the cognitive approach lies in the fact that these different models of cognition allow researchers to make explicit and fine-grained predictions about how humans will perform in particular problem-solving or learning situations. In fact, the ability to measure knowledge and cognitive performances is central to the cognitive tradition. However, rather than being measured directly, knowledge is inferred from observable behaviors connected by models of the mind.

Transfer. Transfer is the use of knowledge in a new situation, different than where it was originally acquired. Traditionally, cognitive approaches to transfer contend that knowledge has to be represented in a suitably abstract manner to be applied in multiple situations (Day & Goldstone, 2012). The similarity between these situations involves mapping features of the original situation to the new context (Reed, 2012). More recently, approaches within the cognitive tradition have noted that these mappings do not involve only static concepts, but also processes and approaches which may be used to solve problems (Day & Goldstone, 2012). Broadly speaking, however, cognitive approaches to transfer focus on how information has been represented within the individual mind, and whether this representation affords the use of this information in new contexts.

Motivation. Within cognitive traditions, motivation involves the internal states and drives that predict whether one approaches or avoids a situation. The theories of motivation that have developed from this perspective are wide-ranging and diverse, but generally share a focus on how an individual feels (about herself, her abilities, about the situation), what the individual desires (her goals, values), how those fit together, and how they respond to environmental characteristics (Wigfield & Eccles, 2000). For example, when students attempt to solve a science problem and succeed, they gain knowledge regarding their ability with respect to that class of problems. Motivation is thus informed by their awareness of how challenging this kind of problem is for them, as well as the likelihood of overcoming that challenge. A key assumption that underlies these theories is that motivation is an individual trait that is tied to existing individual interests, that has some stability, and can be investigated independently from contexts. For example, motivation is commonly measured by surveys or questionnaires that ask respondents to rank their relative agreement with a set of statements. These questionnaires often mention particular contexts (for example, "mathematics" in general, or "math class") but do not examine motivation in relation to those contexts.

Sociocultural Theories

The class of theories that we term "sociocultural" include perspectives that, at their core, consider human activity to be inseparable from the contexts, practices, and histories in which activity takes place. From this perspective, studies of learning must focus beyond the individual to include the context in which the individual is interacting. There are myriad theories that fall into this category, the most well-known of which is called "sociocultural" theory, but also within this category are situated cognition, cultural-historical activity theory, social constructivism, and some versions of distributed cognition. Although there are distinctions among these theories, both in their histories and in their specific foci, they share more commonalities than differences, particularly when contrasted with cognitive theories.

Knowing. Sociocultural perspectives generally take a sociohistorical stance to knowing (Case, 1996; Engeström, 1999), assuming that the origins of knowledge and the processes of engaging knowledge stem from the cultural and historical practices in which the individual is immersed. This means that *how* one comes to know something is inseparable from *what* one ultimately comes to know. Across sociocultural perspectives, few doubt that language, tools, social categories, or histories influence the ways we see and experience the world. Indeed, this assumption is central not only to sociocultural theories, but also to many cognitive theories, seen, for example, in the claim that the structure of schema influences perception of new information. However, the implications of the focus on *inseparability* of person and context is unique to sociocultural theories, leading, for example, to skepticism regarding the generalizability of research that takes place primarily in rarified laboratory environments. Sociocultural theorists argue that each context (including experiments) is unique in its own right, and experiences or findings may not apply to other contexts (Lave, 1980). Contexts are richly theorized and complex places that include histories and cultures that frame what one is expected or entitled to do, the meaning that is made of those actions, and how those actions are mediated by artifacts, people, and motives (Engeström, 1999). The core assumption is that cognition and knowing are a *joint accomplishment* between the individual and the rich context in which she is participating. Furthermore, due to the centrality of activity contexts in explaining knowing and learning, sociocultural theorists believe it is valuable to explore how they came to be, and how they may transform over time.

Transfer. Sociocultural theories of transfer often explicitly recognize the fact that transfer, as defined by cognitive traditions, is really hard to find. However, they note that human activity is full of examples of transfer, as we routinely move from situation to situation with little effort or challenge. Thus, the question becomes one of accounting for this cross-situational fluidity. To answer this question, sociocultural theories of transfer broaden the unit of analysis beyond the individual to include the contexts in which information is engaged. Although specifics differ, theorists who have written about transfer from a sociocultural perspective focus on: (1) the practices that are present in the learning situation; (2) the participation of individuals with those practices; (3) the potential overlap between the transfer context and the learning context (Lobato, 2012). The paired focus on individual participation in relation to context and the overlapping practices in the transfer context are consistent with the different assumptions about learning that are made by sociocultural theories, specifically that whether or what the individual does is only part of the ultimate activity.

Motivation. When exploring motivation, sociocultural theories tend to move away from considering individuals' goals, desires, and confidence independently, and instead consider the ways that activities and practices frame participation and human agency such that people act in more or less motivated ways (Gresalfi, 2009; Nolen, Horn, & Ward, 2015). From this perspective, motivation is seen as both an individual and collective endeavor: the behavior of pursuing or avoiding an activity is co-constructed between the opportunities in the environment and the individual's participation with those practices. Central to these kinds of analyses is the claim that people are not motivated or unmotivated, but rather act in motivated or unmotivated ways in relation to the practices of the context. This shift in the unit of analysis requires examining not whether or how to make *people* more motivated, but rather to consider how to reform practices and contexts to invite engaged and motivated participation. Furthermore, motivation shifts to being a mediator that may shape how people participate instead of simply informing whether they will (Engeström, 1999).

Tensions and Synergy in Theoretical Assumptions

Cognitive and sociocultural theoretical perspectives make different assumptions about the world and human activity. Cognitive perspectives are critiqued for their focus on individual characteristics within experimental contexts, thus missing or ignoring details that reflect the real-world links between individuals and their context. Sociocultural perspectives are critiqued for their focus on context, making it difficult to produce any systematic, actionable, and generalizable results; they often lose the individual due to the focus on the collective. For many, these differences are irreconcilable. However, there are also scholars who view these different theories as an important starting point to build upon the models that are central to cognitive science while also accounting for the importance of context and social-historical issues that are central to sociocultural theorists.

In line with the tension noted above, traditional cognitive approaches typically aim to refine models of how individual students understand and learn core disciplinary concepts. In contrast, sociocultural approaches focus on the social environment that supports and inhibits students' engagement with the discipline. However, the two perspectives can come together in the same work. For example, a project by Enyedy and colleagues (Enyedy, Danish, & Fields, 2011) explored teaching the mathematics of central tendency to traditionally underrepresented middle school students in the Los Angeles, CA area. The authors began with extant models of student cognition related to mean, median, and mode as a starting point (cf. Bakker & Gravemeijer, 2003; Konold & Higgins, 2002; Makar, Bakker, & Ben-Zvi, 2011). Enyedy et al. (2011) also used culturally relevant pedagogy (Ladson-Billings, 1995) as a framework for adapting their instructional design to focus on students' use of mathematics to support argumentation using data that were relevant to their lives (such as the presence of graffiti in their neighborhood, or violence in local parks). Enyedy et al.'s (2011) focus was thus on how students might learn normative mathematical concepts *while* engaging in difficult but locally meaningful questions and arguments. The results are thus tied more closely to the kinds of generalizable mathematical conceptions favored in prior cognitive work, while also attending to important issues of context that are valued by sociocultural theorists.

Data Collection and Methods

The data collection and analytic methods that are leveraged by theorists must be tightly coupled with the questions that are posed, and thus it is often the case that cognitive and sociocultural theorists use different analytical methods. The distinction between the perspectives generally falls along the lines of debates between quantitative and qualitative methodologies, a conversation not revisited here. However, with respect to developing and contributing to theories of learning, the distinctions between the questions posed, the methods used, and, ultimately, the claims that are made, matter, both in terms of what the field learns and, ultimately, what kinds of questions get attention.

Cognitive Approaches

Due to the focus on empirically testable models of cognition and learning, studies within the cognitive tradition frequently contain measures that allow for comparisons between people and across time points such as surveys or standardized assessments. These measures can then be easily quantified so that parametric statistical methods can be employed. In this strand of research, scholars are often interested in making claims about causality, to link specific activities or events to models of cognition and learning. As a result, experimental designs that contrast intervention and control groups are quite common and, indeed, considered by some to be the gold standard of "scientific" research (Feuer, Towne, & Shavelson, 2002).

Sociocultural critiques of these approaches have noted that they are often conducted in settings that lack ecological validity (e.g., laboratory experiments and interviews), and that can oversimplify important interactional and cultural dimensions. The very same assumptions that allow for these kinds of statistical inferences are theoretically problematic in that they do not allow for the messy interactions between individuals and their environment; quantifying results can inadvertently gloss the role of interaction and the research setting in producing those results.

Sociocultural Approaches

In contrast, many sociocultural studies rely heavily upon qualitative methodologies including discourse and interaction analysis, interviewing, and ethnographies. The goal in employing these approaches is to understand learning as continually mediated by the local activity system, which is in turn continually in transition. Key concepts that are treated as "variables" are considered to be dynamic and locally produced within the sociocultural tradition. For example, sociocultural theorists note that culture is not static, and should not be treated as such (Gutiérrez & Rogoff, 2003). Rather, culture is continually created and transformed in the moment, as individuals contribute to and are impacted by their cultural milieu. Likewise, key mediators of activity, including tools, classroom practices, language, and students' relations, are analyzed along with student participation.

These theoretical assumptions therefore frequently lead sociocultural theorists to focus on qualitative analyses which allow for a deeper look at how a specific set of participants engages within their local context. The highly localized nature of these analyses is often what draws critique from those in the cognitive tradition who are skeptical that findings will generalize. Cognitive theorists also frequently note that core theoretical concepts are not effectively operationalized in this tradition, remaining vague and underspecified as a result of looking for them as produced in interaction rather than identifying them a priori.

A Synergy in Methods

In short, cognitive approaches frequently aim to collect systematic, generalizable, and quantifiable data from controlled environments, whereas sociocultural theorists place a higher premium upon ecological validity and rely more heavily upon qualitative data to support exploration of emergent and interactional results. Is it possible to reconcile these tensions and support both experimental, a priori contrasts and analyses of emergent, interactional accomplishments? We believe it is, and that the learning sciences have developed increasingly robust hybrid approaches that reflect the strengths of both traditions (see Dingyloudi & Strijbos, this volume). One example is work by Russ, Lee, and Sherin (2012), who explored the impact of social framing on student answers provided during interviews about science concepts. The authors built on the notion of social frames from interaction analysis to note that, while there were patterns in student cognition within their interviews, those patterns were also heavily influenced by the perceived social frame that the students engaged in with the interviewer. The authors were thus able to incorporate social cues into their model of how individual students presented their knowledge in interaction, accounting for the concepts that students understood as well as how their view of the context shaped their presentation of that understanding.

Designing for Learning

The assumptions we make about how people learn fundamentally drive the ways we design to support that learning. Designing for learning is very broad and can focus on different areas, such as classroom norms and instructional practices, particular disciplinary tools, or broader immersive learning environments. In the sections that follow, we offer overviews of the ways the two perspectives have typically thought about design, and then offer examples of designs that represent the extreme of each perspective. We then follow with two examples of synergistic designs, drawing from our own research, and highlight the contribution of that work.

Designing from a Cognitive Perspective

At the core of the majority of cognitive designs is an explicit awareness of and inclusion of a specific model of cognition. For example, much of the prolific work in the domain of cognitive tutors often builds on Act* model of cognition (see Graesser, Hu, & Sottilare, this volume). Furthermore, cognitive designs usually build on a refined model that is specific to the kinds of cognition that have been observed in experts within the specific content area, such as a model of how students learn new science concepts (White & Frederiksen, 2000) or how they process historical information (Wineburg, 1991). Cognitive models that drive design also frequently include an acknowledgment of previous, common misconceptions that the target population holds. Once these models have been specified, designs in the cognitive tradition are intended to help students to develop the target normative model or schema, addressing common misconceptions on the way.

A long-running program of research that exemplifies this approach is the development of cognitive tutors (Koedinger, Anderson, Hadley, & Mark, 1997). At the core of this approach to computer-assisted instruction is a model of student cognition. In one of the most famous examples, the Algebra Tutor, there are models of how to solve algebra problems as well as common mistakes made by students. Students can attempt to solve problems, and the continually updated model of their performance allows the cognitive tutor software to offer guidance as needed. Thus, the cognitive model is not only an inspiration for this research, but an actual core component of the software system. One of the goals of the cognitive tutor research was to bring cognitive science into the classroom, and it has been quite successful in doing so. As a result, researchers have paid quite a bit of attention to how the tutor might be adapted into local classroom contexts.

Designing from a Sociocultural Perspective

Sociocultural perspectives on design aim to accomplish two things. First, they look beyond the individual to understand the multiple mediators within the local context. As a result, sociocultural designs typically focus on entire activity systems rather than single tools (e.g., the Fifth Dimension projects). This also means that sociocultural theorists are often interested in supporting "authentic" environments that mirror the practices of the discipline and not just the concepts to be learned. Second, as a result, sociocultural perspectives on design tend to question what it means for practices to be authentic, and for whom. These approaches often challenge the status quo, noting how schools, and the disciplines they aim to prepare students for, are frequently not as valuable or well aligned with the goals, experiences, and histories of all students. Inequities within the multiple levels of our societal systems are thus a common focus of these design approaches.

An example of this can be seen in Lee's (1995) well known project that developed a high school curriculum for literary interpretation, drawing on and leveraging practices from the African American community—particularly *signifying*—and incorporating them into classroom activity. Signifying is a form of verbal play in the African American community that involves sophisticated language use, including "irony, double entendre, satire, and metaphorical language" (Lee, 1995, p. 612). A key assumption of this work was that African American students' performance on school-based assessments of literary interpretation did not accurately represent their actual understanding of literary interpretation. Instead, Lee hypothesized that the practices of interpretation of the spoken word, such as signifying, in which students were already central participants, were treated as unrelated to the practices of school English classrooms. As a consequence, students failed to transfer the practices of signifying to the classroom, because, although the underlying skills were equivalent, the contexts of use were notably different. Thus, the intervention involved bringing into alignment the tacit, everyday practice of signifying and the formal, academic practice of interpretation. Lee's work demonstrated that students who participated in the instructional intervention involving connecting everyday signifying to school practices learned twice as much as the control group.

Synergy in Design

The examples above highlight the differences in the ways that theoretical frameworks direct our perception of problems and, relatedly, the solutions that we pose. Taking extreme cases from cognitive science and sociocultural theory, we see work that has demonstrated effectiveness by carefully considering the way individual processing unfolds, and work that has demonstrated effectiveness by theorizing about the nature of the context that shapes individual participation. However, there is nothing inherently incommensurate between these two foci: one can draw on our understanding of the structure of human mental representation while simultaneously acknowledging that this structure is only part of understanding and predicting learning and activity. To highlight the potential to design across theoretical perspectives in design work, we present two examples of our own work, where we explicitly attempted to build on both traditions.

In the BeeSign project, Danish (2014) designed a series of activities intended to help early elementary students engage with complex systems concepts in the context of honeybees collecting nectar. Danish began his design work by exploring more individually focused work that describes the challenges and misconceptions that students face in exploring complex systems concepts. At the same time, Danish aimed to support this individual learning by focusing on designing collective, mediated activities where multiple participants were necessary to help students explore these concepts, and where key new practices were developed or supported. For example, inquiry with the BeeSign software relied upon the teachers' ability to help guide the students through cycles of inquiry, and also built on students' ability to help their peers attend to useful patterns in how bees collect nectar, and to challenge each other's assumptions by running simulated experiments. Student learning was demonstrated in both the changes in students' ongoing collective activity, as well as in individual interviews that took place afterwards. In particular, Danish demonstrated how ideas that were first made visible in collective activity were also seen within the individual interviews, though sometimes in different forms. This is an example, therefore, of how cognitive analyses of individual learning can be synthesized with a focus on collective activity to better understand how the design of collective activity can lead to new forms of interaction as well as individual outcomes.

Similarly, Gresalfi and Barnes (2015) describe a series of design studies that focused on supporting the development of a particular kind of mathematical problem-solving practice, which they call critical engagement. Beginning with research about the development of students' multiplicative and proportional thinking (Lesh, Post, & Behr, 1988; Misailidou & Williams, 2003), they designed an interactive immersive game and focused on the ways the narrative and feedback of the game supported students to consider different possible solutions, and the effectiveness of those solutions. This design framework built on ecological psychology, specifically focusing on the kinds of affordances that are included in designed environments, and how those affordances interact with students' incoming effectivities (prior knowledge, history with mathematics, etc.). Integrating theories of student knowing about ratio with an ecological framework allowed for the development of a set of conjectures about individual student reasoning as it related to and played out in relation to the interactive tools that were a part of the game.

Conclusions

There are many fundamental differences between cognitive and sociocultural theories of learning. As a result, the field has discussed, debated, and taught our students about these differences. One of the most important results of this ongoing work, and a hallmark of the Learning Sciences, has been that both traditions have continually refined their approach, and many scholars have worked to synthesize findings, theories, and designs from both traditions. We do not mean to suggest that we are moving, as a field, to one grand unified theory—while that might be possible, many productive debates and differences still exist. Rather, we believe that the last few decades' worth of pushback, argument, and discussion have led researchers across the Learning Sciences to focus on issues of interest to all of us. Regardless of the theoretical orientations that are taken up, we see more work that is explicitly addressing issues of individual performance and cognition while also focusing on social context and its role in constructing and being constructed by individual cognition. Perhaps even more importantly, we see scholars across the Learning Sciences explicitly recognizing that, in order to unpack the role of context in learning, we have to recognize and begin to address fundamental issues of equity and access that we know are so intertwined with the learning opportunities and experience of students across the world.

Further Readings

diSessa, A., Sherin, B., & Levin, M. (2015). Knowledge analysis: An introduction. In A. diSessa, M. Levin, & N. Brown (Eds.), *Knowledge and interaction: A synthetic agenda for the learning sciences* (pp. 377–402). New York: Routledge.

In this edited volume, efforts to analyze knowledge and interaction are compared, contrasted, and synthesized. The efforts to do so parallel our own in noting how not all of the differences in theoretical camps are irreconcilable, and contributions provide promising next steps for synergy.

Greeno, J. G., Collins, A., & Resnick, L. (1996). Cognition and learning. In D. C. Berliner & R. C. Calfee (Eds.), *Handbook of educational psychology* (pp. 15–46). New York: Routledge.

This classic piece provides a clear breakdown of the core principles within each theoretical framework. While more recent work has moved towards greater synergy, this remains a clear, high-level summary of core differences.

Sfard, A. (1998). On two metaphors for learning and the dangers of choosing just one. *Educational Researcher, 27*(2), 4–13.

This canonical piece helps not only to contrast the two core theoretical approaches, but to highlight the impact of their underlying differences. Sfard also argues compellingly for the danger of focusing too closely on only one approach.

Svihla, V., & Reeve, R. (Eds.). (2016). *Design as scholarship: Case studies from the learning sciences.* New York: Routledge.
This volume provides a rare look into the actual design process within the Learning Sciences, providing the kinds of depth and exploring challenges that rarely fit into a traditional article format. In doing so, it also helps make visible the role of the different theories in informing the design process.

Note

1 We are referring here to the theory proposed by Piaget as opposed to the philosophical approach, although many cognitive theories agree with the philosophical approach.

References

Alibali, M. W., & Nathan, M. (2018). Embodied cognition in learning and teaching: action, observation, and imagination. In F. Fischer, C .E. Hmelo-Silver, S. R. Goldman, & P. Reimann (Eds.), *International handbook of the learning sciences* (pp. 75–85). New York: Routledge.

Anderson, J. R., Reder, L. M., & Simon, H. A. (1996). Situated learning and education. *Educational Researcher, 25*(4), 5–11. doi: 10.3102/0013189x025004005

Bakker, A., & Gravemeijer, K. (2003). Planning for teaching statistics through problem solving. In R. Charles & H. L. Schoen (Eds.), *Teaching mathematics through problem solving: Grades 6–12* (pp. 105–117). Reston, VA: National Council of Teachers of Mathematics.

Case, R. (1996). Changing Views of Knowledge and Their Impact on Educational Research and Practice. In D. R. Olson & N. Torrance (Eds.), *The Handbook of Education and Human Development: New Models of Learning, Teaching and Schooling* (pp. 75–99). Malden, MA: Blackwell Publishers.

Danish, J. A. (2014). Applying an activity theory lens to designing instruction for learning about the structure, behavior, and function of a honeybee system. *Journal of the Learning Sciences, 23*(2), 1–49. doi:10.1080/10508406.2013.856793

Day, S. B., & Goldstone, R. L. (2012). The import of knowledge export: Connecting findings and theories of transfer of learning. *Educational Psychologist, 47*(3), 153–176. doi:10.1080/00461520.2012.696438

Dingyloudi, F., & Strijbos, J. W.(2018). Mixed methods research as a pragmatic toolkit: Understanding versus fixing complexity in the Learning Sciences. In F. Fischer, C. E. Hmelo-Silver, S. R. Goldman, & P. Reimann (Eds.), *International handbook of the learning sciences* (pp. 444–454). New York: Routledge.

Engeström, Y. (1999). Activity theory and individual and social transformation. In Y. Engeström, R. Miettinen, & R.-L. Punamäki (Eds.), *Perspectives on activity theory* (pp. 19–38). New York: Cambridge University Press.

Enyedy, N., Danish, J. A., & Fields, D. (2011). Negotiating the "relevant" in culturally relevant mathematics. *Canadian Journal for Science, Mathematics, and Technology Education, 11*(3), 273–291.

Feuer, M. J., Towne, L., & Shavelson, R. J. (2002). Scientific culture and educational research. *Educational Researcher, 31*(8), 4–14.

Goodwin, C. (1994). Professional vision. *American Anthropologist, 96*(3), 606–633.

Graesser, A. C., Hu, X., & Sottilare, R. (2018). Intelligent tutoring systems. In F. Fischer, C. E. Hmelo-Silver, S. R. Goldman, & P. Reimann (Eds.), *International handbook of the learning sciences* (pp. 246–255). New York: Routledge.

Greeno, J. G. (1997). On claims that answer the wrong questions. *Educational Researcher, 26*(1), 5–17. doi:10.3102/0013189x026001005

Greeno, J. G., Collins, A., & Resnick, L. (1996). Cognition and learning. In D. C. Berliner & R. C. Calfee (Eds.), *Handbook of educational psychology* (pp. 15–46). New York: Routledge.

Greeno, J. G., & Engeström, Y. (2014). Learning in activity. In R. K. Sawyer (Ed.), *The Cambridge handbook of the learning sciences* (2nd ed., pp. 128–147). Cambridge, UK: Cambridge University Press.

Gresalfi, M. S. (2009). Taking up opportunities to learn: Constructing dispositions in mathematics classrooms. *Journal of the Learning Sciences, 18*(3), 327–369.

Gresalfi, M. S., & Barnes, J. (2015). Designing feedback in an immersive videogame: Supporting student mathematical engagement. *Educational Technology Research and Development*, 1–22. doi:10.1007/s11423-015-9411-8

Gutiérrez, K. D., & Rogoff, B. (2003). Cultural ways of learning: Individual traits or repertoires of practice. *Educational Researcher, 32*(5), 19–25.

Koedinger, K. R., Anderson, J. R., Hadley, W. H., & Mark, M. A. (1997). Intelligent tutoring goes to school in the big city. *International Journal of Artificial Intelligence in Education, 8*, 30–43.

Konold, C., & Higgins, T. (2002). Working with data: Highlights of related research. In S. J. Russell, D. Schifter, & V. Bastable (Eds.), *Developing mathematical ideas: Working with data* (pp. 165–201). Parsippany, NJ: Dale Seymour Publications.

Ladson-Billings, G. (1995). Toward a theory of culturally relevant pedagogy. *American Educational Research Journal, 32*(3), 465.

Lave, J. (1980). What's special about experiments as contexts for thinking. *Quarterly Newsletter of the Laboratory of Comparative Human Cognition, 2*(4), 86–91.

Lee, C. D. (1995). A culturally based cognitive apprenticeship: Teaching African American high school students skills in literary interpretation. *Reading Research Quarterly, 30*(4), 608–630.

Lesh, R. A., Post, T., & Behr, M. (1988). Proportional reasoning. In J. Hiebert & M. Behr (Eds.), *Number concepts and operations in the middle grades* (pp. 93–118). Reston, VA: Lawrence Erlbaum & National Council of Teachers of Mathematics.

Lobato, J. (2012). The actor-oriented transfer perspective and its contributions to educational research and practice. *Educational Psychologist, 47*(3), 232–247. doi:10.1080/00461520.2012.693353

Makar, K., Bakker, A., & Ben-Zvi, D. (2011). The reasoning behind informal statistical inference. *Mathematical Thinking and Learning, 13*(1–2), 152–173.

Misailidou, C., & Williams, J. (2003). Children's proportional reasoning and tendency for an additive strategy: The role of models. *Research in Mathematics Education, 5*(1), 215–247.

Nolen, S. B., Horn, I. S., & Ward, C. J. (2015). Situating motivation. *Educational Psychologist, 50*(3), 234–247. doi:10.1080/00461520.2015.1075399

Reed, S. K. (2012). Learning by mapping across situations. *Journal of the Learning Sciences, 21*(3), 353–398. doi: 10.1080/10508406.2011.607007

Russ, R. S., Lee, V. R., & Sherin, B. L. (2012). Framing in cognitive clinical interviews about intuitive science knowledge: Dynamic student understandings of the discourse interaction. *Science Education, 96*(4), 573–599.

Sfard, A. (1998). On two metaphors for learning and the dangers of choosing just one. *Educational Researcher, 27*(2), 4–13.

Sherin, M., & van Es, E. A. (2009). Effects of video club participation on teachers' professional vision. *Journal of Teacher Education, 60*(1), 20–37. doi:10.1177/0022487108328155

Stevens, R., & Hall, R. (1998). Disciplined perception: Learning to see in technoscience. In M. Lampert & M. L. Blunk (Eds.), *Talking mathematics in school: Studies of teaching and learning* (pp. 107–149). Cambridge, UK: Cambridge University Press.

White, B., & Frederiksen, J. R. (2000). Technological tools and instructional approaches for making scientific inquiry accessible to all. In M. J. Jacobson & R. B. Kozma (Eds.), *Innovations in science and mathematics education* (pp. 321–359). Mahwah, NJ: Lawrence Erlbaum.

Wigfield, A., & Eccles, J. S. (2000). Expectancy–value theory of achievement motivation. *Contemporary educational psychology, 25*(1), 68–81.

Wineburg, S. S. (1991). Historical problem solving: A study of the cognitive processes used in the evaluation of documentary and pictorial evidence. *Journal of Educational Psychology, 83*(1), 73–87.

5
Apprenticeship Learning

Julia Eberle

At the beginning of the design process for a complex learning environment, many important decisions have to be made. What knowledge and skills are learners supposed to acquire in this environment? What are appropriate ways to help learners to construct knowledge, internalize scripts, or practice skills? How can we make sure that learners stay motivated throughout the learning process and understand the value of the learning content? What are ways to make sure that learners are able to apply their knowledge and skills later on in appropriate situations?

The answers to these questions and indeed even the questions that are asked reveal assumptions about the nature and processes of learning, and visions of learning environments that might enable such learning to occur. For example, the "acquisition metaphor" for learning (Sfard, 1998) assumes a form of traditional formal schooling in which the learning context is separated from the application context. Learners accumulate bits of knowledge for use some time in the future. Teachers impart these bits of knowledge to learners grouped with others of similar age and/or skill levels. In contrast, having said that, there are also other educational approaches based on different goals for learning and on different assumptions on how learning can best be fostered (Paavola, Lipponen, & Hakkarainen, 2004; Sfard, 1998). One of those, apprenticeship learning, is the focus of the present chapter.

Apprenticeship learning is based on a "participation metaphor" for learning (Sfard, 1998) and comes, in part, with fundamental differences in its implications for how learning needs to be structured and what a learning environment needs to look like. Most Learning Sciences research on the design of learning environments aims at integrating fundamental concepts of both approaches to learning and assumptions about what "travels" and how.

Apprenticeship learning is closely related to research focusing on "situated learning" and "situated cognition." This research was strongly influenced by the rediscovery of Vygotsky's work on sociocultural learning (see Danish & Gresalfi, this volume) in the late 1970s. Resurrecting Vygotsky's approach brought to researchers' attention that context is an essential aspect for understanding how people learn (Hoadley & van Heneghan, 2012).

However, apprenticeship learning is not one thing. Research focusing on apprenticeship learning can be divided into different strands that have developed in parallel and for the most part remain separate. These research strands differ not only in their theoretical focus but also in their core methodological approaches. The first strand focuses on understanding apprenticeship learning in its original meaning, looking at how people learn with the apprenticeship learning approach (and how this differs from the schooling approach). Two perspectives within this research strand are described in the first part of this chapter. The second strand comes from an instructional design perspective and aims at including features of apprenticeship learning in schooling situations to benefit from the strengths of both approaches. This strand is described in the second part of this chapter, exemplified by the cognitive apprenticeship framework.

Apprenticeship Learning: Defining Characteristics

Defining apprenticeship learning and understanding its underlying mechanisms is not trivial and, thus, has led to a huge body of research dedicated to the description of apprenticeship learning situations. This research is dominated by ethnographic studies, in which apprenticeship learning situations are observed and analyzed in detail.

Two distinct sub-strands can be identified: The first line of research investigates apprenticeship learning in the classical sense. Most studies focus on novices in certain occupations, or contexts of customary practices, who learn to master the given practices and become recognized members within the community of practice (Lave & Wenger, 1991). The second line of research transfers the concept of apprenticeship learning metaphorically to child development and children's learning in their surrounding world (Rogoff, 1995). Both perspectives are reviewed and the central elements are then contrasted with the acquisition metaphor of traditional schooling.

Apprenticeship Learning in Communities of Practice

The core foundation for understanding apprenticeship learning is Lave and Wenger's (1991) book *Situated Learning: Legitimate Peripheral Participation*. In this book, the authors describe several ethnographic studies on apprenticeship learning in very different contexts, e.g., midwives in Yucatán, butchers in German butcher shops, and Alcoholics Anonymous, and introduce the underlying learning process as legitimate peripheral participation in communities of practice.

Communities of practice form the social and epistemic context in which apprenticeship learning takes place. In apprenticeship learning, this context has a much more important role than in schooling, where the learning context is created top-down or by chance when a class or group of learners comes together for the first time and dissolves when the course is finished or when the school year ends. A community of practice, in contrast, is based on several epistemic and social aspects (Barab & Duffy, 2000; Barab, MaKinster, & Scheckler, 2003): A common practice and/or mutual enterprise brings people together and there is a mutual interdependence among the members that makes coming together inevitable. As the main reason for the existence of the community of practice is to benefit from each other, community members create opportunities for interactions and participation and respect diverse perspectives and minority views. A community of practice has a long-term perspective, resulting in overlapping (learning) histories, meaningful relationships, as well as shared knowledge, values, and beliefs. The community of practice, finally, goes beyond individual members and, consequently, mechanisms for reproduction appear to secure the community's existence when members leave for some reason.

The concept of communities of practice has received much attention by practitioners and researchers far beyond the Learning Sciences, especially in organizational contexts. Researchers have studied, applied, and differentiated the concept from other types of context in a large number of scientific papers. Consequently, the definition of the community of practice concept is still evolving and not used coherently (Lindkvist, 2005).

However, the apprenticeship learning process—defined as legitimate peripheral participation in such communities of practice—has received less attention. It describes learners' increasing participation in the community of practice as requiring two educationally relevant aspects (Lave & Wenger, 1991). First, legitimate access to the interactions, practices, and knowledge of the community is necessary. Second, learners need opportunities to participate peripherally—on the boundaries—but in ways that are authentic and valued by the community. As they become more able and ready for more active forms of participation, they move from peripheral to more central forms of activity in the community.

Peripheral participation is not only participation as a learner, who listens and observes, but also as a functional member of the community, who contributes to the mutual enterprise and/or executes the practices of the community. As newcomers to a community of practice usually lack the necessary knowledge and skills, they start to participate in a way that is possible with the knowledge and skills they have. There is a common "curriculum" that often begins with the observation of the results of the activities and practices. For example, in the case of midwifery in Yucatán, apprentices are initially directed to the results—a healthy child and mother after midwife-assisted delivery. Based on this desired and valued outcome, the learners are able to participate proactively in the community's practices in ways consistent with their skill levels. Over time and with guided observation, they become more skilled and able to do more and more of the core practices. At the core of apprenticeship learning are guidance by and collaboration with masters, observation of practices by members of the community at varying skill levels, and interactions with other community members of different experience levels.

Researchers have studied many different apprenticeship situations, mostly in occupational contexts, in which novices learn how to become, e.g., a police officer, a nurse, a psychiatrist, or a teacher (e.g. Lambson, 2010; Lave & Wenger, 1991). However, apprenticeship learning has also been studied in other everyday situations such as parents new to home schooling in communities of home schoolers (Safran, 2010), as well as in exotic situations, such as new members of a witch circle (Merriam, Courtenay, & Baumgartner, 2003). Most of this research is ethnographic, often focusing on one apprentice or a very small community of practice. However, there are also recent quantitative studies on the different participation support structures that enable legitimate peripheral participation (Eberle, Stegmann, & Fischer, 2014; Nistor, 2016).

Apprenticeship as a Metaphor for Children's Development

An apprenticeship learning perspective on the development of children emphasizes the interpersonal mechanisms through which children acquire knowledge of the world, learn their language, and develop ways of thinking (Rogoff, 1991; Rogoff, Paradise, Arauz, Correa-Chavez, & Angelillo, 2003). Of interest from this perspective are investigations of how children learn and develop in interaction with the people around them—adults and peers (Rogoff, 1991). The child is considered an apprentice in her surrounding world, a "universal" legitimate peripheral participant (Lave & Wenger, 1991). For that participation to be effective in promoting development, the participation must occur within what Vygotsky defined as the Zone of Proximal Development, with more knowledgeable others guiding the child's participation (Rogoff, Mystry, Göncü, & Mosier, 1993).

Fundamental to much of the research on children as apprentices is a commitment to cultural comparison, contrasting children's learning processes in different cultural contexts. The research often takes the form of ethnographic studies of a small number of children. This method allows researchers to capture, describe, and understand complex situations in deep detail but does not allow inferences about causal relations between different factors in the cultural context (Rogoff et al., 1993). For example, Rogoff and colleagues (1993) observed and interviewed the everyday interactions of caregivers and toddlers (14 families) who were members of four communities that differed in their traditions of schooling versus apprenticeship. Extensive analyses revealed that caregivers in communities in which the apprenticeship approach was dominant viewed the child as having the responsibility for initiating learning. Children participated in their caregivers' everyday and social activities (e.g., household chores). This participation provided them with opportunities to observe and, if they wished, initiate interactions with caregivers around ongoing activities and the objects that were part of them. Caregivers, then, turned their attention to helping the children perform the activities. In other words, the caregivers guided children's participation in activities in which the children showed interest; eventually, children could engage in these activities without guided participation (Rogoff et al., 1993). In contrast, caregivers from cultures with prevalent schooling approaches usually took the initiative for interaction with their children and explicitly created learning situations such as playing games or reading children's books. These experiences were typically apart from the everyday activities that caregivers engaged in when children were not present. When children were present, the center of attention for caregivers was on creating experiences for the child. Rogoff and colleagues also found that caregivers with apprenticeship approaches focused much more on supporting children's observations and attention to the general context and provided far less verbal instruction than caregivers from the schooling-focused cultures. The authors suggested that these differential foci were likely related to differences between the cultures in the goals of children's early learning. Schooling-oriented caregivers aimed to prepare children for academics, such as learning to read, which are hardly observable, whereas apprenticeship-oriented caregivers were preparing their children for the everyday activities (e.g., cooking or weaving).

The differences between the goals of caretakers in apprenticeship-oriented and school-oriented cultures reported by Rogoff and colleagues parallel and reflect contrasting theories of learning. As a theory of learning, apprenticeship relies on social and interpersonal observation: "newcomers to a community of practice advance their skill and understanding through participation with others in culturally organized activities in which apprentices become more responsible participants" (Rogoff, 1995, p. 143). There are essential differences between apprenticeship as a theory of learning and the associationist and behavioral learning theories of Thorndike, Skinner, and Watson that underpin traditional schooling (see, for discussion, Greeno, Collins, & Resnick, 1996). Recently, Collins (2015) summarized these differences; an adapted version of his summary is provided in Table 5.1.

Table 5.1 Differences Between Apprenticeship Learning and Schooling

	Apprenticeship learning	*Traditional schooling*
Epistemic/ functional context	Real situation in which a real problem needs to be solved or a real task needs to be performed	Artificially created for learning purposes
Content of learning	Observable practices	Largely unobservable academic skills "in the head" of the learner
Relevance of learning	Immediate relevance to solve a current problem or to accomplish important tasks	Assumed relevance in a distant future
Social context	Real-life social context	Created for learning purposes
Role of the expert	Coach who guides the learner during a mostly self-regulated learning process	Teacher who structures the learning process and designs learning experiences
Responsibility for learning	The context provides learning opportunities and/or the learner initiates learning activities based on occurring interest and receives or seeks guidance if necessary	Teacher pre-structures and initiates learning experiences for the learner, in which the learner navigates more or less actively
Relationship between expert and learner	Often one-on-one situation or a small group of learners; the "master" knows the learner very well which makes it easier to find tasks in the Zone of Proximal Development and to reduce failure	One teacher for usually more than 15 students; it is challenging for the teacher to find optimal learning tasks for so many learners at the same time
Central form of learning	Observational learning and active doing	Oral instruction and practice
Sequencing of learning tasks	Bottom-up, created either by occurring incidents or from a clear main goal to details; tasks learners can handle	Top-down, based on knowledge structured by experts, mostly from detail to detail, chronologically, etc.

Source: Adapted from Collins (2015).

The contrasts evident in Table 5.1 between apprenticeship learning and traditional schooling reflect the epistemic functions of each and their emergence in response to very different societal needs. These, in turn, drive the nature of what is to be learned and the central learning processes that characterize each approach. Nevertheless, the more active participatory character of apprenticeship learning holds a good deal of appeal to those interested in moving away from acquisition models of schooling. A number of learning researchers began to conceptualize and research instructional designs that were consistent with apprenticeship learning perspectives but that were adapted to learning the subject matter of traditional schooling, in particular language arts and mathematics (e.g., Collins, Brown, & Newman, 1988).

Apprenticeship Learning Goes to School: The Cognitive Apprenticeship Framework

The *cognitive apprenticeship framework* (Collins, 2015; Collins & Kapur, 2014; Collins et al., 1988) proposes essential characteristics of apprenticeship learning that can be applied in the classroom to make the schooling experience more apprenticeship-like. The framework emphasizes four core components: a focus on several types of knowledge, use of a variety of methods to promote learning, specific sequencing of learning activities, and emphasis on the social context of learning. The cognitive apprenticeship framework can be seen as a theoretical umbrella over many strands of current Learning Sciences research so the framework's core components are exemplary connected to recent research in the following sections. This research is methodologically diverse, including many of the research methods and analytic strategies that are included in Section 3 of this handbook.

Types of Knowledge

While schooling usually is based on the idea of knowledge accumulation, cognitive apprenticeship is fundamentally based on the idea of expertise development (Collins & Kapur, 2014; see Reimann & Markauskaite, this volume). Expertise encompasses multiple types of knowledge, not simply declarative and procedural content knowledge. Expertise also draws on knowledge of how to figure out how to perform a task or solve a problem, that is, problem-solving skills for tackling non-routine problems. As well, metacognitive knowledge and strategies are needed to monitor learning and problem-solving processes, evaluate progress, and adjust accordingly, in part by drawing on resources available in the situation or based on prior learning and experiences. Many of these types of knowledge have been investigated by cognitive psychologists, largely from an "in the head" individual perspective. Learning Scientists are making important contributions to broadening this individual-orientation by examining a socially shared perspective on many of these types of knowledge—for example, on metacognition (Garrison & Akyol, 2015) and on problem solving (see Goldman & Brand-Gruwel, this volume).

A Variety of Methods to Promote Learning

The different teaching and learning methods that are part of the cognitive apprenticeship framework serve specific purposes (Collins, 2015): cognitive modeling, coaching, and scaffolding are traditional apprenticeship learning methods that foster learning of a given task. Articulation and reflection, in contrast, foster generalization across the specific learning tasks and contexts. Exploration, finally, aims at giving control and ownership to learners.

The three traditional apprenticeship learning methods have received much attention in the Learning Sciences and instructional design research. Cognitive modeling describes the externalization of internal thinking and problem solving of an expert, so learners can "observe" the invisible cognitive processes. Only a few approaches for providing cognitive modeling in different domains have been explored (e.g., Schoenfeld, 1985) but there is a growing strand of research on worked examples that is closely related to cognitive modeling (e.g. Mulder, Lazonder, & Jong, 2014).

Coaching and scaffolding are central in the Learning Sciences. Coaching refers to ways in which teachers monitor individual students' or groups of students' learning processes and intervene when necessary. Scaffolding refers to tools that support students performing a task and are intended to be withdrawn in a fading process as students develop greater facility and proficiency with a task (Collins, 2015; Collins & Kapur, 2014). In a sense they are useful for creating a Zone of Proximal Development that is expected to change through the use of the tool. Current Learning Sciences research is investigating the role, effects, and detailed use of feedback by experts and peers (e.g. Bolzer, Strijbos, & Fischer, 2015), as well as scaffolding and scripting of individual and collaborative learning situations (e.g. Vogel, Wecker, Kollar, & Fischer, 2016; Kollar, Wecker, & Fischer, this volume; Tabak & Kyza, this volume). A long-standing area of work has been on cognitive tutors (see Graesser, Hu, & Sottilare, this volume). As noted by Collins (2015), technology such as adaptive cognitive tutors have the potential to simulate the close apprentice-master relationship, a relationship that affords optimal tailoring of learning experiences to an individual learner's Zone of Proximal Development. As such, there is the potential for moving to scale even when there are high student to teacher ratios.

Methods for fostering articulation and reflection to support students in externalizing their own thoughts and cognitive processes and looking back at how they came to their current state of knowledge and thinking are also heavily investigated in the Learning Sciences (Collins, 2015; Collins & Kapur, 2014). Research on prompting explores ways to trigger students' articulation and reflection. A number of studies look at prompts that foster self-explanation and metacognitive processes (e.g., Bannert, Sonnenberg, Mengelkamp, & Pieger, 2015; Heitzmann, Fischer, Kühne-Eversmann, & Fischer, 2015; Rau, Aleven, & Rummel, 2015). Other studies focus on fostering the creation of representations to visualize the learning content or the development of argumentation—not only for individual visualization but also as a tool for communication (e.g. Ainsworth, Prain, Vaughan, & Tytler, 2011; Bell, 1997; Schwarz, this volume).

Fostering exploration refers to enabling students to solve problems on their own and to encouraging them to seek out challenging learning and problem-solving opportunities (Collins, 2015; Collins & Kapur, 2014). This idea is closely connected to larger educational approaches, such as problem-based learning or inquiry learning (see de Jong, Lazonder, Pedaste, & Zacharia this volume; Hmelo-Silver, Kapur, & Hamstra, this volume; Linn, McElhaney, Gerard, & Matuk, this volume). However, exploration as the center of an educational approach needs guidance (Kirschner, Sweller, & Clark, 2006). Implementing a phase of exploration within a broader educational design that leaves freedom of exploration to the students is different. Such an unguided phase can be beneficial for learning, even when its outcome is failure instead of success (Kapur & Rummel, 2012).

Sequencing Learning Activities

The sequencing of learning tasks in the cognitive apprenticeship framework differs significantly from sequencing in traditional schooling (see Table 5.1). Collins (Collins, 2015; Collins & Kapur, 2014; Collins et al., 1988) derives three important rules for cognitive apprenticeship from apprenticeship learning research. First, the overall picture must be clear before focusing on detailed skills and tasks necessary to achieve the main goal. When students focus on developing a conceptual model of the overall task and activity and the different components it consists of, they better understand the value of the more detailed aspects and are more able to monitor their own learning progress on the more detailed aspects. Second, it is important to design the learning experience in a way that learners begin with very simple tasks and to increase the complexity in line with learners' developing skills. This approach fosters feelings of success and reduces frustration in learners. Third, cognitive flexibility research (Spiro, Feltovich, Jacobson, & Coulson, 1991) suggests ways to counteract learners' "overfitting" to a limited application context and a limited range of problem solving skills—a possible disadvantage of authentic apprenticeship learning. It is necessary that not only complexity, but also diversity of learning tasks and problems, increase with learners' growing expertise so learners both deepen and broaden their skills and knowledge.

Perhaps there is a limited amount of research on sequencing of learning activities because the three principles seem very obvious from a theoretical point of view. However, practical execution of the three principles is complex. Nevertheless, a few researchers have attempted to investigate sequencing. For example, Loibl and colleagues (Loibl & Rummel, 2015; Loibl, Roll, & Rummel, 2016) analyzed different features and sequencing orders in an exploration phase and in a cognitive modeling phase. They found differences for the process and outcome of the learning process depending on the sequence and features employed. Likewise, in computer-supported learning environments, researchers have looked at the orchestration of different social planes, scaffolds, and learning materials (e.g., Kollar & Fischer, 2013).

Social Context of Learning

The most complex challenge in bringing apprenticeship learning into schooling settings is, however, to build an authentic epistemic and social context for learning. According to the cognitive apprenticeship framework (Collins, 2015; Collins & Kapur, 2014) this means that learning tasks need to be situated in a real-world context in a way that naturally requires the acquisition of certain skills and knowledge to enable learners to solve the problems they face. The environment also needs to foster collaboration among the learners. Working in such a context is assumed to foster a sense of ownership of the problems and intrinsic motivation to solve them. At the same time, working with other learners is intended to motivate the individual learner and the shared learning experiences in joint projects are expected to foster a sense of community among the learners.

Several approaches to building communities of learners in schooling settings have been explored, including collaborative learning in groups of varying sizes. These approaches emphasize authentic peer relations in larger communities, such as a whole class or even several classes, and focus on joint learning experiences (see Chan & van Aalst, this volume; Slotta, Quintana, & Moher, this volume).

The Future of Apprenticeship Learning Research

This chapter presented the foundations of apprenticeship learning and the cognitive apprenticeship framework for supporting the design of learning environments that bring features of apprenticeship learning into schooling. As noted earlier in the chapter, ethnographic studies dominate the research on apprenticeship learning. It would be interesting to explore whether and how other types of methodologies might provide useful ways to address issues of social context and community (e.g., social network analysis as illustrated in Eberle, Stegmann, & Fischer, 2015). Insights from a broader variety of research methods may contribute additional perspectives on the dynamics and mechanisms of apprenticeship learning as it occurs in authentic settings.

Promising new contexts for future research are those that have traditionally been sites for either apprenticeship learning or schooling and that now are attempting to integrate both approaches. Examples are after school clubs, maker spaces (see Halverson & Peppler, this volume), and vocational training, particularly in several countries (e.g., Switzerland and Germany) where the integration of both approaches has a long-standing tradition. These contexts provide opportunities for observing naturally emerging apprenticeship learning situations and instructionally designed opportunities for the cognitive aspects of apprenticeship learning.

Further Readings

Brown, J. S., Collins, A., & Duguid, P. (1989). Situated cognition and the culture of learning. *Educational Researcher, 18*(1), 32–42.
This paper describes how schooling approaches neglect the situatedness of knowledge and how cognitive apprenticeship can be applied in mathematics teaching.

Collins, A., Brown, J. S., & Newman, S. E. (1988). Cognitive apprenticeship: Teaching the craft of reading, writing and mathematics. *Thinking: The Journal of Philosophy for Children, 8*(1), 2–10.
This paper describes the idea of cognitive apprenticeship.

Hod, Y., & Ben-Zvi, D. (2015). Students negotiating and designing their collaborative learning norms: A group developmental perspective in learning communities. *Interactive Learning Environments, 23*(5), 578–594.
This empirical study explores how students in a learning community take responsibility for their community as another core aspect of the cognitive apprenticeship framework, applying a qualitative, discourse oriented research design.

Lave, J., & Wenger, E. (1991). *Situated learning: Legitimate peripheral participation.* Cambridge: Cambridge University Press.
This book is the central source for understanding the foundations of apprenticeship learning.

Loibl, K. & Rummel, N. (2015). Productive failure as strategy against the double course of incompetence. *Learning: Research and Practice, 1*(2), 113–121.
This empirical study explores sequencing of learning activities as one of the core elements of the cognitive apprenticeship framework, applying a quantitative, experimental research design in a field setting.

Rogoff, B., Mystry, J., Göncü, A., & Mosier, C. (1993). *Guided participation in cultural activity by toddlers and caregivers. Monographs of the Society for Research in Child Development, 236.*
This book reports the cross-cultural ethnographic study on guided participation and provides an elaborated explanation of child development from an apprenticeship learning perspective.

NAPLeS Resources

Collins, A., *Cognitive apprenticeship* [Webinar]. In *NAPLeS video series.* Retrieved October 19, 2017, from http://isls-naples.psy.lmu.de/intro/all-webinars/collins/index.html
This webinar explains the cognitive apprenticeship framework for instructional application of apprenticeship learning.

Eberle, J., *Apprenticeship learning* [Video file]. *Introduction and short discussion.* In *NAPLeS video series.* Retrieved October 19, 2017, from http://isls-naples.psy.lmu.de/video-resources/guided-tour/15-minutes-eberle/index.html

Eberle, J., *Apprenticeship learning* [Video file]. *Interview.* In *NAPLeS video series.* Retrieved October 19, 2017, from http://isls-naples.psy.lmu.de/video-resources/interviews-ls/eberle/index.html

Acknowledgments

I would like to express my gratitude for the support I received while writing this chapter: to Susan Goldman for detailed editorial suggestions and to Barbara Rogoff for suggestions on the structure. I am especially thankful to both of them for sharing with me their unique knowledge concerning the history of apprenticeship learning research. Furthermore, I am grateful to my local research community—the Bochum Unicorns and in particular to Malte Elson—for their feedback.

References

Ainsworth, S., Prain, V., & Tytler, R. (2011). Drawing to learn in science. *Science, 333*(6046), 1096–1097.
Bannert, M., Sonnenberg, C., Mengelkamp, C., & Pieger, E. (2015). Short- and long-term effects of students' self-directed metacognitive prompts on navigation behavior and learning performance. *Computers in Human Behavior, 52,* 293–306. doi:10.1016/j.chb.2015.05.038
Barab, S. A., & Duffy, T. M. (2000). From practice fields to communities of practice. In D. H. Jonassen & S. M. Land (Eds.), *Theoretical Foundations of Learning Environments* (pp. 25–55). Mahwah, NJ: Lawrence Erlbaum Associates.
Barab, S. A., MaKinster, J. G., & Scheckler, R. (2003). Designing system dualities: Characterizing a web-supported professional development community. *The Information Society, 19*(3), 237–256. doi:10.1080/01972240309466
Bell, P. (1997). Using argumentation representation to make thinking visible for individuals and groups. In R. Hall, N. Miyake, & N. Enyedy (Eds.), *Proceedings of the 2nd international conference on computer support for collaborative learning* (pp. 10–19). Toronto, ON: International Society of the Learning Sciences.

Bolzer, M., Strijbos, J. W., & Fischer, F. (2015). Inferring mindful cognitive-processing of peer-feedback via eye-tracking: Role of feedback-characteristics, fixation-durations and transitions. *Journal of Computer Assisted Learning, 31*(5), 422–434. doi:10.1111/jcal.12091

Chan, C., & van Aalst, J. (2018). Knowledge building: Theory, design, and analysis. In F. Fischer, C. E. Hmelo-Silver, S. R. Goldman, & P. Reimann (Eds.), *International handbook of the learning sciences* (pp. 295–307). New York: Routledge.

Collins, A. (2015). Cognitive apprenticeship: NAPLeS Webinar. Retrieved from http://isls-naples.psy.lmu.de/intro/all-webinars/collins/index.html

Collins, A., Brown, J. S., & Newman, S. E. (1988). Cognitive apprenticeship: Teaching the craft of reading, writing and mathematics. *Thinking: The Journal of Philosophy for Children, 8*(1), 2–10.

Collins, A., & Kapur, M. (2014). Cognitive apprenticeship. In R. K. Sawyer (Ed.), *The Cambridge handbook of the learning sciences* (2nd ed., pp. 109–127). New York: Cambridge University Press.

Danish, J., & Gresalfi, M. (2018). Cognitive and sociocultural perspective on learning: Tensions and synergy in the Learning Sciences. In F. Fischer, C. E. Hmelo-Silver, S. R. Goldman, & P. Reimann (Eds.), *International handbook of the learning sciences* (pp. 34–43). New York: Routledge.

de Jong, T., Lazonder, A., Pedaste, M., & Zacharia, Z. (2018). Simulations, games, and modeling tools for learning. In F. Fischer, C. E. Hmelo-Silver, S. R. Goldman, & P. Reimann (Eds.), *International handbook of the learning sciences*. New York: Routledge.

Eberle, J., Stegmann, K., & Fischer, F. (2014). Legitimate peripheral participation in communities of practice: Participation support structures for newcomers in faculty student councils. *Journal of the Learning Sciences, 23*(2), 216–244. doi:10.1080/10508406.2014.883978

Eberle, J., Stegmann, K., & Fischer, F. (2015). Moving beyond case studies: Applying social network analysis to study learning-as-participation. *Learning: Research and Practice, 1*(2), 100–112. doi:10.1080/23735082.2015.1028712

Garrison, D. R., & Akyol, Z. (2015). Toward the development of a metacognition construct for communities of inquiry. *The Internet and Higher Education, 24*, 66–71. doi:10.1016/j.iheduc.2014.10.001

Goldman, S. R., & Brand-Gruwel, S. (2018) Learning from multiple sources in a digital society, In F. Fischer, C. E. Hmelo-Silver, S. R. Goldman, & P. Reimann (Eds.), *International handbook of the learning sciences* (pp. 86–95). New York: Routledge.

Graesser, A. C., Hu, X., & Sottilare, R. (2018). Intelligent tutoring systems. In F. Fischer, C. E. Hmelo-Silver, S. R. Goldman, & P. Reimann (Eds.), *International handbook of the learning sciences* (pp. 246–255). New York: Routledge.

Greeno, J. G., Collins, A. M., & Resnick, L. B. (1996). Cognition and learning. In D. C. Berliner & R. C. Calfee (eds.), *Handbook of educational psychology* (pp. 15–46). New York: Simon & Schuster Macmillan.

Halverson, E., & Peppler, K. (2018). The Maker Movement and learning. In F. Fischer, C. E. Hmelo-Silver, S. R. Goldman, & P. Reimann (Eds.). *International handbook of the learning sciences* (pp. 285–294). New York: Routledge.

Heitzmann, N., Fischer, F., Kühne-Eversmann, L., & Fischer, M. R. (2015). Enhancing diagnostic competence with self-explanation prompts and adaptable feedback. *Medical Education, 49*(10), 993–1003. doi:10.1111/medu.12778

Hmelo-Silver, C. E., Kapur, M., & Hamstra, M. (2018) Learning through problem solving. In F. Fischer, C. E. Hmelo-Silver, S. R. Goldman, & P. Reimann (Eds.), *International handbook of the learning sciences* (pp. 210–220). New York: Routledge.

Hoadley, C., & van Heneghan, J. P. (2012). The Learning Sciences: Where they came from and what it means for instructional designers. In R. A. Reiser (Ed.), *Trends and issues in instructional design and technology* (3rd ed., pp. 53–63). Boston, MA: Pearson.

Kapur, M., & Rummel, N. (2012). Productive failure in learning from generation and invention activities. *Instructional Science, 40*(4), 645–650. doi:10.1007/s11251-012-9235-4

Kirschner, P. A., Sweller, J., & Clark, R. E. (2006). Why minimal guidance during instruction does not work: An analysis of the failure of constructivist, discovery, problem-based, experiential, and inquiry-based teaching. *Educational Psychologist, 41*(2), 75–86. doi:10.1207/s15326985ep4102_1

Kollar, I., & Fischer, F. (2013). Orchestration is nothing without conducting—but arranging ties the two together! A response to Dillenbourg (2011). *Computers & Education, 69*, 507–509.

Kollar, I., Wecker, C., & Fischer, F. (2018). Scaffolding and scripting (computer-supported) collaborative learning. In F. Fischer, C. E. Hmelo-Silver, S. R. Goldman, & P. Reimann (Eds.), *International handbook of the learning sciences* (pp. 340–350). New York: Routledge.

Lambson, D. (2010). Novice teachers learning through participation in a teacher study group. *Teaching and Teacher Education, 26*(8), 1660–1668. doi:10.1016/j.tate.2010.06.017

Lave, J., & Wenger, E. (1991). *Situated learning: Legitimate peripheral participation*. Cambridge: Cambridge University Press.

Lindkvist, L. (2005). Knowledge communities and knowledge collectivities: A typology of knowledge work in groups. *Journal of Management Review, 42*(6), 1189–1210.

Linn, M. C., McElhaney, K. W., Gerard, L., & Matuk, C. (2018). Inquiry learning and opportunities for technology. In F. Fischer, C. E. Hmelo-Silver, S. R. Goldman, & P. Reimann (Eds.), *International handbook of the learning sciences* (pp. 221–233). New York: Routledge.

Loibl, K., Roll, I., & Rummel, N. (2016). Towards a theory of when and how problem solving followed by instruction supports learning. *Educational Psychology Review, 26*(4), 435. doi:10.1007/s10648-016-9379-x

Loibl, K., & Rummel, N. (2015). Productive failure as strategy against the double curse of incompetence. *Learning: Research and Practice, 1*(2), 113–121. doi:10.1080/23735082.2015.1071231

Merriam, S. B., Courtenay, B., & Baumgartner, L. (2003). On becoming a witch: Learning in a marginalized community of practice. *Adult Education Quarterly, 53*(3), 170–188.

Mulder, Y. G., Lazonder, A. W., & Jong, T. de. (2014). Using heuristic worked examples to promote inquiry-based learning. *Learning and Instruction, 29*, 56–64. doi:10.1016/j.learninstruc.2013.08.001

Nistor, N. (2016). Quantitative analysis of newcomer integration in MMORPG communities. In Y. Li, M. Chang, M. Kravcik, E. Popescu, R. Huang, Kinshuk, & N.-S. Chen (Eds.), *Lecture Notes in Educational Technology. State-of-the-Art and Future Directions of Smart Learning* (pp. 131–136). Singapore: Springer Singapore. doi:10.1007/978-981-287-868-7_15

Paavola, S., Lipponen, L., & Hakkarainen, K. (2004). Models of innovative knowledge communities and three metaphors of learning. *Review of Educational Research, 74*(4), 557–576. doi:10.3102/00346543074004557

Rau, M. A., Aleven, V., & Rummel, N. (2015). Successful learning with multiple graphical representations and self-explanation prompts. *Journal of Educational Psychology, 107*(1), 30–46. doi:10.1037/a0037211

Reimann, P., & Markauskaite, L. (2018). Expertise. In F. Fischer, C. E. Hmelo-Silver, S. R. Goldman, & P. Reimann (Eds.), *International handbook of the learning sciences* (pp. 54–63). New York: Routledge.

Rogoff, B. (1991). *Apprenticeship in thinking: Cognitive development in social context*. New York: Oxford University Press.

Rogoff, B. (1995). Observing sociocultural activity on three planes: participatory appropriation, guided participation, and apprenticeship. In J. V. Wertsch, P. del Rio, & A. Alvarez (Eds.), *Sociocultural studies of mind* (pp. 139–164). Cambridge: Cambridge University Press.

Rogoff, B., Mystry, J., Göncü, A., & Mosier, C. (1993). *Guided participation in cultural activity by toddlers and caregivers. Monographs of the Society for Research in Child Development, 236*.

Rogoff, B., Paradise, R., Arauz, R. M., Correa-Chavez, M., & Angelillo, C. (2003). Firsthand learning through intent participation. *Annual Review of Psychology, 54*, 175–203. doi:10.1146/annurev.psych.54.101601.145118

Safran, L. (2010). Legitimate peripheral participation and home education. *Teaching and Teacher Education, 26*(1), 107–112. doi:10.1016/j.tate.2009.06.002

Schoenfeld, A. (1985). *Mathematical problem solving*. Burlington: Elsevier Science.

Schwarz, B. B. (2018). Computer-supported argumentation and learning. In F. Fischer, C. E. Hmelo-Silver, S. R. Goldman, & P. Reimann (Eds.), *International handbook of the learning sciences* (pp. 318–329). New York: Routledge.

Sfard, A. (1998). On two metaphors for learning and the dangers of choosing just one. *Educational Researcher, 27*(2), 4–13. doi:10.3102/0013189X027002004

Slotta, J. D., Quintana, R., & Moher, T. (2018). Collective inquiry in communities of learners. In F. Fischer, C. E. Hmelo-Silver, S. R. Goldman, & P. Reimann (Eds.), *International handbook of the learning sciences* (pp. 308–317). New York: Routledge.

Spiro, R. J., Feltovich, P. J., Jacobson, M. J., & Coulson, R. L. (1991). Cognitive flexibility, constructivism, and hypertext: Random access instruction for knowledge acquisition in ill-structured domains. *Educational Technology, 31*(5), 24–33.

Tabak, I., & Kyza., E. (2018). Research on scaffolding in the learning sciences: A methodological perspective. In F. Fischer, C. E. Hmelo-Silver, S. R. Goldman, & P. Reimann (Eds.), *International handbook of the learning sciences* (pp. 191–200). New York: Routledge.

Vogel, F., Wecker, C., Kollar, I., & Fischer, F. (2016). Socio-cognitive scaffolding with computer-supported collaboration scripts: A meta-analysis. *Educational Psychology Review*. Advance online publication. doi:10.1007/s10648-016-9361-7

6
Expertise

Peter Reimann and Lina Markauskaite

Expertise, competence and skillful performance have been studied from many disciplinary perspectives, including philosophy (Collins & Evans, 2007), sociology (Young & Muller, 2014) and cognitive science (Chi, Glaser, & Farr, 1988; Ericsson, 2009; Ericsson, Charness, Feltovich, & Hoffman, 2006). In this chapter, we begin by discussing the study of expert competence and performance from a cognitive perspective in part because of its contributions to defining the results of successful learning and its relevance to education (Bransford, Brown, & Cocking, 2000). We then take up a broader ecological perspective on expertise that stems from a more situated and sociocultural studies and emphasize person-environment interactions. The influence of each perspective can be seen in conceptions of teacher expertise and its development, discussed in the third section of the chapter. In the final section, we briefly turn to some emerging areas of expertise research that are particularly pertinent to the learning sciences.

Cognitive Science Perspective on Expertise

Cognitive science provided seminal contributions to understanding components of knowledge and skills that make up competence in a particular field, such as classical mechanics or genetics. It also contributed to insights into how people who are highly proficient in a field actually solve problems, and what enables them to do so. This research also contributed to debunking the proposition that expertise, in many domains, depended on "talent" or "exceptional intelligence," indicating instead that expertise resulted from the acquisition of knowledge and skills acquired and honed through extensive practice.

General Characteristics of Expertise

Many insights into what constitutes expertise came from expert-novice comparison studies. These studies converged on four key characteristics: (i) expertise is domain-specific, (ii) experts perceive larger perceptual units than novices, (iii) experts' knowledge is organized differently from non-experts, and (iv) experts solve routine problems differently from novices. We briefly summarize these characteristics (for more extensive reviews, see Bransford et al., 2000; Chi et al., 1988; Feltovich, Prietula, & Ericsson, 2006).

Domain and task specificity. The higher the level of performance achieved in one particular domain, the less it is transferable to a new domain. That is, world-class chess players are not necessarily good managers or generals, and vice versa (Ericsson & Lehmann, 1996). This is true even when the domains are seemingly very similar. For instance, when Eisenstadt and Kareev (1979) compared the memory for board positions for expert Go and Gomoku players, they found no transfer of expertise across games, even though these two games are played on the same board.

The general explanation for this finding is that expert behavior is dependent more on knowledge, and knowledge acquisition, than on talent, gift, or general intelligence (Chi et al., 1988; Feltovich et al., 2006; Reimann & Chi, 1989). Consistent with the age-old adage, practice does indeed make the master. Furthermore, children demonstrate domain expertise in areas of interest (e.g., dinosaurs) and have been shown to outperform adults on memory and problem-solving tasks that rely on their specialized domain knowledge (Chi & Koeske, 1983). Rather than being a trait, "the development of expertise is largely a matter of amassing considerable skills, knowledge, and mechanisms that monitor and control cognitive processes to perform a delimited set of tasks efficiently and effectively" (Feltovich et al., 2006, p. 57). For education and instruction, this specificity of expertise is a challenge, because transfer is seen as a hallmark of successful learning. Adaptations to suit different task demands and environments allow the expert to function despite the constraints of the human cognitive architecture, the main ones being limits on attention, working memory, and access to long-term memory (Feltovich et al., 2006). However, there are tradeoffs to be considered between efficiency and reliability of performing a skill, and its flexibility and applicability.

Larger perceptual patterns. Experts literally see important objects in their area of expertise differently than novices. This has been demonstrated many times in many areas of expertise, but was first discovered for chess experts. In a series of classical studies, DeGroot (1965) in the Netherlands, and Chase and Simon (1973) in the US, presented boards with configurations of chess pieces to master-level and less accomplished players for only a few (typically 5) seconds, and let them then reproduce the boards. Chase and Simon (1973) found that the expert players correctly recalled approximately 20 pieces, four to five times the number of pieces recalled by less accomplished players. This compares with the expected 5–7 pieces for normal adults, given known capacity limitations of short-term memory (STM) (Miller, 1956).

The explanation advanced for this finding was "chunking": experts would perceive and store in STM larger patterns (e.g., specific opening moves), rather than each individual chess figure comprising the pattern. In other words, experts remembered "chunks" of pieces organized by board configurations that were meaningful in the game context. These configurations reflected experts' long-term memory for knowledge of chess and chess game patterns, knowledge that was automatically activated when looking at a chessboard. This 'knowledge dependence' theory of experts' perception is further supported by the observation that, when presented with random configurations of pieces on a chessboard, chess masters' recall was not better than that of novices.

Organization around problem-solving principles. The fascinating observation has been that experts not only remember (much) more, they also remember differently from novices, implicating differences in the organization of knowledge in long-term memory. Chase and Simon (Chase & Simon, 1973; Simon & Chase, 1973) studied qualitative aspects of recall by analyzing the characteristics of the chess piece configurations remembered by the experts. They found that these configurations were based largely on strategic aspects of the game, such as threats and opportunities. Studies by Chi, Feltovich, and Glaser (1981) and Glaser and Chi (1988) in physics yielded comparable results: experts sorted mechanics problems by principles of mechanics, such as Newton's Second Law being applicable, whereas novices grouped problems by surface features, such as involving an inclined plane.

The general explanation for this phenomenon is that expertise involves adaptation to the requirements of the task environment. As part of the adaptive process, experts build abstractions that are functional; they build schemas that are optimized for problem solving. Zeitz (1997) calls these "moderately abstracted conceptual representations" and describes how such abstractions aid in problem solving and decision making. Not surprisingly, the findings about knowledge organization in expertise have seen the most intensive application in education, particularly in science education (Chi et al., 1988) and medical education (Schmidt, 1993).

Strong problem-solving methods, forward reasoning, and automaticity. Experts use 'strong' methods of problem solving in their area of expertise, whereas novices use 'weak' ones. They are called 'weak' because they make little use of domain knowledge and it can take a very long time to find a solution and one may not be found. 'Strong' methods, in contrast, make use of domain knowledge, such as former solutions or domain principles.

Both weak and strong problem-solving methods can lead to learning, and when repeatedly practiced for a specific class of problems can become automatized. This kind of learning—the process of everyday skill acquisition—is well understood and has been described by John Anderson (1982) as "proceduralization." Problem-solving behavior becomes fast and reliable, and less vulnerable to interruptions and distractions. To the problem solver, it feels effortless and cognitive resources are freed up and can be used for higher order tasks, other tasks, or reflective learning.

Limitations of the Cognitive Perspective

The general cognitive characteristics of expertise emerged from a line of research that compared experts to novices on specific tasks that are relatively easy to administer in a laboratory setting. Despite the value of this research in elucidating the ways in which experts and novices differ as described above, it neglects the physical, symbolic, and social environment that supports competent behavior under authentic circumstances. The methods used, such as card sorting (Fincher & Tenenberg, 2005) and think-aloud protocols (Someren, Barnard, & Sandberg, 1994), necessitate a concentration on the individual expert, in a highly restricted task environment.

There are also limitations on the application of these cognitive characteristics of expertise to formal education. For one, formal education has neither the goal nor the means to enable novices to become experts. The development of professional expertise and the development of high levels of proficiency require the mobilization of resources that go far beyond what formal education alone can provide. Formal education may be where students first encounter areas in which they later develop expertise. In this sense, it may play an important role in fostering and sustaining interest (Hidi & Renninger, 2006; see also Renninger, Ren, & Kern, this volume).

Expertise from Sociocultural and Situated Perspectives

The cognitive view can be contrasted with studies of expertise "in the wild" (e.g., Hutchins, 1995; Klein, Calderwood, & Clinton-Cirocco, 1985). Such studies, in conjunction with the view of learning as situated (Greeno, 1998), have contributed to a view of expertise as much more distributed over the person and their environment and that expert performance is fundamentally dependent on resources in the experts' environment. As indicated earlier, cognitive-psychological research on expertise explains the development of expert-level behaviour by growth in the extent of content in long-term memory and changes in the organisation of memory content, brought about by abstraction (Zeitz, 1997) and "proceduralization" (Anderson, 1987). The 'domain-specificity' of expertise resides in experts' knowledge structures (Charness, 1976). Sociocultural and situated perspectives provide an additional—and many would argue alternative—view that development of expertise is explainable by changes in the relations between a person and their environment (see Danish & Gresalfi, this volume). From this perspective, learning takes the form of changes in one's relations to physical and social resources. Although, as emphasized in the cognitive view, experts may excel at adapting their knowledge for mastering demands in their area of expertise, they also excel at structuring their environment so that the resources available to them in relevant situations support their activities. For instance, cooks and many other vocational experts use space as a tool (Kirsh, 1995), and, importantly, use all sorts of tools (Hutchins, 1995). And, while the individual expert can be quite inventive in the use of tools, these tools and ways of using them are initially passed to them as cultural knowledge maintained by communities of practice (Greeno & Engeström, 2014).

There is perhaps a meeting of the cognitive and socio-culturally situated perspectives in the way Vygotsky has suggested: socially invented tools get encountered and then internalized, typically in the context of an apprentice-like relationship (Collins, Brown, & Newman, 1989; Lave, 1988). However, the focus of these two perspectives is different. The cognitive perspective focuses on studying the processes of internalization and their products—such as generalizations, abstractions, automatisms—whereas the socio-cultural perspective concentrates on the way knowledge gets socially constructed and used in specific situations, and how it undergoes changes over time (generations) and across different communities of practice.

In the learning sciences, the Brown, Collins, and Duguid (1989) milestone paper was highly influential in introducing and popularizing the socio-cultural view of learning and expertise. This perspective extended the unit of analysis from the individual to the *activity system*, thus providing learning researchers with the conceptual tools to study systems of the kind that matter in educational and professional workplace settings: organizations, units (classrooms), smaller groups and individuals interacting through tools with their environments (Greeno & Engeström, 2014). Further, it brought to the fore issues of curriculum—the conceptual tools reflecting cumulative wisdom—and an interest in studying the differences between the mastery of these tools by students in the classrooms compared to their mastery by expert practitioners in professional settings (Billett, 2001).

In the next section, we consider the specific profession of teaching, as both cognitive and socio-cultural/situated perspectives seem to contribute to characteristics of teaching expertise and its development.

Conceptions of Teacher Expertise and Its Development

Expertise in teaching is of particular relevance to the learning sciences for three reasons beyond simply the design of learning environments. First, research on what constitutes expertise in teaching, how it develops, and how it could be facilitated illustrates the main features of expertise research in many other complex professional domains. Second, in contrast to classical cognitive studies of expertise that investigated chess, diagnostic reasoning, and other well-defined domains in experimental settings, research on teacher expertise has also involved attempts to understand the capabilities involved in effective professional performance in naturalistic settings. Third, as a result, such studies have provided a rich account of expertise development that informs teacher professional education. We discuss three main areas of work that have been shaping thinking about teacher expertise: (1) the scope of teacher expertise; (2) research of teacher expertise; and (3) changing notions about teacher expertise.

Defining Teaching and Teacher Expertise

What does it take to be an expert teacher? Literature has defined the *notion and scope* of "teaching" very differently. Many classical studies of expertise have primarily associated teaching with what happens in the classrooms. Thus, the main focus has been effective teacher's performance during face-to-face teaching time (Leinhardt, 1989). Some studies expanded this notion to include other kinds of work that teachers do such as planning, design, and reflection (Borko & Livingston, 1989). Others studies expanded this notion even further to a whole range of activities that teachers do as professionals, including engaging with parents, participating in communities of practice and in broader society (Darling-Hammond & Bransford, 2005; Shulman & Shulman, 2004). Some accounts have adopted an even broader view, seeing teacher expertise as firmly linked to overall teacher identity that extends to values, dispositions, and other professional qualities (Darling-Hammond & Bransford,

2005; Shulman & Shulman, 2004). While these different notions suggest fundamentally different views of what teaching is, almost all of them acknowledge that teaching involves a complex set of tasks and expertise requires mastering a set of knowledge, skills, and dispositions. What counts as an expert teacher is also highly culture and context dependent (Li & Kaiser, 2011).

Researching Teaching Expertise

Classical studies of teacher expertise usually have focused on what distinguishes expert teachers from novices. Four differences have been commonly investigated: (1) structure of behavior and knowledge; (2) content of knowledge; (3) perception and representation of knowledge; and (4) nature of teacher knowledge.

In early studies, one of the common approaches for investigating expertise was to compare expert *behavior schemas* and *knowledge structures* with the structures and schemas of novices (Borko & Livingston, 1989; Leinhardt & Greeno, 1986). These studies usually attributed differences between experts and novices to the structure of their cognitive schemata, arguing that novices' schemata were less elaborate, interconnected, and accessible, and thus their pedagogical reasoning skills were less developed than schemata and reasoning skills of experts. These studies typically showed that mastery of a large number of powerful scripts and other kinds of structures for effective teaching was the main distinguishing feature of expert teachers.

Later studies have disagreed that expertise in teaching relies solely on schematic kinds of knowledge. Rather, this work argued the importance of the *content of teacher knowledge*, including its breath and depth. They contended that effective teaching relies on rich, well-integrated, and flexible knowledge that includes different domains. For example, Shulman (1987), drawing on his empirical studies, argued that teachers have a distinct knowledge base that included, at a minimum, seven components: content knowledge, general pedagogical knowledge, curriculum knowledge, pedagogical content knowledge, knowledge of learners, knowledge of educational contexts, and knowledge of educational ends or outcome goals. He particularly emphasized pedagogical content knowledge, a combination of pedagogy and subject matter from the perspective of how to support student learning of that content. Shulman regarded this knowledge as a unique kind of knowledge for teaching and that it differentiated teachers from subject experts who did not have expertise in teaching. More recent empirical studies have been suggesting that teacher expertise relies not only on unique forms of pedagogical content knowledge but also specialized forms of content knowledge in the specific discipline being taught (Lachner & Nückles, 2016).

Conceptions of teacher expertise have expanded beyond the structure and content of knowledge to encompass a variety of additional constructs. For example, *professional perception* (Seidel & Stürmer, 2014) and *problem representation* (Chi, 2011) refer to the idea that experts perceive and represent classroom situations differently from novices (Wolff, Jarodzka, van den Bogert, & Boshuizen, 2016). Having defined the problem differently, experts approach it differently than novices.

Yinger and Hendricks-Lee (1993) describe another important aspect of teacher knowledge as that which is "particularly useful to get things accomplished in practical situations" (p. 100). This has been variously labeled in the literature as "working knowledge," "personal practical knowledge," "craft knowledge," or "actionable knowledge." This characteristic of knowledge is increasingly seen as central to professional expertise performance (Markauskaite & Goodyear, 2017).

Three Complementary Views on the Development of Expertise in Teaching

Three complementary views shape current thinking about how teachers develop expertise and ways to facilitate the process: deliberative practice, adaptive expertise, and the role of tacit knowledge. We discuss each in turn.

Teaching as deliberative practice. The notion of expertise in teaching as *deliberative practice* has its origins in a broader view of expertise development that claims that experts who reach an elite status and those who remain non-elite experts spend about the same time practicing. What differentiates the former group of high achievers is their intentional, focused, goal-oriented, and structured efforts to improve their performance (Chi, 2011; Ericsson, 2006). While traditional daily practice and exposure to recurring tasks lead to attainment of a certain level of skill that produces certain levels of efficiency and automatization of stable performance, Ericsson (2006) argued that it is precisely experts' resistance to automaticity that distinguishes elite experts from other experienced performers. Bereiter and Scardamalia (1993) similarly argued that expertise in teaching involves continuous deliberative effort to "surpass oneself." The achievement of a certain level of performance and automaticity is necessary, as this frees up mental resources that can be invested in intentional efforts to improve and move to higher levels of proficiency.

Adaptive expertise. Hatano and Inagaki (1986) introduced a contrast between two kinds of expertise: routine and adaptive. Routine expertise enables efficient performance of procedural skills without necessarily having knowledge of why the procedure works. Adaptive expertise rests on understanding the meaning of the procedures and the nature of the objects on which they operate or towards which they are directed. Adaptive expertise allows efficient execution of procedural skills but with conceptual understanding of why these skills work, how they are related to important principles and constructs in the domain, and how to choose knowledgeably between alternatives. Importantly, it also undergirds invention and flexibility and enables the adaptive expert to act sensibly in new or unfamiliar situations. In teachers' work, these two dimensions are complementary and both are needed for effective, adaptive teaching.

Darling-Hammond and Bransford (2005) extend this notion to teachers' expertise, describing it as a progress along two dimensions: efficiency and innovation. The efficiency dimension involves abilities to perform various teaching related tasks without having to allocate much attention and mental resources for them, such as handing out worksheets while keeping everyone's attention or using a particular teaching technique effortlessly. The innovation dimension involves capacities to move beyond existing routines, accepted ideas, or even values, and to try to solve encountered problems in new ways.

Tacit knowledge. While the notions of deliberative practice and adaptive expertise emphasize the critical role of intentional behavior and articulated conceptual knowledge, teaching is inseparable from *intuition and tacit knowledge* that originate in actions and are deeply rooted in personal experiences. Tacit knowledge is often seen as essential for expert performance, particularly in complex professions that rely not only on cognitive capabilities, but also on motor skills and interpersonal capacities, such as medicine. However, when it comes to teaching, tacit knowledge often attracts negative connotations (Perry, 1965). Research on teacher expertise often argues that people who are untrained in teaching hold strong intuitive conceptions of teaching and learning that they develop through a long "apprenticeship of observing" teaching while they are students (Darling-Hammond & Bransford, 2005; Torff, 1999; Torff & Sternberg, 2001). These observations and experiences are often seen "as superficial trappings of teaching" (Darling-Hammond & Bransford, 2005, p. 367) that lead to serious misconceptions about teaching and are counterproductive for the development of deep professional knowledge. This intuitive knowledge has to be proactively confronted, as it blocks formation of expert knowledge. However, some recent studies suggest that the positive role of intuitive knowledge developed through everyday encounters with teaching and learning is likely to be significantly underestimated (Markauskaite & Goodyear, 2014). Teachers, similarly as experts in other domains, such as physics (Gupta, Hammer, & Redish, 2010), often draw on their "naive" experiences of phenomena in productive ways and these experiences provide essential experiential ground for making conceptual understanding actionable. It may be that the issue of tacit knowledge is more about having access to it; that is, making it explicit and open to inspection so that teachers not only "know-how" but they "know-why" (Bereiter, 2014) and can explain to students and others their thinking and the basis for their actions.

Overall, while the literature offers various suggestions of how expertise in teaching might look like and develop, evidence is far from conclusive. One of the biggest challenges is that resourceful teacher professional thinking and skill are inseparable from embodied, distributed, enacted, and situated performance (Markauskaite & Goodyear, 2017). The cognitive accounts of expertise that build on the mentalist models of human mind are largely inadequate for explaining such performance. In contrast, the accounts of expertise that build on more ecological views of cognition tend to be dispersed across different theoretical traditions, and more synthetic accounts such as that attempted by Darling-Hammond and Bransford (2005) continue to be needed.

Conclusions and Future Directions

Expertise research, as we have shown, is a well-established field to which both cognitive and sociocultural/situated perspectives have made valuable contributions. Over the last three decades, studies have provided various explanations of what kinds of cognitive and social structures and mechanisms underpin expertise. While these explanations are usually tightly linked to a particular domain due to the domain-specificity of expertise, they provide important general-level insights into the nature of experts' thinking, knowledge, actions, and interactions with the people and objects in their environment. Many of these findings have been generated by comparing expert and novice performance, thus providing valuable clues for teaching and instructional design—at least in some domains. Further, our review of expertise in teaching illustrates that expertise in real-world settings, particularly in complex professional domains, cannot be understood in isolation from broader expert cultures that shape what counts as expertise, capabilities to adapt to dynamically changing context, as well as capacities to drive change and innovate (see also Cress & Kimmerle, this volume).

Over the last 25 years an important development in theorizing and researching expertise has been the move away from a psychological bias, by which we mean the emphasis on explanations of human performance solely in terms of internal person characteristics, toward a view that encompasses the person-environment interactions. The enactive nature of human skillfulness, and particularly the constitutive entwinement between human capacities and their environments, cannot be overlooked (Markauskaite & Goodyear, 2017). A full account of expertise needs to incorporate the reality that humans are not only adapted and attuned to their environment, but that they are the *designers and creators* of their environment (Clark, 2011). Nevertheless there is scant systematic study of how experts *structure and use* their environment to achieve their skillful behavior. In this respect the relevant research has been highly fragmented, with studies of relevance to understanding expertise being distributed over a number of research and disciplinary communities that often do not refer to each other systematically (e.g., cognitive science, experimental psychology, educational psychology, human factors, anthropology). Addressing this important question in the future and connecting studies of expertise with research in neuroscience will require interdisciplinary efforts. It is a challenge that the Learning Sciences are particularly well positioned to address.

Further Readings

Bransford, J. D., Brown, A. L., & Cocking, R. C. (Eds.). (2000). *How People Learn: Brain, Mind, Experience and School*. Washington, DC: National Academy Press.
Although almost 20 years old, this volume remains an excellent source for understanding the influence of cognitive-psychological research on student learning and teacher cognition. It is highly readable, and freely available.

Clark, A. (2011). *Supersizing the mind. Embodiment, action, and cognitive extension*. Oxford, UK: Oxford University Press.
This is an excellent introduction to the role of tools in human cognition, and the relation to neuro-scientific research.

Greeno, J. G., & Engeström, Y. (2014). Learning in activity. In R. K. Sawyer (Ed.), *The Cambridge handbook of the learning sciences* (2nd ed., pp. 128–147). New York, NY: Cambridge University Press.
This piece provides a concise account of the notion of activity system and its role in explaining learning and competent performance.

Markauskaite, L., & Goodyear, P. (2017). *Epistemic fluency and professional education: innovation, knowledgeable action and actionable knowledge.* Dordrecht, Netherlands: Springer.
In this book, the authors explore the nature of professional resourcefulness, covering a range of explanatory accounts in depth.

References

Anderson, J. R. (1982). Acquisiton of cognitive skill. *Psychological Review, 89,* 369–406.
Anderson, J. R. (1987). Skill acquisition: Compilation of weak-method problem solutions. *Psychological Review, 94,* 192–210.
Bereiter, C. (2014). Principled practical knowledge: Not a bridge but a ladder. *Journal of the Learning Sciences, 23*(1), 4–17.
Bereiter, C., & Scardamalia, M. (1993). *Surpassing ourselves: An inquiry into the nature and implications of expertise.* Chicago, IL: Open Court.
Billett, S. (2001). Knowing in practice: Re-conceptualising vocational expertise. *Learning and Instruction, 11*(6), 431–452.
Borko, H., & Livingston, C. (1989). Cognition and improvisation: Differences in mathematics instruction by expert and novice teachers. *American Educational Research Journal, 26*(4), 473–498.
Bransford, J. D., Brown, A. L., & Cocking, R. C. (Eds.). (2000). *How people learn: Brain, mind, experience and school.* Washington, DC: National Academy Press.
Brown, J. S., Collins, A., & Duguid, P. (1989). Situated cognition and the culture of learning. *Educational Researcher, 18*(1), 32–42.
Charness, N. (1976). Memory for chess positions: Resistance to interference. *Journal of Experimental Psychology: Human Learning and Memory, 2,* 641–653.
Chase, W. G., & Simon, H. A. (1973). Perception in chess. *Cognitive Psychology, 4,* 55–81.
Chi, M. T. H. (2011). Theoretical perspectives, methodological approaches, and trends in the study of expertise. In Y. Li & G. Kaiser (Eds.), *Expertise in mathematics instruction: An international perspective* (pp. 17–39). Boston, MA: Springer.
Chi, M. T. H., Feltovich, P., & Glaser, P. (1981). Categorization and representation of physics problems by experts and novices. *Cognitive Science, 5,* 121–152.
Chi, M. T. H., Glaser, R., & Farr, M. J. (1988). *The nature of expertise.* Hillsdale, NJ: Lawrence Erlbaum Associates.
Chi, M. T. H., & Koeske, R. (1983). Network representation of a child's dinosaur knowledge. *Developmental Psychology, 19,* 29–39.
Clark, A. (2011). *Supersizing the mind. Embodiment, action, and cognitive extension.* Oxford, UK: Oxford University Press.
Collins, A., Brown, J. S., & Newman, S. E. (1989). Cognitive apprenticeship: Teaching the craft of reading, writing and mathematics. In L. B. Resnick (Ed.), *Knowing, learning and instruction: Essays in honor of Robert Glaser* (pp. 453–494). Hillsdale, NJ: Lawrence Erlbaum Associates.
Collins, H., & Evans, R. (2007). *Rethinking expertise.* Chicago, IL: University of Chicago Press.
Cress, U., & Kimmerle, J. (2018) Defining collective knowledge construction. In F. Fischer, C. E. Hmelo-Silver, S. R. Goldman, & P. Reimann (Eds.), *International handbook of the learning sciences.* New York: Routledge.
Danish, J., & Gresalfi, M. (2018). Cognitive and sociocultural perspective on learning: Tensions and synergy in the Learning Sciences. In F. Fischer, C. E. Hmelo-Silver, S. R. Goldman, & P. Reimann (Eds.), *International handbook of the learning sciences* (pp. 34–43). New York: Routledge.
Darling-Hammond, L., & Bransford, J. (Eds.) (2005). *Preparing teachers for a changing world: What teachers should learn and be able to do.* San Francisco, CA: Jossey-Bass.
DeGroot, A. D. (1965). *Thought and choice in chess.* The Haag, Netherlands: Mouton.
Eisenstadt, M., & Kareev, Y. (1979). Aspects of human problem solving: The use of internal representations. In D. A. Norman & D. E. Rumelhart (Eds.), *Exploration in cognition.* San Francisco, CA: W H. Freeman.
Ericsson, K. A. (2006). The influence of experience and deliberate practice on the development of superior expert performance. In K. A. Ericsson, M. Charness, P. J. Feltovich, & R. R. Hoffman (Eds.), *The Cambridge Handbook of Expertise and Expert Performance* (pp. 683–703). New York, NY: Cambridge University Press.
Ericsson, K. A. (Ed.) (2009). *Development of professional expertise: Toward measurement of expert performance and design of optimal learning environments.* New York, NY: Cambridge University Press.
Ericsson, K. A., Charness, M. N., Feltovich, P. J., & Hoffman, R. R. (Eds.). (2006). *The Cambridge handbook of expertise and expert performance.* New York, NY: Cambridge University Press.
Ericsson, K. A., & Lehmann, A. C. (1996). Expert and exceptional performance: Evidence of maximal adaptation to task constraints. *Annual Review of Psychology, 47,* 273–305.

Feltovich, P. J., Prietula, M. J., & Ericsson, K. A. (2006). Studies of expertise from psychological perspectives. In K. A. Ericsson, M. N. Charness, P. J. Feltovich, & R. R. Hoffman (Eds.), *The Cambridge handbook of expertise and expert performance* (pp. 41–68). New York, NY: Cambridge University Press.

Fincher, S., & Tenenberg, J. (2005). Making sense of card sorting data. *Expert Systems, 22*(3), 89–93.

Glaser, R., & Chi, M. T. H. (1988). Overview. In M. T. Chi, R. Glaser, & M. J. Farr (Eds.), *The nature of expertise* (pp. xv–xxviii). Hillsdale, NJ: Lawrence Erlbaum Associates.

Greeno, J. G. (1998). The situativity of knowing, learning, and research. *American Psychologist, 53*, 5–26.

Greeno, J. G., & Engeström, Y. (2014). Learning in activity. In R. K. Sawyer (Ed.), *The Cambridge handbook of the learning sciences* (2nd ed., pp. 128–147). New York, NY: Cambridge University Press.

Gupta, A., Hammer, D., & Redish, E. F. (2010). The case for dynamic models of learners' ontologies in physics. *Journal of the Learning Sciences, 19*(3), 285–321.

Hatano, G., & Inagaki, K. (1986). Two courses of expertise. In H. A. H. Stevenson & K. Hakuta (Eds.), *Child development and education in Japan* (pp. 262–272). New York, NY: Freeman.

Hidi, S., & Renninger, K. A. (2006). The four-phase model of interest development. *Educational Psychologist, 41*, 111–127.

Hutchins, E. (1995). *Cognition in the wild*. Cambridge, MA: Cambridge University Press.

Kirsh, D. (1995). The intelligent use of space. *Artificial Intelligence, 73*, 31–68.

Klein, G., Calderwood, R., & Clinton-Cirocco, A. (1985). Rapid decision making on the fire ground (KATR-84–41–7). *Yellow Springs, OH: Klein Associates Inc. Prepared under contract MDA903—85-G—0O99 for the US Army Research Institute, Alexandria, VA.*

Lachner, A., & Nückles, M. (2016). Tell me why! Content knowledge predicts process-orientation of math researchers' and math teachers' explanations. *Instructional Science, 44*(3), 221–242.

Lave, J. (1988). *Cognition in practice: Mind, mathematics, and culture in everyday life*. Cambridge, UK: Cambridge University Press.

Leinhardt, G. (1989). Math lessons: A contrast of novice and expert competence. *Journal for Research in Mathematics Education, 20*(1), 52–75.

Leinhardt, G., & Greeno, J. G. (1986). The cognitive skill of teaching. *Journal of Educational Psychology, 77*, 247–271.

Li, Y., & Kaiser, G. (Eds.). (2011). *Expertise in mathematics instruction: an international perspective*. New York, NY: Springer.

Markauskaite, L., & Goodyear, P. (2014). Tapping into the mental resources of teachers' working knowledge: Insights into the generative power of intuitive pedagogy. *Learning, Culture and Social Interaction, 3*(4), 237–251.

Markauskaite, L., & Goodyear, P. (2017). *Epistemic fluency and professional education: innovation, knowledgeable action and actionable knowledge*. Dordrecht, Netherlands: Springer.

Miller, G. A. (1956). The magical number seven, plus minus two. Some limits on our capacity to process information. *Psychological Review, 63*, 81–87.

Perry, L. R. (1965). Commonsense thought, knowledge and judgement and their importance for education. *British Journal of Educational Studies, 13*(2), 125–138.

Reimann, P., & Chi, M. T. H. (1989). Human expertise. In K. J. Gilhooly (Ed.), *Human and machine problem solving* (pp. 161–192). London, UK: Plenum Press.

Renninger, K. A., Ren, Y., & Kern, H. M. (2018). Motivation, engagement, and interest: "In the end, it came down to you and how you think of the problem." In F. Fischer, C. E. Hmelo-Silver, S. R. Goldman, & P. Reimann (Eds.), *International handbook of the learning sciences* (pp. 116–126). New York: Routledge.

Schmidt, H. G. (1993). Foundation of problem-based learning: Some explanatory notes. *Medical Education. 27* (5), 422–432.

Seidel, T., & Stürmer, K. (2014). Modeling and measuring the structure of professional vision in preservice teachers. *American Educational Research Journal, 51*(4), 739–771.

Shulman, L. S. (1987). Knowledge and teaching: Foundations of the new reform. *Harvard Educational Review, 57*(1), 1–22.

Shulman, L. S., & Shulman, J. H. (2004). How and what teachers learn: A shifting perspective. *Journal of Curriculum Studies, 36*(2), 257–271.

Simon, H. A., & Chase, W. G. (1973). Skill in chess. *American Scientist, 61*, 394–403.

Someren, M. W. v., Barnard, Y. F., & Sandberg, J. A. C. (1994). *The think aloud method: A practical guide to modelling cognitive processes*. London, UK: Academic Press.

Torff, B. (1999). Tacit knowledge in teaching: Folk pedagogy and teacher education. In R. J. Sternberg & J. A. Horvath (Eds.), *Tacit knowledge in professional practice: Researcher and practitioner perspectives* (pp. 195–213). Mahwah, NJ: Lawrence Erlbaum Associates.

Torff, B., & Sternberg, R. J. (Eds.). (2001). *Understanding and teaching the intuitive mind: Student and teacher learning*. Mahwah, NJ: Lawrence Erlbaum Associates.

Wolff, C. E., Jarodzka, H., van den Bogert, N., & Boshuizen, H. P. A. (2016). Teacher vision: Expert and novice teachers' perception of problematic classroom management scenes. *Instructional Science, 44*(3), 243–265.

Yinger, R., & Hendricks-Lee, M. (1993). Working knowledge in teaching. In C. Day, J. Calderhead, & P. Denicolo (Eds.), *Research on teacher thinking: Understanding professional development* (pp. 100–123). London, UK: Falmer Press.

Young, M., & Muller, J. (Eds.). (2014). *Knowledge, expertise and the professions*. Abingdon, Oxon: Routledge.

Zeitz, C. M. (1997). Some concrete advantages of abstraction: How experts' representations facilitate reasoning. In P. J. Feltovich, K. M. Ford, & R. R. Hoffman (Eds.), *Expertise in context: Human and machine.* (pp. 43–65). Cambridge, MA: MIT Press.

7
Cognitive Neuroscience Foundations for the Learning Sciences

Sashank Varma, Soo-hyun Im, Astrid Schmied, Kasey Michel, and Keisha Varma

Introduction

Cognitive neuroscience and the learning sciences arose around the same time, in the early 1990s. Both are interdisciplinary fields, and both borrowed the theoretical constructs and experimental paradigms of cognitive psychology during their formative years. Even though they share some common intellectual ancestry, they have diverged considerably over the past 25 years. Each has pursued different research questions, adopted different methods, and incorporated different disciplines into their respective folds.

This chapter considers the relation between these now mature fields. What, if anything, can cognitive neuroscience offer the learning sciences? (The converse question is equally interesting, but beyond the scope of this chapter and handbook.) We begin by reviewing the research methods of cognitive neuroscience, to reveal how they generate facts about the mind and brain. Next, we consider cognitive neuroscience conceptions of the large-scale structure of the brain, emphasizing how thinking and learning can be understood as network phenomena. We then present a case study of what cognitive neuroscience research has revealed about mathematical thinking, an ability that the learning sciences also seeks to understand and foster in students. Finally, we consider the broader philosophical and social barriers to importing cognitive neuroscience findings into the learning sciences and applying them to improve learning in formal and informal contexts.

Neuroscience Methods

The growth of new sciences is often driven by the development of new methods. This has certainly been true of cognitive neuroscience, which has been fueled in particular by the maturation of neuroimaging. Here, we review contemporary neuroimaging methods of greatest relevance to the learning sciences. We refer the interested reader to Gazzaniga, Ivry, and Mangun (2013) for further details about these and other methods.

Neuroimaging methods have driven the rise of cognitive neuroscience because they allow non-invasive measurement of brain function (and also brain structure) in typical adults and children. These methods differ in their spatial resolution, i.e., the volume of tissue whose activity can be individuated and measured. They also differ in their temporal resolution, i.e., the time required to collect an accurate measure of activity in those volumes. There is no perfect method, and cognitive neuroscientists choose the one best matched to the phenomena under study.

Neural firing generates electrical signals that propagate through tissue, and that can be recorded at the scalp using a net of electrodes. These recordings constitute an electroencephalogram (EEG). Electrical signals propagate nearly instantly, and therefore EEG has very good temporal resolution relative to other neuroimaging techniques. However, because there are billions of neurons, each generating an electrical field, and because these fields are distorted as they travel through brain tissue, EEG has very poor spatial resolution. Nevertheless, EEG has proven to be an informative method for investigating questions of interest to learning scientists. For example, it can be used to detect neural responses to stimuli even in the absence of behavioral responses. Using this method, Molfese (2000) found that newborn infants who show aberrant EEG recordings when listening to normal speech sounds are more likely to develop reading difficulties when they enter elementary school.

Magnetic resonance imaging (MRI) is a technique for imaging the structure of the brain. In the 1980s, a variant, functional MRI (fMRI), was developed that enables imaging of the correlates of thinking. The word "correlates" is key. fMRI does not directly image neural computation. Rather, it images the vascular response to this computation. More precisely, when neurons fire, they consume local metabolic resources such as glucose. These resources are resupplied by what is called the hemodynamic response. fMRI can be used to measure this response, specifically local changes in the concentration of oxygen—this is the blood-oxygen level dependent (BOLD) signal. The spatial resolution and temporal resolution of fMRI are both constrained by the properties of the hemodynamic response. They turn out to occupy a "sweet spot" in the investigation of cognition. fMRI images ("volumes") are composed of picture elements ("voxels") encompassing 10 or so cubic millimeters of tissue, and can be acquired every second or so. Given the popularity of fMRI studies, both in the scientific community and in the public media, it is important to remember that this method—like all methods—must be used properly to generate results that are interpretable and replicable. Whether fMRI can address a research question depends on whether its spatial resolution and temporal resolution are appropriate for the research questions at hand, the logic of the experimental design, the soundness of the data analysis, and the proper interpretation of the results. A failure in any of these areas will result in an uninformative—or, worse, a misleading—study.

The Neuroarchitecture of Cognition and Learning

Cognitive neuroscience shares with cognitive psychology the claim that thinking and learning are forms of information processing. However, it differs in the mechanisms and metaphors it offers. Here we present a cognitive neuroscience view of the brain and neural information processing that is compatible with a learning sciences view of how people learn.

Thinking as a Network Activity

Cognitive neuroscience has revealed novel decompositions of cognition, and in this way provided new insights into thinking and learning. There are two broad approaches to mapping cognitive information processing to neural information processing. The classical, localist approach is to understand cognitive task X by looking for "the X area." The more contemporary, distributed approach is to identify the suite of cognitive functions Y_i that are recruited to perform cognitive task X, and then to look for their mapping to the brain areas that compose "the X network."

Many theories of the brain networks that underwrite cognition have been offered over the years (e.g., Luria, 1966). More recent network theories decompose the brain into many areas that are organized hierarchically and interconnected in a many-to-many fashion (Mesulam, 1990). Computational models have revealed how these areas dynamically self-organize into networks capable of performing complex tasks extended in time (Just & Varma, 2007). Current efforts are attempting to identify "the human connectome" (Sporns, 2011).

Given that neural processing is highly distributed, it is perhaps not surprising that neural representations are, as well. The notion of distributed representations is an old one (Lashley, 1950). Early cognitive neuroscience studies found evidence of distributed representations of visual categories (i.e., faces, buildings, furniture) in the ventral visual pathway (Ishai, Ungerleider, Martin, Schouten, & Haxby, 1999). More recent machine-learning approaches are revealing that representations of words and concepts are highly distributed over many brain areas (Just, Cherkassky, Aryal, & Mitchell, 2010). These include perception and motor areas, a result consistent with embodied approaches to cognition (see Alibali & Nathan, this volume, on the role of embodiment in the learning sciences).

Learning as a Network Activity

Plasticity may be the defining feature of the human brain that enables learning, and it takes different forms at different levels. At the cellular and molecular level, long-term potentiation—a long-lasting increase in the synaptic strength between neurons after high-frequency stimulation—is an important mechanism for learning. Of greatest relevance to the learning sciences are conceptions of learning at the level of brain networks. The different memory systems identified by cognitive psychologists are associated with different brain areas. Learning of new episodic memories is associated with the medial temporal lobe, including the hippocampus. Semantic memories are distributed over many areas of cortex (Just et al., 2010), although there is also some degree of specialization. For example, the fusiform gyrus is associated with memories for complex visual categories such as faces, whereas the left angular gyrus (AG) is associated with memories for the phonology of words. Procedural memories, including cognitive skills, are associated with the subcortical nuclei that constitute the basal ganglia, including the caudate and putamen.

Like thinking, learning is also a network activity. In particular, learning a new cognitive ability X requires (1) that the composite cognitive functions Y_i can be performed by their respective brain areas and (2) that these areas are properly connected so that information can be communicated and co-processed. This conception of learning can be formalized in three principles that find support in the cognitive neuroscience literature. The first is that learning a new cognitive ability requires strengthening the communication between the brain areas that support the composite cognitive functions. For example, increased functional connectivity between areas of the visual system (i.e., increased correlation in their activation levels over time) is observed when people learn to associate objects with locations (Büchel, Coull, & Friston, 1999).

The second principle is that when one component of a brain network is impaired, then task performance will be affected. It is sometimes possible, however, to restore the function of the impaired area, and therefore to improve task performance, through instruction. For example, dyslexia is associated with reduced activation in the left AG, an area associated with phonological processing, during word reading (Shaywitz, Lyon, & Shaywitz, 2006). However, when people with dyslexia complete remediation programs and experience behavioral improvements, activation in this area is normalized (Eden et al., 2004).

The third principle is that, when the components of a brain network are under-connected, then task performance will be affected. In this case, the restoration of connectivity sometimes accompanies the successful remediation of behavioral difficulties. For example, dyslexia is also associated with reduced integrity of the white-matter tracts over which anterior and posterior areas of the language network communicate. However, the integrity of these white-matter tracts has been found to improve when people with dyslexia successfully complete remediation programs (Keller & Just, 2009).

Case Study: Mathematical Thinking and Learning

To illustrate what cognitive neuroscience research can tell us about thinking and learning, we consider a subset of neuroscience findings relevant to mathematics education. Our review focuses on the contributions of two components of the brain network that supports number and arithmetic understanding, the intra-parietal sulcus (IPS) and the left AG.

Representations and Strategies

The key finding in understanding the mental representation and neural correlates of number is the *distance effect*: when comparing which of two numbers is greater, people are faster when the numbers are far apart (e.g., 1 vs. 9) than when they are close together (e.g., 1 vs. 2). This has been interpreted as evidence that children and adults possess *magnitude representations* of numbers (Moyer & Landauer, 1967; Sekuler & Mierkiewicz, 1977).

Neuroimaging studies have identified a neural correlate of these magnitude representations, the IPS. The closer the numbers being compared, the greater the activation observed in this area (Pinel, Dehaene, Rivière, & Le Bihan, 2001). This neural distance effect is also found in children (Ansari & Dhital, 2006).

The *problem size effect* is the key finding in studies investigating how people solve arithmetic problems: the larger the operands, the longer the solution times (Groen & Parkman, 1972). One explanation is that because problems with smaller operands are encountered earlier and practiced more frequently in school, the corresponding arithmetic facts are more strongly encoded in memory, and can therefore be directly retrieved—a fast process. By contrast, because problems with larger operands are encountered later and less frequently, they are less strongly encoded in memory, and therefore people must deploy slower counting and calculation strategies (LeFevre, Sadesky, & Bisanz, 1996; Siegler, 1988).

Neuroimaging studies have documented the neural correlates of arithmetic problem solving. A key finding is that when people report solving a problem using direct retrieval, there is greater activation in the left AG, an area associated with the retrieval of verbally coded information. By contrast, when they report using slower non-retrieval strategies, there is greater activation in a broad fronto-parietal network associated with spatial problem solving, one that includes the IPS, which is associated with magnitude representations of numerical operands (Grabner et al., 2009). Over development, children shift from using non-retrieval strategies supported by the fronto-parietal network to using direct retrieval supported by the left AG (Rivera, Reiss, Eckert, & Menon, 2005).

Learning

There have been relatively few neuroscience studies of mathematics learning. This is because the constraints of neuroscience methods make it difficult to continuously measure brain activation during extended learning episodes. For this reason, these studied have adopted pre-post designs, and imaged the effect of different kinds of instruction on brain function. For example, Delazer et al. (2005) evaluated whether, when teaching a mathematical concept, different approaches to instruction lead to different understandings (i.e., different brain networks) or the same understanding (i.e., the same brain network). Adults learned new arithmetic operations using two different kinds of instruction. The memorization group learned to directly associate operands with results. The procedure group learned an algorithm for iteratively transforming operands into results. The instructional parallel would be memorizing math facts versus learning to compute them (e.g., Baroody, 1985). The memorization group organized a brain network for the concept that included the left AG, consistent with direct retrieval of memorized arithmetic facts. By contrast, the procedural group recruited a fronto-parietal network consistent with the use of spatial working memory to store partial products during execution of the algorithm. This study makes the point, obvious to learning scientists, that different instructional approaches lead students to adopt different strategies to solve the same problems. More importantly, it illustrates how neuroscience methods can be used to detect and understand these differences at the level of brain function.

Individual Differences

The size of a person's distance effect can be interpreted as an index of the precision of his or her magnitude representations. Researchers have investigated whether *individual differences* in the precision of magnitude representations are associated with individual differences in mathematical achievement as measured by standardized tests. This is the case for elementary school children when precision is measured behaviorally (De Smedt, Verschaffel, & Ghesquière, 2009). This is also true when precision is measured neurally, by the amount of activation in the left IPS for difficult close-distance comparisons (Bugden, Price, McLean, & Ansari, 2012). This relationship also holds for middle school students, who have moved beyond number and arithmetic to more abstract mathematical concepts (Halberda, Mazzocco, & Feigenson, 2008).

Studies of the neural correlates of individual differences in mathematical achievement have focused on the problem size effect for arithmetic. Recall that one explanation of this effect is that when solving small-operand problems, people use direct retrieval, whereas when solving large-operand problems, they use slower counting and calculation strategies. The prediction is that people of higher mathematical achievement will use direct retrieval even when solving large-operand problems, because they will have memorized these arithmetic facts. Grabner et al. (2007) found that higher mathematical achievement was associated with greater activation in the left AG when solving large-operand multiplication problems, presumably reflecting direct retrieval of the corresponding multiplication facts.

Researchers have also used the precision of number magnitude representations to invest *group differences* in mathematical achievement. People with dyscalculia, who comprise 3–6% of the population, have low mathematical achievement in the context of otherwise normal intelligence, academic achievement, and access to educational resources (Butterworth, Varma, & Laurillard, 2011). Neuroimaging studies of people with dyscalculia have found reduced activation in the IPS, the neural correlate of magnitude representations, when comparing numerosities, i.e., sets of objects (Price, Holloway, Räsänen, Vesterinen, & Ansari, 2007). This has led to the development of computerized learning environments that "train" magnitude representations (Butterworth et al., 2011). For example, in "Rescue Calcularis," the user is shown a number line with only the poles and midpoint labeled (e.g., 0, 50, and 100), given a target number (e.g., 32) or sum (e.g., 27 + 5), and tasked with landing a ship on its approximate location; see Figures 7.1a–7.1c (Kucian et al., 2011). Playing this game improved the behavioral performance of people with dyscalculia, and also normalized the activation in the IPS (Kucian & von Aster, 2015); see Figure 7.1d.

Number, Space, and Instruction

Educators, cognitive scientists, and learning scientists have long understood the strong connection between number and arithmetic concepts on one hand and spatial manipulations on the other (Montessori, 1966). The relationship between numbers, space, and instruction is demonstrated by a series of studies investigating how children and adults understand the integers spanning from the neuroimaging center to the psychological laboratory to the classroom. In a psychological study, Varma and Schwartz (2011) found that when children reason about negative numbers, they employ rules such as "positives are greater than negatives." By contrast, adults have magnitude representations of negative numbers, and they organize these via *symmetry*, as reflections of positive numbers. The surprising role of symmetry was supported by a neuroimaging study finding activation in brain

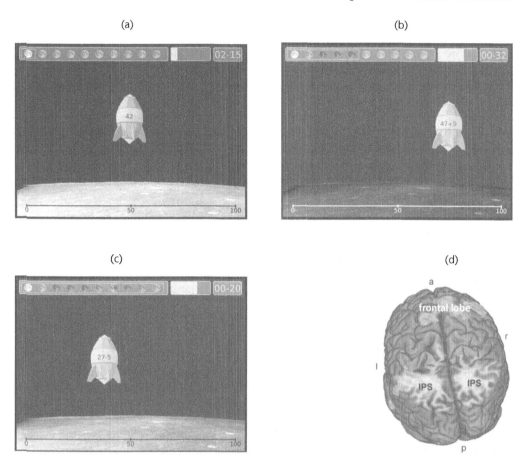

Figure 7.1a–7.1c "Rescue Calcularis"
Source: Kucian et al. (2011).

Figure 7.1d Improved Behavioral Performance for People with Dyscalculia, Normalized Activation in the IPS
Source: Kucian and von Aster (2015).

areas associated with processing visual symmetry (i.e., left lateral occipital cortex) when comparing integer magnitudes (Blair, Rosenberg-Lee, Tsang, Schwartz, & Menon, 2012). The instructional question, then, is whether emphasizing the symmetry of positive and negative integers in mathematics instruction leads to better learning? Tsang, Blair, Bofferding, and Schwartz (2015) addressed this question in a classroom study of elementary school children, finding an advantage for symmetry-based instruction over conventional approaches.

Challenges for Building Cognitive Neuroscience Foundations for the Learning Sciences

Cognitive neuroscience and the learning sciences share some research goals such as understanding thinking and learning. However, they follow different philosophies of science, employ different methods, and adopt different conceptualizations of learning. Here, we consider methodological and social barriers against productive exchange between the disciplines.

Methodological Differences

There are non-trivial methodological barriers to bridging between neuroscience and the learning sciences (Varma, McCandliss, & Schwartz, 2008). One difference between the disciplines concerns the context of learning. The methods of neuroscience target learning in individual brains. This is at odds with the social and collaborative nature of learning as it occurs in formal learning environments such as classrooms and informal learning environments such as museums (see Lyons, this volume). This is also at odds with social-constructivist conceptualizations of learning that are prevalent in the learning sciences (see Danish & Gresalfi, this volume). Another methodological difference between the disciplines concerns the timescale of learning: neuroscience experimental sessions typically last a few hours, whereas instructional studies span multiple days, weeks, or even months.

Although these methodological differences between the disciplines appear to be insurmountable, they are not. The neuroscience studies of greatest relevance to the learning sciences often adopt pre-post designs. This allows learning to take place in naturalistic settings, for it is the *consequences* of learning that are imaged, not the learning process itself. Such studies can investigate questions of interest to both disciplines, as we saw above in the example of the Delazer et al. (2005) study of the consequences of different kinds of instruction for organizing different brain networks in order to understand a new mathematical concept.

Educational Neuromyths

Although neuroscience findings have the potential to reveal why some kinds of instruction work and to inform the design of new educational activities, it is very easy for enthusiasm for neuroscience to slip into belief in *neuromyths*, which are incorrect applications of neuroscience findings (Howard-Jones, 2014). There are three primary sources of neuromyths in education. The first is improper extrapolations from neuroscience research, often conducted on animals, to recommendations about education (Bruer, 1997). This has led to overly strong claims about the importance of enriching preschool environments and widespread beliefs about punctuated critical periods in cognitive development and the validity of teaching to the left vs. right hemispheres (Dekker, Lee, Howard-Jones, & Jolles, 2012).

The second source of neuromyths is the "brain-based learning" industry that has grown over the past 20 years. These commercial interests repackage existing instructional approaches, add a gloss of neuroscience, and sell them to school districts, teachers, and parents. A recent example is the "brain training" industry, which has been criticized by cognitive psychologists and neuroscientists who study memory and learning for making exaggerated claims (Max Planck Institute for Human Development and Stanford Center on Longevity, 2014).

The third source of neuromyths is the seductive allure of neuroscience explanations (SANE): the finding that applications of psychological findings are rated as more credible when they are framed in terms of neuroscience findings (McCabe & Castel, 2008; cf. Farah & Hook, 2013). The SANE effect has recently been extended to the case where the applications concern educational topics (Im, Varma, & Varma, 2017). This might explain the persistence of educational neuromyths, because popular media articles that extrapolate wildly from neuroscience findings to educational recommendations are often accompanied by colorful brain images, which increase their believability (Keehner, Mayberry, & Fischer, 2011).

One explanation for why educational neuromyths arise and persist is that people find the simplicity of reductionist explanations appealing for complex phenomena such as thinking and learning in classroom contexts. They interpret the finding that "learning X 'lights up' area Y" as an *explanation* of what it means to 'learn X,' especially when the verbal finding is accompanied by a colorful image of brain "activation" (Rhodes, Rodriguez, & Shah, 2014).

Educational Neuroethics

It is becoming increasingly possible to apply advances from cognitive neuroscience to improve learning and instruction. But, should we? The ethical, legal, and social implications of applying neuroscience findings to social problems have been generally discussed in the literature (e.g., Farah, 2012). However, questions specific to applications to educational contexts have received comparatively little attention (Stein, Della Chiesa, Hinton, & Fischer, 2011).

How do relevant stakeholders in educational systems reason about the ethical dimensions of these applications? In a study including semi-structured interviews, Ball and Wolbring (2014) reported that parents were more likely to approve the use of safe and effective pharmacological interventions if their children have cognitive impairments. Howard-Jones and Fenton (2012) found that most teachers believe that grades obtained as a result of pharmacological interventions should be revised downward compared to those acquired without them, and that these interventions should not be available for free. Schmied et al. (2016) recently found that pre-service teachers were more appropriately cautious than undergraduate science majors when evaluating the use of pharmacological interventions, brain imaging techniques, and brain stimulation procedures in educational contexts, especially when the applications targeted more vulnerable groups such as low-achieving students and those with learning disabilities.

Conclusion

We are, of course, not the first to consider the relationship between cognitive neuroscience and the learning sciences (and education) (Bruer, 1997; Dubinsky, Roehrig, & Varma, 2013; Goswami, 2006; Howard-Jones et al., 2016; Varma et al., 2008). In this chapter, we first reviewed the foundations of cognitive neuroscience data and theories, sketched the neuroarchitecture of complex cognition, presented a case study of the insights that cognitive neuroscience is providing into mathematical thinking and learning, and considered fundamental barriers and broader obstacles to applying cognitive neuroscience findings to research problems in the learning sciences.

We end on an optimistic note. There is overlap in the research questions of cognitive neuroscience and the learning sciences. This represents a potential source of new methods, phenomena, and theoretical perspectives for each discipline. This potential is already being realized in studies that take their inspiration from, and contribute their results to, both fields (e.g., Tsang et al., 2015). Such transformative research is only possible when cognitive neuroscientists and learning scientists collaborate on questions of mutual interest. Such collaborations, although difficult to orchestrate, have the potential to make important contributions to the learning sciences.

Further Readings

Butterworth, B., Varma, S., & Laurillard, D. (2011). Dyscalculia: From brain to education. *Science, 332*, 1049–1053.
Butterworth et al. review the behavioral and neural correlates of developmental dyscalculia, and the use of games to strengthen number representations in this population.

Eden, G. F., Jones, K. M., Cappell, K., Gareau, L., Wood, F. B., Zeffiro, T. A., et al. (2004). Neural changes following remediation in adult developmental dyslexia. *Neuron, 44*, 411–422.
Eden et al. investigate whether successfully completing a behavioral intervention normalizes the brain activation patterns of people with dyslexia.

Fugelsang, J. A., & Dunbar, K. N. (2005). Brain-based mechanisms underlying complex causal thinking. *Neuropsychologia, 43*, 1204–1213.
Fugelsang and Dunbar investigate the neural correlates of reasoning from causal theories and evaluating experimental results that falsify those theories.

Grabner, R. H., Ansari, D., Reishofer, G., Stern, E., Ebner, F., & Neuper, C. (2007). Individual differences in mathematical competence predict parietal brain activation during mental calculation. *NeuroImage, 38*, 346–356.
Grabner et al. investigate the neural correlates of individual differences in mathematical achievement.

Varma, S., McCandliss, B. D., & Schwartz, D. L. (2008). Scientific and pragmatic challenges for bridging education and neuroscience. *Educational Researcher, 37*, 140–152.
Varma et al. outline the concerns of applying neuroscience findings to educational problems, and why these concerns also represent opportunities for innovative new research.

NAPLeS Resource

Varma, S., *Neurocognitive foundations for the Learning Sciences* [Webinar]. *In NAPLeS video series.* Retrieved October 19, 2017, from http://isls-naples.psy.lmu.de/intro/all-webinars/varma_video/index.html

References

Ansari, D., & Dhital, B. (2006). Age-related changes in the activation of intraparietal sulcus during nonsymbolic magnitude processing: An event-related functional Magnetic Resonance Imaging study. *Journal of Cognitive Neuroscience, 18*, 1820–1828.

Ball, N., & Wolbring, G. (2014). Cognitive enhancement: Perceptions among parents of children with disabilities. *Neuroethics, 7*, 345–364.

Baroody, A. J. (1985). Mastery of basic number combinations: Internalization of relationships or facts? *Journal for Research in Mathematics Education*, 83–98.

Blair, K. P., Rosenberg-Lee, M., Tsang, J. M., Schwartz, D. L., & Menon, V. (2012). Beyond natural numbers: Negative number representation in parietal cortex. *Frontiers in Human Neuroscience, 6*, 17.

Bruer, J. (1997). Education and the brain: A bridge too far. *Educational Researcher, 26*, 4–16.

Büchel, C., Coull, J. T., & Friston, K. J. (1999). The predictive value of changes in effective connectivity for human learning. *Science, 283*, 1538–1541.

Bugden, S., Price, G. R., McLean, D. A., & Ansari, D. (2012). The role of the left intraparietal sulcus in the relationship between symbolic number processing and children's arithmetic competence. *Developmental Cognitive Neuroscience, 2*, 448–457.

Butterworth, B., Varma, S., & Laurillard, D. (2011). Dyscalculia: From brain to education. *Science, 332*, 1049–1053.

De Smedt, B., Verschaffel, L., & Ghesquière, P. (2009). The predictive value of numerical magnitude comparison for individual differences in mathematics achievement. *Journal of Experimental Child Psychology, 103*, 469–479.

Dekker, S., Lee, N. C., Howard-Jones, P. A., & Jolles, J. (2012). Neuromyths in education: Prevalence and predictors of misconceptions among teachers. *Frontiers in Psychology, 3*, 429.

Delazer, M., Ischebeck, A., Domahs, F., Zamarian, L., Koppelstaetter, F., Siedentopf, C. M., et al. (2005). Learning by strategies and learning by drill—evidence from an fMRI study. *NeuroImage, 25*, 838–849.

Dubinsky, J. M., Roehrig, G., & Varma, S. (2013). Infusing neuroscience into teacher professional development. *Educational Researcher, 42*, 317–329.

Eden, G. F., Jones, K. M., Cappell, K., Gareau, L., Wood, F. B., Zeffiro, T. A., et al. (2004). Neural changes following remediation in adult developmental dyslexia. *Neuron, 44*, 411–422.

Farah, M. J. (2012). Neuroethics: The ethical, legal, and societal impact of neuroscience. *Annual Review of Psychology, 63*, 571–591.

Farah, M. J., & Hook, C. J. (2013). The seductive allure of "seductive allure." *Perspectives on Psychological Science, 8*, 88–90.

Gazzaniga, M. S., Ivry, R. B., & Mangun, G. R. (2013). *Cognitive neuroscience: The biology of the mind* (4th ed.) New York: W. W. Norton.

Goswami, U. (2006). Neuroscience and education: From research to practice? *Nature Reviews Neuroscience, 7*, 406–413.

Grabner, R. H., Ansari, D., Koschutnig, K., Reishofer, G., Ebner, F., & Neuper, C. (2009). To retrieve or to calculate? Left angular gyrus mediates the retrieval of arithmetic facts during problem solving. *Neuropsychologia, 47*, 604–608.

Grabner, R. H., Ansari, D., Reishofer, G., Stern, E., Ebner, F., & Neuper, C. (2007). Individual differences in mathematical competence predict parietal brain activation during mental calculation. *NeuroImage, 38*, 346–356.

Groen, G. J., & Parkman, J. M. (1972). A chronometric analysis of simple addition. *Psychological Review*, *79*, 329–343.

Halberda, J., Mazzocco, M. M. M., & Feigenson, L. (2008). Individual differences in non-verbal number acuity correlate with maths achievement. *Nature*, *455*, 665–668.

Howard-Jones, P. A. (2014). Neuroscience and education: Myths and messages. *Nature Reviews Neuroscience*, *15*, 817–824.

Howard-Jones, P. A., & Fenton, K. D. (2012). The need for interdisciplinary dialogue in developing ethical approaches to neuroeducational research. *Neuroethics*, *5*, 119–134.

Howard-Jones, P., Varma, S., Ansari, D., Butterworth, B., De Smedt, B., Goswami, U., et al. (2016). The principles and practices of educational neuroscience: Commentary on Bowers (2016). *Psychological Review*, *123*, 620–627.

Im, S.-h., Varma, K., & Varma, S. (2017). Extending the seductive allure of neuroscience explanations (SANE) effect to popular articles about educational topics. *British Journal of Educational Psychology*, *87*(4), 518–534.

Ishai, A., Ungerleider, L. G., Martin, A., Schouten, J. L., & Haxby, J. V. (1999). Distributed representation of objects in the human ventral visual pathway. *Proceedings of the National Academy of Science USA*, *96*, 9379–9384.

Just, M. A., Cherkassky, V. L., Aryal, S., & Mitchell, T. M. (2010). A neurosemantic theory of concrete noun representation based on the underlying brain codes. *PloS One*, *5*, e8622.

Just, M. A., & Varma, S. (2007). The organization of thinking: What functional brain imaging reveals about the neuroarchitecture of cognition. *Cognitive, Affective, and Behavioral Neuroscience*, *7*, 153–191.

Keehner, M., Mayberry, L., & Fischer, M. H. (2011). Different clues from different views: The role of image format in public perceptions of neuroimaging results. *Psychonomic Bulletin & Review*, *18*, 422–428.

Keller, T. A., & Just, M. A. (2009). Altering cortical connectivity: Remediation-induced changes in the white matter of poor readers. *Neuron*, *64*, 624–631.

Kucian, K., Grond, U., Rotzer, S., Henzi, B., Schönmann, C., Plangger, F., et al. (2011). Mental number line training in children with developmental dyscalculia. *NeuroImage*, *57*, 782–795.

Kucian, K., & von Aster, M. (2015). Developmental dyscalcula. *European Journal of Pediatrics*, *174*, 1–13.

Lashley, K. S. (1950). In search of the engram. *Symposia of the Society for Experimental Biology*, *4*, 454–482.

LeFevre, J. A., Sadesky, G. S., & Bisanz, J. (1996). Selection of procedures in mental addition: Reassessing the problem size effect in adults. *Journal of Experimental Psychology: Learning, Memory, and Cognition*, *22*, 216–230.

Luria, A. R. (1966). *Higher cortical functions in man*. London: Tavistock.

Max Planck Institute for Human Development and Stanford Center on Longevity. (2014, October 20). *A consensus on the brain training industry from the scientific community*. Retrieved from http://longevity3.stanford.edu/blog/2014/10/15/the-consensus-on-the-brain-training-industry-from-the-scientific-community/

McCabe, D., & Castel, A. (2008). Seeing is believing: The effect of brain images on judgments of scientific reasoning. *Cognition*, *107*, 343–352.

Mesulam, M.-M. (1990). Large-scale neurocognitive networks and distributed processing for attention, language and memory. *Annals of Neurology*, *28*, 597–613.

Molfese, D. L. (2000). Predicting dyslexia at 8 years of age using neonatal brain responses. *Brain and Language*, *72*, 238–245.

Montessori, M. (1966). *The discovery of the child* (M. Johnstone, Trans.). Madras, India: Kalakshetra Publications.

Moyer, R. S., & Landauer, T. K. (1967). Time required for judgments of numerical inequality. *Nature*, *215*, 1519–1520.

Pinel, P., Dehaene, S., Rivière, D., & Le Bihan, D. (2001). Modulation of parietal activation by semantic distance in a number comparison task. *NeuroImage*, *14*, 1013–1026.

Price, G. R., Holloway, I., Räsänen, P., Vesterinen, M., & Ansari, D. (2007). Impaired parietal magnitude processing in developmental dyscalculia. *Current Biology*, *17*, R1042–R1043.

Rhodes, R. E., Rodriguez, F., & Shah, P. (2014). Explaining the alluring influence of neuroscience information on scientific reasoning. *Journal of Experimental Psychology: Learning, Memory, and Cognition*, *40*, 1432–1440.

Rivera, S. M., Reiss, S. M., Eckert, M. A., & Menon, V. (2005). Developmental changes in mental arithmetic: Evidence for increased functional specialization in the left inferior parietal cortex. *Cerebral Cortex*, *15*, 1779–1790.

Schmied, A., Varma, S., Im, S.-h., Schleisman, K., Patel, P. J., & Dubinsky, J. M. (2016). *Reasoning about educational neuroethics*. Poster presented at the 2016 Meeting of the International Mind Brain and Education Society (IMBES), Toronto, Canada.

Sekuler, R., & Mierkiewicz, D. (1977). Children's judgements of numerical inequality. *Child Development*, *48*, 630–633.

Shaywitz, B. A., Lyon, G. R., & Shaywitz, S. E. (2006). The role of functional magnetic resonance imaging in understanding reading and dyslexia. *Developmental Neuropsychology*, *30*, 613–632.

Siegler, R. S. (1988). Strategy choice procedures and the development of multiplication skill. *Journal of Experimental Psychology: General, 117,* 258–275.

Sporns, O. (2011). *Networks of the brain.* Cambridge, MA: MIT Press.

Stein, Z., Della Chiesa, B., Hinton, C., & Fischer, K. W. (2011). Ethical issues in educational neuroscience: Raising children in a brave new world. In Illes & Sahakian (Eds.), *Oxford Handbook of Neuroethics* (pp. 803–822). Oxford: Oxford University Press.

Tsang, J. M., Blair, K. P., Bofferding, L., & Schwartz, D. L. (2015). Learning to "see" less than nothing: Putting perceptual skills to work for learning numerical structure. *Cognition and Instruction, 33,* 154–197.

Varma, S., McCandliss, B. D., & Schwartz, D. L. (2008). Scientific and pragmatic challenges for bridging education and neuroscience. *Educational Researcher, 37,* 140–152.

Varma, S., & Schwartz, D. L. (2011). The mental representation of integers: An abstract-to-concrete shift in the understanding of mathematical concepts. *Cognition, 121,* 363–385.

8
Embodied Cognition in Learning and Teaching
Action, Observation, and Imagination

Martha W. Alibali and Mitchell J. Nathan

A central claim of theories of embodied cognition is that cognitive processes are rooted in the actions of the human body in the physical world. In any given environment and setting, a specific individual has a set of potential perceptual experiences and actions. These experiences and actions depend on the specifics of the individual's body shape, morphology, and scale; the body's sensory and perceptual systems; the neural systems involved in planning and producing actions; and the affordances of these sensory, perceptual, and motor systems in that specific environment and setting. Although there is to date no single, unified theory of embodied cognition, the grounding of thought in perception and action is a common theme across a range of theoretical perspectives (Barsalou, 2008; Glenberg, 1997; Shapiro, 2011; Wilson, 2002).

The view that cognition is grounded in action implies that, across domains of reasoning, fundamental concepts and activities are based in actions of the body. This is the case, even for non-physical, non-observable ideas, which are conceptualized through their relations with sensorimotor experiences via metaphor (Lakoff & Johnson, 1999). From this perspective, then, action should matter for learning, and in turn, for instruction. In this chapter, we consider both learning and teaching from the perspective of embodied cognition, with a special focus on the importance of action.

Points of Contact with Cognitivist and Situated-Cognition Views

An embodied perspective on learning contrasts with more traditional, cognitivist perspectives on learning, which draw on the metaphor of the human brain as an information processing system (DeVega, Glenberg, & Graesser, 2008). As such, cognitivist approaches tend to focus on how arbitrary symbol systems and internal, mental representations mediate behavior. Studies in the cognitivist tradition often examine behavior in tasks and settings that are distant from authentic practices and contexts.

The embodied perspective has more in common with the situated cognition perspective and its assumption that thinking is bound to activities that occur in physical, social, and cultural contexts. From a situated cognition perspective (Robbins & Aydede, 2009), cognition is distributed across (or embedded in) cultural tools, inscriptions, and spaces, as exemplified by cognitive "offloading" to the environment (Wilson, 2002), and extended in the sense that sociocultural and physical settings are viewed as part of the cognitive system (Clark & Chalmers, 1998). The embodied perspective, in contrast, takes as central the physical body, the motor system, and the systems involved in sensing and perceiving. As such, the embodied cognition perspective construes cognition as the goal-directed activity of a human with a particular body in a particular physical environment and setting.

Martha W. Alibali and Mitchell J. Nathan

Overview of the Chapter

In this chapter, our focus is on the implications of an embodied cognition perspective for learning and instruction. We focus on three major principles. First, action matters for cognitive performance and learning. Appropriate actions can promote performance and learning, and unaligned actions can interfere. Second, observing others' actions can influence cognitive performance and learning, in the same ways that producing actions does. Third, imagining or simulating actions can also influence cognitive performance and learning. We consider both imagined actions and gestures, which are a form of representational action that manifests simulated actions. Taken together, evidence for these principles converges to highlight the importance of actions—both real and imagined—in cognitive performance, learning, and instruction. A focus on action also has far-reaching implications for central topics in the learning sciences, including instructional design, assessment, and educational technology; we consider these issues in the final section of the chapter.

Our focus on action, observation, and imagination can be illustrated by the forms of behavior that a child produces during a mathematics lesson about numerical equality that uses a pan balance (see Figure 8.1). The child may reason about equality via *action*, as she places cubes in the pans of the pan balance (Panel A); via *observation*, as she looks on while another child places cubes in the pans (Panel B); or via *imagination*, as she thinks about placing the cubes in the pans (Panel C).

Action Matters for Cognition and Learning

The idea that action matters for cognition and learning is rooted in developmental psychological theories, most notably those of Piaget and his successors, and in phenomenology. Both of these theoretical perspectives highlight the sensorimotor origins of knowledge and thought. According to Piaget, action is the foundation of thought; physical operations are internalized and transformed into mental operations, and eventually mental representations (Beilin & Fireman, 2000). Phenomenologists reject this last step and, instead, consider how actions and ways of engaging with the world provide a more direct account of cognition (Dreyfus, 2002). The feasibility of this approach has been demonstrated using mobile robot designs without mental representations (Brooks, 1991).

If cognitive processes are grounded in action, then the nature of the actions that people produce should make a difference for their cognitive performance and learning. Indeed, a growing body of research on language comprehension, problem solving, and mathematical reasoning supports this view. This body of work has revealed that actions that are structurally aligned with key features of the target material can foster comprehension, memory, and learning. In contrast, unaligned actions can interfere with comprehension, memory, and learning.

Language Comprehension

One formulation of the view that language comprehension is grounded in action understanding is that people index words and grammatical structures to real-world experiences (Glenberg & Robertson, 1999). Therefore, comprehending language evokes motor and perceptual affordances. If this is the case, then concurrent motor activity should affect language comprehension—and, indeed, it does. In one compelling demonstration of this effect, participants were asked to read sentences that implied motion away from the body (such as "Close the drawer") or motion towards the body (such as "Open the drawer"). Participants were asked to verify whether the sentences were sensible, either by making arm movements away from their bodies (to press a button farther from the body than the hand-at-rest position) or by making arm movements towards their bodies (to press a button closer to the body than the hand-at-rest position). Participants responded faster when the motion implied in the sentence was aligned with the direction of motion in the response (Glenberg & Kaschak, 2002). Importantly, this phenomenon held, not only for sentences describing concrete, physical

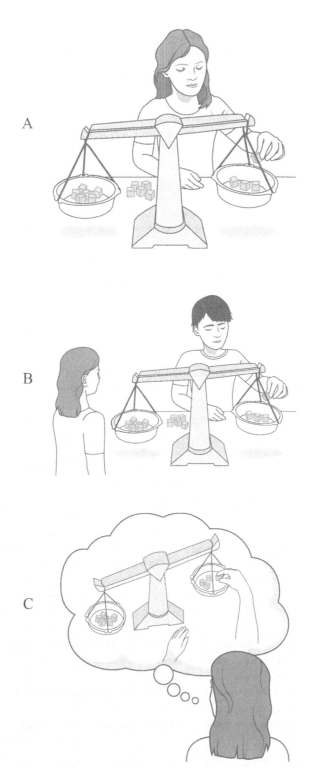

Figure 8.1 A Child Reasoning About Numerical Equality by Producing Actions (Panel A); by Observing Another Child's Actions (Panel B); or by Imagining Actions (Panel C).

Source: Figures available at http://osf.io/ydmrh under a CC-BY4.0 license (Alibali, 2017).

movements, but also for sentences describing the metaphorical movement of abstract entities (such as "Liz told you the story"). Later studies revealed that there is modulation of activity in the hand muscles when comprehending sentences that describe concrete actions and figurative "actions" that involve transfer of information (Glenberg et al., 2008).

Are the same processes involved in more complex contexts of language comprehension and use—for example, in tasks that require more than comprehending simple sentences? Glenberg and colleagues addressed this question in a set of studies on learning from text (Glenberg, Gutierrez, Levin, Japuntich, & Kaschak, 2004). First- and second-grade students were presented with three toy scenarios (a farm, a house, and a gas station) and asked to read brief texts about them. In one condition, children were asked to manipulate the toys (characters and objects, e.g., for the farm scenario, a tractor, a barn, animals, etc.) to correspond to each of the sentences in the text; in another condition, children observed but did not manipulate the toys. Children in the manipulate group showed better memory for the text than children in the observe group, and they were also better at drawing inferences based on the text. Similar findings were obtained when comparable manipulation interventions were implemented for solving mathematical story problems (Glenberg, Jaworski, Rischal, & Levin, 2007) and in small-group settings for reading (Glenberg, Brown, & Levin, 2007). Thus, manipulating relevant objects in text-relevant ways facilitates children's text comprehension, at least for simple passages about concrete objects and situations.

Problem Solving

Action also plays a powerful role in shaping the cognitive processes involved in problem solving. For example, Thomas and Lleras (2007) asked participants to solve a well-known insight problem that had a spatial solution (Duncker's radiation problem). During the solution period, in the guise of an unrelated visual tracking task, participants' eye movements were guided, either in a pattern that aligned with the problem's solution or in a pattern that did not align with the solution. Participants who moved their eyes in a way that aligned with the solution were more likely to solve the insight problem. Importantly, most participants did not suspect that there was a relationship between the tracking task and the insight problem, suggesting that the connection from the eye movements to the spatial solution to the problem was implicit.

Directed actions can also influence problem solving in more academic tasks, such as generating mathematical proofs. Nathan and colleagues (2014) asked undergraduate students to generate proofs for two mathematical statements, after performing either actions that were aligned with the key insights that underlay the proofs or actions that were not aligned with those key insights. Participants who produced aligned actions were more likely to discover those key insights as they worked to generate proofs.

Taken together, these studies help build a case that actions are integral in cognition and learning across a range of tasks, including language comprehension, problem solving, and mathematical reasoning. This view has motivated instructional approaches that involve actions, such as lessons with concrete manipulatives (see Figure 8.1). Indeed, a recent meta-analysis revealed beneficial effects of manipulatives on mathematics learning across a wide range of ages and concepts, although many additional factors moderate these effects (Carbonneau, Marley, & Selig, 2013). In this regard, it is important to consider whether learners must actually *produce* those actions, or whether viewing another's actions or even simply imagining actions might matter, as well. Research suggests that all three—acting, observing, and imagining—influence cognition and learning.

Observing Others' Actions Can Activate Learners' Embodied Knowledge

A large body of research, going back to early social learning theory (e.g., Bandura, 1965), and continuing to the present day (e.g., Chi, Roy & Hausmann, 2008; also see also Rummel & van Gog, this volume) demonstrates that people learn from observing others. Researchers from different theoretical perspectives refer to this process with different terms (e.g., imitative learning, vicarious learning, observational learning) and have offered different views regarding underlying mechanisms. An embodied perspective on this literature highlights the possibility that others' actions may also activate learners' action-based knowledge, and therefore influence their cognition and learning.

Many studies have shown that observing actions activates the same brain regions that are involved in actually executing those actions. Nearly two decades ago, Rizzolatti and colleagues demonstrated that the human motor cortex is activated in similar ways when humans execute motor tasks and when they observe them (Hari et al., 1998; Rizzolatti, Fogassi, & Gallese, 2001). They showed that patterns of brain activation were similar when people manipulated a small object with their hands and when they viewed another person performing that same action. Later research demonstrated the specificity of this effect, showing that viewing actions produced with specific effectors (e.g., the mouth, the hands, or the feet) elicited activation in the corresponding motor areas in people who observed the actions (Buccino et al., 2001). These findings suggest that when people observe an action, they generate an internal simulation of that action in their premotor cortex.

These findings open the possibility that learners need not physically produce actions themselves in order for action to affect their cognition and learning. Viewing others' actions in learning contexts may be sufficient to activate learners' action-based knowledge, or to facilitate learners' generating action-based knowledge.

The claim that individual learning may depend on *others'* actions highlights the importance of considering the learning context as a system involving multiple participants in interaction. The actions produced by a learner's teachers or peers in the learning context have the potential to influence that learner's thinking by activating action-based knowledge. Although many perspectives on learning—including sociocultural and ecological perspectives—highlight the importance of social interactions in learning, an embodied perspective foregrounds others' observed actions as a crucial element of the learning context.

Imagined or Simulated Actions Can Influence Cognition and Learning

Some scholars have highlighted the close ties between imagination and action. Indeed, Nemirovsky and Ferrara (2008) define *imagining* as "entertaining possibilities for action; entertaining (in the sense of 'holding' or 'keeping') a state of readiness for the enactment of possible actions" (p. 159). In our view, imagining involves mentally experiencing actions by engaging in motor imagery or mental simulation of action. This mental experience of action could be triggered in a range of ways, such as by reading or listening to words that describe actions, by planning an intended action, by thinking about performing an action, or by thinking about viewing another person's action.

There is evidence that listening to or reading sentences about actions activates the brain areas involved in producing those actions, suggesting that thinking about actions also engages simulations of actions. In one study (Tettamanti et al., 2005), participants passively listened to sentences involving actions with different effectors ("I bite an apple," "I grasp a knife," "I kick the ball"). The sentences about actions elicited activation in the brain areas involved in action execution and action observation, whereas control sentences about abstract content did not. Moreover, there were additional, specific loci of activation that reflected motor representations of the specific actions (i.e., mouth vs. hand vs. leg actions), suggesting that hearing action-related words led to activation in the corresponding motor regions. In a related study, passively reading the words *lick*, *pick*, and *kick* activated the same brain regions that were activated when participants moved their tongues, fingers, and feet (Hauk, Johnsrude, & Pulvermüller, 2004). Further, words denoting more specific actions (e.g., *wipe*) elicited greater activation in motor areas than words denoting more general actions (e.g., *clean*) (van Dam, Rueschemeyer, & Bekkering, 2010). Taken together, these studies suggest that hearing or reading action words automatically activates the brain regions involved in actually producing those actions.

Indeed, there is evidence suggesting that these forms of "mentally" acting are sufficient for action to influence cognition. In their study of actions and reading comprehension, Glenberg and colleagues (2004) also tested the effect of imagining actions. In one experiment, children practiced imagining rather than actually acting out actions on the task objects described in the text passages. Children who imagined manipulating the objects showed better memory for the text and better inference making than children in a control condition, who read the text twice. Presumably, children who were directed to specifically imagine acting on objects produced richer action simulations that children who simply read the text. Glenberg and colleagues argued that imagining actions conferred benefits for cognition and inference making that were similar to actually performing the actions.

Research in sports psychology has also addressed the relations between imagined action and actual action. Past research in this area focused on whether "mental practice" can influence performance and learning of motor skills. An early meta-analysis demonstrated that imagined actions can indeed influence later performance and learning (Feltz & Landers, 1983). The boundary conditions for such effects remain to be ascertained; however, at a minimum, they highlight the functional similarity—and, in some cases, functional equivalence—of action and action simulations (Beilock & Gonso, 2008).

Simulating action functions like producing action in many key respects, including activating relevant neural circuitry. From this perspective, then, it makes sense that imagining action—either in response to observing another's action or on one's own volition—may affect performance and learning in the same ways that producing action does. In some cases, simulating action may lead to actually producing action: simulations activate premotor and motor areas in the brain, and this activation may give rise to overt action. For example, one could think about tracing the shape of a triangle in the air, and this simulated action might give rise to actual movements that trace a triangle in the air.

Gesture as Simulated Action

The *Gesture as Simulated Action* framework (Hostetter & Alibali, 2008) holds that simulated actions and perceptual states are sometimes manifested in overt behavior as spontaneous gestures, which are a form of action that *represents* actual actions and perceptual states. According to this framework, when the level of activation of an individual's simulation exceeds that individual's "gesture threshold" (which depends on a set of individual, social, and situational factors), that simulation will give rise to a gesture. Gestures typically occur with speech, presumably because producing oral movements for speaking increases overall activation in motor and premotor areas, and this increased activation makes it more likely that activation will exceed the individual's gesture threshold. However, gestures can also occur in the absence of speech. For example, people often gesture without speech when engaging in challenging spatial tasks, such as mentally rotating objects (e.g., Chu & Kita, 2011).

Producing gestures may also feed back to increase the activation level of the simulated actions or perceptual states that gave rise to the gestures. Thus, producing gestures may increase activation on a simulation, and may consequently make an individual more likely to reason with that simulation. For example, in gear movement problems, which involve predicting how one gear will turn if another gear is moved in a particular direction, individuals who produce gestures are more likely to use strategies that involve simulating the movements of each gear (Alibali, Spencer, Knox, & Kita, 2011).

Just as viewing actions has much in common with producing actions, we argue that viewing gestures has much in common with producing gestures—and with producing the actions or experiencing the perceptual states that are represented in those gestures. Thus, viewing others' gestures may evoke simulations of actions in the viewers (Ping, Goldin-Meadow, & Beilock, 2014), and these simulated actions may in turn influence the viewers' thinking and their actions (Cook & Tanenhaus, 2009). Indeed, this may be one of the reasons why teachers' gestures are so influential in affecting student learning (e.g., Cook, Duffy, & Fenn, 2013; Singer & Goldin-Meadow, 2005). Teachers' gestures may evoke simulated actions in their students' thinking, and these simulated actions may in turn give rise to gestures or actions on the part of the students. In fact, students gesture more when their teachers gesture more (Cook & Goldin-Meadow, 2006), lending support to this idea.

This pathway provides a potential mechanism by which demonstrations and instructional gestures may influence learning. In brief, gestures may both manifest simulated actions on the part of the gesture producers (Alibali & Nathan, 2012) and evoke simulated actions on the part of those who observe the gestures (Kita, Alibali, & Chu, 2017).

Implications for Key Topics in the Learning Sciences

Taken together, the lines of work reviewed in this chapter converge to demonstrate the importance of actions—actual, observed, and imagined—in cognition, learning, and instruction. These ideas have implications for several key topics in the learning sciences, including instructional design, assessment, and educational technology.

Instructional Design

An embodied perspective on instructional design highlights the importance of considering students' opportunities to engage in actions, with a focus on "action-concept congruencies" (Lindgren & Johnson-Glenberg, 2013). One strand of research in this vein has focused on designing instructional interventions in which students engage in bodily actions (e.g., Fischer, Moeller, Bientzle, Cress, & Nuerk, 2010; Johnson-Glenberg, Birchfield, Tolentino, & Koziupa, 2013; Nathan & Walkington, 2017) or produce specific gestures (Goldin-Meadow, Cook, & Mitchell, 2009). Another strand of research has focused on concrete manipulatives and the actions they afford (e.g., Martin & Schwartz, 2005; Pouw, van Gog, & Paas, 2014). An emerging approach extends this perspective to digital manipulatives, as touch screen technology also affords actions (Ottmar & Landy, 2017).

An embodied perspective also highlights the potential value of observing others' actions and gestures. A growing body of recent work has sought to characterize the ways in which teachers use gestures in instructional settings and to investigate how teachers' gestures are involved in students' learning (e.g., Alibali et al., 2013; Furuyama, 2000; Richland, 2015). This body of work highlights the role of teachers' gesture in guiding students' simulations of relevant actions and perceptual states (Kang & Tversky, 2016).

Assessment

An embodied perspective on learning also has implications for the assessment of students' knowledge. In this regard, we consider two related issues. First, some action-based knowledge may not be verbally coded. Such knowledge may be readily exhibited in demonstrations or gestures, but it may not be spontaneously expressed in speech or writing. Indeed, many studies have shown that learners often express some aspects of their knowledge uniquely in gestures. Such "mismatching" gestures have been observed across a range of tasks, including conservation of quantity (Church & Goldin-Meadow, 1986), mathematical equations (Perry, Church, & Goldin-Meadow, 1988), and reasoning about balance (Pine, Lufkin, & Messer, 2004), and it has been argued that such gestures reveal the contents of learners' Zones of Proximal Development (Goldin-Meadow, Alibali, & Church, 1993). Given that learners' gestures often reveal information they do not express in speech, it follows that some aspects of learners' knowledge may be inappropriately discounted by assessment methods that consider only knowledge conveyed through speech or writing. Students' actions and gestures can reveal aspects of their knowledge that must be considered if formative assessments are to be accurate. Attending to students' actions and gesture may improve the validity of summative assessments, as well.

Second, assessment practices that prevent learners from producing task-relevant actions and gestures (such as assessment practices that require typing) may actually impair higher-order thinking, such as inference making (Nathan & Martinez, 2015). When assessment practices interfere with the processes involved in embodied thinking, this may compromise the validity of those assessment practices, yielding a false portrayal of what a test taker actually knows. For assessment to be accurate and valid, it may be necessary to use methods that allow learners to engage their bodies.

Educational Technology

As Lee (2015) notes, technological advances support the embodiment of concepts in new ways. The increasing availability of motion capture technologies such as the Kinect™ have made it possible—and practical—to design and implement interventions that elicit, track, and respond to learners' movements. There is an emerging class of embodied learning technologies that expressly uses players' movements to foster learning. Some of the systems in this early wave of embodied designs enlist participants to engage in movements that elicit intended associations and conceptualizations. For example, the GRASP project draws on the gestures and actions that students exhibited during successful reasoning about science phenomena, including gas pressure, heat transfer, and seasons (Lindgren, Wallon, Brown, Mathayas, & Kimball, 2016). Other students can be cued to use such gestures to interact with computer simulations of these phenomena when reasoning about causal mechanisms (Wallon & Lindgren, 2017). As a second example, players of *The Hidden Village* (Nathan & Walkington, 2017) match the actions of in-game agents. These actions are intended to foster players' insights about generalized properties of shapes and space, and these insights in turn may influence their geometry justifications and proofs. The target directed actions were curated from analyses of successful mathematical reasoning.

Embodied learning technologies can also support bottom-up processes that facilitate the generation and exploration of concepts. For example, in the Mathematical Imagery Trainer (Abrahamson & Trninic, 2014), students find the proper pacing of their two hands to represent target proportions (such as 1:2) without overt direction to do so, thereby generating sensorimotor schemes that may foster their conceptual development.

Conclusions

An embodied perspective on learning and instruction highlights the importance of producing, observing, and imagining actions. Actions that are well aligned with target ideas can promote cognitive performance and learning, whereas actions that are not well aligned can interfere. It is not always necessary for learners to produce actions on their own; observing others' actions can also activate action-based knowledge. Imagined or simulated actions can do so, as well, and these simulations may be manifested in gestures, which are a form of representational action. Finally, observing others' actions and gestures may provoke learners to simulate or to produce actions. In these ways, perceiving others' actions and gestures can also influence cognitive performance and learning.

A focus on action requires attention, not only to processes taking place in the learner's own brain and cognitive system, but also to the learner's activities with physical and cultural tools, and to the learner's interactions with other people. The physical and cultural tools that are available to learners both afford and constrain the actions that learners engage in. Learners' interactions with others—for example, in classrooms or in collaborative learning environments—typically involve opportunities for joint action and for observing others' actions. Thus, an embodied perspective on learning and instruction requires a broader view of the learner in physical, cultural, and social context.

In sum, embodied action is a foundational element of learning. As such, embodied action must be viewed as a fundamental construct in the learning sciences, with implications that broadly encompass theories, designs, and practices for assessment, teaching, and learning.

Further Readings

Glenberg, A. M., Gutierrez, T., Levin, J. R., Japuntich, S., & Kaschak, M. P. (2004). Activity and imagined activity can enhance young children's reading comprehension. *Journal of Educational Psychology, 96*(3), 424–436.
A set of empirical studies documenting the implications of action and imagined action in text comprehension and inference making.

Goldin-Meadow, S. (2005). *Hearing gesture: How our hands help us think.* Cambridge, MA: Harvard University Press.
A comprehensive introduction to gesture in reasoning, learning, and teaching.

Hall, R., & Nemirovsky, R. (Eds.). (2012). Modalities of body engagement in mathematical activity and learning [Special Issue]. *Journal of the Learning Sciences, 21*(2).
A compendium of papers representing a range of embodied perspectives on student performance, learning, and teaching of mathematics.

Lakoff, G., & Johnson, M. (1999). *Philosophy in the flesh: The embodied mind and its challenge to Western thought.* New York: Basic Books.
A seminal text that makes a strong case for embodied cognition on philosophical and linguistic grounds.

Shapiro, L. (2011). *Embodied cognition.* New York: Routledge.
An excellent introduction to basic themes and debates regarding embodied cognition.

References

Abrahamson, D., & Trninic, D. (2014). Bringing forth mathematical concepts: Signifying sensorimotor enactment in fields of promoted action. *ZDM, 47*, 295–306. doi:10.1007/s11858-014-0620-0

Alibali, M. W. (2017, February 27). *Action-observation-imagination figures.* Retrieved from http://osf.io/ydmrh

Alibali, M. W., & Nathan, M. J. (2012). Embodiment in mathematics teaching and learning: Evidence from students' and teachers' gestures. *Journal of the Learning Sciences, 21*, 247–286. doi:10.1080/10508406.2011.611446

Alibali, M. W., Nathan, M. J., Wolfgram, M. S., Church, R. B., Jacobs, S. A., Martinez, C. J., & Knuth, E. J. (2013). How teachers link ideas in mathematics instruction using speech and gesture: A corpus analysis. *Cognition & Instruction, 32*, 65–100. doi:10.1080/07370008.2013.858161

Alibali, M. W., Spencer, R. C., Knox, L., & Kita, S. (2011). Spontaneous gestures influence strategy choices in problem solving. *Psychological Science, 22*, 1138–1144. doi:10.1177/0956797611417722

Bandura, A. (1965). Vicarious processes: A case of no-trial learning. In L. Berkowitz (Ed.), *Advances in experimental social psychology* (Vol. 2, pp. 1–55). New York: Academic Press.

Barsalou, L. W. (2008). Grounded cognition. *Annual Review of Psychology, 59*, 617–645. doi:10.1146/annurev.psych.59.103006.093639

Beilin, H., & Fireman, G. (2000). The foundation of Piaget's theories: Physical and mental action. *Advances in Child Development and Behavior, 27*, 221–246. doi:10.1016/S0065-2407(08)60140-8

Beilock, S. L., & Gonso, S. (2008). Putting in the mind versus putting on the green: Expertise, performance time, and the linking of imagery and action. *Quarterly Journal of Experimental Psychology, 61*, 920–932. doi:10.1080/17470210701625626

Brooks, R. A. (1991). Intelligence without representation. *Artificial Intelligence, 47*, 139–159. doi:10.1016/0004-3702(91)90053-M

Buccino, G., Binkofski, F., Fink, G. R., Fogassi, L., Gallese, V., Seitz, R. J., et al. (2001). Action observation activates premotor and parietal areas in a somatotopic manner: an fMRI study. *European Journal of Neuroscience, 13*, 400–404. doi:10.1046/j.1460-9568.2001.01385.x

Carbonneau, K. J., Marley, S. C., & Selig, J. P. (2013). A meta-analysis of the efficacy of teaching mathematics with concrete manipulatives. *Journal of Educational Psychology, 105*, 380–400. doi:10.1037/a0031084

Chi, M. T. H., Roy, M. & Hausmann, R. G. M. (2008). Observing tutorial dialogues collaboratively: Insights about human tutoring effectiveness from vicarious learning. *Cognitive Science, 32*, 301–341. doi:10.1080/03640210701863396

Chu, M., & Kita, S. (2011). The nature of gestures' beneficial role in spatial problem solving. *Journal of Experimental Psychology: General, 140*, 102–116. doi:10.1037/a0021790

Church, R. B., & Goldin-Meadow, S. (1986). The mismatch between gesture and speech as an index of transitional knowledge. *Cognition, 23*, 43–71. doi:10.1016/0010-0277(86)90053-3

Clark, A., & Chalmers, D. (1998). The extended mind. *Analysis, 58*(1), 7–19.

Cook, S. W., Duffy, R. G., & Fenn, K. M. (2013). Consolidation and transfer of learning after observing hand gesture. *Child Development, 84*, 1863–1871. doi:10.1111/cdev.12097

Cook, S. W., & Goldin-Meadow, S. (2006). The role of gesture in learning: Do children use their hands to change their minds? *Journal of Cognition and Development, 7*, 211–232. doi:10.1207/s15327647jcd0702_4

Cook, S. W., & Tanenhaus, M. K. (2009). Embodied communication: Speakers' gestures affect listeners' actions. *Cognition, 113*, 98–104. doi:10.1016/j.cognition.2009.06.006

DeVega, M., Glenberg, A. M., & Graesser, A. C. (Eds.). (2008). *Symbols, embodiment and meaning: A debate*. Oxford, England: Oxford University Press.

Dreyfus, H. L. (2002). Intelligence without representation—Merleau-Ponty's critique of mental representation: The relevance of phenomenology to scientific explanation. *Phenomenology and the Cognitive Sciences, 1*, 367–383. doi:10.1023/A:1021351606209

Feltz, D. L., & Landers, D. M. (1983). The effects of mental practice on motor skill learning and performance: A meta-analysis. *Journal of Sport Psychology, 5*, 25–57. doi:10.1123/jsp.5.1.25

Fischer, U., Moeller, K., Bientzle, M., Cress, U., & Nuerk, H.-C. (2010). Sensori-motor spatial training of number magnitude representation. *Psychonomic Bulletin & Review, 18*, 177–183. doi:10.3758/s13423-010-0031-3

Furuyama, N. (2000). Gestural interaction between the instructor and the learner in origami instruction. In D. McNeill (Ed.), *Language and gesture: Window into thought and action* (pp. 99–117). Cambridge, UK: Cambridge University Press. doi:10.1017/CBO9780511620850.007

Glenberg, A. M. (1997). What memory is for. *Behavioral and Brain Sciences, 20*, 1–55. doi:10.1017/s0140525x97000010

Glenberg, A. M., Brown, M., & Levin, J. R. (2007). Enhancing comprehension in small reading groups using a manipulation strategy. *Contemporary Educational Psychology, 32*, 389–399. doi:10.1016/j.cedpsych.2006.03.001

Glenberg, A. M., Gutierrez, T., Levin, J. R., Japuntich, S., & Kaschak, M. P. (2004). Activity and imagined activity can enhance young children's reading comprehension. *Journal of Educational Psychology, 96*, 424–436. doi:10.1037/0022-0663.96.3.424

Glenberg, A. M., Jaworski, B., Rischal, M., & Levin, J. R. (2007). What brains are for: Action, meaning, and reading comprehension. In D. McNamara (Ed.), *Reading comprehension strategies: Theories, interventions, and technologies* (pp. 221–240). Mahwah, NJ: Lawrence Erlbaum Associates.

Glenberg, A. M., & Kaschak, M. P. (2002). Grounding language in action. *Psychonomic Bulletin & Review, 9*, 558–565. https://doi.org/10.3758/BF03196313

Glenberg, A. M., & Robertson, D. A. (1999). Indexical understanding of instructions. *Discourse Processes, 28*, 1–26. doi:10.1080/01638539909545067

Glenberg, A. M., Sato, M., Cattaneo, L., Riggio, L., Palumbo, D., & Buccino, G. (2008). Processing abstract language modulates motor system activity. *Quarterly Journal of Experimental Psychology, 61*, 905–919. doi:10.1080/17470210701625550

Goldin-Meadow, S., Alibali, M. W., & Church, R. B. (1993). Transitions in concept acquisition: Using the hand to read the mind. *Psychological Review, 100*, 279–297. doi:10.1037/0033-295X.100.2.279

Goldin-Meadow, S., Cook, S. W., & Mitchell, Z. A. (2009). Gesturing gives children new ideas about math. *Psychological Science, 20*, 267–272. doi:10.1111/j.1467-9280.2009.02297.x

Hari, R., Forss, N., Avikainen, S., Kirveskari, E., Salenius, S., & Rizzolatti, G. (1998). Activation of human primary motor cortex during action observation: A neuromagnetic study. *Proceedings of the National Academy of Sciences, 95*, 15061–15065. doi:10.1073/pnas.95.25.15061

Hauk, O., Johnsrude, I., & Pulvermüller, F. (2004). Somatotopic representation of action words in human motor and premotor cortex. *Neuron, 41*, 301–307. doi:10.1016/S0896-6273(03)00838-9

Hostetter, A. B., & Alibali, M. W. (2008). Visible embodiment: Gestures as simulated action. *Psychonomic Bulletin & Review, 15*, 495–514. doi:10.3758/PBR.15.3.495

Johnson-Glenberg, M. C., Birchfield, D. A., Tolentino, L., & Koziupa, T. (2013). Collaborative embodied learning in mixed reality motion-capture environments: Two science studies. *Journal of Educational Psychology, 106*, 86–104. doi:10.1037/a0034008.supp

Kang, S., & Tversky, B. (2016). From hands to minds: Gestures promote understanding. *Cognitive Research: Principles and Implications, 1*, 4. doi:10.1186/s41235-016-0004-9

Kita, S., Alibali, M. W., & Chu, M. (2017). How do gestures influence thinking and speaking? The gesture-for-conceptualization hypothesis. *Psychological Review, 124*, 245–266. doi: 10.1037/rev0000059

Lakoff, G., & Johnson, M. (1999). *Philosophy in the flesh: The embodied mind and its challenge to Western thought*. New York: Basic Books.

Lee, V. R. (Ed.). (2015). *Learning technologies and the body: Integration and implementation in formal and informal learning environments*. New York: Routledge.

Lindgren, R., & Johnson-Glenberg, M. (2013). Emboldened by embodiment: Six precepts for research on embodied learning and mixed reality. *Educational Researcher, 42*(8), 445–452. doi:10.3102/0013189X13511661

Lindgren, R., Wallon, R. C., Brown, D. E., Mathayas, N., & Kimball, N. (2016). "Show me" what you mean: Learning and design implications of eliciting gesture in student explanations. In C.-K. Looi, J. Polman, U. Cress, & P. Reimann (Eds), *Transforming learning, empowering learners: The International Conference of the Learning Sciences (ICLS) 2016* (Vol. 2, pp. 1014–1017). Singapore: International Society of the Learning Sciences.

Martin, T., & Schwartz, D. L. (2005). Physically distributed learning: Adapting and reinterpreting physical environments in the development of fraction concepts. *Cognitive Science, 29*, 587–625. doi.10.1207/s15516709cog0000_15

Nathan, M. J., & Martinez, C. V. J. (2015). Gesture as model enactment: The role of gesture in mental model construction and inference making when learning from text. *Learning: Research and Practice, 1*, 4–37. doi:10.1080/23735082.2015.1006758

Nathan, M. J., & Walkington, C. A. (2017). Grounded and embodied mathematical cognition: Promoting mathematical insight and proof using action and language. *Cognitive Research: Principles and Implications, 2*:9. doi 10.1186/s41235-016-0040-5

Nathan, M. J., Walkington, C., Boncoddo, R., Pier, E., Williams, C. C., & Alibali, M. W. (2014). Actions speak louder with words: The roles of action and pedagogical language for grounding mathematical proof. *Learning and Instruction, 33*, 182–193. doi:10.1016/j.learninstruc.2014.07.001

Nemirovsky, R., & Ferrara, F. (2008). Mathematical imagination and embodied cognition. *Educational Studies in Mathematics, 70*, 159–174. doi:10.1007/s10649-008-9150-4

Ottmar, E., & Landy, D. (2017). Concreteness fading of algebraic instruction: Effects on learning. *Journal of the Learning Sciences, 26*, 51–78.

Perry, M., Church, R. B., & Goldin-Meadow, S. (1988). Transitional knowledge in the acquisition of concepts. *Cognitive Development, 3*, 359–400. doi:10.1016/0885-2014(88)90021-4

Pine, K. J., Lufkin, N., & Messer, D. (2004). More gestures than answers: Children learning about balance. *Developmental Psychology, 40*, 1059–1067. doi:10.1037/0012-1649.40.6.1059

Ping, R., Goldin-Meadow, S., & Beilock S. (2014). Understanding gesture: Is the listener's motor system involved? *Journal of Experimental Psychology: General, 143*, 195–204. doi:10.1037/a0032246

Pouw, W. T. J. L., van Gog, T., & Paas, F. (2014). An embedded and embodied cognition review of instructional manipulatives. *Educational Psychology Review, 26*, 51–72. doi:10.1007/s10648-014-9255-5

Richland, L. E. (2015). Linking gestures: Cross-cultural variation during instructional analogies. *Cognition and Instruction, 33*, 295–321. doi:10.1080/07370008.2015.1091459

Rizzolatti, G., Fogassi, L., & Gallese, V. (2001). Neurophysiological mechanisms underlying the understanding and imitation of action. *Nature Reviews Neuroscience, 2*, 661–670. doi:10.1038/35090060

Robbins, P., & Aydede, M. (2009). A short primer on situated cognition. In P. Robbins & M. Aydede (Eds.), *The Cambridge handbook of situated cognition* (pp. 3–10). Cambridge, England: Cambridge University Press.

Rummel, N., & van Gog, T. (2018). Example-based learning, In F. Fischer, C. E. Hmelo-Silver, S. R. Goldman, & P. Reimann (Eds.), *International handbook of the learning sciences* (pp. 201–229). New York: Routledge.

Shapiro, L. (2011). *Embodied cognition*. New York: Routledge.

Singer, M. A., & Goldin-Meadow, S. (2005). Children learn when their teacher's gestures and speech differ. *Psychological Science, 16*, 85–89. doi:10.1111/j.0956-7976.2005.00786.x

Tettamanti, M., Buccino, G., Saccuman, M. C., Gallese, V., Danna, M., Scifo, P., et al. (2005). Listening to action-related sentences activates fronto-parietal motor circuits. *Journal of Cognitive Neuroscience, 17*, 273–281. doi:10.1162/0898929053124965

Thomas, L. E., & Lleras, A. (2007). Moving eyes and moving thought: On the spatial compatibility between eye movements and cognition. *Psychonomic Bulletin & Review, 14*, 663–668. doi:10.3758/BF03196818

van Dam, W. O., Rueschemeyer, S.-A., & Bekkering, H. (2010). How specifically are action verbs represented in the neural motor system: An fMRI study. *NeuroImage, 53*, 1318–1325. doi:10.1016/j.neuroimage.2010.06.071

Wallon, R. C., & Lindgren, R. (2017). Cued gestures: Their role in collaborative discourse on seasons. In B. K. Smith, M. Borge, E. Mercier, & K. Y. Lim (Eds.), *Making a difference: Prioritizing equity and access in CSCL, 12th International Conference on Computer Supported Collaborative Learning (CSCL) 2017* (Vol. 2, pp. 813–814). Philadelphia, PA: International Society of the Learning Sciences.

Wilson, M. (2002). Six views of embodied cognition. *Psychonomic Bulletin & Review, 9*, 625–636. doi:10.3758/BF03196322

9

Learning From Multiple Sources in a Digital Society

Susan R. Goldman and Saskia Brand-Gruwel

Introduction

Twenty-first-century society relies on digital technology for information relevant to all spheres of life. Yet, the majority of citizens lack knowledge, skills, and practices needed to meet the challenges posed by the ubiquitous array of information and technologies that they have at their fingertips (American College Testing, 2006; Ananiadou, & Claro, 2009; National Center for Education Statistics, 2013; National Research Council, 2012; Organization for Economic Cooperation and Development [OECD], 2013).

This chapter brings together work from two research areas that have focused on the competencies demanded by 21st-century society: information problem solving and multiple source discourse comprehension. Information problem solving has tended to focus on processes of search, evaluation, and selection of task-relevant, useful, and accurate sources of information, typically involving use of the internet. On the other hand, the majority of multiple source comprehension work has tended to focus on processes of sense-making from researcher-provided information resources, often pre-selected to vary along dimensions hypothesized as important to learners' decisions regarding task-relevant information within and across information resources. Both research areas have been concerned with source attributes (e.g., expertise of the author, potential self-interest of the publisher), as well as learner and task characteristics that impact search, evaluation, and integration processes within and across sources. Investigations in both areas rely on experimental and quasi-experimental paradigms and use multiple methodologies and dependent measures, including think-alouds, read-alouds, eye-tracking, navigation logs, rating tasks, and constructed and forced-choice responses. Topics and spheres of life tapped by this research reflect a range from academic topics typical of formal schooling (e.g., in history "Was U.S. intervention in the Panama justified?"; in science "Explain volcanic eruptions") to more informal, personal decisions about environmental and health issues (e.g., climate change and cell phone use).

The integrative effort of this chapter seems particularly apt for this handbook because of the central emphasis in Learning Sciences on authentic tasks and situations as contexts for studying learning and problem solving through iterative design-based research (DBR) (e.g., Cobb, Confrey, diSessa, Lehrer, & Schauble, 2003). Although work on learning and problem solving from multiple sources more typically employs (quasi-)experimental methodologies, there is convergence with research in the DBR tradition on the importance of characteristics and attributes associated with learners, tasks, and the resources and tools available or provided. How and what individuals and groups of learners actualize of the potential affordances of specific multiple source comprehension and information problem-solving situations is

Learning From Multiple Sources

of interest from theoretical, pedagogical, and practical perspectives (Barab & Roth, 2006). Thus, this chapter can inform efforts in the Learning Sciences to design learning contexts, technology-based, or otherwise, that support authentic learning and problem solving from multiple information sources.

In addition, the Learning Sciences draws on multiple disciplines that have stimulated interest in multiple source comprehension and information problem solving. For example, cognitive analyses show that experts and novices in history use different approaches to sense-making with historical documents (e.g., Wineburg, 1991). Scientists adopt different reading goals for articles in their disciplinary specialization compared to articles outside their specializations (Bazerman, 2004). Sociolinguistic and anthropological studies of scientists' in situ comprehension and problem solving reveal spirited argumentation about data, models, and their integration with "accepted theory" (Latour & Woolgar, 1986). The appearance of computer technologies and of hypermedia and webpages challenged discourse and reading comprehension researchers to explore what these meant for sense-making, meaning construction, and literacy (e.g. New London Group, 1996). Finally, those interested in the public understanding of science and civics education have explored laypersons' comprehension of and problem solving with information available outside traditional gatekeeping and review processes (e.g., Bromme & Goldman, 2014).

We begin with an integrative model of multiple source discourse comprehension and information problem solving. We then summarize research findings regarding search and selection of information resources and sense-making from multiple resources. We conclude with suggestions for future research.

An Integrative Model (Framework) for Multiple Source Comprehension and Information Problem Solving

Figure 9.1 depicts a conceptual model as an integrative framework for various elements of multiple source comprehension and information problem solving. It is consistent with the Multiple

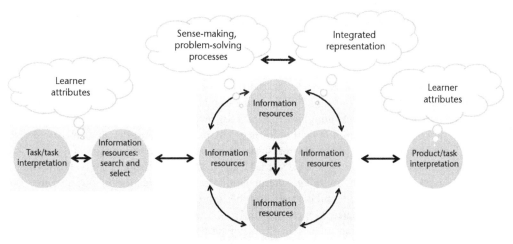

Figure 9.1 Representation of comprehension of and problem solving with multiple information resources.

Note: *Learner attributes* include prior content knowledge, epistemic cognitions and beliefs, attitudes, skills, and dispositions. *Integrated representations* result from the application of *sense-making and problem-solving processes*, including information search and selection, basic (e.g., decoding, word recognition, parsing, resonance, simple inferences) and complex comprehension processes (e.g., analysis within and across resources, interpretation, reasoning; synthesis within and across resources; sourcing processes; critique and evaluation)

Document Task-based Relevance Assessment and Content Extraction (MD-TRACE) model of Rouet and Britt (2011) and with models such as the Information-Problem Solving with Internet (IPS-I) model (Brand-Gruwel, Wopereis, & Vermetten, 2005). This class of models points to a complex of interacting processes, knowledge, and dispositions that are called on when learning and solving problems in multiple source situations, including identifying information needs, locating information sources, evaluating these sources (e.g., in terms of quality, trustworthiness, etc.), extracting, comprehending, and organizing the information from each source, and then synthesizing across sources to address the task as interpreted by the learner.

More specifically, the leftmost column of Figure 9.1 depicts the intersection of learner attributes, task, and source components at the start of problem solving. *Learner attributes* encompass what learners bring to the task, including prior knowledge of domain and topic content, epistemic cognitions and beliefs, dispositions, and attitudes. *Task/task interpretation* refers to what learners are asked to do as well as how they interpret the instructions. Importantly, tasks vary along a continuum of "known-answer" questions for which it is likely a solution can be retrieved (e.g., "What is the capital city of France?") to open-ended inquiry problems for which multiple solutions might be possible. Tasks also differ with respect to the sphere of life and/or sociocultural context to which they apply. For example, tasks may be situated within a discipline (e.g., Asian Studies, Biology) and invoke formal, academic norms and conventions or they may be more informal (e.g., personal decisions). *Information resources* refers to what are more frequently referred to as sources, documents, or texts. We prefer the phrase "information resources" because we are referring to the multiplicity and variety of representational forms that convey ideas that might be relevant to the task and that are accessible digitally or physically. We include traditional texts as well as multiple modalities (e.g., verbal, visual; dynamic, static) and genres (e.g., traditional print books, journals, peer reviewed, blogs, commercial advertisements). In authentic problem solving, resources must be identified through search processes that on the internet produce search results pages (SERPs), the contents of which need to be evaluated and prioritized for examination. When resources are provided, search and selection are more constrained. The two-way arrow between task and resources indicates that task interpretations guide search and selection of relevant resources but that results of search and selection feed back into interpretations and reinterpretations of tasks.

The middle portion of Figure 9.1 reflects the sense-making comprehension and problem-solving processes learners apply to information resources and that produce integrated mental representations of information the learner believes relevant to the task. For sense-making and problem solving to occur, selected resources must be "read." To read, the learner uses basic reading processes plus more complex comprehension and reasoning processes that are involved in evaluating relevance, reliability, and validity of the information in light of learners' task interpretations. Processes associated with "reading" visuals and relating verbal and visual information are also involved in light of their ubiquitous presence in web-based resources (e.g., Stieff, Hegarty, & Deslongchamps, 2011). Furthermore, in multiple resource situations, meaning, interpretation, and representation of one resource can potentially stimulate comparisons and contrasts to other sources (indicated in the arrows among the information resources and integrated representation).

The integrated representation reflects an expansion of the mental representation proposed for single text comprehension (Kintsch, 1998; van Dijk & Kintsch, 1983). In addition to three levels of representation for single texts, i.e., surface level (the specific words and layout that often fades rapidly), textbase (a literal level of what the text says), and situation model (integration of prior knowledge with textbase), two critical elements need to be added to accommodate multiple information resources: source nodes and inter-text links. As initially articulated sourcing processes result in source nodes that convey author/producer, time, place, and purpose of production and that are linked to the specific text to which they refer, producing a Document Model (Perfetti, Rouet, & Britt, 1999). Note that here and in referencing prior work we use the terminology of the models and studies. Links between different sources/texts reflect judgments and decisions based on content comparisons and contrasts, resulting in Intertext Models (Rouet & Britt, 2011). A substantial amount of multiple source discourse research focuses on when and how these additional elements are created and represented (see Bråten, Stadtler, & Salmerón, in press). We propose that evaluative decisions regarding perspective and judged "trustworthiness" or reliability of information are represented, either in source nodes as part of the Document Model, or in links between documents, although to date the representation of such information has not been directly investigated. Finally the two-way arrow between the leftmost and middle portions of Figure 9.1 indicates that sense-making processes sometimes reveal the need for additional or different information resources and can thus reinitiate searches and selection (Anmarkrud, Bråten, & Strømsø, 2014; Goldman, Braasch, Wiley, Graesser, & Brodowinska, 2012).

The results of sense-making are used to create a product that meets the requirements for task completion (Figure 9.1, rightmost portion). The quality of the product depends on the adequacy of source selection and use of the information to meet the demands of the interpreted/reinterpreted task. Furthermore, having completed the task, learner attributes are expected to be altered, although how and with what robustness are open questions.

In Figure 9.1, we explicitly acknowledge the sociocultural context of multiple source comprehension and information problem solving, although as yet there is little research exploring the impact of these more macro-level dimensions. Similarly, there is scant research on the iterative relationships among task interpretation, search and source selection, processing of sources to accomplish tasks, and reinterpretations of tasks as problem solutions proceed. There is, however, a good bit of research investigating subsets of the elements and relationships depicted in Figure 9.1, with the lion's share of attention on sourcing processes, sense-making, and individual differences in these. We turn now to that research.

Sourcing Processes in Comprehension and Problem Solving

Research examining sourcing processes encompasses several types of studies. Some focus on initial search and selection processes particularly in internet-based search engines (e.g. Mason, Boldrin, & Ariasi, 2010; Walraven, Brand-Gruwel, & Boschuizen, 2009). In others, information resources are described as the output of search engines (e.g., Goldman et al., 2012; Strømsø, Bråten, & Britt, 2010). In both situations, questions of interest concern why sites are selected for further examination and use in completing tasks. Another type of study examines selection among sources and selection of specific sections within a source as learners attempt to comprehend and make sense of the information in light of the task requirements (e.g., Stadtler, Scharrer, Skodzik, & Bromme, 2014; Wiley et al., 2009).

Studies of selection among and within sources assume that evaluations are being made of the utility of the information resource in the task context. Hence, selection and evaluation are often used interchangeably in describing the findings. However, other research explicitly asks learners to justify their selections or provide ratings that evaluate information resources. Persistent themes in studies of sourcing processes concern the impact on performance of task and information resource characteristics as well as learner attributes (e.g., Brand-Gruwel, Kammerer, van Meeuwen, & van Gog, 2017; Bråten, Ferguson, Strømsø, & Anmarkrud, 2014; Wiley et al., 2009).

Initial Search and Selection

A typical multiple resource problem-solving situation begins with locating potentially relevant information. This involves an initial interpretation of the task to identify and/or generate keywords or search terms to guide the search. Research indicates that when working on inquiry problems (e.g., "In what ways was economics a factor in the emergence of Europe from the period known as the Dark Ages?") as compared to known-answer questions or fact-based questions (e.g., "What is the capital city of the Netherlands?"), people use more search queries and search terms and are more likely to adapt initial search queries (Barsky & Bar-Ilan, 2012; Singer, Norbisrath, & Lewandowski, 2012).

Inputting keywords to a search engine produces a search engine results page (SERP) with multiple entries. Studies on evaluation of SERPs reveal that the top-ranked search results are most frequently selected for further examination, with little regard for attributes of the listing (e.g. Guan & Cutrell, 2007; Salmerón, Kammerer, & García-Carrión, 2013). For example, think-alouds indicate that students emphasized the rank in the list and the title/summary of the search result as the most important evaluation criteria (Walraven et al., 2009). However, Höchstötter and Lewandowski (2009) pointed out two flaws in the "first-listed" strategy: (1) search engines are not neutral in ordering the listings on the results page, with certain websites being "weighted" to appear in first position on SERPs; and (2) many websites use search engine optimization to improve the likelihood of being listed among the top half-dozen sites on SERPs. Many internet users are uninformed about how the ordering on SERPS is determined. The tendency to go with first-listed site and proceed to inspect sites "in order" occurs both when users generate their own search terms and when SERPs are provided (e.g., Wiley, 2009).

Although the research indicates that spontaneous selections of information resources tend to be based on relatively superficial features (surface match of content words with question) and potentially artificial "relevance" of the site to the task, with minimal prompting or instruction students readily engage additional criteria in the selection of sites on SERPs (Walraven et al., 2009). For example, in one study problem solvers who were explicitly instructed to include several aspects of relevancy made selections reflecting these criteria (Kim, Kazai, & Zitouni, 2013). As well, Gerjets, Kammerer, and Werner (2011) found that the use of explicit instructions to evaluate search results led to a significantly higher number of verbalized evaluation comments relative to those made under standard instructions.

Making Sense of Resources to Accomplish the Task

An issue in multiple resource comprehension and problem-solving situations is how readers make sense of the multiple resources to create a mental model representation that is relevant to accomplishing the task and that does so in a coherent and complete manner. A similar challenge exists even when only a single information resource is in play. However, unlike single texts, where it is reasonable for readers to assume that authors will attempt to present coherent accounts or indicate why and where accounts disagree, no such assumption applies when learners are dealing with multiple texts (see, for discussion, Britt, Rouet, & Braasch, 2013). Hence, it falls to the learner to piece together a coherent whole, much like assembling a jigsaw puzzle but without the benefit of the finished picture to guide the construction process. Of major interest is how learners engage in this activity and differences related to type of task (e.g., description versus argument), topic (e.g., relatively non-controversial versus controversial), consistency of the content in the information resources (e.g., consistent or conflicting), domain expertise (e.g., content and domain epistemology), and, in the case of controversial topics, personal attitudes and beliefs.

Similar to the relatively naive approaches evidenced in search tasks, studies on comprehension and problem solving conducted with adolescents and young adults have revealed little in the way of attention to source attributes or critical analysis, whether in history (e.g., Rouet, Favart, Britt, & Perfetti, 1997; Wineburg, 1991), science (e.g., Walraven et al., 2009; Ward, Henderson, Coveney, & Meyer, 2012; Wiley et al., 2009), or for controversial socioscientific issues (e.g., Strømsø et al., 2010). Think-alouds have shown very infrequent use of source attributes, including author credibility, to guide processing (e.g., Barzilai & Eshet-Alkalai, 2015; Goldman et al., 2012). What does seem to guide processing is content relevance at the surface text level, with useful resources reported to be those that contain a lot of seemingly relevant content and content that is different from that contained in previously processed information resources (e.g., Goldman et al., 2012; Perfetti, Britt, & Georgi, 1995).

Furthermore, readers often fail to notice inconsistent information and/or do not react to it strategically (e.g., Otero & Kintsch, 1992; Perfetti et al., 1995).

Although these studies paint a rather dismal picture of multiple source comprehension, they provide evidence that better learning performance is associated with attending to source attributes, allocating more effortful processing to more reliable resources, and relating information across resources during sense-making predicts better learning performance (Anmarkrud et al., 2014; Goldman et al., 2012; Wiley et al. 2009; Wolfe & Goldman, 2005). Thus, researchers explored characteristics of tasks, information resources, and learners in efforts to encourage attention to resources and cross-resource processing.

Attention to sourcing increases when tasks require participants to make recommendations or come to decisions about controversial topics, generate arguments rather than descriptions, and provide information resources that contain conflicting information, especially if the conflicting information occurs in different information resources (e.g., Braasch, Rouet, Vibert, & Britt, 2012; Gil, Bråten, Vidal-Abarca, & Strømsø, 2010). Furthermore when there are explicit linguistic markers inserted into the language of texts to highlight conflicting information (e.g., *on the contrary, by comparison*), learners are more likely to include conflicting information in essays written from memory (Stadtler et al., 2014).

Interventions that direct learners to the importance of author competence and purpose increase the acceptance likelihood of knowledge claims made by more competent and well-intended authors (e.g., Kammerer, Amann, & Gerjets, 2015; Stadtler, Scharrer, Macedo-Rouet, Rouet, & Bromme, 2016). Furthermore, instruction that emphasized criteria for determining reliability of science information resources increased sensitivity to differences in reliability of information resources about a new and unrelated topic (Graesser et al., 2007; Wiley et al., 2009).

Individual Differences in Multiple Resource Comprehension and Information Problem Solving

Research studies consistently find that prior knowledge and epistemic cognition are significantly and positively related to performance on search, selection, information analysis and synthesis, and evaluation (e.g., trustworthiness and reliability) in multiple source tasks (e.g., Brand-Gruwel et al., 2017; Brand-Gruwel, Wopereis, & Walraven, 2009; Rouet et al., 1997; Strømsø et al., 2010). For example, McCrudden, Stenseth, Bråten, and Strømsø (2016) found that high school students paid less attention to sources for topics that they knew little about compared to topics about which they knew more. Likewise, epistemic cognition, i.e., thinking about the nature of knowledge and how it is generated, shows strong positive relationships to performance on sourcing processes and multiple source tasks (Barzilai & Eshet-Alkalai, 2015; Bråten et al., 2014; Kammerer et al., 2015). Epistemic cognition encompasses thinking about the nature of knowledge claims in the discipline, what counts as evidence to support or refute knowledge claims, criteria governing reliability of evidence and principles, frameworks, and/or disciplinary core ideas that define valid reasoning (Sandoval, Greene, & Bråten, 2016). The answers to these questions differ substantially from discipline to discipline (Goldman et al., 2016). Learners need to be aware of this knowledge *about* the discipline/content domain lest they inappropriately attempt to apply, for example, scientific reasoning and criteria when making sense of a literary work, historical narrative, or mathematical proof. Knowledge *about* the discipline is as critical to selecting relevant and reliable sources as knowledge *of* the disciplinary content. Thus, it is not surprising that both forms of disciplinary knowledge influence multiple components of reading, reasoning and problem solving, including selection, interpretation, and evaluation of tasks, sources, and adequacy of sense-making and task completion.

In addition, research in decision sciences indicates that attitudes, beliefs, and opinions that people bring to comprehension and information problem solving situations are visible in the reasons people search, the keywords they use, and the particular links they follow on SERPs. For example, in retrospective interviews following searches for self-generated yes/no questions participants reported searching to confirm their beliefs (White, 2014). Similarly, Van Strien, Brand-Gruwel, and Boshuizen (2014) found that students with strong prior attitudes were significantly more likely to use biased information to write their essays.

Conclusions and Future Directions

In this brief and perhaps overly ambitious effort to integrate multiple source comprehension and information problem-solving research we have had to largely forgo discussion of the burgeoning work on interventions aimed at enhancing search, selection, and sourcing processes. Bråten et al. (in press) provide an excellent review. However, it is important to acknowledge that many "search" and "sourcing" interventions target only one or two of the components depicted in Figure 9.1, and do so over short periods of time. Future research needs to move beyond these types of interventions to begin to investigate interactions and iterative cycles among the full set of components. We need to better understand synthesis and integration processes and whether there are linguistic cues and/or heuristic metacognitive strategies that would support critical analysis and evaluation of the probity of information. We also need a more nuanced approach to the purpose and value of sourcing processes; identifying the perspective of a particular source is not the "end goal." Perspective is not so much about trustworthiness of sources as it is about how perspective *informs* what learners make of the information with respect to forming interpretations, making decisions, and proposing solutions.

Other important research questions concern the types of cues and signals that reinitiate search, selection, and task reinterpretation. And how do we decide we are done, deadlines aside? These questions need to be pursued in parallel in different disciplinary areas. Contrasts and comparisons within and among disciplines are important for revealing similarities that can potentially serve as leverage points as learners move into unfamiliar disciplines or subdisciplines (Goldman et al., 2016). As well, contrasts can reveal differences in the importance and role of particular source attributes (e.g., date as indicating currency of science information vs. date as contextualizing historical artifacts) and what those attributes imply for interpretations and utility of the information in the disciplinary task.

Finally, in the spirit of design research traditions in the Learning Sciences, we encourage the pursuit of these issues in formal and informal settings and across diverse time spans, learners, and tools. Close examination of the potential affordances of different learning situations and how learner attributes affect, and are affected by, realized affordances will go a long way to revealing mechanisms of robust sense-making and problem solving in a context of increasingly available information resources.

Further Readings

Brand-Gruwel, S., Kammerer, Y., van Meeuwen, L., & van Gog, T. (2017). Source evaluation of domain experts and novices during Web search. *Journal of Computer Assisted Learning, 33*(3), 234–251. doi:10.1111/jcal.12162
Empirical report of experts and novices in the psychology domain evaluating internet sources through think-aloud and cued retrospective methods.

Goldman, S. R., Britt, M. A., Brown, W., Cribb, G., George, M., Greenleaf, C., et al. (2016). Disciplinary literacies and learning to read for understanding: A conceptual framework of core processes and constructs. *Educational Psychologist, 51*, 219–246.
Theoretical framework for multiple source sense-making processes and knowledge about three disciplines (the sciences, history, and literature) needed to engage evidence-based argument in each.

Sandoval, W. A., Greene, J. A., & Bråten, I. (2016). Understanding and promoting thinking about knowledge: Origins, issues, and future directions of research on epistemic cognition. *Review of Research in Education, 40*, 457–496.
A critical review and conceptual meta-analysis of philosophical and empirical perspectives on epistemic cognition with suggestions for moving the conversation forward.

NAPLeS Resources

Brand-Gruwel, S., Information-problem solving [Webinar]. In *NAPLeS video series*. Retrieved October 19, 2017, from http://isls-naples.psy.lmu.de/intro/all-webinars/brand-gruwel/index.html

Chinn, C., Epistemic cognition [Video file] Interview. In *NAPLeS video series*. Retrieved October 19, 2017, from http://isls-naples.psy.lmu.de/video-resources/interviews-ls/chinn/index.html

References

American College Testing (2006). *Reading between the lines: What the ACT reveals about college readiness in reading*. Iowa City, IA; Author.

Ananiadou, K., & Claro, M. (2009). *21st century skills and competences for new millennium learners in OECD countries* (OECD Education Working Papers No. 41). Paris: OECD Publishing.

Anmarkrud, Ø., Bråten, I., & Strømsø, H. I. (2014). Multiple-documents literacy: Strategic processing, source awareness, and argumentation when reading multiple conflicting documents. *Learning and Individual Differences, 30*, 64–76.

Barab, S. A., & Roth, W. M. (2006). Curriculum-based ecosystems: Supporting knowing from an ecological perspective. *Educational Researcher, 35*(5), 3–13.

Barsky, E., & Bar-Ilan, J. (2012). The impact of task phrasing on the choice of search keywords and on the search process and success. *Journal of the American Society for Information Science & Technology, 63*(10), 1987–2005.

Barzilai, S., & Eshet-Alkalai, Y. (2015). The role of epistemic perspectives in comprehension of multiple author viewpoints. *Learning and Instruction, 36*(0), 86–103.

Bazerman, C. (2004). Speech acts, genres, and activity systems: How texts organize activity and people. In C. Bazerman & P. A. Prior (Eds.), *What writing does and how it does it: An introduction to analyzing texts and textual practices* (pp. 309–339). Mahwah, NJ: Erlbaum.

Braasch, J. L. G., Rouet, J.-F., Vibert, N., & Britt, M. A. (2012). Readers' use of source information in comprehension. *Memory & Cognition, 40*(3), 450–465.

Brand-Gruwel, S., Kammerer, Y., van Meeuwen, L., & van Gog, T. (2017). Source evaluation of domain experts and novices during Web search. *Journal of Computer Assisted Learning*. doi:10.111/jcal.12162.

Brand-Gruwel, S., Wopereis, I., & Vermetten, Y. (2005). Information problem solving by experts and novices: analysis of a complex cognitive skill. *Computers in Human Behaviour, 21*, 487–508.

Brand-Gruwel, S., Wopereis, I., & Walraven, A. (2009). A descriptive model of Information Problem Solving while using Internet. *Computers & Education, 53*, 1207–1217.

Bråten, I., Ferguson, L. E., Strømsø, H. I., & Anmarkrud, Ø. (2014). Student working with multiple conflicting documents on a science issue: Relations between epistemic cognition while reading and sourcing and argumentation in essays. *British Journal of Educational Psychology, 84*, 58–85.

Bråten, I., Stadtler, M., & Salmerón, L. (in press). The role of sourcing in discourse comprehension. In M.F. Schober, D. N. Rapp, & M. A. Britt (Eds.), *Handbook of discourse processes* (2nd ed.). New York: Routledge.

Britt, M. A., Rouet, J. F., & Braasch, J. L. G. (2013). Documents experienced as entities: Extending the situation model theory of comprehension. In M. A. Britt, S. R. Goldman, & J. F. Rouet (Eds.), *Reading from words to multiple texts* (pp. 160–179). New York: Routledge.

Bromme, R., & Goldman, S. R. (2014). The public's bounded understanding of science. *Educational Psychologist, 49*, 59–69.

Cobb, P., Confrey, J. diSessa, A., Lehrer, R., & Schauble, L. (2003). Design experiments in educational research. *Educational Researcher, 32*(1), 9–13.

Gerjets, P., Kammerer, Y., & Werner, B. (2011). Measuring spontaneous and instructed evaluation processes during web search: Integrating concurrent thinking-aloud protocols and eye-tracking data. *Learning and Instruction, 21*, 220–231.

Gil, L., Bråten, I., Vidal-Abarca, E., & Strømsø, H. I. (2010). Understanding and integrating multiple science texts: Summary tasks are sometimes better than argument tasks. *Reading Psychology, 31*, 30–68.

Goldman, S. R., Braasch, J. L. G., Wiley, J., Graesser, A. C., & Brodowinska, K. (2012). Comprehending and learning from internet sources: Processing patterns of better and poorer learners. *Reading Research Quarterly, 47*, 356–381.

Goldman, S. R., Britt, M. A., Brown, W., Cribb, G., George, M., Greenleaf, C., et al. (2016). Disciplinary literacies and learning to read for understanding: A conceptual framework of core processes and constructs. *Educational Psychologist, 51*, 219–246.

Guan, Z., & Cutrell, E. (2007). An eye tracking study of the effect of target rank on Web search. In *Proceedings of the SIGCHI Conference on Human Factors in Computing Systems* (pp. 417–420). New York: ACM Press.

Graesser, A. C., Wiley, J., Goldman, S. R., O'Reilly, T., Jeon, M., & McDaniel, B. (2007). SEEK Web tutor: Fostering a critical stance while exploring the causes of volcanic eruption. *Metacognition and Learning, 2* (2–3), 89–105.

Höchstötter, N., & Lewandowski, D. (2009). What users see—Structures in search engine results pages. *Information Sciences, 179*(12), 1796–1812.

Kammerer, Y., Amann, D., & Gerjets, P. (2015). When adults without university education search the internet for health information: The roles of internet-specific epistemic beliefs and a source evaluation intervention. *Computers in Human Behavior, 48*, 297–309.

Kim, J., Kazai, G., & Zitouni, I. (2013). Relevance dimensions in preference-based evaluation. In *Proceedings of the 36th International ACM SIGIR Conference on Research and Development in Information Retrieval* (pp. 913–916). New York: ACM Press.

Kintsch, W. (1998). *Comprehension: A paradigm for cognition*. Cambridge, UK: Cambridge University Press.

Latour, B., & Woolgar, S. (1986). *Laboratory life: The construction of scientific facts*. Princeton, NJ: Princeton University Press.

Mason, L., Boldrin, A., & Ariasi, N. (2010). Searching the web to learn about a controversial topic: Are students epistemically active? *Instructional Science, 38*(6), 607–633.

McCrudden, M. T., Stenseth, T., Bråten, I., & Strømsø, H. I. (2016). The effects of author expertise and content relevance on document selection: A mixed methods study. *Journal of Educational Psychology, 108*, 147–162.

National Center for Education Statistics (2013). *The Nation's Report Card: A First Look: 2013 Mathematics and Reading* (NCES 2014-451). Washington, DC: Institute for Education Sciences.

National Research Council. (2012). *Education for life and work: Developing transferable knowledge and skills in the 21st century* (J. W. Pellegrino and M. L. Hilton, Eds.). Washington, DC: National Academies Press.

New London Group (1996). A pedagogy of multiliteracies: Designing social futures. *Harvard Educational Review, 66*, 60–92.

Organization of Economic Cooperation and Development (2013). *PISA 2012: Results in focus*. Paris: OECD.

Otero, J., & Kintsch, W. (1992). Failures to detect contradictions in a text: What readers believe versus what they read. *Psychological Science, 3*(4), 229–235.

Perfetti, C. A., Britt, M. A., & Georgi, M. C. (1995). *Text-based learning and reasoning: Studies in history*. New York: Psychology Press. (Reissued by Routledge, 2012.)

Perfetti, C. A., Rouet, J. F., & Britt, M. A. (1999). Toward a theory of documents representation. In H. Van Oostendorp & S. R. Goldman (Eds.), *The construction of mental representation during reading* (pp. 99–122). Mahwah, NJ: Erlbaum.

Rouet, J.-F. & Britt, M. A. (2011). Relevance processes in multiple document comprehension. In M. T. McCrudden, J. P. Magliano, & G. Schraw (Eds.), *Relevance instructions and goal-focusing in text learning* (pp. 19–52). Greenwich, CT: Information Age Publishing.

Rouet, J. F., Favart, M., Britt, M. A., & Perfetti, C. A. (1997). Studying and using multiple documents in history: Effects of discipline expertise. *Cognition and Instruction, 15*, 85–106.

Salmerón, L., Kammerer, Y., & García-Carrión, P. (2013). Searching the Web for conflicting topics: page and user factors. *Computers in Human Behavior, 29*, 2161–2171.

Sandoval, W. A., Greene, J. A., & Bråten, I. (2016). Understanding and promoting thinking about knowledge: Origins, issues, and future directions of research on epistemic cognition. *Review of Research in Education, 40*, 457–496.

Singer, G., Norbisrath, U., & Lewandowski, D., (2012). Impact of gender and age on performing search tasks online. In H. Reiterer & O. Deussen (Eds.), *Mensch & Computer 2012: interaktiv informiert—allgegenwärtig und allumfassend!?* (pp. 23–32). Munich: Oldenbourg Verlag.

Stadtler, M., Scharrer, L., Macedo-Rouet, M., Rouet, J. F., & Bromme, R. (2016). Improving vocational students' consideration of source information when deciding about science controversies. *Reading and Writing, 29*(4), 705–729.

Stadtler, M., Scharrer, L., Skodzik, T., & Bromme, R. (2014). Comprehending multiple documents on scientific controversies: Effects of reading goals and signaling rhetorical relationships. *Discourse Processes, 51*(1–2), 93–116.

Stieff, M., Hegarty, M., & Deslongchamps, G. (2011). Coordinating multiple representations in scientific problem solving: Evidence from concurrent verbal and eye-tracking protocols. *Cognition & Instruction, 29*, 123–145.

Strømsø, H. I., Bråten, I., & Britt, M. A. (2010). Reading multiple texts about climate change: The relationship between memory for sources and text comprehension. *Learning and Instruction, 20*, 192–204.

van Dijk, T. A., & Kintsch, W. (1983). *Strategies of discourse comprehension*. New York: Academic Press.

Van Strien, J., Brand-Gruwel, S., & Boshuizen, H. P. A. (2014). Dealing with conflicting information from multiple nonlinear texts: Effects of prior attitudes. *Computer in Human Behavior, 32*, 101–111.

Walraven, A., Brand-Gruwel, S., & Boshuizen, H. P. A. (2009). How students evaluate information and sources when searching the World Wide Web for information. *Computers and Education, 52*(1), 234–246.

Ward, P. R., Henderson, J., Coveney, J., & Meyer, S. (2012). How do south Australian consumers negotiate and respond to information in the media about food and nutrition? The importance of risk, trust and uncertainty. *Journal of Sociology, 48*(1), 23–41.

White, R. W. (2014). Belief dynamics in web search. *Journal of the Association for Information Science and Technology, 65*, 2165–2178.

Wiley, J., Goldman, S. R., Graesser, A. C., Sanchez, C. A., Ash, I. K., & Hemmerich, J. A. (2009). Source evaluation, comprehension, and learning in internet science inquiry tasks. *American Educational Research Journal, 46*, 1060–1106.

Wineburg, S. S. (1991). Historical problem solving: A study of the cognitive processes used in the evaluation of documentary and pictorial evidence. *Journal of Educational Psychology, 83*, 73–87.

Wolfe, M. B. & Goldman, S. R. (2005). Relationships between adolescents' text processing and reasoning. *Cognition & Instruction, 23*, 467–502.

10
Multiple Representations and Multimedia Learning

Shaaron Ainsworth

Consider three diverse learning situations: in the first, a high school student is asked to assess the evidence that the universe is expanding; in the second, a training dentist learns to clean and fill a root canal; and in the third, a family visit to a natural history museum prompts conversation about whether the diversity of animal life is related to how continents are formed. It is easy to see how these situations differ: they involve individual and social learning, occur in formal schooling, professional education and informal contexts, with learners of any age and whose duration ranges from minutes to months. But there is something they all have in common, and that is learning will be mediated by external representations such as pictures, animations, graphs, augmented reality, haptics, as well as text and speech.

Human learning is increasingly (multi-)representational, as we constantly invent new forms of representations whose appearance and interactive possibilities are partly due to technological development. Accordingly, the purpose of this chapter is to review research from diverse branches of learning sciences to trace some of the history of the field, before summarizing what we currently know about the opportunities offered by multi-representational learning, as well as addressing the challenges that it brings. Finally, it will end by predicting future trends and suggesting a focus for new research on multi-representational learning.

Background

The learning sciences approach to multi-representational learning draws together three main themes, each with a distinguishing preoccupation and specific methodological approach. It is the combination of these three that give the learning sciences approach its distinctive flavor.

The first approach is based upon cognitive accounts of instructional psychology and is particularly associated with Multimedia Learning Theory (Mayer, 2014), Cognitive Load Theory (Sweller, Van Merrienboer, & Paas, 1998), and the Integrated Theory of Text and Picture Learning (Schnotz, 2005). These theories share some assumptions. They argue that understanding is enhanced when learners' working memories are not overloaded, and therefore environments should be designed to use representations in ways that minimize their impact on working memory. They assume that there are limited-capacity modality specific processing systems: one focused on verbal, auditory, or descriptive representations (depending upon the theory), and one for visual, pictorial or depictive representations. Thus, learning is more effective when learners actively process representations, by selecting and organizing relevant information from material and integrating it into coherent long-term memory structure(s).

Arguably, these theories provided the main impetus behind the argument that multiple representations can be advantageous for learning: for example, Mayer's multimedia learning principle famously concludes that people learn better from pictures and text than text alone (e.g. Mayer, 2014). A large number of researchers have adopted the approach and together have produced a body of research that can be distilled into guidelines for the design of learning material. Examples include *avoid a split attention effect* (Ayres & Sweller, 2014), so that materials a student needs to mentally integrate are collocated spatially and temporally, or *use spoken not written text with an animation*, so that a learner can direct their visual attention to the pictorial elements (Sweller, 2005). Furthermore, as the theories have developed these claims have become more nuanced, with research focused on identifying the boundary conditions for the principles—the most common condition being that materials designed for learners low in prior knowledge may be less suitable for those with high prior knowledge (Kalyuga, 2007), and vice versa.

Methodologically, the vast majority of this research is experimental. The canonical study recruits from a broad university population to achieve large numbers of participants who are assigned to study variants of material that differ only in representation; for example, an animation which explains how lightning is formed accompanied by either written, spoken, or both forms of text. Assessments are given immediately after a short period of study and test the students' retention of the material and whether they can transfer their understanding to new issues. Given a common theoretical framework, as well as sharing materials and tests, it has proven possible to produce meta-analyses of such studies allowing estimations of the effects of using spoken text with pictures (e.g., Ginns, 2005).

However, there are still relatively few studies that test these principles in more realistic learning situations, with people studying authentic materials for longer periods of time where understanding is tested after some delay; see Eilam and Poyas (2008) for such an example. Consequently, it is not clear the extent to which the guidelines formulated and tested under lab conditions apply in classrooms, professional training, or museums. Nor is it certain that the underlying theoretical explanation, based upon working memory, is sufficient (e.g., Rummer, Schweppe, Furstenberg, Scheiter, & Zindler, 2011). Finally, these cognitive studies typically rest on a taxonomic approach to the representation's form (e.g., picture, text), whereas frameworks such as DeFT (Ainsworth, 2006) argue that a functional analysis that first considers the specific educational purpose is vital.

The second area of research that informs the learning sciences approach to multiple representations develops from a very different perspective, as it starts from understanding expert performance in a domain (see Reimann & Markauskaite, this volume). From radiologists detecting tumors on X-rays (Lesgold et al., 1988) to archaeologists studying soil using a Munsell colour chart (Goodwin, 1994), expertise is seen as fundamentally representational. Historians of science (Gooding, 2004) describe phases of representational construction and invention as scientists make new discoveries. Moreover, working with representations is not the individualized practice common to instructional psychology; instead, representations are at the center of a community of practice where they mediate communication between members, drive explanation, or are martialled in argument (Kozma, Chin, Russell, & Marx, 2000; Latour, 1999). Thus, developing expertise in a domain is judged by increasing proficiency in using representational tools at the heart of cultural practices.

The theoretical frameworks that underpin this approach to multiple representations are more diverse. Clearly, grounded in a sociocultural approach emphasizing how communities develop representations and how membership involves acquiring cultural tools and practices (Säljö, 1999). This approach draws more explicitly on semiotic theorists (e.g., Roth & Bowen, 2001), particularly social semiotics (Kress, 2009). Cognitive theories are also important, although, rather than the information processing models above, representational learning is understood in relation to situated (Tversky, 2005) or distributed (Zhang, 1997) approaches to cognition. Additionally, there are increasingly welcome attempts to combine these sociocultural and cognitive approaches in integrative perspectives (Airey & Linder, 2009; Prain & Tytler, 2012).

Typical studies explore how representations are used in practice. Kozma et al. (2000) spent 64 hours studying chemists in laboratories and found structural diagrams and equations drawn onto flasks, glass hoods, and white boards; in reference books and articles; and numeric and graphic output from instruments. They found that chemists' understanding was inherently multi-representational, as they selected specific representations for particular functions. They recount how scientists coordinate these different representations to justify a particular interpretation of a chemical reaction, where initial disagreement becomes joint understanding as interlocutors draw diagrams, consult instruments, and look up findings. Similarly, Hutchins (1995) describes how practices, such as navigating ships or landing planes, involved coordinating a distinct set of representations. These could include constantly changing visual displays of speed or weight with static representations such as numbers looked up on paper records. Individuals in the team take responsibility for coordinating different actions and must remember different parts of the system. Together, the team uses a combination of external representations, spoken talk, and their memories to safely land airplanes.

These accounts offer a rich picture of how constructing, selecting, and coordinating different representations is fundamental to professional practice. As such they are effective at describing the results of long-term representational learning. However, translating this knowledge into classroom practices is not easy. Whilst we might argue that representational practices are necessary knowledge, students themselves find such learning highly complicated. Many studies have shown that when learning representational systems, students fail to integrate representations (see Ainsworth, 2006 for examples) and that attempts to teach them are often surprisingly unsuccessful.

The final theme in this account is one that is often implicit: the role of technological development in shaping our approach to multi-representational learning. As digital technologies have become commonplace in classrooms, museums, workplaces, and homes, so the representations that support teaching and learning have rapidly changed, although not automatically for the better. This is seen in learning material as familiar as the textbook, which has seen a growth in the variety and number of representations on a typical page, and an increased use of high-fidelity representations such as photographs (Lee, 2010), as well as infographics (Polman & Gebre, 2015). And, of course, "textbooks" are increasingly digital and so now routinely include sound, video, and animation.

In addition, representational technologies that used to be only found in the hands of professionals are increasingly present in classrooms offering opportunities to engage in authentic disciplinary practices. Scientific visualization allows school students to experience the scale of physical phenomena—from atomic interactions to planetary imaging (Gordon & Pea, 1995). Geographic information systems combine satellite images, maps, field data, and aerial photographs in ways that can help students understand the complexity of their local communities (Kerski, 2003).

Finally, some representational technologies offer students participation in digitally simulated experiences that are inaccessible within the physical world. For example, simulations are now commonplace in science and engineering classrooms, as teachers hope to save time, offer safe spaces to acquire skills, allow manipulation of variables that would otherwise be unmanipulable (e.g., alter gravity), and offer students opportunities for increased control of scientific inquiry (Rutten, van Joolingen, & van der Veen, 2012). They also easily permit multiple representations of phenomena (van der Meij & de Jong, 2006). Whilst there is some debate as to whether simulations are "better" or "worse" than physical laboratories, most researchers accept that the two situations have distinct affordances and learning is maximized by sensible combinations of both (de Jong, Linn, & Zacharia, 2013).

Methodologically, much research in this arena consisted of developing systems and pronouncing their success (Dillenbourg, 2008), but learning sciences approaches tend to be more nuanced. They typically involve design-based research studies, iterating through cycles of development and use, to refine the systems themselves or the way they are used in practice (e.g., Barab, Thomas, Dodge, Carteaux, & Tuzun, 2005; Puntambekar, this volume).

Looking backwards, we can see that learning sciences approaches to multi-representational learning have been theoretically and methodologically diverse. I will argue in the next section that this has resulted in a mature field of inquiry with useful insight, as well as more open questions.

Where Is Research on Multiple Representations Now?

The first and most important lesson we have learned about multiple representations is to be skeptical about implicit or explicit claims that *more* is always better: for example, two representations are not always better than one (Kalyuga & Sweller, 2014), three dimensions not better than two (Keller, Gerjets, Scheiter, & Garsoffky, 2006), interactive and dynamic representations are not always better than static (Bétrancourt, 2005). Consequently, one important message is to embrace the affordances that new technologies bring to representational learning, without assuming that this will magically resolve learners' difficulties.

So what are effective multi-representational systems? I want to argue against general principles and suggest the following formulism: well-designed combinations of representations manipulate information to make their key (task-relevant) aspects more accessible to learners for beneficial cognitive, social, and affective processes. This claim makes salient aspects of representational learning that I consider to be the most important. First, it identifies the importance of *task* analysis—representations are not generally good or bad; they are more or less suitable for a particular task for a specific learner. So, when looking at a multi-representational simulation of population density (see Figure 10.1), a graph or (even better) a phase-plot allows for perceptual inferences (is this ecosystem oscillating or moving towards stable equilibrium?), a table supports precise read-off (the number of prey and predators at this point in time), whereas an equation allows precise calculation of expected future states (Larkin & Simon, 1987). Second, this definition also draws attention to the way that this analysis needs to embrace the representational system as *a combination* and not each representation in isolation. Consequently, by combining the table, graph, and equation in a system, these representations complement one another and allow an experienced student to select the representations most appropriate for their specific needs at that time. Alternatively, someone learning to read phase-plots may benefit from the support of the more familiar table or time-series graph to help them interpret

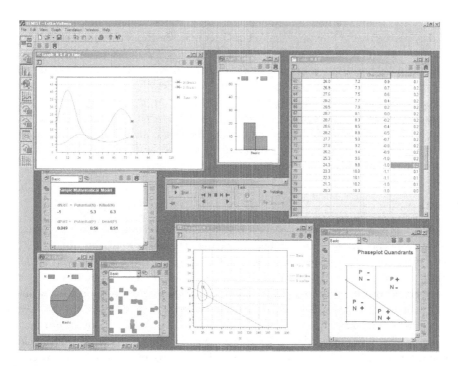

Figure 10.1 An Example of Many Representations From a Predator–Prey Simulation Including Table, Equation, Time-Series Graph, and Phase-Plot

this less familiar form (see Ainsworth, 2006). Hopefully, this example also makes clear that representations should be seen as something that stands to somebody for something in some respect or capacity (Peirce, 1906) and so, when considering their suitability, it is likely that both learner and task factors will be important (Acevedo Nistal, Van Dooren, & Verschaffel, 2013; Kalyuga, 2007).

The third main point of this definition is the equal attention given to the triumvirate of *cognitive, social, and affective* processes. Classical research on multi-representational and multimedia learning focused on individualized cognitive accounts of learning. This is clearly an important consideration for multiple representations. However, the learning sciences traditions of sociocultural and situated accounts of learning and professional practice make clear that representations are developed and used by communities of practice to mediate communication between members. Representations serve multiple social functions as students learn: facilitating communication between peers, and becoming joint resources for subsequent problem solving (White & Pea, 2011); supporting teachers and students to work together in the classroom as well as becoming the focus of what is to be learned (e.g. Prain & Tytler, 2012), and, of course, are vivid in the account of workplace learning and action (Kozma et al., 2000; Latour, 1999). Finally, this draws attention to the affective aspects of learning. Unfortunately, wider claims about affect or motivation have too often just been simplistic and overgeneralized: "video, multimedia, augmented reality, etc. helps children learn as they enjoy it so much." Happily, learning sciences approaches can draw on design studies and conceptual frameworks from game-based learning, which have provided better theorized impetus to study and design for affect and motivation in representational learning (Habgood & Ainsworth, 2011; Virk, Clark, & Sengupta, 2015). Traditional cognitive approaches have also been broadened to focus on motivational components (Moreno & Mayer, 2007).

The final point that I wish to draw attention to is the importance of *processes*. Learning with multiple representations is not a magical activity whereby simply presenting an animation with a picture results in new understanding. There is abundant evidence that learners need to master many complicated aspects of these representations in isolation and in combination for multi-representational learning to be successful. Learners need to understand how representations encode and present information, they need to know how to select or construct representations and, particularly for multiple representations, how they relate to one another (e.g., Ainsworth, 2006). This understanding can take a long time to develop as learners become more familiar with the representations and the roles they play in their communities (Kozma & Russell, 2005). Novices may be misled by features of representational systems that are vivid but not important (Lowe, 2004), or lack metacognitive insight into the need to process representations actively (Salomon, 1984). However, it is not all doom and gloom! There is evidence that that visual representations can encourage the use of effective metacognitive strategies (Ainsworth & Loizou, 2003) and that even younger students have insight into how to design considered representations (diSessa, 2004).

The Future

Certain things are clearly predictable, at least in general, about the future of multi-representational learning. We will continue to learn surrounded by representations as we read textbooks, run simulations, visit museums and play games. It is also highly likely these representations will take forms that we cannot currently imagine. Furthermore, some scenarios seem likely given the current direction of travel. First, we may expect increasing attention to representations that more actively involve the body. These draw upon the development of natural user interfaces (NUIs) whereby the interaction is felt to be "natural" and draws upon increasingly invisible forms of control such as body movements, gesture, or speech; for example, when children visiting an art gallery can interact and animate paintings by moving on the floor (Price, Sakr, & Jewitt, 2016). This is closely related to approaches such as participatory simulations, where students become viruses in an ecosystem by wearing programmable tags (Colella, 2000) or experience and investigate earthquakes in their classrooms (Moher, 2008).

Another approach is to provide representations of somatosensory (touch) information through haptics allowing children to feel viruses (Minogue & Jones, 2006) or training dentists to prepare a cavity (Suebnukarn, Haddawy, Rhienmora, & Gajananan, 2010). Opportunely, at the same time that technology is permitting more body-based representation and interaction, developments in our understanding of the importance of these body-based representations is increasing. This resonates with those who argue for a more embodied approach to cognition and where the importance of gesture for supporting learning is increasingly demonstrated (Kontra, Goldin-Meadow, & Beilock, 2012; Alibali & Nathan, this volume).

A second emerging theme is the importance of learners constructing and even inventing their own representations for learning. This is seen in those who argue for this as a fundamental aspect of representational (Kozma & Russell, 2005) or meta-representational competence (Disessa, 2004). It is also resonant with the resurgence of interest in construction and making (e.g., Halverson & Peppler, this volume). Again, technology progress is now bringing these practices more easily to university classrooms, where tools such as CogSketch (Forbus, Usher, Lovett, Lockwood, & Wetzel, 2011) or beSocratic (Bryfczynski et al., 2015) support learning and assessment in subjects like geology and chemistry. Simpler interfaces can even allow children to engage in model based reasoning through drawing (van Joolingen, Aukes, Gijlers, & Bollen, 2015). However, like others, I would not want to forget the importance of good old-fashioned pen and paper drawing (Ainsworth, Prain, & Tytler, 2011).

Another important arena for research is how best we support learners to work with multiple representations. Unfortunately, far too much research can still mistake the process *of* learning a multi-representational system with learning *with* a multi-representational system. Learners need more time to master the environment and probably more explicit teaching in representational practices if we are to describe successful multi-representational learning. In addition, we need to actively research the best ways to support learning with multiple representations. This might include teaching learners effective ways to engage with standalone learning environments, such as when Stalbovs, Scheiter, and Gerjets (2015) teach learners If-Then plans to integrate text and pictures in multimedia learning or when students are taught specific animation understanding strategies (Kombartzky, Ploetzner, Schlag, & Metz, 2010).

However, compared to the sizable amount of research on learning with representations, there has been relatively little to say about how teachers teach with and about multiple representations (a strong exception to that claim being the work of socio-semioticians, e.g., Kress et al., 2005). Fortunately, this gap is increasingly being filled by researchers exploring what teachers understand about representations (e.g., Eilam, Poyas, & Hashimshoni, 2014), how they can support their students so that their classrooms become sites of representational activities (Prain & Tytler, 2012), and how teachers provide instruction in representational conventions as part of learning (Cromley et al., 2013). Nevertheless, there is still much to explore about the teachers' roles in multi-representational classrooms.

A final theme to explore is the importance of considering assessment. At present formal schooling relies on written and mathematical forms (Yore & Hand, 2010). This is of concern when learning is multi-representational, as it not only may be an unsuitable way to assess students' understanding, but also sends a worryingly message of what knowing in a professional domain actually means (Lemke, 2004). In research, we do see improvement with researchers designing assessment aligned to representational goals (Lowe, Schnotz, & Rasch, 2010) as well as more usage of online processing measures such as eye movements (van Gog & Scheiter, 2010), data mining and verbal protocols (Rau, Michaelis, & Fay, 2015) to relate processing of learning to outcomes. Moreover, representational technologies for assessment are being developed. For example, beSocratic or CogSketch assess students' understanding by asking them to construct visual and multiple forms of representations, and are becoming sufficiently mature to move from proof of concepts studies to large-scale deployment. Games and simulations are also increasing being used as innovative forms of assessment (Clarke-Midura & Dede, 2010).

Conclusion

The future is multi-representational! There seems little doubt that learning will be mediated by a variety of representational forms whose interactive possibilities are increasingly diverse. Learning sciences can contribute to this evolution by offering thoughtful insight into how representations can be designed, taking into account insight into cognitive, affective and social processes of learning, discovering ways that learners can be supported to use these representations effectively in ways appropriate to the contexts and, where required, ensuing assessment is sensitive to the multi-representational learning that has taken place.

Further Readings

Hegarty, M. (2011). The cognitive science of visual-spatial displays: Implications for design. *Topics in Cognitive Science, 3*(3), 446–474. doi:10.1111/j.1756-8765.2011.01150
This paper synthesizes cognitive science approaches to design of visual-spatial displays and an overview of the perceptual and cognitive processes involved in reading such displays

Kozma, R., & Russell, J. (2005). Students becoming chemists: Developing representational competence. In J. K. Gilbert (Ed.), *Visualization in science and education* (pp. 121–146). Dordrecht: Kluwer Academic Publishers.
This chapter summarizes evidence to explore the types of representational practices that expert chemists are proficient in and novices need to acquire. It illustrates how foundational theory and disciplinary knowledge integrate in learning sciences approaches to multi- representational learning.

Mayer, R. E., & Moreno, R. (2003). Nine ways to reduce cognitive load in multimedia learning. *Educational Psychologist, 38*(1), 43–52. doi:10.1207/S15326985EP3801_6
This paper is an excellent example of a paper that illustrates the classic cognitive approach to multimedia learning.

Rau, M. A., Michaelis, J. E., & Fay, N. (2015). Connection making between multiple graphical representations: A multi-methods approach for domain-specific grounding of an intelligent tutoring system for chemistry. *Computers & Education, 82,* 460–485. doi:10.1016/j.compedu.2014.12.009
This paper provides an account of how connection making between multiple representations is important as well as demonstrating how eye tracking and log data can be used to relate the processes of multi-representational learning-to-learning outcomes.

Stieff, M. (2017). Drawing for promoting learning and engagement with dynamic visualizations. In R. Lowe & R. Ploetzner (Eds.), *Learning from dynamic visualization* (pp. 333–356). Cham, Switzerland: Springer.
This reference illustrates a number of themes in this chapter. It is theoretically integrative, addresses the construction of representations, and considers both cognitive and affective processes involved in drawing to support learning from visualization.

NAPLeS Resources

Ainsworth, S., *Multiple representations and multimedia learning* [Video file]. *Interview.* In *NAPLeS video series.* Retrieved October 19, 2017, from http://isls-naples.psy.lmu.de/video-resources/interviews-ls/ainsworth/index.html

References

Acevedo Nistal, A., Van Dooren, W., & Verschaffel, L. (2013). Students' reported justifications for their representational choices in linear function problems: An interview study. *Educational Studies, 39*(1), 104–117. doi:10.1080/03055698.2012.674636

Ainsworth, S. (2006). Deft: A conceptual framework for considering learning with multiple representations. *Learning and Instruction, 16*(3), 183–198. doi:10.1016/j.learninstruc.2006.03.001

Ainsworth, S., & Loizou, A. T. (2003). The effects of self-explaining when learning with text or diagrams. *Cognitive science, 27*(4), 669–681. doi:10.1016/s0364-0213(03)00033-8

Ainsworth, S., Prain, V., & Tytler, R. (2011). Drawing to learn in science. *Science, 333*(6046), 1096–1097. doi:10.1126/science.1204153

Airey, J., & Linder, C. (2009). A disciplinary discourse perspective on university science learning: Achieving fluency in a critical constellation of modes. *Journal of Research in Science Teaching, 46*(1), 27–49.

Alibali, M. W., & Nathan, M. (2018). Embodied cognition in learning and teaching: Action, observation, and imagination. In F. Fischer, C. E. Hmelo-Silver, S. R. Goldman, & P. Reimann (Eds.), *International handbook of the learning sciences* (pp. 75–85). New York: Routledge.

Ayres, P., & Sweller, J. (2014). The split-attention principle in multimedia learning. In R. E. Mayer (Ed.), *The Cambridge handbook of multimedia learning* (Vol. 2, pp. 206–226). New York: Cambridge University Press.

Barab, S., Thomas, M., Dodge, T., Carteaux, R., & Tuzun, H. (2005). Making learning fun: Quest Atlantis, a game without guns. *Educational Technology Research and Development, 53*(1), 86–107. doi:10.1007/bf02504859

Bétrancourt, M. (2005). The animation and interactivity principles. In R. E. Mayer (Ed.), *The Cambridge handbook of multimedia learning* (pp. 287–296). New York: Cambridge University Press.

Bryfczynski, S., Pargas, R. P., Cooper, M. M., Klymkowsky, M., Hester, J., & Grove, N. P. (2015). Classroom uses for besocratic. In T. Hammond, S. Valentine, A. Adler, & M. Payton (Eds.), *The impact of pen and touch technology on education* (pp. 127–136). Cham: Springer International Publishing.

Clarke-Midura, J., & Dede, C. (2010). Assessment, technology, and change. *Journal of Research on Technology in Education, 42*(3), 309–328. doi:10.1080/15391523.2010.10782553

Colella, V. (2000). Participatory simulations: Building collaborative understanding through immersive dynamic modeling. *Journal of the Learning Sciences, 9*(4), 471–500. doi:10.1207/s15327809jls0904_4

Cromley, J. G., Perez, T. C., Fitzhugh, S. L., Newcombe, N. S., Wills, T. W., & Tanaka, J. C. (2013). Improving students' diagram comprehension with classroom instruction. *Journal of Experimental Education, 81*(4), 511–537. doi:10.1080/00220973.2012.745465

de Jong, T., Linn, M. C., & Zacharia, Z. C. (2013). Physical and virtual laboratories in science and engineering education. *Science, 340*(6130), 305–308. doi:10.1126/science.1230579

Dillenbourg, P. (2008). Integrating technologies into educational ecosystems. *Distance Education, 29*(2), 127–140. doi:10.1080/01587910802154939

diSessa, A. A. (2004). Metarepresentation: Native competence and targets for instruction. *Cognition and Instruction, 22*(3), 293–331. doi:10.1207/s1532690xci2203_2

Eilam, B., & Poyas, Y. (2008). Learning with multiple representations: Extending multimedia learning beyond the lab. *Learning and Instruction, 18*(4), 368–378. doi:10.1016/j.learninstruc.2007.07.003

Eilam, B., Poyas, Y., & Hashimshoni, R. (2014). Representing visually: What teachers know and what they prefer. In B. Eilam & J. K. Gilbert (Eds.), *Science teachers' use of visual representations* (pp. 53–83). Dordrecht: Springer.

Forbus, K., Usher, J., Lovett, A., Lockwood, K., & Wetzel, J. (2011). Cogsketch: Sketch understanding for cognitive science research and for education. *Topics in Cognitive Science, 3*, 648–666. doi:10.1111/j.1756-8765.2011.01149.x

Ginns, P. (2005). Meta-analysis of the modality effect. *Learning and Instruction, 15*(4), 313–331. doi:10.1016/j.learninstruc.2005.07.001

Gooding, D. C. (2004). Cognition, construction and culture: Visual theories in the sciences. *Journal of Cognition and Culture, 3*(4), 551–593. doi:10.1163/1568537042484896

Goodwin, C. (1994). Professional vision. *American Anthropologist, 96*(3), 606–633. doi:10.1525/aa.1994.96.3.02a00100

Gordon, D. N., & Pea, R. D. (1995). Prospects for scientific visualisation as an educational technology. *Journal of the learning sciences, 4*(3), 249–279. doi:10.1207/s15327809jls0403_1

Habgood, M. P. J., & Ainsworth, S. E. (2011). Motivating children to learn effectively: Exploring the value of intrinsic integration in educational games. *Journal of the Learning Sciences, 20*(2), 169–206. doi:10.1080/10508406.2010.508029

Halverson, E., & Peppler, K. (2018). The Maker Movement and learning. In F. Fischer, C. E. Hmelo-Silver, S. R. Goldman, & P. Reimann (Eds.), *International handbook of the learning sciences* (pp. 285–294). New York: Routledge.

Hutchins, E. (1995). How a cockpit remembers its speeds. *Cognitive Science, 19*(3), 265–288. doi:10.1207/s15516709cog1903_1

Kalyuga, S. (2007). Expertise reversal effect and its implications for learner-tailored instruction. *Educational Psychology Review, 19*(4), 509–539. doi:10.1007/s10648-007-9054-3

Kalyuga, S., & Sweller, J. (2014). The redundancy principle in multimedia learning. In R. E. Mayer (Ed.), *The Cambridge handbook of multimedia learning* (p. 247). New York: Cambridge University Press.

Keller, T., Gerjets, P., Scheiter, K., & Garsoffky, B. (2006). Information visualizations for knowledge acquisition: The impact of dimensionality and color coding. *Computers in Human Behavior, 22*(1), 43–65. doi:10.1016/j.chb.2005.01.006

Kerski, J. J. (2003). The implementation and effectiveness of geographic information systems technology and methods in secondary education. *Journal of Geography, 102*(3), 128–137. doi:10.1080/00221340308978534

Kombartzky, U., Ploetzner, R., Schlag, S., & Metz, B. (2010). Developing and evaluating a strategy for learning from animations. *Learning and Instruction, 20*(5), 424–433. doi:10.1016/j.learninstruc.2009.05.002

Kontra, C., Goldin-Meadow, S., & Beilock, S. L. (2012). Embodied learning across the life span. *Topics in Cognitive Science, 4*(4), 731–739. doi:10.1111/j.1756-8765.2012.01221.x

Kozma, R., Chin, E., Russell, J., & Marx, N. (2000). The roles of representations and tools in the chemistry laboratory and their implications for chemistry learning. *Journal of the Learning Sciences, 9*(2), 105–143. doi:10.1207/s15327809jls0902_1

Kozma, R., & Russell, J. (2005). Students becoming chemists: Developing representational competence. In J. K. Gilbert (Ed.), *Visualization in science and education* (pp. 121–146). Dordrecht: Kluwer Academic Publishers.

Kress, G. (2009). *Multimodality: A social semiotic approach to contemporary communication*. London: Routledge.

Kress, G., Jewitt, C., Bourne, J., Franks, A., Hardcastle, J., Jones, K., & Reid, E. (2005). *English in urban classrooms: Multimodal perspectives on teaching and learning*. New York: Routledge.

Larkin, J. H., & Simon, H. A. (1987). Why a diagram is (sometimes) worth 10000 words. *Cognitive Science, 11*(1), 65–99. doi:10.1016/S0364-0213(87)80026-5

Latour, B. (1999). *Pandora's hope: Essays on the reality of science studies*. Cambridge, MA: Harvard University Press.

Lee, V. R. (2010). Adaptations and continuities in the use and design of visual representations in U.S. middle school science textbooks. *International Journal of Science Education, 32*(8), 1099–1126. doi:10.1080/09500690903253916

Lemke, J. L. (2004). The literacies of science. In E. W. Saul (Ed.), *Crossing borders in literacy and science instruction: Perspectives on theory and practice* (pp. 33–47). Newark, DE: International Reading Association.

Lesgold, A., Rubinson, H., Feltovich, P., Glaser, R., Klopfer, D., & Wang, Y. (1988). Expertise in a complex skill: Diagnosing x-ray pictures. In M. T. Chi, R. E. Glaser, & M. J. Farr (Eds.), *The nature of expertise* (pp. 311–342). Hillsdale, NJ: Lawrence Erlbaum Associates.

Lowe, R. (2004). Interrogation of a dynamic visualization during learning. *Learning and Instruction, 14*(3), 257–274. doi:10.1016/J.Learninstruc.2004.06.003

Lowe, R., Schnotz, w., & Rasch, T. (2010). Aligning affordances of graphics with learning task requirements. *Applied Cognitive Psychology, 25*(3), 452–459. doi:10.1002/acp.1712

Mayer, R. E. (2014). Cognitive theory of multimedia learning. In R. E. Mayer (Ed.), *The Cambridge handbook of multimedia learning* (pp. 43–71). New York: Cambridge University Press.

Minogue, J., & Jones, M. G. (2006). Haptics in education: Exploring an untapped sensory modality. *Review of Educational Research, 76*(3), 317–348. doi:10.3102/00346543076003317

Moher, T. (2008). Learning and participation in a persistent whole-classroom seismology simulation. *Proceedings of the 8th International Conference for the Learning Science*, Utrecht, Netherlands.

Moreno, R., & Mayer, R. (2007). Interactive multimodal learning environments. *Educational Psychology Review, 19*(3), 309–326. doi:10.1007/s10648-007-9047-2

Peirce, C. S. (1906). Prolegomena to an apology for pragmaticism. *The Monist, 16*, 492–546.

Polman, J. L., & Gebre, E. H. (2015). Towards critical appraisal of infographics as scientific inscriptions. *Journal of Research in Science Teaching, 52*(6), 868–893. doi:10.1002/tea.21225

Prain, V., & Tytler, R. (2012). Learning through constructing representations in science: A framework of representational construction affordances. *International Journal of Science Education, 34*(17), 2751–2773. doi: 10.1080/09500693.2011.626462

Price, S., Sakr, M., & Jewitt, C. (2016). Exploring whole-body interaction and design for museums. *Interacting with Computers, 28*(5), 569–583. doi:10.1093/iwc/iwv032

Puntambekar, S. (2018). Design-based research (DBR). In F. Fischer, C. E. Hmelo-Silver, S. R. Goldman, & P. Reimann (Eds.), *International handbook of the learning sciences* (pp. 383–392). New York: Routledge.

Rau, M. A., Michaelis, J. E., & Fay, N. (2015). Connection making between multiple graphical representations: A multi-methods approach for domain-specific grounding of an intelligent tutoring system for chemistry. *Computers & Education, 82*, 460–485. doi:10.1016/j.compedu.2014.12.009

Reimann, P., & Markauskaite, L. (2018). Expertise. In F. Fischer, C. E. Hmelo-Silver, S. R. Goldman, & P. Reimann (Eds.), *International handbook of the learning sciences* (pp. 54–63). New York: Routledge.

Roth, W. M., & Bowen, G. M. (2001). Professionals read graphs: A semiotic analysis. *Journal for Research in Mathematics Education, 32*(2), 159–194. doi:10.2307/749672

Rummer, R., Schweppe, J., Furstenberg, A., Scheiter, K., & Zindler, A. (2011). The perceptual basis of the modality effect in multimedia learning. *Journal of Experimental Psychology—Applied, 17*(2), 159–173. doi:10.1037/a0023588

Rutten, N., van Joolingen, W. R., & van der Veen, J. T. (2012). The learning effects of computer simulations in science education. *Computers & Education, 58*(1), 136–153. doi:10.1016/j.compedu.2011.07.017

Säljö, R. (1999). Learning as the use of tools. In K. Littleton & P. Light (Eds.), *Learning with computers: Analysing productive interaction* (pp. 144–161). London: Routledge.

Salomon, G. (1984). Television is easy and print is tough—The differential investment of mental effort in learning as a function of perceptions and attributions. *Journal of Educational Psychology, 76*(4), 647–658. doi:10.1037/0022-0663.76.4.647.

Schnotz, W. (2005). An integrated model of text and picture comprehension. In R. Mayer (Ed.), *The Cambridge handbook of multimedia learning* (pp. 49–69). Cambridge: Cambridge University Press.

Stalbovs, K., Scheiter, K., & Gerjets, P. (2015). Implementation intentions during multimedia learning: Using if-then plans to facilitate cognitive processing. *Learning and Instruction, 35*, 1–15. doi:10.1016/j.learninstruc.2014.09.002

Suebnukarn, S., Haddawy, P., Rhienmora, P., & Gajananan, K. (2010). Haptic virtual reality for skill acquisition in endodontics. *Journal of Endodontics, 36*(1), 53–55. doi:http://dx.doi.org/10.1016/j.joen.2009.09.020

Sweller, J. (2005). The redundancy principle in multimedia learning. In R. Mayer (Ed.), *The Cambridge handbook of multimedia learning:* (pp. 159–168). Cambridge: Cambridge University Press.

Sweller, J., van Merrienboer, J. J. G., & Paas, F. G. W. C. (1998). Cognitive architecture and instructional design. *Educational Psychology Review, 10*(3), 251–296. doi:10.1023/a:1022193728205

Tversky, B. (2005). Spatial cognition: Embodied and situated. In P. Robbins & M. Aydede (Eds.), *The Cambridge handbook of situated cognition* (pp. 201–216). Cambridge: Cambridge University Press.

van der Meij, J., & de Jong, T. (2006). Supporting students' learning with multiple representations in a dynamic simulation-based learning environment. *Learning and Instruction, 16*(3), 199–212. doi:10.1016/j.learninstruc.2006.03.007

van Gog, T., & Scheiter, K. (2010). Eye tracking as a tool to study and enhance multimedia learning. *Learning and Instruction, 20*(2), 95–99. doi:10.1016/j.learninstruc.2009.02.009

van Joolingen, W., Aukes, A. V., Gijlers, H., & Bollen, L. (2015). Understanding elementary astronomy by making drawing-based models. *Journal of Science Education and Technology, 24*(2–3), 256–264. doi:10.1007/s10956-014-9540-6

Virk, S., Clark, D., & Sengupta, P. (2015). Digital games as multirepresentational environments for science learning: Implications for theory, research, and design. *Educational Psychologist, 50*(4), 284–312. doi:10.1080/00461520.2015.1128331

White, T., & Pea, R. (2011). Distributed by design: On the promises and pitfalls of collaborative learning with multiple representations. *Journal of the Learning Sciences, 20*(3), 489–547. doi:10.1080/10508406.2010.542700

Yore, L., & Hand, B. (2010). Epilogue: Plotting a research agenda for multiple representations, multiple modality, and multimodal representational competency. *Research in Science Education, 40*(1), 93–101. doi:10.1007/s11165-009-9160-y

Zhang, J. J. (1997). The nature of external representations in problem solving. *Cognitive Science, 21*(2), 179-217. doi: 10.1207/s15516709cog2102_3.

11
Learning Within and Beyond the Disciplines

Leslie R. Herrenkohl and Joseph L. Polman

The learning sciences, as the -s on "sciences" indicates, has its roots in diverse social science disciplines such as psychology, sociology, and applied linguistics, as well as computer science and artificial intelligence (Hoadley & Van Haneghan, 2011; Hoadley, this volume). The signature method that has evolved from this diverse collection of fields is itself a hybrid form called "design based research" (Puntambekar, this volume) and more recently as the scale of research shifts to larger systems "design based implementation research" (Penuel, Fishman, Cheng, & Sabelli, 2011; Fishman & Penuel, this volume). Practitioners in the field have always drawn on multiple disciplinary perspectives to address organizational, social, cognitive, linguistic, and historical aspects of learning and learning environments. So, it could appear ironic to focus a chapter on disciplinary reasoning within a field that transcends typical boundaries. Yet, for many years a primary focus of the learning sciences has been on addressing more effective ways to support and understand teaching and learning in traditional disciplinary contexts, typically in formal educational settings, although there have been exceptions (e.g., Hutchins, 1995). In this chapter we explore the fruitfulness of adopting a disciplinary perspective and important findings about disciplinary learning, which has primarily been carried out in a school-based, formal setting. We then juxtapose this work with research that transcends traditional disciplinary boundaries, which has often been carried out in learning environments that extend beyond formal schooling.

Disciplinary Thinking and Learning

What we now call the learning sciences began to take shape during the 1970s, when information processing became a theoretical explanation for the inner workings of the human mind. Computational models of human thinking were offered from cognitive science. Researchers and scholars of human cognition debated central questions such as: Is human cognition domain general? Is it domain specific? (Perkins & Salomon, 1989; see Stevens, Wineburg, Herrenkohl, & Bell, 2005 for a history of the rise of discipline-specific thinking). Findings from important studies of expertise pointed to a *both/and* conclusion—that the most effective thinking recruits deep local knowledge while using strong general thinking strategies to support effective metacognition, reflection, and ongoing learning (Bransford, Brown, & Cocking, 2000; Bruer, 1993). Dreyfus and Dreyfus (1986) also suggested that expertise goes beyond calculated rationality to engage intuition—a way of effortlessly seeing similarities across experiences that allow one to act effectively and fluidly without calling on a deliberative, rational process. As the field developed, researchers also asked how developments in neuroscience, both basic research and technical tools, could help us better understand and improve the science of learning, or whether this was a bridge too far (Bruer, 1997; Stern, 2016; Varma, Im, Schmied, Michel, & Varma, this volume).

At the same time that the domain-specific/domain-general debate was underway, others were examining the role of culture and context in human cognition (Cole, 1996; Rogoff, 1990; Wertsch, 1985). Drawing on translations of Lev Vygotsky's work that first became available in English in 1962, more scholars began to explore the deeply social, cultural, and discursive dimensions of human learning (Brown, Collins, & Duguid, 1989; Lave & Wenger, 1991; Saxe, 1991). These theoretical approaches focused on situated cognition and the generative nature of cultural tools, their process of production and adaptation, and their critical role in mediating human thinking and learning. Studies of learning focused on the role of language as a tool to mediate human interaction, thinking, and ultimately human learning (Wertsch, 1991). Building on a related set of ideas, Latour (e.g., 1990) and others who studied the practices of scientists in the field and in the laboratory demonstrated how data are assembled into material inscriptions such as tables, drawings, maps, and so on, and how important the assembly, transformation, and social interaction around such inscriptions are to the conduct and advancement of science disciplines. Recognizing the role of culture and cultural tools has contributed to establishing how access to and appropriation of certain cultural tools and discourses confers power and status to individuals and groups (Carlone & Johnson, 2012; Cornelius & Herrenkohl, 2004; Lemke, 1990; O'Connor, Peck, & Cafarella, 2015).

These roots are reflected in learning sciences research that foregrounds cognition in ways that privilege the role that tools of social exchange and cultural engagement play in thinking and understanding. From this perspective, "the disciplines," as educators and academics commonly refer to them, can be seen as cultural legacies coming from particular communities of practice (Lave & Wenger, 1991; Wenger, 1998). Many learning scientists have converged on seeing disciplines as embedded historically and institutionally with specific linguistic and thinking practices acting as signature ways of knowing (Bruner, 1960; Schwab, 1978). Two sets of tools characterize disciplinary thinking. One focuses on *products* like key concepts and theories. The other highlights the *processes* or *practices*, i.e., the sanctioned methods of generating important disciplinary products and evaluating their merit.

Contributions of the Learning Sciences to Disciplinary Learning

Learning sciences scholars have contributed significantly to research on disciplinary learning in many subject matter with the bulk of the work emphasizing disciplines prominent in formal educational settings. Learning scientists, often in collaboration with school-based educational professionals, have identified important domain specific cognitive skills and have elevated the use of metacognition as a critical domain general tool that integrates domain-specific thinking (Bransford et al., 2000; Brown, 1992). In the past several decades, instead of focusing efforts on transmitting knowledge as a body of facts to be memorized, learning scientists and their collaborators have created the conditions for students to experience processes of knowledge creation and evaluation that reflect disciplinary practices, at least to some degree.

An influential example is the work of Wineburg on historical thinking. Wineburg's research (e.g., 1991, 1998) identified specific differences in the ways that experts and novices approach historical reasoning. Experts, but not novices, evidenced careful consideration of sources, used corroboration more meaningfully, and made efforts to historically contextualize ideas. These findings contributed to efforts to make primary source documents and lesson plans available to support teachers to engage students in "thinking like a historian" (Wineburg & Reisman, 2015; see also https://sheg.stanford.edu/). In science learning, Linn and colleagues' web-based Inquiry Science Environment (WISE) supports inquiry science through domain-specific practices of visualizing data while also encouraging knowledge integration through processes focused on eliciting, adding, distinguishing, and reflecting on ideas (e.g., Linn, n.d.; Linn & Eylon, 2011). Tabak, Reiser, and colleagues have shown how discipline-specific scaffolds can contribute to important science learning, such as in the formulation of explanations, as well as the adoption of a productive overall stance toward the discipline (e.g., Tabak & Reiser, 2008; Tabak & Kyza, this volume). In literature learning, Lee (e.g., 2001) demonstrated how high school students could build on cultural resources and practices from outside school to engage in the practices of literary analysis in school contexts. Goldman et al. (2016) brought together research on disciplinary expertise in history, science and literature to create a general framework of core constructs that were then instantiated appropriately within each domain. This process highlighted high level similarities across the three disciplines, but more importantly it led to defining learning goals specific to each discipline. These efforts reflect a general view that learning in the disciplines should reflect the work of actual practitioners of the discipline itself (Engle & Conant, 2002; Shulman & Quinlan, 1996). These approaches to disciplinary learning have contributed to redefining standards for outcomes of schooling in the United States and to a degree in Europe.

Limitations of Disciplinary Learning

Disciplinary learning has important limitations, especially when applied at the grain size typically found in formal schooling settings at elementary, secondary, and pre-university levels. For example, referring to science writ large as a discipline masks substantial variation in practices and norms among physics, chemistry, geology, and astronomy. Classical experimental methods in physics and chemistry are often taken as the prototypes for all science practices, but geology and cosmology (the science of the universe's origin) utilize interpretive and historical methods that have epistemic assumptions and norms that radically differ from experimental science (Frodeman, 1995). Similarly, the "social studies" as formulated in many schools encompass disciplines from history to political science to geography, which have wildly differing methods and norms. Accordingly, learning scientists have taken the differences between disciplines and subdisciplines into account by identifying shared practices such as argumentation and working to understand how arguing is different across subject matter contexts (e.g., Herrenkohl & Cornelius, 2013).

A second problem is more an issue created by the way in which institutions typically structure the school day with specific periods of time devoted to singular subject matter areas. Students progress through school in subject,matter silos corresponding to disciplines. Although there may be "cross-departmental" efforts to integrate content over time (one takes algebra, geometry, pre-calculus, and calculus in that order to build increasingly complex concepts over time), there are often no significant attempts to create opportunities to engage across different subject matter areas at the same time. Typically, interdisciplinarity is evident only in thematically integrated curriculum rather than epistemically oriented cross-cutting concepts and tools that would make meaningful connections to the ways that knowledge is created and justified in different disciplinary contexts (Stevens et al, 2005).

Yet, phenomenologically, people do not experience the world from singular disciplinary perspectives (Stevens, et al., 2005). Researchers have demonstrated how learners develop knowledgeability across time and place in differently organized settings and contexts, from schools to community organizations to museums and parks to homes to online spaces (Leander, Phillips, & Taylor, 2010; Lemke, 2000). For example, researchers in literacy (e.g., Barron, Gomez, Pinkard, & Martin, 2014; Gomez, n.d.) have studied the development of literate practices across online and face-to-face/offline spaces spanning schools and multiple sites. Nasir and Hand (2008) showed how school courses and basketball teams provided differential opportunities for learners to engage in mathematical practices. As a whole, this sort of work demonstrates that, in order to fully understand learning in the disciplines, we must recognize that disciplinary practices and concepts are inextricably tied to and transformed by the varied contexts in which they appear.

Related to the phenomenological challenges, there are longstanding non-dominant cultural practices and forms that demonstrate important differences in epistemological commitments within a domain (Bang, 2015; Bang, Warren, Rosebery & Medin, 2013). Bang's work on Indigenous epistemologies, including the idea of relational epistemologies as they apply to sense-making in ecological domains, is particularly relevant. Her studies demonstrate that our understanding of human knowledge in scientific domains has been constrained by dominant, white Western views (Bang, 2015). She highlights how Indigenous epistemologies position the activity system, including a wider set of non-human biological entities, as possessing agency and that human actors exist in relation to this wider system. Western dominant epistemologies position human actors outside this wider system and endow human beings with agency that is not attributed to other non-human dimensions of the ecological system. Bang found that learning contexts play a critical role in shaping these epistemologies and that the same people can take up different epistemologies in response to the context in which they are asked to activate their knowledge. These stances have important implications for how people think about and respond to the natural world as well as how we engage children who are learning about the world (Bang et al., 2013; Rosebery, Ogonowski, Di Schino, & Warren, 2010). This work highlights the importance of preserving epistemic heterogeneity in order to ensure that our field captures the range of human thinking and sense-making.

A further issue is the difference between the uses of disciplines by insiders and outsiders. Insiders to a discipline participate in the practices of academic knowledge production and professional work while outsiders to that discipline use the knowledge or practices of that discipline in their everyday lives, creating crucial differences (Bromme & Goldman, 2014; Feinstein, 2011). However, in formal educational settings there is scant instructional time devoted to use of disciplinary knowledge in contexts of everyday life decisions and dilemmas. For instance, the model of expert, practicing scientists has dominated school science curricula and assessment internationally, privileging a near-exclusive focus on how science is conceptualized and practiced from the inside (Roberts, 2011). Feinstein, Roberts, and others advocate for an alternative approach to science literacy that emphasizes knowledgeability about science-related situations that people are likely to encounter in everyday life, such as personal choices on treatments for health conditions and public debates about policy related to climate change. Advocates of using "socioscientific dilemmas" in education (e.g., Sadler, 2004), as well as community-based science inquiry and action (e.g., Barton, Tan, & Rivet, 2013; Roth & Lee, 2004), take seriously this perspective; ideas and practices of insiders are threaded through problems consistent with the goals, values, and priorities of "everyday citizens." Onwu and Kyle (2011) point out that a focus on issues like sustainable development can increase the relevance of education and better fulfill goals of equity and democratic citizenship. Although the above examples are drawn from science education, similar arguments can be made in other disciplines. For instance, Gottlieb and Wineburg (2012) explicated how scholars of religion do "epistemic switching" between historical and religious assumptions about the nature of knowledge depending on whether aspects of a text they were reading evoked identification with their academic community or their religious community.

Moving Beyond Disciplinary Learning

Many of the limitations of a disciplinary learning approach are bound up in the constraints of formal educational systems and longstanding debates about the goals and functions of schooling (see discussion in Stokes, 2011) Since its emergence in the early 1990s, the learning sciences has vastly expanded the focus, contexts, and constructs of interest. The field has and continues to shift toward a more human science view (Flyvbjerg, 2001; Penuel & O'Connor, 2010) where the values, purposes, and goals of learning as well as who has the power to decide such matters are critically important to understanding learning. As such, motivation, emotion, values and purposes are considered a part of any design and analysis of learning.

Knowledge in this view is not neutral, nor is it separated from actors and contexts where it is put to use. For example, Herrenkohl and Mertl (2010) examined how fourth-grade students and their teacher used intellectual and social roles to support scientific reasoning. Their analysis demonstrated the complex ways that different students used knowledge and skills to achieve particular intellectual, social, and emotional goals even within a formal educational setting. The structure of knowledge is not unimportant; however, knowledge exists in the hands and mouths of people in specific contexts who breathe life and meaning into it. We expect productive research to result from shifting our focus in learning sciences to *people* who *employ* knowledge and other tools *in hybrid settings* to solve complex problems that involve purposeful collaboration and managing competing values and goals.

Across Space, Time, and Contexts

In order for our field to effectively study learning outside of classroom settings and learners moving across space and time, multiple disciplines and approaches are necessary. Actors in non-school settings often draw on multiple traditions and approaches to learn or take action because the focus of their efforts does not align neatly with just one cell of a disciplinary taxonomy. First, they must identify a problem to be solved, then they build understanding of the current conditions and scope of the problem, and finally they create possible solutions to the problem drawing on whatever resources help them make progress. In this frame, the attention shifts from understanding concepts and processes to *designing solutions* using relevant concepts and processes. Cross (2007) suggests this solution-oriented practice is at the heart of "designerly ways of knowing" and it is a "third culture" (in addition to the sciences and humanities) largely left out of formal education. This culture focuses on the material world and on synthesizing across perspectives to create practical solutions. This approach mirrors the signature method of the learning sciences (design-based research; see Puntambekar, this volume) so it is not a surprise that some of the most exciting contemporary research in the learning sciences is now focused on settings where design is at the center.

The trend of addressing a diverse array of settings for learning has contributed to growing interest in the learning sciences community in "boundary work" and "hybridity" that necessitates breaking out of disciplinary constraints. As a case in point, journalism is a community of practice with a history of crossing boundaries: using social sciences, sciences, history, and rhetoric to inform citizens' lives (Polman & Hope, 2014; Polman, Newman, Saul, & Farrar, 2014; Shaffer, 2006). Journalism involves domain-general and domain-specific understanding, it seeks to understand tools and practices on their own terms and in relationship to human institutions, histories, etc. The best journalism opens up alternative points of view and critical concerns representing multiple perspectives on issues without necessarily resolving valid disputes. And, like any community of practice, it has its own rhetorical and evidentiary norms, which should affect interpretation of the community's work.

Stember (1991) argued that the social sciences would benefit from engaging more than one discipline, which can be accomplished through: *multi*disciplinary work involving people from different disciplines collaborating, with each drawing on their disciplinary knowledge; *inter*disciplinary work involving knowledge integration and synthesis and methods from different disciplines; and *trans*disciplinary work that creates unified frameworks transcending disciplinary perspectives. In learning sciences, scholars who study complex systems (e.g., Jacobson & Wilensky, 2006) draw on multidisciplinary and interdisciplinary perspectives. Additionally, a growing number of authors draw on Star's (2010) notion of boundary objects (e.g., Akkerman & Bruining, 2016; Wenger, 1998), or document the work of brokers, boundary spanners (e.g., Buxton, Carlone, & Carlone, 2005) or boundary crossing (e.g., Engeström, Engeström, & Kärkäinen, 1995) to explain important aspects of learning and development at both individual and communal levels. Such work can be an important means of connecting formal school contexts to wider communities. For instance, Polman and Hope (2014) reported several outcomes at individual and communal levels of authoring science news articles. This activity provided a context for supporting important learning goals for students, while benefitting from interactions between students and an editor of a science news magazine from outside of the school boundary. As well, the activity facilitated the school community engaging with their wider civic community.

A focus on boundary work leads naturally to considering hybrid spaces where discourses from different communities come together (e.g., Barton et al., 2008; Gutiérrez, Baquedano-López, & Tejeda, 1999). Additionally, numerous learning sciences scholars are adapting and developing innovative frameworks and methods that show promise as models for future engaged scholarship: social design experiments (Gutiérrez & Vossoughi, 2010), youth participatory action research (Kirshner, Pozzoboni, & Jones, 2011), and participatory design research (Bang & Vossoughi, 2016) have increased in the past decade as learning scientists seek to create, study, and understand equitable learning opportunities. Such work frequently draws on ideas from cultural, feminist, and queer theory (e.g., Anzaldúa, 1987) or geography (e.g., Soja, 1996).

Of particular interest from the perspective of identity and agency in the context of communal activity is research being conducted from what might be called a youth culture perspective. Taylor (2017) urges the field to consider the role of mobility in youth learning and youth mobility as a context for learning. Her notion of "city science" hybridizes science-literacy-technology to support youth in using location aware and mobile tools to collect, analyze, and argue from data they collect about their daily lives in order to create more equitable and safe transportation options. This initiative supports youth in changing their communities at the city scale, thereby democratizing the urban planning process to include youth voices thoughtfully and intentionally using ubiquitous computing.

In a related vein, Peppler's work focuses on media arts and literacy in the digital age to reimagine opportunities for marginalized youth voices (Peppler, 2010). Peppler also intentionally uses making as a context to better engage girls in STEM fields (Peppler, 2013). Ultimately, she uses these approaches, like Taylor, to move the learning sciences toward engaging larger sets of values for learning in a democracy. Using the toolkit BlockyTalky, Shapiro and colleagues are bringing together multiple disciplines including computer science, music, and design. They study the socially and materially distributed processes of cognition and action among groups of youths engaged in musical instrument design and group performance (Shapiro, Kelly, Ahrens, & Fiebrink, 2016). Rather than specifying particular learning goals from one of the related disciplines of computer science, music, or mathematics, they build capabilities for networked interaction into the toolkit, empirically examine the work of youth jointly programming for performance, and use their examination to refine the designs of tools and organization of activities to refine the performance and composition possibilities. This sort of boundary work across disciplines and creatively situated educational contexts and tools is critically important to advancing the learning sciences.

A Challenging Future

As the learning sciences expands into multiple and varied settings in which learning occurs with multiple goals for the learning, multi-, inter-, and transdisciplinary alliances and collaborations are essential. The research requires critical collaborations with people trained in fields with different norms, values, and approaches. Integrating different stances, tools, lexicons, and routines to create shared understanding and achieve goals is complex and time-consuming. In addition, institutional recognition and support for work that pushes boundaries may be limited. Although many academic institutions encourage cross-disciplinary collaborations, few mechanisms support this kind of work for new scholars. Opportunities to educate people to participate in this type of research are limited, so mid-career scholars may also find it challenging to navigate and reconcile these new intellectual opportunities with material support to pursue them. Yet, this is the history of our field and this moment in time brings new twists on a familiar theme. We will continue to develop and bring together tools within *and* across disciplines to create powerful learning opportunities with practitioners and provide insight into important questions about how people learn.

Further Readings

Bang, M., Warren, B., Rosebery, A. S., & Medin, D. (2013). Desettling expectations in science education. *Human Development, 55(5–6)*, 302–318.
Examines two episodes to "desettle" the relationship between science learning, classroom teaching, and emerging understandings of grounding concepts in scientific fields. Demonstrates how desettling and reimagining core relations between nature and culture shifts possibilities in learning and development, particularly for nondominant students.

Barron, B., Gomez, K., Pinkard, N., & Martin, C. K. (2014). *Digital Youth Network: Cultivating new media citizenship in urban communities*. Cambridge, MA: MIT Press.
Describes the motivations for and findings from the Digital Youth Network, where economically disadvantaged middle school students develop technical, creative, and analytical skills in a learning ecology spanning school, community, home, and online.

Peppler, K. (2010). Media arts: Arts education for a digital age. *Teachers College Record, 112*(8), 2118–2153.
Mixed method study documenting what youth learn through media art making in informal settings, the strengths and limitations of capitalizing on youth culture in media art production, and the distinct contributions that media arts education can make to the classroom environment.

Shapiro, R. B., Kelly, A., Ahrens, M., & Fiebrink, R. (2016) BlockyTalky: A physical and distributed computer music toolkit for kids. In *Proceedings of the 2016 Conference on New Interfaces for Musical Expression*. Brisbane, Australia.
Describes a computer music toolkit for kids called BlockyTalky, which enables users to create networks of sensing devices and synthesizers, and offers findings from research on student learning through programming and performance.

Taylor, K. H. (2017). Learning along lines: Locative literacies for reading and writing the city. *Journal of the Learning Sciences, 26*(4), 533–574. doi:10.1080/10508406.2017.1307198
Contributes to our understanding of learning place-based, digital literacies through urban spaces. The analyses push our field's understanding of digital and physical mobility in conceptualizing and designing new forms of learning locative literacies.

NAPLeS Resources

Gomez, K., *Learning with digital technologies across learning ecologies* [Webinar]. In *NAPLeS video series*. Retrieved October 19, 2017, from http://isls-naples.psy.lmu.de/intro/all-webinars/gomez/

Linn, M. C., *Inquiry learning.* [Webinar]. In *NAPLeS video series*. Retrieved October 19, 2017, from http://isls-naples.psy.lmu.de/intro/all-webinars/linn_all/index.html

Tabak, I., & Reiser, B., *Scaffolding* [Webinar]. In *NAPLeS video series*. Retrieved October 19, 2017, from http://isls-naples.psy.lmu.de/intro/all-webinars/tabak_reiser_all/index.html

References

Akkerman, S., & Bruining, T. (2016). Multilevel boundary crossing in a professional development school partnership. *Journal of the Learning Sciences, 25*(2), 240–284, doi:10.1080/10508406.2016.1147448

Anzaldúa, G. (1987). *Borderlands: La frontera* (Vol. 3). San Francisco: Aunt Lute.

Bang, M. (2015). Culture, learning, and development about the natural world: Advances facilitated by situative perspectives. *Educational Psychologist, 50*(3), 220–233.

Bang, M., & Vossoughi, S. (2016). Participatory design research and educational justice: Studying learning and relations within social change making. *Cognition & Instruction, 34*(3), 173–193.

Bang, M., Warren, B., Rosebery, A. S., & Medin, D. (2013). Desettling expectations in science education. *Human Development, 55(5–6)*, 302–318.

Barron, B., Gomez, K., Pinkard, N., & Martin, C. K. (2014). *Digital Youth Network: Cultivating new media citizenship in urban communities*. Cambridge, MA: MIT Press.

Barton, A. C., Tan, E., & Rivet, A. (2008). Creating hybrid spaces for engaging school science among urban middle school girls. *American Educational Research Journal, 45*(1), 68–103.

Bransford, J. D., Brown, A. L., & Cocking, R. R. (2000). *How people learn: Brain, mind, experience, and school.* Washington, DC: National Academies Press.

Bromme, R. & Goldman, S. R. (2014) The public's bounded understanding of science, *Educational Psychologist, 49 (2)*, 59–69.

Brown, A. L. (1992). Design experiments: Theoretical and methodological challenges in creating complex interventions in classroom settings. *Journal of the Learning Sciences*, 2(2), 141–178.

Brown, J. S., Collins, A., Duguid, P. (1989). Situated cognition and the culture of learning. *Educational Researcher*, 18(1), 32–42.

Bruer, J. T. (1993). *Schools for thought: A science of learning in the classroom*. Cambridge, MA: MIT Press.

Bruer, J. T. (1997). Education and the brain: A bridge too far. *Educational Researcher*, 26(8), 4–16.

Bruner, J. S. (1960). On learning mathematics. *Mathematics Teacher*, 53(8), 610–619.

Buxton, C. A., Carlone, H. B., & Carlone, D. (2005). Boundary spanners as bridges of student and school discourses in an urban science and mathematics high school. *School Science & Mathematics*, 105(6), 302–312.

Carlone, H., & Johnson, A. (2012). Unpacking 'culture' in cultural studies of science education: Cultural difference versus cultural production. *Ethnography and Education*, 7(2), 151–173.

Chinn, C. A., Buckland, L. A., & Samarapungavan, A. L. A. (2011). Expanding the dimensions of epistemic cognition: Arguments from philosophy and psychology. *Educational Psychologist*, 46(3), 141–167.

Cole, M. (1996). *Cultural psychology: A once and future discipline*. Cambridge, MA: Harvard University Press.

Cornelius, L., & Herrenkohl, L.R. (2004). Power in the classroom: How the classroom environment shapes students' relationships with each other and with concepts. *Cognition and Instruction*, 22, 467–498.

Cross, N. (2007). *Designerly ways of knowing*. Berlin: Birkhäuser.

Dreyfus, H. L., & Dreyfus, S. E. (1986). *Mind over machine: The power of human intuition and expertise in the era of the computer*. New York: Free Press.

Engeström, Y., Engeström, R., & Kärkkäinen, M. (1995). Polycontextuality and boundary crossing in expert cognition: Learning and problem solving in complex work activities. *Learning and Instruction*, 5(4), 319–336.

Engle, R. A., & Conant, F. R. (2002). Guiding principles for fostering productive disciplinary engagement: Explaining an emergent argument in a community of learners classroom. *Cognition and Instruction*, 20(4), 399–483.

Feinstein, N. (2011). Salvaging science literacy. *Science Education*, 95(1), 168–185.

Fishman, B., & Penuel, W. (2018). Design-based implementation research, In F. Fischer, C. E. Hmelo-Silver, S. R. Goldman, & P. Reimann (Eds.), *International handbook of the learning sciences* (pp. 393–400). New York: Routledge.

Flyvbjerg, B. (2001). *Making social science matter: Why social inquiry fails and how it can succeed again*. New York: Cambridge University Press.

Frodeman, R. (1995). Geological reasoning: Geology as an interpretive and historical science. *Geological Society of America Bulletin*, 107(8), 960–968.

Goldman, S. R., Britt, M. A., Brown, W., Cribb, G., George, M., Greenleaf, C., et al. (2016). Disciplinary literacies and learning to read for understanding: A conceptual framework for disciplinary literacy, *Educational Psychologist* (May), 1–28.

Gomez, K., *Learning with digital technologies across learning ecologies* [Webinar]. In *NAPLeS video series*. Retrieved October 19, 2017, from http://isls-naples.psy.lmu.de/intro/all-webinars/gomez/

Gottlieb, E., & Wineburg, S. (2012). Between veritas and communitas: Epistemic switching in the reading of academic and sacred history. *Journal of the Learning Sciences*, 21(1), 84–129.

Gutiérrez, K. D., Baquedano-López, P., & Tejeda, C. (1999). Rethinking diversity: Hybridity and hybrid language practices in the third space. *Mind, Culture, and Activity*, 6(4), 286–303.

Gutiérrez, K. D., & Vossoughi, S. (2010). Lifting off the ground to return anew: Mediated praxis, transformative learning, and social design experiments. *Journal of Teacher Education*, 61(1–2), 100–117.

Herrenkohl, L. R., & Cornelius, L. (2013). Investigating elementary students' scientific and historical argumentation. *Journal of the Learning Sciences*, 22(3), 413–461.

Herrenkohl, L. R., & Mertl, V. (2010). *How students come to be, know, and do: A case for a broad view of learning*. New York: Cambridge University Press.

Hoadley, C., (2018) A short history of the learning sciences. In F. Fischer, C. E. Hmelo-Silver, S. R. Goldman, & P. Reimann (Eds.), *International handbook of the learning sciences* (pp. 11–23). New York: Routledge.

Hoadley, C., & Van Haneghan, J. (2011). The learning sciences: Where they came from and what it means for instructional designers. In R. A. Reiser & J. V. Dempsey (Eds.), *Trends and issues in instructional design and technology* (3rd ed., pp. 53–63). New York: Pearson.

Hutchins, E. (1995). *Cognition in the wild*. Cambridge, MA: MIT Press.

Jacobson, M. J., & Wilensky, U. (2006). Complex systems in education: Scientific and educational importance and implications for the learning sciences. *Journal of the Learning Sciences*, 15(1), 11–34.

Kirshner, B., Pozzoboni, K., & Jones, H. (2011). Learning how to manage bias: A case study of youth participatory action research. *Applied Developmental Science*, 15(3), 140–155.

Latour, B. (1990). Drawing things together. In M. Lynch & S. Woolgar (Eds.), *Representation in scientific practice* (pp. 19–68). Cambridge, MA: MIT Press.

Lave, J., & Wenger, E. (1991). *Situated learning: Legitimate peripheral participation.* New York: Cambridge University Press.

Leander, K. M., Phillips, N. C., & Taylor, K. H. (2010). The changing social spaces of learning: Mapping new mobilities. *Review of Research in Education, 34*(1), 329–394.

Lee, C. D. (2001). Is October Brown Chinese? A cultural modeling activity system for underachieving students. *American Educational Research Journal, 38*(1), 97–141.

Lemke, J. L. (1990). *Talking science: Language, learning, and values.* Norwood, NJ: Ablex.

Lemke, J. L. (2000). Across the scales of time: Artifacts, activities, and meanings in ecosocial systems. *Mind, Culture, and Activity, 7*(4), 273–290.

Linn, M. C., Inquiry learning. [Webinar]. In *NAPLeS video series.* Retrieved October 19, 2017, from http://isls-naples.psy.lmu.de/intro/all-webinars/linn/

Linn, M. C., & Eylon, B. S. (2011). *Science learning and instruction: Taking advantage of technology to promote knowledge integration.* New York: Routledge.

Nasir, N. I. S., & Hand, V. (2008). From the court to the classroom: Opportunities for engagement, learning, and identity in basketball and classroom mathematics. *Journal of the Learning Sciences, 17*(2), 143–179.

O'Connor, K., Peck, F. A., & Cafarella, J. (2015). Struggling for legitimacy: Trajectories of membership and naturalization in the sorting out of engineering students. *Mind, Culture, and Activity, 22*(2), 168–183.

Onwu, G. O. M., & Kyle, Jr., W. C. (2011). Increasing the socio-cultural relevance of science education for sustainable development. *African Journal of Research in Mathematics, Science and Technology Education, 15*(3), 5–26.

Penuel, W. R., Fishman, B. J., Cheng, B. H., & Sabelli, N. (2011). Organizing research and development at the intersection of learning, implementation, and design. *Educational Researcher, 40*(7), 331–337.

Penuel, W. R., & O'Connor, K. (2010). Learning research as a human science: Old wine in new bottles. *National Society for the Study of Education, 109*(1), 268–283.

Peppler, K. (2010). Media arts: Arts education for a digital age. *Teachers College Record, 112*(8), 2118–2153.

Peppler, K. (2013). STEM-powered computing education: Using e-textiles to integrate the arts and STEM, *IEEE Computer* (September), 38–43.

Perkins, D. N., & Salomon, G. (1989). Are cognitive skills context-bound? *Educational Researcher, 18*(1), 16–25.

Polman, J. L., & Hope, J. M. (2014). Science news stories as boundary objects affecting engagement with science. *Journal of Research in Science Teaching, 51*(3), 315–341.

Polman, J. L., Newman, A., Saul, E. W., & Farrar, C. (2014). Adapting practices of science journalism to foster science literacy. *Science Education, 98*(5), 766–791.

Puntambekar, S. (2018). Design-based research (DBR). In F. Fischer, C. E. Hmelo-Silver, S. R. Goldman, & P. Reimann (Eds.), *International handbook of the learning sciences* (pp. 383–392). New York: Routledge.

Rogoff, B. (1990). *Apprenticeship in thinking: Cognitive development in social context.* New York: Oxford University Press.

Roberts, D. A. (2011). Competing visions of scientific literacy: Influence of a science curriculum policy image. In C. Linder, L. Östman, D. A. Roberts, P.-O. Wickman, G. Erickson, & A. MacKinnon (Eds.), *Exploring the landscape of scientific literacy* (pp. 11–27). New York: Routledge.

Rosebery, A., Ogonowski, M., Di Schino, M., & Warren, B. (2010). "The coat traps all your body heat": Heterogeneity as fundamental to learning. *Journal of the Learning Sciences, 19*, 322–357.

Roth, W.-M., & Lee, S. (2004). Science education as/for participation in the community. *Science Education, 88*(2), 263–291.

Sadler, T. D. (2004). Informal reasoning regarding socioscientific issues: A critical review of research. *Journal of Research in Science Teaching, 41*(5), 513–536.

Saxe, G. B. (1991). *Culture and cognitive development: Studies in mathematical understanding.* Hillsdale, NJ: Erlbaum.

Schwab, J. J. (1978). Education and the structure of the disciplines. In I. Westbury & N. Wilkof (Eds.), *Science, curriculum and liberal education: Selected essays* (pp. 229–272). Chicago, IL: University of Chicago Press.

Shaffer, D. W. (2006). *How computer games help children learn.* New York: Palgrave Macmillan.

Shapiro, R. B., Kelly, A., Ahrens, M., & Fiebrink, R. (2016) BlockyTalky: A physical and distributed computer music toolkit for kids. In *Proceedings of the 2016 Conference on New Interfaces for Musical Expression.* Brisbane, Australia

Shulman, L. S., & Quinlan, K. M. (1996). The comparative psychology of school subjects. In D. C. Berliner & R. Calfee (Eds.), *Handbook of educational psychology* (pp. 399–422). New York: Macmillan.

Soja, E. (1996). *Thirdspace: Journeys to Los Angeles and other real-and-imagined places.* Oxford, UK: Blackwell.

Star, S. L. (2010). This is not a boundary object: Reflections on the origin of a concept. *Science, Technology, & Human Values, 35*, 601–617.

Stember, M. (1991). Advancing the social sciences through the interdisciplinary enterprise. *Social Science Journal*, *28*(1), 1–14.

Stern, E. (2016, June). *Educational neuroscience: A field between false hopes and realistic expectations.* Keynote presentation at the 12th International Conference of the Learning Sciences, Singapore.

Stevens, R., Wineburg, S., Herrenkohl, L. R., & Bell, P. (2005). Comparative understanding of school subjects: Past, present, and future. *Review of Educational Research*, *75*(2), 125–157.

Stokes, D. E. (2011). *Pasteur's quadrant: Basic science and technological innovation.* Washington, DC: Brookings Institution Press.

Tabak, I., & Kyza., E. (2018) Research on scaffolding in the learning sciences: A methodological perspective. In F. Fischer, C. E. Hmelo-Silver, S. R. Goldman, & P. Reimann (Eds.), *International handbook of the learning sciences* (pp. 191–200). New York: Routledge.

Tabak, I., & Reiser, B. J. (2008). Software-realized inquiry support for cultivating a disciplinary stance. *Pragmatics & Cognition*, *16*(2), 307–355.

Varma, S., Im, S., Schmied, A., Michel, K., & Varma, K. (2018) Cognitive neuroscience foundations for the learning sciences. In F. Fischer, C. E. Hmelo-Silver, S. R. Goldman, & P. Reimann (Eds.), *International handbook of the learning sciences* (pp. 64–74). New York: Routledge.

Wenger, E. (1998). *Communities of practice: Learning, meaning, and identity.* New York: Cambridge University Press.

Wertsch, J. V. (1985). *Vygotsky and the social formation of the mind.* Cambridge, MA: Harvard University Press.

Wertsch, J. V. (1991). *Voices of the mind: A sociocultural approach to mediated action.* Cambridge, MA: Harvard University Press.

Wineburg, S. (1991). Historical problem solving: A study of the cognitive processes used in the evaluation of documentary and pictorial evidence. *Journal of Educational Psychology*, *83*(1), 73–87.

Wineburg, S. (1998). Reading Abraham Lincoln: An expert/expert study in the interpretation of historical texts. *Cognitive Science*, *22*(3), 319–346.

Wineburg, S., & Reisman, A. (2015). Disciplinary literacy in history. *Journal of Adolescent & Adult Literacy*, *58*(8), 636–639.

12

Motivation, Engagement, and Interest

"In the End, It Came Down to You and How You Think of the Problem"

K. Ann Renninger, Yanyan Ren, and Heidi M. Kern

This chapter addresses the similarities and differences between motivation, engagement, and interest and their implications for learning science research. For example, it is possible for a person to be motivated or engaged, but not interested, whereas, when something is of some interest to a person, it is always motivating and engaging. Understanding these variables and the relations among them can contribute to the effective design, facilitation, and evaluation of learning environments as wide-ranging as everyday experiences (e.g., TV programming, family interactions, Facebook), designed settings (e.g., museums, online courses, zoos), and out-of-school programs (e.g., scouting, sports, music lessons).

"Motivation" concerns individuals and their response to their social and cultural circumstances; specifically the will to engage, and the influence of will on individuals' setting goals and working to accomplish them (Hidi & Harackiewicz, 2000; Wigfield, Eccles, Schiefele, Roeser, & Davis-Kean, 2006). "Engagement" typically deals with the context of participation and individuals' cognitive, affective, and behavioral responses to it, which reflect their beliefs about the possibility of their participation (Christenson, Reschly, & Wylie, 2012; Fredricks, Blumenfeld, & Paris, 2004; Shernoff, 2013). Finally, "interest" describes individuals' participation with particular content (e.g., computer science, ballet)—individuals' psychological states during engagement with that content, as well as the likelihood of their independent and voluntary reengagement with it over time (Hidi & Renninger 2006; Renninger & Hidi, 2016).

Consider the case studies of Nasir and Emily (Figure 12.1). They demonstrate how a learner's motivation and level of engagement can change depending on factors such as their interactions with other people and the structure of the task or environment. The cases also show the motivation and level of engagement that can characterize persons in different phases of interest development (Nasir with less, Emily more).

Nasir describes himself as motivated by the open-ended assignments he is given in his computer science (CS) courses, compared to the tight structure of chemistry assignments. He finds that his CS teachers put him in charge of his own learning by asking him to set and achieve realistic goals. He describes having fun working on rich problems alongside his friends. It may also be important that they may all be doing different things in order to accomplish their goals.

In contrast, Emily reports thriving on the structure and discipline of ballet once she understood that the intense practice required to master harder steps put her closer to being a "real" ballerina.

Case 1: Nasir*	Case 2: Emily
Nasir's eyes lit up when asked about his major in computer science (CS). Coming into college, Nasir had no idea what CS was. He assumed that he would be pursuing chemistry and only decided to take an introductory CS course because his friends were signing up for it. He explained, "CS [computer science] feels like art, like drawing." He described the first course, saying, "We all worked on designing a slot machine that worked. In the end, they did, and they were all different. We would look at each other's efforts to build a slot machine and laugh (even when it wasn't working). There wasn't a better or best answer. In the end, it came down to you and how you think of the problem." He explained that the project focus was novel for him and a contrast to the advanced chemistry course in which he was also enrolled. He said, "The [chemistry] lab journal felt unreasonably strict, and everything felt like a procedure," and noted that even though chemistry had been his intended major, he switched to major in CS after taking more classes and doing a summer internship as a software engineer. He described CS as challenging and at the same time doable, and the kind of thing that he and his friends had fun hanging around thinking and talking about.	Emily says that dance is an important part of who she is, but reports that it was not always this way. She started taking ballet classes at the age of 5. She looked up to the older dancers as role models, and worked hard in class to impress her teacher, who challenged and encouraged her. Each year Emily took more classes and performed more roles. Age 10 was a critical year for her. "I remember when I was 10, my teacher moved me into the advanced class, and I felt so out of place. I didn't think I'd ever be as good as the older girls. The steps were too hard for me, and I wanted to quit." She explained that instead of boosting her confidence, being moved into the advanced class lessened her self-efficacy. She left most classes feeling discouraged, but continued to attend them because of her mother's encouragement. That spring, her teacher announced that the following fall Emily would be ready for her first pair of pointe shoes (which she knew everyone got at age 12, not 10). "When my teacher said I was ready for pointe, my whole perception changed. I came to class every day working hard to make her proud and prove that I could do it. With pointe shoes, I could show everyone that I really was a ballerina." Emily persevered through the challenging exercises and continued into company classes and pointe work, earning more lead roles in the annual performances.

Figure 12.1 Case Studies

Note: Pseudonyms have been used for both cases.

She needed to figure out for herself that she could master the movements, even though other indicators might have shown her that this was the case (she was moved into the advanced class; she would be receiving pointe shoes in the fall).

In the sections that follow, we provide working definitions and findings in the study of motivation, engagement, and interest, and use Nasir's and Emily's cases for purposes of illustration. We review methods employed to assess these variables, consider their relation, and use the literature to suggest design principles for learning. We suggest that these variables are central to supporting deeper learning.

Working Definitions and Research

Motivation

Motivation is a broad term that encompasses both engagement and interest, as well as other topics such as: perceptions or beliefs about achievement, capability, or competence; expectancy (likelihood of benefit from one versus another action); value; and choice. These factors are addressed by a person's consideration of the possibility, utility, importance, and benefit of participating and belonging. Motivation can address learners' mindsets about whether learning or being able to understand new content/skills is possible; future time perspective; beliefs about self-efficacy, or the ability to work with a specific task, and/or their self-concept of ability, their sense of their ability to work with the types of tasks that characterize a domain, subject area, or field of study (e.g., science). Studies of motivation may focus on goals that individuals set for themselves and their readiness to take initiative and/or to self-regulate in order to achieve these goals. Topics in motivation may also include consideration of boredom, as well as incentive salience or reward.

Topics in motivation are distinct variables and areas of study, but they often co-occur. In Nasir's and Emily's case material, for example, several topics in motivation are evident.

In his CS class, Nasir wants to (or has the motivation or will to):

- make the slot machine work (an identified goal);
- complete the homework (achievement motivation); and
- be successful in comparison to his classmates (expectancy value, the expectancy that this project is worth the time it is requiring).

Nasir's motivation for CS is informed by his work on the initial assignment to design a slot machine, the appeal of additional CS courses, and opportunities outside of class, such as the summer internship, to develop his ability to program. His motivation is supported by his belief that he had the coding skills needed to build the slot machine, a project with clear outcomes and goals. He kept coding even though he ran into some obstacles in the process, requiring him to take initiative and self-regulate in order to accomplish his goals. At any given point he might have been identified as being in a flow state (Csikszentmihalyi, 1990), and as interested. He did not find the same challenge in chemistry; he found the assignments procedural and constraining, which led him to lose interest.

Nasir's perseverance illustrates how the right level of challenge serves as motivation: he uses his knowledge to solve the problems, gains a sense of accomplishment, and gets ready to solve more complex problems. As a result, Nasir found CS rewarding. His interest in CS had begun to develop; he was both motivated and engaged.

Considering Emily's case along with Nasir's reveals the similarity of their motivation, despite differences in their disciplinary focus and interest level.

In her advanced class, Emily has the motivation to:

- do well (short-term goal);
- become a ballet dancer (long-term goal);
- take on more roles in the annual performance to impress friends and family in the audience (performance goal); and
- master ballet for the sake of art (mastery goal).

Emily's motivation for ballet is influenced by the structure and discipline of ballet, as well as the ballet class she is taking. She has developed her ability to dance in the context of successively challenging classes, and is influenced by role models like her teacher and older classmates. Knowing that the advanced class was for the older and more developed dancers, Emily was very motivated to get into it (achievement motivation). But when she was asked to master steps she did not yet know, participation felt overwhelming, and the challenge affected her self-efficacy. Emily's motivation and her interest in ballet began to wane. Her mother's encouragement was critical to Emily's continued participation in the class, and the promise of pointe shoes signaled that her teacher thought the challenges of the class were within her reach.

Emily's case portrays the mix of goals that underlie learner motivation: short-term and long-term, performance and mastery (Vedder-Weiss & Fortus, 2013). Even though getting pointe shoes made her feel accomplished, it is her mastery of the dance sequences—the development of her knowledge and corresponding value for the dance sequences as integral to ballet—that is motivating. Emily's case also provides insight into how transitional support from other people and possible attainments such as pointe shoes can contribute to continued motivation, even for those with more developed interest.

Nasir's and Emily's cases highlight people's different needs for support in recognizing and engaging with opportunity (e.g., new disciplinary pursuits, advanced coursework). It is through interactions

Engagement

Although many topics either individually or together can be used to describe motivation, engagement is a meta-construct that describes the context of participation (e.g., school, sports team, family) and individuals' cognitive, affective, and behavioral responses to it. Moreover, as Fredricks et al. (2004) point out, the cognitive, emotional, and behavioral components of engagement co-occur and, unlike topics in motivation, overlap.

Cognitive engagement describes how invested people are in a given task, their conscientiousness and/or willingness to exert effort in order to master challenging content and difficult skills. Both Nasir and Emily are cognitively engaged, because they are receptive to support from other people and the structure of the tasks on which they are working. They are able to work with challenge and persevere to set and achieve their goals. Nasir and Emily are emotionally and behaviorally engaged, as well. Their emotional engagement refers to their attitudes about the learning environment, including their feelings that they can engage. Their behavioral engagement includes their continuing participation in CS and ballet, respectively, and their subscribing to rules, expectations, and norms of these learning environments.

Nasir's developing interest in CS and Emily's interest in ballet distinguish them from those who lack motivation and interest, and whose disengagement and ensuing school dropout rates have motivated much of the research on engagement (Reschly & Christenson, 2012). When learners have not yet made a connection to the assignments, tasks, or contexts in or on which they are supposed to be working, they may engage superficially, and can benefit from support to evaluate their situations, rules, and expectations. Their situations differ radically from those of students like Nasir who are meaningfully engaged, even if their interest is not fully developed.

Nasir was excited to engage deeply with the CS assignment. He saw that the assignment was not just an opportunity to earn a grade, but allowed him to learn to code. The learning environment gave him the chance to work with challenging content and collaborate with others for strategies, yet follow his personal design ideas, creating a unique slot machine that he was proud of. Nasir's meaningful engagement is similar to that of the youth Boxerman, Lee, and Olsen (2013) describe, who used video to document science on their outdoor field trip, and those who participated in the structured and collaborative version of Gutwill and Allen's (2012) museum program, *Inquiry Games*.

Interest

Interest describes meaningful participation with particular content: people's psychological state during engagement, as well as the likelihood that they will continue to re-engage that content over time. In their four-phase model of interest development, Hidi and Renninger (2006) describe interest as developing through four phases: triggered situational interest, maintained situational interest, emerging individual interest, and well-developed individual interest (Table 12.1).

Neuroscience has established that triggering, or activation, of interest is associated with reward circuitry in the brain (Berridge, 2012). This means that once interest is triggered and begins to develop, engaging the content of interest becomes its own reward. It also indicates that interest can be triggered, regardless of a person's age, gender, previous experience, and personality. The goal in working with those who have less developed interest or none at all is to make engagement feel rewarding.

Interest can be triggered or introduced by other people (e.g., teachers, coaches, peers), by the tasks and activity of the learning environment, and by a person's own efforts to deepen understanding. The process is one in which attention is piqued, and present knowledge and value is then

Table 12.1 Learner Characteristics, Feedback Wants, and Feedback Needs in Each of the Four Phases of Interest Development

	Less developed (earlier)		More developed (later)	
	PHASE 1 – Triggered Situational Interest	PHASE 2 – Maintained Situational Interest	Phase 3 – EMerging Individual Interest	PHASE 4 – WELL-DEVELOPED INDIVIDUAL INTEREST

Learner characteristics	Learners:
	
	• Attend to content, if only fleetingly
	• Need support to engage content:
	  ◦ From others (e.g., group work, instructional conversation)
	  ◦ Through instructional design (e.g., software)
	• May experience either positive or negative feelings
	• May or may not be reflectively aware of the experience.
		Learners:
		
		• Re-engage content that previously triggered attention
		• Are supported by others to find connections between their skills, knowledge, and prior experience
		• Have positive feelings
		• Are developing knowledge of the content
		• Are developing a sense of the content's value.
			Learners:
			
			• Are likely to independently re-engage content
			• Have curiosity questions that lead them to seek answers
			• Have positive feelings
			• Continue developing knowledge and value for what they understand
			• Are very focused on their own questions
			• May have little value for the canon of the discipline and most feedback.
				Learners:
				
				• Independently re-engage content
				• Have curiosity questions
				• Self-regulate easily to reframe questions and seek answers
				• Have positive feelings
				• Can persevere through frustration and challenge in order to meet goals
				• Recognize others' contributions to the discipline, as well as the presence of additional information/skills/perspectives to be understood
				• Actively seek feedback.

Feedback wants	Learners want: • To have their ideas respected • Others to understand how hard work with this content is • To simply be told how to complete assigned tasks in as few steps as possible.	Learners want: • To have their ideas respected • Concrete suggestions • To be told what to do.	Learners want: • To have their ideas respected • To express their ideas • Not to be told to revise present efforts.	Learners want: • To have their ideas respected • Information and feedback • To balance their personal standards with more widely accepted standards in the discipline.
Feedback needs	Learners need: • To feel genuinely appreciated for the efforts they have made • A limited number of concrete suggestions.	Learners need: • To feel genuinely appreciated for the efforts they have made • Support to explore their own ideas.	Learners need: • To feel that their ideas and goals are understood • To feel genuinely appreciated for their efforts • Feedback that enables them to see how their goals can be more effectively met.	Learners need: • To feel that their ideas have been heard and understood • Constructive feedback • Challenge.

stretched. Novel information can trigger interest by calling attention to gaps in or differences from previous understanding, enabling the development of new knowledge. Triggers for interest can be included in lectures or in discussion, and can also be embedded in assignments, tasks, or activities.

Nasir's and Emily's cases illustrate interest development. In describing his developing interest in CS, Nasir points to support from his friends, the challenge (and doable nature) of the task, differences between the assignments in CS and chemistry, and his feelings about them. Even though Nasir had entered college with an interest in chemistry, this interest had fallen off, at least in part because of the procedural nature of the assignments, and presumably because neither his professor nor peers had supported him to understand the importance or utility of the procedural nature of the tasks, nor had he asked questions about this. In contrast, the connections that Emily had developed to ballet were deep. However, the difficult period that she described is common even for those who have a developed interest. Emily needed support in order to persevere. In Emily's case, her mother's and teacher's encouragement served as critical triggers for her continued study of ballet, and the promise of pointe shoes enabled her to continue to work on what at the time probably felt like the procedural details of the dance steps she needed to master.

The four-phase model describes phases (not stages) because an initial triggering of interest may, or may not, lead to the development of a well-developed interest. If interest is not supported to develop, it will fall off, go dormant, or possibly disappear altogether (Bergin, 1999). Interest continues to develop depending on the quality of a person's interactions with other people and/or the design of tasks in the learning environment (e.g., Linnenbrink, Patall, & Messersmith, 2012; Renninger et al., 2014; Xu, Coates, & Davidson, 2011).

The development of interest is coordinated with the development of other motivational variables and with a person's self-representation, or identity (Renninger, 2009). Those with less developed interest like Nasir may need scaffolding from others and/or their tasks in order to know how to work with new content. They do not identify as persons who pursue the potential interest (e.g., computer scientists); they may not even think that developing an interest is possible. Their self-efficacy and their ability to self-regulate may be low. These people need scaffolding and feedback to enable them to make connections to the content to be learned (Table 12.1).

On the other hand, those with more developed interest, like Emily, are typically able to deepen their knowledge independently, as long as they continue to feel challenged and have confidence. Emily identifies as a ballerina, and has well-developed self-efficacy and self-regulation ability. However, if the tasks on which they are working feel impossible, their situation is not unlike those with less developed interest. They, too, need support to find continued engagement rewarding.

Methods, an Overview

Studies of motivation primarily address basic research questions such as how and why a particular variable works as it does. As such, they tend to focus on one or two motivation topics as independent variables, and analyze their relation to outcomes such as performance on standardized achievement measures. The different topics in motivation are typically targeted for study as though they were distinct from one another, although in practice they co-occur, as Nasir's and Emily's case material indicates. With the exception of studies of engagement and to some extent those addressing interest, motivation research does not usually report on, or include analyses of, the learning environment.

Because studies of engagement often focus on understanding disengagement and how to enable the disengaged to become productive participants, they are concerned with how people engage in the learning environment. The learning environment is typically studied as a dependent variable, and participants' engagement, as reflected in their cognitive, affective, and behavioral responses, is studied in relation to it. Researchers have conducted both survey-based quantitative studies (e.g., Martin et al., 2015) as well as more qualitative studies (e.g., Dhingra, 2003; Rahm, 2008) that provide descriptions of learning environments and participants' cognitive, affective, and behavioral responses to them. Although some researchers investigating engagement do not conduct interventions, the forms of descriptive data that are collected can be used to enable school psychologists and educators to intervene to increase the likelihood of meaningful engagement (e.g., Christenson & Reschly, 2010).

Studies of interest are by definition both studies of continuing motivation and studies of meaningful engagement. As such, they have focused on interest as both a dependent and independent variable. When interest is studied as a dependent variable, the focus is often on topics of interest or the development of interest through interactions with the learning environment. When interest is studied as an independent variable, investigations consider the effect of interest on other variables such as attention, memory, or school performance.

Researchers studying motivation, engagement, and interest often use self-reports from surveys (e.g., Patterns of Adaptive Learning Scales [PALS; Midgley et al., 2000]), Likert-type anonymous questionnaires (Vedder-Weiss & Fortus, 2013), or semi-structured, in-depth interviews (e.g., Azevedo, 2013). These data provide descriptive information about the frequency (or amount) and/or quality of the variable under study, and are often analyzed as influences on the learning environment (as independent variables).

Because self-reports are dependent on how self-aware and reflective participants are, coupling self-report data with other data sources may be important in ensuring the accuracy of self-reports, especially when participants lack motivation, engagement, or interest. Examples of such data sources include: ethnographic data (e.g. Ito et al., 2010); observational or video data (e.g., Barron, Gomez, Pinkard, & Martin, 2014), artifact analysis (e.g., Cainey, Bowker, Humphrey, & Murray, 2012), and experience sampling (e.g., Järvelä, Veermans, & Leinonen, 2008).

As sampling permits, and research questions specify, age, gender, race, and status as a first generation student are likely to be addressed in each of these literatures. As described in the previous section of this chapter, the motivation, engagement, and interest of all individuals have similar characteristics. However, groups of individuals with similar demographic profiles also can vary in the frequency, intensity, or development of their motivational or engagement profiles or topics of interest. Such differences may require different support for learning.

For example, work on competence and achievement indicates that, even at very young ages, children are aware of their performance and care about it (Wigfield et al., 1997). At approximately 8–10 years of age, they begin engaging in self-other comparisons (Harter, 2003) and, as Renninger (2009) points out, these have implications for the kinds of support that they may need to seriously engage content that they have not already mastered (Table 12.1). It is because of this that learners roughly below the age of 8 are more likely and readily able to work in and explore different content areas than are those who are older. This does not mean that those who are older cannot develop interest, but it does affect the nature of the supports that may need to be in place for that to happen.

Gender is another variable that can affect performance, and appears to be related to the context of the tasks provided. Topics such as health and caring for humans have been found to be more girl-friendly. For example, Hoffmann and Häussler (1998) reported that girls' learning is benefited if heart pumps rather than oil pumps are used to provide the context for physics instruction, and that boys work with either context effectively. Another finding from this project was that, given that girls often have little experience with mechanical objects, it made a difference when teachers focused units dealing with force and velocity on cyclists' use of safety helmets (Hoffmann, 2002). Similarly, self-perception (e.g. "I love math") plays a more important role for women than men when they are deciding to pursue a computing-related education (Hong, Wang, Ravitz, & Fong, 2015).

Ethnicity too has been found to affect patterns of engagement and participation (e.g., Huang, Taddese, & Walter, 2000). For example, in the sciences, Latino and Asian families have been identified as communicating stronger and clearer messages about participation than either white or African American families (e.g., Archer et al., 2012).

Participation of underrepresented students and first generation students who are at risk for school success is also positively influenced by utility value, or relevance, interventions (see Hulleman, Kosovich, Barron, & Daniel, 2016). These interventions are relatively simple, involving supporting students to articulate and reflect on their connections to subject matter, frequently through writing. This type of intervention has been repeatedly demonstrated to improve course performance, likely pursuit of additional courses, and increased interest in the subject matter for those who are at risk, with no detriment to those who are not.

Implications: Design Principles for Learning Informed by Motivation, Engagement, and Interest

In the course of reviewing the literature on motivation, engagement, and interest for this chapter, and again in working with Nasir's and Emily's cases, we identified three design principles for learning:

1. Learners need to work with relevant disciplinary content in order for it to become rewarding. They need to work with the language and tasks of the content, and begin to develop an interest in it in order to develop their abilities to work with its challenge and through this extend their current understanding.
2. Support for learners to work with the content can be provided through scaffolding by another person or by the design of tasks and activities. Moreover, learners in earlier and later phases of interest development are likely to need different types of interactions and/or support to engage disciplinary content.
3. The structure of tasks, activities, or the learning environment may need to be adjusted for learners in different phases of interest development to enable them to focus on relevant aspects of the tasks and be challenged to pursue understanding.

We set these design principles out as generalized principles (not "fix-all" step-by-step formulas) and encourage subsequent studies, with replications, of each principle (Makel & Plucker, 2014). We also note the importance of reporting studies with insignificant results, as these data are as critical for moving the field ahead as those that are significant: they can flag questionable assumptions and balance others reporting the same studies (Rothstein, Sutton, & Borenstein, 2006).

Nasir's and Emily's cases illustrate these design principles. Both Nasir and Emily need to make their own connections to content. They also both need support from others or the tasks themselves, despite differences in what they are prepared to work with. The nature of the interactions they have with others, or the tasks and challenges that they are given, need to be aligned to their phase of interest development.

In his Network of Academic Programs in the Learning Sciences (NAPLeS) webinar, Hoadley (2013) describes working to identify design principles that can be generalized across contexts. Each of the questions he identified involve either motivation, engagement, or interest, or would be informed by considering them. It would be a mistake to overlook the centrality of motivation, engagement, and interest in individuals' participation and learning: it could affect whether a project's goals will be achieved, and/or whether a research project is likely to inform practice. Understanding the design implications of research on these terms could significantly improve equity of resource allocation, the quality of support provided to youth, and so forth. Nasir's and Emily's motivation, engagement, and interest were influenced by how each thought about their respective problems, and this influenced their participation and learning. As Nasir observed, "In the end, it came down to you and how you think of the problem."

Acknowledgments

Work on this chapter has been supported by the Swarthmore College Faculty Research Fund, a Howard Hughes Medical Institute Fellowship awarded to Yanyan Ren, and a Eugene M. Lang Summer Research Fellowship to Heidi M. Kern. Editorial support from Melissa Running is also gratefully acknowledged.

Further Readings

Gutwill, J. P., & Allen, S. (2012). Deepening students' scientific inquiry skills during a science museum field trip. *Journal of the Learning Sciences, 21*(1), 130–181. doi:10.1080/10508406.2011.555938
This article describes the creation and study of the museum program *Inquiry Games*, designed to enhance students' inquiry skills at interactive science museums. It reports on the measurement of engagement (*holding time*, or how much time students chose to spend at an exhibit), outcomes of meaningful engagement, and the role of chaperones in a field trip group.

Hidi, S., & Harackiewicz, J. M. (2000). Motivating the academically unmotivated: A critical issue for the 21st century. *Review of Educational Research, 70*(2), 151–179. doi:10.3102/00346543070002151
This article provides a comprehensive review of the literature on goals and interest and explains why reference to dichotomies such as extrinsic and intrinsic motivation, or mastery and performance goals is problematic.

Ito, M. S., Baumer, S., Bittanti, M., boyd, d., Cody, R., Herr Stephenson, B., et al. (2010). *Hanging out, messing around, and geeking out: Kids living and learning with new media.* Cambridge, MA: MIT Press.
This volume reports on the "connected learning" approach, which combines interest-driven learning with interpersonal support and a link to academics, career success, or civic engagement. It also addresses how media can support learning environments to foster connected learning.

Renninger, K. A., & Hidi, S. (2016). *The power of interest for motivation and engagement.* New York: Routledge.
This volume provides an overview of interest research and includes detailed notes for researchers at the end of each chapter. It explains how interest can be supported to develop its measurement, the relation between interest and the development of other motivational variables, studies of interest across in- and out-of-school topic areas, and declining interest.

Vedder-Weiss, D., & Fortus, D. (2013). School, teacher, peers, and parents' goals emphases and adolescents' motivation to learn science in and out of school. *Journal of Research in Science Teaching, 50*(8), 952–988. doi:10.1002/tea.21103
This article reports on students' motivation and engagement: how parents' perceptions of goals predict students' motivation, how school structure influences students' goal setting, and how the peer network affects students' levels of motivation.

NAPLeS Resources

Hoadley, C. *A short history of the learning sciences* [Video file]. In *NAPLeS video series.* Retrieved October 19, 2017, from http://isls-naples.psy.lmu.de/intro/all-webinars/hoadley_video/index.html

Renninger, K. A., Ren, Y., & Kern, H. M. *Motivation, engagement, and interest* [Video file]. *Introduction and discussion.* In *NAPLeS video series.* Retrieved October 19, 2017, from http://isls-naples.psy.lmu.de/video-resources/guided-tour/15-minutes-renninger/index.html

References

Archer, L., DeWitt, J., Osborne, J., Dillon, J., Willis, B., & Wong, B. (2012). Science aspirations, capital, and family habitus how families shape children's engagement and identification with science. *American Educational Research Journal, 49*(5), 881–908. doi:10.3102/0002831211433290

Azevedo, F. S. (2013). Knowing the stability of model rockets: An investigation of learning in interest-based practices. *Cognition and Instruction, 31*(3), 345–374. doi:10.1080/07370008.2013.799168

Barron, B. Gomez, K., Pinkard, N., & Martin, C. K. (Eds.) (2014). *The Digital Youth Network: Cultivating digital media citizenship in urban communities.* Cambridge, MA: MIT Press.

Bergin, D.A. 1999. Influences on classroom interest. *Educational Psychologist*, *34*, 75–85. doi:10.1207/s15326985ep3402_2

Berridge, K. C. (2012). From prediction error to incentive salience: Mesolimbic computation of reward motivation. *European Journal of Neuroscience*, *35*, 1124–1143. doi:10.1111/j.1460-9568.2012.07990.x

Boxerman, J. Z., Lee, V. R., & Olsen, J. (2013, April). *As seen through the lens: Students' encounters and engagement with science during outdoor field trips*. Paper presented at the Annual Meeting of the American Educational Research Association, San Francisco, CA.

Cainey, J., Bowker, R., Humphrey, L., & Murray, N. (2012). Assessing informal learning in an aquarium using pre- and post-visit drawings. *Educational Research and Evaluation*, *18*(3), 265–281.

Christenson, S. L., & Reschly, A. L. (2010). Check & Connect: Enhancing school completion through student engagement. In B. Doll, W. Pfohl, & J. Yoon (Eds.). *Handbook of youth prevention science* (pp. 327–348). New York: Routledge.

Christenson, S. L., Reschly, A. L., & Wylie, C. (Eds.). (2012). *Handbook of research on student engagement*. New York: Springer. doi:10.1007/978-1-4614-2018-7

Csikszentmihalyi, M. (1990). *Flow: The psychology of optimal experience*. New York: Harper & Row.

Dhingra, K. (2003). Thinking about television science: How students understand the nature of science from different program genres. *Journal of Research in Science Teaching*, *40*(2), 234–256. doi:10.1002/tea.10074

Fredricks, J. A., Blumenfeld, P. C., & Paris, A. H. (2004). School engagement: Potential of the concept, state of the evidence. *Review of Educational Research*, *74*(1), 59–109. doi:10.3102/00346543074001059

Gutwill, J. P., & Allen, S. (2012). Deepening students' scientific inquiry skills during a science museum field trip. *Journal of the Learning Sciences*, *21*(1), 130–181. doi:10.1080/10508406.2011.555938

Harter, S. (2003). The development of self-representation during childhood and adolescence. In M. R. Leary & J. P. Tangney (Eds.), *Handbook of self and identity* (pp. 610–642). New York: Guilford.

Hidi, S., & Harackiewicz, J. M. (2000). Motivating the academically unmotivated: A critical issue for the 21st century. *Review of Educational Research*, *70*(2), 151–179. doi:10.3102/00346543070002151

Hidi, S., & Renninger, K. A. (2006). The four-phase model of interest development. *Educational Psychologist*, *41*(2), 111–127. doi:10.1207/s15326985ep4102_4

Hoadley, C. (2013, September 30). *A short history of the learning sciences* [Video file]. In *NAPLeS video series*. Retrieved October 19, 2017, from http://isls-naples.psy.lmu.de/intro/all-webinars/hoadley_video/index.html

Hoffmann, L. (2002). Promoting girls' interest and achievement in physics classes for beginners. *Learning and Instruction*, *12*(4), 447–465. doi:10.1016/s0959-4752(01)00010-x

Hoffmann, L. & Häussler, P. (1998). An intervention project promoting girls' and boys' interest in physics. In L. Hoffmann, A. Krapp, K.A. Renninger, & J. Baumert (Eds.), *Interest and learning* (pp. 301–316). Kiel, Germany: IPN.

Hong, H., Wang, J., Ravitz, J., & Fong, M. Y. L. (2015, February). Gender differences in high school students' decisions to study computer science and related fields. *Symposium on Computer Science Education* (p. 689), Kansas, MO.

Huang, G., Taddese, N. & Walter, E. (2000). *Entry and persistence of women and minorities in college science and engineering education* (Research and Development Report NCES 2000-601). Washington, DC: National Center for Education Statistics. Retrieved January 24, 2018, from https://nces.ed.gov/pubs2000/2000601.pdf

Hulleman, C. S., Kosovich, J. J., Barron, K. E., & Daniel, D. B. (2016). Making connections: Replicating and extending the utility value intervention in the classroom. *Journal of Educational Psychology*, *109*(3), 387–404. doi:10.1037/edu0000146

Ito, M. S., Baumer, S., Bittanti, M., boyd, d., Cody, R., Herr Stephenson, B., et al. (2010). *Hanging out, messing around, and geeking out: Kids living and learning with new media*. Cambridge, MA: MIT Press.

Järvelä, S., Veermans, M., & Leinonen, P. (2008). Investigating learners' engagement in a computer-supported inquiry—A process-oriented analysis. *Social Psychology in Education*, *11*, 299–322.

Linnenbrink-Garcia, L., Patall, E. A., & Messersmith, E. E. (2012). Antecedents and consequences of situational interest. *British Journal of Educational Psychology*, *83*(4), 591–614. doi:10.1111/j.2044-8279.2012.02080.x

Makel, M. C., & Plucker, J. A. (2014). Facts are more important than novelty. *Educational Researcher*, *43*(6), 304–316. doi:10.3102/0013189X14545513

Martin, A., Papworth, B., Ginns, P., Malmberg, L., Collie, R., Calvo, R. (2015). Real-time motivation and engagement during a month at school: Every moment of every day for every student matters. *Learning and Individual Differences*, *38*, 26–35.

Midgley, C., Maehr, M. L., Hruda, L., Anderman, E. M., Anderman, L., Freeman, K. E., et al. (2000). *Manual for the Patterns of Adaptive Learning Scales (PALS)*. Ann Arbor, MI: University of Michigan.

Rahm, J. (2008). Urban youths' hybrid positioning in science practices at the margin: A look inside a school-museum-scientist partnership project and an after school science program. *Cultural Studies of Science Education*, *3*(1), 97–121. doi:10.1007/s11422-007-9081-x

Renninger, K. A. (2009). Interest and identity development in instruction: An inductive model. *Educational Psychologist*, *44*(2), 1–14. doi:10.1080/00461520902832392

Renninger, K. A., Austin, L., Bachrach, J. E., Chau, A., Emmerson, M. S., King, R. B., et al. (2014). Going beyond the "Whoa! That's cool!" of inquiry. Achieving science interest and learning with the ICAN Intervention. In S. Karabenick & T. Urdan (Eds.), *Motivation-based learning interventions*, Advances in Motivation and Achievement series (Vol. *18*, pp. 107–140). United Kingdom: Emerald Publishing. doi:10.1108/S0749-742320140000018003

Renninger, K. A., & Hidi, S. (2016). *The power of interest for motivation and engagement*. New York: Routledge.

Reschly, A. L., & Christenson, S. L. (2012). Jingle, jangle, and conceptual haziness: Evolution and future directions of the engagement construct. *Handbook of research on student engagement* (pp. 3–19). New York: Springer. doi:10.1007/978-1-4614-2018-7_1

Rothstein, H. R., Sutton, A. J., & Borenstein, M. (Eds.). (2006). *Publication bias in meta-analysis: Prevention, assessment and adjustments*. Hoboken, NJ: John Wiley.

Shernoff, D. J. (2013). *Optimal learning environments to promote student engagement*. New York: Springer. doi:10.1007/978-1-4614-7089-2

Vedder-Weiss, D., & Fortus, D. (2013). School, teacher, peers, and parents' goals emphases and adolescents' motivation to learn science in and out of school. *Journal of Research in Science Teaching*, *50*(8), 952–988. doi:10.1002/tea.21103

Wigfield, A., Eccles, J., Schiefele, U., Roeser, R., & Davis-Kean, P. (2006). Development of achievement motivation. In R. Lerner & W. Damon (Series Eds.), N. Eisenberg (Vol. Ed.), *Handbook of child psychology*, Vol. *3*, *Social, emotional, and personality development* (6th ed., pp. 933–1002). New York: Wiley.

Wigfield, A., Eccles, J. S., Yoon, K. S., Harold, R. D., Arbreton, A. J. A., Freedman-Doan, C., & Blumenfeld, P. C. (1997). Change in children's competence beliefs and subjective task value across the elementary school years: A 3-year study. *Journal of Educational Psychology*, *89*, 451–469.

Xu, J., Coats, L. T., & Davidson, M. L. (2011). Promoting student interest in science: The perspectives of exemplary African American teachers. *American Educational Research Journal*, *49*(1), 124–154. doi:10.3102/0002831211426200

13

Contemporary Perspectives of Regulated Learning in Collaboration

Sanna Järvelä, Allyson Hadwin, Jonna Malmberg, and Mariel Miller

Introduction: Self-Regulated Learning Covers a Set of Critical 21st-Century Skills

Successful students actively engage in numerous activities, including planning their learning, utilizing effective strategies, monitoring their progress, and handling the difficulties and challenges associated with their learning tasks (e.g., Zimmerman, 2000). In today's knowledge-based society, the importance of 21st-century skills for learning, creative and critical thinking, collaboration, and strategic use of information and communication technology (ICT) are essential (Beetham & Sharpe, 2013). Knowing how to learn and improve one's learning skills are the keys to individuals' and society's well-being. Decades of self-regulated learning (SRL) research success in solo and collaborative learning tasks requires the development of regulatory learning skills and strategies for working individually or collaboratively (Hadwin, Järvelä, & Miller, 2017). Being able to strategically regulate one's own learning and that of others has great potential for optimizing cognitive, motivational, and emotional behavior throughout life and work. (Zimmerman & Schunk, 2011).

Unfortunately, research consistently shows that learners often fail to plan adequately, use adaptive learning strategies, or leverage technologies for learning, collaborating, and problem solving (cf. Järvelä & Hadwin, 2013; Kirschner & Van Merriënboer, 2013). Regulation of one's own learning is not easy and often needs to be both learned and supported with self-regulation tools and/or environments (e.g., Hadwin, Oshige, Gress, & Winne, 2010). Learners may lack skills or knowledge to direct their own learning, or the motivation to enact successful strategies and processes. Additionally, difficulties regulating at the individual level (self-regulation), are compounded when interacting with peers and teams, (co-regulation and shared regulation). Socially shared regulation is especially critical as many of today's and tomorrow's problems are dependent on teams that can solve complex tasks together. In short, properly planning and strategically adapting one's learning to the challenges encountered during the learning process requires the ability to strategically regulate oneself (i.e., SRL), socio-cultural situations and people (i.e., co-regulated learning; CoRL), and the group collectively (i.e., socially shared regulation of learning; SSRL) (Hadwin et al., 2017). This chapter introduces these concepts and discusses their contribution to the learning sciences. We also describe the design principles and technologies that support regulated learning.

Sanna Järvelä, Allyson Hadwin, Jonna Malmberg, and Mariel Miller

SRL Development During the Past Decades

Early ideas about SRL emphasized the individual, particularly cognitive-constructive aspects of regulation such as cognition, behavior, and motivation (e.g., Winne, 1997). SRL stressed the importance of students taking charge of their own learning (e.g., Zimmerman, 1989). Socio-cognitive perspectives of SRL emphasized triadic reciprocity, whereby cognitive and personal factors, environments, and behaviors interacted in reciprocal ways (Schunk & Zimmerman, 1997). However, the emergence of situated perspectives of learning in the early 1990s (Greeno, 2006) began to challenge the limitations of socio-cognitive models of SRL to explain regulation in highly interactive and dynamically changing learning situations.

Learning situations are increasingly social and interactive, and enriched with various technologies. Regulated learning has therefore come increasingly under consideration in social and collaborative situations, often in computer-supported collaborative learning contexts (Järvelä & Hadwin, 2013; Järvenoja, Järvelä, & Malmberg, 2015). In recent years, researchers have begun exploring where the social and self meet in strategic learning regulation. We ourselves have been striving to define and conceptualize the three forms of regulation (self-regulation, co-regulation, and shared regulation) as central processes in highly interactive and collaborative learning contexts (Hadwin et al., 2017).

Defining the Concepts of SRL, CoRL, and SSRL, and Their Contributions to the Learning Sciences

Three regulation modes are central to collaborative learning: self-regulated learning, socially shared learning regulation, and co-regulated learning. Given the proliferation of these constructs over the past five years, it is important to clarify the meaning of each. To do so, we rely heavily on our most recent review of the field (Hadwin et al., 2017).

Self-regulated learning (SRL) refers to an individual's deliberate and strategic planning, enactment, reflection, and adaptation when engaged in any task in which learning occurs. Individuals self-regulate learning when they learn or refine sports skills, complete academic work such as studying, learn new parenting skills as children develop, or adjust to a new workplace. In SRL, metacognitive monitoring and evaluation drive adaptation, and personal goals and standards set the stage for learner agency. The resulting SRL processes are iterative and recursive, adapting continuously as new metacognitive feedback is generated. Importantly, these metacognitive, adaptive, and agentic processes extend beyond controlling cognition to behavior, motivation, and emotions. This perspective recognizes that metacognitive knowledge and awareness are critical for adaptively responding to a complex set of challenges contributing to learner success. Individual SRL in the service of a group task is necessary for optimal productive collaboration to occur. In other words, evidence of SRL during collaboration is complementary rather than antagonistic to the emergence of shared regulation.

Co-regulated learning (CoRL) in collaboration broadly refers to the processes through which appropriation of strategic planning, enactment, reflection, and adaptation are stimulated or constrained. Co-regulatory affordances and constraints exist in actions and interactions, environmental features, task design, regulation tools or resources, and cultural beliefs and practices that either support or thwart productive regulation. Interpersonal interactions and exchanges may play a role in stimulating this type of transitional, flexible regulation, but other aspects of situation and task may also contribute. This definition of CoRL acknowledges the role of co-regulation in shifting groups toward more productive shared regulation, not just individuals toward self-regulation. Co-regulation also generates affordances and constraints that shape potential for shared regulation, which is why co-regulatory prompts are often found to be embedded within shared regulation episodes (e.g., Grau & Whitebread, 2012). Co-regulation involves group members developing awareness of each other's goals and beliefs and temporarily transferring regulatory support to one another or to technologies and tools that support regulation. The regulator can initiate co-regulation, such as when regulatory support is requested (e.g., asking someone to clarify the task criteria). Alternatively, regulation can be prompted by a peer or group member (e.g., prompting a strategy: "maybe you should review your notes"). Finally, co-regulation can be supported by tools and technologies, such as a digital reminder to check the time, or a goal-setting scaffold embedded in a learning system.

Two important points should be made regarding co-regulation. First, co-regulation is more than merely promoting a regulatory action. It is a temporary shifting or internalization of a regulatory process that enables uptake by the "co-regulated" (Hadwin, Wozney, & Pontin, 2005). Second, CoRL emerges from distributed regulatory expertise across several individuals; it is strategically invoked when necessary, by and for whom it is appropriate. Co-regulation can be difficult to distinguish from shared regulation because consistent and productive co-regulation is likely a necessary condition for shared regulation to emerge.

Socially shared regulation of learning (SSRL) in collaboration refers to a group's deliberate, strategic, and transactive planning, task enactment, reflection, and adaptation. It involves collectively taking control of cognitive, behavioral, motivational, and emotional conditions through negotiation and continual adaptation. Transactivity implies that multiple individual perspectives contribute to the emergence of joint metacognitive, cognitive, behavioral, and motivational states, with meta cognition being central to shared regulation. Metacognitive processes that fuel regulation (monitoring and evaluation) shared among group members, thereby driving negotiated changes, are referred to as large- or small-scale adaptations. Individual SRL provides a critical foundation for collective agency; collective agency depends upon the emergence of joint goals and standards that may be informed by individual goals but do not always replace them. Finally, shared regulation is sociohistorically and contextually situated in both individual and collective beliefs and experiences that together inform joint task engagement and are changed as a result of collaboration.

What Is Regulation in Learning, and What Is It Not?

Over the past 10 years, interest in regulatory constructs, as well as their application to research and practice, has burgeoned well beyond their origins in educational psychology. Increased interest in the topic signals the relevance of regulated learning for understanding the complex, multifaceted processes associated with learning and engagement. However, the construct's emerging uses and elaborations have often been divorced from their psychological underpinnings, leading to inconsistent interpretations, definitions, and operationalization of the constructs, as well as their misuse.

Forms of SRL and social regulation have been studied in educational psychology for more than two decades. This chapter draws from a base of theoretical and empirical research, beginning with early conceptions of SRL (cf. Schunk & Zimmerman, 1997), to emphasize the cognitive, motivational, and metacognitive foundations from which the constructs of self-regulated, co-regulated, and shared regulation of learning originate. We present six guiding themes for conceptualizing all forms of regulation.

First, learning is more than metacognitive control or executive functioning. While metacognitive monitoring, evaluation, and control fuel regulated learning, they should not be treated as the same construct. For researchers adopting a multifaceted view of regulated learning, this means collecting data about the interplay between motivation, behavior, metacognition, and cognition during regulated learning, not just attending to a single facet.

Second, regulation (SRL, CoRL, and SSRL) arises because human beings exercise agency in striving toward goals as part of learning and collaboration. Self-set and collectively generated goals, whether transparent or not, contextualize engagement, strategic action, and interaction. For researchers, this means that data about learner and group intent need to be examined as well as the degree to which that intent matches the task goal or objectives. Without knowledge of learner intent, inferences about observed strategies, behaviors, motivation, or emotions are limited at best.

Third, regulation develops over time and across tasks. This notion of regulation as an adaptive process, rather than a state, is central to both Zimmerman and Schunk's (2011) macro-level phases (forethought, performance or volitional control, and self-reflection) and Winne and Hadwin's (1998) regulation model micro-level COPES architecture (Conditions, Operations, Products, Evaluations, and Standards), which inform phases of SRL. For researchers, this means collecting data about regulation as it temporally unfolds, emerging from and continuing to shape future beliefs, knowledge, and experiences.

Fourth, regulation is situated in personal history. New learning situations are always informed by knowledge, beliefs, and mental models of self, task, domain, and teams and based on past experiences. Learners start from what they know and feel about learning and collaborating, not just their prior knowledge about the domain. As a result, strategic task engagement is always heavily personalized—rooted in past individual and collective experiences. For researchers, this means that data about personal and collective conditions that situated regulation must be collected and observed over time.

Fifth, the mark of regulation is intent or purposeful action in response to situations such as challenges. For example, learners engage in positive self-talk when negative self-efficacy lowers task engagement or performance, or when a situation is anticipated to have lower efficacy. Learners overtly articulate goals when task persistence wanes. The proficiency with which people toggle regulation on and off creates cognitive capacity for complex processing (Hadwin et al., 2017). For researchers, this means that regulation cannot be observed spontaneously at just any time or place. Rather, data collection should be carefully timed to capture situated responses to overt and tacit challenges or situations that simulate self-monitoring and action.

Finally, regulation emerges when learners engage with real learning activities and situations that have personal meaning and create opportunities for them to connect past knowledge and experiences to the situation at hand. It is in these situations that cultural milieu and relationships, as well as interactions, context, and activities, give rise to self-regulation, co-regulation, and shared regulation of learning. We specifically draw from Winne and Hadwin's COPES model of SRL because it models the unfolding of updates to internal, external, and shared conditions within and across task work phases. By so doing, this model acknowledges regulation's situated nature, as well as the ways different modes of regulation (self-, co-, and shared) interact with one another. For researchers, this means being extremely cautious about inferences and interpretations drawn from studies during which learners complete learning or collaboration tasks solely to satisfy the requirements of a research study.

Prompting Metacognitive Awareness with Technologies

Over the past decade, researchers have designed and introduced technologies that support SRL and metacognitive awareness (Azevedo, 2015; Bannert, Reimann, & Sonnenberg, 2014). These technologies offer learning environments guided by the theories of how people typically learn and behave in such environments. For example, MetaTutor (Azevedo, 2015) is a state-of-the-art software that is grounded in SRL theory. MetaTutor gives learners prompts to promote metacognitive awareness based on learners' activities in the environment (Johnson, Azevedo, & D'Mello, 2011). The prompts involve a learning goal (e.g., a science topic to master), learning session sub-goals, and the possibility of communicating with the learning environment (e.g., animated pedagogical agents fostering metacognitive awareness).

Similarly, Bannert, Reimann, and Sonnenberg (2014) developed metacognitive prompts in order to support orientation, planning, goal specification, searching information, monitoring, and evaluation. They prompted learners to exercise metacognitive awareness aiming at supporting regulated learning processes and the learning product. Findings revealed a systematic difference between the occurrence of loops comprising cognitive and metacognitive learning activities in which students received or did not receive prompts.

New technology is often developed and tested under the assumption that students already possess sufficient self-regulatory skills, but are not metacognitively aware of when to spontaneously recall or execute regulated learning (Bannert & Reimann, 2012). Technology has the potential to automatically identify moments in which there is a need to promote metacognitive awareness by prompting cognitive and metacognitive processes. Contemporary research has begun to explore sequential and temporal associations between metacognitive and regulatory processes (Malmberg, Järvelä, & Järvenoja, 2017). However, further research is needed in authentic learning and collaboration contexts.

Why Regulation of Learning Is Relevant to Learning Sciences

The learning sciences are concerned with deep learning that occurs in complex social and technological environments (Sawyer, 2014). Studying and learning in these physical and social contexts introduces new demands for learning. Although the importance of collaboration for deep learning is well established in the learning sciences (e.g., O'Donnell & Hmelo-Silver, 2013), researching and supporting social and collaborative learning requires considering complex interactions between cognitive, social, emotional, motivational, and contextual variables (Thompson & Fine, 1999). To illustrate, Lajoie et al. (2015) examined socio-emotional processes contributing to metacognition and co-regulation used by medical students learning to how to deliver bad news. By coding for metacognitive processes, positive expression of emotions, and negative socio-emotional interactions, they revealed the dynamic relationships between emotions and metacognition in a distributed online problem-based learning environment.

We posit that regulation of learning is the quintessential skill for successful 21st-century learning (Hadwin et al., 2017). However, empirical research consistently indicates that group learning and sharing mental processes in the context of social interaction are challenging. Even when group activity is carefully designed pedagogically, groups can encounter numerous difficulties, including cognitive and socio-emotional challenges (Van den Bossche, Gijselaers, Segers, & Kirschner, 2006). Cognitive challenges can derive from difficulties in understanding each other's thinking or from negotiating multiple perspectives (Kirschner, Beers, Boshuizen, & Gijselaers, 2008). Motivational problems can emerge due to differences in group members' goals, priorities, and expectations (Blumenfeld, Marx, Soloway, & Krajcik, 1996). Addressing these challenges means moving beyond supporting knowledge construction and collaborative interactions alone. Working together means co-constructing shared task representations, goals, and strategies. It also means regulating learning through shared metacognitive monitoring and control of motivation, cognition, and behavior.

Previous studies indicate that students (a) construct shared task perceptions, negotiate their plans and goals together by building upon each other's thinking (Malmberg, Järvelä, Järvenoja, & Panadero, 2015), and (b) equally share their strategic engagement with the task and collectively monitor their learning progress toward their shared goals (Malmberg et al., 2017). For example, Järvelä, Järvenoja, Malmberg, Isohätälä, and Sobocinski (2016) examined groups' cognitive and socio-emotional interactions with respect to three phases of regulation (forethought, performance, and reflection). They studied how self- and shared regulation activities are used in collaboration, as well as whether they are useful for collaborative learning outcomes. Their findings indicated that collaborative planning of regulatory activities became shared in practice. Furthermore, groups that achieved good learning results used several regulatory processes to support their learning, in addition to engaging in shared regulation.

It has also become clear that the tasks used to study and support SRL must be difficult enough to require students to engage in monitoring and control their learning. Hadwin, Järvelä, and Miller (2017) explain that challenging learning situations create SRL opportunities. That is, challenges invite learners to contextualize their regulation strategies in a situation and to put them into practice in order to test whether their SRL processes are conscious or not. When the learning process is effortless, conscious SRL ceases and will not emerge again until a challenge activates the need.

It is our belief that situations designed according to the principles of learning sciences aiming for active learning provide opportunities for training regulation. Furthermore, SRL can be facilitated or constrained by task characteristics (Lodewyk, Winne, & Jamieson-Noel, 2009) or domain (Wolters & Pintrich, 1998). For example, over the last two decades, Perry has examined the quality of classroom contexts that support elementary school children's SRL. In their studies (Perry & VandeKamp, 2000), Perry used classroom observations of teachers and students, work samples, and student interviews to identify factors that encouraged and constrained SRL. The findings revealed that students engage in SRL most often in classrooms where they: (a) have the opportunity to engage in complex, meaningful tasks across multiple sessions; (b) have the opportunity to exercise choice about the task, who to work with, and where to work; (c) can control how challenging the task is; and (d) participate in defining criteria for evaluation, as well as reviewing and reflecting on learning.

Designing collaborative learning tasks with optimal levels of challenge and student responsibility is central for activating students' regulated learning (Malmberg et al., 2015). Collaborative learning research shows that it takes time to progress to productive collaboration (Fransen, Weinberger, & Kirschner, 2013), as it takes time to progress in regulating learning. Malmberg et al.'s (2015) findings indicate that when collaborating groups work on open tasks, the focus of groups' shared regulatory activities shifts over time. At the beginning of the collaboration, groups may focus on regulating external aspects of collaboration (such as time management and environment), whereas in the later stages the focus shifts to cognitive-oriented and motivational issues. In short, groups must be given abundant opportunities to collaborate with each other, complemented by guided opportunities to systematically plan for and reflect on their collaborative progress and challenges.

Learners bring different beliefs and interests to their activity, resulting in varied motivation and engagement, efficacy for success, and the types of support needed for learning (Järvelä & Renninger, 2014). Researchers have identified motivating features of group work, such as the integration of challenging tasks for supporting interest (Järvelä & Renninger, 2014) or individual accountability and interdependence (Cohen, 1994). Others have studied motivational challenges related to group members' different goals, priorities, and expectations toward group activities. Findings in motivation and emotional regulation indicate that, in successful groups, members are aware of socio-emotional challenges and are able to activate socially shared motivation regulation (Järvenoja et al., 2015). As suggested by Järvelä and Renninger (2014), educators could anticipate differences in learners' interest, motivation, and engagement, and include project or problem features in their designs that increase the likelihood that one or another of these features will feel possible to the learner, triggering interest, enabling motivation, and supporting learning regulation. In summary, it appears that design principles for learning must account for differences of interest, motivation, and engagement, and they must do so by (a) supporting content-informed interactions, (b) providing the learners with scaffolding for thinking and working with the content, and (c) providing regulatory support.

Design Principles and Technological Tools for Supporting Regulated Learning

Increasingly, technologies have been used in learning sciences to provide new ways for prompting and supporting learning. For example, computer-supported collaborative learning environments (CSCL) offer learners opportunities to guide and support their own learning, and allow researchers to study the different forms of regulation.

Technology has been used in five ways to support regulation. First, technological tools and environments have been developed for sharing information and co-constructing knowledge as solutions to joint problems (e.g., Scardamalia & Bereiter, 1994). Research has examined the quality and efficiency of the knowledge construction processes and outcomes within these knowledge-building environments (e.g., Fischer, Kollar, Stegmann, & Wecker, 2013).

Second, group awareness and sociability have been supported with the goal of positively affecting social and cognitive performance (Kirschner, Strijbos, Kreijns, & Beers, 2004). Three core CSCL elements (sociability, social space, and social presence; Kreijns, Kirschner, & Vermeulen, 2013), along with their relationships with group members' mental models, social affordances, and learning outcomes, have been implemented in tools and widgets (e.g., Janssen, Erkens, & Kirschner, 2011; see also Bodemer, Janssen, & Schnaubert, this volume).

Third, adaptive tools and agents have been developed to support SRL and metacognitive processes (cf. Azevedo & Hadwin, 2005). Computer-based pedagogical tools are designed to support learners to activate existing SRL skills as needed. Adaptive systems have the potential to react "on the fly" to learner activity, providing tailored and targeted SRL support (Azevedo, Johnson, Chauncey, & Graesser, 2011). Pedagogical tools can vary from being relatively short-term reminders to goal-setting planning tools that depend on the learning phase (Bannert & Reimann, 2012).

Fourth, developing awareness and understanding of self and other when working together on a task over a period of time has been supported with awareness and visualization tools (Kreijns, Kirschner, & Jochems, 2002). Derived from computer-supported collaborative work (CSCW) (e.g., Dourish & Bellotti, 1992), these tools focus on achieving optimal coordination between and within loose- and tightly knit group activities, both between and within collaboration. In CSCL, tools applied ideas of history awareness and group awareness (Kreijns, Kirschner, & Jochems, 2002). Mirroring tools collect, aggregate, and reflect data back to the users about individual and collective interaction and engagement (Buder & Bodemer, 2008).

Finally, SSRL processes have been promoted and sustained by developing regulatory planning, enacting, and monitoring supports (Järvelä, Kirschner, Hadwin, et al., 2016). These supports are grounded in findings that learners seldom recognize opportunities for socially shared regulation and often require support in order to enact these processes (Järvelä, Järvenoja, Malmberg, & Hadwin, 2013). For example, individual and group planning tools have been integrated directly into complex collaborative tasks (Miller & Hadwin, 2015). Comparisons of levels of support (individual vs. group, high vs. low) indicate that, regardless of the individual support level, a high level of group support promotes transactive planning discussions; these, in turn, lead to the construction of more accurate shared task perceptions that capitalize on individuals' task perceptions.

Future Trends and Developments

Despite theoretical and conceptual progress regarding the social aspects of SRL theory, future research may focus on the development of tools that make the intangible mental regulation processes and their accompanying social and contextual reactions more concrete for researchers and learners. Self-reports and subjective coding of video and/or verbal protocols alone are not sufficient to examine how regulation develops and adapts over time. Self-reports are based on students' perceptions of how they would or did enact certain processes; these perceptions often do not align with what actually occurs during learning (Zimmerman, 2000). Subjective coding of observation data is also weak due to the coders' interpretations of observed behaviors. The results lack generalizability, as they are content-specific, time-dependent, and individualistic.

A current trend in research about regulated learning includes (a) collecting rich multimodal data, (b) using data-driven analytical techniques (e.g., learning analytics), and (c) aggregating these data sources to guide learners to strategically regulate individual and group cognition, motivation, and emotion (Roll & Winne, 2015). *Multimodal data* comprises objective and subjective data from different channels, simultaneously tracing a range of cognitive and non-cognitive processes (Reimann, Markauskaite, & Bannert, 2014). While multimodal data collection in SRL research is in its early stages, multichannel data triangulation can provide a fundamentally new approach that captures critical SRL phases as they occur in challenging learning situations.

The progress of SRL research benefits learning scientists who are actively designing and implementing innovative methods for teaching and learning in various contexts, as well as testing their interventions in design-based research (Sawyer, 2014). Researchers and instructors can inspect the extent to which their interventions change the learning processes, in addition to the material that is learned and learners' motivational and affective states. However, in reality, this is seldom the case. Interventions do not determine how learners engage with tasks. Rather, interventions are affordances that agentic learners absorb along with other elements in the learning context, as they perceive it, to regulate learning (Roll & Winne, 2015). Focusing on the online learning processes and collecting data about learning traces can make these complex regulatory processes visible and contribute to better learning design. Multidisciplinary collaboration in the SRL field is promising for producing more efficient tools and models for the learning sciences.

Further Readings

Hadwin, A. F., Järvelä, S., & Miller, M. (2017). Self-regulation, coregulation and shared regulation in collaborative learning environments. In D. Schunk & J. Greene (Eds.), *Handbook of self-regulation of learning and performance* (2nd ed., pp. 65–86). New York: Routledge.
This book chapter provides the most recent overview of the definitions and mechanisms of the three modes of regulated learning.

Järvelä, S., Malmberg, J., & Koivuniemi, M. (2016). Recognizing socially shared regulation by using the temporal sequences of online chat and logs in CSCL. *Learning and Instruction, 42,* 1–11. doi:10.1016/j.learninstruc.2015.10.006
This article provides empirical evidence for how self- and shared regulation activities are used and whether they are useful for collaborative learning outcomes.

Miller, M., & Hadwin, A. (2015). Scripting and awareness tools for regulating collaborative learning: Changing the landscape of support in CSCL. *Computers in Human Behavior, 52,* 573–588. doi:10.1016/j.chb.2015.01.050
This paper addresses the need to apply a theoretical framework of self-regulation, co-regulation, and socially shared regulation to design tools for supporting regulation in CSCL.

Winne, P. H., & Azevedo, R. (2014). Metacognition. In R. K. Sawyer (Ed.), *The Cambridge handbook of the learning sciences* (2nd ed., pp. 63–87). Cambridge, UK: Cambridge University Press.
This chapter explains the basic principles of metacognition and its role in productive self-regulation.

NAPLeS Resources

Järvelä, S., *Shared regulation in CSCL* [Webinar]. In *NAPLeS video series.* Retrieved October 19, 2017, from http://isls-naples.psy.lmu.de/intro/all-webinars/jaervelae/index.html

References

Azevedo, R. (2015). Defining and measuring engagement and learning in science: Conceptual, theoretical, methodological, and analytical issues. *Educational Psychologist, 50*(1), 84–94. doi:10.1080/00461520.2015.1004069
Azevedo, R., & Hadwin, A. F. (2005). Scaffolding self-regulated learning and metacognition – Implications for the design of computer-based scaffolds. *Instructional Science, 33*(5–6), 367–379. doi:10.1007/s11251-005-1272-9
Azevedo, R., Johnson, A., Chauncey, A. & Graesser, A. (2011). Use of hypermedia to convey and assess self-regulated learning. In B. Zimmerman & D. Schunk (Eds.), *Handbook of self-regulation of learning and performance* (pp. 102–121). New York: Routledge.
Bannert, M. & Reimann, P. (2012). Supporting self-regulated hypermedia learning through prompts. *Instructional Science, 40*(1), 193–211. doi:10.1007/s11251-011-9167-4
Bannert, M., Reimann, P., & Sonnenberg, C. (2014). Process mining techniques for analyzing patterns and strategies in students' self-regulated learning. *Metacognition and Learning, 9,* 161–185. doi:10.1007/s11409-013-9107-6

Beetham, H., & Sharpe, R. (2013). *Rethinking pedagogy for a digital age: Designing for 21st century learning* (2nd ed.). New York: Routledge.

Blumenfeld, P. C., Marx, R. W., Soloway, E., & Krajcik, J. (1996). Learning with peers: From small group cooperation to collaborative communities. *Educational Researcher, 25*(8), 37–39. doi:10.3102/0013189X025008037

Bodemer, D., Janssen, J., & Schnaubert, L. (2018). Group awareness tools for computer-supported collaborative learning. In F. Fischer, C. E. Hmelo-Silver, S. R. Goldman, & P. Reimann (Eds.), *International handbook of the learning sciences* (pp. 351–358). New York: Routledge.

Buder, J., & Bodemer, D. (2008). Supporting controversial CSCL discussions with augmented group awareness tools. *International Journal of Computer-Supported Collaborative Learning, 3*(2), 123–139. doi:10.1007/s11412-008-9037-5

Cohen, E. G. (1994). Restructuring the classroom: Conditions for productive small groups. *Review of Educational Research, 64*(1), 1–35. doi:10.3102/00346543064001001

Dourish, P., & Bellotti, V. 1992. *Awareness and coordination in shared workspaces*. Proceedings of the ACM Conference on Computer-Supported Cooperative Work CSCW '92 (pp. 107–114), Toronto, Canada. New York: ACM.

Fischer, F., Kollar, I., Stegmann, K., & Wecker, C. (2013). Toward a script theory of guidance in computer-supported collaborative learning. *Educational Psychologist, 48*(1), 56–66.

Grau, V., & Whitebread, D. (2012). Self and social regulation of learning during collaborative activities in the classroom: The interplay of individual and group cognition. *Learning and Instruction, 22*(6), 401–412. doi:10.1016/j.learninstruc.2012.03.003

Greeno, J. G. (2006). Learning in activity. In R. Sawyer (Ed.), *The Cambridge handbook of the learning sciences* (pp. 79–96). Cambridge, NY: Cambridge University Press.

Fransen, J., Weinberger, A., & Kirschner, P. A. (2013). Team effectiveness and team development in CSCL. *Educational Psychologist, 48*(1), 9–24. doi:10.1080/00461520.2012.747947

Hadwin, A. F., Järvelä, S., & Miller, M. (2017). Self-regulation, co-regulation and shared regulation in collaborative learning environments. In D. Schunk & J. Greene (Eds.), *Handbook of self-regulation of learning and performance* (2nd ed., pp. 65–86). New York: Routledge.

Hadwin, A. F., Oshige, M., Gress, C. L. Z., & Winne, P. H. (2010). Innovative ways for using gStudy to orchestrate and research social aspects of self-regulated learning. *Computers in Human Behavior, 26*(5), 794–805. doi:10.1016/j.chb.2007.06.007

Hadwin, A. F., Wozney, L., & Pontin, O. (2005). Scaffolding the appropriation of self-regulatory activity: A socio-cultural analysis of changes in teacher-student discourse about a graduate student portfolio. *Instructional Science, 33*(5–6), 413–450. doi:10.1007/s11251-005-1274-7

Janssen, J., Erkens, G., & Kirschner, P. A. (2011). Group awareness tools: It's what you do with it that matters. *Computers in Human Behavior, 27*(3), 1046–1058. doi:10.1016/j.chb.2010.06.002

Järvelä, S., & Hadwin, A. F. (2013). New frontiers: Regulating learning in CSCL. *Educational Psychologist, 48*(1), 25–39. doi:10.1080/00461520.2012.748006

Järvelä, S., Järvenoja, H., Malmberg, J., & Hadwin, A. F. (2013). Exploring socially shared regulation in the context of collaboration. *Journal of Cognitive Education and Psychology, 12*(3), 267–286. doi:10.1891/1945-8959.12.3.267

Järvelä, S., Järvenoja, H., Malmberg, J., Isohätälä, J., & Sobocinski, M. (2016). How do types of interaction and phases of self-regulated learning set a stage for collaborative engagement? *Learning and Instruction 43*, 39–51. doi:10.1016/j.learninstruc.2016.01.005

Järvelä, S., Kirschner, P. A., Hadwin, A., Järvenoja, H., Malmberg, J. Miller, M., & Laru, J. (2016). Socially shared regulation of learning in CSCL: Understanding and prompting individual- and group-level shared regulatory activities. *International Journal of Computer Supported Collaborative Learning, 11*(3), 263–280. doi:10.1007/s11412-016-9238-2

Järvelä, S., & Renninger, K. A. (2014). Designing for learning: Interest, motivation, and engagement. In K. Sawyer (Ed.), *Cambridge handbook of the learning sciences* (2nd ed., pp. 668–685). New York: Cambridge University Press.

Järvenoja, H., Järvelä, S., & Malmberg, J. (2015). Understanding the process of motivational, emotional and cognitive regulation in learning situations. *Educational Psychologist, 50*(3), 204–219.

Johnson, A. M., Azevedo, R., & D'Mello, S. K. (2011). The temporal and dynamic nature of self-regulatory processes during independent and externally assisted hypermedia learning. *Cognition and Instruction, 29*(4), 471–504. doi:10.1080/07370008.2011.610244

Kirschner, P. A., Beers, P. J., Boshuizen, H. P. A., & Gijselaers, W. H. (2008). Coercing shared knowledge in collaborative learning environments. *Computers in Human Behavior, 24*, 403–420.

Kirschner, P. A., Strijbos, J.-W., Kreijns, K., & Beers, P. J. (2004). Designing electronic collaborative learning environments. *Educational Technology Research and Development, 52*(3), 47–66.

Kirschner, P. A. & Van Merriënboer, J. G. (2013). Do learners really know best? Urban legends in education. *Educational Psychologist, 48*(3), 169–183.

Kreijns, K., Kirschner, P. A., & Jochems, W. M. G. (2002). The sociability of computer-supported collaborative learning environments. *Educational Technology & Society, 5*(1), 8–22.

Kreijns, K., Kirschner, P. A., & Vermeulen, M. (2013). Social aspects of CSCL environments: A research framework. *Educational Psychologist, 48*(4), 229–242. doi:10.1080/00461520.2012.750225

Lajoie, S. P., Lee, L., Poitras, E., Bassiri, M., Kazemitabar, M., Cruz-Panesso, et al. (2015). The role of regulation in medical student learning in small groups: Regulating oneself and others' learning and emotions. *Computers in Human Behavior, 52*, 601–616. doi:10.1016/j.chb.2014.11.073

Lodewyk, K. R., Winne, P. H., & Jamieson-Noel, D. L. (2009). Implications of task structure on self-regulated learning and achievement. *Educational Psychology, 29*(1), 1–25. doi:10.1080/01443410802447023

Malmberg, J., Järvelä, S., & Järvenoja, H. (2017). Capturing temporal and sequential patterns of self-, co- and socially shared regulation in the context of collaborative learning. *Contemporary Journal of Educational Psychology*. doi:10.1016/j.cedpsych.2017.01.009

Malmberg, J., Järvelä, S., Järvenoja, H., & Panadero, E. (2015). Socially shared regulation of learning in CSCL: Patterns of socially shared regulation of learning between high and low performing student groups. *Computers in Human Behavior, 52*, 562–572. doi:10.1016/j.chb.2015.03.082

Miller, M., & Hadwin, A. (2015). Scripting and awareness tools for regulating collaborative learning: Changing the landscape of support in CSCL. *Computers in Human Behavior, 52*, 573–588. doi:10.1016/j.chb.2015.01.050

O'Donnell, A. M., & Hmelo-Silver, C. E. (2013). Introduction: What is collaborative learning? An overview. In C. Hmelo-Silver, A. O'Donnell, C. Chan, & C. Chin (Eds.), *The international handbook of collaborative learning*, (pp. 1–15). New York: Routledge.

Perry, N. E., & VandeKamp, K. J. (2000). Creating classroom contexts that support young children's development of self-regulated learning. *International Journal of Educational Research, 33*(7), 821–843.

Reimann, P., Markauskaite, L., & Bannert, M. (2014). e-Research and learning theory: What do sequence and process mining methods contribute? *British Journal of Educational Technology, 45*(3), 528–540. doi:10.1111/bjet.12146

Roll, I., & Winne, P. H. (2015). Understanding, evaluating, and supporting self-regulated learning using learning analytics. *Journal of Learning Analytics, 2*, 7–12.

Sawyer, R. K. (2014). Conclusion: The future of learning: Grounding educational innovation in the learning sciences. In Sawyer, R. K. (Ed.), *The Cambridge handbook of the learning sciences*, (2nd ed., pp. 726–746). New York: Cambridge University Press. doi:10.1017/CBO9781139519526

Scardamalia, M., & Bereiter, C. (1994). Computer Support for Knowledge-Building Communities. *Journal of the Learning Sciences, 3*(3), 265–283. doi: 10.1207/s15327809jls0303_3

Schunk, D. H., & Zimmerman, B. J. (1997). Social origins of self-regulatory competence. *Educational Psychologist, 32*(4), 195–208. doi:10.1207/s15326985ep3204_1

Thompson, L., & Fine, G. A. (1999). Socially shared cognition, affect and behavior: A review and Integration. *Personality and Social Psychology Review, 3*(4), 278–302.

Van den Bossche, P., Gijselaers, W. H., Segers, M., & Kirschner, P. A. (2006). Social and cognitive factors driving teamwork in collaborative learning environments. *Small Group Research, 37*, 490–521. doi: 10.1177/1046496406292938

Winne, P. (1997). Experimenting to bootstrap self-regulated learning. *Journal of Educational Psychology, 89*(3), 397–410.

Winne, P. H. (2011). A cognitive and metacognitive analysis of self-regulated learning. In B. Zimmerman & D. Schunk (Eds.), *Handbook of self-regulation of learning and performance* (pp. 15–32). New York: Routledge.

Winne, P. H., & Hadwin, A. F. (1998). Studying as self-regulated engagement in learning. In D. Hacker, J. Dunlosky, & A. Graesser (Eds.), *Metacognition in Educational Theory and Practice* (pp. 277–304). Hillsdale, NJ: Lawrence Erlbaum.

Wolters, C. A., & Pintrich, P. R. (1998). Contextual differences in student motivation and self-regulated learning in mathematics, English, and social studies classrooms. *Instructional Science, 26*(1–2), 27–47. doi:10.1023/A:1003035929216

Zimmerman, B. J. (1989). A social cognitive view of self-regulated academic learning. *Journal of Educational Psychology, 81*(3), 329. doi:10.1037//0022-0663.81.3.329

Zimmerman, B. J. (2008). Investigating self-regulation and motivation: Historical background, methodological developments, and future prospects. *American Educational Research Journal, 45*(1), 166–183.

Zimmerman, B. J., & Schunk, D. H. (2011). Self-regulated learning and performance: An introduction and an overview. In B. J. Zimmerman & D. H. Schunk (Eds.), *Handbook of self-regulation of learning and performance* (pp. 1–12). New York: Routledge.

14

Collective Knowledge Construction

Ulrike Cress and Joachim Kimmerle

Defining Collective Knowledge Construction

Collective knowledge construction refers to a process in which people create new knowledge and new content collaboratively. It is an interpersonal activity that may take place in relatively small groups of people or in large communities where huge numbers of people are involved. The underlying idea is that learning is a social and collective activity rather than a solitary one (Bereiter, 2002; van Aalst & Chan, 2007). Collective knowledge construction is based on (1) knowledge that people introduce into the collaborative process on the basis of their own individual prior knowledge and their personal expertise, as well as (2) knowledge that is already part of the communication and that has previously been shared in the group or community.

Often, the people involved use technologies and shared digital artifacts that facilitate their interaction and collaboration, enabling them to share content and form communities. Knowledge construction is an *emergent* process, that is, it may result in a new group product that could not be predicted from the previous knowledge available in the group. In knowledge construction, individuals do not merely contribute additively but refer to each other and take up each other's arguments in such a way that the group as a whole may arrive at new insights. Thus, knowledge construction requires that people introduce their own knowledge, opinions, and perspectives into the discussion and are willing to take those of others into account.

Examples of the emergence of new knowledge have been described for groups and communities of different sizes. Dyads may arrive at new insights by considering each other's suggestions (Roschelle & Teasley, 1995), or a class of students may become a community through knowledge-building activities in which they collaboratively create ideas to solve problems (Scardamalia & Bereiter, 1999). An example for knowledge construction comes from our own research with individuals who worked with a wiki text about schizophrenia (Kimmerle, Moskaliuk, & Cress, 2011). Initially, individual users had only the information that schizophrenia was caused by genetic factors. Then they encountered information in a wiki that schizophrenia has social causes. In the course of the interaction, several participants came up with the improved idea that an interplay between genetic and social factors might cause schizophrenia (which is also the state-of-the-art opinion in clinical psychology).

Knowledge construction might also take place on an organizational level—for example, when enterprises (and their members) drive innovations. In their prominent model, Nonaka and Takeuchi (1995) describe knowledge creation as a continuous transformation between implicit and explicit knowledge (Polanyi, 1966). Through successive processes of socialization (implicit to implicit knowledge), externalization (implicit to explicit), combination (explicit to explicit), and internalization (explicit to implicit), individual knowledge can be expanded and made usable on higher organizational levels (Nonaka, 1994). An example would be when workers of an organization observe practices of others, describe differences among various practices, and discuss how to generate the most efficient method. It is a synergetic process in a way that the practices of all people are combined to reach a new level that describes the abilities of the collective. We will refer back to these examples of knowledge construction at various points throughout this text to demonstrate the relative value of the different concepts we introduce in this chapter.

In the following sections, we first describe how research on collective knowledge construction has been developing in the past. We highlight two different research traditions—the cognitive and the sociocultural—that represent the individual and the social sides of learning and knowledge construction. In the state-of-the-art section, we present not only several approaches that have been developed in the Learning Sciences but also theories from sociology. Those sociological theories build a conceptual background for sociocultural considerations and describe the nature of emergent social processes. The state-of-the-art section concludes with the presentation of a model that brings together cognitive, sociocultural, and sociological considerations in the context of technology-supported collective knowledge construction. We then describe research methods that have typically been applied in those different research traditions. In the last section, we discuss potential future developments in methodology and theory.

How Research on Collective Knowledge Construction Has Been Developing

Since the 1990s, the research area of *computer-supported collaborative learning* (CSCL) has been developing as a strand within the Learning Sciences. It focuses on collaborative learning and on how information and communication technology can support the development of new knowledge (for an early portrayal of the field, see e.g., Koschmann, 1996). This research area comprises two traditions that differ with regard both to their theoretical assumptions and their methodology (Jeong & Hartley, this volume). These two research traditions may be referred to as cognitive and sociocultural approaches respectively (Danish & Gresalfi, this volume).

The *cognitive approach* emphasizes the individual and deals with individual information processing. Accordingly, this approach considers knowledge production explicitly and exclusively as taking place in the individual mind. In this tradition, which for the most part is upheld by cognitive psychologists, learning is primarily seen as the acquisition, extension, and development of mental structures. Knowledge is conceptualized as *internal* mental representations that people have about the world, whether about facts, events, or their social environment (e.g., Freyd, 1987; Smith, 1998).

People's cognitive processes may also be seen as constructive activities (e.g., Piaget, 1977). The constructivist conceptualization of cognitive processes has had a very strong impact in the Learning Sciences on describing and explaining issues of learning, understanding, knowledge, and memory (Packer & Goicoechea, 2000; Smith, Disessa, & Roschelle, 1994). It says that humans understand the world around them based on their internal knowledge. People aim to make sense of incoming information in terms of their existing cognitive structures and processes, that is, they assimilate new information into their cognitive structures. They are able to build more advanced knowledge from their previous way of understanding. Thus, prior cognitive conceptions can be considered to be resources for cognitive development (Smith et al., 1994). If people's current mental structures cannot adequately deal with new experiences, accommodation processes take place, in which individuals adapt their mental structures to their experiences. In a constructivist sense, all learning is a process of *individual* knowledge construction.

In line with the focus on the individual, researchers who first systematically addressed knowledge-related aspects of cooperation and collaboration in educational settings (e.g., Johnson & Johnson,

1989; Slavin, 1990) considered learners' interactive activities primarily to be a means of fostering *individual* knowledge acquisition. When people learn collaboratively, they have to externalize their own previous knowledge, reflect upon it, explain it to others, and ask questions. These activities, in turn, lead to learning in each individual. Studies in this tradition focused on varying the social situation in controlled settings in order to examine if and how this situation influenced individual cognitive processes. Slavin (1990), for example, studied which kind of reward system (individual grades vs. group grades) led to the best individual performances; King (1990) examined which reciprocal peer questions led to deeper comprehension and more sophisticated individual decisions. In those cases in which these studies investigated knowledge and learning at the group level, they typically aggregated the individual learning measures and regarded them as group measures.

However, emergent aspects of collective knowledge construction come into play (see below for more details on emergence) when people interact in such a way that they refer to each other, take up each other's statements, opinions, and arguments, and integrate them into their own line of reasoning (e.g., Chinn & Clark, 2013). This collective process is key to the second research tradition, which focuses on the group (with all of its members, tools, settings, and activities) and considers the collective to be the relevant unit of analysis for knowledge construction. This *sociocultural approach* has a broad and prolific history, ranging from early approaches by Leontiev (1981) or Vygotsky (1978) to those of contemporary theorists like Engeström (2014). In contrast to the cognitive approach, this tradition proposes that knowledge is not something a person owns or acquires. It is rather something that is embedded into people's activities and cultural practices. In Scardamalia and Bereiter's (e.g., 1994) work, it is the whole group that builds (i.e., constructs) knowledge; knowledge cannot simply be attributed to the processes going on in each individual (Chan & van Aalst, this volume). In this tradition, knowledge is never related to individuals in isolation; instead, knowledge occurs in shared activities and is influenced by the context and the culture of learning (Brown, Collins, & Duguid, 1989; Rogoff, 1990).

State of the Art

All theories having to do with collective knowledge construction must come to terms with its multilevel structure, that is, with the interplay of the individual and the collective levels. Theorizing must aim to lead to an understanding of how people collaborate in such a way that they not only acquire individual knowledge but also develop new knowledge on the group level. Research also needs to lead to explanations of how the development of new knowledge can be described as an inter-individual process. In this section, we introduce three lines of research that address these issues. First, we present several approaches that have been developed within the Learning Sciences and CSCL research. Next, we introduce sociological approaches that deal with the relationship between the individual and the collective and that may be informative for understanding the phenomenon of knowledge emergence in this respect. Finally, we discuss a model that brings together considerations from various research traditions.

Collective Knowledge Construction in the Learning Sciences

To describe the development of new knowledge as an inter-individual process, Stahl (2005, 2006) uses the term *group cognition*. His model describes the way in which artifacts, utterances, and interactions of the group members start to become a network of mutual references during collaboration. This network enables "cognition" at the level of the group, that is, shared understandings. The network allows for meaning-making by the group. The meaning itself is not attached to any single part of this network, neither to particular words nor to artifacts or persons. Instead, meaning emerges from the reciprocity of references. Through this kind of reciprocal reference, it then becomes possible for a group to arrive at new insights, as in the example described above where participants came to the new insight that schizophrenia is possibly caused by the interaction of genetic and social factors. The group develops inter-subjective knowledge that occurs in and arises from the group discourse. Trausan-Matu (2009) refers to this intricate process by using the metaphor of *polyphonic collaborative learning*. Relying on the dialogic model of Bakhtin (1981) and using methods from lexical, semantic, and discourse analysis, he describes knowledge construction in terms of different individual voices that together create a melody. This process allows other voices to establish dissonances, to join in the chorus, and to attune to each other.

Scripting procedures and scaffolding tools (Kollar, Wecker, & Fischer, this volume; Tabak & Kyza, this volume) are often applied as catalysts for collaborative learning and collective knowledge construction (Kollar, Fischer, & Hesse, 2006). They provide guidelines for shaping people's contributions and support the learners in their interactions—for example, by providing explicit social rules or implicit affordances that guide a member's attention to the activities going on in the group. In this way, scripts and tools can intertwine the activities of different people who are involved in the process of coming to a successful collaboration, during which people consider each other's contributions and integrate them into the discussion that is going on in the group.

In the process of collaboration, the availability of knowledge-related artifacts and tools is considered to be highly relevant. For example, the *knowledge creation metaphor* of learning (Paavola & Hakkarainen, 2005) points to the importance of artifacts for knowledge construction. The *trialogical* approach to learning (Paavola & Hakkarainen, 2014) refers to a form of collaborative learning where the people involved create knowledge-related artifacts and develop knowledge-related practices. This approach emphasizes that collaboration among people not only occurs through direct communication (which would be dialogical) but also through common development of shared artifacts or practices which constitutes a continually evolving process. Artifacts present knowledge in a materialized form that exists independently of their creators. These artifacts make people's contributions manifest in the collaborative space. The wiki text that people used to interact with each other in the schizophrenia case described above is an example of such a shared artifact, where the collective knowledge that people developed had materialized and was available to everyone in the community for future reference.

Sociological Approaches

As outlined above, collective knowledge construction is particularly characterized by knowledge emergence. Generally speaking, the term *emergence* refers to the manifestation of occurrences or structures at a higher level of a system that cannot be fully understood by considering the features of its lower-level elements in isolation (Johnson, 2001). In the context of collective knowledge construction, the concept of emergence points to the phenomenon that in a community new knowledge may develop that has not been part of the individual community members' knowledge structures prior to the collaborative activities (Nonaka & Nishiguchi, 2001). So, any theory that deals with the phenomenon of emergence needs to take into consideration this relationship of parts to the whole. Theories dealing with the social aspects of emergence have been proposed in sociology (see also Sawyer, 2005). For a better understanding of the epistemological issues of collective knowledge construction in terms of knowledge emergence, we argue that the Learning Sciences may benefit from taking sociological approaches into account, of which we have selected three for presentation: structuration theory as outlined by Giddens (1984); critical realism in the tradition of Archer (1996); and social systems theory as conceptualized by Luhmann (1995; for a comparative overview, see Elder-Vass, 2007a, 2007b). All of these theories are very comprehensive, and we therefore can provide just a very selective presentation, singling out those aspects that we consider relevant for describing knowledge emergence.

There is a long sociological tradition of theorizing about how social structures (macro level) and individual agency (micro level) are related. Very generally speaking, we might say that there are objectivist and subjectivist social theories, with the objectivist theories emphasizing structures (and largely ignoring individual agency) and the subjectivist theories focusing on actors (while neglecting social frame conditions). One of the most prominent attempts to bring those two approaches together has been made by Giddens (1984) in his *structuration theory*. He suggests that they are only seemingly contradictory to each other, so he aims to reconcile them. He argues that the relationship between individual and collective needs to be reconsidered insofar as both methodological individualism and methodological collectivism on their own fall short of explaining this relationship. To this end, he presents a theory that connects structure and agency by introducing a *duality of structure* that refers to an interaction of agents and structure, a process that allows for the development of social systems. The general idea is that social structures, such as formal rules in institutions, have only a limited impact on behavior, because they may be interpreted very differently by different individual actors. But, of course, individuals also have to take structures into account, and thus, structures enable, support, or hinder particular activities. In their activities, actors rely on and refer to existing social structures. Actors control their activities and thereby regulate their surroundings. In this way, actors produce and reproduce social structures. Accordingly, social structures are both the means and the result of activities.

Emergentist theories of *critical realism* also focus on the interplay between the individual and the social structures, but they emphasize other aspects of this interaction and, accordingly, have taken different concepts into account. Archer (2003) has criticized Giddens' approach of conflating social structures and human agency and considering them to be ontologically inseparable. Archer aims instead at linking culture, social structure, and agency without any ontological reduction or conflation. She rejects the proposal of a duality of agency and structure. She points to the sovereignty of social structures, while, on an analytical level, she also distinguishes structures from the practices of agents that generate or modify them. As a result, social structures and individual activities, and the associations between them, do not blend together, and thus can and should be analyzed distinctly. In addition, critical realism appears to be more appropriate than structuration theory for explaining *causality*. Critical realism assumes that it is human agency that mediates the causal power of social forms (see also Bhaskar, 2014). From the perspective of emergentist ontology, individual agency is identified with the emergent causal powers of people as agents (Elder-Vass, 2010). Individuals are part of a hierarchy of elements, "each with emergent causal powers of their own, including in this case both the biological parts of human beings and the higher level social entities composed (at least in part) of human beings" (Elder-Vass, 2007b, p. 326). In other words, the consequences of individual activities and interactions are combined with the causal power of social structures (see also Sawyer, 2005).

Regarding the schizophrenia example, both structuration theory and the critical realism approach would tend to provide a similar description of the knowledge emergence. Critical realism would allow for a more detailed examination due to an analytical distinction between the individual contributions of the wiki authors and the conditions of the wiki environment. With respect to innovation in organizations, Nonaka's considerations are in line with the critical realism approach insofar as organizational settings (e.g., bureaucratic guidelines) provide social structures for individual activities that allow for the transfer as well as the internalization of knowledge.

A quite different approach as to what constitutes social systems and the phenomenon of emergence can be found in Luhmann (1995). The basic issue of Luhmann's systems theory is the concept of "systems". Systems are defined by their self-referential activities. Cognitive systems come into existence through thinking. Thoughts are based on previous thoughts; hence, thinking is a self-referential process. Social systems are generated through communication. This is also a self-referential process, as utterances can only be understood based on previous utterances. Both cognitive and social systems are meaning-making systems. Meaning only exists within a system at the very moment when it is being realized. The meaning of a concrete concept is created by the interdependency with the meanings of other concepts, whether in a cognitive or in a social system. That is to say, the meaning of an utterance in communication is developed by its relationships to other communications (see Elder-Vass, 2007a). The meaning of an individual perception is developed by its relationships to other mental representations. Meaning arises from how these relations allow a system to represent issues of the outside world: What is relevant is how the relations to other concepts can be reproduced and sustained within the system.

Co-Evolution of Cognitive and Social Systems

The co-evolution model of individual learning and collective knowledge construction (Cress & Kimmerle, 2008) aims to bring together several of the concepts introduced so far, with the further goal of addressing the emergent character of collective knowledge construction. In particular, this model takes up Luhmann's perspective, which posits a theory of self-referential systems wherein everything that is part of the system results from self-production (Luhmann, 2006; see also Maturana & Varela, 1987). Understanding collective knowledge construction entails focusing on the interplay between cognitive and social meaning-based systems. Both cognitive and social systems construct meaning through their own particular operations (i.e., cognition and communication respectively; Luhmann, 1995). Each individual as well as each social system is unique in its knowledge construction process and, thus, each may come to quite different results. Even though systems are closed with regard to their operations, they are open in the sense that they select elements from their environment and operate on them. Wikipedia supplies an example of this selection in a social system in that Wikipedia selects only verified and referenced information for further processing (Oeberst, Halatchliyski, Kimmerle, & Cress, 2014). Social systems constitute the environment for cognitive systems and vice versa. Hence, both types of systems can make meaning from stimuli coming from the respective other. Both the cognitive and the social system externalize and internalize knowledge, and both can assimilate and accommodate, all of which makes knowledge-related dynamics possible.

The example of individuals working on the wiki text about the origin of schizophrenia demonstrates these processes. In a wiki, a person may describe her personal knowledge about the origin of schizophrenia, such as her knowledge that genetic dispositions are relevant. Her interpretation and how she makes meaning from these findings are based on her individual considerations. The considerations she chooses to externalize constitute the environment for the communication in the wiki community, which is the basis for the social system. The social system may ignore her contributions or take them up and start to make meaning out of them. This meaning-making of the social system happens through linking specific contributions to certain previous contributions. In the wiki there may be dominant positions about the origin of schizophrenia (e.g., social aspects like psycho-social stress). There may also be a particular way in which certain arguments are described and valued. These characteristics result from the specific composition of the group members as a whole, from the aim of the wiki, or from dominant members. Over time an individual contribution may become interwoven with other contributions in such a way that it does not fit the original interpretation of the individual contribution any more. The meaning-making of the group may differ from the meaning-making of the individual—both processes belong to separate systems. However, both systems can change and develop dynamically through processes of internalization and externalization. Both systems can "irritate" the other system and lead to assimilation or accommodation of the systems, resulting in new insights about the origin of schizophrenia. These may become manifest as *individual learning* (development of the cognitive system) and/or *collective knowledge construction* (development of the social system). In sum, the co-evolution model describes the interaction of individuals with the help of shared artifacts that allow for the manifestation of communication in a social system (for an overview, see Kimmerle, Moskaliuk, Oeberst, & Cress, 2015).

Research Methods

As pointed out above, cognitive and sociocultural approaches apply different research methods. The dominant research method in cognitive psychology is the controlled randomized laboratory experiment, where potentially disruptive factors can optimally be controlled. However, this method requires reducing the range of potentially relevant variables to a very limited number and largely excluding influences of surrounding factors. Accordingly, controlled experiments mainly deal with the processes within cognitive systems. Sociocultural approaches, in contrast, take into account people's surroundings as well as social and cultural aspects. It follows that sociocultural research settings aim to examine individuals and groups in their everyday (inter)activities. Therefore, those approaches in the Learning Sciences and in CSCL research that view themselves as part of the sociocultural tradition often use ethnomethodological methods (Garfinkel, 1984) and conversation analysis (Sacks & Jefferson, 1995; Koschmann, this volume; Lund & Suthers, this volume). Both methods examine social interactions, including verbal and non-verbal behavior, and deliver precise descriptions of how groups and communities produce social reality and make meaning in a given situation.

When people use a wiki, for example, in order to construct new knowledge collaboratively, conversation analysis allows for examining linguistically how people take up each other's information, how their individual contributions relate to each piece of information, and how they develop further each other's considerations until new ideas emerge. This enables researchers to understand at which point and under what conditions the key incidents of collective knowledge construction take place and how a new insight (such as the interaction of genetic and social aspects for the onset of schizophrenia) is activated. However, this method has weaknesses in determining causalities and hardly allows for concrete predictions.

The spectrum of methods for analyzing collective knowledge construction is broad and still broadening. Further methods include, for example, social network analysis. This method allows for examining social relationships by representing networked structures as nodes (e.g., signifying actors) and ties (signifying interactions; Wassermann & Faust, 1994). Halatchliyski, Moskaliuk, Kimmerle, and Cress (2014) used this method to analyze and predict collective knowledge construction based on the features of individuals.

Future Trends and Developments

Future developments in research on collective knowledge construction will likely be related to both methodology and theory building. With regard to new methods, one future trend will be that "big data" are used increasingly to analyze inter-individual processes and activities in large networks (Wise & Shaffer, 2015). Learning platforms, massive open online courses (MOOCs), internet forums, wikis, blogs, or Twitter provide a wealth of data that allow researchers to trace people's contributions and to see how they change over time. Many of these tools provide extensive databases for analyzing individual and collective dynamics and describing how they interact, making possible better analysis of collective knowledge construction in large groups. Semantic technology will also become more and more useful for analyzing the processes of collective knowledge construction, by making possible the analysis of large group discourse and identifying relevant topics that emerge from the conversations, along with how they develop over time and spread among different people (Chen, Vorvoreanu, & Madhavan, 2014). Finally, computer simulations may be applied more often in the future to simulate multi-agent systems that will allow researchers to identify higher-level knowledge-related behavior that emerges from lower-level components (Nasrinpour & Friesen, 2016).

What is important with respect to methodology in general, and should be taken into account even more in the future, is that the research methods applied should be appropriate for the multi-level structure of complex, knowledge constructing systems (De Wever & Van Keer, this volume).

With regard to theory building, we call for research that connects some of the different approaches described in this chapter. Even though the gaps between cognitive, sociocultural, and sociological theories are fairly wide in many respects, we claim that their respective shortcomings can be compensated for. We see a need for new theoretical developments in social theory that deal more explicitly with the emergence of knowledge and that allow for empirical examination. Including more research into complex systems more consistently in the Learning Sciences (e.g., Jacobson & Wilensky, 2006) would contribute greatly to the understanding of knowledge emergence.

Further Readings

Kimmerle, J., Moskaliuk, J., Oeberst, A., & Cress, U. (2015). Learning and collective knowledge construction with social media: A process-oriented perspective. *Educational Psychologist, 50*, 120–137.
In this review article the authors propose a systemic-constructivist approach for the examination of the co-evolution of cognitive and social systems and discuss implications for educational design.

Luhmann, N. (2006). System as difference. *Organization, 13*, 37–57.
This article presents the basic concepts of Luhmann's general systems theory and explains major components such as the distinction between system and environment and the modes of operation of different systems.

Nonaka, I. (1994). A dynamic theory of organizational knowledge creation. *Organization Science, 5*, 14–37.
In this article Nonaka develops a framework for the management of the dynamics of organizational knowledge creation that takes place through a constant interchange between tacit and explicit knowledge.

Paavola, S., & Hakkarainen, K. (2005). The knowledge creation metaphor: An emergent epistemological approach to learning. *Science & Education, 14*, 535–557.
This conceptual article introduces the "trialogical" approach to learning as a collaborative process in which people develop shared knowledge-related artifacts and common practices.

Stahl, G. (2005). Group cognition in computer-assisted collaborative learning. *Journal of Computer Assisted Learning, 21*, 79–90.
By outlining a social theory of collaborative knowing, Stahl explores how computer-supported collaboration may support "group cognition", a process that goes beyond individual cognition.

References

Archer, M. S. (1996). *Culture and agency: The place of culture in social theory*. Cambridge, UK: Cambridge University Press.
Archer, M. S. (2003). *Structure, agency and the internal conversation*. Cambridge, UK: Cambridge University Press.
Bakhtin, M. (1981). *The dialogic imagination: Four essays*. Austin: University of Texas Press.
Bereiter, C. (2002). *Education and mind in the knowledge age*. Mahwah, NJ: Lawrence Erlbaum Associates.
Bhaskar, R. (2014). *The possibility of naturalism: A philosophical critique of the contemporary human sciences* (4th ed.). London: Routledge.
Brown, J. S., Collins, A., & Duguid, P. (1989). Situated cognition and the culture of learning. *Educational Researcher, 18*, 32–42.
Chan, C. K. K., & van Aalst, J. (2018). Knowledge building: Theory, design, and analysis. In F. Fischer, C. E. Hmelo-Silver, S. R. Goldman, & P. Reimann (Eds.), *International handbook of the learning sciences* (pp. 295–307). New York: Routledge.
Chen, X., Vorvoreanu, M., & Madhavan, K. (2014). Mining social media data for understanding students' learning experiences. *IEEE Transactions on Learning Technologies, 7*, 246–259.
Chinn, C. A., & Clark, D. B. (2013). Learning through collaborative argumentation. In C. E. Hmelo-Silver, C. A. Chinn, C. K. K. Chan, & A. M. O'Donnell (Eds.), *International handbook of collaborative learning* (pp. 314–332). New York: Taylor & Francis.
Cress, U., & Kimmerle, J. (2008). A systemic and cognitive view on collaborative knowledge building with wikis. *International Journal of Computer-Supported Collaborative Learning, 3*, 105–122.
Danish, J. A., & Gresalfi, M. (2018). Cognitive and sociocultural perspective on learning: Tensions and synergy in the Learning Sciences. In F. Fischer, C. E. Hmelo-Silver, S. R. Goldman, & P. Reimann (Eds.), *International handbook of the learning sciences* (pp. 34–43). New York: Routledge.
De Wever, B., & Van Keer, H. (2018). Selecting statistical methods for the learning sciences and reporting their results. In F. Fischer, C. E. Hmelo-Silver, S. R. Goldman, & P. Reimann (Eds.), *International handbook of the learning sciences* (pp. 532–541). New York: Routledge.
Elder-Vass, D. (2007a). Luhmann and emergentism: Competing paradigms for social systems theory? *Philosophy of the Social Sciences, 37*, 408–432.
Elder-Vass, D. (2007b). Reconciling Archer and Bourdieu in an emergentist theory of action. *Sociological Theory, 25*, 325–347.

Elder-Vass, D. (2010). *The causal power of social structures.* Cambridge, UK: Cambridge University Press.

Engeström, Y. (2014). *Learning by expanding: An activity-theoretical approach to developmental research.* Cambridge, MA: Cambridge University Press.

Freyd, J. J. (1987). Dynamic mental representations. *Psychological Review, 94,* 427–438.

Garfinkel, H. (1984). *Studies in ethnomethodology.* Cambridge, UK: Polity Press.

Giddens, A. (1984). *The constitution of society: Outline of the theory of structuration.* Cambridge, UK: Polity Press.

Halatchliyski, I., Moskaliuk, J., Kimmerle, J., & Cress, U. (2014). Explaining authors' contribution to pivotal artifacts during mass collaboration in the Wikipedia's knowledge base. *International Journal of Computer-Supported Collaborative Learning, 9,* 97–115.

Jacobson, M. J., & Wilensky, U. (2006). Complex systems in education: Scientific and educational importance and implications for the learning sciences. *Journal of the Learning Sciences, 15,* 11–34.

Jeong, H., & Hartley, K. (2018). Theoretical and methodological frameworks for computer-supported collaborative learning. In F. Fischer, C. E. Hmelo-Silver, S. R. Goldman, & P. Reimann (Eds.), *International handbook of the learning sciences* (pp. 330–339). New York: Routledge.

Johnson, D. W., & Johnson, R. T. (1989). *Cooperation and competition: Theory and research.* Edina, MN: Interaction Book Company.

Johnson, S. (2001). *Emergence: The connected lives of ants, brains, cities, and software.* New York: Scribner.

Kimmerle, J., Moskaliuk, J., & Cress, U. (2011). Using wikis for learning and knowledge building: Results of an experimental study. *Educational Technology & Society, 14*(4), 138–148.

Kimmerle, J., Moskaliuk, J., Oeberst, A., & Cress, U. (2015). Learning and collective knowledge construction with social media: A process-oriented perspective. *Educational Psychologist, 50,* 120–137.

King, A. (1990). Enhancing peer interaction and learning in the classroom through reciprocal questioning. *American Educational Research Journal, 27,* 664–687.

Kollar, I., Fischer, F. & Hesse, H. W. (2006). Collaboration scripts: A conceptual analysis. *Educational Psychology Review, 18,* 159–185.

Kollar, I., Wecker, C., & Fischer, F. (2018). Scaffolding and scripting (computer-supported) collaborative learning. In F. Fischer, C. E. Hmelo-Silver, S. R. Goldman, & P. Reimann (Eds.), *International handbook of the learning sciences* (pp. 340–350). New York: Routledge.

Koschmann, T. (Ed.). (1996). *CSCL: Theory and practice of an emerging paradigm.* Mahwah, NJ: Lawrence Erlbaum.

Koschmann, T. (2018). Ethnomethodology: Studying the practical achievement of intersubjectivity. In F. Fischer, C. E. Hmelo-Silver, S. R. Goldman, & P. Reimann (Eds.), *International handbook of the learning sciences* (pp. 465–474). New York: Routledge.

Leontiev, A. N. (1981). *Problems of the development of the mind.* Moscow: Progress.

Luhmann, N. (1995). *Social systems.* Stanford, CA: Stanford University Press.

Luhmann, N. (2006). System as difference. *Organization, 13,* 37–57.

Lund, K., & Suthers, D. (2018). Multivocal analysis: Multiple perspectives in analyzing interaction. In F. Fischer, C. E. Hmelo-Silver, S. R. Goldman, & P. Reimann (Eds.), *International handbook of the learning sciences* (pp. 455–464). New York: Routledge.

Maturana, H. R., & Varela, F. J. (1987). *The tree of knowledge: The biological roots of human understanding.* Boston, MA: New Science Library/Shambhala Publications.

Nasrinpour, H. R., & Friesen, M. R. (2016). An agent-based model of message propagation in the Facebook electronic social network. arXiv:1611.07454.

Nonaka, I. (1994). A dynamic theory of organizational knowledge creation. *Organization Science, 5,* 14–37.

Nonaka, I., & Nishiguchi, T. (2001). *Knowledge emergence: Social, technical, and evolutionary dimensions of knowledge creation.* New York: Oxford University Press.

Nonaka, I., & Takeuchi, H. (1995). *The knowledge-creating company: How Japanese companies create the dynamics of innovation.* New York: Oxford University Press.

Oeberst, A., Halatchliyski, I., Kimmerle, J., & Cress, U. (2014). Knowledge construction in Wikipedia: A systemic-constructivist analysis. *Journal of the Learning Sciences, 23,* 149–176.

Paavola, S., & Hakkarainen, K. (2005). The knowledge creation metaphor: An emergent epistemological approach to learning. *Science & Education, 14,* 535–557.

Paavola, S., & Hakkarainen, K. (2014). Trialogical approach for knowledge creation. In S. C. Tan, H. J. So, & J. Yeo (Eds.), *Knowledge creation in education* (pp. 53–73). Singapore: Springer.

Packer, M. J., & Goicoechea, J. (2000). Sociocultural and constructivist theories of learning: Ontology, not just epistemology. *Educational Psychologist, 35,* 227–241.

Piaget, J. (1977). *The development of thought: Equilibration of cognitive structures.* New York: Viking Press.

Polanyi, M. (1966). *The tacit dimension.* Chicago, IL: University of Chicago Press.

Rogoff, B. (1990). *Apprenticeship in thinking: Cognitive development in social context.* New York: Oxford University Press.

Roschelle, J., & Teasley, S. D. (1995). The construction of shared knowledge in collaborative problem solving. In C. O'Malley (Ed.), *Computer-supported collaborative learning* (pp. 69–197). Berlin: Springer.

Sacks, H., & Jefferson, G. (1995). *Lectures on conversation.* Oxford: Blackwell.

Sawyer, R. K. (2005). *Social emergence. Societies as complex systems.* Cambridge, UK: Cambridge University Press.

Scardamalia, M., & Bereiter, C. (1994). Computer support for knowledge-building communities. *Journal of the Learning Sciences, 3,* 265–283.

Scardamalia, M., & Bereiter, C. (1999). Schools as knowledge building organizations. In D. Keating & C. Hertzman (Eds.), *Today's children, tomorrow's society: The developmental health and wealth of nations* (pp. 274–289). New York: Guilford.

Slavin, R. E. (1990). *Cooperative learning: Theory, research, and practice.* Englewood Cliffs, NJ: Prentice-Hall.

Smith, E. R. (1998). Mental representation and memory. In D. T. Gilbert, S. T. Fiske, & G. Lindzey (Eds.), *The handbook of social psychology* (pp. 391–445). New York: McGraw-Hill.

Smith, J. P., III, Disessa, A. A., & Roschelle, J. (1994). Misconceptions reconceived: A constructivist analysis of knowledge in transition. *Journal of the Learning Sciences, 3,* 115–163.

Stahl, G. (2005). Group cognition in computer-assisted collaborative learning. *Journal of Computer Assisted Learning, 21,* 79–90.

Stahl, G. (2006). *Group cognition: Computer support for building collaborative knowledge.* Cambridge, MA: MIT Press.

Tabak, I., & Kyza, E. A. (2018). Research on scaffolding in the learning sciences: A methodological perspective. In F. Fischer, C. E. Hmelo-Silver, S. R. Goldman, & P. Reimann (Eds.), *International handbook of the learning sciences* (pp. 191–200). New York: Routledge.

Trausan-Matu, S. (2009). The polyphonic model of hybrid and collaborative learning. In F. L. Wang, J. Fong, & R. C. Kwan (Eds.), *Handbook of research on hybrid learning models: Advanced tools, technologies, and applications* (pp. 466–486). Hershey, PA: Information Science Publishing.

van Aalst, J., & Chan, C. K. (2007). Student-directed assessment of knowledge building using electronic portfolios. *Journal of the Learning Sciences, 16,* 175–220.

Vygotsky, L. S. (1978). *Mind in society: The development of higher psychological processes.* Cambridge, MA: Harvard University Press.

Wassermann, S., & Faust, K. (1994). *Social network analysis: Methods and application.* New York: Cambridge University Press.

Wise, A. F., & Shaffer, D. W. (2015). Why theory matters more than ever in the age of big data. *Journal of Learning Analytics, 2*(2), 5–13.

15

Learning at Work
Social Practices and Units of Analysis

Sten Ludvigsen and Monika Nerland

Introduction

In the knowledge economy, working life in many professions and organization has become increasingly complex in most countries. The current discussion about the automatization of many types of work seems both realistic and sometimes more like futurist hype (Autor, Levy, & Murmane, 2003; Susskind & Susskind, 2015). However, what is obvious is that participants in the labor market who perform what we will call knowledge-intensive work need to develop new competencies. Some of these competencies are related to the understanding and use of new forms of data that are automatically generated. The interfaces between humans and technologies can and do change, which means that participants will experience new divisions of labor that will create new mediational processes between social and cognitive processes and new infrastructures and tools. As a consequence, many fields of knowledge must contribute to our understanding of workplaces and the development of skills and competences needed in the labor market. We argue here that the learning sciences should be placed at the heart of such contributions.

Different Positions in Learning at Work

In one strand of research, studies focus on individuals' learning and skill development over time, what skills and competencies are required, and how individuals adapt and learn from changing environments (e.g., Carbonell, Stalmeijer, Könings, Segers, & van Merriënboer, 2014; Gijbels, Raemdonck, Vervecken, & van Herck, 2012; and the Organisation for Economic Co-operation and Development [OECD], which initiated The Programme for the International Assessment of Adult Competencies [PIAAC]). Through survey instruments and knowledge tests, certain research designs can measure how participants perceive themselves within an organization or, more generally, in the labor market. These studies provide insights into and overviews of structural features and aggregated results that society, organizations, and clusters of organizations need. Depending on the research design, we can also obtain results about the efficiency of training participants in particular skills in organizations. Outcome studies of training represent a valuable tradition at the intersection of fields such as workplace studies and education (Billett, Harteis, & Gruber, 2014). Although these studies provide important insights into how organizations and workplaces function, few focus on work and learning processes, as such. Moreover, many scholars in this tradition have emphasized the development of vertical expertise. *Vertical expertise* is what we will refer to here as individual development, and these competences are what actors bring to the microgenetic process and collaborative efforts (see also Reimann & Markauskaite, this volume).

Billett (2001) discusses three forms of knowledge that are important for the individual participant: propositional knowledge, procedural knowledge, and dispositions. *Propositional knowledge* is what participants know in the domain, such as facts, information, concepts, and ways of reasoning and solving problems. To become part of a profession or to be seen as an expert in a community, deep knowledge is often required. *Procedural knowledge* is needed to execute actions and activities. Procedural knowledge can mean executing standard operations that are seen as routines. The foundation for the development of such knowledge is connected to a high degree of repetition. The variation in the tasks is limited. If the tasks become more complex, the procedural knowledge becomes more complex. The complexity depends on the degree of new framing every time a task is executed. One conceptualization for this is moving from routine problems to non-routine problems. *Non-routine problems* involve a higher degree of cognitive capacity. After years of experience, professionals and participants develop deeper cognitive structures and social norms for activating particular forms of knowledge in action sequences (see also Danish & Gresalfi, this volume, for similar theoretical stances). The distinction between routine and non-routine problem solving has been important in the field of learning at work and the development of expertise.

In this chapter, we mainly emphasize knowledge-intensive work, which is non-routine work. However, in many professions, the distinction between routine and non-routine is not straightforward. Often, framing the problem in dialogues with other professionals is the most important aspect of the work; for example, interpreting the test results in medical encounters is routine. Moreover, knowledge-intensive work often involves problems that need to be solved by groups of participants—what is conceptualized as horizontal expertise. Horizontal expertise is dependent on the vertical expertise in collective problem solving. Horizontal expertise is part of the configurations that include task distribution, division of labor, leadership, and norms and values. No single participant can achieve and justify what a team of experts can do together. The justification takes place through a gradual increase in the repertoires of the experts and the materialization of new tools and standards.

The terms *adaptive expertise* and *preparation for future learning* (Bransford & Schwartz, 1999) point us in similar directions. Adaptivity and future orientation involve the capacity to transfer different forms of knowledge when engaging with a new task. The transfer of knowledge is seen as a classical problem in the learning sciences that needs to be conceptualized in relation to individual dispositions and knowledge but also in relation to how knowledge becomes activated in the interaction in a context (Greeno, 2006; Greeno & Engeström, 2013). Adaptivity and preparation for future learning act as mechanisms to integrate different forms of cognitive dimensions (propositional and procedural knowledge and dispositions). However, they are more limited when it comes to a differentiated analysis of the social and cultural context.

The Social Practice Stance

Our stance is defined as a practice perspective on human activity and learning. The concept of social practices creates a specific analytical starting point for understanding how and what people learn in work settings. The roots of this tradition go back to classical work in many of the social sciences, which highlights the interplay between individuals' actions and the social environment in processes of learning and knowledge construction (e.g., Engeström & Sannino, 2010; Ludvigsen & Nerland, 2014; Nicolini, 2013). More recently, however, researchers in the social sciences have focused on practice as the level of human organization where processes for change and learning reside. This focus has led to what is described as a practice turn in social theory (Nicolini, 2013; Schatzki, Knorr Cetina, & von Savigny, 2001) and to renewed interest in understanding the semiotic and emergent dimensions of work. We follow this line of reasoning and view the turn toward practice as a theoretical turn. At the same time, the concept of practice has been the subject of considerable discussion and has given rise to different lines of theoretical development. Different positions within this turn use different analytic concepts. What they have in common is an interest in the emerging relationships between humans and between humans and their material or semiotic tools as entry points to studying learning and change.

Within the learning sciences, the influential work by Engeström and others in cultural-historical activity theory (CHAT) and, more broadly, the socio-cultural perspective, can also be seen as specifications of the turn toward social practices as analytical premises for understanding learning at work (Danish & Gresalfi, this volume; Engeström & Sannino, 2012; Ludvigsen & Nerland, 2014; Scribner, 1984). When using social practice as a key concept, we can choose to specify this stance by concepts that are important for the empirical studies we will perform. We describe different perspectives on learning at work; social interaction around artifacts emphasizing language and talk, expansive learning following from object-oriented work within or across activity systems (Engeström, 2001), and the work within science and technology studies in which scholars offer analytical concepts devoted to revealing further specification of the epistemic dimensions of practice and the role knowledge plays in learning (Knorr Cetina, 2001). The concepts of epistemic practices and epistemic objects are used to describe differences in how fields of expertise are composed and what it means to participate in relevant ways within distinct knowledge regimes and practices of knowledge production.

When taking social practice as an analytical stance, we can include socio-cultural layers and psychological levels, it implies a functional approach to learning and cognition in work settings. A functional approach means studying cognition and learning in a specific context and explains how and what participants learn in specific settings and over time. The functional approach makes it possible to identify mechanisms that can be generalized across settings. The mechanism that can explain learning and cognition in work settings is seen as a conceptual system that describes and explains how practices are constituted and become transformed or emerge.

Tracing Learning in Knowledge-Intensive Work Across Layers of Development

The phrase knowledge-intensive work refers to processes that involve the systematic use of formalized knowledge in fields of expertise (see Reimann & Markauskaite, this volume). Such knowledge is often inscribed in and mediated by advanced cultural tools. Moreover, knowledge-intensive work is often part of larger infrastructures of knowledge and networked configurations (Engeström, 2001; Knorr Cetina, 2007). Different forms of technology continuously create new challenges regarding the kinds of knowledge participants have access to, what this means for the division of labor in the organization, and whether norms and values are contested. Technology creates new contingencies between participants and tools in the workplace. In this chapter we use key concepts from practice theory and learning theory that are coherent with our epistemological stance. This stance can be considered a socio-cultural perspective for the study of learning, cognition, and development (Lave & Wenger, 1991; Ludvigsen, 2012; Scribner, 1984). The socio-cultural perspective provides the required variation in units of analysis and levels of description (Ludvigsen, 2012). The premise for understanding how and what participants learn in the workplace should begin with an analysis of microgenetic processes. These processes are often labeled social interactions with mediational means.

Action, Activity, and Expansive Learning

As mentioned above, one can conceptualize learning at work as the vertical development of expertise and the horizontal development of collective expertise (Engeström, 1987; Engeström & Sannino, 2010). Until the late 1980s, most learning theories addressed vertical or individual development and used the individual as the unit of analysis. Lave and Wenger's (1991) contribution was very influential at the time and changed the discourse within the learning sciences. Lave and Wenger's core concepts of legitimate peripheral participation and communities of practice opened the way for concepts such as transparency and access and moved the unit of analysis away from the individual and to the participation in settings, activities, and communities. Changes in participation in communities and in discourse became an important aspect of understanding learning. Greeno (2006) connected social and cognitive processes as a unit of analysis. The term *distributed cognition* (Hutchins, 1995) is part of conceptual shift in learning theory. The concept of *distributed cognition* gives awareness to the fact that intelligence is built into tools and instruments, and the human cognitive process is dependent on the use of the knowledge inscribed in—for example, a cockpit or the dashboard of a ship. In advanced environments in workplaces, accumulated knowledge is inscribed and must be seen as part of the cognitive work needed to perform tasks and solve problems. In the turn toward social practice as a foundational concept for understanding and explaining learning, these contributions have shifted in terms of the unit of analysis—what needs to be included in the analysis.

Within the learning sciences, CHAT has been one prominent approach for understanding and explaining the development of horizontal expertise. Horizontal expertise is the expansion and change of objects within and between activity systems. Through concepts based on the practice perspective, we can understand the relation between changes at the micro, meso, and macro levels, which we see as intertwined.

Construction of a New Object—Illustration One

In this example, we describe how a group of researchers that worked on studying infectious diseases with a statistician developed a new capture-recapture estimation (CRE) method to count children hospitalized with influenza (Hall & Horn, 2012). This method was originally developed for estimating the size of an animal population and further developed for use in hospitals. However, the research team thought that it was better for capturing the number of children hospitalized with influenza. The idea was to combine an active screening procedure (using DNA, which is expensive) with a passive screening of hospital charts (less expensive). Presented below is a conversation between Alberto (a junior doctor/researcher) and Ted (the statistician), in which they try to understand the problem. Alberto first describes what he thinks the problem is, and then he is challenged by Ted (data reused from Hall & Horn, 2012, pp. 243–244).

1 *Ted:* Why not use the seven day numbers here (pointing at manuscript table)
2 *Alberto:* Mm hm
3 *Ted:* and the four day numbers there? As far as the method's concerned, I mean the . . . it doesn't care about the fact that one method has a smaller probability of capture than the other. (Three-second pause) I mean there's no reason why you can't.
4 *Alberto:* Mm (head on hands, five-second pause)
5 *Ted:* And I would . . . I would think that would be uh, it's gonna increase your numbers for one thing.
6 *Alberto:* Mm hm.
7 *Ted:* And uh, it's in some sense simpler. (Pointing at manuscript table) You're gonna get a higher number here [than you] would get there.
8 *Alberto:* [Right . . .]
9 *Ted:* But you know that that's going to be the case because you know that you're only sampling for four days here and you're sampling for seven days there. (Pointing at manuscript table)
10 *Alberto:* But how could we interpret the results after that estimation?
(Ted then explains that CRE does not assume equal probability of capture, 98 sec)

In the talk between Alberto and Ted, we see that they do not agree about the sampling of days that should be included in the analysis. Alberto and some of the senior research colleagues were skeptical about the move from sampling data once a week or four days a week to sampling screening data on all weekdays. Ted, in his argument, used a concrete example that showed that using his sampling procedure would lead to smaller confidence intervals; this is more informative for the hospital when estimating the number of children hospitalized with influenza. The discussion between Ted and the research team went on for several months before they were able to find the final solution. After six months, the research team published a new system for influenza screening. The result was that the expensive screening could be used once a week and that the less expensive screening should be used on all weekdays. This representation changed the participants' understanding of the counting involved and of how to obtain an accurate count of children hospitalized with influenza. The participant defined the sampling procedure in a new way.

The whole process can be described as a microgenetic process in which the participants talked and negotiated about what should be included in the sampling and how the results should be interpreted. The different screening tools were reconceptualized to construct the object; namely, the new sampling procedure. The problem space was changed, and the new sampling procedure was stabilized through the new tool. The new tool supported the construction of a new object: the number of hospitalized children with influenza. The outcome was an improved overview of the patient population. The individual development for members of the research team is seen as part of the reconceptualization of the sampling procedure. In addition, we can say that the rules for how to work with data, which types of expertise were needed, and how the labor was divided between the research team and the statistician became important in the object construction and the transformation of the sampling procedure from being the object to becoming a new tool in the activity system. Herein, the transformation of the object to a tool is conceptualized as steps in a process of expansive learning, in which the participants can transform their practice and provide their community with a new tool; this could be considered part of the development of the knowledge infrastructure in which the work is performed.

Epistemic Practices and Objects in Knowledge—Intensive Work

A related yet different approach takes specific interest in the practices through which knowledge is generated and evaluated in distinct ways in different fields of expertise and in the role of material representations in this regard. Expansive processes are related to epistemic dimensions of practice and to the way in which knowledge unfolds and branches out in an activity. Thus, the analytic resources allow for investigation of the very practices of inquiry in which experts or professionals engage and for examining learning as intrinsic to (collaborative) knowledge construction.

A core concept in this regard is epistemic objects. Expert communities are typically object-centered, in the sense that they are oriented toward exploring, developing, and mobilizing knowledge objects (Knorr Cetina, 2001). However, such objects are not understood as separate, material things. Instead, the objects may be described as complex amalgams of material and symbolic resources that constitute knowledge about a problem and, through their inherent complexity, activate a set of opportunities when the objects are approached (Nerland & Jensen, 2012). Moreover, following Knorr Cetina's (2001) definition, epistemic objects are characterized by their unfolding and question-generating character. In the context of work and learning, epistemic objects could be models for medical treatment, computer programs, legal texts, and complex representations of financial markets. Such objects are created in expert cultures and further developed as people in different settings attend to the objects, explore their complexity, and materialize their potential in local activities.

While epistemic objects give focus to the "what" dimension of work and knowledge construction, the concept of epistemic practices brings attention to the specific ways in which knowledge is approached, developed, and shared in an enacted culture (Knorr Cetina, 2007). Epistemic practices embody the methodological principles and ways of working that are distinctive to the expert culture and, thus, they are fundamental for the procedural side of professional expertise. This concept is relevant in the learning sciences as an approach to specify the types of activities involved in inquiry learning; for example, in science education (Kelly, 2011; Sandoval, 2005) and in research on professional learning (Markauskaite & Goodyear, 2016). Epistemic practices play a critical role in making knowledge applicable and in making expert performance transparent for learners. Moreover, the enactment of epistemic practices is critical for constructing meaning, as they are the means through which connections are made, knowledge and meaning are translated, and representations of a given object are produced and maintained in experts' work.

Developing and Aligning Forms of Knowledge in Architectural Design—Illustration Two

This second example illustrates how knowledge construction in non-routine work evolves around the participants' construction and exploration of epistemic objects. We use a study by Ewenstein and Whyte (2009) to illustrate how participants with different professional backgrounds work together on a shared object based on joint explorative activities. The researchers followed an architectural design process, in this case designing a herbarium and library extension in a botanic garden, which involved architects, engineers, and representatives from the client. Work was driven forward through a series of interactions around visual representations, such as photographs, drawings, calculations, and 3-D sketches. These visual representations took the function of epistemic objects in the activity, as participants used them to explore features of what they represented to move forward in time to imagine the results of the design process and to jointly refine and rework these representations. The visual representations also served to align different forms of knowledge, such as technical and aesthetic knowledge.

The visual representations formed a substantial part of a report that should be presented to the board of trustees during the process. The following conversation, taken from Ewenstein and Whyte (2009, pp. 20–21), illustrates how the participants collaborated to construct and refine this object:

1	*Structural engineer:*	I think it'd be useful if you send me my two pages, because at the moment I think we should either make it smaller or just sort of make it more compatible to the sort of the style that you've got in the rest of the report. I think I can probably work them up again. Last Wednesday, I didn't give myself lots of time. I can add a bit more to that section.
2	*Lead architect:*	And then we need to sort out how we're going to put the information for this new bit of work together.
3	*Services engineer:*	To me it sounds like we should be doing combined commentary.
4	*Lead architect:*	Yes, that's what it feels like.
		The service engineer states that this is quite a complicated bit of design, but the structural engineer points out that there are three options for the basic building shape, each of which could be described with "a few bullet points from each of us." The lead architect then suggests going back to the site:
5	*Lead architect:*	So, I'd say the first thing that we need to do is to all go out there again together, and look, walk around the buildings, the existing bits.
6	*Service engineer:*	Yes.
7	*Structural engineer:*	Yes.
8	*Lead architect:*	And we can talk, and talk and walk.
9	*Service engineer:*	Yes.
10	*Lead architect:*	And take pics, and . . .
11	*Structural engineer:*	Could do it together . . .
12	*Lead architect:*	What's your time like this week? So that we have to do that sooner rather than . . .

In this excerpt, we see how the participants work to combine different inputs and forms of expertise by working tightly together on the same object. Adding the participants' preparations of material to the report, the work involves a range of epistemic practices, such as modelling, testing out, analyzing, providing explanations, and aligning different contributions to the report with other elements. The visual representations take shifting functions in this regard, sometimes serving as objects of inquiry and sometimes mediating the exploration and negotiation of other issues. When the various contributions are to be combined, the need for new and joint explorations arises (line 5 above). In this phase, the participants move between the physical site, where the new building is to be realized, and the drawings and other representations they develop in the design process. The visual representations served to integrate the contributions from different experts in the team and kept the overall design idea in motion as an unfolding object.

The social practice is truly a collaborative one, in which the participants need to find new ways of working together to realize the design process. At the same time, the excerpt above illustrates a division of labor in which the lead architect initiates the further steps. What is striking in this conversation is also the lack of professional concepts in the dialogue, with a possible exception for the phrase *doing combined commentary*. This may be due to the inter-professional setting, in which everyday language forms more of a common ground for joint discussions. However, it may also reflect the heavy dependence on visual representations in this work process. Learning is intrinsic to the way different forms of knowledge are combined and represented in the process and to the achievement of shared understanding in the inter-professional team. Equally important, however, is the creative process of designing something not-yet-existing by exploring unfulfilled possibilities and modifying scenarios by moving back and forth in time and between the abstract and the concrete. The concepts of epistemic objects and practices are productive for conceptualizing forms of expansive learning that arise from practices in which knowledge generation and processes of materialization are at the core.

Conclusion

We started out with the aim of contributing a theoretical and methodological stance on how we can understand how and what participants learn in work settings. In this final section, we return to this issue by discussing the implications for units of analysis in research, as well as the implications for how learning at work can be facilitated. Before we elaborate our stance, we would also like to emphasize that different stances to learning exist and that they explain different aspects of the phenomenon. In this chapter, we have mainly focused on the microgenetic level in order to show learning processes involved in learning at work.

First, when we take social practices as the unit of analysis, it implies that the participants' actions and activities with tools are included. We have emphasized that the current studies on practice are interdisciplinary, since the practice perspective is inspired by a set of different traditions taken from the social sciences, humanities, and socio-cultural psychology. The stance we have taken is that the participants' learning is embedded in the tasks and the use of tools, the division of labor in the microstructures at work, through collaboration, and the norm values and mechanisms of justification involved. The practice perspective entails the use of analytic concepts to study what is at stake for the participants.

Studying historical and situated contingencies means zooming in and zooming out (Little, 2012; Nicolini, 2013). Zooming in on the details in practices is important when we want to understand how participants can learn from new tasks and in collaboration with others. When a new digital tool is introduced in the diagnostic assessment of a patient, a medical professional can experience gaps in his or her understanding between the previous procedures without the tool and the new procedures. Zooming in highlights microgenetic processes and how individuals display their agencies. By zooming out, we can analyze how the practice is influenced by other practices and is part of social and technological infrastructures. This also refers to how the knowledge produced is justified beyond the actions level. Through the connection between zooming in and zooming out, we can understand and explain learning in the workplace. Here, we can study the learning, development, and execution of horizontal expertise.

The stance that learning at work is intrinsic to work practices also means that the tools, infrastructures, and epistemic resources available in these practices frame opportunities for knowledge exploration and determine ways of knowing. Hence, the organization of knowledge, tasks, and epistemic resources at work are important for the conditions for learning. At the same time, the social interactions among participants and their enacted forms of discourse are critical for participants' meaning making and ways of taking advantage of the available resources. It is through social interaction that knowledge resources are invoked, made sense of, explored, and further developed. In order to facilitate productive environments for learning at work, we suggest that work organizations should attend to the types of practices offered to participants, the organization of knowledge in relation to these practices, and the ways in which participants are provided access to learning-conducive tasks and mediating tools.

In the context of professional work, Simons and Ruijters (2014) argue that three ways of learning can occur in professional practices: learning through inquiring, through practicing (skillful performance), and through creating. In the above discussions, we have shown how such processes are stimulated through engagement with shifting concepts, objects, and technologies in knowledge-intensive work. However, these three modes do not connect automatically. Different types of knowledge-related actions are necessary to move between, as well as reflect on, how these activity types relate. Actions such as exploration, translation, expanding, and externalizing are examples through which participants orient to and engage with knowledge. These correspond to what we have termed epistemic practices in the second example.

In order to facilitate work-based learning, we suggest that attention should be given in work organizations to the ways in which work practices call for inquiring, practicing, and creating knowledge and/or knowledgeable performances, as well as to the tools and resources that participants can access and utilize in these activities. Moreover, the division of labor and distribution of responsibilities in organizations are important in securing productive learning opportunities for participants. At the same time, learning needs to be carried out by participants who have an awareness of the learning opportunities and resources available and who deliberately engage with their own learning as part of their work performance.

Further Readings

Engeström, Y. & Sannino, A.L. (2010). Studies of expansive learning: Foundations, findings and future challenges. *Educational Research Review, 5*, 1–24.
Gives an overview of expansive learning as a theory of collective change, central concepts, and empirical findings over the last 20 years.

Little, J. W. (2012). Understanding data use practice among teachers: The contribution of micro-process studies, *American Journal of Education, 118*(2), 143–166.
Gives a theoretical account and rich illustration of how teachers work and how teachers use data in everyday activities.

Ludvigsen, S. R., & Nerland, M. (2014). Knowledge sharing in professions: Working creatively with standards in local settings. In A. Sannino & V. T. Ellis (Eds.), *Learning and collective creativity: Activity-theoretical and sociocultural studies* (pp. 116–131). New York: Routledge.
Gives a theoretical contribution of how professional develop their expertise in practices and what standards means in their work.

Markauskaite, L., & Goodyear, P. (2016). *Epistemic fluency and professional education: Innovation, knowledgeable action and actionable knowledge.* Dordrecht: Springer.
Gives an overview of recent theoretical developments in the learning sciences, especially related to expertise, professional development, and learning in higher professional education.

Simons, P. R.-J., & Ruijters, M. C. P. (2014). The real professional is a learning professional. In S. Billett et al. (Eds.), *International handbook of research in professional and practice-based learning* (pp. 955–985). Dordrecht: Springer Science & Business Media.

Gives an overview of how the term "professional" has been understood, and argues for a reconceptualization that sees professionalism as a self-chosen characteristic closely related to learning.

NAPLeS Resources

Greeno, J., & Nokes-Malach, T., *Situated cognition* [Webinar]. In *NAPLeS video series*. Retrieved October 19, 2017, from http://isls-naples.psy.lmu.de/intro/all-webinars/greeno-nokes-malach/index.html

Ludvigsen, S., *15 minutes about workplace learning with digital resources* [Video file]. In *NAPLeS video series*. Retrieved October 19, 2017, from http://isls-naples.psy.lmu.de/video-resources/guided-tour/15-minutes-ludvigsen/index.html

References

Autor, D., Levy, F., & Murnane, R. J. (2003). The skills content of recent technological change: An empirical exploration. *Journal of Economics, 118*(4), 1279–1334.

Billett, S. (2001). Knowing in practice: Re-conceptualising vocational expertise. *Learning and Instruction, 11*(6), 431–452.

Billett, S., Harteis, C., & Gruber, H. (Eds.). (2014). *International handbook of research in professional and practice-based learning*. Dordrecht: Springer Science & Business Media.

Bransford, J. D.. & Schwartz, D. L. (1999). Rethinking transfer: A simple proposal with multiple implications. *Review of Research in Education, 24*, 61–100.

Carbonell, K. B., Stalmeijer, R. E., Könings, K. D., Segers, M., & van Merriënboer, J. J. G. (2014). How experts deal with novel situations: A review of adaptive expertise. *Educational Research Review, 12*, 14–29.

Danish, J., & Gresalfi, M. (2018). Cognitive and sociocultural perspective on learning: tensions and synergy in the Learning Sciences. In F. Fischer, C. E. Hmelo-Silver, S. R. Goldman, & P. Reimann (Eds.), *International handbook of the learning sciences* (pp. 34–43). New York: Routledge.

Engeström, Y. (1987). *Learning by expanding: An activity-theoretical approach to developmental research*. Helsinki: Orienta–Konsultit.

Engeström, Y. (2001). Expansive learning at work: Toward an activity theoretical reconceptualization. *Journal of Education and Work, 14*(1), 133–156.

Engeström, Y.. & Sannino, A.L. (2010). Studies of expansive learning: Foundations, findings and future challenges. *Educational Research Review, 5*, 1–24.

Ewenstein, B., & Whyte, J. (2009). Knowledge practices in design: The role of visual representations as "epistemic objects." *Organization Studies, 30*(1), 7–30.

Gijbels, D., Raemdonck, I., Vervecken, D., & van Herck, J. (2012). Understanding work-related learning: The case of ICT workers. *Journal of Workplace Learning, 24*(6), 416–429.

Greeno, J. (2006). Authoritative, accountable positioning and connected, general knowing: Progressive themes in understanding transfer. *Journal of the Learning Sciences, 15*(4), 539–550.

Greeno, J., & Engeström, Y. (2013). Learning in activity. In R. K. Sawyer (Ed.), *The Cambridge handbook of the learning sciences* (2nd ed., pp. 128–150). Cambridge, UK: Cambridge.

Hall, R., & Horn, I. S. (2012). Talk and conceptual change at work: Adequate representation and epistemic stance in a comparative analysis of statistical consulting and teacher work groups. *Mind, Culture, and Activity, 19*(3), 240–258.

Hutchins, E. (1995). *Cognition in the wild*. Cambridge, MA: MIT Press.

Kelly, G. J. (2011). Scientific literacy, discourse, and epistemic practices. In C. Linder, L. Östman, D. A. Roberts, P. Wickman, G. Erikson, & A. McKinnon (Eds.), *Exploring the landscape of scientific literacy* (pp. 61–73). New York: Routledge.

Knorr Cetina, K. (2001). Objectual practice. In T. Schatzki, K. Knorr Cetina, & E. von Savigny (Eds.), *The practice turn in contemporary theory* (pp. 175–188). London: Routledge.

Knorr Cetina, K. (2007). Culture in global knowledge societies: Knowledge cultures and epistemic cultures. *Interdisciplinary Science Reviews, 32*(4), 361–375.

Lave, J., & Wenger, E. (1991). *Situated learning. Legitimate peripheral participation*. Cambridge, MA: Cambridge University Press.

Little, J. W. (2012). Understanding data use practice among teachers: The contribution of micro-process studies, *American Journal of Education, 118*(2), 143–166.

Ludvigsen, S. (2012). What counts as knowledge: Learning to use categories in computer environments. *Learning, Media & Technology, 37,* 40–52.

Ludvigsen, S. R., & Nerland, M. (2014). Knowledge sharing in professions: Working creatively with standards in local settings. In A. Sannino & V. T. Ellis (Eds.), *Learning and collective creativity: Activity-theoretical and sociocultural studies* (pp. 116–131). New York: Routledge.

Markauskaite, L., & Goodyear, P. (2016). *Epistemic fluency and professional education: Innovation, knowledgeable action and actionable knowledge.* Dordrecht: Springer.

Nerland, M., & Jensen, K. (2012). Epistemic practices and object relations in professional work. *Journal of Education and Work, 25*(1), 101–120.

Nicolini, D. (2013). *Practice theory, work and organization.* Oxford, UK: Oxford University Press.

Reimann, P., & Markauskaite, L. (2018). Expertise. In F. Fischer, C. E. Hmelo-Silver, S. R. Goldman, & P. Reimann (Eds.), *International handbook of the learning sciences* (pp. 54–63). New York: Routledge.

Sandoval, W. A. (2005). Understanding students' practical epistemologies and their influence on learning through inquiry. *Science Education, 89,* 634–656.

Schatzki, T., Knorr Cetina, K., & von Savigny, E. (Eds.). (2001). *The practice turn in contemporary theory.* London: Routledge.

Scribner, S. (1984). Studying working intelligence. In Rogoff and Lave (Eds.), *Everyday cognition: Development in social context* (pp. 9–40). Cambridge, MA: Harvard University Press.

Susskind, R., & Susskind, D. (2015). *The future of the professions: How technology will transform the work of human experts.* Oxford, UK: Oxford University Press.

16
Complex Systems and the Learning Sciences
Implications for Learning, Theory, and Methodologies

Susan A. Yoon

> *I think the next century will be the century of complexity.*
>
> —Stephen Hawking[1]

In the quote above, theoretical physicist Stephen Hawking was acknowledging that, although the scientific community has described basic laws governing matter under normal conditions, there is still much to learn about the potential emergent effects when systems evolve. What does this mean? The key to understanding this idea is the concept of *emergence*. *Complex systems* can be described as collective behavior that emerges from the interactions of individual units but cannot be anticipated simply by investigating the behavior of the units in isolation (Sherrington, 2010). We live in continuously evolving conditions with catastrophic events like hurricanes and drought exacerbated by human-generated activity. Scientists have focused recent efforts on setting research agendas to investigate and manage issues related to complex systems that impact our lives, such as the spread of disease, power grid robustness, and biosphere sustainability (National Academies, 2009). The goal of this work is to identify the limits, optimal states, and weaknesses within systems such that interventions could be applied to enhance stability in the face of perturbations.

Mirroring activities in the scientific community, science education researchers have recognized the importance of teaching and learning about complex systems. The recently enacted science education standards in the US—*Next Generation Science Standards* (NGSS; NGSS Lead States, 2013)—defined a central role for systems learning in the *Crosscutting Concepts*, which feature topics such as *Systems and System Models, Energy and Matter: Flows, Cycles, and Conservation*, and *Stability and Change*. In their efforts to translate these standards into classroom practice, learning scientists have studied what makes complex systems challenging to learn and have constructed frameworks, technological tools, and interventions to support learning.

In this chapter, I first outline why learning about complex systems is important to science and society. Next, I discuss what learning researchers have discovered about why complex systems are challenging to learn. The discussion then turns to theoretical frameworks that have been used to shape curriculum and instruction, highlighting similarities and differences in these approaches. This is followed by a review of technological tools and interventions that learning scientists have developed. I conclude with suggestions for future development and growth of the field of complex systems in the learning sciences.

Susan A. Yoon

Why Students Need to Learn About Complex Systems

Figuring out how complex systems exist and change is critical to addressing some of the most challenging issues we face today. The study of complex systems occurs in many domains of knowledge due to their ubiquity. They exist when multiple parts or individuals interact within bounded space (e.g., families in a house or forest ecosystems) or through common purposes (educational systems or banks). Because the parts or individuals are interconnected, the results of their interactions are often nonlinear, meaning that information is not simply transferred one part or individual to the next on a one-to-one basis but rather from one to many (Bar-Yam, 2016). This nonlinearity makes it difficult to know the exact path that information will travel, and the resulting pattern of interactions and impact is considered to be emergent. Furthermore, in both social and natural systems, there is continual development and growth, which perturbs the state of equilibrium that systems strive to reach. Through continual feedback, systems self-organize to adapt to shifts in their environment and between the parts or individuals in order to maintain stability. In their seminal work *Order out of Chaos* (1984), Prigogine and Stengers discussed remarkable processes that illustrate how instability often gives rise to spontaneous self-organization. For example, as the temperature increases in a pot of hot water, at a macro-level, we see turbulence or boiling; however, at the micro-level, millions of molecules have formed hexagonal shapes called *Benard* cells that operate in coherent motion. Similarly, in the realm of biology, Kauffman (1995, p. 112) marveled at the striking order that is revealed in natural phenomena that results from lower level interactions:

> Yet who seeing the snowflake ... who seeing the potential for the crystallization of life in swarms of reacting molecules, who seeing the stunning order for free in networks linking tens upon tens of thousands of variables, can fail to entertain a central thought: if ever we are to attain a final theory in biology we will surely, surely have to understand the commingling of self-organization and selection.

Multidisciplinary organizations such as the Santa Fe Institute and the New England Complex Systems Institute are devoted to investigating the hidden order of systems that researchers hypothesize are universal in nature across different domains (West, 2014).

What We Know About Student Understanding of Complex Systems

Complex systems have intricate dependencies, multiple causes and effects, and behaviors and structures that exist at different scales (Bar-Yam, 2016). The fact that order is hidden in complex systems is precisely why learning about such systems is challenging. Learning scientists have shown that students fail to comprehend these complexities, and instead tend to hold naive ideas or misconceptions about complex systems and the mechanisms that fuel them. Grotzer and colleagues found that students tend to reason about immediate effects rather than cascading or indirect effects. For example, students fail to realize that a change in one population can have impacts on populations that are not directly linked and this can happen through domino-like or cyclic complex causal relationships (Grotzer & Basca, 2003; Grotzer et al., 2015). Grotzer and Tutwiler (2014) outlined a number of characteristics of complex systems that may contribute to these learning challenges. For example, some phenomena, such as climate change, occur across large spatial scales that involve distance between causes and effects. This makes covariation relationships difficult to understand because a learner may attend to an effect that was caused temporally at a different time or spatially in a different place.

Chi and colleagues (Chi, 2005; Chi, Roscoe, Slotta, Roy, & Chase, 2012) have hypothesized that learning difficulties about complex causality and nonlinear dynamics may be due to how students learn to write and communicate about event sequences. They suggested that we commonly learn narratives that follow a typical linear progression—for example, the introduction of triggering events, followed by a protagonist's response, followed by a series of overt actions that are logically related with causal relations, and so on. They further suggested that everyday events follow similar scripts that are causally related in a linear sequence, such as feeling hungry, and then going to a restaurant, and then sitting down and ordering food from the menu. In both of these characterizations, initiating events and goal-directed behaviors are controlled or dictated by the agent. Although such direct causal schema are often adequate for making sense of students' lives, in most social and natural systems there is no single initiating or triggering event, behavior is not intentional with respect to the goal, and there is no central control. To illustrate this concept, Resnick (1994) and Chi et al. (2012) both used the example of a long line of ants moving between a food source and the ants' colony. The ants marching directly from their colony to the food is a macro-level pattern that emerges only after the ants roamed around randomly, accidentally bumped into the food, and laid down a pheromone for other ants to follow. Thus, the pattern of marching ants emerges from lower level actions of individual ants following simple foraging rules. Understanding how global or macro-level patterns (e.g., lines of marching ants) form from local or micro-level interactions (e.g., individual ants foraging randomly) is what Chi et al. (2012) called *emergent schema*, and this type of schema is a great deal more difficult to comprehend than direct-causal schema.

Indeed, learning scientists have documented a variety of student learning challenges that stem from a lack of understanding about emergence and other system characteristics, including causality, scale, and self-organization. Hmelo-Silver, Marathe, and Liu (2007) found that novices (middle school students and pre-service teachers) attend only to the superficial structural components of a system rather than the mechanisms that drive global system patterns that experts (biology researchers in their study) more easily recognize. Jacobson (2001) and Yoon (2008) found that students in general are predisposed to understanding systems as linear, centrally controlled, and predictable or nonrandom. Other learning scientists have shown that students have trouble accurately describing complex systems as nonlinear and cyclical (e.g., Eilam, 2012; Jordan et al., 2009), decentralized (e.g., Danish, Peppler, Phelps & Washington, 2011), and nondeterministic or probabilistic (e.g., Wilkerson-Jerde & Wilensky, 2015). In developing an assessment system to diagnose students' progression in understanding complex scientific phenomena like food webs, Gotwals and Songer (2010) found that students lacked an understanding of the components comprising an ecosystem (e.g., what algae were) and therefore couldn't reason accurately about relationships between them.

Conceptual Frameworks for Building Curriculum and Instruction to Support Learning

In addition to investigating learning challenges, researchers have developed a number of conceptual frameworks to inform interventions. I detail four frameworks that learning scientists have used to support their research and illustrate similarities and differences among them: a systems thinking approach; a structure, behavior, and function (SBF) approach; a clockwork-versus-complex systems approach; and organizational approaches.

Systems Thinking

Systems thinking emphasizes the identification of components and processes of a system, the dynamic relationships between the components, and the ability to organize the components and processes within a framework of relationships (Assaraf & Orion, 2010; Assaraf & Orpaz, 2010). These qualities of systems thinking are used as scaffolds to support student problem solving with domain-specific content. For example, Ben-Zvi Assaraf, Dodick, and Tripto (2013) studied students' understanding of the human body as a system after participating in a 10th-grade biology unit that used a systems thinking approach. They investigated increases in students' abilities to identify components of the body (e.g., ventricles, nerves, lungs) and how they are related (e.g., cells compose the organs). They also studied how well students were able to identify dynamic relations (e.g., there is digestion of proteins in the stomach), and how well students understood the hierarchical nature of systems (e.g., cells comprise all the systems). The focus of these studies was to help students recognize large-scale connections and structural components that enable the body to function. Noting that students have difficulties identifying mechanisms that drive systems, they hypothesized that emphasizing structures and processes without making explicit connections to mechanisms can prevent students from fully comprehending the focal system of study.

Structure, Behavior, Function (SBF)

Similar to a systems thinking approach, SBF is a theoretical framework used in learning sciences research on complex systems that is based in understanding the components, connections, and behaviors of systems. The SBF framework originated from the field of systems engineering and artificial intelligence (e.g., Bhatta & Goel, 1997), where SBF is used to construct models that represent functional, causal, and compositional aspects of device designs. A systems understanding follows from a hierarchical knowledge of system characteristics. The components or structures (e.g., hybrid or electric motor) and behaviors (e.g., energy consumption) must first be understood in order to work with the system to achieve the desired output or function (e.g., how far a car can travel).

Over more than a decade of research, Hmelo-Silver and colleagues have used the SBF framework to conduct studies with students and teachers to explore differences in how experts and novices understand systems as well as to develop domain-specific concepts and promote process-based reasoning. They found that experts recognize the integrated nature of SBF components, and use the latter two (i.e., behavior and function) as deep principles that organize their knowledge of the system and of the content domain in general (Hmelo-Silver & Pfeffer, 2004; Hmelo-Silver et al. 2007). Novices, on the other hand, only reason about structures and largely ignore behaviors and functions of systems. Building on research that points to misconceptions students hold when reasoning about specific science phenomena such as ecosystems, SBF researchers have attempted to overcome the tendency in instruction to focus mainly on macro-level structural components (e.g., trees, animals, oxygen). Instead, they advocate first teaching a process- or mechanism-oriented curriculum (e.g., photosynthesis, carbon cycle) (Jordan, Brooks, Hmelo-Silver, Eberbach, & Sinha, 2014). Results from this research suggest that such a conceptual shift is not easy to achieve through hand-drawn modeling tasks. However, other work using interactive computer simulations for engaging conceptual change has shown preliminary but promising outcomes for students in middle school and as young as early elementary school to develop enhanced understanding of system behaviors and functions (Danish, 2014; Vattam et al., 2011).

In both *systems thinking* and *SBF* frameworks, systems understanding is derived using an inductive process that examines how particular systems uniquely operate. These frameworks by and large do not aim to identify patterns across different systems. The next two frameworks posit generalized mechanisms and organizations that can be applied across systems in different domains.

Clockwork-Versus-Complex Systems Understanding

As this framework has been used in the learning sciences, it pertains to learners' beliefs about how the world operates. A *clockwork orientation* views the world from a Cartesian perspective (Capra, 1982), seeing the world and its constituents as machines. It is based on a method of analytic thinking that involves breaking up complex phenomena into pieces to understand the behavior of the whole from the properties of its parts. This is in contrast to a *complex systems* view, according to which the essential properties of an organism (or complex system with a constant influx of energy) are properties of the whole—properties that none of the parts have on their own. A system's central properties arise from the interactions and relationships among the parts, which is the dynamic process of *emergence* that I described at the beginning of the chapter. This framework investigates the processes that fuel emergence and change in systems from micro to macro levels. At its core, a complex systems orientation presupposes mechanisms that are believed to govern many different and seemingly unrelated domains.

Similar to the study of experts using the SBF framework, Jacobson (2001) compared expert and novice understanding of eight different complex systems mechanisms: random effects, nonlinearity, multiple causes, self-organization, decentralized control, nondeterminism, ongoing dynamics, and emergence from simple rules. He found that experts almost always reason from a complex systems perspective, while novices often reason from a clockwork perspective. Later, Yoon (2008) found that building curriculum and instruction using a framework that involves public display of students' ideas, interaction between students, and examination and selection of the best ideas can help to move learners from a clockwork to complex systems understanding. For example, students come to understand that complex issues have multiple causes and that, based on those multiple causes, the system can be influenced in a number of ways that become self-organized and decentralized.

Organizational Approaches

A fourth theoretical framing that learning scientists have embraced examines the idea of emergence but posits that intermediate levels can be formed between micro and macro levels of phenomena. Levy and Wilensky (2009) discussed several social and biological phenomena in which small groups, clusters, or packs are formed between the agent-based and aggregate levels in the process of emergence. For example, they described a situation in which a student explains how rumors spread by initially reaching three or four people in a person's immediate clique, which is a mid-level structure. Then individuals in the clique spread the rumor to three or four more people until the whole population learns about the rumor. Students who tended to reason using a mid-level, bottom-up perspective on system organization scored high on the complex systems selected mechanisms described in Jacobson (2001). The study points out that investigating middle levels, also known as meso-levels, is common in scientific investigations.

Other studies in the learning sciences have also investigated complex systems using organizational approaches that emphasize levels. Each advances the notion that when students are able to access resources either from their own everyday reasoning or through agent-based modeling tools, their ability to connect micro and macro levels improves, which in turn enables learners to demonstrate advanced understanding in a variety of content areas. The concept of *complementarity* is hypothesized to enable students to toggle between dynamic, agent-based organization and dynamic, aggregate-level organization, each of which afford different frames of reference that scaffold deeper knowledge construction in topics like how diseases spread (Stroup & Wilensky, 2014). Levy and Wilensky (2009) offered a useful conceptual framework for supporting student learning in chemistry through model exploration that identifies macroscopic and microscopic phenomena as well as multiple representations, such as symbols and mathematical equations, that must be connected for improved sense-making. In a follow-up study, Levy and Wilensky (2011) further demonstrated that requiring students to construct symbolic representations can provide important connections between students' conceptual and mathematical forms of knowledge.

Tools and Interventions from the Learning Sciences

Learning scientists have constructed tools and interventions to support complex systems learning. These learning supports reveal the hidden dynamics and interactions among system constituents and display phenomena that emerge and exist at different scales. In this section, I describe several learning supports that have been designed or used by learning scientists.

Susan A. Yoon

Concept Maps

A number of the systems thinking and SBF studies discussed in this chapter relied on concept maps to both support and evaluate student learning (e.g., Assaraf & Orpaz, 2010; Hmelo-Silver et al., 2007). Much has been written about the use of concept maps in education (e.g., Novak, 1990), and I will not repeat that material here. However, it is useful to examine how the concept map structure can support the development of complex systems understanding. Concept maps enable learners to identify elements or concepts in a bounded system (e.g., humans, fossil fuels, climate change), which are called nodes. The nodes are connected through interaction qualities or characteristics (e.g., overuse) to form propositions (e.g., humans *overuse* fossil fuels, which leads to climate change) that illustrate mechanisms and states of the system. Learners are able to investigate and make connections between concepts that they previously hadn't considered because the visual display of chains or pathways, loops, and multiple causes shows different influences within the system. As learners add more nodes and connections to the map, the visual characteristics of the concept map change with new and evolving relationships between variables, which can help to restructure the learner's understanding. This emphasis on knowledge integration has been hypothesized to encourage deeper learning (Assaraf & Orpaz, 2010). Concept maps have also been used to evaluate levels of knowledge by investigating the sophistication of the propositions (e.g., "medulla controls ventral medial aspects" versus "brain controls heart"), the number of nodes and connections present, and whether behaviors or functions are illustrated among the propositions identified (Hmelo-Silver et al., 2007).

Agent-Based Models

Agent-based computational models also provide visual displays for the learner, but they are more dynamic and interactive than concept maps. Agent-based models consist of multiple agents interacting with each other and with the environment within specific content domains. StarLogo and NetLogo are agent-based modeling platforms that have had a long history of development in the learning sciences (Colella, Klopfer, & Resnick, 2001; Klopfer, Scheintaub, Huang, Wendel, & Roque, 2009; Wilensky & Rand, 2015). With these platforms, the user can manipulate models with buttons that determine what variables are present in the system (e.g., rabbits and grass), sliders that determine the values of variables, and parameters that apply conditions to the variables (e.g., lifespan). Buttons and sliders are connected to computational procedures that define how the variables behave in the system as they interact over time. These interactions are often captured in graphical displays alongside the simulation space so that users can observe quantitative fluctuations in variables in relation to each other. A key feature of the StarLogo and NetLogo platforms is that users can go "under the hood" to investigate the code that drives system mechanisms. For example, users can observe how feedback loops, random effects, and ongoing dynamics are programmed. Users also learn that individual agents often operate under simple rules. Users can also witness the emergence of patterns that are self-organized and decentralized on the screen. These modeling tools also afford users the ability to build their own models, which provides an added layer of complex systems analysis and learning where users can program system behaviors that they can immediately visualize on the computer screen.

Other visualization tools such as system dynamics models have been explored by learning scientists (e.g., Thompson & Reimann, 2010) to investigate what students learn about systems through explicit representation of dynamical processes such as feedback loops and stock and flow diagrams that model the input/output rate of systems.

Simulation Units and Games

Learning sciences research in complex systems has also focused on embedding learner experiences within whole curricular units and games. The Web-Based Inquiry Science Environment (WISE;

Linn, Clark, & Slotta, 2003) connects simulations with inquiry-based pedagogy to develop content knowledge and scientific practices. WISE units have examined complex systems concepts that highlight the reciprocal effects between human activity and the environment. For example, Varma and Linn (2012) reported on a curriculum module that simulates the effects of human production of greenhouse gases and the effects on global climate change. They showed that by integrating scientific content and processes that are illustrated in dynamic visualization tools, students develop more sophisticated mental models of the phenomenon. Similarly, Yoon (2011) used social network graphs generated from student discussions about system complexities in a unit on genetic engineering. The graphs showed whom each student talked to over time and where those individuals stood on the issue of whether genetic engineering is good or bad for the environment. The graphs served as an important visualization tool to help students diversify whom they talked to in order to gain more information on the topic, which ultimately helped them to understand it in a more complex way.

Capitalizing on the popularity of educational games, other researchers have combined games with complex systems content to support learning. For example, Klopfer et al. (2009) discussed learning about complexity through a model called the *Simulation Cycle*. This model combines games, agent-based simulations, engineering design, and inquiry to scaffold greater conceptual sophistication. Through this model, students are able to grasp challenging scientific content such as random fluctuation and variation in emergent patterns based on initial conditions.

Future Development and Growth for Learning Sciences Research

It is evident that learning scientists have been productive in conducting research and building frameworks and resources for developing complex systems understanding in K-12 students. But there is room to grow when comparing these efforts to applications of real-world complex systems research or that meets the goals of the Next Generation Science Standards. In this section, I frame comments about next steps in learning sciences research around the studies that have been reviewed in this chapter. Because many of the authors referenced are among the most highly cited in complex systems in learning sciences research, these studies are a reasonably good representation of the field as a whole.

In real-world complex systems research, scientists focus on understanding processes of equilibrium, system stability, and robustness. These practical applications of complexity are essential to the sustainability of life on Earth and arguably should be a central focus in school science curricula as well. However, of the 27 empirical studies presented here, three discussed the concept of randomness, two discussed the concept of equilibrium, and only one discussed the concept of fluctuations. No study addressed the concept of robustness (e.g., how to fortify a system to absorb negative perturbations). Furthermore, complex systems research is done in many different knowledge domains. However, in the sample of learning sciences studies, there is an overwhelming bias toward topics in biology, ecology, and earth science (23 studies), whereas topics in chemistry and physics are represented in only four studies. This fact has some ramifications for how to teach NGSS crosscutting concepts that feature systems in all domains of science learning. For example, there may not be readily available curriculum and instructional resources for teachers to use in the less represented subjects.

Learning scientists have historically concerned themselves with situated and sociocultural characteristics that may impact learning–what works for whom and under what conditions (e.g., Penuel & Fishman, 2012). Variability in gender, ethnicity, and socioeconomic status, for example, can play a big role in whether the intervention is the best one for a particular population and whether it can be scaled up to meet the needs of a diverse number of groups. However, ethnicity is reported in only about half of the studies, with eight studies using only the words "mixed" or "a range" to describe the racial/ethnic characteristics of the group. Similarly, with respect to gender, 18 studies do not report a gender ratio, and over half of the studies (14) do not report any information about the socioeconomic status of the group or the location of the research (e.g., urban, suburban). Without this information, it is difficult to determine what works for whom and under what conditions. Research on teacher learning and professional development using complex systems curricula is also rare in learning sciences research (for further discussion on this issue, see Yoon et al., 2017).

In terms of research design, 18 out of the 27 studies are single-group non-comparative, and only nine studies tested their interventions with populations larger than 100. With respect to methodology, all studies can be categorized either as early stage or exploratory (12) or as design and development (15). No studies in this sample belong in the categories of effectiveness, replication, or efficacy (Institute of Education Sciences/National Science Foundation, 2013). Thus, we are not able to determine whether particular interventions actually produced learning gains that were significantly better than comparable interventions. This information would be helpful for teachers and schools when making decisions about what interventions would be the best ones to adopt.

Future work in complex systems research in the learning sciences should attend to these four areas. Specifically, we need research that investigates mechanisms that align with real-world scientific exploration, that explores different domains of knowledge, and that investigates situated and sociocultural characteristics. Furthermore, the design and methodology of that research needs to test interventions at larger scales with a variety of populations. The last point, which may be the most important one to support systems learning goals in educational policies, will require learning scientists to reconcile different conceptual frameworks and what are the core features to be learned about complex systems.

Further Readings

Davis, B., & Sumara, D. (2006). *Complexity and education: Inquiries into learning, teaching, and research.* Mahwah, NJ: Lawrence Erlbaum.
This book provides an excellent overview of the definition, mechanisms, and general applications of complex systems and relates these concepts to educational applications and research.

DeBoer, G. E., Quellmalz, E. S., Davenport, J. L., Timms, M. J., Herrmann-Abell, C. F., Buckley, B. C., & Flanagan, J. C. (2014). Comparing three online testing modalities: Using static, active, and interactive online testing modalities to assess middle school students' understanding of fundamental ideas and use of inquiry skills related to ecosystems. *Journal of Research in Science Teaching, 51*(4), 523–554. doi:10.1002/tea.21145
This article compares three modalities of learning about ecosystems and tests the interventions with a larger sample size of students compared to existing learning sciences research.

Repenning, A., Ioannidou, A., Luhn, L., Daetwyler, C., & Repenning, N. (2010). Mr. Vetro: Assessing a collective simulation framework. *Journal of Interactive Learning Research, 21*(4), 515–537.
This is a rare comparative study in a less researched content domain of physiology using a participatory simulations framework.

Stavrou, D., & Duit, R. (2014). Teaching and learning the interplay between chance and determinism in non-linear systems. *International Journal of Science Education, 36*(3), 506–530. doi:10.1080/09500693.2013.802056
This article aims at understanding student learning of physical systems and represents a rare study that investigates the complex systems concepts of instability and equilibrium.

Yoon, S. A., Koehler-Yom, J., Anderson, E., Lin, J., & Klopfer, E. (2015). Using an adaptive expertise lens to understand the quality of teachers' classroom implementation of computer-supported complex systems curricula in high school science. *Research in Science and Technology Education, 33*(2), 237–251. doi:10.1080/02635143.2015.1031099
To contribute to research needed on teacher learning and instruction in complex systems, this study investigates three characteristics of adaptive expertise that teachers should develop for working with complex systems curricula and technology tools in classrooms.

Note

1 Quoted in "'Unified Theory' is getting closer, Hawking predicts," *San Jose Mercury News,* January 23, 2000, p. 29A.

References

Assaraf, O. B., & Orion, N. (2010). System thinking skills at the elementary school level. *Journal of Research in Science Teaching, 47*(5), 540–563. doi:10.1002/tea.20351

Assaraf, O. B., & Orpaz, I. (2010). The "Life at the Poles" study unit: Developing junior high school students' ability to recognize the relations between earth systems. *Research in Science Education, 40*(4), 525–549. doi:10.1007/s11165-009-9132-2

Bar-Yam, Y. (2016). From big data to important information. *Complexity, 21*(S2), 73–98. doi:10.1002/cplx.21785

Ben-Zvi Assaraf, O., Dodick, J., & Tripto, J. (2013). High school students' understanding of the human body system. *Research in Science Education, 43*, 33–56. doi:10.1007/s11165-011-9245-2

Bhatta, S. R., & Goel, A. (1997). Learning generic mechanisms for innovative strategies in adaptive design. *Journal of the Learning Sciences, 6*(4), 367–396. doi:10.1207/s15327809jls0604_2

Capra, F. (1982). *The turning point*. New York: Bantam Books.

Chi, M. T. H. (2005). Commonsense conceptions of emergent processes: Why some misconceptions are robust. *Journal of the Learning Sciences, 14*(2), 161–199. doi:10.1207/s15327809jls1402_1

Chi, M. T. H., Roscoe, R., Slotta, J., Roy, M., & Chase, M. (2012). Misconceived causal explanations for "emergent" processes. *Cognitive Science, 36*, 1–61. doi:10.1111/j.1551-6709.2011.01207.x

Colella, V., Klopfer, E., & Resnick, M. (2001). *Adventures in modeling: Exploring complex, dynamic systems with StarLogo*. New York: Teachers College Press.

Danish, J. A. (2014). Applying an activity theory lens to designing instruction for learning about the structure, behavior, and function of a honeybee system. *Journal of the Learning Sciences, 23*(2), 100–148. doi:10.1080/10508406.2013.856793

Danish, J. A., Peppler, K., Phelps, D., & Washington, D. (2011). Life in the hive: Supporting inquiry into complexity within the zone of proximal development. *Journal of Science Education and Technology, 20*(5), 454–467. doi:10.1007/s10956-011-9313-4

Eilam, B. (2012). System thinking and feeding relations: Learning with a live ecosystem model. *Instructional Science, 40*(2), 213–239. doi:10.1007/s11251-011-9175-4

Gotwals, A. W., & Songer, N. B. (2010). Reasoning up and down a food chain: Using an assessment framework to investigate students' middle knowledge. *Science Education, 94*(2), 259–280. doi:10.1002/sce.20368

Grotzer, T., & Basca, B., (2003). How does grasping the underlying causal structures of ecosystems impact students' understanding? *Journal of Biological Education, 38*(1), 16–29. doi:10.1080/00219266.2003.9655891

Grotzer, T. A., Powell, M. M., Derbiszewska, K. M. Courter, C. J., Kamarainen, A. M., Metcalf, S. J., & Dede, C. J. (2015). Turning transfer inside out: The affordances of virtual worlds and mobile devices in real world contexts for teaching about causality across time and distance in ecosystems. *Technology, Knowledge, and Learning, 20*, 43–69. doi:10.1007/s10758-014-9241-5

Grotzer, T. A., & Tutwiler, M. S. (2014). Simplifying causal complexity: How interactions between modes of causal induction and information availability lead to heuristic-driven reasoning. *Mind, Brain, and Education, 8*(3), 97–114. doi:10.1111/mbe.12054

Hmelo-Silver, C., Marathe, S., & Liu, L. (2007). Fish swim, rocks sit, and lungs breathe: Expert-novice understanding of complex systems. *Journal of the Learning Sciences, 16*(3), 307–331. doi:10.1080/10508400701413401

Hmelo-Silver, C., & Pfeffer, M. G. (2004). Comparing expert and novice understanding of a complex system from the perspective of structures, behaviors, and functions. *Cognitive Science, 28*(1), 127–138. doi:10.1016/S0364-0213(03)00065-X

Institute of Education Sciences/National Science Foundation. (2013). *Common guidelines for education research and development: A report from the Institute of Education Sciences, U.S. Department of Education and the National Science Foundation*. Washington, DC: Authors.

Jacobson, M. (2001). Problem solving, cognition, and complex systems: Differences between experts and novices. *Complexity, 6*(3), 41–49. doi:10.1002/cplx.1027

Jordan, R. C., Brooks, W. R., Hmelo-Silver, C., Eberbach, C., & Sinha, S. (2014). Balancing broad ideas with context: An evaluation of student accuracy in describing ecosystem processes after a system-level intervention. *Journal of Biological Education, 48*(2), 57–62. doi:10.1080/00219266.2013.821080

Jordan, R., Gray, S., Demeter, M., Lui, L., & Hmelo-Silver, C. (2009). An assessment of students' understanding of ecosystem concepts: Conflating ecological systems and cycles. *Applied Environmental Education and Communication, 8*(1), 40–48. doi:10.1080/15330150902953472

Kauffman, S. (1995). *At home in the universe*. New York: Oxford University Press.

Klopfer, E., Scheintaub, H., Huang, W., Wendel, D., & Roque, R. (2009). The simulation cycle: Combining games, simulations, engineering and science using *StarLogo TNG*. *E-Learning and Digital Media, 6*(1), 71–96. doi:10.2304/elea.2009.6.1.71

Levy, S. T., & Wilensky, U. (2009). Students' learning with the connected chemistry (CC1) curriculum: Navigating the complexities of the particulate world. *Journal of Science Education and Technology, 18*(3), 243–254. doi:10.1007/s10956-009-9145-7

Levy, S. T., & Wilensky, U. (2011). Mining students' inquiry actions for understanding of complex systems. *Computers & Education, 56*(3), 556–573. doi:10.1016/j.compedu.2010.09.015

Linn, M. C., Clark, D., & Slotta, J. D. (2003). WISE design for knowledge integration. *Science Education, 87*, 517–538. doi:10.1002/sce.10086

National Academies, The. (2009). *Keck Futures Initiative: Complex systems: Task group summaries*. Washington, DC: The National Academies Press.

NGSS Lead States. (2013). *Next generation science standards: For states, by states*. Washington, DC: The National Academies Press.

Novak, J. D. (1990). Concept maps and Vee diagrams: Two metacognitive tools for science and mathematics education. *Instructional Science, 19*, 29–52. doi:10.1007/BF00377984.

Penuel, W., & Fishman, B. (2012). Large scale science education research we can use. *Journal of Research in Science Teaching, 49*, 281–304. doi:10.1002/tea.21001

Prigogine, I., & Stengers, I. (1984). *Order out of chaos*. New York, NY: Bantam Books.

Resnick, M. (1994). *Turtles, termites and traffic jams: Explorations in massively parallel microworlds*. Cambridge, MA: MIT Press.

Sherrington, D. (2010). Physics and complexity. *Philosophical Transactions of the Royal Society of London Series A - Mathematical Physical and Engineering Sciences, 368*(1914), 1175–1189. doi:10.1098/rsta.2009.0208

Stroup, W. M., & Wilensky, U. (2014). On the embedded complementarity of agent-based and aggregate reasoning in students' developing understanding of dynamic systems. *Technology, Knowledge and Learning: Learning Mathematics, Science and the Arts in the Context of Digital Technologies, 19*(1–2), 19–52. doi:10.1007/s10758-014-9218-4

Thompson, K., & Reimann, P. (2010). Pattern of use of an agent-based model and a system dynamics model: The application of patterns of use and the impacts on learning outcomes. *Computer & Education, 54*, 392–403. doi:10.1016/j.compedu.2009.08.020

Varma, K., & Linn, M. C. (2012). Using interactive technology to support students' understanding of the greenhouse effect and global warming. *Journal of Science Education and Technology, 21*(4), 453–464. doi:10.1007/s10956-011-9337-9

Vattam, S. S., Goel, A. K., Rugaber, S., Hmelo-Silver, C., Jordan, R., Gray, S., & Sinha, S. (2011). Understanding complex natural systems by articulating structure-behavior-function models. *Educational Technology & Society, 14*(1), 66–81.

West, G. B. (2014). A theoretical physicists journey into biology: From quarks and strings to cells and whales. *Physical Biology, 11*(5), 1–6. doi:10.1088/1478-3975/11/5/053013

Wilensky, U., & Rand, W. (2015). *An introduction to agent-based modeling: Modeling natural, social, and engineered complex systems with NetLogo*. Cambridge, MA: MIT Press.

Wilkerson-Jerde, M. H., & Wilensky, U. J. (2015). Patterns, probabilities, and people: Making sense of quantitative change in complex systems. *Journal of the Learning Sciences, 24*(2), 204–251. doi:10.1080/10508406.2014.976647

Yoon, S. A. (2008). An evolutionary approach to harnessing complex systems thinking in the science and technology classroom. *International Journal of Science Education, 30*(1), 1–32. doi:10.1080/09500690601101672

Yoon, S. A. (2011). Using social network graphs as visualization tools to influence peer selection decision-making strategies to access information about complex socioscientific issues. *Journal of the Learning Sciences, 20*(4), 549–588. doi:10.1080/10508406.2011.563655

Yoon, S., Anderson, E., Koehler-Yom, Evans, C., Park, M., J., Sheldon, J., et al.. (2017). Teaching about complex systems is no simple matter: Building effective professional development for computer-supported complex systems instruction. *Instructional Science, 45*(1), 99–121.

Section 2
Learning Environments
Designing, Researching, Evaluating

17
4C/ID in the Context of Instructional Design and the Learning Sciences

Jeroen J. G. van Merriënboer and Paul A. Kirschner

Introduction

Nowadays, the four-component instructional design (4C/ID) model receives a lot of attention (e.g., Maggio, ten Cate, Irby, & O'Brien, 2015; Postma & White, 2015) because it nicely fits current trends in education: (a) a focus on the development of complex skills or professional competencies, (b) increasing emphasis on the transfer of what is learned in school to new situations including the workplace, and (c) the development of self-directed and self-regulated learning skills and information literacy skills that are important for lifelong learning. The 4C/ID model has been extensively described in two books, *Training Complex Cognitive Skills* (van Merriënboer, 1997) and *Ten Steps to Complex Learning* (van Merriënboer & Kirschner, 2018), as well as in numerous articles and book chapters.

The aim of this chapter is to give a concise description of the 4C/ID model, its development, and its application in an international context. First, a description will be given of the background of the model and how it is positioned in the fields of instructional design and the learning sciences. Second, the model itself is described, including its four components, its "10 steps to complex learning" for designing programs based on the four components, and its applications in an international context. Third, future trends and developments are described. The chapter ends with a short discussion section focusing on underlying research.

Background of the 4C/ID Model

The 4C/ID model was originally developed in the late 1980s, in the Dutch tradition of "onderwijskunde" which is best translated as "applied educational sciences." Its first description can be found in van Merriënboer, Jelsma, and Paas (1992). The context in which the model was developed is foremost a Dutch and European one. In this context, there were and still are few pronounced differences between related fields such as instructional design, educational technology, learning sciences, educational psychology, and so forth. This is quite different from the situation in the United States, where instructional design is traditionally seen as a discipline of its own. Indeed, the differences and commonalities between instructional design and the learning sciences are an ongoing topic of debate in the United States, and some researchers and practitioners make a plea for better integration of the two (Hoadley, 2004; Reigeluth, Beatty, & Myers, 2016). In Europe, such a discussion is largely absent because instructional design is not seen as a separate discipline, but as a diffuse scientific and practical field in which researchers—and often practitioners—from many different disciplines meet each other.

This is not to say that the development of 4C/ID was unaffected by the international debate. In the late 1980s and early 1990s, there was a heated discussion on how to design better education, because students often experienced their educational program as a disconnected set of topics and courses, with implicit relationships between them and unclear relevance to their future profession (Merrill, Li, & Jones, 1990). There was a call for a paradigm shift, which took two different forms. First, fueled by the learning sciences, which were upcoming at that time (the *Journal of the Learning Sciences* was first published in 1991), there was a call for a paradigm shift from "objectivism" towards "constructivism" (e.g., Jonassen, 1991). According to this view, the dominant objectivist approach, where knowing and learning are seen as processes for representing and mirroring reality, had to be replaced by a social constructivist approach where knowing and learning are seen as processes of actively interpreting and constructing knowledge representations—often situated and in collaboration with others. Second, fueled by the field of instructional design, there was a call for a paradigm shift from "objectives-driven models" to "models for integrative goals" (e.g., Gagne & Merrill, 1990). According to this view, the traditional atomistic approach, where complex contents and tasks are reduced into simpler elements up to a level where the single elements can be specified as "objectives" and be transferred to learners through presentation and/or practice, had to be replaced by a holistic approach where complex contents and tasks are taught from simple-to-complex wholes in such a way that relationships between elements are retained.

Both the call for social constructivism from the field of the learning sciences and the call for integrative goals from the field of instructional design affected the development of 4C/ID. In line with both calls, 4C/ID stresses using meaningful, whole learning tasks as the driving force for learning. In the field of learning complex skills, 4C/ID was one of the first models to replace the prevailing part-task approach with a whole-task approach; rather than working from part-tasks to the whole task, simple-to-complex versions of the whole task are used to set up the complete educational program. The 4C/ID model shares this perspective with other whole-task models such as cognitive apprenticeship learning (Brown, Collins, & Duguid, 1989; see Eberle, this volume), goal-based scenarios (Schank, Berman, & MacPerson, 1999), and first principles of instruction (Merrill, 2002, 2012; for a critical comparison of whole-task models, see Francom & Gardner, 2014).

Given its strong focus on whole tasks as the driving force for learning, 4C/ID has both constructivist and objectivist features. According to 4C/ID, schema construction by inductive learning from concrete learning tasks, and by elaboration of new information by connecting it to knowledge already available in memory, are basic learning processes. These processes are under strategic control of the learners: They, thus, actively construct meaning and/or new cognitive schemas that allow for deep understanding and complex task performance. Yet, the 4C/ID model also has "objectivist" features. These are readily visible in the provision of how-to instructions and corrective feedback for learning routine aspects of learning tasks in a process of rule formation, and in part-task practice for routines that need to be developed to a very high level of automaticity in a process of strengthening. It is assumed that social constructivist and objectivist approaches rest on a common psychological basis and should best complement each other (see Danish & Gresalfi, this volume; Eberle, this volume). Thus, the 4C/ID model aims to combine the best of both worlds.

Description of the 4C/ID Model

4C/ID is an instructional design approach for complex learning; that is, learning aimed at integrative goals where knowledge, skills, and attitudes are developed simultaneously in order to acquire complex skills and professional competencies. It provides guidelines for the analysis of real-life tasks and the transition into a blueprint for an educational program. It is typically used for designing and

4C/ID

developing substantial educational programs ranging in length from several weeks to several years and/or that entail a substantial part of a curriculum.

The Four Components

A basic assumption of the 4C/ID model is that educational programs for complex learning can always be described in terms of four basic components—namely, (a) learning tasks, (b) supportive information, (c) procedural information, and (d) part-task practice (see Figure 17.1). Learning tasks provide the backbone of the educational program; they provide learning from varied experiences and explicitly aim at the transfer of learning. The three other components are connected to this backbone.

Component 1: Learning Tasks

Learning tasks are treated as the backbone of an educational program (see the large circles in Figure 17.1). A learning task can be a case, project, professional task, problem, or assignment that learners work on, and so forth. Learners perform these tasks in a simulated task environment and/or a real-life task environment (e.g., the workplace). Simulated task environments can vary from very low fidelity, for example, a "paper-and-pencil" case ("Suppose you are a doctor and a patient comes into your office") or a role play or project in the classroom to very high fidelity; for example, a high-fidelity flight simulator for training professional pilots or an emergency room for training trauma care teams. Learning tasks are preferably based on whole tasks that appeal to knowledge, skills, and attitudes needed to perform tasks in one's future profession or daily life. In addition, the tasks require carrying out both non-routine skills such as problem solving, reasoning, and decision making, as well as routine skills which are always performed in the same way (van Merriënboer, 2013). Learning tasks drive a basic learning process that is known as inductive learning—students learn while doing and by being confronted with concrete experiences.

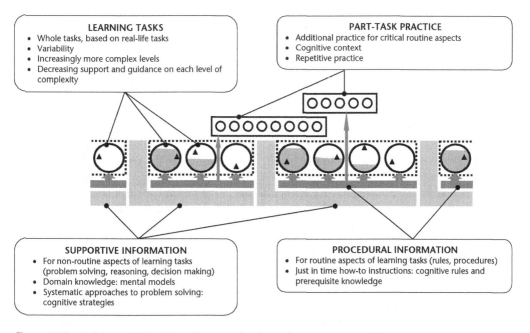

Figure 17.1 A Schematic Training Blueprint for Complex Learning and the Main Features of Each of the 4C/ID Components

Effective inductive learning is only possible when there is *variability* over learning tasks (indicated by the small triangles in the learning tasks in Figure 17.1) (Paas & van Merriënboer, 1994; Schilling, Vidal, Ployhart, & Marangoni, 2003); that is, learning tasks must be different from each other on all dimensions on which tasks in the later profession or in daily life are also different from each other. Only then will it be possible for students to construct those cognitive schemas that generalize away from the concrete experiences; such schemas are critical for reaching transfer of learning.

To prevent cognitive overload, students will typically begin working on relatively simple learning tasks and, as their expertise increases, work on more and more complex tasks (van Merriënboer & Sweller, 2010). There are, thus, *levels of complexity* with equally complex tasks (see the dotted lines encompassing a set of equally complex learning tasks in Figure 17.1). But, there must be variability of practice on each level of complexity. At the first level, students are confronted with learning tasks based on the least complex tasks a professional might encounter; at the highest level of complexity, students are confronted with the most complex tasks a beginning professional must be able to handle, and additional levels of complexity in between enable a gradual increase of complexity over levels.

Students will often receive *support and guidance* when working on the learning tasks (see the filling of the large circles in Figure 17.1). When students start to work on more complex tasks, thus progressing to a higher level of complexity, they will initially receive much support and guidance. Within each particular level of complexity, the support and guidance gradually decreases in a process known as "scaffolding"—as an analogy of a scaffold that is broken down as a building is constructed (Renkl & Atkinson, 2003; Tabak & Kyza, this volume). When students are able to independently perform the final learning tasks at a particular complexity level, thus without support or guidance (i.e., "empty" learning tasks without any filling in Figure 17.1), they are ready to progress to a next level of complexity. There, the process of scaffolding starts again, yielding a sawtooth pattern of support and guidance throughout the whole educational program.

Component 2: Supportive Information

Learning tasks typically make an appeal on both non-routine and routine skills, which are often performed simultaneously. Supportive information (indicated by the extended L-shaped forms in Figure 17.1) helps students with performing the non-routine aspects of learning tasks which require problem solving, reasoning, and/or decision making. This is what teachers often call "the theory" (i.e., the concepts and theories underlying the tasks at hand). This supportive information is typically presented in study books, lectures, and online resources. It describes how the task domain is organized and how problems in the domain can be approached in a systematic fashion.

The organization of the task domain is represented by the learner in cognitive schemas known as mental models. In the medical domain, for example, it pertains to knowledge of symptoms of particular diseases (i.e., conceptual models—what is this?), knowledge of the structure of the human body (i.e., structural models—how is this built?), and knowledge of the working of the structures or organ systems (i.e., causal models—how does this work?). The organization of one's own actions in the task domain is represented by the learner in cognitive schemas known as cognitive strategies. Such strategies identify the subsequent phases in a systematic problem-solving process (e.g., diagnostic phase–treatment phase–follow-up phase) as well as the heuristics that can be helpful for successfully completing each phase.

Supportive information provides the link between what students already know (i.e., their prior knowledge) and what they need to know to perform the non-routine aspects of learning tasks. Instructional methods for the presentation of supportive information facilitate the construction of cognitive schemas in a process of elaboration; that is, the information is presented in a way that helps learners establish meaningful relationships between newly presented information elements and their prior knowledge. This is a form of deep processing, yielding rich cognitive schemas (i.e., mental models and cognitive strategies) that enable the learner to understand new phenomena and approach unfamiliar problems (Kirschner, 2009). Providing cognitive feedback plays an important role in this process. This feedback stimulates learners to critically compare their own mental models and cognitive strategies with those of others, including experts, teachers, and peer learners.

The supportive information is identical for all learning tasks at the same level of complexity, because these tasks appeal to the same knowledge base. This is why the supportive information in Figure 17.1 is not connected to individual learning tasks but to levels of complexity; it can be presented before learners start to work on the learning tasks ("first the theory and only then start to practice") and/or it can be consulted by learners who are already working on the learning tasks ("only consult the theory when needed"). The supportive information for each next level of complexity allows students to perform more complex tasks that they could not previously complete.

Component 3: Procedural Information

Procedural information (in Figure 17.1, the beam with arrows pointing upwards to the learning tasks) helps students with performing the routine aspects of learning tasks, that is, aspects that are always performed in the same fashion. Procedural information is also called just-in-time information because it is best provided during the performance of particular learning tasks exactly when it is needed. It typically has the form of "how-to" instructions given to the learner by a teacher, quick reference guide, or computer program, telling how to perform the routine aspects of the task while doing it. The advantage of a teacher over most other media is that the teacher can act as an "assistant looking over your shoulder" and give instructions and corrective feedback at precisely the moment it is needed by the learner to correctly perform routine aspects of the task. Procedural information for a particular routine aspect is preferably presented to the learner the first time (s)he must perform this aspect as part of a whole learning task. For subsequent tasks, the presentation of procedural information is faded because the need for it diminishes as the learner slowly masters the routine.

Procedural information is always specified at a basic level that can be understood by the lowest ability learners. Instructional methods for the presentation of procedural information aim at a learning process known as rule formation: Learners use how-to instructions to form cognitive rules that couple particular—cognitive—actions to particular conditions (e.g., *If* you work on an electrical installation, *then* first switch the circuit breakers off). Rule formation is facilitated when knowledge prerequisite to the correct use of how-to instructions is presented together with those instructions (e.g., prerequisite knowledge for the presented rule is: "You can find the circuit breakers in the meter box").

Component 4: Part-Task Practice

Learning tasks appeal to both non-routine and routine aspects of a complex skill or professional competency; as a rule, they provide enough practice for learning the routine aspects. Part-task practice of routine aspects (the small circles in Figure 17.1) is only needed when a very high level of automaticity is needed, and when the learning tasks do not provide the required amount of practice. Familiar examples of part-task practice are practicing the multiplication tables of 1 to 10 in primary school (in addition to whole arithmetic tasks, such as paying in a shop or measuring the area of a floor), practicing the musical scales when playing an instrument (in addition to whole tasks, such as playing musical pieces), or practicing physical examination skills in a medical program (in addition to whole tasks, such as patient intake).

Instructional methods for part-task practice aim at strengthening cognitive rules by extensive repetitive practice. Strengthening is a basic learning process that ultimately leads to fully automated cognitive schemas (Anderson, 1993). It is important to start part-task practice in a fruitful cognitive context, that is, after learners have been confronted with the routine aspect in the context of a whole, meaningful learning task. Only then will the learners understand how practicing the routine aspects helps them improve their performance on the whole tasks. The procedural information specifying how to perform the routine aspect can be presented in the context of whole learning tasks, but in addition can be presented again during part-task practice (in Figure 17.1, see the long upward-pointing arrow from procedural information to part-task practice). Part-task practice is best mixed with work on the learning tasks (*intermix training*; Schneider, 1985), yielding a highly integrated knowledge base.

10 Steps to Complex Learning

Part of the research related to 4C/ID aims to better support designers in their application of the model. Van Merriënboer and Kirschner (2018) describe 10 steps to complex learning which specify the whole design process typically employed by a designer to produce effective, efficient, and appealing programs for complex learning (see Table 17.1 and Figure 17.2). The four blueprint components directly correspond with four design steps: the design of learning tasks (Step 1), of supportive information (Step 4), of procedural information (Step 7), and of part-task practice (Step 10).

The other six steps are auxiliary and are only performed when necessary. Step 2, where assessment instruments are developed, specifies performance objectives including standards for acceptable performance. Such standards are needed to assess student performance, monitor progress over tasks/time, and provide useful feedback on progress. Assessment results can also be used for developing adaptive instruction in which students follow individualized learning trajectories. Step 3, in which levels of complexity are defined, organizes learning tasks from simple to complex that ensure that students gradually develop complex skills or professional competencies by working on tasks that begin simply and smoothly increase in complexity. Steps 5, 6, 8, and 9, finally, pertain to in-depth cognitive task analysis (CTA; Clark, Feldon, van Merriënboer, Yates, & Early, 2008). CTA needs to be performed when to-be-presented supportive and procedural information is not yet available in existing instructional materials, job aids, manuals, quick reference guides, and so forth. It should be noted that real-life design projects are never a straightforward progression from Step 1 to Step 10. New findings and decisions will often require the designer to reconsider previous steps, causing an iterative zigzag design process.

Applications in an International Context

Practical applications of the 4C/ID model can be found around the world, and books and articles on the model have been translated in several languages, including Chinese, Dutch, German, Korean,

Table 17.1 10 steps to complex learning

Blueprint components	10 Steps to complex learning
Learning tasks	1 Design learning tasks
	2 Develop assessment instruments
	3 Sequence learning tasks
Supportive information	4 Design supportive information
	5 Analyze cognitive strategies
	6 Analyze mental models
Procedural information	7 Design procedural information
	8 Analyze cognitive rules
	9 Analyze prerequisite knowledge
Part-task practice	10 Design part-task practice

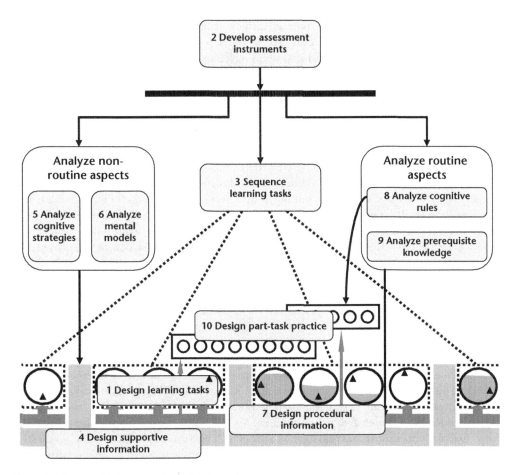

Figure 17.2 The 10 Steps to Complex Learning

Portuguese, and Spanish. The majority of its practical applications are not well described in the international literature, but instead described in local publications or not published at all. Yet, we will provide some recent examples to give an impression of its use on different continents. In the Netherlands, Belgium, and Germany, 4C/ID is probably the most popular instructional design model and used in all educational sectors, ranging from primary education to adult learning. An edited book describes a broad set of these applications (Hoogveld, Janssen, & van Merriënboer, 2011). In Ghana, 4C/ID is used to develop technical training at the junior and senior vocational level. Results indicate that a 4C/ID learning environment is more effective than the conventional approach and, moreover, that this can also be realized with minimal use of technology, which is especially important for developing countries in Africa and Asia (Sarfo & Elen, 2007). In Indonesia, 4C/ID is used for teaching communication skills in nursing programs. The 4C/ID model was successfully used to develop interprofessional training programs for improving communication between nurses, doctors, and patients, which is not evident in the culturally hierarchical context of Indonesia (Claramita & Susilo, 2014). In Brazil, 4C/ID is used to develop online educational programs for learning database management. In a learner-controlled online course, graphics in combination with spoken text proved to be superior to graphics in combination with written text for the presentation of procedural information, as predicted by 4C/ID (De Oliveira Neto, Huang, & De Azevedo Melli, 2015). In the United States, 4C/ID is used to develop educational programs in evidence-based medicine (EBM). Medical educators are suggested to adopt 4C/ID for designing, modifying, and implementing EBM training programs in classrooms and clinical settings to increase transfer of learning (Maggio et al., 2015). As a final example, in Australia, 4C/ID is used to develop guided simulations for the training of naval officers. In these programs, special attention is given to reflection tools for cognitive feedback (Harper & Wright, 2002).

Future Developments

Future developments in 4C/ID relate to acquiring self-regulated and self-directed learning skills, the nature of the learning environments in which complex learning takes place, and non-cognitive factors that affect complex learning. Adaptive education has been an important research topic for 4C/ID where, assessment results are used to develop individualized learning trajectories for learners with different needs. Nowadays, the focus is shifting from "selecting optimal tasks for individual learners" to "helping learners (learn to) select their *own* learning tasks" in a process of self-directed learning. In a process of "second order scaffolding," learners can be taught how to self-assess their performance and select suitable tasks (Kostons, van Gog, & Paas, 2012). A similar approach can be followed for the three other components of the 4C/ID model. For example, information literacy skills can be taught when learners must search for and find supportive information, procedural information, and part-task practice relevant for performing learning tasks. The key issue under study is then how to intertwine the teaching of domain-specific skills and domain-general skills, such as self-directed learning and information literacy.

A second development relates to the nature of learning environments. 4C/ID typically employs simulation-based task environments and nowadays these take the form of serious games. An example in Dutch senior vocational education is CRAFT, a game-facilitated curriculum based on 4C/ID in the field of mechatronics, a multidisciplinary field of science that includes mechanical engineering, electronics, computer engineering, telecommunications engineering, systems engineering, and control engineering (Van Bussel, Lukosch, & Meijer, 2014). CRAFT contains a simulated workplace with virtual machines that students use to build a variety of mechatronic products, and an amusement park where they build attractions from these products; these attractions can be shared with friends and family. CRAFT, however, is not just a serious game but a tool to provide a game-facilitated curriculum setting learning tasks that students perform: in the simulated workplace in the game, on real machines in the school setting, and as interns at the workplace.

A third and final trend concerns an increasing interest in non-cognitive factors, dealing with emotions, affect, and motivation. So far, the focus of research on 4C/ID has been on cognitive outcomes (i.e., performance, cognitive load, transfer), but there are strong indications that working on real-life learning tasks is often associated with emotions that affect or mediate both cognitive and non-cognitive outcomes. For example, Fraser et al. (2014) report research on the emotional and cognitive impact of unexpected patient death in simulation-based training of medical emergency skills. They found that the unexpected death of the mannequin yielded more negative emotions, higher cognitive load, and poorer learning outcomes. Obviously, these findings have direct implications for the design of learning tasks—more research is needed on how to best use emotional experiences during learning in simulated as well as real task environments.

Discussion

This chapter provided a short description of the 4C/ID model, including its background and future developments. The model has a very strong research base. There have been hundreds of experimental studies providing support for the principles prescribed by the 4C/ID model, such as for the use of different types of learning tasks (e.g., worked examples, completion tasks), variability of practice, simple-to-complex sequencing, support/guidance and their scaffolding, timing of information presentation, individualization, feedback, and so forth. Many of those distinct principles were first studied in the context of cognitive load theory (van Merriënboer & Sweller, 2010), which deals with the design of relatively short instructional events, before being included in the 4C/ID model, which aims at the design of substantial educational programs. Research into these principles has been largely experimental, including both true experiments conducted under controlled conditions and quasi-experiments conducted in externally valid settings.

Yet, 4C/ID is more than a loosely coupled set of evidence-informed principles. It is justified by the way the principles are organized in one consistent theoretical framework. Research on the value of this framework has been largely design-based and is conducted by teachers and designers in close collaboration with researchers. In this design-based research (e.g. Vandewaetere et al., 2015; see Puntambekar, this volume, for an overview of design-based research), 4C/ID is used to conceptualize new educational programs or to redesign existing programs, which are then iteratively implemented in the natural learning setting. Each iteration provides information on the value of the model and is used to further improve it. For example, the formulation of the "10 steps" as a systematic design approach has been largely based on design-based research showing, for example, that a stepwise participatory approach may help to combine the perspectives of designers, students, and teachers (Könings, Brand-Gruwel, & van Merriënboer, 2005). To conclude, over the last 25 years, both experimental and design-based research contributed to the ongoing development of the 4C/ID model from a rather rigid model focusing on the design of technical training for adults to a highly flexible model that is nowadays applied in all educational sectors.

Further Readings

Francom, G. M., & Gardner, J. (2014). What is task-centered learning? *TechTrends, 58*(5), 28–36.
In this article, the 4C/ID model is critically compared with cognitive apprenticeship learning, elaboration theory, and first principles of instruction.

Merrill, M. D. (2002). First principles of instruction. *Educational Technology Research and Development, 50*(3), 43–59.
In this article, Merrill presents his first principles of instruction, which are based on a collection of task-centered models, including 4C/ID. See also Merrill's (2012) book with the same title.

van Merriënboer, J. J. G. (1997). *Training complex cognitive skills.* Englewood Cliffs, NJ: Educational Technology Publications.
This book provides the first full description of the 4C/ID model, with a focus on its psychological basis.

van Merriënboer, J. J. G., & Kirschner, P. A. (2018). *Ten steps to complex learning* (3rd Rev. Ed.). New York: Routledge.
This book provides the most recent description of the 4C/ID model, with a focus on 10 steps for systematically designing educational programs based on the four components.

Vandewaetere, M., Manhaeve, D., Aertgeerts, B., Clarebout, G., van Merriënboer, J. J. G., & Roex, A. (2015). 4C/ID in medical education: How to design an educational program based on whole-task learning (AMEE Guide No. 93). *Medical Teacher, 37*, 4–20.
This article reports on a study in which 4C/ID was used to develop a double-blended educational program for general medical practitioners in training.

References

Anderson, J. R. (1993). Problem solving and learning. *American Psychologist, 48*(1), 35–44.
Brown, J. S., Collins, A., & Duguid, P. (1989). Situated cognition and the culture of learning. *Educational Researcher, 18*, 32–42.
Claramita, M., & Susilo, A. P. (2014). Improving communication skills in the Southeast Asian health care context. *Perspectives on Medical Education, 3*, 474–479.
Clark, R. E., Feldon, D. F., van Merriënboer, J. J. G., Yates, K. A., & Early, S. (2008). Cognitive task analysis. In J. Spector, M. Merrill, J. van Merriënboer, & M. Driscoll (Eds.), *Handbook of research on educational communications and technology* (3rd Ed., pp. 577–594). New York: Routledge.

Danish, J., & Gresalfi, M. (2018). Cognitive and sociocultural perspective on learning: Tensions and synergy in the Learning Sciences. In F. Fischer, C. E. Hmelo-Silver, S. R. Goldman, & P. Reimann (Eds.), *International handbook of the learning sciences* (pp. 34–43). New York: Routledge.

De Oliveira Neto, J. D., Huang, W. D., & De Azevedo Melli, N. C. (2015). Online learning: Audio or text? *Educational Technology Research and Development, 63*, 555–573.

Eberle, J. (2018). Apprenticeship learning. In F. Fischer, C. E. Hmelo-Silver, S. R. Goldman, & P. Reimann (Eds.), *International handbook of the learning sciences* (pp. 34–43). New York: Routledge.

Francom, G. M., & Gardner, J. (2014). What is task-centered learning? *TechTrends, 58*(5), 28–36.

Fraser, K., Huffman, J., Ma, I., Sobczak, M., McIlwrick, J., Wright, B., & McLaughlin, K. (2014). The emotional and cognitive impact of unexpected simulated patient death. *Chest, 145*, 958–963.

Gagné, R. M., & Merrill, M. D. (1990). Integrative goals for instructional design. *Educational Technology Research and Development, 38*, 23–30.

Harper, B., & Wright, R. (2002). Designing simulations for complex skill development. *Proceedings of the Educational Multimedia and Hypermedia & Telecommunications* (pp. 713–718). Waynesville, NC: AACE.

Hoadley, C. H. (2004, May–June). Learning and design: Why the learning sciences and instructional systems need each other. *Educational Technology Magazine, 44*(3), 6–12.

Hoogveld, B., Janssen, A., & van Merriënboer, J. J. G. (2011) (Eds.). *Innovatief onderwijs ontwerpen in de praktijk: Toepassing van het 4C/ID model* [Designing innovative education in practice: Applications of the 4C/ID model]. Groningen, Netherlands: Noordhoff.

Jonassen, D. H. (1991). Objectivism versus constructivism: Do we need a new philosophical paradigm? *Educational Technology Research and Development, 39*, 5–14.

Kirschner, P. A. (2009). Epistemology or pedagogy, that is the question. In S. Tobias & T. M. Duffy. (Eds.), *Constructivist instruction: Success or failure?* (pp. 144–157). New York: Routledge.

Könings, K. D., Brand-Gruwel, S., & van Merriënboer, J. J. G. (2005). Towards more powerful learning environments through combining the perspectives of designers, teachers, and students. *British Journal of Educational Psychology, 75*, 645–660.

Kostons, D., van Gog, T., & Paas, F. (2012). Training self-assessment and task-selection skills: A cognitive approach to improving self-regulated learning. *Learning and Instruction, 22*, 121–132.

Maggio, L. A., ten Cate, O., Irby, D., & O'Brien, B. (2015). Designing evidence-based medicine training to optimize the transfer of skills from the classroom to clinical practice: Applying the four-component instructional design model. *Academic Medicine, 90*, 1457–1461.

Merrill, M. D. (2002). First principles of instruction. *Educational Technology Research and Development, 50*(3), 43–59.

Merrill, M. D. (2012). *First principles of instruction*. San Francisco, CA: Pfeiffer.

Merrill, M. D., Li, Z., & Jones, M. K. (1990). Second generation instructional design. *Educational Technology, 30*(2), 7–14.

Paas, F., & van Merriënboer, J. J. G. (1994b). Variability of worked examples and transfer of geometrical problem-solving skills: A cognitive-load approach. *Journal of Educational Psychology, 86*, 122–133.

Postma, T. C., & White, J. G. (2015). Developing clinical reasoning in the classroom—Analysis of the 4C/ID-model. *European Journal of Dental Education, 19*(2), 74–80.

Puntambekar, S. (2018). Design-based research (DBR). In F. Fischer, C. E. Hmelo-Silver, S. R. Goldman, & P. Reimann (Eds.), *International handbook of the learning sciences* (pp. 383–392). New York: Routledge.

Reigeluth, C. M., Beatty, B. J., & Myers, R. D. (Eds.) (2016). *Instructional-design theories and models*, Vol. 4, *The learner-centered paradigm of education*. New York: Routledge.

Renkl, A., & Atkinson, R. K. (2003). Structuring the transition from example study to problem solving in cognitive skill acquisition: A cognitive load perspective. *Educational Psychologist, 38*(1), 15–22.

Sarfo, F. K., & Elen, J. (2007). Developing technical expertise in secondary technical schools: The effect of 4C/ID learning environments. *Learning Environments Research, 10*(3), 207–221.

Schank, R. C., Berman, T. R., & MacPerson, K. A. (1999). Learning by doing. In C. M. Reigeluth (Ed.), *Instructional design theories and models: A new paradigm of instructional theory* (Vol. 2, pp. 161–181). Mahwah, NJ: Erlbaum.

Schilling, M. A., Vidal, P., Ployhart, R. E., & Marangoni, A. (2003). Learning by doing something else: Variation, relatedness, and the learning curve. *Management Science, 49*, 39–56.

Schneider, W. (1985). Training high-performance skills: Fallacies and guidelines. *Human Factors, 27*, 285–300.

Tabak, I., & Kyza., E. (2018). Research on scaffolding in the learning sciences: A methodological perspective. In F. Fischer, C. E. Hmelo-Silver, S. R. Goldman, & P. Reimann (Eds.), *International handbook of the learning sciences* (pp. 191–200). New York: Routledge.

Van Bussel, R., Lukosch, H., & Meijer, S. A. (2014). Effects of a game-facilitated curriculum on technical knowledge and skill development. In S. A. Meijer & R. Smeds (Eds.), *Frontiers in gaming simulation* (pp. 93–101). Berlin: Springer.

van Merriënboer, J. J. G. (1997). *Training complex cognitive skills*. Englewood Cliffs, NJ: Educational Technology Publications.
van Merriënboer, J. J. G. (2013). Perspectives on problem solving and instruction. *Computers and Education, 64*, 153–160.
van Merriënboer, J. J. G., Jelsma, O., & Paas, F. (1992). Training for reflective expertise: A four-component instructional design model for complex cognitive skills. *Educational Technology Research and Development, 40*, 23–43.
van Merriënboer, J. J. G., & Kirschner, P. A. (2018). *Ten steps to complex learning* (3rd Rev. Ed.). New York: Routledge.
van Merriënboer, J. J. G., & Sweller, J. (2010). Cognitive load theory in health professional education: Design principles and strategies. *Medical Education, 44*, 85–93.
Vandewaetere, M., Manhaeve, D., Aertgeerts, B., Clarebout, G., van Merriënboer, J. J. G., & Roex, A. (2015). 4C/ID in medical education: How to design an educational program based on whole-task learning (AMEE Guide No. 93). *Medical Teacher, 37*, 4–20.

18
Classroom Orchestration

Pierre Dillenbourg, Luis P. Prieto, and Jennifer K. Olsen

The Concept of Orchestration

The metaphor of "orchestration" has been used in various educational situations (Hazel, Prosser, & Trigwell, 2002; Meyer, 1991; Watts, 2003) in which multiple "voices" (learning resources or activities, conceptions of learning, learning styles, etc.) needed to be integrated by the learner or teacher. In the Learning Sciences community, the multi-tasking nature of the teacher role has received more attention in the last decade, especially with collaborative learning technologies in genuine classrooms. To understand the relevance of this concept, we briefly review the evolution of computer-supported collaborative learning (CSCL) (Dillenbourg & Fischer, 2007).

Initially, "computer-supported" meant that digital networks enabled distant interactions. However, it quickly became clear that technology shapes interactions among learners. Therefore, collaboration technologies could be "designed" in a way to promote desirable interactions (Roschelle & Teasley, 1995; Suthers, Weiner, Connelly, & Paolucci, 1995). However, these mediation effects were sometimes too subtle, and hence stronger interventions were developed, namely, collaboration scripts. The goal of collaboration scripts is to structure the tasks or tools assigned to teams in a way that maximizes the probability that productive interactions emerge (Dillenbourg & Hong, 2008). Productive interactions are interactions that produce learning effects such as conflict solving, explanation, argumentation, or mutual regulation. Two types of scripts have been developed, both requiring teacher intervention:

- A micro-script governs the sequence of (verbal) interactions in a team. For instance, if learner A formulates a proposition to her or his peer, a micro-scripted communication tool invites learner B to refute A's proposition, then prompts learner A to reject B's refutation and so forth (Weinberger, Stegmann, Fischer, & Mandl, 2007). The aim is for learners to internalize this argumentation script and at some point be able to follow this protocol without any prompts. To favor internalization, the teacher needs to fade out the scaffolding as learners become more fluent and conversely to increase scaffolding if learners face difficulties. Hence, orchestration refers to adapting the level of scaffolding to learner needs.
- A macro-script (also called "classroom script") fosters productive interactions in an indirect way (Dillenbourg & Hong, 2008). Let us take again the example of argumentation. We can identify within a class two learners, A and B, with opposite opinions about topic X, and ask A and B to make a joint decision about X. A macro-script is a pedagogical scenario that learners

are expected to follow but not internalize. Macro-scripts require teacher intervention to run the sequence, which often includes individual activities, team activities, and class-wide activities (such as lectures). Reusing Vygostky's terms, we refer to these levels of activities as social planes (Dillenbourg, 2015) and denote them π_1 (individual), π_2 (team), and π_3 (class).

Orchestration for micro-scripts is internal to one activity (e.g., argumentation), while orchestration for macro-scripts concerns the sequence and, specifically, the transitions between activities. We define orchestration as the real-time management of multi-plane scenarios under multiple constraints. Let us develop these two elements, multiple constraints and multiple planes.

- *Multiple Constraints.* Educational design is constrained by the cognitive mechanisms of learning, the limitations of the human brain, the epistemology of each field, etc. In addition, formal education generates constraints that are extrinsic to learning, such as "finish the lesson by 10:00." A contribution of classroom orchestration research has been to emphasize the importance these extrinsic constraints have on learning, as explained later.
- *Multiple Planes.* A mono-plane scenario (e.g., a sequence of individual activities) does require some orchestration, but the difficulty of orchestration occurs especially when activity transitions include plane shifts. For instance, if output from an individual task is needed as input for a collaborative task, how does the teacher cope with the cases where some learners do not complete the individual task and the rest of the class is waiting? Similarly, what does an orchestra conductor do if one musician is slight offbeat (Kollar & Fischer, 2013)? Plane switching is a concrete problem that raises rich research questions (Study 4).

If switching social planes raises orchestration issues, why don't we design instructional scenarios that remain within a single plane? The first reason is to enrich scenarios with learning activities inspired by multiple learning theories such as mastery learning, constructivism, or socio-cultural approaches. Even if they can be considered as antagonists at the theoretical level, these theories are not mutually exclusive in practice. For instance, individual or team discovery learning activities are more effective when followed by a debriefing lecture: there is a "time for telling," as pointed out Schwartz and Bransford (1998). The lack of theoretical ecumenism in education research comes from our methods: if a treatment includes both activities inspired by behaviorist principles and activities based on constructivist principles, to which learning processes can its outcomes be attributed to? In daily practice, though, a commonsense intuition is that a diversity of learning activities is a simple way to cope with the diversity of learners.

The second motivation for including multiple planes is to engineer "pedagogical tricks." For instance, in the macro-script "ArgueGraph" (Jermann & Dillenbourg, 1999), an individual questionnaire is used for collecting the learners" opinions, which allows for making teams of learners with opposite opinions. The transfer of data from one activity to another belongs to the process of orchestration. In some cases, this workflow can be automated as in the peer grading mechanisms in massive open online courses (MOOCs). We will not further develop the notion of workflow here, since it is specific to one approach to orchestration (Dillenbourg, 2015).

Study 1: Empowering Teachers (π_2/π_3)

TinkerLamp is an augmented reality environment for vocational education for learning logistics. Teams of apprentices lay out a warehouse by placing miniature shelves on the table, whose position is detected by a camera. The warehouse mock-up is then augmented by digital information projected by a beamer to display forklifts that carry boxes from shelves to truck platforms, and vice versa. We conducted controlled studies that showed that these physical manipulations of shelves outperformed their virtual counterpart doing the same task on a multi-touch table (Schneider, Jermann, Zufferey, & Dillenbourg, 2011). The activity per se was engaging and effective, but we failed to show in real classroom environments that it outperformed a similar scenario conducted by drawing the warehouse layout on paper (Do-Lenh Jermann, Cuendet, Zufferey, & Dillenbourg 2010).

When analyzing the log files of team activities, we found that teams that learned well had a low ratio between the number of manipulations (moving quickly some shelves) and the amount of discussion of the warehouse layout. In a nutshell, lower learning teams manipulated the tangibles a lot and did not discuss their designs much. While, in many learning tasks, teachers have to encourage learners to engage more with the learning tasks, here it was the opposite: the technology inverted the teacher role to encouraging learners to disengage occasionally to reflect on results they obtained or to try to predict what they would obtain.

We, therefore, developed a simple orchestration tool—paper cards with two sides. The first card had a "block simulation" side and an "enable simulation" side. Initially, the simulation was disabled and the teams had to signal the teacher when they wanted to run the simulation. The teacher would then come to their table and ask them what they expected to see in the next simulation (i.e., whether the performance would increase or decrease). Teams would usually have difficulties expressing hypotheses and justifying their predictions. The teacher would then let them think more and would return when they had a better answer. When the teacher was satisfied by their answer, he or she would simply show the interactive lamp the other side of the card, which allowed learners to run the simulation (see Figure 18.1, left). Predicting outcomes is a great way of reflecting. These paper cards were used in addition to a dashboard (see Figure 18.1, right) where the teacher could see the number of manipulations performed by each team, depicted by a colour code ranging from yellow (few manipulations) to red (many manipulations).

Another way to encourage reflection was to ask students to compare the warehouse layouts developed by different teams. Typically, a layout with many shelves could maximize storage surfaces, but if the alleys are so narrow that forklifts could not cross each other, this would reduce the overall performance. This comparison was difficult to orchestrate, since the teacher needed to get the attention of all learners busy manipulating their shelves. Therefore, we added a radical orchestration card: when shown to any interactive lamp in the classroom, it would turn all displays to white, blocking any interaction. This would enable the teacher to obtain the attention of all learners without having to ask multiple times. In terms of social planes, this tool supports plane switching from π_2 to π_3. As with the previous card, this orchestration card had a dashboard (see Figure 18.1, right) that would facilitate the comparison.

While the use of paper cards was very natural to teachers, we were concerned that the dashboard would generate too much cognitive load for the teachers. However, a new study revealed the added

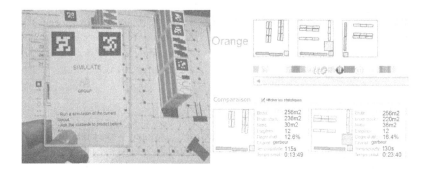

Figure 18.1 Orchestration Card (left) and Dashboard (right)

Source: Dillenbourg (2013).

value of this interactive environment as compared to paper (Do-Lenh, Jermann, Legge, Zufferey, & Dillenbourg, 2012). This convinced us that facilitating orchestration is a condition of the effectiveness of rich learning environments.

Classroom Usability (π_3)

One way to rephrase Study 1 could be to state that, while the learning environment had a high level of usability at π_1 (individual manipulations of shelves) and at π_2 (supporting team argumentation), the technology only "worked well" when we took into consideration usability at π_3 (how teacher regulated four teams of four apprentices). This third plane of usability, or classroom usability (Dillenbourg et al., 2011), enables articulating the notion of orchestration with the concept of usability in human–computer interaction (HCI). We therefore compare here our approach with several HCI methods for maximizing usability.

HCI scholars captured good design by expressing design principles (Nielsen, 1999; Norman, 1984). Two examples include consistency across interactions (e.g., the cancel button is always on the right of the OK button) and easy reversal of any user action. Similarly, design principles for orchestration have been proposed by Cuendet, Bonnard, Do-Lenh, and Dillenbourg (2013):

- Empowerment: giving the classroom control to the teacher, as illustrated by the orchestration cards.
- Awareness of the learner states: for instance, with the tangible interface for logistics, the teacher could see what is happening at every classroom table more easily than if learners were using tablets.
- Flexibility: the ability to change the pedagogical scenario on the fly.
- Minimalism: applies to both information provided to teachers and functionalities (e.g., with a dashboard that can be read at a glance).
- Integration: a workflow that carries data across planes and activities (e.g., the team warehouse designs are collected in the teacher dashboard).

Even though design principles can inspire designers, they do not constitute an algorithm that generates solutions, among other reasons because they are partially contradictory. For instance, the visibility principle recommends providing teacher with information about learner activities, but the minimalism principle recommends providing only critical information. Usability, at any plane, remains an art, a quest to find the sweet spot in design trade-offs.

Another HCI approach, contextual inquiry, broadens the scope of design; it does not only derive software functionalities from the task that users must perform but also from observing how this task is performed in an authentic context. In orchestration, a contextual approach reveals the extrinsic constraints (as defined in the introductory paragraphs above) associated with the context of formal education (Dillenbourg & Jermann, 2010). These constraints include:

- Time: learning has to occur within a time budget allocated to a specific goal within a curriculum (e.g., 5 hours for adding fractions), and this time budget is often divided into slices of 50 minutes.
- Discipline (and safety): a collaborative discovery learning scenario can hardly unfold without some level of noise and movement in the classroom, but if the class turns into chaos, there won't be any learning either (and the teacher will be blamed).
- Space: for instance, if the course is taught in a lecture theatre with a slope, it's difficult to switch from π_3 to π_2.
- Teacher energy: the teacher's energy is not unlimited; hence, scenarios that require long preparation or grading time are not sustainable.

Nussbaum and Diaz (2013) referred to these constraints as "classroom logistics," a term that distinguishes orchestration from theories of learning. Despite the rise for situated learning theories, extrinsic constraints have been underexplored in the Learning Sciences. For instance, even when empirical studies are conducted in authentic classroom contexts, it often occurs that some constraints are partly waived (e.g., the teacher may allocate a bit more time to the topic than usual, invest a bit more energy). We hypothesize that this small reduction of extrinsic constraints may account for the difficulty in obtaining the same results in large-scale studies that were collected at smaller scales.

Finally, the main HCI method for usability is usability testing (i.e., to design prototypes, starting with low-fidelity mock-ups, to test their usability and iterate the process until a satisfactory level of usability is reached). In the Learning Sciences, a similarly iterative design process is central to design-based research (DBR). The logistics environment presented in Study 1 was developed during six years of DBR, leading to the conceptual framework of orchestration. While DBR seems to be the natural approach to investigate orchestration processes, the following studies investigate classroom orchestration with various computational approaches.

Study 2: Orchestration Load ($\pi_1/\pi_2/\pi_3$)

One way HCI has operationalized the notion of usability is by measuring the user's cognitive load, which is the mental effort needed by a human to perform a certain task (Paas, Renkl, & Sweller, 2004). Well-designed, usable user interfaces should impose little or no extraneous cognitive load. Cognitive load can be measured in many ways, from subjective measures ("Was doing the task difficult?") to direct or indirect physiological measures (e.g., brain imaging, involuntary heart rate, or pupillary dilation) (Brunken, Plass, & Leutner, 2003). However, HCI has most often studied cognitive load in relatively short and simple individual tasks (π_1). When studying orchestration, teachers perform multiple tasks that stretch over several planes of interaction. Can we then operationalize a measure of orchestration load for such situations?

This is the question we pursued in a series of experiments done over two years, in which we recorded teachers while they orchestrated real (or realistic) classroom situations, using mobile eye-trackers (Figure 18.2). These devices allow freedom of movement and interaction with students while recording the teacher's field of view and eye movements (as multiple studies relate certain eye-tracking parameters with cognitive load). However, unlike other studies, our studies were not done in controlled lab conditions, but rather in noisy, messy classrooms. The potential lack of reliability could be in part ameliorated by triangulating between multiple eye-tracking metrics, but we were unsure whether such physiological data would capture anything more than noise. Hence, in a first stage, we merely wanted to see whether patterns emerged, in terms of what classroom episodes tended to be "high load" or "low load." As detailed in Prieto, Wen, Caballero, Sharma, and Dillenbourg (2014), by looking at those episodes in which all the considered measures agreed load should be high (or low), some patterns did emerge. Several such patterns were later confirmed by additional studies with different teachers and in different kinds of classrooms: the orchestration of class-wide episodes (π_3) tended to appear much more often in the "high load" category than small-group or individual ones. Also, looking at students' faces tended to appear much more often in the "high load" category, while looking at the technology display tended to appear in the "low load" category.

These empirical operationalizations of orchestration load are still in their infancy. However, initial validation of these eye-tracking methods using "reasonable assumptions" of ground truth are providing promising results (Prieto, Sharma, Kidzinski, & Dillenbourg, 2017). Using orchestration process variables like the current orchestration activity or the social plane of interaction (manually coded by researchers) along with these physiological measures, we built statistical models of how different classroom context factors (e.g., receiving help from a second teacher, using a familiar technology) affect orchestration load. Also, these methods provide additional confirmatory evidence for some of the "orchestrable technology" design principles (e.g., the high load of class-wide monitoring supports the needs outlined by the "awareness" design principle).

Figure 18.2 Teacher Wearing a Mobile Eye-Tracker in a Multi-Tabletop Classroom Situation
Note: Teacher pictured is one of the authors.

Study 3: Extracting Orchestration Graphs (π_2/π_3)

We developed a formal model of orchestration, "orchestration graphs" (Dillenbourg, 2015): a pedagogical scenario is described by a graph in which the vertices represent activities at social planes and the edges encompass the conditional and operational relationship between activities. The conditional relationship captures the probability to succeed in the second activity after succeeding on the first. The operational relationship addresses data manipulation that transforms the output of an activity into the input of the next activity. Since this theory has not yet reached the maturity required to appear in a handbook, we present instead two studies (Studies 3 and 4) that connect the concept orchestration with learning analytics.

In Study 3, multimodal analytics are used to determine at which plane the teacher is acting, by using machine learning techniques (Prieto, Sharma, Dillenbourg, & Rodríguez-Triana, 2016). In a context very similar to that of Study 2 (multi-tabletop mathematics lessons), we recorded the teacher's orchestration behaviors using multiple wearable sensors (a mobile eye-tracker, a mobile EEG device, plus accelerometer data from a mobile phone in the teacher's pocket). Would we be able to extract automatically the classroom events (i.e., the orchestration graph) of the lessons from the sensor data?

The orchestration graph planned by the teacher (Figure 18.3, top) was identical for all the analyzed sessions. The horizontal lines represent the two planes of interaction involved in this experiments (π_2 and π_3). The colors of the bars represent the different teacher activities (monitoring, distributing tasks, lecturing, etc.). The graph in the middle of Figure 18.3 shows the actual enactment of one of the lessons as captured by a human observer: note how the general structure is more or less the same, but transitions between planes and activities are more fluid, and timing is not closely followed. Naturally, the orchestration graphs of the other enacted lessons were also different from the designed graphs.

Our computational models were able to predict, from relatively simple sensor data features, what actually happened during the lesson, represented by the orchestration graph at the bottom of Figure 18.3. As we can see, the resemblance with the human-coded orchestration graph is striking, although far from perfect (accuracies were around 66% for the teacher activity, and close to 90% for the social plane of interaction). However, they were certainly closer to reality, when compared

Pierre Dillenbourg, Luis P. Prieto, and Jennifer K. Olsen

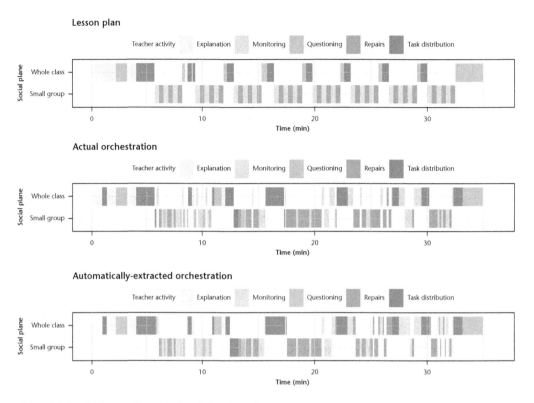

Figure 18.3 Orchestration Graphs of the Same Lesson

Note: Initial lesson plan (top); actual enacted orchestration, as coded by a human (middle); and graph extracted automatically from wearable sensor data (bottom).

with the initial lesson plan. And, more importantly, quite simple models using a limited number of features of the audio-video feeds were able to achieve comparable accuracies (i.e., the more expensive sensors had a relatively low added value). This opens the door to applications of these methods to everyday practices.

Studies 2 and 3 illustrate that research on classroom orchestration goes beyond the art of designing orchestrable learning technologies. It also includes the development of methods to model the processes of classroom orchestration.

Study 4: Plane Switching (π_1/π_2)

Studies 2 and 3 pave the road towards using computational models to support orchestration. If such computational models can be developed as research instruments, they can be applied to automate—or at least support—some aspects of orchestration.

In Study 4 (Olsen, Rummel, & Aleven, 2017) involved integrated team activities within an intelligent tutoring system (ITS) on fractions. ITSs historically have been developed primarily for individual exercises and support activity selection for students by tracking students' skills to support mastery learning (VanLehn, 2006). When combining the collaborative and individual activities within the ITS, the system needs additional support for plane switching (i.e., changing from individual to team activities, and vice versa). This suggests that the learner model (the model tracking student skills) needs to take into account this plane switching and that there is tension between mastery learning (often used within ITSs) and the orchestration of plane switching.

First, learner models need to account for the learning context that includes orchestrating plane switching. Within ITSs, student learning is often tracked in real time through Bayesian Knowledge Tracing (Corbett & Anderson, 1995) and through offline modeling, such as Additive Factor Models (Cen, Koedinger, & Junker, 2007). Olsen, Aleven, and Rummel (2015) enhanced the Additive Factors Model to account for the different individual learning rates that can occur when students are working individually compared to team activities, as well as the added learning that can occur when observing/helping a partner on a tutor step. They found that the enhanced statistical models were a better fit for the collaborative and individual data than the original models. By accounting for the plane on which the students are working, we can more accurately model and predict individual student learning. Statistical models of more accurate student learning within each plane may provide better support for orchestrating plane switching by tracing when students are taking advantage of the context of a particular social plane and being able to automate the switching process.

Second, when the use of the collaborative and individual ITS is extended from a lab setting to a real classroom, tensions emerged around the points of plane switching (Olsen et al., 2017). What if learner A has to be paired for activity 2 with learner B, who has not yet completed activity 1? There is a tension between individual activities, in which individual pace is foundational to mastery learning and to collaborative activities that require synchronicity. This makes it important to orchestrate how the timing of an activity can be changed and optimized such that a balance is found between completing a task and learning in a restricted-time environment. In our example, the system could offer learner B another task to be done while waiting for a partner, but if learner A's marginal benefit to finishing the last few problems in activity 1 is small, it may be better to move them onto activity 2. As this shows, a core property of ITSs (i.e., individual pacing) can be in conflict with key plane switching points that need to be resolved through orchestration.

Discussion

In summary, the notion of orchestration was first investigated in the context of designing technologies with high "classroom usability." In recent years, this branch of research evolved towards modelling the process of classroom orchestration with novel analytics methods. However, tackling the problems of real-time management of classroom activities and the design of technologies that have this coordination in mind, this is not the only aspect of orchestration worthy of research.

Researchers in the Learning Sciences have also studied other aspects of orchestration, from very different perspectives (Prieto, Holenko-Dlab, Adbulwahed, Gutiérrez, & Balid, 2011; Roschelle, Dimitriadis, & Hoppe, 2013). These include, for instance, the search for the most effective ways to structure the classroom activities, so as to maximize the chances of learning (e.g., the aforementioned micro- and macro-scripts, and the work towards a "theory of orchestration"; see Fischer et al., 2013). Others have looked into "logistics of the classroom" and the ways to support teachers in introducing novel learning interactions, not only by designing technologies that are tailored to the local constraints (e.g., the "silent collaboration" described in Rosen, Nussbaum, Alario-Hoyos, Readi, & Hernandez, 2014), but also looking in a more systemic way at the interplay between such new technologies, their appropriation by teachers, professional development, and even school policy (Looi, So, Toh, & Chen, 2011). There have also been pedagogy-specific approaches to orchestration in which the coordination load is reduced by subverting the usual division of labor in the classroom (e.g., in the context of inquiry-based learning—see Sharples, 2013), to name just a few.

The notion of classroom orchestration is not new, but it took a special flavor in the Learning Sciences: viewing teachers as orchestra conductors may be a misleading metaphor (all metaphors are somehow misleading), but conveys some recognition of their critical role in learning. Retrospectively, one may consider that the rise of interest for classroom orchestration is a side effect of the quest for increasing the impact on learning technologies on formal educational systems by empowering teachers (Roschelle et al., 2013). We often hear that the poor spread of learning technologies is due to either teachers' resistance to technologies or lack of proper teacher training. While the latter is certainly true, another explanation is that any rational person—teacher or not—does not like tools that make his or her job more difficult. It has often been the case that learning technologies make classroom orchestration more difficult. We don't think that any learning scientist would deny the importance of the teacher role, but this role has for the most part been neglected in learning technologies. It is common to state that the designed learning environment should be learner-centric. It would be an overstatement to claim that it should instead be teacher-centric, especially if this was understood as synonymous to "more lecturing." But, as Study 1 illustrated, learner-centric technology also has to be teacher-centric to be used effectively in formal education.

The dominant voice in the Learning Sciences promoted a shift of the teacher role, from a role of information provider to a role of facilitator. A widespread slogan, "From a sage on the stage to a guide on the side," illustrates this voice. In Study 1, the teacher was not presenting knowledge, he was not a "sage on the stage," learners were engaged in constructivist activities, but the study revealed that learning does not occur if the teacher was too much "on the side." Some confusion exists between two dimensions of the teacher role: the pedagogical style (instructionism versus socio-constructionism) and the teacher "drive." Constructivist activities also require a powerful driver, require teacher agency. Learning Sciences do not explicitly minimize the role of teachers, but this misunderstanding has been latent in the design of many learning technologies. Repairing this misunderstanding is a condition for scaling up rich learning environments.

Further Readings

Classroom orchestration [Special section]. *Computers & Education* (2013). *69*, 485–526
As an overview, we recommend this special issue of *Computers & Education*, which includes a section with 11 short papers on the same theme.

Fischer, F., Kollar, I., Stegmann, K., & Wecker, C. (2013). Toward a script theory of guidance in computer-supported collaborative learning. *Educational Psychologist, 48*(1), 56–66.
In this chapter, we refer to a slightly different understanding of orchestration developed in the team of Frank Fischer and explained in this paper.

Olsen, J. K., Rummel, N., & Aleven. V. (2017). Learning alone or together? A combination can be best! In Smith, B. K., Borge, M., Mercier, E., and Lim, K. Y. (Eds.), *Making a difference: Prioritizing equity and access in CSCL, 12th International Conference on Computer Supported Collaborative Learning (CSCL)* (Vol. 1, pp. 95–102). Philadelphia, PA: International Society of the Learning Sciences.
Summarized in Study 4, this paper describes the integration of pair activities in technologies traditionally focused on individual learning.

Prieto, L. P., Sharma, K., Kidzinski, L., & Dillenbourg, P. (2017). Orchestration load indicators and patterns: In-the-wild studies using mobile eye-tracking. *IEEE Transactions on Learning Technologies*. doi:10.1109/TLT.2017.2690687
Summarized in Study 2, this paper describes a model and method to quantify orchestration load, using physiological data and human-coded behavioral information.

NAPLeS Resources

Dillenbourg, P., *15 minutes about orchestrating CSCL* [Video file]. In *NAPLeS video series*. Retrieved October 19, 2017, from http://isls-naples.psy.lmu.de/video-resources/guided-tour/15-minutes-dillenbourg/index.html
Dillenbourg, P., *Interview about orchestrating CSCL* [Video file]. In *NAPLeS video series*. Retrieved October 19, 2017, from http://isls-naples.psy.lmu.de/video-resources/interviews-ls/dillenbourgh/index.html
Dillenbourg, P., *Orchestrating CSCL* [Webinar]. In *NAPLeS video series*. Retrieved October 19, 2017, from http://isls-naples.psy.lmu.de/intro/all-webinars/dillenbourg_all/index.html

References

Brunken, R., Plass, J. L., & Leutner, D. (2003). Direct measurement of cognitive load in multimedia learning. *Educational Psychologist, 38*(1), 53–61.

Cuendet, S., Bonnard, Q., Do-Lenh, S., & Dillenbourg, P. (2013). Designing augmented reality for the classroom. *Computers & Education, 68*, 557–569.

Cen, H., Koedinger, K. R., & Junker, B. (2007). Is over practice necessary?—Improving learning efficiency with the Cognitive Tutor through educational data mining. In R. Luckin, K. R. Koedinger, & J. Greer (Eds.), *Proceedings of 13th International Conference on Artificial Intelligence in Education* (pp. 511–518).

Corbett, A. T. & Anderson, J. R. (1995). Knowledge tracing: Modeling the acquisition of procedural knowledge. *User Modeling and User-Adapted Interaction, 4*, 253–278.

Dillenbourg, P. (2015). *Orchestration graphs: Modeling scalable education*. Lausanne, Switzerland: EPFL Press.

Dillenbourg, P., & Fischer, F. (2007). Computer-supported collaborative learning: The basics. *Zeitschrift für Berufs-und Wirtschaftspädagogik, 21*, 111–130.

Dillenbourg, P., & Hong, F. (2008). The mechanics of CSCL macro scripts. *International Journal of Computer-Supported Collaborative Learning, 3*(1), 5–23.

Dillenbourg, P., & Jermann, P. (2010). Technology for classroom orchestration. In M. S. Kline & I. M. Saleh (Eds.), *New science of learning: Cognition, computers, and collaboration in education* (pp. 525–552). New York: Springer.

Dillenbourg, P., Zufferey, G., Alavi, H., Jermann, P., Do-Lenh, S., Bonnard, Q. & Kaplan, F. (2011). Classroom orchestration: The third circle of usability. *CSCL2011 Proceedings, 1*, 510–517.

Do-Lenh, S., Jermann, P., Cuendet, S., Zufferey, G., & Dillenbourg, P. (2010). Task performance vs. learning outcomes: A study of a tangible user interface in the classroom. In *European Conference on Technology Enhanced Learning* (pp. 78–92). Berlin/Heidelberg: Springer.

Do-Lenh, S., Jermann, P., Legge, A., Zufferey, G., & Dillenbourg, P. (2012). TinkerLamp 2.0: Designing and evaluating orchestration technologies for the classroom. *European Conference on Technology Enhanced Learning* (pp. 65–78). Berlin/Heidelberg: Springer.

Fischer, F., Slotta, J., Dillenbourg, P., Tchounikine, P., Kollar, I., Wecker, C., et al. (2013). Scripting and orchestration: Recent theoretical advances. *Proceedings of the International Conference of Computer-Supported Collaborative Learning (CSCL2013)* (Vol. 1, pp. 564–571).

Jermann, P., & Dillenbourg, P. (1999, December). An analysis of learner arguments in a collective learning environment. *Proceedings of the 1999 Conference on Computer Support for Collaborative Learning* (Article 33). International Society of the Learning Sciences.

Hazel, E., Prosser, M., & Trigwell, K. (2002). Variation in learning orchestration in university biology courses. *International Journal of Science Education, 24*(7), 737–751.

Kollar, I., & Fischer, F. (2013). Orchestration is nothing without conducting—But arranging ties the two together!: A response to Dillenbourg (2011). *Computers & Education, 69*, 507–509. doi:10.1016/j.compedu.2013.04.008

Looi, C. K., So, H. J., Toh, Y., & Chen, W. (2011). The Singapore experience: Synergy of national policy, classroom practice and design research. *International Journal of Computer Supported Collaborative Learning, 6*(1), 9–37.

Meyer, J. H. F. (1991). Study orchestration: The manifestation, interpretation and consequences of contextualised approaches to studying. *Higher Education, 22*(3), 297–316.

Nielsen, J. (1999). *Designing web usability: The practice of simplicity*. Thousand Oaks, CA: New Riders Publishing.

Norman, D. A. (1984). *Cognitive engineering principles in the design of human–computer interfaces. Human computer interaction*. Amsterdam: Elsevier Science.

Nussbaum, M., & Diaz, A. (2013). Classroom logistics: Integrating digital and non-digital resources. *Computers & Education, 69*, 493–495.

Olsen, J. K., Aleven, V., & Rummel, N. (2015). Predicting student performance in a collaborative learning environment. In O. C. Santos et al. (Eds.), *Proceedings of the 8th International Conference on Educational Data Mining* (pp. 211–217). Worcester, MA: Educational Data Mining Society.

Olsen, J. K., Rummel, N., & Aleven, V. (2017). Learning alone or together? A combination can be best! In B. K. Smith, M. Borge, E. Mercier, & K. Y. Lim (Eds.), *Making a difference: Prioritizing equity and access in CSCL, 12th International Conference on Computer Supported Collaborative Learning (CSCL)* (Vol. 1, pp. 95–102). Philadelphia, PA: International Society of the Learning Sciences.

Paas, F., Renkl, A., & Sweller, J. (2004). Cognitive load theory: Instructional implications of the interaction between information structures and cognitive architecture. *Instructional Science, 32*(1), 1–8.

Prieto, L. P., Holenko-Dlab, M., Abdulwahed, M., Gutiérrez, I., & Balid, W. (2011). Orchestrating Technology Enhanced Learning: a literature review and a conceptual framework. *International Journal of Technology-Enhanced Learning (IJTEL), 3*(6), 583–598.

Prieto, L. P., Sharma, K., Dillenbourg, P., & Rodríguez-Triana, M. J. (2016). Teaching analytics: Towards automatic extraction of orchestration graphs using wearable sensors. *Proceedings of the Sixth International Conference on Learning Analytics & Knowledge* (pp. 148–157). ACM.

Prieto, L. P., Sharma, K., Kidzinski, L. & Dillenbourg, P. (2017). Orchestration load indicators and patterns: In-the-wild studies using mobile eye-tracking. *IEEE Transactions on Learning Technologies.* doi:10.1109/TLT.2017.2690687.

Prieto, L. P., Wen, Y., Caballero, D., Sharma, K., & Dillenbourg, P. (2014). Studying teacher cognitive load in multi-tabletop classrooms using mobile eye-tracking. *Proceedings of the Ninth ACM International Conference on Interactive Tabletops and Surfaces* (pp. 339–344). ACM.

Roschelle, J., Dimitriadis, Y., & Hoppe, U. (2013). Classroom orchestration: Synthesis. *Computers & Education, 69*, 523–526.

Roschelle, J., & Teasley S. D. (1995). The construction of shared knowledge in collaborative problem solving. In C. E. O'Malley (Ed.), *Computer-supported collaborative learning* (pp. 69–197). Berlin: Springer-Verlag.

Rosen, T., Nussbaum, M., Alario-Hoyos, C., Readi, F., & Hernandez, J. (2014). Silent collaboration with large groups in the classroom. *IEEE Transactions on Learning Technologies, 7*(2), 197–203.

Schneider, B., Jermann, P., Zufferey, G., & Dillenbourg, P. (2011). Benefits of a tangible interface for collaborative learning and interaction. *IEEE Transactions on Learning Technologies, 4*(3), 222–232.

Schwartz, D. L., & Bransford, J. D. (1998). A time for telling. *Cognition and Instruction, 16*(4), 475–522.

Sharples, M. (2013). Shared orchestration within and beyond the classroom. *Computers & Education, 69*, 504–506.

Suthers, D., Weiner, A., Connelly J., & Paolucci, M. (1995). Belvedere: Engaging students in critical discussion of science and public policy issues. In. J. Greer (Ed.), *Proceedings of the International Conference in Artificial Intelligence in Education* (pp. 266–273). Washington, August 16–19.

VanLehn, K. (2006). The behavior of tutoring systems. *International Journal of Artificial Intelligence in Education, 16*(3), 227–265.

Watts, M. (2003). The orchestration of learning and teaching methods in science education. *Canadian Journal of Math, Science & Technology Education, 3*(4), 451–464.

Weinberger, A., Stegmann, K., Fischer, F., & Mandl, H. (2007). Scripting argumentative knowledge construction in computer-supported learning environments. In F. Fischer, I. Kollar, H. Mandl, & J. M. Haake (eds.), *Scripting computer-supported collaborative learning* (pp. 191–211). New York: Springer.

19
Research on Scaffolding in the Learning Sciences
A Methodological Perspective

Iris Tabak and Eleni A. Kyza

Introduction

In the learning sciences, the study of learning and design for learning go hand in hand. It is not surprising, therefore, that *scaffolding* (Greenfield, 1984; Wood, Bruner, & Ross, 1976), the titrated support that enables learners to perform an action that would be outside their independent activity, and that fades as learners gain competence, is a central concept and area of research in the learning sciences (Reiser & Tabak, 2014; Sawyer, 2014). Tailoring instruction to learners' needs is inherently a design task, regardless of whether it is a parent or teacher tacitly designing their words to provide guidance and feedback, similar to recipient design (Sacks, Schegloff, & Jefferson, 1974), or whether it refers to a computer-based interface designed to decrease the constraints imposed on learners in response to their growing proficiency. In addition to designing environments that facilitate learning, learning scientists are interested in examining what forms of learning arise, and what set of supports and interactions can explain how this learning occurred (Barab, 2014; Brown, 1992; Design Based Research Collective, 2003).

In this chapter, we focus on how scientific knowledge on scaffolding is created. We open with a brief overview of the essence of scaffolding in the learning sciences, then discuss different approaches to the study of scaffolding. We conclude with open questions and suggestions for the future of learning sciences research on scaffolding.

Scaffolding in the Learning Sciences

What Is Scaffolding?

Scaffolding refers to an approach to supporting learners in gaining greater proficiency in a task. In scaffolding, learners perform a task, and the scaffolding enables them to perform those aspects of the task that they are unable to perform unaided. As learners continue to perform the task and gain competence, the assistance is gradually removed (fading), until the learner is completely autonomous. A typical example is training wheels for learning to ride a bicycle. The training wheels enable learners to ride a bicycle, even if they are unable to balance the bicycle by themselves. Over time, the children learn how to balance, and the training wheels are gradually removed (faded) until the learners ride their bicycles without training wheels.

The scaffolding approach derives from Vygotskian theory (Wertsch, 1979). The concepts of the *Zone of Proximal Development* (ZPD) and *internalizing external interactions* are particularly central in scaffolding. ZPD refers to the range of activity that, unaided, would be outside a person's reach but with support is within reach. Internalizing the external refers to the idea that development occurs through a process of transforming what occurs through interaction with others, and internalizing it so that it guides future action (Pea, 2004). This motivates the notion that learners can perform complex activities that seem outside their immediate reach, and that mediated activity can advance learners' knowledge and skills.

A seminal article by Wood et al. (1976) introduced the metaphor of scaffolding as a way to conceive of the tutoring interactions between an adult and a child, helping children to master a task that they could not achieve alone. Ethnographic studies of apprenticeship (Greenfield, 1984; Jordan, 1989) revealed similar interactions. These studies have been influential in the emphasis on joint activity and participation as a facet of scaffolding. Rogoff's (1990) research brings the Vygotskyan and apprenticeship traditions together (see Eberle, this volume).

Scaffolding supports all aspects of performance: cognitive, metacognitive, and affective. Generally, scaffolding is intended to serve six functions (Wood et al., 1976): (1) modeling idealized ways to perform the task; (2) reducing the complexity of the task so that it is within the learner's current range of ability, while (3) maintaining a good balance of risk and avoiding frustration; (4) recruiting interest in the task; and (5) sustaining motivation to continue to pursue the goal of the activity; (6) pointing the learner to key differences between the current performance and ideal performance. Modeling and monitoring, the first and last functions listed, connect between the learner's performance and idealized performance, and are key to learners performing independently in the future.

One theoretical assumption of scaffolding is that it enables learners to experience a complete task rather than isolated components of a task. It, thus, avoids problems encountered by learners who master isolated task components but then, often with little guidance, have to figure out how to integrate all of the sub-skills into a full performance. Recognizing the demands of a full task and integrating sub-skills is difficult, and often results in problems of transfer and inert knowledge (Bransford, Brown, & Cocking, 2000). In contrast, if the learning consists of repeated experience with the full task, then integrating and coordinating different task components and sub-skills is an ongoing part of the learning process, and has stronger potential to result in useable rather than inert knowledge.

Computational and Embedded Scaffolding in the Learning Sciences

The learning sciences emerged against the backdrop of the co-evolution of cognitive studies and computational technologies, and with time were also strongly influenced by situated and sociocultural approaches (Hoadley & Van Haneghan, 2011; Kolodner, 2004). This legacy helps to explain the tight coupling between scaffolding and the design of computational artifacts (Pea, 1985; Quintana et al., 2004) and software-realized scaffolding (Guzdial, 1994; Tabak & Reiser, 2008). Software menu labels and the visual representation of tasks can help learners identify when and under what circumstances particular actions make sense to perform. These environments can enable learners to work on complex tasks, such as census data analysis or scientific modeling, that are outside their independent ability, by structuring the sequence and coordination of actions, and by having some aspects of the task—for example, sophisticated computations—handled by the software. Similarly, software prompts, such as prompts that ask learners to evaluate whether they are meeting the goals of the task, can serve a function similar to the guiding questions of a human tutor.

Learning scientists design computational tools to fulfill some of the roles that a human tutor or parent might fulfill, especially where situational demands create challenges of management, that is, a single teacher cannot provide support to all students all the time and do so in a way that optimizes the support the individual learner receives. There are two reasons for this. First, an individual learner's needs change over the course of instruction and it is difficult for a single teacher to monitor these changes closely enough to match these changes precisely. Second, different learners require different levels of support due to their individual ZPDs (Herrenkohl, Palincsar, DeWater, &

Kawasaki, 1999). For a teacher charged with supporting a class of 25 or 30 individuals, it is simply unrealistic to deal with each individual all the time. However, embedding scaffolding in computational or other material tools can address both of these issues, although dynamic adaptation is a challenge designers are still tackling (Krajcik, Blumenfeld, Marx, & Soloway, 2000; Lehrer, Carpenter, Schauble, & Putz, 2000).

Conceptualizations of Scaffolding in the Learning Sciences

Learning scientists consider relationships between different forms of scaffolding, such as teacher prompts and computational prompts that co-occur in a setting to support the same set of learning goals. Such scaffolding systems are referred to as distributed scaffolding (Puntambekar & Kolodner, 2005). Distributed scaffolding can be configured in multiple ways (Tabak, 2004): *differentiated* scaffolding refers to cases where different tools and representations support different needs. *Redundant* scaffolding refers to cases where different tools and representations support the same need but at different times. *Synergistic* scaffolding refers to different tools or agents that support the same skill in different ways. Each form of support may not be sufficient, but when they operate in tandem they can help learners progress to more expert practices.

Distributed scaffolding offers a way to provide support for a range of different learners in a way that might not be possible for a single instructor with numerous learners. For example, if some learners require more support than others they can benefit from the availability of redundant scaffolds, making use of every instance of the supports. Other learners, who do not require this level of support, can ignore the redundant instances of the support. Ignoring or removing scaffolds from a distributed scaffolding system can be a form of titrating support to individual needs, and a form of fading (Chen, Chung, & Wu, 2013; Wecker, Kollar, Fischer, & Prechtl, 2010).

Another emerging approach in computational scaffolding that can provide ongoing support in one-to-many (or no instructor) settings, while also providing individualized and adaptive support, is the design of automated adaptive guidance (Gerard, Matuk, McElhaney, & Linn, 2015). According to Bell and Kozlowski, "adaptive guidance . . . is designed to augment the interpretation process with future-oriented information that will enhance self-regulation in learning. It provides information that helps trainees not only interpret the meaning of their past performance, but also determine what they should be studying and practicing to achieve mastery" (Bell & Kozlowski, 2002, p. 269).

Automated adaptive guidance may use learning analytics (Graesser, Hu, & Sottilare, this volume) to analyze the computerized actions that learners take (e.g., Roll & Winne, 2015). These analyses can be used to identify sequences of actions and patterns that can be compared to expert actions, to known associations between such patterns and learning outcomes, and to the learners' past actions. This information can further be used to infer learners' knowledge, skill, or areas of difficulty, which, in turn, can be used to offer personalized guidance.

How Knowledge on Scaffolding Is Created in the Learning Sciences

Empirical studies of scaffolding in the learning sciences tend to fall into three main categories of study: experimental studies, usually in laboratory settings; quasi-experiments, often in classroom settings; and qualitative and mixed method studies, also predominantly in classroom settings. Increasingly, there is research on scaffolding in higher education (e.g., Demetriadis, Papadopoulos, Stamelos, & Fischer, 2008; Van Zoest & Stockero, 2008), and in informal settings, such as museums (e.g., Kuhn, Cahill, Quintana, & Soloway, 2010; Yoon, Elinich, Wang, Steinmeier, & Tucker, 2012). There are more studies on learning in science, technology, engineering, and mathematics (STEM) at first occurrence than other topics, and on the use of technology and specialized materials, but this is a reflection of the field overall, and not necessarily specific to the study of scaffolding.

Experimental Studies

Laboratory-based experiments are not prevalent in the learning sciences, but are part of the tradition of the study of scaffolding, offering the opportunity to identify the relative effects of different forms of support. Wood and colleagues conducted a number of studies that isolated components of scaffolding or compared scaffolding to other approaches. For example, in one study (Wood, Wood, & Middleton, 1978), 3- to 4-year-olds' performance on a puzzle task was compared between groups that received either scaffolding, modeling, verbal instructions, or an interleaving of modeling with encouragement. Children who received scaffolding were better able to perform the task following instruction, and they were also more efficient.

Studies in the learning sciences have compared performance when a particular scaffold is present or absent, or between different configurations of components of scaffolding. In these studies, participants typically complete a pre-test, engage in learning tasks, and complete a post-test. The instruments used usually examine conceptual knowledge, or skill-related written questions, such as identifying faults in a description of an investigation as an index of inquiry skills. Additional instruments employed in these studies include, but are not limited to, self-reports of learning strategies and dispositions such as satisfaction or confidence in learning.

In one typical study of this type, Azevedo, Moos, Greene, Winters, and Cromley (2008) examined middle and high school students (n = 128) using a hypermedia tool to learn about the circulatory system. In the experimental condition, a human tutor provided learners with dynamic and adaptive regulation prompts (external regulation) according to a specified set of intervention rules, whereas those in the "control" group had no such support. The group that was aided by the tutor developed stronger models of the content, as well as exhibited a broader range of self-monitoring activities. In another study (Wu & Looi, 2012), prompts were provided by a computational agent. The participants in this study were to teach material they had just learned in the first phase of the study to a computational agent tutee. The study compared generic prompts that focus on metacognitive strategies to domain-specific prompts to no prompts. Both types of prompts supported higher learning gains than no prompts, but only the general prompts were effective in supporting transfer to another domain.

Quasi-Experiments

Quasi-experimental studies (Shadish, Cook, & Campbell, 2002) compare conditions in naturalistic settings, such as assigning different conditions to different classrooms. Even though it is not possible to isolate the effects of a component with the same precision as laboratory experiments, these studies offer a more accurate depiction of learning as it unfolds in real contexts, which is valued in the learning sciences. These studies have two main goals: elaborating the theoretical understanding of the nature of learning and scaffolding; and identifying more efficacious designs.

Quasi-experiments of scaffolding study intact classrooms over extended periods of time, and employ mixed methods (Dingiloudy & Strijbos, this volume) that include qualitative analyses of process, as described in the next section. In some studies, a design created in the research context is compared to an "off the shelf" alternative (Kyza, Constantinou, & Spanoudis, 2011; Roschelle et al., 2010), other studies compare design alternatives within (Chang & Linn, 2013) or between (Davis, 2003; Davis & Linn, 2000) research iterations. They use the same type of instruments in a pre/post format as those used in experimental studies, and analyses of variance to test differences between the compared groups. Designs of this type effectively nest individuals within classroom/teacher and analytic and statistical treatment methods need to take these design constraints into account in determining the significance of differences between groups. Sometimes non-parametric tests may be appropriate (Siegel & Castellan, 1988).

Quasi-experimental studies of scaffolding in the learning sciences have compared scaffolded to non-scaffolded environments (Demetriadis et al., 2008; Roll, Holmes, Day, & Bonn, 2012; Simons &

Klein, 2007), and have examined the relative efficacy of different forms of prompts and guiding questions. These studies sought to understand the relative merit of general versus specific prompts, and persistent versus faded prompts (Bulu & Pedersen, 2010; McNeill & Krajcik, 2009; Raes, Schellens, De Wever, & Vanderhoven, 2012), or of providing guidance versus prompting learners to take particular actions such as critiquing (Chang & Linn, 2013), or raising counterarguments (e.g., Kyza, 2009). Process support receives considerable attention (Belland, Glazewski, & Richardson, 2011; Kyza et al., 2011; Roll et al., 2012; Simons & Klein, 2007); in particular, collaboration scripts also specify roles to accomplish these task processes (Kollar, Fischer, & Slotta, 2007; Kollar et al., 2014; Kollar, Wecker, & Fischer, this volume).

Although research may focus on a particular comparison, the investigated learning environments usually include a set of scaffolds consistent with the ideas of distributed scaffolding. Some studies specifically test the differentiated or synergistic effects of the scaffolds. For example, Raes et al. (2012) compared the single contribution of teacher scaffolding or software scaffolding with the combined contribution of teacher and software scaffolding. They found that for domain learning and for metacognitive learning, the teacher and software, respectively, had differentiated effects. The combined scaffolding was just as effective for metacognitive learning as software alone. Interestingly, the combined scaffolding was most beneficial for domain learning, but only for girls. Kollar et al. (2014) examined the differentiated, crossover, and combined effects of collaboration scripts on the development of social discursive mathematical argumentation skills and of worked example problems (van Gog & Rummel, this volume) on problem-solving skills. Although the combination of these scaffolds yielded stronger results than each support in isolation, there were no significant interactions to support a synergy between them. Thus, the empirical support for synergistic scaffolding is still sparse, and it might be necessary to devise specialized methods to study it more effectively.

The prevalence of software-realized scaffolding in the learning sciences has gone a long way in providing ongoing individualized support in the one-to-many setting of classrooms, but has also been criticized for lacking the key feature of fading (Pea, 2004). The lack of fading has to do, in part, with the state-of-the-art of technology, and in part, with the relatively short duration of research in the context of ambitious learning goals (Reiser, 2004). Short duration studies may be insufficient for learners to develop targeted performances that they can sustain in faded or unaided circumstances (van de Pol, Volman, & Beishuizen, 2010). However, recently, findings from a number of studies that investigated the fading of scaffolding reported more robust learning for groups that experienced fading, as compared to persistent or no scaffolding (Bulu & Pedersen, 2010; McNeill, Lizotte, Krajcik, & Marx, 2006; Raes et al., 2012). For instance, McNeill et al. (2006) compared a group that received persistent prompts on making claims, supporting them with evidence and justifying their reasoning, to a group that received fading prompts on successive problems. The faded scaffolding group composed stronger explanations in a non-scaffolded post-test than the group that received persistent prompts.

Qualitative and Mixed Method Studies

A third form of scholarship aims to understand how scaffolding works by empirically depicting a sequence of learner interactions in the context of scaffolding (e.g., Kyza, 2009), or by illuminating how participants experience scaffolding (e.g., Kim & Hannafin, 2011). For example, such analyses might examine how learners make use of sentence-starter prompts or similar scaffolds to construct high-quality explanations. Examining such processes alongside changes in learner products (e.g., explanations or problem solutions) helps to identify whether and how learners' competencies develop (e.g., Pifarre & Cobos, 2010). This research can be situated in single or multiple case studies (Yin, 2013), and is sometimes coupled with quantified analyses of aggregate pre-/post-testing or other forms of outcome measures, or with the quantification of interactional data (Chi, 1997).

The types of data collected in qualitative research on scaffolding include: observations (open-ended or following a structured observation protocol); audio/video recordings of students' discourse and interactions; semi-structured interviews at different milestones; brief ad hoc interviews; and learner artifacts. The crux of analysis in these studies codes transcripts of interaction (e.g., Kim & Hannafin, 2011; Kyza, 2009; Pifarre & Cobos, 2010). The coded interactions are used to establish frequency counts of different types of interaction, and to identify patterns of interaction. This type of analysis can also include statistical tests of correlations and group comparisons. For further information, see Chapter 48 (this volume), and resources on the NAPLeS website (see "NAPLeS Resources" below).

In mixed methods research, qualitative observations of learners' interactions with scaffolding embodied in the designed materials, technological tools, or enacted by the teacher are related to measured outcomes from the quantitative analysis. This relationship reveals the functions of scaffolding (Goldman, Radinsky, & Rodriguez, 2007; Kim & Hannafin, 2011; Radinsky, 2008), and how the scaffolding gives rise to outcomes. The product of such research is a process-explanation of learning through scaffolding (Sherin, Reiser, & Edelson, 2004).

Conclusions and Future Directions

Research on scaffolding in the learning sciences employs a range of empirical methods, and emphasizes the design of scaffolds that are distributed and embedded in material and social means. This research has yielded design insights, such as the circumstances under which specific prompts are more advantageous to learning than generic prompts. There are many remaining open questions.

One question relates to findings from studies that compared software-realized scaffolding to teacher support. These studies found an advantage to the complementary supports of the teacher (Azevedo et al., 2008; Gerard et al., 2015; Raes et al., 2012). Do these findings reflect the need for more dynamic and adaptive software-realized scaffolding, or is material and technological scaffolding inherently limited because it lacks the affective component of human scaffolding? This motivates further research into the affective dimensions of scaffolding (Tabak & Baumgartner, 2004), as well as into dynamic adaptive guidance.

Additional questions relate to dynamic adaptive guidance and to how we can better tailor technological scaffolds to individual needs. It seems that it is easier to provide automatic adaptive guidance on well-structured rather than ill-structured tasks, though work on the latter is increasing. Some open research areas are: the degree of autonomy required to achieve self-regulation, the possible mediation of individual characteristics, who benefits more from which type of automated guidance (Bell & Kozlowski, 2002), as well as the optimal synergies of automated and traditional types of guidance (Gerard et al., 2015).

The hope is that these lines of research will be influential not only for the design of learning environments, but also for advancing our understanding of the nature of scaffolding. At present, we can provide a rich step-by-step account of micro-longitudinal changes for a small number of learners, or we can show that conditions that received scaffolding performed better beyond chance when compared to a non-scaffolded condition. However, these empirical results provide us with an aggregate account, often obscuring what has occurred at the individual level. Perhaps learning analytics techniques will enable us to garner generalizable results that provide both individual and aggregate depictions. This, in turn, will enable us to better understand minute changes in individual or group development, across a large dataset and over larger periods of time, which can advance our understanding of scaffolding and of learning.

Further Readings

Gerard, L., Matuk, C., McElhaney, K., & Linn, M. C. (2015). Automated, adaptive guidance for K-12 education. *Educational Research Review, 15*, 41–58. doi:10.1016/j.edurev.2015.04.001
This review discusses developments on automated adaptive guidance systems. The article provides an overview of comparisons between automated adaptive guidance and teacher-led instruction, as well as of studies that isolated the effects of specific adaptive guidance features.

Quintana, C., Reiser, B. J., Davis, E. A., Krajcik, J., Fretz, E., Duncan, R. G., et al. (2004). A scaffolding design framework for software to support science inquiry. *Journal of the Learning Sciences, 13*(3), 337–386.
This article provides a number of examples of how scaffolding has been conceptualized in software tools, as well as a more design-principled way of approaching this task.

Reiser, B. J., & Tabak, I. (2014). Scaffolding. In R. K. Sawyer (Ed.), *Cambridge handbook of the learning sciences* (pp. 44–62). New York: Cambridge University Press.
This chapter provides more background than the current chapter on the theoretical foundations and research traditions that inform the takeup of scaffolding in the learning sciences.

van de Pol, J., Volman, M., Oort, F., & Beishuizen, J. (2014). Teacher scaffolding in small-group work: An intervention study. *Journal of the Learning Sciences, 23*(4), 600–650. doi:10.1080/10508406.2013.805300
This article provides a good example of scaffolding through teacher moves in the classroom, which was also the target of a careful empirical study.

Wood, D., Bruner, J. S., & Ross, G. (1976). The role of tutoring in problem solving. *Journal of Child Psychology and Psychiatry, 17*, 89–100.
This is the classic reference that coined the term "scaffolding." It is often cited, but also sometimes without sufficiently attending to its actual content; therefore, it is a key reading.

NAPLeS Resources

Tabak, I., & Reiser, B. J., *Scaffolding* [Webinar]. In *NAPLeS video series*. Retrieved October 19, 2017, from http://isls-naples.psy.lmu.de/intro/all-webinars/tabak_reiser_all/index.html

Tabak, I., & Reiser, B. J., *15 minutes about scaffolding* [Video file]. In *NAPLeS video series*. Retrieved October 19, 2017, from http://isls-naples.psy.lmu.de/video-resources/guided-tour/15-minutes-tabak-reiser/index.html

Tabak, I., & Reiser, B. J., *Interview about scaffolding* [Video file]. In *NAPLeS video series*. Retrieved October 19, 2017, from http://isls-naples.psy.lmu.de/video-resources/interviews-ls/tabac-reiser/index.html

References

Azevedo, R., Moos, D. C., Greene, J. A., Winters, F. I., & Cromley, J. G. (2008). Why is externally-facilitated regulated learning more effective than self-regulated learning with hypermedia? *Educational Technology Research and Development, 56*(1), 45–72. doi:10.1007/s11423-007-9067-0

Barab, S. (2014). Design-based research: A methodological toolkit for engineering change. In R. K. Sawyer (Ed.), *The Cambridge handbook of the learning sciences* (2nd ed., pp. 151–170). New York: Cambridge University Press.

Bell, B. S., & Kozlowski, S. W. J. (2002). Adaptive guidance: Enhancing self-regulation, knowledge, and performance in technology-based training. *Personnel Psychology, 55*(2), 267–306. doi:10.1111/j.1744-6570.2002.tb00111.x

Belland, B. R., Glazewski, K. D., & Richardson, J. C. (2011). Problem-based learning and argumentation: Testing a scaffolding framework to support middle school students' creation of evidence-based arguments. *Instructional Science, 39*(5), 667–694. doi:10.1007/s11251-010-9148-z

Bransford, J. D., Brown, A., & Cocking, R. R. (Eds.). (2000). *How people learn: Brain, mind, experience and schools*. Washington DC: National Academy Press.

Brown, A. L. (1992). Design experiments: Theoretical and methodological challenges in creating complex interventions in classroom settings. *Journal of the Learning Sciences, 2*(2), 141–178.

Bulu, S. T., & Pedersen, S. (2010). Scaffolding middle school students' content knowledge and ill-structured problem solving in a problem-based hypermedia learning environment. *Educational Technology Research and Development, 58*(5), 507–529. doi:10.1007/s11423-010-9150-9

Chang, H.-Y., & Linn, M. C. (2013). Scaffolding learning from molecular visualizations. *Journal of Research in Science Teaching, 50*(7), 858–886. doi:10.1002/tea.21089

Chen, C. H., Chung, M. Y., & Wu, W. C. V. (2013). The effects of faded prompts and feedback on college students' reflective writing skills. *Asia-Pacific Education Researcher, 22*(4), 571–583. doi:10.1007/s40299-013-0059-z

Chi, M. T. H. (1997). Quantifying qualitative analyses of verbal data: A practical guide. *Journal of the Learning Sciences, 6*(3), 271–315.

Davis, E. A. (2003). Prompting middle school science students for productive reflection: Generic and directed prompts. *Journal of the Learning Sciences, 12*(1), 91–142.

Davis, E. A., & Linn, M. C. (2000). Scaffolding students' knowledge integration: Prompts for reflection in KIE. *International Journal of Science Education, 22*(8), 819–837.

Demetriadis, S. N., Papadopoulos, P. M., Stamelos, I. G., & Fischer, F. (2008). The effect of scaffolding students' context-generating cognitive activity in technology-enhanced case-based learning. *Computers & Education, 51*(2), 939–954. doi:10.1016/j.compedu.2007.09.012

Design Based Research Collective. (2003). Design-based research: An emerging paradigm for educational inquiry. *Educational Researcher, 32*(1), 5–9.

Dingyloudi, F., & Strijbos, J. W. (2018). Mixed methods research as a pragmatic toolkit: Understanding versus fixing complexity in the Learning Sciences. In F. Fischer, C. E. Hmelo-Silver, S. R. Goldman, & P. Reimann (Eds.), *International handbook of the learning sciences* (pp. 444–454). New York: Routledge.

Eberle, J. (2018). Apprenticeship learning. In F. Fischer, C. E. Hmelo-Silver, S. R. Goldman, & P. Reimann (Eds.), *International handbook of the learning sciences* (pp. 44–53). New York: Routledge.

Gerard, L., Matuk, C., McElhaney, K., & Linn, M. C. (2015). Automated, adaptive guidance for K–12 education. *Educational Research Review, 15*, 41–58. doi:10.1016/j.edurev.2015.04.001

Goldman, S., Radinsky, J., & Rodriguez, C. (2007). *Teacher interactions with small groups during investigations: Scaffolding the sense-making process and pushing students to construct arguments with data*. Paper presented at the Annual Meeting of the American Educational Research Association (AERA), Chicago, IL.

Graesser, A. C., Hu, X., & Sottilare, R. (2018). Intelligent tutoring systems. In F. Fischer, C. E. Hmelo-Silver, S. R. Goldman, & P. Reimann (Eds.), *International handbook of the learning sciences* (pp. 246–255). New York: Routledge.

Greenfield, P. M. (1984). A theory of teacher in the learning activities of everyday life. In B. Rogoff & J. Lave (Eds.), *Everyday cognition: Its development in social context* (pp. 117–138). Cambridge, MA: Harvard University Press.

Guzdial, M. (1994). Software-realized scaffolding to facilitate programming for science learning. *Interactive Learning Environments, 4*, 1–44.

Herrenkohl, L. R., Palincsar, A. S., DeWater, L. S., & Kawasaki, K. (1999). Developing scientific communities in classrooms: A sociocognitive approach. *Journal of the Learning Sciences, 8*(3–4), 451–493.

Hoadley, C., & Van Haneghan, J. (2011). The learning sciences: Where they came from and what it means for instructional designers. In R. A. Reiser & J. V. Dempsey (Eds.), *Trends and issues in instructional design and technology* (3rd ed., pp. 53–63). New York: Pearson.

Jordan, B. (1989). Cosmopolitical obstetrics: Some insights from the training of traditional midwives. *Social Science & Medicine, 28*(9), 925–937.

Kim, M., & Hannafin, M. (2011). Scaffolding 6th graders' problem solving in technology-enhanced science classrooms: A qualitative case study. *Instructional Science, 39*(3), 255–282. doi:10.1007/s11251-010-9127-4

Kollar, I., Fischer, F., & Slotta, J. D. (2007). Internal and external scripts in computer-supported collaborative inquiry learning. *Learning and Instruction, 17*(6), 708–721. doi:10.1016/j.learninstruc.2007.09.021

Kollar, I., Ufer, S., Reichersdorfer, E., Vogel, F., Fischer, F., & Reiss, K. (2014). Effects of collaboration scripts and heuristic worked examples on the acquisition of mathematical argumentation skills of teacher students with different levels of prior achievement. *Learning and Instruction, 32*, 22–36. doi:10.1016/j.learninstruc.2014.01.003

Kollar, I., Wecker, C., & Fischer, F. (2018). Scaffolding and scripting (computer-supported) collaborative learning. In F. Fischer, C. E. Hmelo-Silver, S. R. Goldman, & P. Reimann (Eds.), *International handbook of the learning sciences* (pp. 340–350). New York: Routledge.

Kolodner, J. L. (2004). The learning sciences: Past, present, future. *Educational Technology: The Magazine for Managers of Change in Education, 44*(3), 37–42.

Krajcik, J. S., Blumenfeld, P., Marx, R., & Soloway, E. (2000). Instructional, curricular, and technological supports for inquiry in science classrooms. In J. Minstrell & E. H. v. Zee (Eds.), *Inquiring into inquiry learning and teaching science* (pp. 283–315). Washington, DC: American Association for the Advancement of Science.

Kuhn, A., Cahill, C., Quintana, C., & Soloway, E. (2010). Scaffolding science inquiry in museums with Zydeco. *CHI 2010—The 28th Annual CHI Conference on Human Factors in Computing Systems, Conference Proceedings and Extended Abstracts, 3373–3378.*

Kyza, E. A. (2009). Middle-school students' reasoning about alternative hypotheses in a scaffolded, software-based inquiry investigation. *Cognition and Instruction, 27*(4), 277–311. doi:10.1080/07370000903221718

Kyza, E. A., Constantinou, C. P., & Spanoudis, G. (2011). Sixth graders' co-construction of explanations of a disturbance in an ecosystem: Exploring relationships between grouping, reflective scaffolding, and evidence-based explanations. *International Journal of Science Education, 33*(18), 2489–2525.

Lehrer, R., Carpenter, S., Schauble, L., & Putz, A. (2000). Designing classrooms that support inquiry. In J. Minstrell & E. H. v. Zee (Eds.), *Inquiring into inquiry learning and teaching science* (pp. 80–99). Washington, DC: American Association for the Advancement of Science.

McNeill, K. L., & Krajcik, J. (2009). Synergy between teacher practices and curricular scaffolds to support students in using domain-specific and domain-general knowledge in writing arguments to explain phenomena. *Journal of the Learning Sciences, 18*(3), 416–460. doi:10.1080/10508400903013488

McNeill, K. L., Lizotte, D. J., Krajcik, J., & Marx, R. W. (2006). Supporting students' construction of scientific explanations by fading scaffolds in instructional materials. *Journal of the Learning Sciences*, *15*(2), 153–191.

Pea, R. D. (1985). Beyond amplification: Using the computer to reorganize mental functioning. *Educational Psychologist*, *20*(4), 167.

Pea, R. D. (2004). The social and technological dimensions of scaffolding and related theoretical concepts for learning, education, and human activity. *Journal of the Learning Sciences*, *13*(3), 423–451. doi:10.1207/s15327809jls1303_6

Pifarre, M., & Cobos, R. (2010). Promoting metacognitive skills through peer scaffolding in a CSCL environment. *International Journal of Computer-Supported Collaborative Learning*, *5*(2), 237–253. doi:10.1007/s11412-010-9084-6

Puntambekar, S., & Kolodner, J. L. (2005). Toward implementing distributed scaffolding: Helping students learn science from design. *Journal of Research in Science Teaching*, *42*(2), 185–217. doi:10.1002/tea.20048

Quintana, C., Reiser, B. J., Davis, E. A., Krajcik, J., Fretz, E., Duncan, R. G., et al. (2004). A scaffolding design framework for software to support science inquiry. *Journal of the Learning Sciences*, *13*(3), 337–386.

Radinsky, J. (2008). Students' roles in group-work with visual data: A site of science learning. *Cognition and Instruction*, *26*(2), 145–194. doi:10.1080/07370000801980779

Raes, A., Schellens, T., De Wever, B., & Vanderhoven, E. (2012). Scaffolding information problem solving in web-based collaborative inquiry learning. *Computers & Education*, *59*(1), 82–94.

Reiser, B. J. (2004). Scaffolding complex learning: The mechanisms of structuring and problematizing student work. *Journal of the Learning Sciences*, *13*(3), 273–304.

Reiser, B. J., & Tabak, I. (2014). Scaffolding. In R. K. Sawyer (Ed.), *Cambridge handbook of the learning sciences* (pp. 44–62). New York: Cambridge University Press.

Rogoff, B. (1990). *Apprenticeship in thinking: Cognitive development in social context*. New York: Oxford University Press.

Roll, I., Holmes, N. G., Day, J., & Bonn, D. (2012). Evaluating metacognitive scaffolding in guided invention activities. *Instructional Science*, *40*(4), 691–710.

Roll, I., & Winne, P. H. (2015). Understanding, evaluating, and supporting self-regulated learning using learning analytics. *2015*, *2*(1), 6. doi:10.18608/jla.2015.21.2

Roschelle, J., Rafanan, K., Bhanot, R., Estrella, G., Penuel, B., Nussbaum, M., & Claro, S. (2010). Scaffolding group explanation and feedback with handheld technology: Impact on students' mathematics learning. *Educational Technology Research and Development*, *58*(4), 399–419. doi:10.1007/s11423-009-9142-9

Sacks, H., Schegloff, E. A., & Jefferson, G. (1974). A simplest systematics for the organization of turn-taking for conversation. *Language*, *50*(4), 696–735. doi:10.2307/412243

Sawyer, R. K. (2014). Introduction: The new science of learning. In R. K. Sawyer (Ed.), *The Cambridge handbook of the learning sciences* (2nd ed., pp. 1–20). New York: Cambridge University Press.

Shadish, W. R., Cook, T. D., & Campbell, D. T. (2002). *Experimental and quasi-experimental designs for generalized causal inference*. Boston, MA: Houghton, Mifflin.

Sherin, B., Reiser, B. J., & Edelson, D. (2004). Scaffolding analysis: Extending the scaffolding metaphor to learning artifacts. *Journal of the Learning Sciences*, *13*(3), 387–421. doi:10.1207/s15327809jls1303_5

Siegel, S., & Castellan, N. (1988). *Nonparametric statistics for the behavioral sciences* (2nd ed.). New York: McGraw-Hill.

Simons, K. D., & Klein, J. D. (2007). The impact of scaffolding and student achievement levels in a problem-based learning environment. *Instructional Science*, *35*(1), 41–72. doi:10.1007/s11251-006-9002-5

Tabak, I. (2004). Synergy: A complement to emerging patterns of distributed scaffolding. *Journal of the Learning Sciences*, *13*(3), 305–335.

Tabak, I., & Baumgartner, E. (2004). The teacher as partner: Exploring participant structures, symmetry and identity work in scaffolding. *Cognition and Instruction*, *22*(4), 393–429.

Tabak, I., & Reiser, B. (2008). Software-realized inquiry support for cultivating a disciplinary stance. *Pragmatics & Cognition*, *16*(2), 307–355.

van de Pol, J., Volman, M., & Beishuizen, J. (2010). Scaffolding in teacher–student interaction: A decade of research. *Educational Psychology Review*, *22*(3), 271–296. doi:10.1007/s10648-010-9127-6

van Gog, T., & Rummel, N. (2018). Example-based learning. In F. Fischer, C. E. Hmelo-Silver, S. R. Goldman, & P. Reimann (Eds.), *International handbook of the learning sciences* (pp. 201–209). New York: Routledge.

Van Zoest, L. R., & Stockero, S. L. (2008). Synergistic scaffolds as a means to support preservice teacher learning. *Teaching and Teacher Education*, *24*(8), 2038–2048. doi:http://dx.doi.org/10.1016/j.tate.2008.04.006

Vogel, F., & Weinberger, A. (2018). Quantifying qualities of collaborative learning processes. In F. Fischer, C. E. Hmelo-Silver, S. R. Goldman, & P. Reimann (Eds.), *International handbook of the learning sciences* (pp. 500–510). New York: Routledge.

Wecker, C., Kollar, I., Fischer, F., & Prechtl, H. (2010). *Fostering online search competence and domain-specific knowledge in inquiry classrooms: Effects of continuous and fading collaboration scripts*. Paper presented at the 9th International Conference of the Learning Sciences.

Wertsch, J. V. (1979). From social interaction to higher psychological processes: A clarification and application of Vygotsky's theory. *Human Development, 22*, 1–22.

Wood, D., Bruner, J. S., & Ross, G. (1976). The role of tutoring in problem solving. *Journal of Child Psychology and Psychiatry, 17*, 89–100.

Wood, D., Wood, H., & Middleton, D. (1978). An experimental evaluation of four face-to-face teaching strategies. *International Journal of Behavioral Development, 1*(2), 131–147.

Wu, L., & Looi, C.-K. (2012). Agent prompts: Scaffolding for productive reflection in an intelligent learning environment. *Educational Technology & Society, 15*(1), 339–353.

Yin, R. K. (2013). *Case study research: Design and methods*. Thousand Oaks, CA: Sage.

Yoon, S. A., Elinich, K., Wang, J., Steinmeier, C., & Tucker, S. (2012). Using augmented reality and knowledge-building scaffolds to improve learning in a science museum. *International Journal of Computer-Supported Collaborative Learning, 7*(4), 519–541. doi:10.1007/s11412-012-9156-x

20
Example-Based Learning

Tamara van Gog and Nikol Rummel

Example-based learning is a form of observational learning and can be defined as learning by following a demonstration of how to perform a to-be-learned task or skill. Observational learning is a very natural way of learning that even young infants display (Bandura, 1977, 1986). It can be contrasted with learning by doing, defined as attempting to perform the task or skill oneself without assistance or instruction. In contemporary learning environments example-based learning is often facilitated by technology; for instance, video examples are widely implemented in e-learning environments and (adaptive) instructional systems. Research on example-based learning is conducted by researchers reflecting the variety of disciplines that constitute the learning sciences, including cognitive science, educational psychology, instructional design, and computer science. Research on example-based learning is typically (though not necessarily) experimental in nature.

What Is Example-Based Learning?

By limiting itself to acquiring to-be-learned tasks or skills through observation, the term "example-based learning" is more specific than observational learning, which can apply to the acquisition of all kinds of skills, attitudes, or behaviors (including negative ones) from observing others (Bandura, 1986). Furthermore, although both involve observational learning, example-based learning is different from vicarious learning, which refers to learning by observing someone else being taught (e.g., observing a fellow student while that person is interacting with a teacher or tutor in front of the classroom or on a video; e.g., Chi, Roy, & Hausmann, 2008). Thus, in contrast to example-based learning, where the observer is given a demonstration of the to-be-learned task or skill, the observer is not directly addressed in vicarious learning.

The demonstrations of how to perform a to-be-learned task or skill in examples can take different forms that originate from different research traditions (Renkl, 2014; van Gog & Rummel, 2010). *Worked examples*, which originate from cognitive research, consist of a fully written-out account of how to perform the task, and typically provide students with a didactical solution procedure (e.g., Sweller & Cooper, 1985). *Modeling examples*, which originate from social-cognitive research, comprise a live or video demonstration by a human model (i.e., an expert, teacher, tutor, or peer student) and may also show natural behavior, for instance, a peer student struggling with a task (e.g., Schunk, Hanson, & Cox, 1987). However, most contemporary video modeling examples that are used in online learning environments to support teaching or homework (e.g., www.khanacademy.org) are also didactical examples. In example-based learning, worked examples and (video) modeling examples are typically—though not necessarily—alternated with practicing the task or skill oneself (van Gog, Kester, & Paas, 2011). Regardless of whether worked examples or modeling examples are used, example-based learning has proven to be more effective and efficient for novice learners than learning by doing (Renkl, 2014; Sweller, Ayres, & Kalyuga, 2011; van Gog & Rummel, 2010).

Why Is Example-Based Learning Effective?

As mentioned earlier, research on worked examples and modeling examples has developed in parallel in two different research traditions (van Gog & Rummel, 2010). Therefore, theoretical accounts of the cognitive mechanisms underlying the effectiveness of example-based learning relate to different bodies of literature; nevertheless, the principles that can be deduced from these different bodies of work overlap to a large extent (see also Renkl, 2014).

Research on *worked examples*, inspired by cognitive theories like Cognitive Load Theory (CLT; Sweller et al., 2011) and ACT-R (*a*daptive *c*ontrol of *t*hought—*r*ational; Anderson, 1993) has experimentally established the effectiveness of learning from worked examples, mainly in comparison to a learning by doing (i.e., unassisted practice problems) control condition. Research inspired by CLT has shown that replacing a substantial part of conventional practice problems with worked examples is more effective and/or efficient for novices' learning. For instance, many experimental studies conducted in lab, school, or professional training settings, have demonstrated that for novices, studying examples only or example–problem pairs leads to *higher* learning outcomes than problem-solving practice (i.e., higher performance on a posttest), and this is often attained with *less* investment of study time or mental effort during the learning phase. This has become known as "the worked example effect" (Sweller et al., 2011).

The explanation for the greater effectiveness and efficiency of example study or example study alternated with practice problem solving as compared to unassisted practice lies in the cognitive processes involved. In the early stages of skill acquisition, conventional practice problems force novices to resort to general, weak problem-solving strategies (e.g., trial and error, means–ends analysis), as they have not yet learned effective specific procedures for solving such problems. These general, weak strategies take a substantial amount of effort (i.e., they impose a high load on working memory) and time, but are not very effective and efficient for learning. That is, learners may succeed in solving the problem eventually, but as a consequence of the high working memory load they often don't remember what moves were actually effective, and therefore learning (i.e., knowing how to solve similar problems in the future) progresses very slowly (cf. Sweller & Levine, 1982). In contrast, by showing learners how to solve such a problem, worked examples obviate the need for general, weak problem-solving strategy use and instead allow the learner to devote all available working memory capacity to learning; that is, to constructing a cognitive schema of the solution procedure (Sweller et al., 2011).

Following criticism that conventional practice problem solving is a "lousy control condition", because learners do not receive any assistance during practice whatsoever (Koedinger & Aleven, 2007; see also Schwonke et al., 2009), more recent research has compared example-based learning to tutored problem-solving practice conditions. In tutored problem solving, learners can request hints when they get stuck (i.e., do not know how to proceed with a problem-solving step) and get feedback when they make errors (Koedinger & Aleven, 2007). Findings reviewed by Salden, Koedinger, Renkl, Aleven and McLaren (2010) show that adding worked examples to tutored problem-solving environments does not consistently lead to better learning outcomes compared to tutored problem solving only, but does have substantial efficiency benefits (i.e., same learning outcomes reached with less study time). Recently, a study in which a direct comparison was made between studying worked examples only and tutored problem solving only also showed a large efficiency benefit of worked example study of up to almost 60% (McLaren, van Gog, Ganoe, Karabinos, & Yaron, 2016). These findings suggest that the benefit of example study indeed lies in enabling (comparatively) rapid schema construction.

In contrast to worked examples, the effectiveness of *modeling examples* is not usually explained in comparison with other types of learning but more in terms of the general cognitive processes that need to take place for observational learning to be effective. In his well-known social-cognitive learning theory, Bandura postulated that observers acquire a cognitive (symbolic) representation (cf. cognitive schema) of the model's behavior that outlasts the modeling situation and thus enables learners to exhibit the observed and novel behavior at later occasions (e.g., Bandura, 1977, 1986). To acquire this representation, learners must pay attention to the relevant aspects of the modeled behavior (Bandura, 1986). The learner's attention is influenced both by the salience of those aspects and by the characteristics of the model. The information that the learner attended to then needs to be retained in memory, which requires encoding this information. Rehearsal (i.e., imitation), either mentally or physically, is considered to play an important role in retention, as well as in improvement of performance. However, learners may not always be able to produce the observed behaviors themselves. Whether or not they are able to do so depends on the quality of the cognitive representation they have acquired and on the extent to which they master the component skills. Finally, motivational processes determine whether or not the learner will actually exhibit the behavior that was learned through observation.

In light of debates on "cognitivism vs. constructivism," it is interesting to note that both Sweller and colleagues (Sweller & Sweller, 2006), as well as Bandura (1977, 1986), stress that example-based learning does not involve a one-on-one mapping of the observed information to the memory of the learner; rather, it is a constructive process during which information is actively (re)organized and integrated with the existing knowledge of the learner. Similarly, both emphasize the need for focused processing of the demonstrated task or skill.

When and for Whom Is Example-Based Learning Effective?

Example-based learning has been successfully used with many different kinds of learners, ranging from primary school children to university students to workplace-learners and aging adults (see van Gog & Rummel, 2010). Modeling examples have also been successfully implemented for learners with special needs resulting from psychological disorders (e.g., Biederman & Freedman, 2007).

It should be noted, though, that example study is particularly effective when learners have little if any prior knowledge of the demonstrated task or skill (Kalyuga, Chandler, Tuovinen, & Sweller, 2001)—provided that the task or skill that is demonstrated is not too complex for the learner—as this may result in fragmentary learning (Bandura, 1986). The high level of instructional guidance provided by examples, which fosters learning when no cognitive schemata are available yet, has been shown to be ineffective or even detrimental for learning when students have already developed cognitive schemata that can guide their problem solving (Kalyuga et al., 2001). There are indications that the rate at which the guidance provided by examples becomes obsolete may differ for highly structured as compared to less or ill-structured tasks. For the latter, examples may even be effective for students with higher prior knowledge (e.g., Ibiapina, Mamede, Moura, Elói-Santos, & van Gog, 2014; Nievelstein, van Gog, van Dijck, & Boshuizen, 2013). Indeed, for some complex tasks, a certain amount of expertise may even be required to be able to recognize subtle aspects of performance that are not noticed by novices (Bandura, 1986). Thus, learners who are very advanced in a domain may benefit from examples as long as these examples demonstrate a task or skill they have not yet (fully) mastered. As yet research has not systematically addressed the influence of other individual differences than prior knowledge on the effectiveness of example-based learning, such as intelligence or working memory capacity/span (for an exception, see Schwaighofer, Bühner, & Fischer, 2016).

For What Kind of Tasks and Skills Is Example-Based Learning Suitable?

Both worked examples and modeling examples can be used for teaching a wide variety of tasks and skills, although whether worked examples or modeling examples are more suitable may depend on the nature of the task or skill (e.g., if it cannot easily be communicated in writing, then modeling examples are a more sensible choice). Most research on *worked examples* has been conducted with highly structured problem-solving tasks in STEM-domains (see Renkl, 2014; Sweller et al., 2011; van Gog & Rummel, 2010). However, several studies have shown that worked examples can also be effective with less structured tasks, such as learning argumentation skills (Schworm & Renkl, 2007), recognizing designer styles (Rourke & Sweller, 2009) or reasoning about legal cases (Nievelstein et al., 2013).

Like worked examples, (video) *modeling examples* can also be used for teaching highly structured problem-solving tasks like math (Schunk & Hanson, 1985). However, they have more frequently been used for teaching motor skills (e.g., Blandin, Lhuisset, & Proteau 1999) and less structured cognitive skills, including argumentative writing (e.g., Braaksma, Rijlaarsdam, & van den Bergh, 2002), poetry writing and collage making (Groenendijk, Janssen, Rijlaarsdam, & van den Bergh, 2013), assertive communication skills (e.g., Baldwin, 1992), collaboration skills (e.g., Rummel & Spada, 2005; Rummel, Spada & Hauser, 2009), and *meta*cognitive skills such as self-regulation (e.g., Kostons, van Gog, & Paas, 2012; Zimmerman & Kitsantas, 2002).

Some examples qualify as *double-content* examples (Berthold & Renkl, 2009); that is, when modeling argumentation (Schworm & Renkl, 2007), collaboration (Berthold & Renkl, 2009), self-regulation (Kostons et al., 2012), or reflective reasoning (Ibiapina et al., 2014) skills, the examples will simultaneously convey knowledge about the task the model is working on (e.g., the biology/math/medical problem) while demonstrating the skills.

What Should Be Considered When Designing Examples?

It is important to carefully consider the example design and—in case of (video) modeling examples—who the model is, as this may affect learning outcomes. Following early studies (e.g., Cooper & Sweller, 1987; Sweller & Cooper, 1985), it was soon discovered that studying worked examples was not always more effective for learning than problem solving, and that the design of the examples played a crucial role in this (Tarmizi & Sweller, 1988). Research on this issue led to important design guidelines, such as *avoid split-attention* by integrating mutually referring information sources such as text and picture/diagram. This can also be accomplished by providing spoken rather than written text with pictorial information in the example (Mousavi, Low, & Sweller, 1995). Another design guideline, *avoid redundancy*, indicates that multiple sources of information should only be presented when they are *both* necessary for comprehension. If they can be easily understood in isolation, one of the sources is redundant and should be left out (Chandler & Sweller, 1991).

In video modeling examples, split-attention and redundancy are also important considerations, but other issues come into play as well. Modern technology provides a myriad of possibilities for designing video modeling examples. For instance, in "lecture-style" examples, the model may stand next to a screen on which he or she is writing or projecting slides that visualize each step in the task completion process, or only the slides/writing may be shown with a voice-over. In examples in which the model is manipulating (virtual) objects, that person can be visible in the video entirely, partly, or not at all (e.g., in screen-recording examples that show the model clicking, or dragging objects on a computer screen).

This myriad of design possibilities raises questions that are specific to video-modeling examples. For instance:

- *Does it matter for learning whether the model or the model's face is visible in the video?* This does not seem to hamper learning outcomes (Hoogerheide, Loyens, & van Gog, 2014; van Gog, Verveer, & Verveer, 2014).
- *Do 'direct' gaze cues (the model looking at the task) and 'indirect' gaze cues (a cursor displaying what the model is looking at) help learners attend to the right information at the right time?* They do, though effects

on learning outcomes are inconsistent, as in other cueing studies (Jarodzka, van Gog, Dorr, Scheiter, & Gerjets, 2013; Ouwehand van Gog, & Paas, 2015; van Marlen, van Wermeskerken, Jarodzka, & van Gog, 2016).
- *Does it matter if the video is shot from a first-person or third-person perspective?* First-person seems to result in better learning outcomes (Fiorella, van Gog, Hoogerheide, & Mayer, 2017).

Last but not least, a question that garnered attention in early research on modeling (see Schunk, 1987) is also relevant for video modeling examples:

- *Do model characteristics like gender, age, or (perceived) competence affect self-efficacy and learning outcomes?* According to the model–observer similarity hypothesis, the effects of modeling examples on students' self-efficacy and learning outcomes may depend on how similar to the model students perceive themselves to be (Bandura, 1994). However, studies of model–observer similarity have produced inconsistent findings. For example, findings regarding effects of mastery models (displaying faultless performance from the start) compared to coping models (whose performance includes errors that are corrected and expressions of uncertainty that are gradually reduced) on self-efficacy and learning were mixed (Schunk & Hanson, 1985; Schunk et al., 1987). The same applies to observing weak vs. competent models (Braaksma et al., 2002; Groenendijk et al., 2013).

Findings regarding model observer similarity in terms of gender or age were also inconsistent (Schunk, 1987). However, one issue in many model–observer similarity studies is that not only the model characteristics, but also the content of the examples (i.e., what was demonstrated or explained), differed across conditions, which may be a partial explanation for the mixed findings. Keeping the content of video modeling examples equal across conditions, recent studies found no evidence that the model's gender affected learning (Hoogerheide, Loyens, & van Gog, 2016), but the model's age did. In a study by Hoogerheide, van Wermeskerken, Loyens, and van Gog (2016), secondary education students who had observed an adult model rated the quality of the model's explanation as being higher, and showed better post-test performance than students who had observed a peer model, even though the examples had the exact same content.

How Can We Promote Active Learning from Examples?

Research has examined several design features intended to stimulate more active processing of examples or emphasize important aspects of procedures. Both have been hypothesized to improve students' learning of problem-solving procedures in conjunction with their understanding of underlying structures and rationales for the procedures. Learning with understanding is necessary to be able to solve slightly novel problems (i.e., transfer).

One well-known feature is asking students to give *self-explanations*, which has often been compared with providing *instructional explanations*. Having learners self-explain the principles behind the worked-out solution steps seems an effective way to improve deep learning (e.g., Chi, Bassok, Lewis, Reimann, & Glaser, 1989; Renkl, 1997, 2002). However, a precondition is that students are capable of providing high quality self-explanations, which is not always the case (see Berthold & Renkl, 2009; Chi et al., 1989; Lovett, 1992; Renkl, 1997). If this precondition is not met, providing high quality *instructional explanations* may sometimes enhance learning from examples (Lovett, 1992). However, a meta-analysis suggests that instructional explanations generally seem to have minimal benefit for example-based learning. If anything, the benefit lies in the acquisition of conceptual, not procedural knowledge (Wittwer & Renkl, 2010). This may be due to the instructional explanations becoming redundant relatively quickly, at which point they need to be faded out or they may start to hamper procedural learning (van Gog, Paas, & van Merriënboer, 2008).

Other options for promoting active processing are, for instance, to ask students to:

- *compare different problem solutions* (e.g., Große & Renkl, 2006; Rittle-Johnson & Star, 2007; Rittle-Johnson, Star, & Durkin, 2009);
- *compare correct and erroneous examples* (e.g., Baldwin, 1992; Blandin & Proteau, 2000; Durkin & Rittle-Johnson, 2012; Große & Renkl, 2007; Kopp, Stark, & Fischer, 2008);
- *imagine* or *cognitively rehearse* the observed task (e.g., Cooper, Tindall-Ford, Chandler, & Sweller, 2001; Ginns, Chandler, & Sweller, 2003; Leahy & Sweller, 2008);[1]
- *complete steps* in partially worked-out examples (Paas, 1992; van Merriënboer, Schuurman, de Croock, & Paas, 2002). Such "completion problems" are central to the *completion* or *fading strategy*, in which learners start by studying a fully worked-out example, and then progress via completion problems with increasingly more steps for the learner to complete, to solving conventional problems in which they have to complete all steps themselves without any support (Renkl & Atkinson, 2003).

Summary and Conclusions

Example-based learning is a powerful and very natural way of learning that has been described from different theoretical perspectives. Two major research traditions that have been concerned with investigating mechanisms of example-based learning are CLT (Sweller et al., 2011) and social-cognitive learning theory (Bandura, e.g., 1977, 1986). While the example-demonstrations originating from the different research traditions have taken different forms (worked examples vs. modeling examples), the principles of example-based learning that can be deduced from these different bodies of work overlap to a large extent (see also Renkl, 2014; van Gog & Rummel, 2010). In this chapter we have provided a selective overview of relevant research from both traditions, with the goals of highlighting important insights into the cognitive mechanisms and preconditions of example-based learning, and of summarizing findings pointing at relevant principles for instructional design.

One important direction for future research on example-based learning is to start addressing the effects over time, in real classroom contexts. Most studies on the effectiveness of example-based learning have been of the highly controlled single session variety in a lab or school. There are some classroom studies spanning multiple sessions over different days, with promising results in terms of efficiency. That is, students achieve the same level of learning outcomes in up to 60% less study time (McLaren et al., 2016; Salden et al, 2010; van Loon-Hillen, van Gog, & Brand-Gruwel, 2012; Zhu & Simon, 1987). Thus, pursuing these intriguing findings would seem a worthwhile endeavor.

Acknowledgment

While writing this chapter Tamara van Gog was funded by a Vidi grant (#452-11-006) from the Netherlands Organization for Scientific Research.

Further Readings

Bandura, A. (1965). Influence of model's reinforcement contingencies on the acquisition of imitative responses. *Journal of Personality and Social Psychology, 1*, 589–595.
A classic study on modeling: Bandura's Bobo-doll experiment.

Hoogerheide, V., van Wermeskerken, M., Loyens, S. M. M., & van Gog, T. (2016). Learning from video modeling examples—Content kept equal, adults are more effective models than peers. *Learning and Instruction, 44*, 22–30.
A recent study on learning from video modeling examples.

Renkl, A., (2014). Towards an instructionally-oriented theory of example-based learning. *Cognitive Science, 38,* 1–37.
Overview of research on example-based learning from a cognitive science perspective.

Sweller, J., & Cooper, G. A. (1985). The use of worked examples as a substitute for problem solving in learning algebra. *Cognition and Instruction, 2,* 59–89.
A classic study on learning with worked examples.

van Gog, T., & Rummel, N. (2010) Example-based learning: Integrating cognitive and social-cognitive research perspectives. *Educational Psychology Review,* 22(2), 155–174.
Overview of research on example-based learning from an educational psychology perspective.

NAPLeS Resources

Rummel, N., *Example-based learning* [Webinar]. In *NAPLeS video series.* Retrieved October 19, 2017, from http://isls-naples.psy.lmu.de/intro/all-webinars/rummel/index.html

Note

1 Note that the effectiveness of some of those strategies (e.g., comparing different solutions or correct and erroneous examples, imaging/cognitive rehearsal) may depend on learners' prior knowledge.

References

Anderson, J. R. (1993). *Rules of the mind.* Hillsdale: Erlbaum.
Baldwin, T. T. (1992). Effects of alternative modelling strategies on outcomes of interpersonal-skills training. *Journal of Applied Psychology, 77,* 147–154.
Bandura, A. (1977). *Social learning theory.* Englewood Cliffs, NJ: Prentice Hall.
Bandura, A. (1986). *Social foundations of thought and action: A social cognitive theory.* Englewood Cliffs, NJ: Prentice Hall.
Bandura, A. (1994). Self-efficacy. In V. S. Ramachaudran (Ed.), *Encyclopedia of human behavior* (Vol. 4, pp. 71–81). New York: Academic Press.
Berthold, K., & Renkl, A. (2009). Instructional aids to support a conceptual understanding of multiple representations. *Journal of Educational Psychology, 101,* 70–87.
Biederman, G. B., & Freedman, B. (2007). Modeling skills, signs and lettering for children with Down syndrome, autism and other severe developmental delays by video instruction in classroom setting. *Journal of Early and Intensive Behavior Intervention, 4,* 736–743.
Blandin, Y., Lhuisset, L., & Proteau, L. (1999). Cognitive processes underlying observational learning of motor skills. *Quarterly Journal of Experimental Psychology, 52A,* 957–979.
Blandin, Y., & Proteau, L. (2000). On the cognitive basis of observational learning: Development of mechanisms for the detection and correction of errors. *Quarterly Journal of Experimental Psychology, 53A,* 846–867.
Braaksma, M. A. H., Rijlaarsdam, G., & van den Bergh, H. (2002). Observational learning and the effects of model–observer similarity. *Journal of Educational Psychology, 94,* 405–415.
Chandler, P., & Sweller, J. (1991). Cognitive load theory and the format of instruction. *Cognition and Instruction, 8,* 293–332.
Chi, M. T. H., Bassok, M., Lewis, M. W., Reimann, P., & Glaser, R. (1989). Self-explanations: How students study and use examples in learning to solve problems. *Cognitive Science, 13,* 145–182.
Chi, M. T. H., Roy, M., & Hausmann, R. G. M. (2008). Observing tutorial dialogues collaboratively: Insights about human tutoring effectiveness from vicarious learning. *Cognitive Science, 32,* 301–341.
Cooper, G., & Sweller, J. (1987). The effects of schema acquisition and rule automation on mathematical problem-solving transfer. *Journal of Educational Psychology, 79,* 347–362.
Cooper, G., Tindall-Ford, S., Chandler, P., & Sweller, J. (2001). Learning by imagining. *Journal of Experimental Psychology. Applied, 7,* 68–82.
Durkin, K., & Rittle-Johnson, B. (2012). The effectiveness of using incorrect examples to support learning about decimal magnitude. *Learning and Instruction, 22,* 206–214.
Fiorella, L., van Gog, T., Hoogerheide, V. & Mayer, R. (2017). It's all a matter of perspective: Viewing first-person video modeling examples promotes learning of an assembly task. *Journal of Educational Psychology, 109*(5), 653.

Ginns, P., Chandler, P., & Sweller, J. (2003). When imagining information is effective. *Contemporary Educational Psychology, 28,* 229–251.

Groenendijk, T., Janssen, T., Rijlaarsdam, G., & van den Bergh, H. (2013). Learning to be creative. The effects of observational learning on students' design products and processes. *Learning and Instruction, 28,* 35–47.

Große, C. S., & Renkl, A. (2006). Effects of multiple solution methods in mathematics learning. *Learning and Instruction, 16,* 122–138.

Große, C. S., & Renkl, A. (2007). Finding and fixing errors in worked examples: Can this foster learning outcomes? *Learning and Instruction, 17,* 612–634.

Hoogerheide, V., Loyens, S. M. M., & van Gog, T. (2014). Comparing the effects of worked examples and modeling examples on learning. *Computers in Human Behavior, 41,* 80–91.

Hoogerheide, V., Loyens, S., & van Gog, T. (2016). Learning from video modeling examples: Does gender matter? *Instructional Science, 44,* 69–86.

Hoogerheide, V., van Wermeskerken, M., Loyens, S. M. M., & van Gog, T. (2016). Learning from video modeling examples—Content kept equal, adults are more effective models than peers. *Learning and Instruction, 44,* 22–30.

Ibiapina, C., Mamede, S., Moura, A., Elói-Santos, S., & van Gog, T. (2014). Effects of free, cued and modelled reflection on medical students' diagnostic competence. *Medical Education, 48,* 796–805.

Jarodzka, H., van Gog, T., Dorr, M., Scheiter, K., & Gerjets, P. (2013). Learning to see: Guiding students' attention via a model's eye movements fosters learning. *Learning and Instruction, 25,* 62–70.

Kalyuga, S., Chandler, P., Tuovinen, J., & Sweller, J. (2001). When problem solving is superior to studying worked examples. *Journal of Educational Psychology, 93,* 579–588.

Koedinger, K. R., & Aleven, V. (2007). Exploring the assistance dilemma in experiments with cognitive tutors. *Educational Psychology Review, 19,* 239–264.

Kopp, V., Stark, R., & Fischer, M. R. (2008). Fostering diagnostic knowledge through computer-supported, case-based worked examples: Effects of erroneous examples and feedback. *Medical Education, 42,* 823–829.

Kostons, D., van Gog, T., & Paas, F. (2012). Training self-assessment and task-selection skills—A cognitive approach to improving self-regulated learning. *Learning and Instruction, 22,* 121–132.

Leahy, W., & Sweller, J. (2008). The imagination effect increases with an increased intrinsic cognitive load. *Applied Cognitive Psychology, 22,* 273–283.

Lovett, M. C. (1992). Learning by problem solving versus by examples: The benefits of generating and receiving information. *Proceedings of the 14th annual conference of the Cognitive Science Society* (pp. 956–961). Hillsdale: Erlbaum.

McLaren, B. M., van Gog, T., Ganoe, C., Karabinos, M., & Yaron, D. (2016). The efficiency of worked examples compared to erroneous examples, tutored problem solving, and problem solving in computer-based learning environments. *Computers in Human Behavior, 55,* 87–99.

Mousavi, S. Y., Low, R., & Sweller, J. (1995). Reducing cognitive load by mixing auditory and visual presentation modes. *Journal of Educational Psychology, 87,* 319–334.

Nievelstein, F., van Gog, T., van Dijck, G., & Boshuizen, H. P. A. (2013). The worked example and expertise reversal effect in less structured tasks: Learning to reason about legal cases. *Contemporary Educational Psychology, 38,* 118–125.

Ouwehand, K., van Gog, T., & Paas, F. (2015). Designing effective video-based modeling examples using gaze and gesture cues. *Educational Technology & Society, 18,* 78–88.

Paas, F. (1992). Training strategies for attaining transfer of problem-solving skill in statistics: A cognitive load approach. *Journal of Educational Psychology, 84,* 429–434.

Renkl, A. (1997). Learning from worked-out examples: A study on individual differences. *Cognitive Science, 21,* 1–29.

Renkl, A. (2002). Learning from worked-out examples: Instructional explanations supplement self-explanations. *Learning and Instruction, 12,* 149–176.

Renkl, A. (2014). Towards an instructionally-oriented theory of example-based learning. *Cognitive Science, 38,* 1–37.

Renkl, A., & Atkinson, R. K. (2003). Structuring the transition from example study to problem solving in cognitive skills acquisition: A cognitive load perspective. *Educational Psychologist, 38,* 15–22.

Renkl, A., Hilbert, T., & Schworm, S. (2009). Example-based learning in heuristic domains: A cognitive load theory account. *Educational Psychology Review, 21,* 67–78.

Rittle-Johnson, B., & Star, J. R. (2007). Does comparing solution methods facilitate conceptual and procedural knowledge? An experimental study on learning to solve equations. *Journal of Educational Psychology, 99,* 561–574.

Rittle-Johnson, B., Star, J. R., & Durkin, K. (2009). The importance of prior knowledge when comparing examples: Influences on conceptual and procedural knowledge of equation solving. *Journal of Educational Psychology, 101,* 836–852.

Rourke, A., & Sweller, J. (2009). The worked-example effect using ill-defined problems: Learning to recognize designers' styles. *Learning and Instruction, 19*, 185–199.

Rummel, N., & Spada, H. (2005). Learning to collaborate: An instructional approach to promoting collaborative problem-solving in computer-mediated settings. *Journal of the Learning Sciences, 14*, 201–241.

Rummel, N., Spada, H., & Hauser, S. (2009). Learning to collaborate while being scripted or by observing a model. *International Journal of Computer-Supported Collaborative Learning, 4*, 69–92.

Salden, R. J. C. M., Koedinger, K. R., Renkl, A., Aleven, V., & McLaren, B. M. (2010). Accounting for beneficial effects of worked examples in tutored problem solving. *Educational Psychology Review, 22*(4), 379–392.

Schunk, D. H. (1987). Peer models and children's behavioral change. *Review of Educational Research, 57*, 149–174.

Schunk, D. H., & Hanson, A. R. (1985). Peer models: Influence on children's self-efficacy and achievement. *Journal of Educational Psychology, 77*, 313–322.

Schunk, D. H., Hanson, A. R., & Cox, P. D. (1987). Peer-model attributes and children's achievement behaviors. *Journal of Educational Psychology, 79*, 54–61.

Schwaighofer, M., Bühner, M., & Fischer, F. (2016). Executive functions as moderators of the worked example effect: When shifting is more important than working memory capacity. *Journal of Educational Psychology, 108*, 982–1000.

Schwonke, R., Renkl, A., Krieg, C., Wittwer, J., Aleven, V., & Salden, R. J. C. M. (2009). The worked-example effect: Not an artefact of lousy control conditions. *Computers in Human Behavior, 25*, 258–266.

Schworm, S., & Renkl, A. (2007). Learning argumentation skills through the use of prompts for self-explaining examples. *Journal of Educational Psychology, 99*, 285–296.

Sweller, J., Ayres P. L., & Kalyuga, S. (2011). *Cognitive load theory*. New York: Springer.

Sweller, J., & Cooper, G. A. (1985). The use of worked examples as a substitute for problem solving in learning algebra. *Cognition and Instruction, 2*, 59–89.

Sweller, J., & Levine, M. (1982). Effects of goal specificity on means–ends analysis and learning. *Journal of Experimental Psychology. Learning, Memory, and Cognition, 8*, 463–474.

Sweller, J., & Sweller, S. (2006). Natural information processing systems. *Evolutionary Psychology, 4*, 434–458.

Tarmizi, R., & Sweller, J. (1988). Guidance during mathematical problem solving. *Journal of Educational Psychology, 80*, 424–436.

van Gog, T., Kester, L., & Paas, F. (2011). Effects of worked examples, example–problem, and problem–example pairs on novices' learning. *Contemporary Educational Psychology, 36*, 212–218.

van Gog, T., Paas, F., & van Merriënboer, J. J. G. (2008). Effects of studying sequences of process-oriented and product-oriented worked examples on troubleshooting transfer efficiency. *Learning and Instruction, 18*, 211–222.

van Gog, T., & Rummel, N. (2010) Example-based learning: Integrating cognitive and social-cognitive research perspectives. *Educational Psychology Review, 22*, 155–174.

van Gog, T., Verveer, I., & Verveer, L. (2014). Learning from video modeling examples: Effects of seeing the human model's face. *Computers & Education, 72*, 323–327.

van Loon-Hillen, N. H., van Gog, T., & Brand-Gruwel, S. (2012). Effects of worked examples in a primary school mathematics curriculum. *Interactive Learning Environments, 20*, 89–99.

van Marlen, T., van Wermeskerken, M. M., Jarodzka, H., & van Gog, T. (2016). Showing a model's eye movements in examples does not improve learning of problem-solving tasks. *Computers in Human Behavior, 65*, 448–459.

van Merriënboer, J. J. G., Schuurman, J. G., de Croock, M. B. M., & Paas, F. (2002). Redirecting learners' attention during training: Effects on cognitive load, transfer test performance and training. *Learning and Instruction, 38*, 11–39.

Wittwer, J., & Renkl, A. (2010). How effective are instructional explanations in example-based learning? A meta-analytic review. *Educational Psychology Review, 22*, 393–409.

Zhu, X., & Simon, H. A. (1987). Learning mathematics from examples and by doing. *Cognition and Instruction, 4*, 137–166.

Zimmerman, B. J., & Kitsantas, A. (2002). Acquiring writing revision and self-regulatory skill through observation and emulation. *Journal of Educational Psychology, 94*, 660–668.

21
Learning Through Problem Solving

Cindy E. Hmelo-Silver, Manu Kapur, and Miki Hamstra

Learning through problem solving has long been championed by some in the educational sphere (e.g., Dewey, 1938; Kilpatrick, 1918). The latest re-emergence of learning through problem solving coincides with findings in cognitive and educational sciences research showing that transmission models of learning rarely support transfer and application of previously learned information even when learners demonstrate mastery through relatively immediate recitation "tests" of learning. These findings harken back to Whitehead's (1929) discussion of the problem of inert knowledge and lack of transfer or application of relevant prior knowledge in new and appropriate contexts. Learning through problem solving can be contrasted with other types of problem solving research in which the focus is on how people solve problems, including activation of known concepts and procedures (see Greeno, Collins, & Resnick, 1996). In learning through problem solving approaches, the focus is in how people construct new knowledge as they engage in solving problems. These kinds of approaches are important in the learning sciences because they are all theoretically guided designs and provide opportunities to study learning as it unfolds in authentic settings as learners engage with meaningful tasks.

A variety of learning through problem-solving approaches have been developed to address the "inert knowledge problem" (CTGV, 1992; Hmelo-Silver, 2004). For example, research on transfer-appropriate processing examined how people could retrieve problem solutions and ideas based on how they were encoded (e.g., Adams et al., 1988). Building on this, Bransford and Schwartz (1999) argued that early problem-solving experiences could enable learners to see similarities across situations and thus prepare themselves to learn in new situations through both application and adaptation of knowledge (Schwartz, Chase, & Bransford, 2012). Others stressed how knowledge and practices should be framed to foster the expectation that what students learn will be useful in other settings (Engle, 2006). In short, learning through problem-solving approaches promote transfer by helping learners see the relevance of their prior knowledge, preparing them for new learning, and framing that learning as broadly applicable.

Learning through problem solving has been instantiated in a variety of instructional models and designs, including problem- and project-based learning, productive failure, inquiry learning, and design-based learning (Blumenfeld et al., 1991; Kolodner et al., 2003; Linn, McElhaney, Gerard, & Matuk, this volume). Support for learning is embedded in all of these designs in a range of forms, including scaffolds, sequenced and carefully designed tasks, levels, and styles of intervention. Learning through problem-solving instructional designs share two critical features that are advantageous for learning. First, they build integrated conceptual understanding while simultaneously developing problem-solving and self-regulated learning skills (Savery, 2015). When learners confront an ill-structured problem, they must analyze the problem, identify and resolve knowledge deficiencies through self-directed research, and evaluate their proposed solutions (Hmelo-Silver, 2004). In this way, they engage in deep content learning while developing strategies for future learning. Second, engaging in problem solving also provides motivational advantages. Ill-structured, but well-designed complex problems create situational interest. Problems that are complex yet manageable, realistic and relevant, and offer sufficient choice and control, foster intrinsic motivation (Deci, Koestner, & Ryan, 2001; Schmidt, Rotgans, & Yew, 2011). Furthermore, because learning through problem-solving designs is inherently collaborative, learners may also be more motivated to participate (Blumenfeld et al., 1991). Indeed, students attribute increased motivation to enjoying the social interaction as well as to perceived pressure to be contributing to the group (Wijnia, Loyens, & Derous, 2011).

In this chapter we focus on two specific approaches to learning through problem solving that share a number of commonalities: problem-based learning (PBL) and productive failure (PF). Both are stable, empirically supported pedagogical models that emphasize the social nature of learning and focus on robust learning, not short-term performance proficiency. They assume that problem solving is an iterative process that requires time, persistence, and patience. Students learn as they analyze the problem and make solution attempts, initially relying on their existing knowledge and then augmenting that with additional instructional resources. Although PBL and PF have often been contrasted, we consider how these instructional approaches are complementary, drawing on Kapur's (2016) notion of productive success and productive failure.

Productive Success and Failure: A Framework for Understanding Learning Through Problem Solving

Previous research has used instructional designs for learning through problem solving that can be cross-classified along two continua. One refers to the extent to which they are intended to maximize performance in the initial learning task, the other, to the extent to which they maximize learning in the long-term (Kapur, 2008). The four outcomes of this cross-classification are productive success, productive Failure, unproductive success, and unproductive failure. Productive success describes designs that maximize performance in the shorter term as well as maximize learning in the longer term. PF designs may not maximize performance in the short term, but seek to maximize sustained learning in the longer term. Unproductive success results from designs that maximize performance in the shorter term (e.g. through rote memorization) but do not maximize learning in the longer term. Finally, designs that maximize neither short- nor longer-term learning (e.g. unsupported discovery learning) reflect unproductive failure. As indicated above, in this chapter we focus on productive success as exemplified in PBL instructional designs and productive failure designs. Both have shown promise for facilitating transfer for future learning.

Productive Approaches

There is intentionality in the design of the two types of productive approaches to learning through problem solving. The designs differ in terms of the intended short-term problem-solving outcome.

Productive Success. Where success is intended, as in PBL, scaffolding, and facilitation play a central role in the design of the initial problem solving and are seen as critical to outcomes that are productive for future learning (Ertmer & Glazewski, in press; Hmelo-Silver & Barrows, 2008). Scaffolding reduces the cognitive load associated with confronting complex problems and helps students manage the complexity while also learning from their engagement in collaborative problem-solving activities (Blumenfeld et al., 1991). Facilitation refers to kinds of supports that a teacher provides that help guide the learning process (Hmelo-Silver & Barrows, 2008). As students gain expertise and skills, scaffolding then fades as it is needed less (Puntambekar & Hübscher, 2005; Schmidt, Loyens, van Gog, & Paas, 2007).

In addition, the PBL tutorial cycle provides a loose script that helps communicate the PBL process, shown in Figure 21.1 (Collins, 2006; Derry, Hmelo-Silver, Nagarajan, Chernobilsky, & Beitzel, 2006; see also Kollar, Wecker, & Fischer, this volume). PBL facilitators provide key just-in-time guidance through questioning and suggestions (soft scaffolding) or tools to support student inquiry (hard scaffolding), such as graphic organizers and whiteboards that can guide their thinking (Hmelo-Silver, 2004; Puntambekar & Kolodner, 2005). This type of support maximizes both productive success in the initial learning activity, as well as learning in the longer term.

Productive Failure. In contrast to PBL and other success designs, PF does not provide significant initial learning support. Instead students are purposefully engaged in problems requiring concepts they have not yet learned. In this initial problem-solving stage, students explore affordances and constraints of possible solutions. Although they fail in the short term to generate correct solutions, the process of failing prepares them for learning from subsequent instruction and support, provided in the consolidation phase. Consolidation is thought to foster long-term learning by providing opportunities for comparing and contrasting, assembling, and organizing solutions into canonical solutions (Kapur & Bielaczyc, 2012; Schwartz & Martin, 2004).

Learning Mechanisms

Despite the differences in the timing of support for PBL and productive failure, the learning mechanisms are similar. Information processing theory provided the earliest theoretical explanation for the benefits of learning through problem solving—namely, the activation of prior knowledge and transfer-appropriate processing (Adams et al., 1988; Schmidt, 1993). In PBL, learners connect new learning to prior knowledge when they engage in their initial discussion about a problem, thus preparing them for new learning. Transfer-appropriate processing occurs when knowledge encoded in problem-solving contexts is retrieved to solve new problems. In both PBL and PF, students are learning content in the context of a problem-solving situation, and transfer-appropriate processing theory suggests that learners would be more likely to retrieve this knowledge in relevant problem situations.

In PF, the process of differentiating relevant prior knowledge when students generate sub-optimal or incorrect solutions is critical (DeCaro & Rittle-Johnson, 2012; Schwartz, Chase, Oppezzo, & Chin, 2011). It allows students to: (a) notice inconsistencies and draw their attention to gaps in their prior

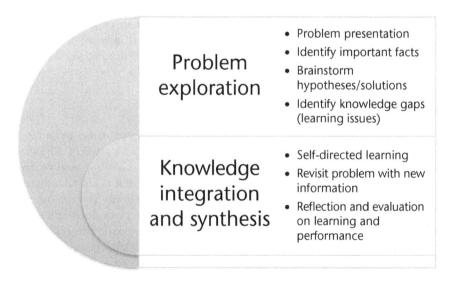

Figure 21.1 PBL Small Group Tutorial Cycle

knowledge (DeCaro & Rittle-Johnson, 2012; Loibl & Rummel, 2014; Ohlsson, 1996); and (b) compare and contrast their generated solutions and the correct solutions during subsequent instruction, enabling students to better encode critical features of the new concept (Kapur, 2014; Schwartz et al., 2011).

More recently, sociocultural theory has broadened our understanding of learning through problem solving. Social constructivist perspectives emphasize active engagement in meaningful learning tasks and the importance of tools in mediating learning (Collins, 2006; Danish & Gresalfi, this volume). As students engage with ill-structured, authentic problems, their learning is supported through facilitator scaffolding of tasks within learners' Zones of Proximal Development (Vygotsky, 1978). This fosters cognitive apprenticeship where students gain problem-solving competencies under the guidance of mentors (Collins, 2006; Quintana et al., 2004; Eberle, this volume). By using conceptual tools, such as language in authentic situations, individuals begin to fully understand their function and complexities. Language itself is a tool learners use to construct meaning and progress in becoming participants in their communities of practice. PBL discourse enables learners to appropriate new disciplinary concepts, vocabulary, and reasoning to engage with other members of the community (Brown et al., 1993; Chernobilsky, DaCosta, & Hmelo-Silver, 2004). As students share what they know, discover what they still need to learn, and form explanations and arguments, their collective thinking becomes visible to the group, inviting discussion and revision.

Research Methods and Findings in PBL and PF

The research literature on PBL is extensive and has used a variety of methods that range from qualitative examinations of curriculum, facilitation, and student development (Bridges, Green, Botelho, & Tsang, 2015; Evensen, Salisbury-Glennon, & Glenn, 2001; Hmelo-Silver & Barrows, 2008) to experimental and quasi-experimental studies of PBL outcomes (e.g., Walker, Leary, & Lefler, 2015). Qualitative methods have included ethnographic, ethnomethodological, and content analysis. Quantitative methods have generally used a range of content and strategy measures. In contrast, PF research methodologies have been largely quantitative. Productive failure research has largely relied on three sequential, yet reinforcing quantitative methodologies. The early work employed design-based research utilizing multiple iterations of design, implementation, and iteration across various contexts and samples to stabilize its design. PF was then tested in classrooms through quasi-experimental investigations. Finally, experimental work was conducted to enable a surgical examination of specific design features and mechanisms (Kapur, 2016). Now that it has achieved a stable design, future qualitative examinations may be employed to better understand the experience of individual learners while engaged in productive failure.

PBL Findings

Because PBL was initially designed to better develop medical reasoning and long-term learning among medical students, much of the early scholarship on its learning gains focused on comparing PBL with direct instruction in medical education. Meta-analytic studies provide mixed findings (e.g., Albanese & Mitchell, 1993; Vernon & Blake, 1993). Vernon and Blake (1993) found that PBL instruction better prepared participants to transfer medical knowledge to solve problems in practice, but was less effective than direct instruction for the acquisition of more basic medical knowledge. However, this advantage may be limited as other researchers have found it to vanish after the second year of medical school (Dochy, Segers, Van den Bossche, & Gijbels, 2003). In addition, when measures of knowledge application are used, PBL demonstrates greater effects over direct instruction (Gijbels, Dochy, Van den Bossche, & Segers, 2005).

PBL has been effectively broadened to include other domains and age groups, including pre-service teachers (Derry et. al., 2006), MBA students (Capon & Kuhn, 2004), and secondary school (CTGV, 1992; Mergendoller, Maxwell, & Bellisimo, 2006), with all of these variations showing positive effects.

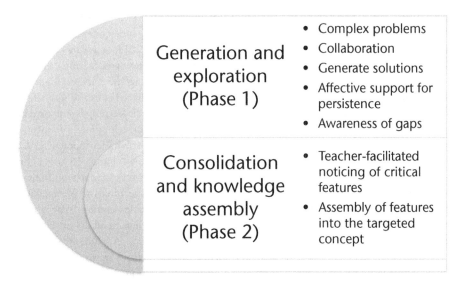

Figure 21.2 Productive Failure Design

A meta-analysis by Walker et al. (2015) found that research across different educational levels and disciplines demonstrated an overall positive but small effect size, but with considerable variability across studies. Moderate effects of PBL were demonstrated when knowledge was assessed at an application level and for strategic performance and design problems. Despite the benefits that aforementioned studies and meta-analysis demonstrated, the effectiveness of PBL continues to be debated. Some researchers warn that the mixed findings across domains suggest that more research is needed to understand the effects of diverse PBL implementation, problem types used, and types of outcomes assessed (Walker et al., 2015). Although the meta-analyses and quantitative studies of PBL tend to show advantages, they are not informative regarding learning processes and facilitation in PBL.

To understand how learning unfolds in PBL discussions, other studies have taken a qualitative turn (cf. Green & Bridges, this volume). For example, Hmelo-Silver and Barrows (2008) demonstrated how medical students engaged in collaborative knowledge building in exploring a complex problem. The facilitator scaffolded the group through asking open-ended questions that helped focus student attention, encourage explanation, causal reasoning, and justification as well as supported groups in monitoring their group dynamics. Students appropriated some of these facilitation functions as they also asked many questions and integrated their self-directed learning. This study also elucidated the important role of shared representations in mediating discussion. Similarly, in teacher professional development, Zhang, Lundeberg, and Eberhardt (2011) studied several PBL groups and their facilitators, finding that facilitators used questioning, revoicing, and a repertoire of strategies to support collaborative advancement of ideas. Yew and Schmidt (2009) found similar evidence of collaborative, self-directed and constructive activity in a polytechnic university. Although much qualitative work in PBL has been conducted with adult learners, Brush and Saye (2008) examined PBL in secondary history classes, finding evidence of students' constructive activity with both cognitive and emotional engagement in historical inquiry. These qualitative studies demonstrate how learning, facilitation, and scaffolding unfold in PBL.

PF Findings

To date, PF research has tended to utilize complex conceptual problems presented to secondary and college-level mathematics and science students in their classroom settings (e.g., DeCaro & Rittle-Johnson, 2012; Kapur, 2012, 2014; Loibl & Rummel, 2014; Schwartz et al., 2011). Findings suggest that students who generate solutions to novel problems before receiving instruction perform significantly better on measures of conceptual understanding and transfer than students who receive instruction prior to engaging in problem solving (Kapur, 2014; Loibl, Roll, & Rummel, 2016). Additionally, PF students who generated multiple solutions performed better on procedural knowledge, conceptual understanding, and transfer items on the posttest. Kapur (2015) called it the "solution generation effect:" the greater the number of solutions generated, the better the learning as measured by performance on the posttest. Kapur argued that the solution generation effect indexes the prior knowledge activation mechanism of PF; that is, when students generate more solutions, relevant prior knowledge gets activated, and such activation prepares students to notice, encode, and retrieve critical features of the targeted concept.

Overall, PBL and PF share a consistent trend in their findings. Compared to direct instruction, both approaches have small or negative effects on basic knowledge acquisition, but positive effects on conceptual understanding and transfer. It is this transfer of prior learning that makes PBL and PF productive pedagogies for robust learning.

Designing for PBL and Productive Failure

As mentioned earlier, the fundamental difference between productive success (exemplified in PBL) and productive failure is in the design of initial problem solving, and where support is provided in the overall learning cycle. Both approaches start with collaborative problem solving, but then the approaches diverge. Although various adaptations of PBL have modified its design in practice, some general design principles of the PBL tutorial process (see Figure 21.1) are key. First, exploration is focused around an ill-structured problem without a single correct answer set in an authentic context. Second, students work in small collaborative groups to identify the problem and design the solution. Students co-construct meaning through integrated exploration across disciplines. Third, teachers act as guides by providing sufficient scaffolding to support student-driven exploration. Finally, reflection and assessment activities are built into the PBL cycle to encourage self-regulated learning (Savery, 2015). Master PBL facilitators employ a range of scaffolds to support students through this learning process (Hmelo-Silver & Barrows, 2008). In the classic PBL model, a whiteboard is used to help students to structure their problem solving and learning by recording facts, ideas, learning issues, and an action plan.

There are several design principles for effective PBL problems (Barrett, Cashman, & Moore, 2011; Hmelo-Silver, 2004; Jonassen & Hung, 2015). First, problems must be sufficiently complex and ill-structured as well as personally relevant to the learners to foster motivation. Second, problems must provide sufficient feedback so learners can evaluate their learning and performance. Good problems promote constructive discussion (Koschmann, Myers, Feltovich, & Barrows, 1994). Jonassen and Hung (2015) examined four problem types typically used in PBL: diagnostic, decision-making, situated cases/policy, and design problems. Design problems can be particularly fruitful, engaging learners in constructing an artifact based on functional specifications (e.g., Jordan & McDaniel, 2014; Kolodner et al., 2003). Design problems may be effective because of the feedback such problems afford.

Problems for the PF approach tend to be sufficiently complex to afford multiple representations and solution methods by drawing on various formal and informal resources. They can be both well-structured (e.g., Kapur, 2014; Kapur & Bielaczyc, 2012) or ill-structured (e.g., Kapur, 2008), but need to be designed with an intuitive and affective hook, embody multiple contrasts that help students notice critical features, and use variant–invariant relations so that student-generated solutions do not lead to successful solutions. This failure then provides the opportunity to compare and contrast the affordances and constraints of their failed and sub-optimal solutions. Ill-structured PBL problems can foster problem-relevant collaborative discussion; however, groups may need more support to make this interaction productive (Kapur & Kinzer, 2007).

Prior knowledge represents the most critical design component of PF, as it primes learners for future learning. The four core mechanisms are: "a) activation and differentiation of prior knowledge, b) attention to critical conceptual features of the targeted concepts, c) explanation and elaboration of these features, and d) organization and assembly of the critical conceptual features into the targeted concepts" (Kapur & Bielaczyc, 2012, p. 75). Students engage in this process through two phases (Figure 21.2): a problem-solving phase (Phase 1) followed by a consolidation phase (Phase 2). In Phase 1, students generate and explore the affordances and constraints of multiple representations and solutions. They then organize and assemble their generated solutions into canonical solutions in Phase 2.

Technology in Learning Through Problem Solving

Some of the challenges for learning through problem solving have involved creating rich problem contexts, scaffolding, providing access to information resources, and communication modalities. Technology has played a role in addressing these challenges. It can be used to provide context, scaffolds, information resources, and spaces for visualizing and co-constructing ideas. In a learning sciences course for preservice teachers, Derry et al. (2006) created an online PBL system that included problem contexts with scenarios that included both a videocase of a student or classroom as well as a problem statement that set the students' goal to redesign the lesson or design a similar one, based on learning sciences principles. They created scaffolds through the use of an eight-step activity structure and prompts to organize the group problem solving in an online whiteboard. The cases included links to a learning sciences hypermedia as a starting point to help guide learners to productive learning issues. In a data analysis course, Holmes, Day, Park, Bonn, and Roll (2014) created the Invention Support Environment to provide contrasting cases and guidance for the invention phase of productive failure.

Other technological tools such as interactive whiteboards can help groups visualize their thinking and organize their process as students generate and refine their solutions (Bridges et al., 2015; Lu, Lajoie, & Wiseman, 2010; Green & Bridges, this volume). Likewise, facilitators could benefit from technological tools that automate some scaffolding, provide learning analytics data, and enable faster feedback to students (Blumenfeld et al., 1991; Rosé, this volume). Understanding the roles for technology is an emerging area for both PBL and PF.

Implications for the Learning Sciences

We gain many insights from jointly considering productive success and productive failure under the umbrella of learning through problem solving. Both approaches help learners understand how knowledge and practices can be tools for thinking, problem solving, and participating in a community of practice. Neither approach focuses on the final solution to the problem, but on how the process of working through carefully designed problems prepares learners for future learning. Although learning through problem-solving approaches has been critiqued for being minimally guided and increasing cognitive load (Kirschner, Sweller, & Clark, 2006), the participant structures, routines, and scaffolding in learning through problem solving help support productive success (Hmelo-Silver, Duncan, & Chinn, 2007).

In many ways, the two approaches to learning through problem solving are two sides of the same coin. Both involve having learners begin with a problem they do not know how to solve but must learn new content, skills, and disciplinary practices along the way. Social practices and complex learning environments are part and parcel of both of these pedagogical designs. In PBL, the goal is to scaffold the students towards productive success, whereas in productive failure, the failure that occurs is an opportunity for learning. In PBL the opportunity for learning comes from identifying knowledge gaps. In the end, we argue that the similarities among different models of learning through problem solving are just as important as the distinctions. Indeed, "productive success could well be conceived as a design that embodies iterative cycles of productive failure" (Kapur, 2016, p. 297). We argue that it is critical to be intentional in the design of the task and the timing of support with an eye towards learning goals. In future design and research efforts, we need to be exploring under what circumstances and for whom different approaches to learning through problem solving are effective and why. How does the designed system function to support learning? Learning through problem-solving approaches provide effective designs for learning, but they also provide contexts to study complex learning in action, providing important implications for research and practice. Future approaches to learning through problem solving will depend on the ability of researchers to examine adaptations of the models in practice within different disciplinary foci, and across problem solvers that reflect diverse demographic characteristics (e.g., age, linguistic and experiential backgrounds, socioeconomic status, and rural–urban geographies).

Further Readings

Blumenfeld, P. C., Soloway, E., Marx, R. W., Krajcik, J. S., Guzdial, M., & Palincsar, A. (1991). Motivating project-based learning: Sustaining the doing, supporting the learning. *Educational Psychologist, 26*(3–4), 369–398.
This article overviews motivational and instructional issues inherent in project-based learning and provides suggestions for how technology may be used to support them. The authors contend that student interest in projects can be bolstered by variety, student control, opportunities for collaboration, and teacher scaffolding.

Hmelo-Silver, C. E. (2004). Problem-based learning: What and how do students learn? *Educational Psychology Review, 16*(3), 235–266.
This article provides an overview of goals, core components, and scholarship of problem-based learning. This article also addresses the PBL scholarship that has provided mixed learning results and calls for additional research in diverse settings beyond the medical school context.

Hmelo-Silver, C. E., & Barrows, H. S. (2008). Facilitating collaborative knowledge building. *Cognition and Instruction, 26*(1), 48–94.
This empirical study of student and facilitator discourse examines how expertly facilitated PBL can develop knowledge-building practices among students. Attention to core conditions of knowledge building, such as student engagement in constructing and transforming knowledge as well as student control over learning, can be supported.

Kapur, M. (2016). Examining productive failure, productive success, unproductive failure, and unproductive success in learning. *Educational Psychologist, 51*(2), 289–299.
This article reviews four theoretical categories of success and failure in learning: productive failure, productive success, unproductive success, and unproductive failure. While clear definitions, empirical evidence, and specific design considerations are provided for each, abandoning a strict, dichotomous understanding of the four categories; attention to cognitive, social, and cultural mechanisms; and assessing students' pre-existing knowledge are encouraged when designing for learning.

Walker, A., Leary, H., Ertmer, P. A., & Hmelo-Silver, C. (2015). Epilogue: The future of PBL. In A. Walker, H. Leary, C. Hmelo-Silver, & P. A. Ertmer (Eds.), *Essential readings in problem-based learning: Exploring and extending the legacy of Howard Barrows* (pp. 373–376). West Lafayette, IN: Purdue University Press.
This chapter provides an overview of how past and present conceptions of problem-based learning will inform its future iterations.

NAPLeS Resources

Hmelo-Silver, C. E. *Problem-based learning* [Webinar]. In *NAPLeS video series*. Retrieved October 19, 2017, from http://isls-naples.psy.lmu.de/intro/all-webinars/hmelo-silver/index.html

References

Adams, L., Kasserman, J., Yearwood, A., Perfetto, G., Bransford, J., & Franks, J. (1988). The effect of fact versus problem oriented acquisition. *Memory & Cognition, 16*, 167–175.
Albanese, M., & Mitchell, S. (1993). Problem-based learning: A review of the literature on its outcomes and implementation issues. *Academic Medicine, 68*, 52–81.

Barrett, T., Cashman, D., & Moore, S. (2011). Designing problems and triggers in different media. In T. Barrett & S. Moore (Eds.), *New approaches to problem-based learning: Revitalising your practice in higher education* (pp. 18–35). New York: Routledge.

Blumenfeld, P. C., Soloway, E., Marx, R. W., Krajcik, J. S., Guzdial, M., & Palincsar, A. (1991). Motivating project-based learning: Sustaining the doing, supporting the learning. *Educational Psychologist, 26*(3–4), 369–398.

Bransford, J. D., & Schwartz, D. (1999). Rethinking transfer: A simple proposal with multiple implications. *Review of Research in Education, 24*, 61–100.

Bridges, S., Green, J., Botelho, M., & Tsang, P. C. (2015). Blended learning and PBL: An interactional ethnographic approach to understanding knowledge construction in situ In A. Walker, H. Leary, C. Hmelo-Silver, & P. A. Ertmer (Eds.), *Essential readings in problem-based learning: Exploring and extending the legacy of Howard S. Barrows* (pp. 107–130). West Lafayette, IN: Purdue University Press.

Brown, A. L., Ash, D., Rutherford, M., Nakagawa, K., Gordon, A., & Campione, J. C. (1993). Distributed expertise in the classroom. In G. Salomon (Ed.), *Distributed cognitions* (pp. 188–228). New York: Cambridge University Press.

Brush, T., & Saye, J. (2008). The effects of multimedia-supported problem-based inquiry on student engagement, empathy, and assumptions about history. *Interdisciplinary Journal of Problem-Based Learning, 2*(1), 4.

Capon, N., & Kuhn, D. (2004). What's so good about problem-based learning? *Cognition and Instruction, 22*(1), 61–79.

Chernobilsky, E., DaCosta, M. C., & Hmelo-Silver, C. E. (2004). Learning to talk the educational psychology talk through a problem-based course in educational psychology. *Instructional Science, 32*, 319–356.

Collins, A. (2006). Cognitive apprenticeship. In K. Sawyer (Ed.), *Cambridge handbook of the learning sciences* (pp. 47–60). New York: Cambridge University Press.

CTGV. (1992). The Jasper series as an example of anchored instruction: Theory, program description and assessment data. *Educational Psychologist, 27*, 291–315.

Danish, J., & Gresalfi, M. (2018). Cognitive and sociocultural perspective on learning: tensions and synergy in the Learning Sciences. In F. Fischer, C. E. Hmelo-Silver, S. R. Goldman, & P. Reimann (Eds.), *International handbook of the learning sciences* (pp. 34–43). New York: Routledge.

DeCaro, M. S., & Rittle-Johnson, B. (2012). Exploring mathematics problems prepares children to learn from instruction. *Journal of Experimental Child Psychology, 113*, 552–568.

Deci, E. L., Koestner, R., & Ryan, R. M. (2001). Extrinsic rewards and intrinsic motivation in education: Reconsidered once again. *Review of Educational Research, 71*, 1–27.

Derry, S. J., Hmelo-Silver, C. E., Nagarajan, A., Chernobilsky, E., & Beitzel, B. (2006). Cognitive transfer revisited: Can we exploit new media to solve old problems on a large scale? *Journal of Educational Computing Research, 35*, 145–162.

Dewey, J. (1938). *Experience and education* (Kappa Delta Phi Lecture Series) New York: Macmillan.

Dochy, F., Segers, M., Van den Bossche, P., & Gijbels, D. (2003). Effects of problem-based learning: A meta-analysis. *Learning and Instruction, 13*(5), 533–568.

Eberle, J. (2018). Apprenticeship learning. In F. Fischer, C. E. Hmelo-Silver, S. R. Goldman, & P. Reimann (Eds.), *International handbook of the learning sciences* (pp. 44–53). New York: Routledge.

Engle, R. A. (2006). Framing interactions to foster generative learning: A situative explanation of transfer in a community of learners classroom. *Journal of the Learning Sciences, 15*(4), 451–498.

Ertmer, P. A., & Glazewski, K. D. (in press). Scaffolding in PBL environments: Structuring and problematizing relevant task features. To appear in W. Hung, M. Moallem, N. Dabbagh (Eds.) *Wiley handbook of problem-based learning*.

Evensen, D. H., Salisbury-Glennon, J., & Glenn, J. (2001). A qualitative study of 6 medical students in a problem-based curriculum: Towards a situated model of self-regulation. *Journal of Educational Psychology, 93*, 659–676.

Gijbels, D., Dochy, F., Van den Bossche, P., & Segers, M. (2005). Effects of problem-based learning: A meta-analysis from the angle of assessment. *Review of Educational Research, 75*(1), 27–61.

Green, J. L., & Bridges, S.M. (2018) Interactional ethnography. In F. Fischer, C. E. Hmelo-Silver, S. R. Goldman, & P. Reimann (Eds.), *International handbook of the learning sciences* (pp. 475–488). New York: Routledge.

Greeno, J. G., Collins, A. M., & Resnick, L. B. (1996). Cognition and learning. In D. Berliner & R. Calfee (Eds.), *Handbook of educational psychology* (pp. 15–46). New York: Macmillan.

Hmelo-Silver, C. E. (2004). Problem-based learning: What and how do students learn? *Educational Psychology Review, 16*(3), 235–266.

Hmelo-Silver, C. E., & Barrows, H. S. (2008). Facilitating collaborative knowledge building. *Cognition and Instruction, 26*(1), 48–94.

Hmelo-Silver, C. E., Duncan, R. G., & Chinn, C. (2007). Scaffolding and achievement in problem-based and inquiry learning: A response to Kirschner, Sweller, and Clark (2006). *Educational Psychologist, 42*, 99–107.

Holmes, N., Day, J., Park, A. H., Bonn, D., & Roll, I. (2014). Making the failure more productive: Scaffolding the invention process to improve inquiry behaviors and outcomes in invention activities. *Instructional Science*, *42*(4), 523–538.

Jonassen, D., & Hung, W. (2015). All problems are not created equal: Implications for problem-based learning. In A. Walker, H. Leary, C. Hmelo-Silver, & P. A. Ertmer (Eds.), *Essential readings in problem-based learning: Exploring and extending the legacy of Howard S. Barrows* (pp. 69–84). West Lafayette, IN: Purdue University Press.

Jordan, M. E., & McDaniel, Jr, R. R. (2014). Managing uncertainty during collaborative problem solving in elementary school teams: The role of peer influence in robotics engineering activity. *Journal of the Learning Sciences*, *23*(4), 490–536.

Kapur, M. (2008). Productive failure. *Cognition and Instruction*, *26*(3), 379–424.

Kapur, M. (2012). Productive failure in learning the concept of variance. *Instructional Science*, *40*, 651–672.

Kapur, M. (2014). Productive failure in learning math. *Cognitive Science*, *38*(5), 1008–1022.

Kapur, M. (2015). Learning from productive failure. *Learning: Research and Practice*, *1*(1), 51–65.

Kapur, M. (2016). Examining productive failure, productive success, unproductive failure, and unproductive success in learning. *Educational Psychologist*, *51*(2), 289–299.

Kapur, M., & Bielaczyc, K. (2012). Designing for productive failure. *Journal of the Learning Sciences*, *21*(1), 45–83.

Kapur, M., & Kinzer, C. K. (2007). Examining the effect of problem type in a synchronous computer-supported collaborative learning (CSCL) environment. *Educational Technology Research and Development*, *55*(5), 439–459.

Kilpatrick, W. H. (1918). The project method. *Teachers College Record*, *19*, 319–335.

Kirschner, P. A., Sweller, J., & Clark, R. E. (2006). Why minimal guidance during instruction does not work. *Educational Psychologist*, *41*, 75–86.

Kollar, I., Wecker, C., & Fischer, F. (2018). Scaffolding and scripting (computer-supported) collaborative learning. In F. Fischer, C. E. Hmelo-Silver, S. R. Goldman, & P. Reimann (Eds.), *International handbook of the learning sciences* (pp. 340–350). New York: Routledge.

Kolodner, J. L., Camp, P. J., Crismond, D., Fasse, B., Gray, J., Holbrook, J., Puntambekar, S., & Ryan, M. (2003). Problem-based learning meets case-based reasoning in the middle-school science classroom: Putting learning by design(tm) into practice. *Journal of the Learning Sciences*, *12*(4), 495–547.

Koschmann, T. D., Myers, A., Feltovich, P. J., & Barrows, H. S. (1994). Using technology to assist in realizing effective learning and instruction: A principled approach to the use of computers in collaborative learning. *Journal of the Learning Sciences*, *3*(3), 227–264.

Linn, M. C., McElhaney, K. W., Gerard, L., & Matuk, C. (2018). Inquiry learning and opportunities for technology. In F. Fischer, C. E. Hmelo-Silver, S. R. Goldman, & P. Reimann (Eds.), *International handbook of the learning sciences* (pp. 221–233). New York: Routledge.

Loibl, K., Roll, I., & Rummel, N. (2016). Towards a theory of when and how problem solving followed by instruction supports learning. *Educational Psychology Review*, *29*(4), 693–715. doi:10.1007/s10648-016-9379-x

Loibl, K., & Rummel, N. (2014). Knowing what you don't know makes failure productive. *Learning and Instruction*, *34*, 74–85.

Lu, J., Lajoie, S. P., & Wiseman, J. (2010). Scaffolding problem-based learning with CSCL tools. *International Journal of Computer Supported Collaborative Learning*, *5*(3), 283–298.

Mergendoller, J. R., Maxwell, N. L., & Bellisimo, Y. (2006). The effectiveness of problem-based instruction: A comparative study of instructional method and student characteristics. *Interdisciplinary Journal of Problem-based Learning*, *1*, 49–69.

Ohlsson, S. (1996). Learning from performance errors. *Psychological Review*, *103*, 241–262.

Puntambekar, S., & Hubscher, R. (2005). Tools for scaffolding students in a complex learning environment: What have we gained and what have we missed? *Educational Psychologist*, *40*(1), 1–12.

Puntambekar, S., & Kolodner, J. L. (2005). Toward implementing distributed scaffolding: Helping students learn science from design. *Journal of Research in Science Teaching*, *42*(2), 185–217.

Quintana, C., Reiser, B. J., Davis, E. A., Krajcik, J., Fretz, E., Duncan, R. G., et al. (2004). A scaffolding design framework for software to support science inquiry. *Journal of the Learning Sciences*, *13*(3), 337–386.

Rosé, C. P. (2018). Learning analytics in the Learning Sciences. In F. Fischer, C. E. Hmelo-Silver, S. R. Goldman, & P. Reimann (Eds.), *International handbook of the learning sciences* (pp. 511–519). New York: Routledge.

Savery, J. R. (2015). Overview of problem-based learning: Definitions and distinctions. In A. Walker, H. Leary, C. Hmelo-Silver, & P. A. Ertmer (Eds.), *Essential readings in problem-based learning: Exploring and extending the legacy of Howard S. Barrows* (pp. 5–16). West Lafayette, IN: Purdue University Press.

Schmidt, H. G. (1993). Foundations of problem-based learning: Some explanatory notes. *Medical Education*, *27*, 422–432.

Schmidt, H. G., Loyens, S. M., Van Gog, T., & Paas, F. (2007). Problem-based learning is compatible with human cognitive architecture: Commentary on Kirschner, Sweller, and Clark (2006). *Educational Psychologist*, *92*, 91–97.

Schmidt, H. G., Rotgans, J. I., & Yew, E. H. (2011). The process of problem-based learning: What works and why. *Medical Education, 45*(8), 792–806.

Schwartz, D., Chase, C. C., & Bransford, J. (2012). Resisting overzealous transfer: Coordinating previously successful routines with needs for new learning. *Educational Psychologist, 47*(3), 204–214.

Schwartz, D., Chase, C. C., Oppezzo, M. A., & Chin, D. B. (2011). Practicing versus inventing with contrasting cases: The effects of telling first on learning and transfer. *Journal of Educational Psychology, 103*, 759–775.

Schwartz, D., & Martin, T. (2004). Inventing to prepare for future learning: The hidden efficiency of encouraging original student production in statistics instruction. *Cognition and Instruction, 22*, 129–184.

Vernon, D. T. A., & Blake, R. L. (1993). Does problem-based learning work? A meta-analysis of evaluative research. *Academic Medicine, 68*, 550–563.

Vygotsky, L. S. (1978). *Mind in society: The development of higher mental process.* Cambridge, MA: Harvard University Press.

Walker, A., Leary, H., & Lefler, M. (2015). A meta-analysis of problem-based learning: Examination of education levels, disciplines, assessment levels, problem types, and reasoning strategies. In A. Walker, H. Leary, C. Hmelo-Silver, & P. A. Ertmer (Eds.), *Essential readings in problem-based learning: Exploring and extending the legacy of Howard S. Barrows* (pp. 303–330). West Lafayette, IN: Purdue University Press.

Whitehead, A. N. (1929). *The aims of education and other essays.* New York: Macmillan.

Wijnia, L., Loyens, S. M., & Derous, E. (2011). Investigating effects of problem-based versus lecture-based learning environments on student motivation. *Contemporary Educational Psychology, 36*(2), 101–113.

Yew, E. H., & Schmidt, H. G. (2009). Evidence for constructive, self-regulatory and collaborative processes in problem-based learning. *Advances in Health Sciences Education, 14*, 251–273.

Zhang, M., Lundeberg, M., & Eberhardt, J. (2011). Strategic facilitation of problem-based discussion for teacher professional development. *Journal of the Learning Sciences, 20*, 342–394.

22
Inquiry Learning and Opportunities for Technology

Marcia C. Linn, Kevin W. McElhaney, Libby Gerard, and Camillia Matuk

Whether in history, science, journalism, economics, or other disciplines, inquiry activities engage learners in exploring meaningful problems, testing conjectures about relationships among variables, comparing alternative explanations (often by building and testing models), using evidence to refine ideas, and developing arguments for promising solutions (Furtak, Seidel, Iverson, & Briggs, 2012). Inquiry instruction can exploit the multiple, often conflicting ideas that students have about personal, societal, and environmental dilemmas, and help them to sort out these ideas to address challenges such as economic disparity or health decision making (Donnelly, Linn, & Ludvigsen, 2014; Herrenkohl & Polman, this volume). Technologies such as natural language processing, interactive simulations, games, collaborative tools, and personalized guidance can support students to become autonomous learners (Quintana et al., 2004; Tabak & Kyza, this volume). Logs of student activities can capture class performance and inform teachers of student progress (Gerard, Matuk, McElhaney, & Linn, 2015).

Autonomous learners identify gaps in arguments and independently seek evidence to select among alternatives. Learning environments can promote autonomous efforts to sort out, link, and connect cultural, social, economic and scientific ideas (see Figure 22.1). Autonomous capabilities empower all citizens to take charge of their lives and strengthen democratic decision making.

This chapter integrates advances in theory, instructional design, and technology concerning interdisciplinary inquiry learning. We highlight autonomous learning from (a) student-initiated investigations of thorny, contemporary problems using modeling and computation tools, (b) design projects featuring analysis of alternatives, testing prototypes, and iteratively refining solutions in complex disciplines, and (c) reflection activities that encourage gathering and synthesizing evidence from multiple sources and using automated, personalized guidance to revise.

Historical Trends Culminating in Impact of Learning Sciences

Inquiry instruction has roots in the experiential learning philosophies of Rousseau and Dewey. Ideals of inquiry were often inspired by images of learners, benevolently guided by a skilled tutor, independently making startling insights. For example, Rousseau (1979) describes a fictitious child who, while playing with a kite, uses the shadow of the kite to infer its position. This image of hands-on, discovery learning implies that autonomy is inherent, when it is actually cultivated through well-designed instruction. Calls for hands-on investigations or active learning tend to come from experts who are already autonomous learners. Teachers, left with the task of guiding

Marcia C. Linn, Kevin W. McElhaney, Libby Gerard, and Camillia Matuk

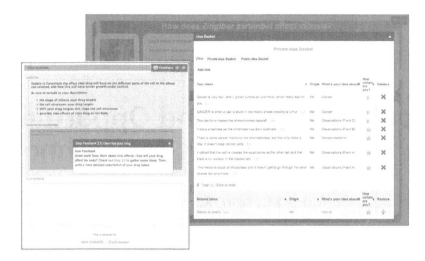

Figure 22.1 Screenshots from the Web-Based Inquiry Science Environment (WISE) Cell Division Unit

Note: This figure shows technologies designed to guide middle school students' inquiry into cancer treatment. The Idea Manager (left) supports students' self-monitoring and collaborative learning as they document, organize, share, and integrate their ideas (Matuk et al., 2016). Automated scoring (right) supports continuous formative assessment and personalized guidance to help students refine their arguments.

Source: Gerard et al. (2015).

students in inquiry, are often challenged to create the classroom structures, scaffolds for autonomous investigation, and student guidance necessary to convert hands-on activities into learning opportunities.

Historically, when open-ended inquiry activities failed, they were often replaced by abstract images of the scientific method accompanied by step-by-step exercises that resonated with emerging behaviorist theories in the 1930s. For example, when students failed to derive Hooke's Law by using the scientific method to experiment with springs, designers attempted to help students identify potential confounds by providing them with explicit, step-by-step instructions. This solution made classroom implementation of experimentation easier while downplaying autonomous investigation. It also generally left students with fragmented ideas because they were not encouraged to distinguish between their own ideas and those promoted by the instruction (Linn & Eylon, 2011).

In the 1980s, spurred by government funding in Europe and the United States, and building on research illustrating how scientific reasoning is entwined in and advances with disciplinary knowledge (Brown, Collins, & Duguid, 1989), disciplinary experts, learning scientists, technology experts, and classroom teachers established partnerships to improve inquiry-oriented curriculum materials and evaluate their effectiveness. For example, the U.S. American National Science Foundation (NSF) funded individual research programs and centers that required multidisciplinary partnerships. These partnerships tackled the challenge of designing instruction that coupled inquiry about realistic, complex problems with guidance to support autonomous investigations. In addition, researchers clarified learners' autonomous inquiry capabilities as a set of interacting practices that develop in concert with disciplinary knowledge. These practices include developing and refining models, evaluating and testing simulations, analyzing and interpreting data, and forming and critiquing explanations (e.g., National Research Council, 2012).

In the 1990s, research in the learning sciences incorporated an emphasis on coherent understanding and researched autonomous learning capabilities such as metacognition and collaboration among a broad and diverse population. Furthermore, learning scientists developed learning environments that log student interactions with the goal of documenting, interpreting, and supporting students' inquiry in a wide range of disciplines. In addition, designers created innovative activities that could be embedded in inquiry learning environments, including concept maps, drawings, essays, hands-on examinations, critiques of experiments, and portfolios. These activities encourage students to integrate and apply the ideas they encounter in instruction while at the same time documenting student progress. Analysis of student trajectories during inquiry helps clarify how inquiry instruction can promote coherent, robust, and durable understanding of complex topics along with the autonomous capability to conduct investigations of new topics (see WISE, Figure 22.1).

Researchers have created culturally responsive curriculum materials featuring personally relevant problems such as contested historical events (e.g., the Spanish–American war) or localized environmental stewardship. They have tested and refined ways to design personalized guidance, facilitate classroom discourse, help students to deal with multiple conflicting ideas, guide interpretation of historical documents, and negotiate cultural expectations. They have studied instructional patterns for guiding students to develop and articulate coherent explanations. The knowledge integration framework emerged to guide the design of learning environments, instruction, assessment, and collaborative tools with the goal of helping students express, refine, and integrate their ideas to construct coherent, causal descriptions of scientific phenomena (Linn & Eylon, 2011). Constructionism, another constructivist view, emerged to guide learning from the making of artifacts (Papert, 1993). Such innovations inform design of instruction that promotes autonomy and prepare learners to use inquiry to tackle new and meaningful problems.

At the same time, the audience for inquiry learning broadened to include all citizens, not just students with professional aspirations. Learning scientists responded to this broadening of participation by incorporating identity and sociocultural learning perspectives into instruction. They began to investigate ways to develop students' identities as intentional, autonomous, lifelong learners (Raes, Schellens, & De Wever, 2013). Researchers developed ways to respect and build on the diverse and culturally rich experiences students bring to inquiry activities. To address stereotypes about who can succeed in specific fields, designers featured role models from the communities of the students and dilemmas relevant to their lives. The focus on relevant dilemmas has accompanied a blurring of boundaries across disciplines and between in-school and out-of-school learning. Investigators have identified ways to motivate learners to intentionally seek to make sense of the world, solve personally relevant problems, build on community knowledge, and participate in a community of learners (Sharples et al., 2015; Danish & Gresalfi, this volume; Slotta, Quintana, & Moher, this volume; see nQuire, Figure 22.2). Studies show that personalizing inquiry in the context of historical games, practical challenges such as designing low-cost e-textiles, and meaningful questions such as how to design a cancer-fighting drug can help students envision themselves as capable of solving novel, relevant challenges (Kafai et al., 2014; Renninger, Kern, & Ren, this volume).

Research methods to support investigation of the complex, systemic, and sociocultural aspects of inquiry learning have evolved with advances in the learning sciences (Hoadley, this volume). Design-based research methods that emphasize iterative refinement, informed by theory, reveal ways to improve outcomes from inquiry learning (Chinn & Sandoval, this volume; Puntambekar, this volume). Inquiry activities embedded in learning environments enable researchers to apply learning analytics to log files to reveal patterns in students' collaborative and inquiry processes (Rosé, this volume). Well-designed, technology-enhanced learning environments make it possible to utilize multiple, robust measures of student progress and implement them as part of learning rather than interrupting learning with assessments that do not themselves advance student understanding (see Pellegrino, this volume). In this chapter we discuss illustrative technological and instructional advances and identify crucial elements of successful instruction that are essential for the success of inquiry instruction and the development of autonomous learning capabilities (Linn, Gerard, Matuk, & McElhaney, 2016).

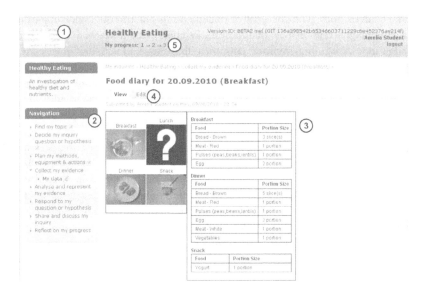

Figure 22.2 Screenshot of an nQuire (Sharples et al., 2015) Activity From an Investigation on Healthy Eating

Note: Numbers illustrate types of supports for learners monitoring their own progress through an investigation: (1) visual representation of the inquiry process, (2) hierarchical panel to navigate between activities, (3) current activity content, (4) tabs for toggling between activity viewing and editing, and (5) progression through temporal stages of the inquiry process.

Modeling, Computational Thinking, and Inquiry

Findings from the learning sciences have motivated educators to design instruction that makes explicit the mechanisms behind scientific and social phenomena. Models are representations (often computer-based) of a phenomenon or system whose central features are explicit and visible. The explanatory and predictive nature of models allows learners to investigate contemporary problems as part of inquiry into the natural or the designed world (de Jong, Lazonder, Pedaste, & Zacharia, this volume). Technology advances (such as visualization tools, programming environments, and the computing power to handle large datasets) have made modeling a central practice in professional inquiry. Natural scientists use models to explain complex systems and to make predictions about newly observed phenomena. Engineers and designers use models to develop, and refine prototypes prior to implementing full design solutions. Social scientists use models to characterize and predict human behavior, such as outcomes of elections or sporting events.

Research syntheses of inquiry instruction that incorporates interactive modeling tools have identified design principles for promoting deep conceptual learning (McElhaney, Chang, Chiu, & Linn, 2014). These studies show that effective design of instructional scaffolds contribute substantially to the value of models for promoting inquiry learning. For instance, supporting autonomous inquiry with models, rather than giving step-by-step instructions, encourages learners to test their own ideas, resulting in better conceptual or mechanistic understanding of a phenomenon. Prompts for learners to engage in self-monitoring and reflection help students achieve more coherent understanding of the phenomenon being modeled.

The emergence of computation (alongside theory and experiment) in science and engineering is promoting the inclusion of computational thinking into K–12 STEM curricular programs. By developing computational models, learners can better understand the mathematical and epistemological foundations of models and make their own decisions about what aspects of a model to include and how to specify their relationships. To meet this need, computational model building environments (such as NetLogo or AgentSheets) make the mathematical relationships that underpin complex phenomena explicit for learners (Reppening et al., 2015; Wilensky & Reisman, 2006). For example, these computational environments enable learners to simulate emergent phenomena that result from relatively straightforward rules such as the spread of disease. Computational modeling thus offers a powerful way for learners to engage in complex inquiry that integrates multiple disciplines.

A significant barrier for pre-college learners is the need for teachers and their students to learn the programming skills required to develop and test a computational model. For students to build their own artifacts or models requires more classroom time than is typically available in a science, mathematics, or history course. Learning scientists are actively exploring ways to make both model building and model exploration more accessible to all learners (Basu et al., 2016).

Future Directions: Models as Assessments of Inquiry Learning

Students' interactions with models can provide a wealth of information with which to assess students' proficiency with inquiry practices. For example, analyses of data logs from Energy3D (Xie, Zhang, Nourian, Pallant, & Hazzard, 2014) and other modeling environments reveal learning processes that can inform the design of adaptive guidance. Many games support model-based assessment by requiring learners to explore and understand variable relationships to achieve goals (see Fields & Kafai, this volume). Modeling activities embedded in games have the additional benefits of providing learners with continuous feedback on their completion of modeling challenges and engaging learners in a wide range of disciplines. For example, SimCity and Civilization connect a compelling narrative to underlying economic models to support causes and remedies for economic disparity (DeVane, Durga, & Squire, 2010; Nilsson & Jakobsson, 2011).

Design and Inquiry

Design is a growing part of inquiry instruction (e.g., Kolodner et al., 2003; Xie et al., 2014). Design offers realistic, ill-defined challenges through which students can apply such core practices as defining problems, making predictions, building arguments, experimentation, and iterative refinement (e.g., Blikstein, 2013). It also highlights the various facets of inquiry, including its collaborative, practical, and disciplinary natures, and its social value. For instance, in using design to address contemporary real-world problems, whether this involves designing solar ovens to replace the burning of soft coal or a campaign to promote social change, learners must develop both a deep understanding of disciplinary ideas and an ability to empathize with the people and communities for whom they design. Moreover, part of a successful design process involves reasoning about the constraints and trade-offs of contexts and available resources, and considering and integrating diverse perspectives through collaboration and ongoing reflection.

Inquiry learning through design is valued because it goes beyond teaching content and practices to also teaching dispositions, including risk-taking, tinkering, and persistence, that foster autonomous learning. It encourages learners to pursue interest-driven projects that engage them with disciplinary ideas and practices, as in how student-driven e-textile projects can be used to introduce programming concepts to high school students and to broaden their perceptions of the role of computer science in society (Kafai et al., 2014; Fields & Kafai, this volume). Design also promotes learners' abilities to define problems, seek help, and use failure productively. Floundering while seeking creative design solutions can frustrate students, but with the proper scaffolding and guidance to surmount vexing challenges, these experiences can also be opportunities to develop self-monitoring, an aspect of autonomy (Järvelä, Hadwin, Malmberg, & Miller, this volume).

Research on the potential of design activities to support inquiry learning arises from the integration of increasingly accessible fabrication technologies and constructionist learning perspectives. For instance, the increasing affordability of digital fabrication technologies has enabled a Maker

Movement to emerge which has stimulated extensive learning activities that traverse disciplines as well as in- and out-of-school contexts (Halverson & Peppler, this volume). Recognition of the universal value of engineering practices has furthermore spurred the creation of new engineering-related curriculum materials in pre-college education.

Future Directions: Providing Structure, Promoting Ownership

Questions remain about how to effectively guide design activities that promote disciplinary learning, while at the same time allowing learners to develop agency and pursue personal interests through taking ownership over problems. Researchers observe several successful guidelines. For instance, teachers can bring attention to productive attitudes toward risk and failure, create opportunities for learners to learn from their peers' mistakes, such as through public tests of designs, and generate authentic motivation for documenting and refining ideas (Sadler, Coyle, & Schwartz, 2000). Technology environments might empower learners with tools for creating complex artifacts, for learning productively from peers, for gaining timely access to resources, and for offering adaptive support for their design reasoning. Technology might also highlight evidence of learning from open-ended design challenges, revealing patterns that teachers can monitor and use to guide progress.

Reflection, Guidance, and Inquiry

Students gather a multitude of new ideas through modeling, design, and other inquiry experiences. However, these ideas often remain distinct from, or conflict with, one another and the ideas students have gathered from their prior experiences. Given opportunities for reflection, students can compare ideas, grapple with inconsistencies, identify gaps, and build connections among their diverse ideas. Students benefit from personalized guidance during reflection (Gerard et al., 2015; Lazonder & Harmsen, 2016).

Learning scientists have developed learning environments to guide students toward productive reflection during inquiry. They embed opportunities for students to construct arguments and explanations as they progress through an inquiry project. These activities encourage students to make their ideas visible, compare their ideas to those of their peers, and connect their ideas to prior instruction and experiences (Quintana et al., 2004). Recent technologies, including argument structuring tools and automated guidance, guide students as they sort out their often disparate views. The tools help students to refine their understanding and achieve coherent views of inquiry topics. Such technologies can also make student ideas immediately available to teachers, enabling them to help students advance their own understanding. By encouraging students to see themselves as investigators rather than accumulators of facts and to see their teachers as guides rather than authorities on knowledge, the technologies guide students to develop autonomous use of inquiry practices (Gerard et al., 2015).

Argument Structuring Tools

Refining arguments during inquiry involves the processes of critique and revision, particularly in professional practice. Studies reveal that students seldom revise their ideas. Like learners of all types, students often misconstrue contradictory evidence to align it with their view, or ignore alternative views and assert their own perspective (Berland & Reiser, 2011). Supporting argumentation in formal learning settings as a process of evidence-based revision calls for a fundamental shift in classroom culture. Rather than directing students toward the "correct" answer, argumentation emphasizes the integration of evidence and continual revision of the connections among ideas in support of integrated understanding. Students reconcile inconsistencies in their views by revisiting evidence, such as a computer model or designed artifact, and consulting resources such as peers. Learning how to use evidence and resources to construct and revise one's argument is central to becoming an autonomous learner.

Argument structuring tools enable students to distinguish and organize evidence they gather during the course of inquiry to create coherent perspectives on complex issues (e.g. using the Idea Manager, Figure 22.1; see Schwarz, this volume). Contemporary argument structuring tools enable students to incorporate visual evidence such as photos, screenshots, or graphs; create concept maps; and add their peers' documented ideas to their own collection of evidence. Research finds that, while students identify relevant ideas during inquiry, they struggle to distinguish and integrate them. Argument structuring tools encourage students to sort their ideas into categories and refine the criteria for categories as new evidence is encountered.

Students who categorize their ideas, as opposed to only accumulating ideas, as they progress through an investigation, form a more lasting understanding of the inquiry issue. For example, students' ability to categorize the ideas they had recorded in the Idea Manager during an investigation of the chemistry of recycling predicted the coherence of their understanding more strongly than the number of ideas they had added (McElhaney, Matuk, Miller, & Linn, 2012). Students who used Belvedere to organize their ideas about natural phenomena were more likely to delineate connections among their ideas than students who composed solely written arguments (Toth, Suthers, & Lesgold, 2002).

Analyses of how students developed argument structures enabled learning scientists to design scaffolds that guide students to distinguish among alternatives within an argument that are likely to otherwise go unnoticed (Reiser, 2004). For example, students provided with some predefined conceptual categories to help them sort the evidence they gathered during an investigation were more likely to generate additional categories based on science principles, whereas students who had to generate all categories on their own were less likely to create categories to effectively distinguish among ideas (Bell & Linn, 2000). Other tools structure peer interactions to encourage students to diversify their ideas as they develop arguments. Students using the Idea Manager who added contrasting ideas from a peer to their collection of ideas, rather than adding ideas similar to their own, developed more robust arguments and science knowledge (Matuk & Linn, 2015).

Automated, Personalized Guidance for Student Explanations

Explanations, like arguments, constitute an inquiry-based artifact that compels learners to synthesize ideas from multiple sources. Providing students with guidance for their explanations can encourage students to revisit evidence and refine the connections among their ideas. This guidance is typically crafted by expert teachers, who, based on their students' explanations, distinguish the promising ideas from those that may hinder reasoning (van Zee & Minstrell, 1997). Automated scoring technologies can support teachers in providing students with personalized guidance for their explanations during inquiry. Natural language processing tools and diagram-scoring algorithms assess the coherence and accuracy of students' explanations with scoring reliability approaching that of human scorers (Liu, Rios, Heilman, Gerard, & Linn, 2016). The resulting computer-generated score can be used to provide students with immediate, personalized guidance. Even with accurate scoring, this guidance must be carefully designed in order to encourage students to make revisions that focus on connections among ideas rather than adding disparate ideas. Current research efforts examine how best to design such guidance (Gerard et al., 2015).

Research across disciplines suggests that encouraging students to critique and refine their explanations is more beneficial than providing feedback on the correctness of explanations. Comparison studies suggest that automated guidance that encourages students to distinguish and clarify the gaps in their reasoning leads to higher quality revisions during an inquiry project and higher pretest to posttest learning gains, compared to giving a specific hint or right/wrong feedback. This finding is most pronounced for students who encounter the activity with low prior knowledge, evidenced in studies of learning in the language arts (Franzke, Kintsch, Caccamise, Johnson, & Dooley, 2005), geometry (Razzaq & Heffernan, 2009), and inquiry science (Tansomboon, Gerard, Vitale, & Linn, 2017). These findings are further supported by a meta-analysis of effect sizes drawn from comparison studies of automated guidance for student-generated artifacts in K–12 classrooms (Gerard et al., 2015). Guidance that provided individualized hints to help students strengthen their revision strategies, in addition to disciplinary support, was most successful for long-term learning. We conjecture this is because guidance that targets students' ideas about the discipline, as well as ideas about how to refine one's understanding, encourages students to develop the inquiry strategies essential to becoming autonomous learners.

Future Directions: Guidance to Promote Autonomous Learning

Technological advancements coupled with learning sciences research points to the next direction: how to design guidance that promotes students' autonomous use of inquiry practices? One approach is for technology to help students form stronger connections between their everyday language and the language of the discipline. Bridging these two linguistic spheres may encourage learners to better see themselves as a participant in inquiry, leverage their existing knowledge and experiences, and subsequently develop more autonomous use of inquiry practices. For example, situating reflection activities in students' natural activities, such as a peer dialogue, could elicit and capture students' articulation of inquiry issues in their everyday language. Natural language processing can be used on this dataset to identify students' expressions, and this language can be used to tailor guidance.

Another promising direction is using automated analysis of students' reflections to provide teachers and school leaders with rich assessment information. This approach converts meaningful learning activities such as student written or diagrammed arguments, into powerful assessments. Using both written and diagrammatic assessments increases scoring validity and provides language learners an alternative method to express their views (Ryoo & Linn, 2015). Drawing on embedded data provides stakeholders with data on student trajectories, as opposed to student performance at a fixed point in time. This focuses assessment on both learners' disciplinary understanding and their use of inquiry practices to strengthen their views.

Conclusions

Inquiry skills can promote lifelong learning and active participation in society. Researchers are exploiting new technologies to broaden the scope of inquiry, enable teachers to localize instruction, and help students develop autonomous learning capabilities that are essential for addressing contemporary issues, thereby improving their own and others' lives.

Promoting Autonomy

Preparing learners to address complex problems through inquiry requires carefully designed curriculum materials that take advantage of innovative learning technologies while also building on learners' ideas. Research has found promise in various technologies for supporting students in generating and testing their own ideas by guiding them in developing models, designing solutions, and constructing evidence-based explanations. These technologies prompt students to continually monitor their progress and evaluate and refine their own inquiry artifacts. Personalized guidance based on automated analysis of logged data can encourage learners to assume greater responsibility for their own progress, rather than view their teacher as the singular authority on knowledge. These technologies show promise in supporting inquiry across disciplines, including language arts, history, science, mathematics, engineering, and economics.

Inquiry-based materials can emphasize issues that students will find relevant at the individual, community, and/or global levels. Selecting relevant contexts can engage diverse students, and promote their agency and identities as inquiring citizens. Inquiry curricula that also highlight connections among historical, social, scientific, and mathematical domains, and make connections to out-of-school learning opportunities, can reveal the connections between classroom learning and everyday life (see Lyons, this volume).

Opportunities for Continuous Assessment

Technology-enhanced inquiry instruction provides opportunities to reconceptualize assessment as a continuous, formative process integrated with instruction instead of a summative, standardized process sequestered from instruction. Continuous assessment offers students more varied and authentic ways to express their ideas than typical summative assessments. Technology-enhanced learning environments can take advantage of learning analytics approaches to measure student trajectories during the course of instruction. Coupling these embedded assessments with guidance tightly integrates instruction and assessment and promotes students' self-monitoring, self-evaluation, and iterative refinement that are central to autonomous inquiry learning.

Technology-enhanced environments, whether investigation-based or game-based, can use data logs to make student progress visible to teachers. Dashboards of student learning can inform teachers' individual or whole class guidance. These reports not only represent learning at discrete points in time, but also provide valuable information about students' learning trajectories over time.

Synthesis of Technological Innovation

To promote more efficient, collaborative progress, the NSF Task Force on Cyberlearning (2008) encouraged researchers to build on, rather than reinvent existing solutions. A learning environment (e.g., Figures 22.1 and 22.2) open to all designers and users could support such efforts, and dramatically accelerate both research on and scalability of inquiry learning. It would offer instructors a repository of tested and refined curriculum materials, and could readily evolve to incorporate emerging technologies such as automated scoring, computational modeling tools, collaborative features, and games (Linn, 2016a, 2016b, 2016c).

Such an environment could support comparative research on the diverse theoretical perspectives and contexts currently used, and help to synthesize our collective understanding of ways to support inquiry learning (e.g., how to balance the quality and amount of guidance for developing autonomous inquiry skills). It could also provide continuous assessment of learners' progress within and across their learning experiences. Information from such assessments could be used to tailor instruction to individuals based on their past experiences in multiple disciplines. Guidance could build on student insights from prior instruction and related courses, ensuring that each learner is appropriately challenged.

Finally, this environment could provide a common platform for teachers to seamlessly implement series of inquiry activities, collaborate on customizing materials, share strategies for enactment, and collaborate with researchers to create and test new learning innovations.

Acknowledgment

This material is based on work supported by the National Science Foundation under NSF Projects 1451604, DRL-1418423, 1119270, 0822388, 0918743, and Department of Education Project DOE R305A11782. Any opinions, findings, and conclusions or recommendations expressed in this material are those of the authors and do not necessarily reflect the views of the National Science Foundation.

Further Readings

Furtak, E. M., Seidel, T., Iverson, H., & Briggs, D. C. (2012). Experimental and quasi-experimental studies of inquiry-based science teaching: A meta-analysis. *Review of Educational Research, 82*(3), 300–329.

This meta-analysis synthesizes 37 studies of inquiry published between 1996 and 2006. Consistent with prior research, the mean effect size was .50 favoring inquiry. The effect size for studies featuring teacher-led inquiry was .40 higher than the effect size for student-led inquiry. These results underscore the importance of guidance to realize the benefit of inquiry instruction.

Gerard, L., Matuk, C., McElhaney, K., & Linn, M. C. (2015). Automated, adaptive guidance for K–12 education. *Educational Research Review, 15*, 41–58.

This meta-analysis synthesized 24 independent comparisons between automated adaptive guidance and guidance provided during typical teacher-led instruction, and 29 comparisons between enhanced adaptive guidance and simple adaptive guidance. Adaptive guidance demonstrated a significant advantage over each of the activities used by the researchers to represent "typical classroom instruction." Enhanced guidance that diagnosed content gaps and encouraged autonomy was more likely to improve learning outcomes than guidance that only addressed content.

McElhaney, K. W., Chang, H. Y., Chiu, J. L., & Linn, M. C. (2015). Evidence for effective uses of dynamic visualisations in science curriculum materials. *Studies in Science Education, 51*(1), 49–85.

This meta-analysis synthesizes 76 empirical design comparison studies testing the effectiveness of instructional scaffolds for modeling tools in science. Each of the 76 studies isolates a single design feature of the tool (or the supporting instruction for the tool) by comparing learning outcomes from a typical version of the tool to an enhanced version. Inquiry-based scaffolds found to be most successful include interactive modeling features and prompts for promoting sense-making, self-monitoring, and reflection.

Raes, A., Schellens, T., & De Wever, B. (2013). Web-based collaborative inquiry to bridge gaps in secondary science education. *Journal of the Learning Sciences, 23*(3), pp. 316–347. doi:10.1080/10508406.2013.836656

This experimental study investigates the impact of a web-based, collaborative inquiry unit on diverse high school students in 19 classes. The results show that inquiry is effective and has advantages for students who are not typically successful in science or are not enrolled in a science track. Furthermore, the unit gives low-achieving students and general-track students an opportunity to develop science practices and confidence in learning science.

Xie, C., Zhang, Z., Nourian, S., Pallant, A., & Hazzard, E. (2014). A time series analysis method for assessing engineering design processes using a CAD tool. *International Journal of Engineering Education, 30*(1), 218–230.

This study describes the use of a computer-aided design environment for assessing students' engineering design processes. Students' interactions with the environment are continuously logged while students engage with a design challenge. A classroom study examined the logs of high school engineering students working on a solar urban design challenge.

NAPLeS Resources

Linn, M., *15 minutes about inquiry learning* [Video file]. In *NAPLeS video series*. Retrieved October 19, 2017, from http://isls-naples.psy.lmu.de/video-resources/guided-tour/15-minutes-linn/index.html

Linn, M., *Interview about inquiry learning* [Video file]. In *NAPLeS Video Series*. Retrieved October 19, 2017, from http://isls-naples.psy.lmu.de/video-resources/interviews-ls/linn/index.html

Linn, M., *Inquiry learning* [Webinar]. In *NAPLeS video series*. Retrieved October 19, 2017, from http://isls-naples.psy.lmu.de/intro/all-webinars/linn_video/index.html

References

Basu, S., Biswas, G., Sengupta, P., Dickes, A., Kinnebrew, J. S., & Clark, D. (2016). Identifying middle school students' challenges in computational thinking-based science learning. *Research and Practice in Technology Enhanced Learning, 11*(1), 1–35.

Bell, P., & Linn, M. C. (2000). Scientific arguments as learning artifacts: Designing for learning from the Web with KIE. *International Journal of Science Education, 22*(8), 797–817.

Berland, L. K., & Reiser, B. J. (2011). Classroom communities' adaptations of the practice of scientific argumentation. *Science Education, 95*(2), 191–216.

Blikstein, P. (2013). Digital fabrication and "making" in education: The democratization of invention. In J. Walter-Herrmann & C. Büching (Eds.), *FabLabs: Of machines, makers and inventors*. Bielefeld, Germany: Transcript.

Brown, J. S., Collins, A., & Duguid, P.(1989). Situated cognition and the culture of learning. *Educational Researcher, 18*(1), 32–42.

Chinn, C., & Sandoval, W. (2018). Epistemic cognition and epistemological development. In F. Fischer, C. E. Hmelo-Silver, S. R. Goldman, & P. Reimann (Eds.), *International handbook of the learning sciences* (pp. 24–33). New York: Routledge.

Danish, J., & Gresalfi, M. (2018). Cognitive and sociocultural perspective on learning: Tensions and synergy in the Learning Sciences. In F. Fischer, C. E. Hmelo-Silver, S. R. Goldman, & P. Reimann (Eds.), *International handbook of the learning sciences* (pp. 34–43). New York: Routledge.

de Jong, T., Lazonder, A., Pedaste, M., & Zacharia, Z. (2018). Simulations, games, and modeling tools for learning. In F. Fischer, C. E. Hmelo-Silver, S. R. Goldman, & P. Reimann (Eds.), *International handbook of the learning sciences*. New York: Routledge.

DeVane, B., Durga, S., & Squire, K. (2010). "Economists Who Think Like Ecologists": Reframing systems thinking in games for learning. *E-Learning and Digital Media, 7*(1), 3–20.

Donnelly, D. F., Linn, M. C., & Ludvigsen, S. (2014). Impacts and characteristics of computer-based science inquiry learning environments for precollege students. *Review of Educational Research, 20*(10), 1–37. doi:10.3102/0034654314546954

Fields, D. A., & Kafai, Y. B. (2018). Games in the learning sciences: Reviewing evidence from playing and making games for learning. In F. Fischer, C. E. Hmelo-Silver, S. R. Goldman, & P. Reimann (Eds.), *International handbook of the learning sciences* (pp. 276–284). New York: Routledge.

Franzke, M., Kintsch, E., Caccamise, D., Johnson, N., & Dooley, S. (2005). Summary Street®: Computer support for comprehension and writing. *Journal of Educational Computing Research, 33*, 53–58.

Furtak, E. M., Seidel, T., Iverson, H., & Briggs, D. C. (2012). Experimental and quasi-experimental studies of inquiry-based science teaching: A meta-analysis. *Review of Educational Research, 82*(3), 300–329. doi:10.3102/0034654312457206

Gerard, L., Matuk, C., McElhaney, K., & Linn, M.C. (2015). Automated, adaptive guidance for K–12 education. *Educational Research Review, 15*, 41–58.

Halverson, E., & Peppler, K. (2018). The Maker Movement and learning. In F. Fischer, C. E. Hmelo-Silver, S. R. Goldman, & P. Reimann (Eds.), *International handbook of the learning sciences* (pp. 285–294). New York: Routledge.

Herrenkohl, L. R., & Polman J. L. (2018). Learning within and beyond the disciplines. In F. Fischer, C. E. Hmelo-Silver, S. R. Goldman, & P. Reimann (Eds.), *International handbook of the learning sciences* (pp. 106–115). New York: Routledge.

Hoadley, C. (2018). A short history of the learning sciences. In F. Fischer, C. E. Hmelo-Silver, S. R. Goldman, & P. Reimann (Eds.), *International handbook of the learning sciences* (pp. 11–23). New York: Routledge.

Järvelä, S., Hadwin, A., Malmberg, J., & Miller, M. (2018). Contemporary perspectives of regulated learning in collaboration. In F. Fischer, C. E. Hmelo-Silver, S. R. Goldman, & P. Reimann (Eds.), *International handbook of the learning sciences* (pp. 127–136). New York: Routledge.

Kafai, Y. B., Lee, E., Searle, K., Fields, D., Kaplan, E., & Lui, D. (2014). A crafts-oriented approach to computing in high school: Introducing computational concepts, practices, and perspectives with electronic textiles. *ACM Transactions on Computing Education (TOCE), 14*(1), 1–20.

Kolodner, J. L., Camp, P. J., Crismond, D., Fasse, B., Gray, J., Holbrook, J., Puntambekar, S., & Ryan, M. (2003). Problem-based learning meets case-based reasoning in the middle-school science classroom: Putting learning by design (tm) into practice. *Journal of the Learning Sciences, 12*(4), 495–547.

Lazonder, A. W., & Harmsen, R. (2016). Meta-analysis of inquiry-based learning effects of guidance. *Review of Educational Research, 86*(3), 681–718. doi:10.3102/0034654315627366

Linn, M. (2016a). *15 minutes about inquiry learning* [Video file]. In *NAPLeS video series*. Retrieved October 19, 2017, from http://isls-naples.psy.lmu.de/video-resources/guided-tour/15-minutes-linn/index.html http://isls-naples.psy.lmu.de/video-resources/guided-tour/15-minutes-linn/index.html

Linn, M. (2016b). *Interview about inquiry learning* [Video file]. In *NAPLeS video series*. Retrieved October 19, 2017, from http://isls-naples.psy.lmu.de/video-resources/interviews-ls/linn/index.html

Linn, M. (2016c). *Inquiry learning* [Webinar]. In *NAPLeS video series*. Retrieved October 19, 2017, from http://isls-naples.psy.lmu.de/intro/all-webinars/linn_video/index.html

Linn, M. C., & Eylon, B.-S. (2011). *Science learning and instruction: Taking advantage of technology to promote knowledge integration*. New York: Routledge.

Linn, M. C., Gerard, L. F., Matuk, C. F., & McElhaney, K. W. (2016). Science education: From separation to integration. *Review of Research in Education, 40*(1), 529–587. doi:10.3102/0091732X16680788

Liu, L., Rios, J., Heilman, M., Gerard, L., & Linn, M. (2016). Validation of automated scoring of science assessments. *Journal of Research in Science Teaching, 53*(2), 215–233.

Lyons, L. (2018). Supporting informal STEM learning with technological exhibits: An ecosystemic approach. In F. Fischer, C. E. Hmelo-Silver, S. R. Goldman, & P. Reimann (Eds.), *International handbook of the learning sciences* (pp. 234–245). New York: Routledge.

Matuk, C. F., & Linn, M. C. (2015). Examining the real and perceived impacts of a public idea repository on literacy and science inquiry. *CSCL'15: Proceedings of the 11th International Conference for Computer Supported Collaborative Learning* (Vol. 1, pp. 150–157). Gothenburg, Sweden: International Society of the Learning Sciences.

McElhaney, K. W., Chang, H. Y., Chiu, J. L., & Linn, M. C. (2015). Evidence for effective uses of dynamic visualisations in science curriculum materials. *Studies in Science Education*, *51*(1), 49–85.

McElhaney, K. W., Matuk, C. F., Miller, D. I., & Linn, M. C. (2012). Using the Idea Manager to Promote Coherent Understanding of Inquiry Investigations. In J. van Aalst, K. Thompson, M. J. Jacobson, & P. Reimann (Eds.), *The future of learning: Proceedings of the 10th International Conference of the Learning Sciences* (Vol. 1, Full papers, pp. 323–330). Sydney, NSW, Australia: International Society of the Learning Sciences.

National Research Council (NRC) (2012). *A framework for K–12 science education: Practices, crosscutting concepts, and core ideas.* doi:10.17226/13165

NSF Task Force on Cyberlearning (2008). *Fostering learning in the networked world: The cyberlearning opportunity and challenge* (C. L. Borgman, Ed.). Retrieved from http://www.nsf.gov/pubs/2008/nsf08204/index.jsp

Nilsson, E. M., & Jakobsson, A. (2011). Simulated sustainable societies: Students' reflections on creating future cities in computer games. *Journal of Science Education and Technology*, *20*(1), 33–50.

Papert, S. (1993). *The children's machine: rethinking school in the age of the computer.* New York: Basic Books.

Pellegrino, J. W. (2018). Assessment of and for learning. In F. Fischer, C. E. Hmelo-Silver, S. R. Goldman, & P. Reimann (Eds.), *International handbook of the learning sciences* (pp. 410–421). New York: Routledge.

Puntambekar, S. (2018). Design-based research (DBR). In F. Fischer, C. E. Hmelo-Silver, S. R. Goldman, & P. Reimann (Eds.), *International handbook of the learning sciences* (pp. 383–392). New York: Routledge.

Quintana, C., Reiser, B. J., Davis, E. A., Krajcik, J., Fretz, E., Golan, R. D., et al. (2004). A scaffolding design framework for software to support science inquiry. *Journal of the Learning Sciences*, *13*(3), 337–386. doi:10.1207/s15327809jls1303_4

Raes, A., Schellens, T., & De Wever, B. (2013). Web-based collaborative inquiry to bridge gaps in secondary science education. *Journal of the Learning Sciences*, *23*(3), pp. 316–347. doi:10.1080/10508406.2013.836656

Razzaq, L., & Heffernan, N. T. (2009). To tutor or not to tutor: That is the question. In M. Dimitrova, du Boulay, & Graesser (Eds.), *Proceedings of the Conference on Artificial Intelligence in Education* (pp. 457–464). Amsterdam, Netherlands: IOS Press.

Reiser, B. J. (2004). Scaffolding complex learning: The mechanisms of structuring and problematizing student work. *Journal of the Learning Sciences*, *13*(3), 273–304

Renninger, K. A., Ren, Y., & Kern, H. M. (2018). Motivation, engagement, and interest: "In the end, It came down to you and how you think of the problem." In F. Fischer, C. E. Hmelo-Silver, S. R. Goldman, & P. Reimann (Eds.), *International handbook of the learning sciences* (pp. 116–126). New York: Routledge.

Repenning, A., Webb, D. C., Koh, K. H., Nickerson, H., Miller, S. B., Brand, C., et al. (2015). Scalable game design: A strategy to bring systemic computer science education to schools through game design and simulation creation. *ACM Transactions on Computing Education (TOCE)*, *15*(2), 1–34. doi:10.1145/2700517

Rosé, C. P. (2018). Learning analytics in the Learning Sciences. In F. Fischer, C. E. Hmelo-Silver, S. R. Goldman, & P. Reimann (Eds.), *International handbook of the learning sciences* (pp. 511–519). New York: Routledge.

Rousseau, J.-J. (1979). *Emile, or on education* (Allan Bloom, Trans.). New York: Basic Books.

Ryoo, K., & Linn, M.C. (2015). Designing and validating assessments of complex thinking in science. *Theory into Practice*, *54*(3), 238–254.

Sadler, P., Coyle, H. A., & Schwartz, M. (2000). Successful engineering competitions in the middle school classroom: Revealing scientific principles through design challenges. *Journal of the Learning Sciences*, *9*(3), 299–327.

Schwarz, B. B. (2018). Computer-supported argumentation and learning. In F. Fischer, C. E. Hmelo-Silver, S. R. Goldman, & P. Reimann (Eds.), *International handbook of the learning sciences* (pp. 318–329). New York: Routledge.

Sharples, M., Scanlon, E., Ainsworth, S., Anastopoulou, S., Collins, T., Crook, C., et al. (2015). Personal inquiry: Orchestrating science investigations within and beyond the classroom. *Journal of the Learning Sciences*, *24*(2), 308–341. doi:10.1080/10508406.2014.944642

Slotta, J. D., Quintana, R., & Moher, T. (2018). Collective inquiry in communities of learners. In F. Fischer, C. E. Hmelo-Silver, S. R. Goldman, & P. Reimann (Eds.), *International handbook of the learning sciences* (pp. 308–317). New York: Routledge.

Tabak, I., & Kyza., E. (2018). Research on scaffolding in the learning sciences: A methodological perspective. In F. Fischer, C. E. Hmelo-Silver, S. R. Goldman, & P. Reimann (Eds.), *International handbook of the learning sciences* (pp. 191–317). New York: Routledge.

Tansomboon, C., Gerard, L., Vitale, J., & Linn, M. C. (2017). Designing automated guidance to promote productive revision of science explanations. *International Journal of Artificial Intelligence in Education*, 27(4), 729–757. doi:10.1007/s40593-017-0145-0

Toth, E., Suthers, D., & Lesgold, A. M. (2002). Mapping to know: The effects of representational guidance and reflective assessment on scientific inquiry skills. *Science Education*, 86(2), 264–286.

van Zee, E., & Minstrell, J. (1997). Using questioning to guide student thinking. *Journal of the Learning Sciences*, 6(2), 227–269. doi:10.1207/s15327809jls0602_3

Wilensky, U., & Reisman, K. (2006). Thinking like a wolf, a sheep or a firefly: Learning biology through constructing and testing computational theories—An embodied modeling approach. *Cognition and Instruction*, 24(2), 171–209.

Xie, C., Zhang, Z., Nourian, S., Pallant, A., & Hazzard, E. (2014). A time series analysis method for assessing engineering design processes using a CAD tool. *International Journal of Engineering Education*, 30(1), 218–230.

23

Supporting Informal STEM Learning with Technological Exhibits

An Ecosystemic Approach

Leilah Lyons

Informal learning settings have been a primary site for the learning sciences to explore educational designs that exploit the transformative potential of technology. An important lesson from these efforts has been the need to consider the settings as ecosystems, including how the culture and organizations have evolved over time in each setting. This chapter applies an ecosystemic perspective to three examples of a particular subclass of informal learning environments, informal STEM institutions (ISIs), so named because they focus on science, technology, engineering, or mathematics: natural history museums, science centers, and zoos and aquaria. Despite some overlap in STEM content coverage, their different evolutionary histories as institutions have fostered different learning practices.

An ecosystemic perspective permits educational designers to think strategically about how their designs will challenge and expand existing "ecosystem services," by which ecologists mean the useful functions an ecosystem provides for its denizens. In the ISI context, ecosystem services are the learning practices made possible by different exhibit designs. Learning practices are supported by "affordance networks," a term given to the synergistic collection of "possibilities for action" provided by the various material, social, and cultural properties of a context (Barab & Roth, 2006). Exhibit design can thus be thought of as a process of marshaling collections of affordances to support learning practices. Technological innovations by definition add new affordances to a learning environment, but designers must carefully consider how to align new affordances with existing affordance networks for the innovations to succeed and be adopted.

The main vehicle ISIs use for supporting visitor learning is the exhibit. The different evolutionary histories of natural history museums, science centers, and zoos and aquaria produced three different archetypes for exhibits: information delivery, phenomenological exploration, and affective spectacle exhibits. The chapter discusses the historical origin stories of these three archetypes, connecting them to their (explicit and implicit) pedagogies and existing attempts to integrate educational technologies into each. The final section discusses future directions for technology designs and research, including issues of broadening access to ISIs.

Natural History Museums and Information-Delivery Exhibits

Natural history museums came of age during the enlightenment era of the 1700s and 1800s, when science was preoccupied with the morphology and categorization of specimens. They were primarily sites of active scholarly research and, when the lay public was grudgingly admitted, they were expected to learn by visually inspecting carefully preserved specimens in glass vitrines and reading labels crafted by curators (Conn, 1998). The design of many collection-based exhibits still rests on a transmission model of learning, which fails to recognize or explain why a visitor might find an exhibit inscrutable (L. C. Roberts, 1997).

Most modern theories of learning acknowledge that learners have some interior model of the world that is amended upon the acquisition of new information. However, for this to occur learners have to be able to connect the new information to what they already know. Herein lies a major limitation of information delivery exhibits: it is patently impossible to create an interpretive experience that caters to the prior knowledge of *all* learners. Moreover, although museums are supposedly environments where visitors can discover interests they did not know they had (Rounds, 2006), legacy vitrine-and-label-style exhibits do not invite casual inspection. Technology offers several opportunities for supporting knowledge-building and motivating discovery with information-delivery exhibits.

Tailoring Information Delivery to Support Knowledge-Building

Natural history museums have attempted to meet the diverse informational needs of visitors via both customization strategies (where the visitor specifies their interests or needs) and personalization strategies (where the visitor's characteristics or actions provide evidence of interests and needs) (Bowen & Filippini-Fantoni, 2004). Digital labels provide more information that the typical static label, allowing visitors to access as much or as little information as they want. Researchers have begun to more systematically explore how design factors affect visitors' use of digital labels (J. Roberts, Banerjee, Matcuk, McGee, & Horn, 2016). One issue with digital labels, however, is that visitors can be overwhelmed when there is too much information, thus limiting the degree of manual customization possible.

Customization can be made less overwhelming with context-aware information delivery, where the system selects information for the learner based on its understanding of the current situation. For example, mobile interpretation systems (Raptis, Tselios, & Avouris, 2005) typically use the visitor's physical location in a museum to deliver information about nearby artifacts. Thus, the "context" is defined by the museum's layout, not the visitor's needs. Several digital guide systems (e.g., Kruppa & Aslan, 2005) deliver customized media after probing visitors about their interests, but it can be hard for a learner to absorb information too far outside their "Zone of Proximal Development" (Vygotsky, 1978) regardless of their interest in the topic. Other systems ask visitors to reflect on their informational needs, as in the mobile field-trip support app, Zydeco (Cahill et al., 2011). However, such systems require students to develop inquiry questions in advance, often requiring extensive teacher support. Visitor knowledge could be assessed on-site more quickly via quizzes, but activities that smack of formal classrooms are often anathema to ISI practitioners, harkening back to Oppenheimer's assertion that "Museums . . . can relieve . . . the tensions which make learning in school ineffective or even painful. No one ever 'fails' in a museum" (Oppenheimer & Cole, 1974).

Personalization entails less visitor effort, but it requires more information about the visitors. The rub for artifactual museums is that much of what learners do is hard to discern (e.g., does a long pause at a digital label indicate deep reading or distraction?). Gaze tracking might be a way of profiling visitors, but this work is in its infancy (Mokatren, Kuflik, & Shimshoni, 2016). Another possible source of visitor information may be social media—via user-generated taxonomies, dubbed "folksonomies," that emerge as visitors create descriptive "tags" for artifacts (Thom-Santelli, Cosley, & Gay, 2010), or via Instagram narratives of their visit experiences (Weilenmann, Hillman, & Jungselius, 2013). Both user contributions and consumption patterns provide evidence of the current state of visitor knowledge and interests. If visitors are using their own mobile devices in ISIs, there is also the controversial possibility of tapping into records of their daily online activity.

Motivating Visitors' Discovery of Information-Delivery Exhibits

Artifactual ISIs were not initially created for learners, which may be why so many digital interventions developed for artifactual ISIs attempt to motivate visitors to discover exhibits. Some museums have experimented with using "gamification" approaches that substitute game-driven motivations for intrinsic interest in artifacts. These games are often take the form of mobile "treasure hunts" (e.g., Ceipidor, Medaglia, Perrone, Di Marsico, & Di Romano, 2009), borrowing from the classic paper worksheets often used by school groups.

The motivational structures used in treasure-hunt games are exogenous, meaning that there is an arbitrary relationship between the motivational embellishments and the content. Ideally, the reason for inspecting objects should be endogenous, i.e., tied more tightly to the learning goals. To illustrate, another mobile application uses endogenous motivation when it asks teams of visitors to virtually link objects on the basis of shared traits. Once linked, the objects are removed as "playable objects" that other visitor teams can use, pushing the teams to find more creative object–object connections (Yiannoutsou, Papadimitriou, Komis, & Avouris, 2009). This type of game can help learners create a richer mental model by endogenously motivating visitors to find similarities between seemingly different objects, similar to how learners might make object comparisons out of their own inquiry-driven curiosity.

Science Centers and Phenomenological Inquiry Exhibits

The Franklin Museum in Philadelphia and the Boston Museum of Science began creating exhibits that allowed visitors to manipulate scientific phenomena like electricity early in the 20th century, a trend that accelerated during the countercultural 1960s and 1970s when the Exploratorium in San Francisco began publishing inquiry exhibit "cookbooks." This shift—from presenting the authentic physical *artifacts* of science to presenting the authentic *phenomena* of science, and allowing visitors to engage in the authentic *practices* of science-like inquiry—represented a sea change in how ISIs regarded learners, from passive receivers of information to active constructors of knowledge. This new "hands-on" pedagogy straddled the Deweyan idea of "education through experience" and the Piagetian notions that learners construct their own understandings by manipulating the world around them, and that learner inquiry is piqued by discrepant events. While sound in theory, in practice visitors struggle with engaging in inquiry with hands-on exhibits (Gutwill, 2008). This, combined with concern that very open-ended exhibits could lead to misconceptions, and with the practical challenge of taming real scientific phenomena for presentation (Gutwill, 2008), led to the genesis of so-called "planned discovery" exhibits (Humphrey, Gutwill, & Exploratorium APE Team, 2005).

Planned discovery exhibits were developed by progressively isolating a phenomenon, tuning the presentation of that phenomenon to be as surprising as possible, and refining how-to instructions to support visitor inquiry. Sometimes exhibit designers would find the original scientific phenomena too unruly or hard to observe, and would instead produce "analogy-based exhibits" (Afonso & Gilbert, 2007), where something like ping-pong balls might substitute for air molecules. But science centers were victims of their own success. While most visitors could easily produce the desired effect with planned discovery exhibits, a whole generation of visitors was enculturated to expect to be told what to do rather than formulating their own questions. Isolating a reliable phenomenon resulted in few options to explore, and while many analogy-based exhibits could support a wider range of interactions, visitors often struggled with connecting the exhibit to the source domain (Afonso & Gilbert, 2007). Even when visitors went "off script" to pursue their own questions, the effects they could generate were essentially guaranteed to not be as interesting as what the exhibit was tuned to produce on first use (Gutwill, 2008). Exhibit designers had unintentionally transformed their constructivist inquiry exhibits into behaviorist opportunities for reinforcement.

Museum researchers began exploring what changes could be made to remedy visitor disinterest in inquiry. One was to relax the goal from supporting *pure inquiry into* a phenomenon to instead supporting *prolonged engagement with* that phenomenon (Humphrey et al., 2005). Permitting this relaxation was a change in how researchers regard misconceptions, from pernicious learning barriers that must be avoided to resources that learners can productively reorganize and apply (Smith, diSessa, & Roschelle, 1993). So, rather than being tasked with discovering the right answer for *why* something happened, visitors can explore the *bounds* of the phenomenon: which effects happen often, which are rare, and which are impossible to produce.

Researchers also explored methods to "retrain" visitors to engage in inquiry practices. In one project, inquiry practices were "reified" in the form of physical, laminated cards that visitors use as conversation starters, resulting in a dramatic improvement in the amount of observed inquiry activity as compared to control groups (Gutwill & Allen, 2010). Physical exhibit designs can also promote different inquiry behaviors via "framings," i.e., the social, affective, and epistemological expectations people bring to situations. One discovery was that when an open-ended exhibit is framed as having a singular goal, visitors tend to treat it like a school lesson (Atkins, Velez, Goudy, & Dunbar, 2009), with parents dominating and reserving the "important conceptual roles" for themselves (Schauble et al., 2002).

Supporting Phenomenological Exploration

The most obvious technology to support phenomenological inquiry is simulation. A simulated world can be as rich or as simple as needed, and can address phenomena otherwise impossible to display because of issues of scale or time. While inquiry learning with simulations has been studied extensively in schools, contextual differences across formal and informal ecosystems preferentially reinforce certain aspects of designed learning experiences while repressing other aspects, so different strategies are likely needed for integrating simulations into science centers. In simulations set in classrooms, learners have often relied on user interface scaffolding or explicit tutorial guidance to engage in productive inquiry. If science center visitors are tasked with exploring the *bounds* of phenomena, rather than enacting pure inquiry, other forms of *support* and *guidance* might be useful.

One thing we know is that visitors enjoy—and find it beneficial—to watch how other visitors interact with exhibits (vom Lehn, Heath, & Hindmarsh, 2001). When digital phenomenological exhibits are designed so that the connections between visitor actions and the phenomenological effects are clearly visible, visitors can monitor one another's activities, so that struggling visitors can echo more successful visitors' actions (a passive form of support), and more successful visitors can diagnose their companions' problems and provide suggestions (an active form of support), as was seen in the Oztoc tabletop exhibit (Lyons et al., 2015). HCI (human–computer interaction) researchers would categorize this kind of spectator experience as being "expressive," meaning that both the users' manipulations and those manipulations' effects are clearly revealed (Reeves, Benford, O'Malley, & Fraser, 2005). With phenomenological exhibits, however, sometimes the cause producing an effect is not clear, becoming what HCI researchers call a "magical" spectator experience (Reeves et al., 2005). Such experiences can encourage visitors to develop "superstitious" behaviors, much like roulette players. For example, some children thought running in circles around a tornado exhibit caused the funnel to form. Onlookers began imitating the behavior despite it being both conceptually and operationally incorrect. By allowing for more expressive spectator experiences, digital technologies can leverage other visitors to support more productive phenomenological exploration.

Guiding phenomenological exploration is trickier—borrowing features from video game culture, like "collecting gems," to drive exploration directions can end up distracting visitors from the phenomenon (Horn, Weintrop, & Routman, 2014). One possible solution is to provide automated guidance not by monitoring learner *comprehension*, as intelligent tutors do, but instead by monitoring learner *exploration patterns* to push visitors towards "unseen" areas of the phenomenological space (Mallavarapu, Lyons, Shelley, & Slattery, 2015).

Authenticity in Simulations

The abstractions inherent to simulations can help visitors focus on processes (Eberbach & Crowley, 2005), but visitors might miss the link to the source domain (Afonso & Gilbert, 2007) and respond negatively to the exhibit because it is not "real" science (L. C. Roberts, 1997). One way around the dilemma of presenting phenomena in an authentic but comprehensible fashion is to use the *representations* of phenomena used by scientists. For example, the DeepTree tabletop exhibit introduces visitors to the phenomena of inheritance by having visitors interact with phylogenetic tree representations (Horn et al., 2016).

Another strategy is to use augmented reality (AR), a family of technologies that allow digital visualizations to be superimposed upon real objects and settings. By augmenting traditional phenomenological exhibits with digital representations like electron flow paths, visitors gain access to abstract phenomena without sacrificing authenticity (Yoon, Elnich, Wang, Van Schooneveld, & Anderson, 2013). Yet another way to dodge questions of authenticity is to literally embed visitors in a simulated phenomenon, allowing them to control it in a first-person fashion. Single-user exhibits like MEteor, wherein visitors act out the trajectories of meteors moving past gravitational bodies as they run across a floor, have visitors *role-play* a phenomenon (Tscholl & Lindgren, 2016). In large-group exhibits like Connected Worlds, visitors are tasked with water distribution and forestry management in order to attain a sustainable ecosystem (Uzzo, Chen, & Downs, 2016). Visitors *generate* emergent phenomena (like droughts, foods, population explosions) via their collective actions (see Figure 23.1).

Situating Scientific Phenomena in a Broader Sociocultural Context

When people think about scientific practices, they most often think of scientific inquiry, but a singular focus on inquiry produced the less-than-effective planned discovery exhibits. Science intersects with many aspects of modern life and culture, giving exhibit designers a number of ways to use technology to create opportunities for visitors to experience the larger sociocultural context of science (see Figure 23.1).

Role-Playing and Decision-Making

One of the easiest ways to embed learners in the sociocultural context of science is to ask them to make decisions that have relatable human consequences. In Sickle Cell Counselor (Bell, Bareiss, & Beckwith, 1993) a visitor acts as a genetic counselor advising "patients," introducing visitors to the experience of applying scientific knowledge to real-world problems, and in Malignancy (Lyons, 2009), visitors role-play as different oncology specialists collaborating to treat a "patient," introducing them to the dilemmas faced by STEM professionals. Other exhibits have explored role play and decision-making at a higher, policy level. The Youtopia tabletop game (Antle et al., 2013) and the Mine Games participatory simulation (Pedretti, 2004) engage groups of visitors in making decisions about what infrastructure to install in communities, and reflecting upon the impacts on stakeholders. Designers of issues-based exhibits just need to be careful that the scientific phenomenon is more than an afterthought, by ensuring that understanding how the phenomenon works is key to negotiating solutions.

Scaffolding Social Interactions at Exhibits

Technology provides interesting opportunities for mediating visitor–visitor interactions, but it is wise to examine existing social practices first, to determine how the tool can be aligned with existing positive practices, and if it can help intercede with existing negative practices (Lyons, Becker, & Roberts, 2010). For example, technology can be used to mediate parents' natural inclination to

Figure 23.1 The Connected Worlds Exhibit at the New York Hall of Science

Note: Large groups of up to 30 visitors make decisions about how to distribute water and manage wildlife in simulated biomes via full-body interaction.

Photo by David Handschuh and DesignIO.

guide their children by supplying location-dependent prompts (McClain & Toomey Zimmerman, 2016) or exhibit-state-dependent prompts (Tscholl & Lindgren, 2016). Inquiry prompts could also be a vehicle to counteract less beneficial parental behaviors. For example, parents (unintentionally) supply more explanations to boys than to girls (Crowley, Callanan, Tenenbaum, & Allen, 2001) and tend to take charge of decision-making when engaged in open-ended explorations (Schauble et al., 2002). Inquiry prompts could be augmented with reminders to engage with each child in turn or suggestions for how to distribute inquiry tasks (Lyons et al., 2010). Apart from explicit scaffolding, technology designers must be conscious of how designs unintentionally evoke social framings—for example, visitors acculturated to competitive computer-based games need only the slightest cues to begin competing (Lyons, 2009).

Zoos and Aquaria and Affective Spectacle Exhibits

Zoos and aquaria began in the same enlightenment era as natural history museums, as a public form of wealthy nobles' menageries. While most supported some level of scientific endeavor, their main competitors for audiences were circuses and trained animal performances. Thus, historically zoos and aquaria have always been prone to catering to public spectacle. When science centers shifted focus from authentic objects to authentic scientific phenomena, zoos and aquaria also shifted focus from presenting animals to presenting animals within authentic natural contexts. This "immersive design" movement had the goal of both supporting the animals' welfare and encouraging animals to engage in more authentic behaviors for the edification of the visitors, although sometimes these goals were subsumed by stagecraft that catered to visitors' perceptions of authenticity (Schwan, Grajal, & Lewalter, 2014). While spectacles like animal feedings draw visitors, the visitor learning is supported via staff interactions and the addition of other media (labels, signs, artifacts, and interactives). Unlike other ISIs, these educational interventions must be designed around infrastructure and weather, and will always be in competition for visitor attention with the animals themselves.

Zoos and aquaria are unique among ISIs in that they take a strong activist stance, desiring to produce changes in visitors' attitudes and behaviors towards conservation (Dierking, Burtnyk, Buchner, & Falk, 2002). Helping visitors make connections between human activities, habitats, and animals (i.e., engaging them in systems thinking) is one known strategy for impacting visitor conservation attitudes and behaviors (Ballantyne & Packer, 2005). A more common way for zoos and aquaria to evoke attitudinal and behavioral change is by cultivating visitor empathy for animals, drawing on conservation psychology and theories of emotion (Dierking et al., 2002).

Systems Thinking via Technologically Augmented Connection-Building

The need to conserve species means that even "immersive" exhibits disconnect the presented animals from most of their systemic relationships (with habitats, with predators or prey, with human activities, etc.), which is why some zoos and aquaria have turned to technology to support systems thinking. One mobile app reflectively encouraged systems thinking by asking visitors to compare the leg morphologies seen in different animals, and connect them to their habitats (Suzuki et al., 2009). More constructively, another app asked visitors to role-play as an escaped zoo animal trying to return to its enclosure, where the paths visitors walked impacted them positively (if passing an exhibit with a potential food source for the avatar animal) or negatively (if passing a potential predator) (Perry et al., 2008). Both apps support making connections between animals and environments, but leave out any connection to human activities. By contrast, a tabletop game in an aquarium used both constructive and reflective strategies by allowing visitors to play as fishermen and reflect on how their decisions impacted animal populations (D'Angelo, Pollock, & Horn, 2015).

Evoking Empathy Through Perspective-Taking

Zoos and aquaria often rely on "charismatic animals" and stagecrafted animal encounters like feedings to evoke visitors' empathy. Technology can support empathy by connecting visitors to animals even more intimately. For example, mobile devices can supply audio clips of an animal's heartbeat, or display videos showing how an animal sees the world around them (Ohashi, Ogawa, & Arisawa, 2008). Sensing technologies build bridges between animals and visitors, supporting augmented reality annotations like displaying an animal's name and information when visitors direct their mobile phone camera at it (Karlsson, ur Réhman, & Li, 2010). Perspective-taking can be encouraged via role-playing strategies as well, as in the aforementioned escaped zoo animal game (Perry et al., 2008), or in a full-body interaction game that invited visitors to feel the toll climate change is taking on polar bears via the increased effort needed to hunt (Lyons, Slattery, Jimenez Pazmino, Lopez Silva, & Moher, 2012) (Figure 23.2).

Future Directions for Technology Integration

Future directions for integrating technology into these three exhibit archetypes, using an ecosystemic perspective, may involve *enhancing* existing ecosystem services (i.e., existing learning practices) by adding new affordances to the network, *evolving* existing ecosystem services to draw more strongly on existing affordances in the network, or *adding* wholly new ecosystem services to embrace new learning activities.

Information-Delivery Exhibits

The main ecosystem services provided by information delivery exhibits, helping visitors build knowledge and allowing visitors to discover new content, could be enhanced by filling gaps in their existing affordance networks. Unfortunately, these two goals are fundamentally opposed: one seeks to increase depth, while the other seeks to expand the breadth of visitors' explorations. So, as researchers begin to collect data on visitors to personalize visit experiences, they should seek a

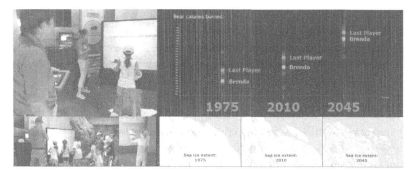

Figure 23.2 Perspective-Taking via Role-Playing Strategies: Full-Body Interaction Game

Note: Players walk (by stomping in place) and swim (by waving their arms while wearing weighted plush gloves) to feel the increased effort a polar bear would expend to hunt through three different decades of a simulated arctic environment (maps, middle right). The top right image depicts the graphical representation of visitors' calorie expenditures across the decades, with the most recent two visitors' data highlighted.

Source: Lyons et al. (2012). Photo by Leilah Lyons.

median between depth and breadth. For example, a treasure-hunt game could use visitor data to make sure that some of the targeted exhibits cater to the player's interests, and that the clues are tailored to the player's existing knowledge, so that the activity supports knowledge consolidation along with discovery. (It would also encourage repeat visits.) Identity exploration is another ecosystem service provided by artifactual museums, and technology could help evolve it from a solo to a social pursuit, facilitating visitors' exploration of artifacts to help them also learn more about each other (J. Roberts, Lyons, Cafaro, & Eydt, 2014).

Phenomenological Exhibits

Simulations seem like the inevitable evolution of phenomenological exhibits, but a new ecosystem service is needed to guide learners' exploration of the bounds of the simulated phenomena, and its affordance network will likely need to span multiple contexts (physical, social, and cultural). For example, exhibits designed to prolong visitor engagement weren't fully effective at producing more inquiry behaviors—the combination of the physical exhibits and the inquiry reacculturation via cards produced the largest gains (Gutwill & Allen, 2010). Similarly, visitors experienced a drop in playful, exploratory interactions when augmented reality exhibits were framed with instructional labels (Yoon et al., 2013), an example of how physical and cultural affordances can combine (here, negatively). Designers usually look to physical affordances for motivating visitors to explore a phenomenon (Gutwill, 2008), but digital technologies can recruit existing social and cultural motivational affordances into the network. For example, using mobile devices to channel into more productive directions parents' latent inclination to guide their children (Lyons et al., 2010; McClain & Toomey Zimmerman, 2016), or designing exhibits so that they are fully "expressive" spectator experiences (Reeves et al., 2005) so that visitors can be inspired by witnessing each others' exploration (Lyons et al., 2015).

Another motivational strategy is to situate scientific phenomena within a sociocultural issues-based framing (Antle et al., 2013; Pedretti, 2004; Uzzo et al., 2016). When digital technology is used to engage learners in a participatory fashion, visitor themselves become part of the phenomenon, suggesting a need for new ecosystem services that can provide feedback on emergent collective behaviors, like flagging "tragedy of the commons" outcomes (D'Angelo et al., 2015). Moreover, borrowing strategies evolved within zoos and aquaria for inducing empathy, like perspective-taking, might be useful for helping visitors understand the full sociocultural implications of issues-based science. Exhibit designs can afford perspective-taking by emphasizing how choices affect the enacted entity physically or emotionally, and how those physical and emotional impacts in turn affect future choices. Augmented reality, virtual reality, immersive experiences, and full-body interaction designs all provide affordances for imparting the physical and emotional aspects of perspective-taking.

Affective Spectacle Exhibits

Technology has shown promise for helping visitors connect animals to each other, to environments, and to human activities, but more can be done to support systems thinking in zoo and aquarium visitors. Traditionally, zoos and aquaria do this by having interpretive staff lead visitors through "explanatory chains" that connect human activities to environmental impacts on animals—technology like mobile support tools could be used to enhance this existing ecosystem service (Lyons et al., 2012). In unguided experiences, it's especially hard to integrate human activities unless they are presented via a standalone activity like a video game, but parents show reluctance to have their children engage with games in zoos and aquaria (D'Angelo et al., 2015), often saying that the kids play enough games at home. Because these ISIs encourage spectacle, technological interventions might need to be stagecrafted experiences that visitors don't encounter in their everyday lives to gain acceptance. Alternatively, it could be that highly engaging digital exhibits violate the expectation that a visit should be spent viewing real, live animals. Some researchers have begun exploring how technology might facilitate animal–human interactions at ISIs (Webber, Carter, Smith, & Vetere, 2017). This might eventually result in exhibits that, for example, allow visitors to play games against orangutans, a new ecosystem service which would turn on its head the idea of technology being a distractor from animals, and have profound impact on visitor empathy.

Conclusion

In concluding, it is important to emphasize that visitors from traditionally underserved populations *do* feel that they "flunk" museums (Dawson, 2014), and do not feel like ISIs are "for them." Learning scientists need to attend to what new ecosystem services might need to be put into place to support non-traditional visitor populations, and what methodologies should be used to discover these visitor needs. Acculturation will certainly play a part, as the new visitors may need guidance in learning how to take advantage of ISIs as learning experiences, and ISI staff may need guidance in understanding how to support learners who come from very different cultural backgrounds.

Fortunately, ecosystems evolve—it is their defining trait. Examining ISIs from an ecosystemic perspective can help designers identify the new kinds of ecosystem services and affordance networks that need to be in place, or flag existing ecosystem elements for adaptation or cross-pollination. Innovation within an established setting is a balancing act: if the designer tries to change too much too fast, stakeholders (in our case, both learners and ISI practitioners) often reject the innovation because they struggle to recognize and use the new "possibilities for action." Information delivery exhibits, phenomenological exhibits, and affective spectacle exhibits each evolved within a distinct context. Education technology designers who acknowledge the full breadth of physical, social, historical, and cultural affordance networks within ISIs can make strategic decisions in how to evolve the ecosystem services to better support STEM learning.

Further Readings

Bell, P., Lewenstein, B., Shouse, A. W., & Feder, M. A. (Eds.). (2009). *Learning science in informal environments: People, places, and pursuits*. Washington, DC: National Academies Press.
A National Academies report that overviews informal learning research and is useful for getting acquainted with how science learning has been explored in informal settings.

Falk, J. H., & Dierking, L. D. (2000). *Learning from museums: Visitor experiences and the making of Meaning*. Walnut Creek, CA: AltaMira Press.

This frequently cited book presents a framework that highlights how physical, sociocultural, and personal contexts shape the learning experiences of visitors, and complements its accessible theoretical perspective with research results and illustrative anecdotes.

Humphrey, T., Gutwill, J. P., & Exploratorium APE Team. (2005). *Fostering active prolonged engagement: The Art of creating APE exhibits*. San Francisco: Exploratorium.

An example of the exhibit design guides that the Exploratorium science center has published. These guides are useful for understanding the design concerns involved in creating interactive exhibits.

Lord, B., & Piacente, M. (Eds.). (2014). *Manual of museum exhibitions* (2nd ed.). Lanham, MD: Rowman & Littlefield.

This handbook provides guidance on pragmatic concerns and best practices (e.g., height, access, lighting) with respect to exhibit design. Researchers are highly encouraged to study these kinds of guides before rediscovering known design recommendations.

References

Afonso, A., & Gilbert, J. (2007). Educational value of different types of exhibits in an interactive science and technology center. *Science Education, 91*(6), 967–987.

Antle, A., Wise, A., Hall, A., Nowroozi, S., Tan, P., Warren, J., et al. (2013). Youtopia: A collaborative, tangible, multi-touch, sustainability learning ability. *Proceedings of Conference on Interaction Design for Children (IDC '13)* (pp. 565–568). New York: ACM Press.

Atkins, L., Velez, L., Goudy, D., & Dunbar, K. N. (2009). The unintended effects of interactive objects and labels in the science museum. *Science Education, 93*(1), 161–184.

Ballantyne, R., & Packer, J. (2005). Promoting environmentally sustainable attitudes and behaviour through free-choice learning experiences: What is the state of the game? *Environmental Education Research, 11*(3), 281–295.

Barab, S. A., & Roth, W.-M. (2006). Curriculum-based ecosystems: Supporting knowing from an ecological perspective. *Educational Researcher, 35*(5), 3–13.

Bell, B., Bareiss, R., & Beckwith, R. (1993). Sickle cell counselor: A prototype goal-based scenario for instruction in a museum environment. *Journal of the Learning Sciences, 3*(4), 347–386.

Bowen, J. P., & Filippini-Fantoni, S. (2004). Personalization and the web from a museum perspective. In D. Beaumont and J. Trant (Eds.), *Museums and the web 2004: Selected papers from an international conference*, Arlington, VA, March 31–April 3.

Cahill, C., Kuhn, A., Schmoll, S., Lo, W.-T., McNally, B., & Quintana, C. (2011). Mobile learning in museums. *Proceedings of Conference on Interaction Design for Children (IDC '11)* (pp. 21–28). New York: ACM Press.

Ceipidor, U., Medaglia, C., Perrone, A., De Marsico, M., & Di Romano, G. (2009). A museum mobile game for children using QR-codes. *Proceedings of Conference on Interaction Design for Children (IDC '09)* (pp. 282–283). New York: ACM Press.

Conn, S. (1998). *Museums and American intellectual life, 1876-1926*. Chicago: University of Chicago Press.

Crowley, K., Callanan, M., Tenenbaum, H., & Allen, E. (2001). Parents explain more often to boys than to girls during shared scientific thinking. *Psychological Science 12*(3), 258–261.

D'Angelo, S., Pollock, D., & Horn, M. (2015). *Fishing with Friends*: Using tabletop games to raise environmental awareness in aquariums. *Proceedings of Conference on Interaction Design for Children (IDC '15)* (pp. 29–38). New York: ACM Press.

Dawson, E. (2014). "Not designed for us": How science museums and science centers socially exclude low-income, minority ethnic groups. *Science Education, 98*(6), 981–1008.

Dierking, L., Burtnyk, K., Buchner, K., & Falk, J. (2002). *Visitor learning in zoos and aquariums*. Silver Spring, MD: American Zoo and Aquarium Association.

Eberbach, C., & Crowley, K. J. (2005). From Living to Virtual: Learning from Museum Objects. *Curator, 48*(3), 317–338.

Gutwill, J. (2008). Challenging a common assumption of hands-on exhibits: How counterintuitive phenomena can undermine inquiry. *Journal of Museum Education 33*(2), 187–198.

Gutwill, J. P., & Allen, S. (2010). Facilitating family group inquiry at science museum exhibits. *Science Education, 94*(4), 710–742.

Horn, M., Phillips, B., Evans, E., Block, F., Diamond, J., & Shen, C. (2016). Visualizing biological data in museums: Visitor learning with an interactive tree of life exhibit. *Journal of Research in Science Teaching, 53*(6), 895–918.

Horn, M., Weintrop, D., & Routman, E. (2014). Programming in the pond: A tabletop computer programming exhibit. *Proceedings of CHI EA '14: Extended abstracts on human factors in computing systems* (pp. 1417–1422). New York: ACM Press.

Humphrey, T., Gutwill, J., & Exploratorium APE Team. (2005). *Fostering active prolonged engagement*. San Francisco: Exploratorium.

Karlsson, J., ur Réhman, S., & Li, H. (2010). Augmented reality to enhance visitors experience in a digital zoo. *Proceedings of the Eighth International Conference on Mobile and Ubiquitous Media (MUM '10)* (pp. 1–4). New York: ACM Press.

Kruppa, M., & Aslan, I. (2005). Parallel presentations for heterogenous user groups—An initial user study. *Proceedings for the First International Conference on Intelligent Technologies for Interactive Enyertainment (INTETAIN '05)* (pp. 54–63). Berlin: Springer.

Lyons, L. (2009). Designing opportunistic user interfaces to support a collaborative museum exhibit. *Proceedings of the Ninth International Conference on Computer Supported Collaborative Learning (CSCL '09)* (pp. 375–384). London: ISLS.

Lyons, L., Becker, D., & Roberts, J. (2010). Analyzing the affordances of mobile technologies for informal science learning. *Museums & Social Issues*, 5(1), 89–104.

Lyons, L., Slattery, B., Jimenez Pazmino, P., Lopez Silva, B., & Moher, T. (2012). Don't forget about the sweat: Effortful embodied interaction in support of learning. In *Proceedings of the Sixth International Conference of Tangible, Embedded and Embodied Interaction (TEI '12)* (pp. 77–84). New York: ACM Press.

Lyons, L., Tissenbaum, M., Berland, M., Eydt, R., Wielgus, L., & Mechtley, A. (2015). Designing visible engineering: Supporting tinkering performances in museums. *Proceedings of Conference on Interaction Design for Children (IDC '15)* (pp. 49–58). New York: ACM Press.

Mallavarapu, A., Lyons, L., Shelley, T., & Slattery, B. (2015). Developing computational methods to measure and track learners' spatial reasoning in an open-ended simulation. *Journal of Educational Data Mining*, 7(2), 49–82.

McClain, L., & Toomey Zimmerman, H. (2016). Integrating mobile technologies into outdoor education to mediate learners' engagement with nature. In L. Avraamidou & W.-M. Roth (Eds.), *Intersections of Formal and Informal Science* (pp. 122–137). New York: Routledge.

Thom-Santelli, J., Cosley, D., & Gay, G. (2010). What do you know? Experts, novices and territoriality in collaborative systems. In *Proceedings of the SIGCHI Conference on Human Factors in Computing Systems (CHI '10)* (pp. 1685–1694). New York: ACM Press.

Ohashi, Y., Ogawa, H., & Arisawa, M. (2008). Making new learning environment in zoo by adopting mobile devices. *Proceedings of the 10th International Conference on Human Computer Interaction with Mobile Devices and services (HCI '08)*. New York: ACM Press.

Oppenheimer, F., & Cole, K. (1974). The Exploratorium: A participatory museum. *Prospects*, 4(1), 1–10.

Pedretti, E. G. (2004). Perspectives on learning through research on critical issues-based science center exhibitions. *Science Education*, 88(S1), S34–S47.

Perry, J., Klopfer, E., Norton, M., Sutch, D., Sandford, R., & Facer, K. (2008). AR gone wild: Two approaches to using augmented reality learning games in Zoos. *Proceedings of Conference on Interaction Design for Children (IDC '08)* (pp. 322–329). New York: ISLS.

Raptis, D., Tselios, N., & Avouris, N. (2005). Context-based design of mobile applications for museums: A survey of existing practices. *Proceedings of the Seventh International Conference on Human Computer Interaction with Mobile a Device s and Services (HCI '05)* (pp. 153–160). New York: ACM Press.

Reeves, S., Benford, S., O'Malley, C., & Fraser, M. (2005). Designing the spectator experience. *Proceedings of the Seventh International Conference on Human Computer Interaction with Mobile a Device s and Services (HCI '05)* (pp. 741–750). New York: ACM Press.

Roberts, J., Banerjee, A., Matcuk, M., McGee, S., & Horn, M. S. (2016). *Uniting big and little data to understand visitor behavior*. Paper presented at the Visitor Studies Association Conference (VSA '16), Boston, MA.

Roberts, J., Lyons, L., Cafaro, F., & Eydt, R. (2014). Interpreting data from within: Supporting human–data interaction in museum exhibits through perspective taking. *Proceedings of Conference on Interaction Design for Children (IDC '14)* (pp. 7–16). New York: ACM Press.

Roberts, L. C. (1997). *From knowledge to narrative: Educators and the changing museum*. Washington, DC: Smithsonian Institution Press.

Rounds, J. (2006). Doing identity work in museums. *Curator*, 49(2), 133–150.

Schauble, L., Gleason, M., Lehrer, R., Bartlett, K., Petrosino, A., Allen, A., et al. (2002). Supporting science learning in museums. In G. Leinhardt, K. Crowley, & K. Knutson (Eds.), *Learning conversations in museums* (pp. 333–356). Mahwah, NJ: Lawrence Erlbaum.

Schwan, S., Grajal, A., & Lewalter, D. (2014). Understanding and engagement in places of science experience: Science museums, science centers, zoos, and aquariums. *Educational Psychologist*, 49(2), 70–85.

Smith, J., diSessa, A., & Roschelle, J. (1993). Misconceptions reconceived: A constructivist analysis of knowledge in transition. *Journal of the Learning Sciences*, *3*(2), 115–163.

Suzuki, M., Hatono, I., Ogino, T., Kusunoki, F., Sakamoto, H., Sawada, K. et al. (2009). LEGS system in a zoo: Use of mobile phones to enhance observation of animals. *Proceedings of Conference on Interaction Design for Children (IDC '09)* (pp. 222–225). New York: ACM Press.

Thom-Santelli, J., Cosley, D., & Gay, G. (2010). What do you know? Experts, novices and territoriality in collaborative systems. *Proceedings of the SIGCHI Conference on Human Factors in Computing Systems (CHI '10)* (pp. 1685–1694). New York: ACM Press.

Tscholl, M., & Lindgren, R. (2016). Designing for learning conversations: How parents support children's science learning within an immersive simulation. *Science Education*, *100*(5), 877–902.

Uzzo, S., Chen, R., & Downs, R. (2016). Connected worlds: Connecting the public with complex environmental systems. *AGU Fall Meeting Abstracts*, San Francisco.

vom Lehn, D., Heath, C., & Hindmarsh, J. (2001). Conduct and collaboration in museums and galleries. *Symbolic Interaction*, *24*(2), 189–216.

Vygotsky, L. S. (1978). *Mind in society: The development of higher mental processes*. Cambridge, MA: Harvard University Press.

Webber, S., Carter, M., Smith, W., & Vetere, F. (2017). Interactive technology and human–animal encounters at the zoo. *International Journal of Human–Computer Studies*, *98*, 150–168.

Weilenmann, A., Hillman, T., & Jungselius, B. (2013). Instagram at the museum: Communicating the museum experiences through social photo sharing. In *Proceedings of the SIGCHI Conference on Human Factors in Computing Systems (CHI '13)* (p. 1843). New York: ACM Press.

Yiannoutsou, N., Papadimitriou, I., Komis, V., & Avouris, N. (2009). "Playing with" museum exhibits: Designing educational games mediated by mobile technology. *Proceedings of Conference on Interaction Design for Children (IDC '09)* (pp. 230–233). New York: ACM Press.

Yoon, S., Elnich, K., Wang, J., Van Schooneveld, J., & Anderson, E. (2013). Scaffolding informal learning in science museums: How much is too much? *Science Education*, *97*(6), 848–877.

24
Intelligent Tutoring Systems

Arthur C. Graesser, Xiangen Hu, and Robert Sottilare

We define an intelligent tutoring system (ITS) as a computer learning environment that helps students master knowledge and skills by implementing intelligent algorithms that adapt to students at a fine-grained level and that instantiate complex principles of learning (Graesser, Hu, Nye, & Sottilare, 2017). An ITS normally works with one student at a time because students differ on many dimensions and the goal is to be sensitive to the idiosyncrasies of individual learners. That being said, pairs of students may benefit from jointly preparing responses to an ITS. It is also possible to have an automated tutor or mentor interact with small teams of learners in collaborative learning and problem-solving environments.

ITS environments can be viewed as a generation beyond conventional computer-based training (CBT). CBT systems also adapt to individual learners, but they do so at a coarse-grained level with simple learning principles. In a prototypical CBT system, the learner (a) studies material presented in a lesson, (b) gets tested with a multiple-choice test or another objective test, (c) gets feedback on the test performance, (d) re-studies the material if the performance in (c) is below threshold, and (e) progresses to a new topic if performance exceeds threshold. The order of topics typically follows a predetermined order, such as ordering on complexity (simple to complex) or ordering on prerequisites. ITSs enhance CBT with respect to the adaptability, grain size, and the power of computerized learning environments. An ITS tracks the knowledge, skills, and other psychological attributes and adaptively responds to the learner by applying computational models in artificial intelligence and cognitive science (VanLehn, 2006; Woolf, 2009). For CBT, interaction histories can be identical for multiple students and the interaction space is small. In contrast, for many ITSs, every tutorial interaction is unique and the space of possible interactions is extremely large, if not infinite.

ITSs have frequently been developed for mathematics and other computationally well-formed topics. For example, the *Cognitive Tutors* (Aleven, McLaren, Sewall, & Koedinger, 2009; Ritter, Anderson, Koedinger, & Corbett, 2007; see the NAPLeS webinar by Aleven) and *ALEKS* (Falmagne, Albert, Doble, Eppstein, & Hu, 2013) together cover basic mathematics, algebra, geometry, statistics, and more advanced quantitative skills. ITSs exist in other STEM areas, such as physics (VanLehn et al., 2005), electronics (Dzikovska, Steinhauser, Farrow, Moore, & Campbell, 2014), and information technology (Mitrovic, Martin, & Suraweera, 2007).

Some ITSs handle knowledge domains that have a stronger verbal foundation as opposed to mathematics and precise analytical reasoning (Johnson & Lester, 2016). *AutoTutor* and its descendants (Graesser, 2016; Nye, Graesser, & Hu, 2014) help college students learn about computer literacy, physics, biology, scientific reasoning, and other STEM topics by holding conversations in natural language. Other successful ITSs with natural language interaction include *Tactical Language and Culture System* (Johnson & Valente, 2009), *iSTART* (Jackson & McNamara, 2013), and *My Science Tutor* (Ward et al., 2013).

Reviews and quantitative meta-analyses confirm that ITS technologies frequently improve learning over reading text and traditional teacher-directed classroom teaching. These meta-analyses normally report effect sizes (d) to convey differences between the ITS condition and a control condition in standard deviation units. A difference of one *d* is approximately a letter grade in a course. The reported meta-analyses show positive effect sizes that vary from 0.05 (Dynarsky et al., 2007) to 1.08 (Dodds & Fletcher, 2004), but most hover between 0.40 and 0.80 (Kulik & Fletcher, 2015; Ma, Adesope, Nesbit, & Liu, 2014; Steenbergen-Hu & Cooper, 2013, 2014; VanLehn, 2011). Our current best meta-meta estimate from all of these meta-analyses is 0.60. This performance is comparable to human tutoring which varies between 0.20 and 0.80 (Cohen, Kulik, & Kulik, 1982; VanLehn, 2011), depending on the expertise of the tutor. Human tutors have not varied greatly from ITSs in direct comparisons between ITSs and trained human tutors (Graesser, 2016; VanLehn, 2011).

The subject matter being tutored is important to consider when analyzing learning gains. It is difficult to obtain high effect sizes for literacy and numeracy because these skills are ubiquitous in everyday life and habits are automatized. In contrast, when the student starts essentially from ground zero, then effect sizes are expected to be more robust. As a notable example, the *Digital Tutor* (Fletcher & Morrison, 2012; Kulik & Fletcher, 2015) improves information technology by an effect size as high as 3.70 for knowledge and 1.10 for skills. The students' knowledge of digital literacy was minimal before being tutored so there was much room to improve.

The remainder of this chapter has two sections. We first identify components of ITSs that are frequently incorporated in most applications. We next identify major challenges in building ITSs, some of their limitations, and promising future directions.

Components of Intelligent Tutoring Systems

ITSs vary in their affordances and learning principles, but they all require some form of active student learning rather than resorting to the mere delivery of information through lectures, films, and books (Chi, 2009). The following affordances always occur in ITSs:

Interactivity. The system systematically responds to actions of the student.

Adaptivity. The system presents information that is contingent on the behavior, knowledge, and characteristics of the student.

Feedback. The system immediately gives feedback to the student on the quality of the student's performance and how the quality could be improved.

The following affordances frequently occur in ITSs but not always:

Choice. The system gives students options on what to learn to encourage self-regulated learning.

Nonlinear access. The technology allows the student to select or receive learning activities in an order that deviates from a rigid scripted order.

Linked representations. The system provides quick connections between representations that emphasize different conceptual viewpoints, pedagogical strategies, and media.

Open-ended learner input. The system allows the students to express themselves through natural language, drawing pictures, and other forms of open-ended communication.

One affordance that is rare in ITSs is *communication with other people*, where the student communicates with one or more other people who are either peers or experts on the subject matter. However, there are exceptions. For example, www.tutor.com has 3,500 individuals available for chat interactions while students are having difficulties interacting with ITSs on STEM topics. There have been several million of these embedded tutoring chats that are currently being analyzed with the long-term goal of automating the exchanges (Rus, Maharjan, & Banjade, 2015). The chapters on learning analytics (Rosé, this volume) and MOOCs (G. Fischer, this volume) discuss how communication with peers and instructors can be coordinated with digital learning technologies.

Pedagogical Interactions in Intelligent Tutoring Systems

ITSs are designed to provide tutoring so one worthwhile source of information in the design of ITSs is to explore how humans tutor. Consequently, many ITSs have been influenced by systematic analyses of the discourse and pedagogical strategies of human tutors who vary from novices to experts (Chi, Siler, Yamauchi, Jeong, & Hausmann, 2001; D'Mello, Olney, & Person, 2010; Graesser, D'Mello, & Person, 2009). Part of tutoring involves the tutor delivering information, but a more fundamental part involves co-constructing responses in specific tasks, such as solving problems, answering challenging questions, and creating artifacts. The *outer loop* of VanLehn's (2006) analysis of ITSs consists of the selection of major tasks for the tutor and student to work on. The *inner loop* consists of the steps and dialogue interactions to manage the interaction within these major tasks.

The inner loop of most ITSs follows systematic mechanisms of interaction. One of these is the *five-step dialogue frame* (Graesser et al., 2009). Once a problem or difficult main question is selected to work on, the five-step tutoring frame is launched: (1) the tutor or student present a task, (2) the student generates an initial attempt to handle the task, (3) the tutor gives short feedback on the quality of the answer, (4) the tutor and student improve the quality of the answer through interaction, and (5) the tutor assesses whether the student understands the correct answer and follows up if necessary. Interestingly, classroom teaching normally (not always!) involves the first three steps (Sinclair & Coulthart, 1975) but not steps 4 and 5.

So how does step 4 evolve? In most ITSs (as well as human tutors), *expectation and misconception tailored dialogue* guides the micro-adaptation of the inner loop at step 4. The ITS typically has a list of *expectations* (anticipated good answers, steps in a procedure) and a list of anticipated *misconceptions* (errors or bugs) associated with each task. The ITS guides the student in generating the expectations through a number of dialogue moves: *pumps, hints,* and *prompts* for the student to fill in missing information. A pump is a move to get the student to provide more information, such as "What else?" or "Tell me more." Hints and prompts are selected by the ITS to get the student to make decisions, generate content at a step, or articulate missing words, phrases, and propositions in ITS with natural language. As the learner produces information over many interactional turns and the tutor fills in missing content, the list of expectations is eventually covered and the task is completed. The ITS keeps track of how well the student does in covering expectations and not having misconceptions. An ITS does this immediately and responds intelligently. The selection of the next task depends on the student's profile in the prior history of performance on tasks in tutorial sessions.

In addition to being interactive and adaptive, the intelligent tutors give feedback to the student on their performance. There is both short feedback (positive, negative, or neutral) and qualitative explanations that justify correct versus incorrect answers. A typical ITS turn has three components:

Tutor Turn → Short Feedback + Dialogue Advancer + Floor Shift

The short feedback addresses the quality of the student's prior turn, whereas the dialogue advancer gives either qualitative feedback or alternatively moves the tutoring agenda forward with hints, prompts, requests, or information to fill gaps. The third component has some cue to indicate that it is the student's turn to take an action or do the talking.

Most ITSs attempt to accommodate a mixed-initiative dialogue by allowing students to ask questions, ask for help, and select tasks to work on. However, it is difficult or impossible for an ITS to anticipate all of the student questions, requests for help, or tasks to work on. This difficulty also applies to human tutoring, which explains why the frequency of student questions and requests for help is low, even when students are encouraged to take the initiative (Graesser et al., 2009).

Student Modeling

One of the hallmarks of ITSs is the detailed tracking of the student on their knowledge on a topic, their skills, and various psychological attributes, including personality, motivation, and emotions (Baker & Ocumpaugh, 2014; Calvo & D'Mello, 2010). All of the behaviors of the learner are logged and classified into various functional categories. For example, *knowledge components* are the primitive content units in many ITSs, particularly those of the Cognitive Tutors (Koedinger, Corbett, & Perfetti, 2012). Knowledge components have a specific scope, such as one of Newton's laws of motion in physics. A task has a set of knowledge components (KCs) and performance on these KC is tracked during the tutorial interaction. There are dozens to hundreds of KCs in an ITS on a subject matter. Mastery of a KC may be manifested in many ways, such as verbally articulating one of Newton's laws (e.g., force equals mass times acceleration), making a decision on a question that involves the KC, or performing an action that presupposes mastery of the KC. These all are tracked in an ITS so one can assess the level of performance and stability of each KC. For example, students can sometimes articulate one of Newton's laws and do so at the right time, but make errors when applying the knowledge to particular problems.

It is important to acknowledge that this fine-grained knowledge tracking is very different from global assessments that measure how well students are performing overall. A single performance score on a lesson is presumably not that helpful compared with detailed feedback on particular KCs. Content feedback that explains reasoning is also more important than a simple "Yes/No" feedback after a student's turn or a global score on a major unit. ITSs are designed for the detailed tracking of the student model, quick adaptive responses, and qualitative feedback.

The student model can store other psychological characteristics of the learner. Generic skills, abilities, and interests may be inferred from the log files with interaction profiles. Examples of these include numeracy, verbosity (number of words or idea units per student turn), fluency (speed of responding to requests), and self-regulated learning (such as asking questions, seeking help, and initiating new topics) (Aleven, McLaren, Roll, & Koedinger, 2016; Biswas, Jeong, Kinnebrew, Sulver, & Roscoe, 2010). Some of the characteristics tracked by ITSs are problematic. These include gaming the system with help abuse (i.e., quickly asking for hints and help, but avoiding learning; Baker, Corbett, Roll, & Koedinger, 2008), disengagement and off-task behavior (Arroyo, Muldner, Burleson, & Woolf, 2014), and wheel spinning (performing the same actions repeatedly without progressing; Beck & Gong, 2013).

Emotions and affective states are tracked in addition to knowledge components and generic cognitive abilities. The most frequent learning-centered affective states that require attention are frustration, confusion, and boredom/disengagement (Baker, D'Mello, Rodrigo, & Graesser, 2010; D'Mello, 2013). These affective states can be detected by analyzing multiple channels of communication and behavior including facial expressions, speech intonation, body posture, physiological responses, natural language, and the patterns and timing of tutorial interaction (Arroyo et al., 2014; Calvo & D'Mello, 2010; D'Mello & Graesser, 2012; McQuiggan & Lester, 2009). D'Mello and Graesser (2012) have reported results that support the claim that an affect sensitive ITS (AutoTutor) facilitates learning in low-knowledge students compared to an adaptive ITS that is only sensitive to cognitive states.

Arthur C. Graesser, Xiangen Hu, and Robert Sottilare

Generalized Intelligent Framework for Tutoring (GIFT)

The Generalized Intelligent Framework for Tutoring (GIFT; www.gifttutoring.org; Sottilare, Brawner, Goldberg, & Holden, 2012) is a framework that articulates the frequent practices, pedagogical and technical standards, and computational architectures for developing ITSs. The goal is to scale up ITS development for schools, the military, industry, and the public. GIFT has evolved through an annual meeting with 20–30 experts in ITS research and development. These ITS experts, who change from year to year, write chapters in a book series that can be downloaded for free (www.gifttutoring.org). Each year there is a particular thematic focus: Learner modeling (2013), instructional strategies (2014), authoring tools (2015), domain modeling (2016), assessment (2017), and team learning (2018). Over 100 ITS experts have contributed chapters, whereas there are over 700 GIFT users. Moreover, these experts come from many countries (Austria, Canada, China, Germany, Netherlands, New Zealand, Taiwan, United Kingdom, United States), branches of the U.S. Department of Defense, academia, and the corporate sector.

GIFT specifies the characteristics of a number of ITS components. The *Learner Record* automatically stores all of the information in the student model as the student interacts with the ITS. The Learner Record Store can accommodate a rich history of interactions in log files, performance scores, mastery of particular KCs, and both cognitive and noncognitive learner characteristics. There are four components that have representations and algorithms that follow particular technical standards. The *Sensor* module accommodates recordings of physiological states, geographical location, facial expressions, speech, gestures, and other multimodal input. The *Learner* module records the raw ITS–student interaction and also various codes that are either expected theoretically or are based on prior research using discovery-oriented data mining. The *Pedagogical* module is the set of tutoring strategies and tactics. The *Domain* module includes subject-matter representations, KCs, and procedures. There can be *External Applications* from third-party learning resources that communicate with GIFT via a *Gateway*. These external applications range from Microsoft PowerPoint presentations to 3-D virtual reality environments. Sottilare et al. (2012) discuss each of the GIFT components in more detail.

Now that GIFT has evolved for over five years, many ITSs have adopted the framework and suite of software tools to develop ITSs for use at scale. There is also an expanded GIFT architecture that incorporates team learning, but it is beyond the scope of this chapter to consider team learning and collaborative problem solving (see Looi & Wong, this volume).

GIFT has been designed to increase quality but simultaneously decrease development costs; that is, tutoring can be developed for one subject matter and then ported to a second application with similar content. Modularity allows GIFT to use the same suite of authoring tools across multiple domains and learning environments. The instructional designers who develop content with GIFT and authoring tools may vary in expertise, ranging from computer scientists to curriculum developers who have limited computer technology skills. Instructional support is needed to assist developers with a wide range of expertise.

The representation of pedagogical strategies in GIFT often consists of IF <state> THEN <action> production rules, a standard representation for strategically selecting instructional strategies. Rule-based tutoring strategies have a long history in ITSs (Anderson, Corbett, Koedinger, & Pelletier, 1995; Woolf, 2009). The system watches over the landscape of current states in the Sensor and Learner modules, which are akin to working memory. If particular states exist or reach some threshold of activation, a production rule is fired probabilistically. Contemporary rules need not be brittle, but rather are activated to some degree and probabilistically. Moreover, the scope of a pedagogical strategy can vary from being domain general to applying to very specific subject matters.

GIFT is a framework and a suite of computational tools to guide instructional designers to build ITSs that embrace the generative power of ITS. Once an ITS is developed, it automatically interacts with students, accumulates information in the Learner Record, and generates feedback and adaptive intelligent responses that guide the students in ways that hopefully improve their learning.

Challenges, Limitations, and Future Directions

The development and use of ITS have had challenges and limitations. This section identifies the major obstacles that have prevented ITS from reaching several millions of learners throughout the globe. Perhaps these obstacles will be circumvented by GIFT and attempts to integrate ITS with other types of learning environments.

One obvious obstacle is that it has historically taken a large amount of time and money to build ITSs because of the complexity of the mechanisms. The GIFT community has been attempting to reduce the cost and time through modularity, standards, and better authoring tools. However, the cost is still measured in millions of dollars and the time is measured in years. That may be necessary for building systems that promote deeper learning (Graesser, Lippert, & Hampton, 2017), as opposed to the shallow learning that is provided by most computer-based training and educational games. Moreover, self-regulated learning is limited because the vast majority of learners have underdeveloped skills of metacognitive monitoring and self-regulated learning (Azevedo, Cromley, Moos, Greene, & Winters, 2011; Goldman, 2016; Graesser et al., 2009; Winne, 2011).

A second obstacle is that it is difficult to develop authoring tools that can be productively used by individuals without expertise in computer science (see GIFT volume 3 on authoring tools; www.gifttutoring.org/documents/56). The ideal author would have the perfect combination of skills in the subject-matter knowledge, analytical computation, and pedagogy, but that is a rare combination of skills to find in anyone. Interestingly, *Authorware* was developed a few decades ago with the vision of instructors creating content to meet their content and learning objectives. That approach was not successful, although such tools are used by persistent, adventuresome instructors. The ITS community is not unique in facing this challenge, but the complexity of the ITS mechanisms aggravates the problem.

A third obstacle is the uncertainty on how much human intervention is needed to contextualize and scaffold the value of an ITS. Instructors are increasing turning to *blended learning* environments in which the instructor attempts to weave in these sophisticated technologies like ITSs (Siemens, Gasevic, & Dawson, 2015), but the existing professional development is not meeting the required needs to bridge humans and computers. That is, many computer environments are available, such as computer-assisted instruction, repetitive skill training, hypertext and hypermedia, simulations (de Jong, Lazonder, Pedaste, & Zacharia, this volume), serious games (Fields & Kafai, this volume), intelligent tutoring systems, massively open online courses (MOOCs; see G. Fischer, this volume), and virtual reality. But instructors need to be trained how to integrate them with human-led instruction. The technology runs the risk of collecting dust without the human element. More generally, the e-learning enterprise is currently exploring how much human intervention and scaffolding is needed to provide a sufficient context for students to effectively use and continue to use computer learning environments (Means, Peters, & Zheng, 2014).

A fourth obstacle to ITS is that they have limited mixed-initiative dialogue and self-regulated learning. The systems cannot provide support for all questions, topics, problems, or tasks that students initiate unless the ITS has already developed them. The learning leans toward being instruction-centered or tutor-centered more than student-centered. Self-regulated leaning is not the emphasis in ITSs, even though self-regulated learning is tracked in some systems (Aleven et al., 2016; Biswas et al., 2010) and directly trained in others (Azevedo et al., 2011). The limited affordances for self-regulated learning is of course applicable to most learning environments, not just ITSs.

A fifth obstacle is the uncertainty of whether to follow the path of human tutoring strategies or ideal pedagogical strategies. Instructors and tutors have many blind spots and misconceptions of what makes tutoring effective (Graesser et al., 2009). Which path should the ITS designer follow? Experienced instructors? Or scientific principles of learning that teachers rarely follow or are trained to follow (Pomerance, Greenberg, & Walsh, 2015)? For example, listed below are several tutoring moves that an ideal tutor would implement, but are rarely implemented by human tutors, even ideal human tutors (Graesser et al., 2009).

(1) *Request student summary.* Instead of the tutor giving the summary of a good answer, the tutor could first request that the student summarize the answer.
(2) *Don't trust the student's accuracy of meta-comprehension.* The tutor performs dialogue moves that troubleshoot the student's understanding, such as a follow-up question or a request for student summary.
(3) *Explore the foundations of student's wrong answers.* Tutors can be on the lookout for opportunities to launch dialogues on the epistemological foundations of the student's mindset.
(4) *Ground referring expressions and quantities.* The tutor does not assume that the tutor and student are on the same page (i.e., common ground, Clark, 1996), so the tutor offers precise specification of ideas and asks confirmation questions to verify common ground.
(5) *Request more explanations behind student answers.* The tutor requests that the student justify the reasoning with an explanation (e.g., ask why, how, or why not?).
(6) *Plant seeds for cognitive disequilibrium.* The tutor challenges the student by disagreeing with the student, presenting contradictions, or expressing ideas that clash with common knowledge.
(7) *Monitor student emotions.* Frustration, confusion and boredom may be productively managed with tutoring moves that promote learning.

ITSs have been designed to implement the above tutoring tactics that are not routinely made by human tutors. Some of these ideal strategies have produced learning gains beyond human strategies, but more empirical studies are needed to sort out the relative impact on learning from the normal versus ideal strategies. If the ideal strategies outstrip the normal, the possibility of an ITS outperforming human tutors is on the horizon.

Aside from making these tactical improvements, the ITS community is continuing to improve enhanced personalization that accommodates a broad diversity of student personalities, abilities and affective states. There are important differences between passive students who wait for the tutor to guide them and the students who are prone to take initiative with self-regulated learning. Students also vary in emotional temperament, motivation to learn, and interests in different subject matters. ITS researchers aspire for the learning environment to be personalized so that it offers the right learning resource to the right person at the right time in the right context. As ITS applications accumulate in the future, the growing repertoire of resources has the potential to help students beyond the mastery of specific subject matters and skills. Lifelong learning could be supported by recommending learning resources on various topics at varying levels of depth and sophistication.

Further Readings

Aleven, V., McLaren, B. M., Roll, I., & Koedinger, K. R. (2016). Help helps, but only so much: Research on help seeking with intelligent tutoring systems. *International Journal of Artificial Intelligence in Education, 26*(1), 205–223.
The article identifies ways to help students become better self-regulated learners by seeking help when they need to in ITS.

Baker, R. S. J. d., D'Mello, S. K., Rodrigo, M. M. T., & Graesser, A. C. (2010) Better to be frustrated than bored: The incidence, persistence, and impact of learners' cognitive-affective states during interactions with three different computer-based learning environments. *International Journal of Human–Computer Studies, 68*(4), 223–241.
The detection of affective states such as confusion, frustration, and boredom, are important to detect and respond to in ITS.

Koedinger, K. R., Corbett, A. C., & Perfetti, C. (2012). The Knowledge–Learning–Instruction (KLI) framework: Bridging the science–practice chasm to enhance robust student learning. *Cognitive Science, 36*(5), 757–798.
This article describes how ITS are motivated by theories in the cognitive and learning sciences.

Kulik, J. A., & Fletcher, J. D. (2015). Effectiveness of intelligent tutoring systems: A meta-analytic review. *Review of Educational Research, 85,* 171–204.
This article reports meta-analyses that assesses the impact of ITS on the learning of different subject matters and skills.

Nye, B. D., Graesser, A. C., & Hu, X. (2014). AutoTutor and family: A review of 17 years of natural language tutoring. *International Journal of Artificial Intelligence in Education, 24*, 427–469.
This article covers the history of AutoTutor and similar systems that use intelligent conversational agents to help students learn STEM topics.

NAPLeS Resources

Aleven, V. *Cognitive tutors* [Webinar]. In *NAPLeS video series*. Retrieved October 19, 2017, from http://isls-naples.psy.lmu.de/intro/all-webinars/aleven/index.html

Goldman, S. R. *Cognition and metacognition* [Webinar] In *NAPLeS video series*. Retrieved October 19, 2017, from http://isls-naples.psy.lmu.de/intro/all-webinars/goldman_video/index.html

Hoadley, C. *A short history of the learning sciences* [Webinar]. In *NAPLeS video series*. Retrieved October 19, 2017, from http://isls-naples.psy.lmu.de/intro/all-webinars/hoadley_video/index.html

References

Aleven, V., McLaren, B. M., Roll, I., & Koedinger, K. R. (2016). Help helps, but only so much: Research on help seeking with intelligent tutoring systems. *International Journal of Artificial Intelligence in Education, 26*(1), 205–223.

Aleven, V., McLaren, B. M., Sewall, J., & Koedinger, K. (2009). A new paradigm for intelligent tutoring systems: Example-tracing tutors. *International Journal of Artificial Intelligence in Education, 19*, 105–154.

Anderson, J. R., Corbett, A. T., Koedinger, K. R., & Pelletier, R. (1995). Cognitive tutors: Lessons learned. *Journal of the Learning Sciences, 4*, 167–207.

Arroyo, I., Muldner, K., Burleson, W., & Woolf, B. P. (2014). Adaptive interventions to address students' negative activating and deactivating emotions during learning activities. In R. Sottilare, A. Graesser, X. Hu, & B. Goldberg (Eds.), *Design recommendations for intelligent tutoring systems: Instructional management* (Vol. 2, pp. 79–91). Orlando, FL: Army Research Laboratory.

Azevedo, R., Cromley, J. G., Moos, D. C., Greene, J. A., & Winters, F. I. (2011). Adaptive content and process scaffolding: A key to facilitating students' self-regulated learning with hypermedia. *Psychological Test and Assessment Modeling, 53*, 106–140.

Baker, R. S. J. d., Corbett, A. T., Roll, I., & Koedinger, K. R. (2008) Developing a generalizable detector of when students game the system. *User Modeling and User Adapted Interaction, 18*(3), 287–314.

Baker, R. S. J. d., D'Mello, S. K., Rodrigo, M. M. T., & Graesser, A. C. (2010) Better to be frustrated than bored: The incidence, persistence, and impact of learners' cognitive-affective states during interactions with three different computer-based learning environments. *International Journal of Human–Computer Studies, 68*(4), 223–241.

Baker, R. S. J. d., & Ocumpaugh, J. (2014) Interaction-based affect detection in educational software. In R. A. Calvo, S. K. D'Mello, J. Gratch, & A. Kappas (Eds.), *The Oxford Handbook of Affective Computing* (pp. 233–245). Oxford, UK: Oxford University Press.

Beck, J. E., & Gong, Y. (2013). Wheel-spinning: Students who fail to master a skill. In H. C. Lane, K. Yacef, J. Mostow, & P. Pavlik (Eds.), *Proceedings of the 16th International Conference on Artificial Intelligence in Education* (pp. 431–440). Berlin: Springer.

Biswas, G., Jeong, H., Kinnebrew, J., Sulcer, B., & Roscoe, R. (2010). Measuring self-regulated learning skills through social interactions in a teachable agent environment. *Research and Practice in Technology-Enhanced Learning, 5*, 123–152.

Calvo, R. A., & D'Mello, S. K. (2010). Affect detection: An interdisciplinary review of models, methods, and their applications. *IEEE Transactions on Affective Computing, 1*, 18–37.

Chi, M. T. H. (2009). Active–Constructive–Interactive: A conceptual framework for differentiating learning activities. *Topics in Cognitive Science, 1*, 73–105.

Chi, M. T. H., Siler, S., Yamauchi, T., Jeong, H., & Hausmann, R. (2001). Learning from human tutoring. *Cognitive Science, 25*, 471–534.

Clark, H. H. (1996). *Using language*. Cambridge, UK: Cambridge University Press.

Cohen, P. A., Kulik, J. A., & Kulik, C. C. (1982). Educational outcomes of tutoring: A meta-analysis of findings. *American Educational Research Journal, 19*, 237–248.

de Jong, T., Lazonder, A., Pedaste, M., & Zacharia, Z. (2018). Simulations, games, and modeling tools for learning. In F. Fischer, C. E. Hmelo-Silver, S. R. Goldman, & P. Reimann (Eds.), *International handbook of the learning sciences*. New York: Routledge.

D'Mello, S. K. (2013). A selective meta-analysis on the relative incidence of discrete affective states during learning with technology. *Journal of Educational Psychology, 105,* 1082–1099.

D'Mello, S. K., & Graesser, A. C. (2012). AutoTutor and affective AutoTutor: Learning by talking with cognitively and emotionally intelligent computers that talk back. *ACM Transactions on Interactive Intelligent Systems, 23,* 1–38.

D'Mello, S., Olney, A. M., & Person, N. (2010). Mining collaborative patterns in tutorial dialogues. *Journal of Educational Data Mining, 2*(1), 1–37.

Dodds, P. V. W., & Fletcher, J. D. (2004). Opportunities for new "smart" learning environments enabled by next generation web capabilities. *Journal of Education Multimedia and Hypermedia, 13,* 391–404.

Dynarsky, M., Agodina, R., Heaviside, S., Novak, T., Carey, N., Campuzano, L., et al. (2007). *Effectiveness of reading and mathematics software products: Findings from the first student cohort.* Washington, DC: Institute of Education Sciences.

Dzikovska, M., Steinhauser, N., Farrow, E., Moore, J., & Campbell, G. (2014). BEETLE II: Deep natural language understanding and automatic feedback generation for intelligent tutoring in basic electricity and electronics. *International Journal of Artificial Intelligence in Education, 24,* 284–332.

Falmagne, J., Albert, D., Doble, C., Eppstein, D., & Hu, X. (2013). *Knowledge spaces: Applications in education.* Berlin-Heidelberg: Springer.

Fields, D. A., & Kafai, Y.B. (2018). Games in the learning sciences: reviewing Evidence from playing and making games for learning. In F. Fischer, C. E. Hmelo-Silver, S. R. Goldman, & P. Reimann (Eds.), *International handbook of the learning sciences* (pp. 276–284). New York: Routledge.

Fischer, G. (2018) Massive open online courses (MOOCs) and rich landscapes of learning: a learning sciences perspective. In F. Fischer, C. E. Hmelo-Silver, S. R. Goldman, & P. Reimann (Eds.), *International handbook of the learning sciences* (pp. 368–380). New York: Routledge.

Fletcher, J. D., & Morrison, J. E. (2012). *DARPA Digital Tutor: Assessment data* (IDA Document D-4686). Alexandria, VA: Institute for Defense Analyses.

Goldman, S. R., (2016). *Cognition and metacognition* [Webinar] In *NAPLeS video series.* Retrieved October 19, 2017, from http://isls-naples.psy.lmu.de/intro/all-webinars/goldman_video/index.html

Graesser, A.C. (2016). Conversations with AutoTutor help students learn. *International Journal of Artificial Intelligence in Education, 26,* 124–132.

Graesser, A. C., D'Mello, S. K., & Person, N., (2009). Meta-knowledge in tutoring. In D. J. Hacker, J. Dunlosky, & A. C. Graesser (Eds.), *Metacognition in educational theory and practice* (pp. 361–382). Mahwah, NJ: Erlbaum.

Graesser, A. C., Hu, X., Nye, B., & Sottilare, R. (2016). Intelligent tutoring systems, serious games, and the Generalized Intelligent Framework for Tutoring (GIFT). In H. F. O'Neil, E. L. Baker, and R. S. Perez (Eds.), *Using games and simulation for teaching and assessment* (pp. 58–79). Abingdon, UK: Routledge.

Graesser, A. C., Lippert, A. M., & Hampton, A. J. (2017). Successes and failures in building learning environments to promote deep learning: The value of conversational agents. In J. Buder & F. Hesse (Eds.), *Informational environments: Effects of use, effective designs* (pp. 273–298). New York: Springer.

Jackson, G. T., & McNamara, D. S. (2013). Motivation and performance in a game-based intelligent tutoring system. *Journal of Educational Psychology, 105,* 1036–1049.

Johnson, W. L., & Lester, J. C. (2016). Twenty years of face-to-face interaction with pedagogical agents. *International Journal of Artificial Intelligence in Education, 26*(1), 25–36.

Johnson, W. L., & Valente, A. (2009). Tactical Language and Culture Training Systems: Using AI to teach foreign languages and cultures. *AI Magazine, 30,* 72–83.

Koedinger, K. R., Corbett, A. C., & Perfetti, C. (2012). The Knowledge–Learning–Instruction (KLI) framework: Bridging the science–practice chasm to enhance robust student learning. *Cognitive Science, 36*(5), 757–798.

Kulik, J. A., & Fletcher, J. D. (2015). Effectiveness of intelligent tutoring systems: A meta-analytic review. *Review of Educational Research, 85,* 171–204.

Looi, C.-K., & Wong, L.-H. (2018) Mobile computer-supported collaborative learning. In F. Fischer, C. E. Hmelo-Silver, S. R. Goldman, & P. Reimann (Eds.), *International handbook of the learning sciences* (pp. 359–367). New York: Routledge.

Ma, W. Adesope, O. O., Nesbit, J. C., & Liu, Q. (2014). Intelligent tutoring systems and learning outcomes: A meta-analytic survey. *Journal of Educational Psychology, 106,* 901–918.

McQuiggan, S., & Lester, J. (2009). Modelling affect expression and recognition in an interactive learning environment. *International Journal of Learning Technology, 4,* 216–233.

Means, B., Peters, V., & Zheng, Y. (2014). *Lessons from five years of funding digital courseware: Postsecondary success portfolio review, full report.* Menlo Park, CA: SRI Education.

Mitrovic, A., Martin, B., & Suraweera, P. (2007). Intelligent tutors for all: The constraint-based approach. *IEEE Intelligent Systems, 22,* 38–45.

Nye, B. D., Graesser, A. C., & Hu, X. (2014). AutoTutor and family: A review of 17 years of natural language tutoring. *International Journal of Artificial Intelligence in Education, 24*, 427–469.

Pomerance, L., Greenberg, J., & Walsh, K. (2015). *Learning about learning: Do textbooks deliver on what every new teacher needs to know.* National Council on Teacher Quality. Retrieved form www.nctq.org/dmsView/Learning_About_Learning_Report

Ritter, S., Anderson, J. R., Koedinger, K. R., & Corbett, A. (2007) Cognitive Tutor: Applied research in mathematics education. *Psychonomic Bulletin & Review, 14*, 249–255.

Rus, V., Maharjan, N., & Banjade, R. (2015). Unsupervised discovery of tutorial dialogue modes in human-to-human tutorial data. *Proceedings of the Third Annual GIFT Users Symposium* (R. Sottilare & A. M. Sinatra, Eds.) (pp. 63–80), Army Research Lab, June 2015.

Siemens, G., Gasevic, D., & Dawson, S. (Eds.) (2015). *Preparing for the digital university: A review of the history and current state of distance, blended, and online learning.* Alberta, Canada: MOOC Research Initiative.

Sinclair, J. & Coulthart, M. (1975) *Towards an analysis of discourse: The English used by teachers and pupils.* London: Oxford University Press.

Sottilare, R. A., Goldberg, B. S., Brawner, K. W., & Holden, H. K. (2012). A modular framework to support the authoring and assessment of adaptive computer-based tutoring systems (CBTS). In *The Interservice/Industry Training, Simulation & Education Conference* (I/ITSEC) (Vol. *2012*, No. 1, pp. 1–13). Orlando, FL: National Training Systems Association.

Steenbergen-Hu, S., & Cooper, H. (2013). A meta-analysis of the effectiveness of intelligent tutoring systems on K–12 students' mathematical learning. *Journal of Educational Psychology, 105*, 971–987.

Steenbergen-Hu, S., & Cooper, H. (2014). A meta-analysis of the effectiveness of intelligent tutoring systems on college students' academic learning. *Journal of Educational Psychology, 106*, 331–347.

VanLehn, K. (2006) The behavior of tutoring systems. *International Journal of Artificial Intelligence in Education, 16*(3), 227–265.

VanLehn, K. (2011). The relative effectiveness of human tutoring, intelligent tutoring systems and other tutoring systems. *Educational Psychologist, 46*, 197–221.

VanLehn, K., Lynch, C., Schultz, K., Shapiro, J. A., Shelby, R., Taylor, L., et al. (2005). The Andes physics tutoring system: Lessons learned. *International Journal of Artificial Intelligence and Education, 15*(3), 147–204.

Ward, W., Cole, R., Bolaños, D., Buchenroth-Martin, C., Svirsky, E., & Weston, T. (2013). My Science Tutor: A conversational multimedia virtual tutor. *Journal of Educational Psychology, 105*, 1115–1125.

Winne, P. H. (2011). Cognitive and metacognitive factors in self-regulated learning. In B. J. Zimmerman and D. H. Schunk (Eds.), *Handbook of Self-Regulation of Learning and Performance* (pp.15–32). New York: Routledge.

Woolf, B. P. (2009). *Building intelligent tutoring systems.* Burlington, MA: Morgan Kaufman.

25
Simulations, Games, and Modeling Tools for Learning

Ton de Jong, Ard Lazonder, Margus Pedaste, and Zacharias Zacharia

Introduction

Learning in an active way is regarded as a necessary condition for acquiring deep knowledge and skills (e.g., Freeman et al., 2014). Experiential and inquiry learning are specific forms of learning in which students make active choices (choosing the next step in performing an action, changing the value of a variable), experience the consequences of their own actions, and are stimulated to adapt their knowledge and skills in response to these experiences. Experiential and inquiry learning can take place in real environments (a "wet" lab or a practical) but are nowadays increasingly enabled by technologies such as games, simulations, and modeling environments. In this chapter, we first give an overview of these technologies and discuss how they can be used in a diversity of educational settings. We then explain why smart design and careful combination with other instructional approaches and support are necessary. We conclude our chapter by trying to give a glimpse of the future.

Technologies for Experiential and Inquiry Learning

Simulations

Simulations have a long history and come in a great variety of forms (de Jong, 2016). Simulations are very often created for science topics such as physics, chemistry, and biology, but they also exist for the behavioral sciences. For example, many psychology courses now offer a set of online simulations that students can explore (e.g., a simulation in which students can practice conditioning with a virtual dog; Hulshof, Eysink, & de Jong, 2006). Figure 25.1 shows an example of a simulation (or virtual lab) for the physics domain of electricity in which students can create and test their own electrical circuits.

Computer simulations for learning basically consist of two components: an underlying computational model that simulates a process or phenomenon and an interface that enables student to interact with this model (de Jong & van Joolingen, 2008). Across simulations, the underlying models can cover very different domains and levels of complexity and the interfaces can differ widely in appearance and facilities offered. Interfaces can range from simple and functional to displaying complex and realistic representations, sometimes even including realistic haptic input (Han & Black, 2011) or 3-D virtual reality (Bonde et al., 2014). Simulations may also offer augmentations for embodied input facilities (Lindgren, Tscholl, Wang, & Johnson, 2016), or provide features that cannot be seen

Simulations, Games, and Modeling Tools

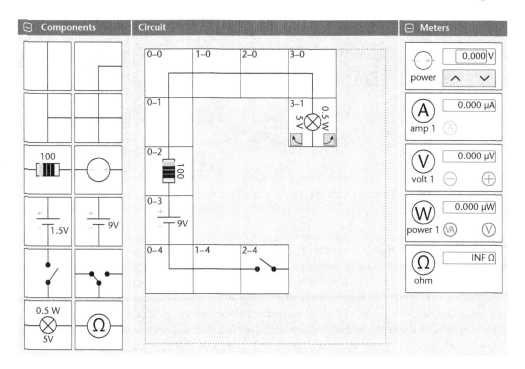

Figure 25.1 Electrical Circuit Laboratory From the Go-Lab Collection (www.golabz.eu)

(Ibáñez, Di Serio, Villarán, & Delgado Kloos, 2014) or felt in the real world (Bivall, Ainsworth, & Tibell, 2011). Bivall et al. (2011, p. 700), for example, added a haptic device to a chemistry simulation allowing students to "feel the interactions (repulsive and attractive) between molecules as forces." These authors found that students acquired better conceptual knowledge with the help of the haptic device. Han and Black (2011) found similar results when a joystick providing haptic feedback about forces was added to a simulation with which students could experiment with gears. An overview of these types of features is given in Zacharia (2015).

Nowadays, simulations can be found online in many (dedicated) repositories. Some well-known examples are PhET (Wieman, Adams, & Perkins, 2008), Amrita (Achuthan et al., 2011), and ChemCollective (Yaron, Karabinos, Lange, Greeno, & Leinhardt, 2010). Some of these repositories offer instructional material around the simulations, while other repositories (e.g., Ing-ITS; Gobert, Baker, & Wixon, 2015) offer simulations in a sequence of activities including adaptive tools. Still other repositories include authoring facilities to create learning environments around a simulation. Examples of these are WISE (Slotta & Linn, 2009) and Go-Lab (de Jong, Sotiriou, & Gillet, 2014).

Lab exercises have always been part of science education (Hofstein & Lunetta, 2004; Waldrop, 2013). These lab exercises traditionally focus on the learning of a skill (e.g., operating a piece of equipment or making a medical diagnosis). What makes the use of technology interesting here (apart from practical issues such as maintenance costs of real labs, etc.) is that many different situations can be offered systematically, so that the skills can be practiced in various circumstances. Figure 25.2 shows an example of a VR (virtual reality) simulation on how to follow the correct protocols in a laboratory (Bonde et al., 2014). Learning a skill from a simulation has received a prominent place in the field of medicine (see, for example, McGaghie, Issenberg, Petrusa, & Scalese, 2010 for an overview). Extant research reveals that simulations are being used to enhance doctors' clinical skills in controlled and safe practice environments, as well as to support future research, practice, and innovation in medicine (Fincher & Lewis, 2002).

Ton de Jong, Ard Lazonder, Margus Pedaste, and Zacharias Zacharia

Figure 25.2 Learning Chemistry Laboratory Skills in a VR Simulation (Labster)
Source: For Labster, see Bonde et al. (2014).

Other advantages of simulations compared to "wet" labs include that computer technology permits the fast manipulation of many variables and immediate feedback from the simulation, the introduction of situations that are not possible in a school lab, and the use of augmented reality (de Jong, Linn, & Zacharia, 2013). This also means that there can be a shift in emphasis from learning the more procedural knowledge in a lab (how to perform the experiment) to (also) acquiring deep conceptual knowledge about the underlying domain. This focus on acquiring conceptual knowledge is central to what is called inquiry learning (Rönnebeck, Bernholt, & Ropohl, 2016; Linn, McElhaney, Gerard, & Matuk, this volume).

Games

Some types of games (those with an underlying computational model) are closely related to simulations; these games add features to a simulation such as competition, goal setting, rules and constraints, rewards, role playing and/or surprise (Leemkuil & de Jong, 2011; see Fields & Kafai, this volume). Examples of such games are Electric Field Hockey (Miller, Lehman, & Koedinger, 1999) or Supercharged! (Anderson & Barnett, 2013). Games can be used like simulations in the sense that students manipulate variables and learn from observing the consequences of their manipulations, but with the added motivational aspects of the game characteristics. As with simulations, one type of games focuses on the acquisition of skills, such as how to do the job of a knowledge engineer in a large company (Leemkuil & de Jong, 2012) or how to act in emergency situations (van der Spek, van Oostendorp, & Meyer, 2013).

Modeling Tools

A third technology useful for active learning is modeling tools. In learning by modeling, students themselves create the models underlying simulations (de Jong & van Joolingen, 2008). During this process, they develop an understanding of the phenomenon or system being modeled. A modeling process may involve several steps: making systematic observations and collecting data about the phenomenon or system at hand; developing a (computer) model of the phenomenon or system based on those observations and data modeling; evaluating the model according to its degree of resemblance to the actual phenomenon or system, its predictive power and its explanatory adequacy; and revising and applying the model (Louca & Zacharia, 2015; Windschitl, Thompson, & Braaten, 2008). These steps keep repeating, not necessarily in this order, through iterative cycles of evaluation and refinement, until the model reaches a state that provides "insights" into the phenomenon or system being represented (National Research Council, 2012, p. 57). For example, Papaevripidou and Zacharia (2015) involved physics and science education graduates in a modeling-based learning activity concerning 1-D elastic collisions of two masses. The participants observed videos on 1-D elastic collisions of two masses colliding (stimulus) and then used a computer modeling environment to construct a model that could potentially represent all possible scenarios where two masses collide elastically in one dimension. The study showed how the students' models progressed after passing through several cycles of evaluation and revision, until they developed a generalized model with optimal predictive power. Computer-based modeling tools use different modeling principles and languages (e.g., system dynamics; Wetzel et al., 2016), which in turn correspond to different interfaces ranging from more textual to graphical or even drawing-based representations (Bollen & van Joolingen, 2013).

How Can We Make Experiential and Inquiry Learning Effective?

Just providing students with an open simulation, game, or modeling tool may not be productive. In the next two sections, we discuss two design principles that help to create effective learning experiences.

Sequencing of Experiences

One specific characteristic of the type of environments discussed in this chapter is that they allow students to decide their own route. However, students may not know where to start so that they may begin with elements of the domain that are, at that point, beyond their reach. So, within the freedom of open learning environments, structuring of the domain is needed, just as in each curriculum. This structuring can be done in a number of ways. First, the domain can be gradually opened up through successively adding more variables (model progression). This means that students do not immediately see the simulation in its full complexity; it is built up gradually (first only velocity present, then acceleration added, etc.; see, e.g., White & Frederiksen, 1990). In games this is a well-accepted principle. Once having completed a level in a game, the players (learners) move to the next level, which offers greater complexity and challenges. Playing at the right level means that students work at the edge of their current knowledge and skill level. They can play the game at that level without being frustrated, and can anticipate moving to a higher level when the current level is mastered. Lazonder and Kamp (2012) compared a group of (young) students learning with a complete simulation on the reverberation time of a gong, with a group of students who were working through the (four) variables one after each other. They found that the latter group performed more systematic experiments and gained more knowledge. In the case of modeling, Mulder, Lazonder, and de Jong (2011) compared two modes of model progression (by successively introducing more variables or by making the relations between variables gradually more specific) and found that students in both model progression groups created better models than students from a group without model progression. Another way to sequence progress through the domain is to provide students with assignments that stimulate them to explore different parts of the domain or to practice increasingly complex skills. In this case, the underlying model stays the same during the learning activity, but by considering different situations or investigating different variables that are indicated in the assignment, students are guided in covering all aspects of the domain in an appropriate sequence. An example of this is described in Trout, Lee, Moog, and Rickey (2008), who outlined in detail how inquiry learning can be used in chemistry education and listed a dedicated set of assignments to guide students through the inquiry process. In a game on learning about propositional reasoning, ter Vrugte et al. (2015) not only used different levels of difficulty in the game, but also introduced a number of sub-games, each one containing situations and assignments dedicated to a specific sub-skill in the game.

Ton de Jong, Ard Lazonder, Margus Pedaste, and Zacharias Zacharia

Instructional Support

Most research on the effectiveness of instructional support revolves around the design of software tools that either aim to alleviate specific difficulties students encounter during the learning process (Zacharia et al., 2015) or guide students through all stages of the activity (Pedaste et al., 2015). De Jong and Lazonder (2014) classified the types of instructional support embedded within simulation-based learning environments, and their typology applies just as well to games and modeling tools.

The least specific type of support, referred to as process constraints, essentially matches the sequencing principle described in the previous section. Direct presentation of information is the most explicit type of support. It can be delivered through the learning environment before and during the learning process—for instance, in help files or by a pedagogical agent. A typical example can be found in *Shakshouka Restaurant*, a business simulation game for children to strengthen their financial and math skills (Barzilai & Blau, 2014). To avoid children developing only an implicit understanding of the central concepts of cost, price, and profit, the game contained an online study unit that children could consult before playing the game. The children who completed this unit scored higher on a math problem posttest than the children who only played the game.

Support at intermediate levels of specificity includes prompts, heuristics, and scaffolds. An example of scaffolding was described by ter Vrugte and colleagues, who embedded faded worked-out examples in a math game about proportional reasoning (ter Vrugte et al., 2017). Vocational education students who played this game received a partly filled-out template of the solution steps and could solve the problem by completing the missing steps. As students progressed through the game's levels, the number of filled-out steps was gradually reduced. This scaffolding proved to be effective: students who had access to worked examples during game play scored significantly higher on a proportional reasoning posttest than students who played the game without them (see Tabak & Kyza, this volume, for an overview on scaffolding).

Prompts and heuristics are offered during the learning process to remind students of certain actions and the way they could be performed. Prompts and heuristics should be carefully designed in order to minimize disruption of the flow of the activity, which makes them perhaps somewhat less appropriate for use in games. In simulations, however, both types of support have been successfully applied to promote self-explanations, which, in turn, enhance learning outcomes. In one study (Johnson & Mayer, 2010), a simple question prompt ("Provide an explanation for your answer in the space below") was added to a simulation of an electric circuit, whereas Berthold, Eysink, and Renkl (2009) used a fill-in-the-blanks format to encourage students to self-explain the to-be-learned principles of probability. In both studies, prompted students acquired more topic knowledge than their counterparts who interacted with a simulation without prompts.

Combining open, active, learning environments with direct instruction (concurrent or sequential) may also help students in the learning process. Overall, for both games and simulations, studies report positive effects of this combination, with an overall tendency to favor presentation of the information before the game or simulation compared to after it (Arena & Schwartz, 2014; Barzilai & Blau, 2014; Brant, Hooper, & Sugrue, 1991; Wecker et al., 2013) with some studies reporting superior effects of presenting the information both before and concurrently with the simulation (Lazonder, Hagemans, & de Jong, 2010).

Meta-analyses show that interventions like those above that combine support with inquiry enhance learning compared to unassisted inquiry (d'Angelo et al., 2014; Lazonder & Harmsen, 2016) and lead to better knowledge acquisition compared to expository forms of instruction (Alfieri, Brooks, Aldrich, & Tenenbaum, 2011). Guidance is not typically added to games, from the fear that adding guidance may interfere with the flow of the game, but a recent review study showed that games that were augmented with support outperformed games without that support (Clark, Tanner-Smith, & Killingsworth, 2016; see Fields & Kafai, this volume). Despite the overall conclusion that support works, little is known about the relative effectiveness of different types of support. Recently, however, Lazonder and Harmsen (2016) concluded from their meta-analysis that students create better products (e.g., models, concept maps, lab reports) during an inquiry if they receive more specific support. Although the specificity of the support did not show an effect on learning outcomes, it did impact learning activities, such that young, less experienced learners tended to benefit more from specific support than older, more experienced learners.

Learning Sciences and Research Methods

The urge to include technology-based simulations and modeling tools in education is based on a number of theoretical premises. The first is that following a scientific inquiry cycle possibly includes encountering a cognitive conflict between existing ideas and data that come from experiments (Limón, 2001). Based on such conflicts, students would then be stimulated to adapt their existing knowledge. This theoretical notion is closely related to the cognitive theories of schema development and adaptation (see e.g., Chi, Glaser, & Farr, 1988). A second theoretical notion underlying the impetus for simulation-based learning is that simulations often use multiple representations. These different representations (graphs, animations, equations, tables, etc.) are dynamic and must be connected by students, which leads to processes of knowledge abstraction (Ainsworth, 2006; Ainsworth, this volume), as also explained by Mayer's multimedia theory (Mayer, 2009). A third underlying principle is that in simulation-based learning students are in charge of their own learning process, which, according to theories of social learning, leads to higher motivation and especially to intrinsic motivation (Ryan & Deci, 2000), while the students get control over the learning process by planning, monitoring, and reflecting about it. In this way, simulation and modeling also support self-regulated learning (Zimmerman, 1990). A fourth relevant theoretical approach is constructionism (e.g., Kafai & Resnick, 1996). According to this theory, students learn through the process of identifying and representing the components that comprise a phenomenon. These components include *objects* (e.g., particles), *processes* (e.g., free fall), *entities* (e.g., acceleration), and *interactions* (e.g., how entities interact with objects or processes). In other words, the learner strips down the phenomenon into its components (an analysis process) and then builds up the phenomenon in a modeling environment (a synthesis process). However, the underlying premises behind each of these approaches could be overly optimistic; for example, sometimes students do not adapt their knowledge in response to anomalous data (Chinn & Brewer, 1993) or they fail to connect representations (van der Meij & de Jong, 2006). In these cases, instructional support is needed for successful learning.

The research methods used in this field can be divided into: design-based research that basically focuses on the design of a specific simulation or modeling tool through iterative rounds of improvement (Wang & Hannafin, 2005; see Puntambekar, this volume, for an introduction to design-based research); experimental studies in which different variants of a learning environment are compared or in which a modeling- or simulation-based learning environment is compared to another type of instruction (for an overview of these type of studies, see d'Angelo et al., 2014); and studies that focus on the learning process and have a qualitative character, using, for example, thinking aloud techniques (Furberg, Kluge, & Ludvigsen, 2013) or using student interaction data to collect information on the learning process or to predict students' developing knowledge (e.g., Käser, Hallinen, & Schwartz, 2017).

Future Developments

Most of the cited studies used a "one-size-fits-all" approach, in that one particular type of instructional support was available to all students throughout the entire learning process. Evidence is now accumulating that a more sophisticated approach is needed that takes the interplay between student characteristics and the provided guidance into account; specific scaffolds may be more or less effective depending on the student's prior knowledge. Future efforts to investigate and employ these technologies should therefore be more sensitive to individual differences and at the same time address ways to embed them in the curriculum.

Authoring facilities offered by repositories such as WISE and Go-Lab enable teachers to adapt instructional support to student characteristics up-front. Teachers can configure a learning environment such that 10th graders receive more background information than 12th graders, or the less knowledgeable students in the class receive a partially specified model whereas the more knowledgeable ones must build this model from scratch (cf. Mulder, Bollen, de Jong, & Lazonder, 2016). Possibilities like these are likely to grow in the near future, and research must provide teachers with specific guidelines on which levels of support are appropriate for which groups of learners.

Going beyond differentiation based on learner characteristics, support should also adapt to what students create or do during the learning process. Teachers can offer this support on the spot while watching their students, but developments in learning analytics (see Rosé, this volume, for an overview on learning analytics and educational data mining) have greatly advanced the possibilities for teachers and software agents to work in synergy. For example, the Inq-ITS platform offers physics simulations that monitor middle school students' actions and provides automatic support right when it is needed, thus enabling the teacher to resolve more specific issues through personal coaching and feedback (Gobert, Sao Pedro, Raziuddin, & Baker, 2013). Likewise, Ryoo and Linn (2016) developed an automated scoring system that assessed the qualities of student-created concept maps and generated guidance based on assessment outcomes. Research and development of such automated support facilities can play a major role in the future development of effective experiential and inquiry-based learning methods.

Conclusion

Modern, web-based technology enables the use of simulations and modeling environments worldwide and increases the accessibility of online labs, simulations, and modeling tools. This increases their usage and with appropriate instructional embedding, simulations, games, and modeling tools become very effective technologies for learning. These technologies and the associated types of learning fit into a trend that pedagogical approaches in schools are gaining versatility and flexibility and are offered to students in all types of combinations. Learning with these technologies can be very nicely integrated with collaborative learning, learning in real (wet) labs and in flip-the-classroom set-ups in which data from interactions or products from the virtual experiences can be the input for class discussions. Issues for attention are the technology skills that teachers need and new pedagogies that should go along with these new developments.

Acknowledgment

This work was partly prepared in the context of the Next-Lab project, which has received funding from the European Union's Horizon 2020 Research and Innovation Programme under Grant Agreement No. 731685. This publication reflects the authors' views only, and the European Commission is not responsible for any use that may be made of the information it contains.

Further Readings

Clark, D. B., Tanner-Smith, E. E., & Killingsworth, S. S. (2016). Digital games, design, and learning: A systematic review and meta-analysis. *Review of Educational Research, 86*, 79–122.
This meta-analysis synthesized the results of 57 primary studies on learning with digital games, including simulation games. Main findings indicate that digital games significantly enhanced student learning relative to non-game control conditions. The meta-analysis also compared augmented game designs to standard game

designs. Across 20 studies, the learning effect of the augmented designs was higher than that of standard designs. Together these findings indicate that digital games can be beneficial for learning, and that their educational effectiveness depends more on the instructional design than on the affordances of computer technology per se.

d'Angelo, C., Rutstein, D., Harris, C., Bernard, R., Borokhovski, E., & Haertel, G. (2014). *Simulations for STEM learning: Systematic review and meta-analysis.* Menlo Park, CA: SRI International.

This meta-analysis had two main research questions: Do students learn more from computer simulations than from other forms of instruction; and do students learn more from plain simulations or from simulations that are combined with other instructional measures? Including 42 studies, the first meta-analysis showed, in a consistent pattern, a clear advantage for simulation-based instruction over non-simulation instruction. Based on 40 comparisons, the second meta-analysis showed that enhanced simulations are more effective than non-enhanced ones, with the largest effects for the enhancements of scaffolding and adding special representations.

McGaghie, W. C., Issenberg, S. B., Petrusa, E. R., & Scalese, R. J. (2010). A critical review of simulation-based medical education research: 2003–2009. *Medical Education, 44,* 50–63.

This literature review addresses the research on simulation-based medical education (SBME) up until 2009. Its authors identify and reflect upon the features and best practices of SBME that medical educators should take into account for optimizing learning through the use of simulations. Each of these features is discussed according to the empirical evidence available in the domain of SBME. The authors conclude that research and development on SBME has improved substantially and that SBME has a lot to offer for the training of future doctors.

Windschitl, M., Thompson, J., & Braaten, M. (2008). Beyond the scientific method: Model-based inquiry as a new paradigm of preference for school science investigations. *Science Education, 92,* 941–967.

The authors discuss findings from a series of five studies and offer an alternative vision for investigative science—model-based inquiry. The aim of model-based inquiry is to engage learners more deeply in the learning process. They focus on the learners' ideas and show that, by modeling, these ideas should be testable, revisable, explanatory, conjectural, and generative. These ideas are also applicable when using simulations that are controllable by the students and where the learning process could be supported by the teacher or specifically designed worksheets.

Zacharia, Z. C. (2015). Examining whether touch sensory feedback is necessary for science learning through experimentation: A literature review of two different lines of research across K–16. *Educational Research Review, 16,* 116–137.

This review aimed at examining whether touch sensory feedback offered through physical or virtual (with haptic feedback) manipulatives affects students' science learning differently, compared to virtual manipulatives that do not offer touch sensory input (no use of haptic devices). The evidence came from two different lines of research; namely, from studies focusing on comparing physical and virtual manipulatives (with no haptic feedback), and from studies comparing virtual manipulatives that do or do not offer touch sensory feedback through haptic devices. The meta-analysis of these studies revealed that touch sensory feedback is not always a prerequisite for learning science through experimentation.

NAPLeS References

Jacobson, M., *15 minutes on complex systems* [Video file]. In *NAPLeS video series.* Retrieved October 19, 2017, from http://isls-naples.psy.lmu.de/video-resources/guided-tour/15-minutes-jacobson/index.html

Linn, M., *15 minutes on inquiry learning* [Video file]. In *NAPLeS video series.* Retrieved October 19, 2017, from http://isls-naples.psy.lmu.de/video-resources/guided-tour/15-minutes-linn/index.html

References

Achuthan, K., Sreelatha, S. K., Surendran, S. K., Diwakar, S., Nedungadi, P., Humphreys, S., et al. (2011, October–November). *The VALUE @ Amrita virtual labs project: Using web technology to provide virtual laboratory access to students.* Paper presented at the 2011 IEEE Global Humanitarian Technology Conference (GHTC).

Ainsworth, S. (2006). Deft: A conceptual framework for considering learning with multiple representations. *Learning and Instruction, 16,* 183–198.

Ainsworth, S. (2018). Multiple representations and multimedia learning. In F. Fischer, C. E. Hmelo-Silver, S. R. Goldman, & P. Reimann (Eds.), *International handbook of the learning sciences* (pp. 96–105). New York: Routledge.

Alfieri, L., Brooks, P. J., Aldrich, N. J., & Tenenbaum, H. R. (2011). Does discovery-based instruction enhance learning? *Journal of Educational Psychology, 103,* 1–18.

Anderson, J. L., & Barnett, M. (2013). Learning physics with digital game simulations in middle school science. *Journal of Science Education and Technology, 22*, 914–926.

Arena, D. A., & Schwartz, D. L. (2014). Experience and explanation: Using videogames to prepare students for formal instruction in statistics. *Journal of Science Education and Technology, 23*, 538–548.

Barzilai, S., & Blau, I. (2014). Scaffolding game-based learning: Impact on learning achievements, perceived learning, and game experiences. *Computers & Education, 70*, 65–79.

Berthold, K., Eysink, T. H. S., & Renkl, A. (2009). Assisting self-explanation prompts are more effective than open prompts when learning with multiple representations. *Instructional Science, 37*, 345–363.

Bivall, P., Ainsworth, S., & Tibell, L. A. E. (2011). Do haptic representations help complex molecular learning? *Science Education, 95*, 700–719.

Bollen, L., & van Joolingen, W. R. (2013). SimSketch: Multiagent simulations based on learner-created sketches for early science education. *IEEE Transactions on Learning Technologies, 6*, 208–216.

Bonde, M. T., Makransky, G., Wandall, J., Larsen, M. V., Morsing, M., Jarmer, H., & Sommer, M. O. A. (2014). Improving biotech education through gamified laboratory simulations. *Nature Biotechnology, 32*, 694–697.

Brant, G., Hooper, E., & Sugrue, B. (1991). Which comes first, the simulation or the lecture? *Journal of Educational Computing Research, 7*, 469–481.

Chi, M. T. H., Glaser, R., & Farr, M. (Eds.). (1988). *The nature of expertise*. Hillsdale, NJ: Lawrence Erlbaum.

Chinn, C. A., & Brewer, W. F. (1993). The role of anomalous data in knowledge acquisition: A theoretical framework and implications for science instruction. *Review of Educational Research, 63*, 1–51.

Clark, D. B., Tanner-Smith, E. E., & Killingsworth, S. S. (2016). Digital games, design, and learning: A systematic review and meta-analysis. *Review of Educational Research, 86*, 79–122.

d'Angelo, C., Rutstein, D., Harris, C., Bernard, R., Borokhovski, E., & Haertel, G. (2014). *Simulations for STEM learning: Systematic review and meta-analysis*. Menlo Park, CA: SRI International.

de Jong, T. (2016). Instruction based on computer simulations and virtual labs. In R. E. Mayer & P. A. Alexander (Eds.), *Handbook of research on learning and instruction* (2nd ed., pp. 1123–1167). New York: Routledge.

de Jong, T., & Lazonder, A. W. (2014). The guided discovery principle in multimedia learning. In R. E. Mayer (Ed.), *The Cambridge handbook of multimedia learning* (2nd ed., pp. 371–390). Cambridge, UK: Cambridge University Press.

de Jong, T., Linn, M. C., & Zacharia, Z. C. (2013). Physical and virtual laboratories in science and engineering education. *Science, 340*, 305–308.

de Jong, T., Sotiriou, S., & Gillet, D. (2014). Innovations in STEM education: The Go-Lab federation of online labs. *Smart Learning Environments, 1*, 1–16.

de Jong, T., & van Joolingen, W. R. (2008). Model-facilitated learning. In J. M. Spector, M. D. Merill, J. v. Merriënboer, & M. P. Driscoll (Eds.), *Handbook of research on educational communication and technology* (3rd ed., pp. 457–468). New York: Lawrence Erlbaum.

Fields, D. A., & Kafai, Y. B. (2018). Games in the learning sciences: Reviewing evidence from playing and making games for learning. In F. Fischer, C. E. Hmelo-Silver, S. R. Goldman, & P. Reimann (Eds.), *International handbook of the learning sciences* (pp. 276–284). New York: Routledge.

Fincher, R. M. E., & Lewis, L. A. (2002). Simulations used to teach clinical skills. In G. R. Norman, C. P. M. van der Vleuten, & D. J. Newble (Eds.), *International handbook of research in medical education* (pp. 499–535). Dordrecht, Netherlands: Kluwer Academic.

Freeman, S., Eddy, S. L., McDonough, M., Smith, M. K., Okoroafor, N., Jordt, H., & Wenderoth, M. P. (2014). Active learning increases student performance in science, engineering, and mathematics. *Proceedings of the National Academy of Sciences, 111*, 8410–8415.

Furberg, A., Kluge, A., & Ludvigsen, S. (2013). Student sensemaking with science diagrams in a computer-based setting. *International Journal of Computer-Supported Collaborative Learning, 8*, 41–64.

Gobert, J. D., Baker, R. S., & Wixon, M. B. (2015). Operationalizing and detecting disengagement within online science microworlds. *Educational Psychologist, 50*, 43–57.

Gobert, J. D., Sao Pedro, M. A., Raziuddin, J., & Baker, R. S. (2013). From log files to assessment metrics: Measuring students' science inquiry skills using educational data mining. *Journal of the Learning Sciences, 22*, 521–563.

Han, I., & Black, J. B. (2011). Incorporating haptic feedback in a simulation for learning physics. *Computers & Education, 57*, 2281–2290.

Hofstein, A., & Lunetta, V. N. (2004). The laboratory in science education: Foundations for the twenty-first century. *Science Education, 88*, 28–54.

Hulshof, C. D., Eysink, T. H. S., & de Jong, T. (2006). The ZAP project: Designing interactive computer tools for learning psychology. *Innovations in Education & Teaching International, 43*, 337–351.

Ibáñez, M. B., Di Serio, Á., Villarán, D., & Delgado Kloos, C. (2014). Experimenting with electromagnetism using augmented reality: Impact on flow student experience and educational effectiveness. *Computers & Education, 71*, 1–13.

Johnson, C. I., & Mayer, R. E. (2010). Applying the self-explanation principle to multimedia learning in a computer-based game-like environment. *Computers in Human Behavior, 26*, 1246–1252.

Kafai, Y. B., & Resnick, M. (Eds.). (1996). *Constructionism in practice: Designing, thinking, and learning in a digital world*. Mahwah, NJ: Lawrence Erlbaum.

Käser, T., Hallinen, N. R., & Schwartz, D. L. (2017). *Modeling exploration strategies to predict student performance within a learning environment and beyond*. Paper presented at the Proceedings of the Seventh International Learning Analytics & Knowledge Conference.

Lazonder, A. W., Hagemans, M. G., & de Jong, T. (2010). Offering and discovering domain information in simulation-based inquiry learning. *Learning and Instruction, 20*, 511–520.

Lazonder, A. W., & Harmsen, R. (2016). Meta-analysis of inquiry-based learning: Effects of guidance. *Review of Educational Research, 86*, 681–718.

Lazonder, A. W., & Kamp, E. (2012). Bit by bit or all at once? Splitting up the inquiry task to promote children's scientific reasoning. *Learning and Instruction, 22*, 458–464.

Leemkuil, H., & de Jong, T. (2011). Instructional support in games. In S. Tobias & D. Fletcher (Eds.), *Can computer games be used for instruction?* (pp. 353–369). Charlotte, NC: Information Age Publishers.

Leemkuil, H., & de Jong, T. (2012). Adaptive advice in learning with a computer-based strategy game. *Academy of Management Learning & Education, 11*, 653–665.

Limón, M. (2001). On the cognitive conflict as an instructional strategy for conceptual change: A critical appraisal. *Learning and Instruction, 11*, 357–380.

Lindgren, R., Tscholl, M., Wang, S., & Johnson, E. (2016). Enhancing learning and engagement through embodied interaction within a mixed reality simulation. *Computers & Education, 95*, 174–187.

Linn, M. C., McElhaney, K. W., Gerard, L., & Matuk, C. (2018). Inquiry learning and opportunities for technology. In F. Fischer, C. E. Hmelo-Silver, S. R. Goldman, & P. Reimann (Eds.), *International handbook of the learning sciences* (pp. 221–233). New York: Routledge.

Louca, L. T., & Zacharia, Z. C. (2015). Examining learning through modeling in K–6 science education. *Journal of Science Education and Technology, 24*, 192–215.

Mayer, R. E. (2009). *Multimedia learning* (2nd ed.). New York: Cambridge University Press.

McGaghie, W. C., Issenberg, S. B., Petrusa, E. R., & Scalese, R. J. (2010). A critical review of simulation-based medical education research: 2003–2009. *Medical Education, 44*, 50–63.

Miller, C. S., Lehman, J. F., & Koedinger, K. R. (1999). Goals and learning in microworlds. *Cognitive Science, 23*, 305–336.

Mulder, Y. G., Bollen, L., de Jong, T., & Lazonder, A. W. (2016). Scaffolding learning by modelling: The effects of partially worked-out models. *Journal of Research in Science Teaching, 53*, 502–523.

Mulder, Y. G., Lazonder, A. W., & de Jong, T. (2011). Comparing two types of model progression in an inquiry learning environment with modelling facilities. *Learning and Instruction, 21*, 614–624.

National Research Council. (2012). *A framework for K–12 science education: Practices, crosscutting concepts, and core ideas*. Washington, DC: National Academies Press.

Papaevripidou, M., & Zacharia, Z. C. (2015). Examining how students' knowledge of the subject domain affects their process of modeling in a computer programming environment. *Journal of Computers in Education, 2*, 251–282.

Pedaste, M., Mäeots, M., Siiman, L. A., de Jong, T., van Riesen, S. A. N., Kamp, E. T., et al. (2015). Phases of inquiry-based learning: Definitions and inquiry cycle. *Educational Research Review, 14*, 47–61.

Puntambekar, S. (2018). Design-based research (DBR). In F. Fischer, C. E. Hmelo-Silver, S. R. Goldman, & P. Reimann (Eds.), *International handbook of the learning sciences* (pp. 383–392). New York: Routledge.

Rönnebeck, S., Bernholt, S., & Ropohl, M. (2016). Searching for a common ground—A literature review of empirical research on scientific inquiry activities. *Studies in Science Education*, 1–37. doi:10.1080/03057267.2016.1206351

Rosé, C. P. (2018). Learning analytics in the Learning Sciences. In F. Fischer, C. E. Hmelo-Silver, S. R. Goldman, & P. Reimann (Eds.), *International handbook of the learning sciences* (pp. 511–519). New York: Routledge.

Ryan, R. M., & Deci, E. L. (2000). Self-determination theory and the facilitation of intrinsic motivation, social development, and well-being. *American Psychologist, 55*, 68–78.

Ryoo, K., & Linn, M. C. (2016). Designing automated guidance for concept diagrams in inquiry instruction. *Journal of Research in Science Teaching, 53*, 1003–1035.

Slotta, J. D., & Linn, M. C. (2009). *WISE science: Web-based inquiry in the classroom*. New York: Teachers College Press.

Tabak, I., & Kyza., E. (2018). Research on scaffolding in the Learning Sciences: A methodological perspective. In F. Fischer, C. E. Hmelo-Silver, S. R. Goldman, & P. Reimann (Eds.), *International handbook of the learning sciences* (pp. 191–200). New York: Routledge.

ter Vrugte, J., de Jong, T., Wouters, P., Vandercruysse, S., Elen, J., & van Oostendorp, H. (2015). When a game supports prevocational math education but integrated reflection does not. *Journal of Computer Assisted Learning, 31*, 462–480.

ter Vrugte, J., de Jong, T., Wouters, P., Vandercruysse, S., Elen, J., & van Oostendorp, H. (2017). Computer game-based mathematics education: Embedded faded worked examples facilitate knowledge acquisition. *Learning and Instruction, 50*, 44–53.

Trout, L., Lee, C. L., Moog, R., & Rickey, D. (2008). Inquiry learning: What is it? How do you do it? In S. L. Bretz (Ed.), *Chemistry in the national science education standards: Models for meaningful learning in the high school chemistry classroom* (2nd ed., pp. 29–43): Washington, DC: American Chemical Society.

van der Meij, J., & de Jong, T. (2006). Supporting students' learning with multiple representations in a dynamic simulation-based learning environment. *Learning and Instruction, 16*, 199–212.

van der Spek, E. D., van Oostendorp, H., & Meyer, J. J. C. (2013). Introducing surprising events can stimulate deep learning in a serious game. *British Journal of Educational Technology, 44*, 156–169.

Waldrop, M. M. (2013). The virtual lab. *Nature, 499*, 268–270.

Wang, F., & Hannafin, M. J. (2005). Design-based research and technology-enhanced learning environments. *Educational Technology Research and Development, 53*, 5–23.

Wecker, C., Rachel, A., Heran-Dörr, E., Waltner, C., Wiesner, H., & Fischer, F. (2013). Presenting theoretical ideas prior to inquiry activities fosters theory-level knowledge. *Journal of Research in Science Teaching, 50*, 1180–1206.

Wetzel, J., VanLehn, K., Butler, D., Chaudhari, P., Desai, A., Feng, J., et al. (2016). The design and development of the dragoon intelligent tutoring system for model construction: Lessons learned. *Interactive Learning Environments*, 1–21. doi: 10.1080/10494820.2015.1131167

White, B. Y., & Frederiksen, J. R. (1990). Causal model progressions as a foundation for intelligent learning environments. *Artificial Intelligence, 42*, 99–157.

Wieman, C. E., Adams, W. K., & Perkins, K. K. (2008). PhET: Simulations that enhance learning. *Science, 322*, 682–683.

Windschitl, M., Thompson, J., & Braaten, M. (2008). Beyond the scientific method: Model-based inquiry as a new paradigm of preference for school science investigations. *Science Education, 92*, 941–967.

Yaron, D., Karabinos, M., Lange, D., Greeno, J. G., & Leinhardt, G. (2010). The chemcollective—Virtual labs for introductory chemistry courses. *Science, 328*, 584–585.

Zacharia, Z. C. (2015). Examining whether touch sensory feedback is necessary for science learning through experimentation: A literature review of two different lines of research across K–16. *Educational Research Review, 16*, 116–137.

Zacharia, Z. C., Manoli, C., Xenofontos, N., de Jong, T., Pedaste, M., van Riesen, S. A. N., et al. (2015). Identifying potential types of guidance for supporting student inquiry in using virtual and remote labs: A literature review. *Educational Technology Research & Development, 63*, 257–302.

Zimmerman, B. J. (1990). Self-regulated learning and academic achievement: An overview. *Educational Psychologist, 25*, 3–17.

26
Supporting Teacher Learning Through Design, Technology, and Open Educational Resources

Mimi Recker and Tamara Sumner

Introduction

Several strands of Learning Sciences research have conceptualized teaching as a design activity and examined resulting instructional implications. Participating in design processes can help teachers learn new content and skills (Davis & Krajcik, 2005), while also serving to support and sustain curricular innovations (Fishman & Krajcik, 2003). In a seminal review, Remillard (2005) argues that curriculum materials can be viewed as models that teachers engage with in order to design activities for their students. Studies show that teachers who engage closely with curriculum and other high-quality materials to design instructional activities can significantly enhance their students' learning (NRC, 2007; Penuel, Gallagher, & Moorthy, 2011).

Engaging teachers in design has been explored from several perspectives within the Learning Sciences (see Kali, McKenney, & Sagy, 2015). For example, one strand has examined the kinds of knowledge teachers need to engage in design, while another investigated different trajectories for engaging teachers in design (for example, collaborative design). A third perspective studies technological supports for design processes (Davis & Krajcik, 2005; Kali, Kenney, & Sagy, 2015; Matuk, Linn, & Eylon, 2015; Voogt, Almekindes, van den Akker, & Moonen, 2005).

New technological arrangements and their affordances to support or enhance learning have been broadly termed *cyberlearning*. Here, the role of open educational resources (OERs) have been specifically called out as an important component (Borgman et al., 2008). OERs are teaching and learning resources that reside in the public domain or have been released under licensing schemes that allow their free use or customization by others (Atkins, Brown, & Hammond, 2007; Borgman et al., 2008). They encompass multiple media and resource types, such as animations, videos, scientific data, maps, images, games, simulations, and complete textbooks. OERs can be created by scientific institutions such as NASA, by publishing companies, by university faculty, by K–12 teachers, or by learners of all ages. OERs can be found on the world wide web and in dedicated repositories such as the National Science Digital Library, OERs Commons, or YouTube's education channel, which contains tens of thousands of educational resources.

In this chapter, we focus on the role technology can play in supporting teachers as they engage in design processes using OERs. As teachers work with OERs, teachers engage in a rich set of instructional and design practices: they adapt and combine resources to create new learning materials, share their labor with others by posting new or remixed resources, and contribute metadata about OERs such as ratings, tags, comments, and reviews. Teachers' prior experiences, their perceptions of student needs and assessments, and their own design capacity all influence how these instructional materials and teaching processes are implemented, refined, customized, and improved.

Developing and studying new technological supports for designing with OERs is increasingly critical. OERs have become a global phenomenon that is reshaping traditional relationships between teachers, instructional materials, and curriculum, and even disrupting traditional teaching arrangements (Fishman & Dede, 2016). At the same time, the growing availability of digital material online has raised new questions about how to frame teaching as a design activity, as teachers increasingly turn to the web and specialized repositories of open educational resources.

The purpose of this chapter is to examine two different types of software tools supporting teachers to design and customize instructional activities and curriculum with OERs. We first describe cyberlearning conceptual frameworks informing the *teaching as design* research agenda. Then, two technology models are presented that support these processes and we describe their implementation in two software tools, the Instructional Architect and the Curriculum Customization Service. The chapter summarizes findings from studies of teacher use of these two tools, and concludes by discussing implications of this perspective for research and practice.

Conceptual Framing

The technology models drew on elements of three approaches for framing teaching as design, each of which is briefly discussed in this section.

Pedagogical Design Capacity

As noted above, Learning Science research has argued that the way teachers use curricula is naturally a process of design. For example, a teacher might add or subtract planned classroom activities based on formative assessment results, personalize instruction for students who performed well on an assessment, or modify text-heavy instruction for English language learners.

However, teachers vary considerably in their ability to identify, sequence, and make principled customizations of curriculum materials in order to design instructional activities for their students, a skill called *pedagogical design capacity* (Brown, 2009). According to this perspective, as teachers engage with instructional materials, their pedagogical design capacity, prior experience, and knowledge all interact with curricular affordances to influence resulting customizations and classroom enactments. In characterizing these customization practices, Brown (2009) found that teachers at different times might "offload" the design onto the curriculum itself (thus offering few customizations), "adapt" it with small changes, or completely "improvise" by relying on their own design and minimally on the provided curriculum.

In addition, research has shown that many teachers are unsupported or unprepared to make productive customizations. As a result, they may unintentionally undermine the intent of curricular innovations (Remillard, 2005). For example, a teacher might choose to demonstrate a science experiment instead of supporting students in conducting the investigation (Fogleman, McNeil, & Krajcik, 2011). However, professional development focused on preparing teachers to make principled customizations to materials—that is, ones that preserve coherence and align to underlying rationales for the organization of materials—can serve to enhance their pedagogical design capacity (Penuel et al., 2011).

Peer Production

Peer production is a set of practices that rely on self-organizing groups of individuals: distributed groups of people connected by networked technologies work together to accomplish tasks in ways

Supporting Teacher Learning

that can be more effective, and more efficient than working alone (Benkler, 2006). Peer production has been productively applied in several domains, for example encyclopedia editing (Wikipedia) and collaborative software design (Linux).

When applied to education, peer production can support teachers in small, iterative cycles of design and sharing of instructional artifacts, which then has the potential to support more incremental and scalable improvement in instructional quality. The rapidly shifting technological landscape, coupled with the increasing availability of OERs, is making possible new trajectories for engaging teachers in design. For instance, access to OERs repositories, coupled with new tools, enables teachers to create, iteratively adapt, and share instructional activities using OERs. These activities, when supported and amplified by networked technologies, enable distributed groups of teachers to build on their peers' work to best serve the needs of their students (Porcello & Hsi, 2013). When combined with a large number of teachers over a period of time, the aggregated impacts of many small changes can support cycles of iterative improvement in which the enhanced quality of teacher designs can lead to improved instructional quality in a mutually reinforcing way (Morris & Hiebert, 2011).

Use Diffusion

While peer production offers a compelling 21st-century view of teaching using OERs, much existing research argues that novel practices and tools in many fields are not uniformly and seamlessly adopted by intended users, including teachers. *Use diffusion* models such adoption behavior in terms of two dimensions, frequency of use and variety of feature use, thus leading to different typologies of usage patterns (Shih & Venkatesh, 2004). For example, one user may be a frequent user of just a few functions (high usage; low variety), while another may use the full range of functions frequently (high usage; high variety). Other typologies are low usage; high variety, and low usage; low variety. Different usage typologies can be revealed by mining the digital traces left by users interacting with online systems. The use diffusion framework may be useful for capturing systematic pattern variation in practices as teachers choose to adopt, design with, and customize OERs.

Two Models and Software Implementations

Elements of these three frameworks have been informative in developing models for providing technological support for teaching as design. In particular, this section describes two models for helping educators use OERs to design and customize teaching. As described below, both models aim to help teachers engage in productive design by enhancing their pedagogical design capacity by leveraging peer production processes. The two models have been instantiated in two software tools – the Instructional Architect (IA) (Recker et al., 2007) and the Curriculum Customization Service (CCS) (Sumner & CCS Team, 2010).

These tools have been deployed for a number of years in different educational contexts, and their use has been studied with a range of teachers. Research examining tool use employed a range of methods, including established approaches such as quasi-experimental and survey designs, as well as qualitative methods such as case studies, interviews, and field observations (Maull et al., 2011; Recker, Yuan, & Ye, 2014; Sumner et al., 2010; Ye, Recker, Walker, Leary, & Yuan, 2015).

In addition, like most online systems, the IA and CSS automatically record all user online interactions. The increasing availability of these datasets, coupled with increased computing power as well as emerging "big data" techniques, offer unparalleled opportunities for research on understanding teaching and learning in online environments (Baker & Siemens, 2014). This has led to a new field of educational research known as educational data mining (EDM). Results presented below apply EDM techniques to examine teacher design patterns through several lenses, including use diffusion.

Case 1: The Instructional Architect

The Instructional Architect (IA.usu.edu) is a free, web-based authoring tool that enables teachers to find and use OERs to create web-based learning activities for their students (Recker et al., 2007). To use the IA, teachers must first create a free account, which enables them to use the IA in several ways. The "My Resources" area of the IA allows teachers to directly search for and save OERs from linked educational digital libraries and the wider web and add them to their list of saved resources.

In the "My Projects" area, teachers can create webpages (called IA projects, themselves a kind of OER) in which they link to selected OERs and provide accompanying instructional text. Finally, teachers can choose to "publish" (or share) their resulting IA projects with their students, or publicly with the wider web. IA users can view public IA projects and make a copy of the ones they like. This copy is then added to that user's personal collection of saved resources for further editing and reuse.

A key use case for the IA is as follows. First, a teacher uses a wizard-like authoring tool to create an IA project using OERs that she has found by searching OERs collections or the IA repository of public IA projects, or by browsing the internet. After teachers have created their IA projects, they can then choose to share these with their students or with the wider public via the IA repository. A teacher can also view IA projects contributed by other teachers to the IA repository. If a teacher especially likes an IA project, she can decide to copy it to her personal collection for further editing, adaptation, and reuse. In this way, teachers can design learning activities around OERs, featuring the discovered OERs as a primary instructional resource or adapting it for a planned instructional activity.

IA usage. Since 2005, the IA has had over 8,200 registered users, who have gathered approximately 78,000 OERs and created over 18,000 IA projects. Since August 2006, public IA projects have been viewed over 5 million times. While this is small activity in terms of enterprise web-based systems, it is large enough to ask interesting educational questions about teacher use.

Study 1. This study examined the activities of 200 users of the IA who created their IA account during one calendar year (see Recker et al., 2014). Analyses of IA usage data (see Table 26.1) suggest that teachers made varied use of the IA. Large standard deviations relative to means suggest a skewed distribution where some teachers made heavy use of particular features, while others made much lighter use. Such usage skew is fairly typical in analyses of online system usage and is consonant with the use diffusion framework.

In addition, teachers, on average, chose to share almost two-thirds of their created IA projects, while a smaller proportion (15%) of their IA projects were copied from IA projects created by other IA users. Finally, on average, teachers preferred to view IA projects rather than to completely copy them. This suggests that teachers may have been browsing for ideas or finding only a smaller set of IA projects that completely met their needs. These findings thus reveal some evidence of peer production in that these teachers can be seen as both contributors to as well as consumers of OERs in the IA community, as they create, copy, and adapt IA projects.

Table 26.1 Descriptive Statistics of Users' (N=200) Activity and Their IA Project Features (Data Collected Over a One-Year Period)

	IA activity	Mean	Median	SD	Min	Max
Teacher activities (N=200)	# of logins	10.38	7	10.59	1	57
	# of OER used in all IA projects	16.82	10	24.02	0	217
	# of IA projects created	2.60	2	2.04	1	10
	# of public IA projects created	1.73	1	1.95	0	10
	# of IA projects copied from others	0.58	0	1.46	0	9
	# of IA projects viewed	12.98	7	17.44	0	134

Table 26.2 Total Activities of Active Users (N=547) During a Nine-Month Period

IA Activity	Total	Mean
# of logins	3,440	6.28
# of OER collected	6,509	11.89
# of IA projects viewed	4,827	8.82
# of IA projects copied (%)	422 (22%)	0.77
# of IA projects created	1,890	3.45
# of IA projects shared back to IA repository (%)	1,194 (63%)	2.18

Study 2. In contrast to a focus on new users, this second study examined the usage activities of 547 active users over a nine-month school year period (see Recker et al., 2014). As shown in Table 26.2, for these users the most frequent activity was collecting OERs and viewing public IA projects. Less common was completely copying an existing IA project, as was found in Study 1. In addition, these users created almost 2,000 IA projects. Of these, users chose to share two-thirds of these projects with the IA user community. These results suggest that teachers find OERs and IA projects useful. They again provide some evidence of peer production, as teachers are building upon other teachers' work in a decentralized way and perhaps also leveraging the pedagogical design capacity of others.

Moreover, in examining the content of IA projects created by these teachers, the majority of IA projects displayed low levels of pedagogical design capacity, or what might be termed an "offload." That is, these IA projects often contained links to OERs, with little evidence of added teacher design or adaptation. Far fewer IA projects showed evidence of "improvisation," in which the teacher engaged in substantial design around OERs. However, since only the content was examined, evidence as to how these IA projects were subsequently enacted in classrooms and other teaching contexts is missing. Since anyone with internet access is free to create an account and use the IA, teachers come from a wide variety of contexts. As a result, an IA project could be designed to serve a broad range of purposes: formal learning, informal learning, homework, homebound students, extra help, etc. Without knowing more about the intent of an IA project's author and the wider enactment context, it is hard to make strong inferences about a teacher's pedagogical design capacity.

These cautions on interpretation notwithstanding, the findings from the two studies suggest that the IA supports multiple ways for teachers to participate in a peer production community—as a contributor with perhaps higher levels of pedagogical design capacity, and as a consumer with perhaps lower levels of pedagogical design capacity.

Case 2: The Curriculum Customization Service

The Curriculum Customization Service (CCS) is a web-based application designed to support groups of teachers within partner school districts in customizing instructional materials, including OERs, in order to implement differentiated instruction for their students. In a typical CCS use case, a teacher can search for and select OERs from online repositories and from his school district curriculum materials, as well as from OERs contributed by other teachers who are using the CCS. In this way, the teacher can leverage the work of other teachers in his school district. A teacher can then align these with different learning goals to create customized lesson plans and activities for his students (Sumner et al., 2010). The teacher can then contribute the customization back to the CCS commons.

The CCS has been developed over several years using iterative and participatory design processes in which, over time, a number of features have been added to better support teachers' work. For example, the CCS has features that allow teachers to match OERs and school district curricula to learning goals, store a personalized set of preferred OERs for later use, and build customized sequences of instructional materials from stored OERs that they can access while teaching.

Social software features also enable teachers to assign and view star ratings and descriptive tags for OERs, share their newly created customizations with other teachers in their school district, see the number of people who have stored a particular OERs, and view an activity stream indicating usage of materials by other teachers. In this way, the CCS features help support a networked professional learning community within a particular school district. The individual teacher can benefit by using the CCS to leverage the work and pedagogical design capacity of other teachers in order to customize content for his students. The school district benefits because the CCS becomes a repository of shared customizations that are aligned to the district's curricular framework.

In contrast to the IA, the CCS is designed as a closed environment. Accounts are created for teachers within partner school districts. This enables teachers to access their school-district-specific learning materials (e.g., district curricula, learning goals) as well as relevant OERs. This also means that teacher customizations are shared only with teachers within the same school district. As a result, these customizations are more closely aligned to the school district's instructional context and thus are more likely to be relevant and useful to teachers.

Study 1. In this study, the CCS was deployed to all middle and high school earth science teachers (N=124) in a large urban district for an academic year. The purpose of this study was to examine how and why educators integrated OERs into their instructional practice, with an emphasis on the role the CCS played in the integration process.

Survey findings showed that the CCS helped teachers integrate OERs into their teaching practices with more confidence, frequency, and effectiveness than was the case prior to the introduction of the CCS. Teachers integrated OERs in order to improve student engagement, address misconceptions about key concepts, offer alternative representations of scientific concepts or phenomena, and differentiate instruction according to student differences such as reading ability and language proficiency. In addition, social network analysis showed that use of the CCS helped teachers become more aware of other teachers' practices, thereby supporting an online community of teachers (Maull, Salvidar, & Sumner, 2011).

Study 2. A second study involved 73 high school science teachers drawn from five school districts who used the CCS over the course of one year with over 2,000 students (Ye et al., 2015). The study examined how teachers with different skills and backgrounds chose to integrate the CCS into their teaching. The study also examined the impact of these integration strategies on their students' learning outcomes.

Data analysis occurred in two parts. The first part, the *impact* study, included all teachers, while the second part, the *active user* study, examined the usage patterns of the 43 teachers who used the CCS more actively over the course of the year.

First, teachers in the *impact* study reported significant increases in their awareness of other science teachers' practices and in their frequency of using OERs in their instruction. Thus, the CCS served its goal of helping increase awareness of others' teaching practices and increasing their use of OERs in teaching. However, findings showed no strong relationship between how teachers used the CCS and their students' learning outcomes.

In the *active user* study, teacher usage data were clustered based on use diffusion. In particular, a "variety of usage" metric was defined using CCS usage log data to produce a user typology to help us understand and classify users, the common tasks they performed, and the details of their online behaviors. This metric was used to partition usage patterns into four quadrants, labelled: (1) Feature Explorer (high usage; low variety), (2) Power User (high usage; high variety), (3) Specialist (low usage; high variety), and (4) Lukewarm (low usage; low variety). Figure 26.1 shows the resulting typology for teachers, based on their CCS usage patterns.

Supporting Teacher Learning

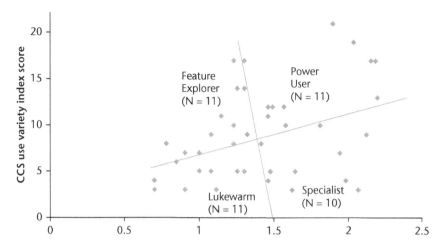

Figure 26.1 Four Typologies of Teacher Users (according to Ye et al., 2015)

Note: The X axis represents frequency of use and the Y axis represents variety of use of the different CCS features.

In contrast to the *impact* study, a teacher's usage pattern showed no relationship with changes in attitudes and teaching practices (as self-reported on the surveys). However, in terms of student learning outcomes, results showed that students of Feature Explorers and Power Users had the highest learning gains (Ye et al., 2015). The common trait between these two typologies is that these teachers showed the most variety in their use of the CCS and this usage positively influenced student learning outcomes.

Conclusion

This chapter argues that conceptualizing teaching as design is an increasingly important lens due to rapid increases in the availability of digital curriculum materials and OERs. It describes two software tools, the Instructional Architect (IA) and the Curriculum Customization Service (CCS), developed to help teachers integrate and customize OERs to design learning activities for their students. The processes behind teachers engaging with curriculum materials and designing activities for students have long been acknowledged as an important component of effective teaching (Remillard, 2005). The two tools described in this chapter, although occupying different parts of the design space, aim to support teachers through leveraging peer production processes around teachers' varying levels of pedagogical design capacity.

In particular, the IA supports teachers in finding relevant resources (including OERs from the wider web and existing IA projects) to use in their teaching. Many teachers also chose to contribute their created IA projects back to the community. This is done in a decentralized way, in that there is no curricular framework unifying teachers' IA projects. Drawing upon peer production processes, other IA users can then leverage from these contributions in opportunistic ways to find and customize "just-in-time" solutions to teaching problems. At the same time, an examination of IA projects showed low levels of pedagogical design capacity, as many IA projects could be characterized as "offloads."

The CCS, on the other hand, is designed to support and sustain a community of teachers within a school district. Through social software features, teachers are made aware of colleagues' contributions (in the form of contributed curricular customization and relevant OERs), including those from teachers with perhaps more pedagogical design capacity. In this way, district curriculum materials can be progressively customized, shared, and improved by the teacher community. These progressive refinements also provide a means for addressing persistent teaching problems that may exist within particular school communities.

Note that the core use case for both the IA and CCS was to support teachers during instructional planning and not necessarily during classroom enactment. However, evidence suggests that some teachers appropriated these tools in varied ways, including in direct support of classroom instruction.

The IA and CCS, as software tools, are of course not the only ways to create opportunities for teaching as design. Instead, we propose these tools as theoretical lenses for examining how teachers use cyberlearning technology and OER to leverage their pedagogical design capacity. This work also sheds light on examining teacher usage patterns and sets the stage for future work exploring how teachers appropriate different aspects of tools and how social software supports teacher learning from vast networks of online information resources and peers in online communities.

Acknowledgments

The author would like to thank our collaborators (Heather Leary, Victor Lee, Ye Lei, Keith Maull, and Andrew Walker), as well as study participants. This material is based upon work supported by the National Science Foundation under Grant Nos. 0554440, 1043660, 0840745, 0937630, and 1043858. Any opinions, findings, and conclusions or recommendations expressed in this material are those of the author(s) and do not necessarily reflect the views of the National Science Foundation.

Further Readings

Davis, E. A., Beyer, C., Forbes, C. T., & Stevens, S. (2011). Understanding pedagogical design capacity through teachers' narratives. *Teaching and Teacher Education, 27*(4), 797–810.
This article presents two case studies of how two elementary schools adapted curriculum materials and reflects on their respective pedagogical design capacity.

Fishman, B., & Dede, C. (2016). Teaching and technology: New tools for new times. In D. Gitomer & C. Bell (Eds.), *Handbook of research on teaching* (5th ed., pp 1269–1335). Washington, DC: AERA.
A comprehensive review of the use of educational technology for teaching and teachers.

Kali, Y., McKenney, S., & Sagy, O. (2015). Teachers as designers of technology enhanced learning. *Instructional Science, 43*(2), 173–179.
The introduction to a special issue of *Instructional Science* examining "teachers as designers of technology."

Porcello, D., & Hsi, S. (2013). Crowdsourcing and curating online education resources. *Science, 341*(6143), 240–241.
This article describes four key issues for supporting the effective development, access, and use of online education resources. These are: converging toward common ways to describe OER, or metadata; enabling both experts and community members to voice their input on what constitutes OER quality; building ways for the community users to be full participants; and developing systems that use common technical standards are thus are interoperable.

NAPLeS Resources

Recker, M. *Teacher learning and technology* [Webinar]. In *NAPLeS video series*. Retrieved October 19, 2017, from http://isls-naples.psy.lmu.de/intro/all-webinars/recker_all/index.html
Recker, M. *15 minutes about teacher learning and technology* [Video file]. In *NAPLeS video series*. Retrieved October 19, 2017, from http://isls-naples.psy.lmu.de/video-resources/guided-tour/15-minutes-recker/index.html
Recker, M. *Interview about teacher learning and technology* [Video file]. *In NAPLeS video series*. Retrieved October 19, 2017, from http://isls-naples.psy.lmu.de/video-resources/interviews-ls/recker/index.html

References

Atkins, D. E., Brown, J. S., & Hammond, A. L. (2007). A review of the open educational resources (OER) movement: Achievements, challenges, and new opportunities. Retrieved January 9, 2013, from www.hewlett.org/Programs/Education/OER/OpenContent/Hewlett+OER+Report.htm

Baker, R., & Siemens, G. (2014). Educational data mining and learning analytics. In R. K. Sawyer (Ed.), *Cambridge handbook of the learning sciences* (2nd ed., pp. 253–274). New York: Cambridge University Press.

Benkler, Y. (2006). *The wealth of networks: How social production transforms markets and freedom*. New Haven, CT: Yale University Press.

Borgman, C., Abelson, H., Dirks, L., Johnson, R., Koedinger, K., Linn, M., et al. (2008). *Fostering learning in the networked world: The cyberlearning opportunity and challenge, a 21st century agenda for the National Science Foundation* (Report of the NSF Task Force on Cyberlearning). Retrieved from www.nsf.gov/pubs/2008/nsf08204/nsf08204.pdf

Brown, M. W. (2009). Toward a theory of curriculum design and use: Understanding the teacher–tool relationship. In J. T. Remillard, B. A. Herbel-Eisenman, & G. M. Lloyd (Eds.), *Mathematics teachers at work: Connecting curriculum materials and classroom instruction* (pp. 17–36). New York: Routledge.

Davis, E. A., & Krajcik, J. S. (2005). Designing educative curriculum materials to promote teacher learning. *Educational Researcher, 34*(3), 3–14.

Fishman, B., & Dede, C. (2016). Teaching and technology: New tools for new times. In D. Gitomer & C. Bell (Eds.), *Handbook of research on teaching*, (5th ed., pp. 1269–1335). Washington, DC: AERA.

Fishman, B. J., & Krajcik, J. (2003). What does it mean to create sustainable science curriculum innovations? A commentary. *Science Education, 87*(4), 564–573.

Fogleman, J., McNeil, K., & Krajcik, J. (2011). Examining the effect of teachers' adaptations of a middle school science inquiry-oriented curriculum unit on student learning. *Journal of Research in Science Teaching, 48*(2), 149–169.

Kali, Y., McKenney, S., & Sagy, O. (2015). Teachers as designers of technology enhanced learning. *Instructional Science, 43*(2), 173–179.

Matuk, C. F., Linn, M. C., & Eylon, B. S. (2015). Technology to support teachers using evidence from student work to customize technology-enhanced inquiry units. *Instructional Science, 43*(2), 229–257.

Maull, K. E., Saldivar, M. G., & Sumner, T. (2011, June). Understanding digital library adoption: A use diffusion approach. In *Proceedings of the 11th annual international ACM/IEEE joint conference on digital libraries* (pp. 259–268). New York: ACM Press.

Morris, A. K., & Hiebert, J. (2011). Creating shared instructional products: An alternative approach to improving teaching. *Educational Researcher, 40*(1), 5–14.

National Research Council (NRC) (2007) *Taking science to school: Learning and teaching science in Grades K–8*. Washington, DC: National Academy Press. Retrieved from www.nap.edu/catalog.php?record_id=11625

Penuel, W. R., Gallagher, L. P., & Moorthy, S. (2011). Preparing teachers to design sequences of instruction in Earth science: A comparison of three professional development programs. *American Educational Research Journal, 48*(4), 996–1025.

Porcello, D., & Hsi, S. (2013). Crowdsourcing and curating online education resources. *Science, 341*(6143), 240–241.

Recker, M., Walker, A., Giersch, S., Mao, X., Palmer, B., Johnson, D., et al. (2007). A study of teachers' use of online learning resources to design classroom activities. *New Review of Multimedia and Hypermedia, 13*(2), 117–134.

Recker, M., Yuan, M., & Ye, L. (2014). CrowdTeaching: Supporting teachers as designers in collective intelligence communities. *The International Review in Open and Distance Learning, 15*(4).

Remillard, J. (2005). Examining key concepts in research on teachers' use of mathematics curricula. *Review of Educational Research, 75*(2), 211–246.

Shih, C., & Venkatesh, A. (2004). Beyond adoption: Development and application of use-diffusion model. *Journal of Marketing, 68*(1), 59–72.

Sumner, T., & CCS Team. (2010). Customizing science instruction with educational digital libraries. In *Proceedings of the Joint Conference on Digital Libraries* (pp.353–356). New York: ACM.

Voogt, J., Almekinders, M., van den Akker, J., & Moonen, B. (2005). A blended in-service arrangement for classroom technology integration: impacts on teachers and students. *Computers in Human Behavior, 21*(3), 523–539.

Ye, L., Recker, M., Walker, A., Leary, H., & Yuan, M. (2015). Expanding approaches for understanding impact: Integrating technology, curriculum, and open educational resources in science education. *Educational Technology Research and Development, 63*(3), 355–380.

27

Games in the Learning Sciences
Reviewing Evidence From Playing and Making Games for Learning

Deborah A. Fields and Yasmin B. Kafai

Introduction

In reviewing the research about play, games and learning, it is only in the past decade that the educational value of videogames and virtual worlds has received significant attention in the learning sciences. Gee's (2003) publication launched discussions about the educational value of videogames and coincided with the launch of the Serious Games movement, games that focus on more than entertainment. He argued that good videogames provided complex learning environments that foster collaboration, problem solving, and literacy. Since then researchers and educators have begun to examine and design educationally motivated games and to create environments (online, local, and connected) that support learning by *playing* games.

More recently, the perspective of *making* games for learning has become an equally prominent part of these discussions, partially fueled by the growth of the Maker and Coder Movements (see Halverson & Peppler, this volume) that promote the design of digital artifacts, on and off the screen for learning. Kafai (1995) provided an early example of how programming your own game could become a rich context for learning not just programming but also mathematics. Various efforts have been dedicated to the development of game making tools that help novice designers (Gee & Tran, 2015), and numerous research studies have documented successful learning while making games inside and outside of school. The research has grown enough to support recent syntheses that have reviewed the impact of making games for learning (e.g., Kafai & Burke, 2015).

Both approaches—*playing* or *making* games for learning—make use of principles of good game design outlined by Gee (2003), including helping players appreciate the design of the game, providing a level of challenge that is at the outer edge of players' competencies, and encouraging multiple legitimate routes to success within the game. Moreover, many of the principles elaborated by Gee were drawn from learning sciences research (or research that has strongly influenced the learning sciences, such as cognitive science and motivation research), and thus it should not be a surprise that many learning scientists have been among those researching and designing games for learning. In Gee's view, videogames would not survive if good learning was not built into the design of the games.

In this chapter we outline major approaches to playing and making games and take stock on where the field is regarding learning with games. In particular, we draw on several recent meta-analyses and meta-syntheses that have examined evidence of classroom, after-school, and online implementations. In the discussion, we address connections between playing and making games for learning, review connected gaming designs, outline new directions for online gaming, examine opportunities in the independent gaming movement, and consider opportunities for future research.

Learning by Playing Games

In their overview of videogames for learning Steinkuehler and Squire (2016) outline three primary reasons that educators support playing games for learning: for content, for "bait," and as assessment. The first, playing for content regards the idea of playing games in order to learn core content, such as playing the simulation game *Civilization* in order to learn about geographical underpinnings of history (see Squire, 2011). The second, playing as bait, uses games as a motivating factor for engaging in learning more generally. For instance, playing some popular games can motivate students to read game-related texts that may be far beyond their assessed reading ability (Steinkuehler, 2012). Third, playing as assessment, concerns the "data exhaust" produced from playing games—all the digital clicks, moves, and words that players type—that can be used to assess to what degree players engage in various activities that might be relevant to learning or that provide feedback game designers can use for improvements. This trifold explanation covers some of the reasons why educators turn to videogames to study and promote learning. Below we consider the results of research on videogames and learning and on the contexts that support good play.

Meta-Analyses of Playing Games for Learning

A report by the U.S. National Research Council (2011) that reviewed evidence of learning while playing science games provided a first official acknowledgment about videogames' growing importance in the educational sector. Several additional meta-analyses (e.g., Clark, Tanner-Smith, & Killingsworth, 2016; Young et al., 2012) have since reviewed the academic and motivational impacts of videogames. Overall these meta-analyses show positive effects on learning and motivational outcomes when participants play games rather than learning in a more traditional manner, though there are some nuances between studies. Each meta-analysis had a different focus, whether simulation games (Sitzmann, 2011; see also de Jong, Lazonder, Pedaste, & Zacharia, this volume), serious games played by adults (Wouters, van Nimwegen, von Oostendorp & van der Spek, 2013) or K–12 students (Young et al., 2012), or games played by K–16 students (Clark et al., 2016). At least four meta-analyses found significant improvements during game conditions (Clark et al., 2016; Sitzmann, 2011; Vogel et al., 2006; Wouters et al., 2013). The analysis by Young and colleagues (2012) saw general improvement in game-based learning situations in the areas of literacy and history but not in math or science. Their study, which included qualitative as well as quantitative analyses, provided more cautious recommendations about the contexts of game play, the importance of well-crafted narratives and backstories for games, and the use of games in the contexts of good teaching.

Clark and colleagues' (2016) meta-analysis focused on the broadest swath of games so long as they were played by elementary, secondary, or college students in educational contexts, and covered research published between 2000 and 2012. They also looked at the ways in which game design mechanics (e.g., visual effects, narrative, anthropomorphism) and the conditions of the studies (e.g., collaboration configurations, teacher presence) served as moderators of the game effects on learning outcomes, which they defined as cognitive (cognitive processes and strategies, knowledge, and creativity), intrapersonal (intellectual openness, work ethic, positive self-evaluation), or interpersonal (teamwork, collaboration, and leadership) learning. Overall, the findings revealed that students who played digital games performed better on learning assessments than students who did not play digital games.

Games that provided a wider range of player activities led to greater learning gains than games with limited actions for players. To illustrate, *Tetris* is a familiar game with one core action that lets players rotate and move geometric objects so that they fall in a close configuration. This is very different from a game like *SimCity* that allows players to manage buildings, streets, economies and taxes as a mayor of a city. Learning outcomes were stronger in the games with medium to large varieties of game actions. Looking at the conditions of game play, they found that single, noncompetitive games led to stronger gains than single competitive or collaborative team competition games. However, within the competitive games, students performed better when they were collaborative. In addition, multiple game sessions (playing more than one session) also resulted in positive gains for students compared to the alternative conditions. Unlike other meta-analyses (Sitzmann, 2011; Wouters et al., 2013), Clark and colleague's research identified that extra curriculum did not make a difference in the overall performance, but the presence of a teacher during game play or enhanced scaffolding designed into the game itself did improve results.

Overall, the outcomes of these meta-analyses show that playing games positively impact learning, and there are clear indicators from moderator analyses that the design of a game as well as contextual factors of game play have a significant impact on learning outcomes. However, understanding the situated nature of game play, which depends on the type of game as well as the context of game play, is not something that can be addressed by meta-analyses which depend on tests of individual achievement and surveys of psychological measures. Thus, while the meta-analyses discussed above generally demonstrate the proof-of-concept that games can support learning and may often be better than other, more traditional forms of learning (e.g., lecture, worksheets), it is time to move beyond the basic proof-of-concept of games and learning and investigate when, how, what types of, and under what conditions games can support deeper learning, and to understand how learning is taking place in social, distributed, and collective forms. Thankfully a substantial body of research has been emerging over the past decade that explores these aspects.

Broader Contexts of Game Play

As suggested by the recent meta-analyses of playing videogames for learning, the broader contexts of play are important for the quality of learning in games. In general, videogames are highly social, even when they are ostensibly single player. In their research studying children playing games at home, Stevens, Satwicz, and McCarthy (2008) identified the ways that children used others (i.e., friends, siblings, parents) as resources for their play, having friends study and problem solve gaming techniques, using a sibling to support the flow of play, and asking others for help. This social dimension of videogaming is extended when one considers networked games (playing with computers on a shared server or networked together), internet and gaming cafés (Lindtner & Dourish, 2011), massive online multi-player games, and multi-player games with built-in competition from friends.

Beyond the immediate situation of active play are the many extended communities of play that take place in affinity, or interest-driven, spaces both online and locally. In these affinity spaces, often sanctioned by game developers and companies, players share knowledge about the games they play in the forms of tips, walkthroughs, character builds (i.e., the best combinations of statistics for certain types of players), and problem solving. Consider *World of Warcraft (WoW)*, a massive online multi-player game that draws in millions of players to collaborate on quests in a magical fantasy world. In *WoW* players decide what kind of character to become, with each character having different abilities (i.e., healer, long-range fighters, short-range fighters). These abilities can also be customized each time a player levels up. For instance, one could decide to strengthen the character's agility instead of strength, defense, or magic attacks. The subject of figuring out how to best customize each type of character is a frequent topic of discussion on *WoW* forums, where Steinkuehler and Duncan (2008) found rich forms of scientific and mathematical thinking as players laid out their reasoning about which character builds were better. Affinity spaces like these are more than just spaces where tips, tricks or cheats are shared. They are knowledge spaces for sharing cultural understanding, building community, celebrating fandom, and reading and writing challenging texts (Steinkuehler, 2012; see also Slotta, Quintana, & Moher, this volume). Understanding and facilitating learning in video games must take account of these surrounding spaces bursting with active learning and collaboration.

Learning by Making Games

While most efforts have focused on having students play commercial or instructional games for learning, there might be equal benefit in having students themselves make games for learning (Kafai & Burke, 2015). In many commercial games players already find opportunities for customizing their online representations such as features of avatars, adding new content to gaming levels, or generating their own modifications. Extending game making to educational contexts can range from *modding* (or modifying) features (Gee & Tran, 2016) in game environments such as Gamestar Mechanic (Games, 2010) to programming all interactions and content in games using tools such as Logo, Scratch, or Alice. Over the last two decades, hundreds of studies have taken up these two instructional approaches to game making and examined student learning. Kafai and Burke (2015) recently completed a substantive review of the literature on learning by making games, and identified three main areas of benefits: learning about computing, other academic content, and problem solving.

Meta-Synthesis of Making Games for Learning

Although programming games is not the only way to make them, most of the efforts to support making games have been code-based, using some programming platform (whether Scratch, Alice, Agentsheets or another platform) to create them. Not surprisingly, 44% of the studies in the literature focused on whether students developed computational strategies or coding knowledge in the process of making games. For instance, one study conducted an analysis of hundreds of games made in Agentsheets (a platform used to make simulation games) by over 10,000 college and middle school students during an eight-week long course. The analysis revealed that both groups of students improved over time in their computational thinking patterns using more complex programming concepts and practices (Repenning et al., 2015).

Overall, a number of studies have demonstrated that children can learn challenging computational concepts by making games on various platforms. For instance, an examination of the 221 games created by 325 middle school students using Storytelling Alice in classes and after-school clubs (Denner, Werner, Campe, & Ortize 2014) revealed not only the use of simple programming constructs but also more complex constructs such as student-created abstractions, concurrent execution, and event handling—all indicative of higher-order thinking. Likewise, a review of 108 games created by 59 middle school girls with Stagecast Creator showed the use of key computational concepts such as loops, variables, and conditionals, though only with moderate usability and low levels of code organization and documentation (Denner, Werner, & Ortiz, 2012). The Globaloria online platform, through which thousands of students design videogames as part of curricular activities in their schools or clubs, demonstrated learning of key programming concepts using Flash (Reynolds & Caperton, 2011).

Aside from learning about computer science concepts and skills discussed in the above section, making games can also focus on learning academic content such as mathematical and writing skills, though this aspect has been less investigated. For instance, recent studies of making math games in Scratch confirmed that students activated their everyday mathematical experiences and understanding (Ke, 2014). Moving beyond traditional STEM content, game-making activities have also connected to literacy studies (Buckingham & Burn, 2007), the arts, and language arts (Robertson, 2012). In a comparative study of 18 fourth-grade classes, researchers found that 186 Canadian students involved in game making classes (as compared to 125 students in control classes reading and writing about the same content) demonstrated significantly better logical sentence construction skills in addition to better retaining content, comparing and contrasting information resources, and integrating of digital resources (Owston, Wideman, Ronda, & Brown, 2009). These examples illustrate that game making activities can focus not only on learning computer science content and skills but also promote other K-12 academic content.

Researchers have also studied whether making games can support learning problem solving or similar skills, and whether making games or playing games (where players do have to solve problems) provide a better scenario for this learning. In one of the few experimental studies that pitched playing versus making games, Vos, van der Meijden, and Denessen (2011) found that Dutch students who engaged in making a game versus a second group of students who just played a similar game, demonstrated significantly deeper engagement in their learning and strategy use which involved system analysis, decision making, and troubleshooting. Similarly, a study comparing two summer camp groups in Turkey indicated that the group involved in game making (versus playing) also produced measurable improvements in problem solving (Akcaoglu & Koehler, 2014). While 20 students (the experimental group) learned problem-solving skills through designing and testing their own videogame using Microsoft Kodu, 24 students (the control group) simply practiced their problem-solving skills by playing already-created games in Kodu. At the end of the intervention, the students who designed their own videogames significantly outperformed students on the validated assessment, the *Program for International Student Assessment* (OECD, 2003) in terms of questions related to the three problem types: system analysis and design, troubleshooting, and decision making.

Broader Contexts of Making Games for Learning

As with playing games, relatively little focus has been on the social situations that support learning by making games. An exception is research using pair programming activities to help support children's making of games. Originally implemented among college-level students as a learning technique, pair programming—sometimes referred to as "peer programing"—is rooted in the belief that learning is an inherently social activity. Working in pairs at a single computer, students code together, with one student taking the role of "driver" and generating the code, while the other student takes the role of "navigator," reviewing each line of code for accuracy. Denner and colleagues (2014) took the pair-programming premise and found that 126 middle school girls (ages 10–14) participating in a summer program were not only more successful in their capacity to program their own games in Flash but also significantly more able to articulate when they had found a problem and subsequently used their partner to help debug the issue. This in turn, made the girls more likely to persist in programming before asking the instructor for external help or even giving up altogether. These results were conceptually replicated two years later with middle school girls programming their own videogames using Storytelling Alice rather than Flash (Werner, Denner, Bliesner, & Rex, 2009). Beyond small group collaborations, sites like Scratch or GameStar Mechanic provide online, interest-driven communities where creators' game designs can be shared, critiqued, built on, and highlighted. Furthermore, organized initiatives like the annual National STEM Video Game Challenge in the United States can provide a public forum and competition through which students can share and celebrate their game designs.

Yet other social situations of supporting kids making games have not been well studied. Cases of successful teams such as the "Gray Bear" collective in the Scratch online community (Kafai & Burke, 2014) have illustrated that youth can self-organize collaborative game making activities, but these types of groups can be hard to replicate in school or after-school club settings. Few studies have made an effort to explicate and theorize about the social fabric that sustains game making learning environments. Fields, Vasudevan and Kafai (2015) considered the role of multiple nested forms of collaboration and audience in one elective class where students participated in an online Scratch design challenge. Not only were students in small groups to construct their entries, but they also worked at the classroom level to provide constructive criticism and motivation, and the online community level where they viewed other projects as well as providing and receiving constructive criticism there. This brings up the importance of creating in community, something taken for granted in learning-by-making settings. Grimes and Fields (2015) point out that, in online sites that support making, whether games or other DIY (do-it-yourself) media, relatively few support sharing and commenting on each other's work. Much more could be done to develop and research collaborations beyond pairs and small groups in the design of classroom and online environments.

New Directions in Designing and Researching Games for Learning

In this chapter we considered the overarching claims and evidence for learning in playing games, including the often-overlooked environments in which games are made. There now exists an extensive body of research and examples of implementation in the field of games and learning, far more than we can attend to in the length allowed in this chapter. Here, we focused on efforts, especially within the learning sciences, to understand and design for games and learning, though we encourage readers to consider research and writings from the humanities, games studies, literacies and other areas in their further reading. Next, we examine new directions for design and research.

If we examine potential overlaps between playing and making games for learning, we can see that the two pedagogical approaches indeed have much in common. First, both approaches provide direct and immediate feedback on whether ideas work or not. The computer plays a pivotal role in providing speedy responses to the design features that students implement in their code or in guiding game play and interactions. Second, both approaches involve players and designers in solving problems, whether it is in facing challenges or in generating design ideas. Third, both require time and practice in developing competencies as players or designers. Game makers spend extensive time in figuring out how they want to make things work and look on the screen, as do game players when games are customizable (e.g., the narrative trajectories in extensive role-playing games). Finally, both game playing and game making are social activities, disrupting the stereotype of game players and makers often portrayed in the media. Both often involve sharing and comparing work (whether creations or problem-solving approaches) and can involve extensive discussion forums and fandoms. All of these ideas are reflected in the principles that Gee (2003) laid out for good videogames, but apply beyond game play to game making as well. In the end, successful game playing and game making simply reflects good learning designs.

Ultimately learning environments that combine playing and making games under one umbrella should become the focus of new developments. Nowhere is the merger of instructional approaches more evident than in the success of *Minecraft*, a videogame where players craft materials and create buildings as part of relatively unstructured game play. On common servers, groups of players could play in the same virtual space, sharing their creations or even creating new game play scenarios for others. In other words, in *Minecraft*, playing is making and making is playing (Kafai & Burke, 2016). A large community of educators have begun to utilize *Minecraft* to let students create and play various scenarios designed by both teachers and/or students, integrating it into science, math, history and other courses as well as clubs after or out of school (e.g., Dikkers, 2015). *Minecraft* is one example of a "sandbox" game where players are given wide rein in how to develop their play, not being restricted to a single, linear path (i.e., like children playing in a sandbox). Educational researchers and designers have begun to develop sandbox games that integrate the worlds of game playing with making (Holbert & Wilensky, 2014). In these sandbox games, students are provided with microworlds (Papert, 1980) in which they have access as designers and players to experiment with ideas and phenomena such as a frictionless world that they may not easily encounter in their regular textbooks or classroom lessons.

We also see great potential in expanding game playing and making approaches into mobile and independent, or indie, games, two new genres that are gaining importance in the commercial sector. As smartphones have become accessible even with younger students, gaming has moved out of contained spaces (e.g., homes, classrooms, internet cafés) into more expanded spaces (e.g., neighborhoods, parks, streets, cities) through mobile devices. As learning sciences designers are attentive to learning contexts, it makes sense to utilize not only the ability to think within screen-based games but also beyond the screen as well. Augmented reality games are one area where some are exploring the opportunities to layer physically based experiences with screen-based representations and narratives that traverse both (see Yoon & Klopfer, 2015). More recently, the ability to develop such games has become more accessible with platforms such as ARIS (Holden et al., 2015), which allow teachers and students to develop their own augmented reality games and scavenger hunts.

The growing popularity of the indie gaming movement means that games created by smaller groups with limited funding can be well received and popular. Indie games, *Minecraft* is a prominent example, are made independently without major game developers backing them. Many indie (including educator-created) games have begun to explore new game genres in ways that could be relevant to relatively unexplored learning scenarios, for instance, making ethical decisions part of game play. That indie games are gaining wide popularity and recognition broadens the field regarding who can create an excellent videogame and provides models for legitimate lower tech games that enable creators and players to think more critically about the medium of the game. Good games do not need to look fancy or involve millions of dollars in development to make an impact.

Finally, while designing new games to play or tools for making games remain critical issues if we want these activities to be accessible to many learners and content areas, equally important is research into implementing these environments in classrooms and after-school contexts. Research has indicated the critical role that teachers play in augmenting the educational experience of students, in playing or making games for learning. Further research is needed on how teachers can work with different game genres, find ways to make them part of their classroom activities, use games for assessment purposes, and even think about adopting games to extend connected learning experiences beyond the school day.

In this chapter we reviewed the evidence on playing and making games for learning. We argued that both making and playing games embody many of the very same learning principles that Gee (2003) saw in well-designed videogames—namely, providing highly responsive contexts for complex problem solving that motivate learners' engagement with the game, content and other people. These connections between game playing and making provide new directions for design and research in the learning sciences. To this end, research on game playing and making should look at intersecting contexts of play and the role of designed technology in individual, collaborative, and collective learning. Methods may include digital data from a game design or online environment but should be brought into conversation with other methods that can illuminate context, interpersonal interactions, and human–game interaction. We are promoting a vision of educational gaming that considers both playing and making games, develops thoughtful social environments to support game-based interventions, and critically explores when and where games can promote deeper learning in and out of school, across disciplines and connected with multiple audiences. As our review of meta-analyses of games showed, games clearly have the potential to support learning; it is time to tap that potential and develop greater sophistication in the ways that we design, study, and implement games for learning.

Further Readings

Clark, D. B., Tanner-Smith, E., & Killingsworth, S. (2016). Digital games, design, and learning: A systematic review and meta-analysis. *Review of Educational Research, 86*(1), 79–122.
A recent meta-analysis of video games and learning covering the greatest breadth of implementations as well as some variables regarding game and intervention qualities.

Gee, E. R., & Tran, K. (2015). Video game making and modding. In B. Guzzetti & M. Lesley (Eds.), *Handbook of research on the societal impact of digital media* (pp. 238–267). Hershey, PA: IGI Global.
Provides an overview of different tools developed for game making and modding activities and reviews empirical evidence of how these tools have been used by students inside and outside of schools for learning technical and academic content.

Kafai, Y. B. (1995). *Minds in play: Computer game design as a context for children's learning*. Mahwah, NJ: Lawrence Erlbaum.

The first study in which students programmed games to teach younger students in their school about fractions. Includes case studies and a quasi-experimental analysis comparing learning of programming concepts and skills and fraction concepts in game design and traditional instruction scenarios.

Kafai, Y. B., & Burke, Q. (2015). Constructionist gaming: Understanding the benefits of making games for learning. *Educational Psychologist, 50*(4), 313–334.

Examined 55 research studies that engaged over 9,000 youth in making games for learning. Shows the landscape of what has been researched and what has been left out.

Steinkuehler, C., & Squire, K. (2016). Videogames and learning. In R. K. Sawyer (Ed.), *The Cambridge handbook of the learning sciences* (2nd ed., pp. 377–396). New York: Cambridge University Press.

This review covers the breadth of the games and learning movement up to present times as well as challenges facing the movement today.

NAPLeS Resources

Yoon, S., & Klopfer, E., *Augmented reality in the learning sciences* [Webinar]. In *NAPLeS video series*, Retrieved October 19, 2017, from http://isls-naples.psy.lmu.de/intro/all-webinars/yoon_klopfer_video/index.html

References

Akcaoglu, M., & Koehler, M. J. (2014). Cognitive outcomes from the game-design and learning (GDL) after-school program. *Computers & Education, 75*, 72–81.

Buckingham, D. & Burn, A. (2007). Game literacy in theory and practice. *Journal of Educational Multimedia and Hypermedia, 16*(3), 323–349.

Clark, D. B., Tanner-Smith, E., & Killingsworth, S. (2016). Digital games, design, and learning: A systematic review and meta-analysis. *Review of Educational Research, 86*(1), 79–122.

de Jong, T., Lazonder, A., Pedaste, M., & Zacharia, Z. (2018). Simulations, games, and modeling tools for learning. In F. Fischer, C. E. Hmelo-Silver, S. R. Goldman, & P. Reimann (Eds.), *International handbook of the learning sciences*. New York: Routledge.

Denner, J., Werner, L., Campe, S. & Ortiz, E. (2014). Pair programming: Under what conditions is it advantageous for middle school students? *Journal of Research on Technology in Education, 46*(3), 277–296.

Denner, J., Werner, L., & Ortiz, E. (2012). Computer games created by middle school girls: Can they be used to measure understanding of computer science concepts? *Computers & Education, 58*(1), 240–249.

Dikkers, S. (2015). *Teachercraft: How teachers learn to use Minecraft in their classrooms*. Pittsburgh, PA: ETC Press.

Fields, D. A., Vasudevan, V., Kafai, Y. B. (2015). The programmers' collective: Fostering participatory culture by making music videos in a high school Scratch coding workshop. *Interactive Learning Environments, 23*(5), 1–21.

Games, I. A. (2010). Gamestar Mechanic: Learning a designer mindset through communicational competence with the language of games. *Learning, Media and Technology, 35*(1), 31–52.

Gee, E. R., & Tran, K. (2015). Video Game Making and Modding. In B. Guzzetti and M. Lesley (Eds), *Handbook of research on the societal impact of digital media* (pp. 238–267). Hershey, PA: IGI Global.

Gee, J. P. (2003). *What videogames have to teach us about learning and literacy*. New York: Palgrave Macmillan.

Grimes. S. M & Fields, D. A. (2015). Children's media making, but not sharing: The potential and limitations of child-specific DIY media websites for a more inclusive media landscape. *Media International Australia, 154*, 112–122.

Halverson, E., & Peppler, K. (2018). The Maker Movement and learning. In F. Fischer, C. E. Hmelo-Silver, S. R. Goldman, & P. Reimann (Eds.), *International handbook of the learning sciences* (pp. 285–294). New York: Routledge.

Holbert, N. R., & Wilensky, U. (2014). Constructible authentic representations: Designing video games that enable players to utilize knowledge developed in-game to reason about science. *Technology, Knowledge and Learning, 19*, 53–79.

Holden, C., Dikkers, S., Martin, J., Litts, B., et al. (Eds.). (2015). *Mobile media learning: Innovation and inspiration*. Pittsburgh, PA: ETC Press.

Kafai, Y. B. (1995). *Minds in play: Computer game design as a context for children's learning*. Mahwah, NJ: Lawrence Erlbaum.

Kafai, Y. B., & Burke, Q. (2014). *Connected code: Why children need to learn programming*. Cambridge, MA: MIT Press.

Kafai, Y. B., & Burke, Q. (2015). Constructionist gaming: Understanding the benefits of making games for learning. *Educational Psychologist, 50*(4), 313–334.

Kafai, Y.B. & Burke, Q. (2016). *Connected gaming: What making video games can teach us about learning and literacy.* Cambridge, MA: MIT Press.

Ke, F. (2014). An implementation of design-based learning through creating educational computer games: A case study on mathematics learning during design and computing. *Computers and Education, 73*, 26–39.

Lindtner, S., & Dourish, P. (2011). The promise of play: A new approach to productive play. *Games and Culture, 6*(5), 453–478.

National Research Council (2011). *Learning Science through Simulations and Games.* Washington, DC: National Academies Press.

OECD (2003). *PISA 2003 Assessment framework: Mathematics, reading, science and problem solving.* Paris, France: OECD.

Owston, R., Wideman, H., Ronda, N. S., & Brown, C. (2009). Computer game development as a literacy activity. *Computers & Education, 53*, 977–989.

Papert, S. (1980). *Mindstorms: Children, computers, and powerful ideas.* New York: Basic Books.

Repenning, A., Webb, D. C., Koh, K. H., Nickerson, H., Miller, S. B., Brand, C., et al. (2015). Scalable game design: A strategy to bring systemic computer science education to schools through game design and simulation creation. *ACM Transactions on Computing Education (TOCE), 15*(2), 11.

Reynolds, R., & Caperton, I. H. (2011). Contrasts in student engagement, meaning-making, dislikes, and challenges in a discovery-based program of game design learning. *Educational Technology Research and Development, 59*(2), 267–289.

Robertson, J. (2012). Making games in the classroom: Benefits and gender concerns. *Computers and Education, 59*(2), 385–398.

Sitzmann, T. (2011). A meta-analytic examination of the instructional effectiveness of computer-based simulations games. *Personnel Psychology, 64*(2), 489–528.

Slotta, J. D., Quintana, R., & Moher, T. (2018). Collective inquiry in communities of learners. In F. Fischer, C. E. Hmelo-Silver, S. R. Goldman, & P. Reimann (Eds.), *International handbook of the learning sciences* (pp. 308–317). New York: Routledge.

Squire, K. (2011). *Video games and learning: Teaching and participatory culture in the digital age.* New York: Teachers College Press.

Steinkuehler, C. (2012). The mismeasure of boys: Reading and online videogames. In W. Kaminski & M. Lorber (Eds.), *Proceedings of Game-based Learning: Clash of Realities Conference* (pp. 33–50). Munich: Kopaed Publishers.

Steinkuehler, C., & Duncan, S. (2008). Scientific habits of mind in virtual worlds. *Journal of Science Education Technology, 17*(6), 530–543.

Steinkuehler, C., & Squire, K. (2016). Videogames and learning. In R. K. Sawyer (Ed.), *The Cambridge handbook of the learning sciences* (2nd ed., pp. 377–396). New York: Cambridge University Press.

Stevens, R., Satwicz, T., & McCarthy, L. (2008). In-game, in-room, in-world: Reconnecting video game play to the rest of kids' lives. In K. Salen (Ed.), *The ecology of games: Connecting youth, games, and learning* (pp. 41–66). Cambridge, MA: MIT Press.

Vogel, J. J., Vogel, D. S., Cannon-Bowers, J., Bowers, C. A., Muse, K., & Wright, M. (2013). Computer gaming and interactive simulations for learning: a meta-analysis. *Journal of Educational Computing Research, 34*(3), 229–243.

Vos, N., van der Meijden, H., & Denessen, E. (2011). Effects of constructing versus playing an educational game on student motivation and deep learning strategy use. *Computers & Education, 56*(1), 127–137.

Werner, L., Denner, J., Bliesner, M., & Rex, P. (2009, April). Can middle-schoolers use Storytelling Alice to make games? Results of a pilot study. In *Proceedings of the 4th International Conference on Foundations of Digital Games* (pp. 207–214). New York: ACM Press.

Wouters, P., van Nimwegen, C., von Oostendorp, H., & van der Spek, E. D. (2013). A meta-analysis of the cognitive and motivational effects of serious games. *Journal of Educational Psychology*, 1–17.

Yoon, S. & Klopfer, E. (2015). *Augmented reality in the learning sciences.* Retrieved June 26, 2017, from http://isls-naples.psy.lmu.de/intro/all-webinars/yoon_klopfer_video/index.html

Young, M. F., Slota, S., Cutter, A. R., Jalette, G., Mullin, G., Lai, B., et al. (2012). Our princess is another castle: A review of trends in serious gaming for education. *Review of Educational Research, 82*(1), 61–89.

28

The Maker Movement and Learning

Erica Halverson and Kylie Peppler

This chapter focuses on the range of teaching, learning, and design practices associated with the Maker Movement in education and how these practices are intimately connected to how the learning sciences conceptualizes, studies, and designs for learning and knowing. The Maker Movement is defined as "the growing number of people who are engaged in the creative production of artifacts in their daily lives and who find physical and digital forums to share the processes and products with others" through analog and digital practices such as woodworking, soldering, cooking, programming, painting, or crafting (Halverson & Sheridan, 2014, p. 496). As this broad definition suggests, there are many inroads into education research that are made possible through the study of the Maker Movement. Here, we suggest that research on the Maker Movement can contribute to three core topics in the learning sciences:

- The Maker Movement contributes to our *theories of how people learn* by merging a constructionist perspective (Martinez & Stager, 2013) with other big theoretical ideas in the learning sciences, including distributed cognition (Halverson, Lakind, & Willett, 2017), embodied cognition (Peppler & Gresalfi, 2014), new materialisms (Wohlwend, Peppler, & Keune, 2017), and the new literacies (Litts, 2015).
- Makerspaces provide opportunities for the *design of learning environments*, particularly to rethink the disconnect between learning in and out of schools (Peppler & Bender, 2013; Peppler, Halverson, & Kafai, 2016).
- Conceptualizing who counts as makers pushes us to think about issues of *equity and diversity*, focusing on deep connections between meaningful content and processes and cultural, historical, material, and social movements in education (Blikstein & Worsley, 2016; Buechley, 2016; Vossoughi, Hooper, & Escudé, 2016).

This chapter provides a roadmap through the emerging field of making and education while focusing on how the field is connected to the learning sciences. We begin with an overview of research on the Maker Movement and then turn to how people learn both in and through making. We conclude by offering reflections on how the three key issues we identify above are central to understanding making and learning specifically, and the learning sciences more broadly.

Erica Halverson and Kylie Peppler

An Introduction to Research on the Maker Movement

While people have been "making things" forever, research on the Maker Movement, FabLabs, and DIY (do-it-yourself) culture in education has exploded over the past decade. From an educational research perspective, the larger movement encompasses *learning activities* that engage people in the creative production of artifacts, *communities of practice* where making activities occur, and the *identities of participation* that people take on as they engage in making (Halverson & Sheridan, 2014). Across these three domains of research, the Maker Movement is characterized by a "do-it-yourself" ethos in a range of domains including textile crafts, electronics, advanced robotics, and traditional woodworking (Peppler & Bender, 2013). Typically, work in these domains includes some focus on the use of new production technologies such as 3-D printers, laser cutters, and microcomputers as well as the sharing of ideas via the internet (Peppler et al., 2016).

While some research on making and learning has identified the ways in which making results in improved outcomes in sanctioned schooling practices (e.g., Peppler & Glosson, 2013), others have identified Maker Movement-specific practices that are worthy of understanding, including: a focus on interest-driven learning (Peppler et al., 2016); tinkering as a valid form of knowing and doing (Wilkinson, Anzivino, & Petrich, 2016); and the importance of sharing and audience for learning (Sheridan et al., 2014). These practices are not often present in the more defined disciplinary practices of schooling.

A Brief History

Over the past decade, there has been an interest in studying creation, sharing, and learning with new technologies within what was initially referred to as online "Do-It-Yourself (DIY) communities" (Guzzetti, Elliot & Welsch, 2010; Kafai & Peppler, 2011) and which have later been reframed by the larger discourse around the Maker Movement as popularized through Make magazine and Maker Media. Both the DIY and the "Maker Movement" share the spirit of self-produced and original projects. What was once solely the domain of in-person "hobby clubs" is now accelerated by the rise of social media, where distributed communities can congregate and share projects and techniques on sites like makezine.com or instructables.com, where members have posted hundreds of thousands of videos on virtually any topic (Torrey, McDonald, Schilit, & Bly, 2007). In some cases, these communities follow the open-source movement and have developed networks around the use of a particular programming language, such as Processing (Reas, 2006; see also processing.org) or Scratch (Brennan, Resnick, & Monroy-Hernandez, 2010). In other cases, these communities have developed around the use and development of an open-source construction kit, Arduino, which hobbyists around the world use to design projects, such as their own laser printers. In the instance of the LilyPad Arduino kit (Buechley, Eisenberg, Catchen, & Crockett, 2008), textile productions now can include sensors and LED lights to be programmed for informative feedback and artistic purposes. While these communities have much of the flair of exclusive clubs found among earlier programmers, their growing presence also signals a larger trend.

Dougherty (2011) states that the "most important thing about DIY [or making] is that it portrays the idea that you can learn to do anything." Furthermore, the resurgence of interest in DIY forms of making stems from the range of what can be produced due to the rapid rise in the number and availability of new digital fabrication technologies, like 3-D printers, wearable computers, and other tools that merge digital and physical materials (Gershenfeld, 2005). FabLabs, for example, are now becoming widespread both in and out of schools, allowing youth and adults alike access to digital fabrication tools that at one time were only accessible to industries (Blikstein, Martinez, & Pang, 2016). The success of these new makerspaces and events like Maker Faires is largely based on exposure: The driving force of the culture is about being inspired when you see someone else making, compelling you to want to make it yourself. This emergent set of artistic and technological practices, compounded by the desire for those online to seek recognition for the work they do, is at the heart of interest-driven learning today, largely existing at the fringes of traditional education and schooling.

Research on Making

One way to explore the affordances of the Maker Movement in education is through the study of *making activities*, "designing, building, modifying, and/or repurposing material objects for playful or useful ends, oriented toward a product of some sort that can be used, interacted with, or demonstrated" (Martin, 2015, p. 31). This definition accomplishes several functions. As "a set of activities" it reminds us that making is something you *do*; multiple identities are possible for participation and multiple kinds of learning environments can support making activities. This also has implications for research; the study of activities in the learning sciences has a rich history that can be leveraged in our study of making. The creation of objects "for playful or useful ends" reminds us that making can be practical, whimsical, or both. As such, it can serve goals often associated with STEM fields—solving problems, innovating products; it can also serve its own goals—creating playful artifacts. Finally, the outcomes of making "can be used, interacted with, or demonstrated," further reminding us that making does not serve a singular purpose but must be judged based on the goal toward which the making activity is aimed.

Research on Makerspaces

Makerspaces present unique contexts for study that are different from formal classroom settings in several key ways. Making in a makerspace is guided by a different set of pedagogical practices than hands-on learning in a traditional classroom context; for example, cross-age learning is common (and often required!) and projects are not guided by predetermined learning outcomes. Current research on makerspaces sees these differences as endemic to the learning environments themselves. In one of the first empirical studies on makerspaces, Sheridan et al. (2014) defined makerspaces as "informal sites for creative production in art, science, and engineering where people of all ages blend digital and physical technologies to explore ideas, learn technical skills, and create new products" (p. 505). This definition suggests some reasons why makerspaces ought to be treated independently from making activities. Since schools are by definition formal, makerspaces as informal spaces struggle to integrate seamlessly into schooling structures (Lacy, 2016). Furthermore, schools resist multi-aged learning, age-segregating students by year and rarely acknowledging anyone as an expert in the classroom other than the adult in charge. Finally, it is important to note the multiple, simultaneous goals that makerspaces support; participants are likely exploring different goals while in the space together. A key feature of makerspaces is that not everyone learns the same thing at the same time—a big challenge for classrooms.

Makerspaces can be broken down into three primary categories: P–16 education, out-of-school spaces, and online spaces. As alluded to above, makerspaces in schools have been difficult to develop, implement, and study, with a few notable exceptions. While there are numerous "how to" trade books available for the development of makerspaces in schools (e.g., Blikstein, Martinez, & Pang, 2016; Martinez & Stager, 2013), there is little empirical research on how we can understand makerspaces as school-based learning environments. Lacy's (2016) dissertation chronicles the implementation of a FabLab at a Midwestern ex-urban high school, finding that the FabLab continues to reify divisions between those in the technical careers track who maintained their focus on the practical skills afforded by the school's shops and college-bound students who saw the FabLab as a natural extension of their AP STEM courses. By contrast Puckett, Gravel, and Vizner (2016) found that the addition of a makerspace to a large comprehensive high school meant that students experienced fewer status distinctions, fewer gendered practices, and came together from a range of academic tracks as opposed to traditional STEM and shop courses. In higher education, colleges and universities have also begun to develop makerspaces for their campuses for a variety of educational contexts, ranging from studying learning through making, to hands-on approaches to various disciplines, including chemistry, engineering, design, and biology (e.g., Fields & Lee, 2016).

Case studies of makerspaces in out-of-school learning environments include museums, libraries, FabLabs, and independent for profit and non-profit organizations. These case studies describe unique features of makerspaces as learning environments, such as side-by-side multidisciplinarity and a diverse set of learning arrangements that distinguish makerspaces from other informal learning environments and participatory cultures (Peppler et al., 2016; Sheridan et al., 2014). As mentioned earlier, online places for engaging in and talking about making are also a robust part of the Maker Movement. Case studies of online makerspaces describe how internet technologies extend face-to-face communities of practice as well as create and support new communities of makers (Peppler et al., 2016).

Research on Makers

Many scholars argue that the Maker Movement, like other areas of the learning sciences, has neglected issues of culture and history (Ames & Rosner, 2014; Blikstein & Worsley, 2016) and devalued the contributions of women of all ages and communities of color (Buechley, 2016; Kafai, Fields, & Searle, 2014; Vossoughi et al., 2016). As a branded arm of the Maker Movement, the MAKE organization overwhelmingly puts White men and their sons on the covers of their magazine (Buechley, 2016) and 89% of the magazine's authors self-identify as male (Brahms & Crowley, 2016). Blikstein and Worsley (2016) argue that the "hacker culture" roots of the Maker Movement "only works for a small elite group of high end students" (p. 66) and prevents equitable access to the kind of deep learning that making can engender. As a result, a third line of research has emerged that explores "makers" as identities of participation.

Research on makers as identities of participation explicitly attends to who gets to be a maker, how we construct the maker identity, and what we can do to both broaden participation and to expand what counts as a maker identity. Efforts to focus on gender equity and making have resulted in design-based research projects that help girls develop maker identities through their interests in fashion and sewing (Erete, Pinkard, Martin, & Sandherr, 2015; Kafai et al., 2014) and through projects that explicitly connect them to their local community (Holbert, 2016). Vossoughi et al. (2016) seek to make a range of maker identities possible by embracing the historical and cultural roots of making that include longstanding making practices such as weaving and the teaching of aesthetic and pragmatic skills across generations within a community.

Other research on makers as identities of participation tries to answer the question, "What makes a maker?" The Agency by Design project describes maker identity as a set of dispositions that learners develop including the discovery of personal passions, the capacity to develop those passions, and the confidence and resourcefulness that results from learning with and from others (Ryan, Clapp, Ross, & Tishman, 2016). Rusk (2016) finds that makers display self-determination as a core characteristic of their identity. Sheridan and Konopasky (2016) provide linguistic evidence of young people's development of resourcefulness as an identity outcome of extended participation in makerspaces.

Learning in and Through Making

In order to understand how the learning sciences contributes to the emerging field of making, we ask: "How do people learn to make?" This question seeks to identify theories of learning that inform how we document, study, and represent process. Here, learning is a dynamic, ongoing process rather than a fixed entity to be evaluated. The work resonates with how the learning sciences constructs knowing, learning, teaching, and design. Learning theorists may also be interested in what making can tell us about theories of learning, about what is learned through participation in an innovative set of practices. Toward this interest, we ask: "What do people learn through making?" Asking questions about making and learning is tricky because "making" can be seen as either a set of instrumental practices that serve as a gateway to already established, school-based disciplinary practices, or as a discipline in and of itself. In both cases, the learning sciences is fundamentally interested in the cognitive and sociocultural mechanisms by which learners engage with content, process, and practice.

How Do People Learn to Make?

The most influential learning theory that has informed learning in and through making is constructionism (see, for discussion, Harel & Papert, 1991; Kafai, 2006; Papert, 1980). The influence of constructionism is discussed in almost every text that explores the research and practice of making and learning (cf. Litts, 2015; Martinez & Stager, 2013; Peppler et al., 2016). In addition to the obvious emphasis on the making and sharing of artifacts, learning to make also includes a focus on ideation, iteration, and reflection; the use of portfolio systems for assessment; and technologies to support powerful ideas—all components of the constructionist tradition. Learning to make is more, however, than just a "leveling up" of constructionism to include the latest technologies and tools. Making also draws on a multiliteracies perspective as a way to understand how people learn (cf. Cope & Kalantzis, 2000; New London Group, 1996). Multiliteracies is built on the idea that knowledge is embedded in social, historical, cultural, and physical contexts, and that learning happens at the intersection of these, primarily through the design and sharing of representations. Although making has primarily been associated with the STEM disciplines, writing can also be considered a form of making (Cantrill & Oh, 2016) and the composition process is closely associated with a multiliteracies perspective on learning. Resnick and Rosenbaum (2013) describe how making, writing, and coding are all fundamentally about "the idea that people create meaning through the things they create" (p. 231). Understanding the process of learning to make, then, is not too different from understanding the process of learning to write, or any other creatively interpretive act, such as generating a novel or a creative "read" of a canonical work.

Proponents of both problem and project-based learning will see echoes of learning to make in their framing of how people learn (see Hmelo-Silver, Kapur, & Hamstra, this volume). Grounded in learner interest, dependent on relationships among people and tools, and fundamentally multidisciplinary, these approaches also embrace the core ideas of both a constructionist and a new literacies perspective. Similarly, scholars who take an embodied cognition perspective on the relationship between mind, body, and tools will also recognize the core ideas of constructionism in their research (see Alibali & Nathan, this volume). Finally, scholars who value a distributed cognition perspective on learning, where knowledge is stretched across people, tools, and time, can also see their research situated within making and learning (Halverson et al., 2017). This is in large part because learning to make requires that not everyone learns and does the same things at the same time. In fact, successful making is often dependent on individualized participation trajectories that converge around the creation of an artifact or artifacts that are shared with an external audience. As with distributed cognition, the answer to "Where is the learning?" in making is at the intersection of people, tools, and space.

All of these connections indicate that making sits squarely within the learning sciences as a way to articulate how people know, learn, and act. Furthermore, making contributes to our understanding of how people learn through explicit connections among theoretical strands that do not often talk to one another in research or in practice. An understanding of learning through making demonstrates that:

- Learning processes are *design processes;*
- The *creation and sharing of artifacts* is essential for learning;
- Learning requires attending to both *process and product as outcomes of participation.*

What Do People Learn Through Making?

Learning theorists are also interested in the second question—"What do people learn through making?"—in order to extend our understanding of how people learn across a range of disciplines and spaces. Most of the work on learning through making has linked making to STEM disciplines. Specifically, Martin and Dixon (2016) argue that making activities can be a gateway to K–12 engineering, and others demonstrate that hybrid digital/physical maker activities contribute to students' knowledge of computer programming, particularly when they are working at the intersection of digital programming and physical tools (Berland, 2016; Shapiro, Kelly, Ahrens, & Fiebrink, 2016). Learning arts practices are also linked to making; Peppler (2016) shows how a focus on physical computing enables young digital media artists to expand their reach into coding, electronics, and craftsmanship. Perhaps the most well-researched set of making activities are those that involve circuitry. Research has shown that students as young as third grade learn basic circuitry concepts through participation in making activities, as demonstrated both through the successful creation of operational, closed circuits and through more traditional pre/post tests identifying more abstract circuitry knowledge (Peppler & Glosson, 2013; Qi, Demir, & Paradiso, 2017).

While all of these studies are encouraging in terms of legitimizing making as practices that might be embraced in formal learning environments, it is perhaps more interesting to ask what people learn through making qua making, rather than in service of exogenous learning goals. *Tinkering* is a core component of making that is often devalued in formal educational settings (Resnick & Rosenbaum, n.d.). Bevan, Gutwill, Petrich, and Wilkinson (2015) identify "tinkering" as a set of practices unique to making that are learned through sustained engagement in making activities: "At the heart of tinkering is the generative process of developing a personally meaningful idea, becoming stuck in some aspects of physically realizing the idea, persisting through the process, and experiencing breakthroughs as one finds solutions to problems" (p. 99). While tinkering can be associated with success in STEM disciplines, the practices described in their research are not often measured or valued in formal learning environments, including playfulness, iteration, failure, experimentation, and the freedom to change course and explore new paths (Resnick & Rosenbaum, n.d.).

Learning outcomes are often determined by what is being made, rather than the abstract set of learning outcomes that are often found in school-based versions of making. For example, young people involved in a "build a bike" challenge at their local makerspace demonstrate learning through the successful creation of a bicycle that they can take with them (Sheridan & Konopasky, 2016). Unsurprisingly, most of the research on making outcomes is done in out-of-school learning spaces, primarily museums and libraries, that are interested in understanding how participants learn but have the freedom to broaden conceptions of what counts as learning outcomes. Less commonly explored are the ways that learning dispositions cultivated during making, including the familiarization with productive failure and experimentation, can help youth develop the social/affective/interpersonal learning (including "grit") sought after in recent national schooling initiatives.

So is making a discipline unto itself? Research on what and how people learn in making has identified roots in STEM, the arts, design, and entrepreneurship (themselves emerging disciplines that combine more established fields). There is evidence that incorporating making into traditional learning environments could connect students more effectively to disciplinary content that they would otherwise struggle with. But "disciplining" making also means acknowledging that practices like tinkering do not mesh well with the standardized outcome structures of traditional schooling. Furthermore, as research on makerspaces and makers points out, the promise of making as a set of democratizing practices will not be realized if we simply drop making into already established social and cultural schooling routines. Rather, this will become another set of practices reserved for those who look the part (Lacy, 2016; Vossoughi et al., 2016).

The Maker Movement in the Learning Sciences

We want to return now to the question of how research and practice within the Maker Movement connect to and advance the learning sciences as a field by exploring theories of how people learn, the design of learning environments, and issues of equity and diversity.

Theoretically, making serves as a bridging construct among theories of learning that do not always communicate with one another. As indicated earlier, learning through making embraces constructionist, multiliteracies, and embodied and distributed cognition perspectives to arrive at three core principles: designing as learning, creating and sharing artifacts, and attending to process and product as outcomes. While most public schooling reforms revolve around standardizing learning sequences and assessment measures as a pathway to equity and access, maker pedagogy is built on individualization—of interest, of skill development, and of participation in a process. This tension between individualization and standardization is of great interest to scholars in the learning sciences, who see the potential of constructionist pedagogy while eschewing the "hacker mindset" that often results from highly individualized learning environments (Blikstein & Worsley, 2016).

Learning research is often caught in the longstanding divide between formal and informal learning environments. However, research on making offers opportunities to stretch across the divide; design experiments often feature partnerships among organizations and the inherent interdisciplinarity encourages each group to bring their expertise to the table. Museum educators, for example, may encourage teachers to embrace tinkering practices, while teachers may help museum educators to link their goals to more measurable, standardized schooling outcomes. While learning theorists are not troubled by the bridging, designers of learning experiences often take this divide as foundational to their practice frequently building solely for in-school or out-school settings rather than tools that have the potential to be used across settings for a variety of purposes (Peppler et al., 2016). By starting with making and not with the setting, research on making can remind us to take a learning-first perspective on research and practice.

Making is a paradigmatic context for attending to issues of equity and diversity in research, practice, and design within the learning sciences (Esmonde & Booker, 2017), given the promise of the Maker Movement to democratize access to means of production and audiences for work (Vossoughi et al., 2016). But democratization is laden with values that are often not shared by the diverse communities we aim to reach with our teaching and learning reform efforts. There is disagreement, for instance, over whether making in its current form is dependent on the use of new technologies and online distribution; What counts as making? Are traditional crafting practices making? Does the presence of a 3-D printer in a space automatically mean that making is happening? These questions prompt many researchers who care deeply about the democratization of teaching and learning practices to criticize the Maker Movement for its lack of attention to the cultural and historical ways that marginalized communities are permitted to participate (Vossoughi et al., 2016). Dropping new ideas into already existing systems can reify inequitable access (Lacy, 2016). Likewise, adopting one-size-fits-all identities of participation often leads to resistance on the part of marginalized communities, whose identities are then not valued. Critical scholars who study making and learning remind us to design for a range of places and identities where making can happen successfully.

Further Readings

Halverson, E. R., & Sheridan, K. (2014). The Maker Movement in education. *Harvard Educational Review, 84*(4), 495–504.
This essay provides an overview of the role the Maker Movement plays in education. Furthermore, the authors offer a framework for conducting research at the intersection of making and learning by focusing on making as a set of learning activities, makerspaces as communities of practice, and makers as identities of participation.

Martin, L. (2015). The promise of the maker movement for education. *Journal of Pre-College Engineering Education Research, 5*(1), 30–39.

In this article, Martin offers three key components of making that are necessary for the design of maker-based learning environments: digital tools, community infrastructure, and the maker mindset. Martin argues that "good making" is well aligned with what learning scientists understand as "good learning."

Peppler, K., Halverson, E., & Kafai, Y. (Eds.). (2016). *Makeology* (Vols. 1 & 2). New York: Routledge.
This two-volume text series is a set of empirical studies by leading scholars in the field of making and learning. The volumes include studies of making activities, cases of formal, informal, and online makerspaces, as well as research on learner identities as makers.

Sheridan, K., Halverson, E. R., Brahms, L., Litts, B., Owens, T., & Jacobs-Priebe, L. (2014). Learning in the making: A comparative case study of three makerspaces. *Harvard Educational Review, 84*(4).
In this piece, Sheridan and colleagues analyze three cases of makerspaces representing the range of environments that focus on making and learning: a museum-based makerspace, a community makerspace, and an adult-oriented member makerspace. They describe features of the makerspaces as well as how participants learn and develop through complex design and making practices.

Vossoughi, S., Hooper, P. K., & Escudé, M. (2016). Making through the lens of culture and power: Toward transformative visions for educational equity. *Harvard Educational Review, 86*(2), 206–232.
This article offers a critique of the Maker Movement and presents a vision for "making" as grounded in social, historical, and cultural practices. The authors argue that by ignoring the contributions of historically marginalized communities in the Maker Movement, educators reify the risks of turning a potentially radical pedagogy into another tool for establishment communities to thrive.

References

Alibali, M. W., & Nathan, M. (2018). Embodied cognition in learning and teaching: action, observation, and imagination. In F. Fischer, C. E. Hmelo-Silver, S. R. Goldman, & P. Reimann (Eds.), *International handbook of the learning sciences* (pp. 75–85). New York: Routledge.

Ames, M., & Rosner, D. (2014). From drills to laptops: Designing modern childhood imaginaries. *Information, Communication & Society, 17*(3), 357–370.

Berland, M. (2016). Making, tinkering, and computational literacy. In K. Peppler, E. Halverson, & Y. Kafai (Eds.), *Makeology: Makers as learners* (pp. 196–205). New York: Routledge.

Bevan, B., Gutwill, J. P., Petrich, M., & Wilkinson, K. (2015). Learning through STEM-rich tinkering: Findings from a jointly negotiated research project taken up in practice. *Science Education, 99*, 98–120.

Blikstein, P., Martinez, S. L., & Pang, H. A. (2016). *Meaningful making: Projects and inspirations for Fab Labs and makerspaces*. Torrance, CA: Constructing Modern Knowledge Press.

Blikstein, P. & Worsley, M. (2016). Children are not hackers: Building a culture of powerful ideas, deep learning, and equity in the maker movement. In K. Peppler, E. Halverson, & Y. Kafai (Eds.), *Makeology: Makerspaces as learning environments* (pp. 64–80). New York: Routledge.

Brahms, L., & Crowley, K. (2016). Making sense of making: Defining learning practices in MAKE magazine. *Makeology: Makers as Learners, 2*, 13–28.

Brennan, K., Resnick, M., & Monroy-Hernandez, A. (2010). Making projects, making friends: Online community as a catalyst for interactive media creation. *New Directions for Youth Development, 2010*(128), 75–83.

Buechley, L. (2016). Opening address. *FabLearn Conference*, Stanford University, Palo Alto, CA, October 14–16. Retrieved from https://edstream.stanford.edu/Video/Play/a33992cc9fb2496488c1afa9b6204a571d

Buechley, L., Eisenberg, M., Catchen, J., & Crockett, A. (2008). The LilyPad Arduino: Using computational textiles to investigate engagement, aesthetics, and diversity in computer science education. *CHI '08: Proceedings of the SIGCHI Conference on Human Factors in Computing Systems*, 423–432.

Cantrill, C., & Oh, P. (2016). The composition of making. In K. Peppler, E. Halverson, & Y. Kafai (Eds.), *Makeology: Makerspaces as learning environments* (pp. 107–120). New York: Routledge.

Cope, B., & Kalantzis, M. (2000). *Multiliteracies: literacy learning and the design of social futures*. London: Routledge.

Dougherty, D. (2011). *The Maker Movement: Young makers and why they matter* [Video file]. Retrieved from www.youtube.com/watch?v=lysTo7-VVg0

Dougherty, D. (2013). The maker mindset. In M. Honey & D. E. Kanter (Eds.), *Design, make, play: Growing the next generation of STEM innovators* (pp. 7–11). New York: Routledge.

Erete, S., Pinkard, N., Martin, C., Sandherr, J. (2015) Employing narratives to trigger interest in computational activities with inner-city girls. *Proceedings of the First Annual Research on Equity and Sustained Participation in Engineering, Computing, and Technology (RESPECT) Conference*, Charlotte, NC, August 14–15, 2015.

Esmonde, I. & Booker, A. N. (Eds.) (2017). *Power and privilege in the learning sciences*. New York: Routledge.

Fields, D., & Lee, V. (2016). Craft Technologies 101: Bringing making to higher education. In K. Peppler, E. Halverson, & Y. Kafai (Eds.), *Makeology: Makerspaces as learning environments* (pp. 121–138). New York: Routledge.

Gershenfeld, N. (2005). *Fab: The coming revolution on your desktop—From personal computers to personal fabrication*. Cambridge, MA: Basic Books.

Guzzetti, B., Elliot, K., & Welsch, D. (2010). *DIY media in the classroom: New literacies across content areas*. New York: Teachers College Press.

Halverson, E. R., Lakind, A., & Willett, R. (2017). The Bubbler as systemwide makerspace: A case study of how making became a core service of the public libraries. *International Journal of Designs for Learning, 8*(1), 57–68.

Halverson, E. R., & Sheridan, K. (2014). The Maker Movement in education. *Harvard Educational Review, 84*(4), 495–504.

Harel, I., & Papert, S. (1991). *Constructionism*. Norwood, NJ: Ablex Publishing.

Hmelo-Silver, C. E., Kapur, M., & Hamstra, M. (2018) Learning through problem solving. In F. Fischer, C. E. Hmelo-Silver, S. R. Goldman, & P. Reimann (Eds.), *International handbook of the learning sciences* (pp. 210–220). New York: Routledge.

Holbert, N. (2016). Leveraging cultural values and "ways of knowing" to increase diversity in maker activities. *International Journal of Child–Computer Interaction. 9–10*, 33–39.

Kafai, Y. B. (2006). Constructionism. In K. Sawyer (Ed.), *Cambridge handbook of the learning sciences* (pp. 35–46). Cambridge, MA: Cambridge University Press.

Kafai, Y. B., Fields, D. A., and Searle, K. A. (2014). Electronic textiles as disruptive designs in schools: Supporting and challenging maker activities for learning. *Harvard Educational Review, 84*(4), 532–556.

Kafai, Y., & Peppler, K. (2011). Youth, technology, and DIY: Developing participatory competencies in creative media production. In V. L. Gadsden, S. Wortham, & R. Lukose (Eds.), *Youth cultures, language and literacy: Review of Research in Education, 35*(1), 89–119.

Lacy, J. E. (2016). *A case study of a high school Fab Lab*. University of Wisconsin, Madison. Unpublished doctoral dissertation.

Litts, B. K. (2015). *Making learning: Makerspaces as learning environments*. University of Wisconsin-Madison. Unpublished doctoral dissertation.

Martin, L. (2015). The promise of the maker movement for education. *Journal of Pre-College Engineering Education Research, 5*(1), 30–39.

Martin, L. & Dixon, C. (2016). Making as a pathway to engineering and design. In K. Peppler, E. Halverson, & Y. Kafai (Eds.), *Makeology: Makers as learners* (pp. 183–195). New York: Routledge.

Martinez, S. L., & Stager, G. S. (2013). *Invent to learn: Making, tinkering, and engineering in the classroom*. Torrance, CA: Constructing Modern Knowledge Press.

New London Group. (1996). A pedagogy of multiliteracies: Designing social futures (C. Cazden, B. Cope, N. Fairclough, J. P. Gee, M. Kalantzis, G. Kress, A. Luke, et al., Eds.). *Harvard Educational Review, 66*(1), 60–92.

Papert, S. (1980). *Mindstorms: Children, computers, and powerful ideas*. Hemel Hempstead, UK: Harvester Press.

Peppler, K. (2016). ReMaking arts education through physical computing. In K. Peppler, E. Halverson, & Y. Kafai (Eds.), *Makeology: Makers as learners* (pp. 206–226), New York: Routledge.

Peppler, K., & Bender, S. (2013). Spreading innovation: Leveraging lessons from the Maker Movement. *Phi Delta Kappan, 95*(3), 22–27.

Peppler, K. & Gresalfi, M. (2014). Re-crafting mathematics education: Designing tangible manipulatives rooted in traditional female crafts. National Science Foundation.

Peppler, K. & Glosson, D. (2013). Stitching circuits: Learning about circuitry through e-textile materials. *Journal of Science Education and Technology, 22*(5), 751–763.

Peppler, K., Halverson, E. & Kafai, Y. (Eds.). (2016). *Makeology* (Vols. 1 & 2). New York: Routledge.

Puckett, C., Gravel, B., & Vizner, M. (2016). *Vocational vestiges: Detracking, choice, and STEM education in the new comprehensive high school*. Paper presented at the American Educational Researchers Association Annual Meeting, April 2016, Washington, DC.

Qi, J., Demir, A., & Paradiso, J. A. (2017, May). Code collage: Tangible programming on paper with Circuit stickers. *Proceedings of the 2017 CHI Conference Extended Abstracts on Human Factors in Computing Systems* (pp. 1970–1977). New York: ACM Press.

Reas, C. (2006). Media literacy: Twenty-first century arts education. *AI & Society, 20*(4), 444–445.

Resnick, M., Eidman-Aadahl, E., & Dougherty, D. (2016). Making-writing-coding. In K. Peppler, E. Halverson, & Y. Kafai (Eds.), *Makeology: Makers as learners* (pp. 229–240). New York: Routledge.

Resnick, M., & Rosenbaum, E. (2013). Designing for tinkerability. Design, make, play: Growing the next generation of STEM innovators. In M. Honey, & D. E. Kanter (Eds.), *Design, make, play: Growing the next generation of STEM innovators* (pp. 163–181). New York: Routledge.

Rusk, N. (2016). Motivation for making. In K. Peppler, E. Halverson, & Y. Kafai (Eds.), *Makeology: Makers as learners* (pp. 85–108). New York: Routledge.

Ryan, J. O., Clapp, E., Ross, J., & Tishman, S. (2016). Making, thinking, and understanding: A dispositional approach to maker-centered learning. In K. Peppler, E. Halverson, & Y. Kafai (Eds.), *Makeology: Makers as learners* (pp. 29–44). New York: Routledge.

Shapiro, R. B., Kelly, A. S, Ahrens, M. S., & Fiebrink, R. (2016) BlockyTalky: A physical and distributed computer music toolkit for kids. *Proceedings of the 2016 Conference on New Interfaces for Musical Expression. Brisbane, Australia.*

Sheridan, K., Halverson, E. R., Brahms, L., Litts, B., Owens, T., & Jacobs-Priebe, L. (2014). Learning in the making: A comparative case study of three makerspaces. *Harvard Educational Review, 84*(4) 505–531.

Sheridan, K. & Konopasky, A. (2016). Designing for resourcefulness in a community-based makerspace. In K. Peppler, E. Halverson & Y. Kafai (Eds.), *Makeology: Makerspaces as learning environments*, New York, NY: Routledge, 30–46.

Sheridan, K. & Konopasky, A. (2016). Designing for resourcefulness in a community-based makerspace. In K. Peppler, E. Halverson & Y. Kafai (Eds.), *Makeology: Makerspaces as learning environments* (pp. 30–46). New York: Routledge.

Torrey, C., McDonald, D. W., Schilit, B. N. & Bly, S. 2007. How-to pages: Informal systems of expertise sharing. *Proceedings of the 10th European Conference on Computer Supported Cooperative Work. ECSCW '07,* 391–410.

Vossoughi, S., Hooper, P. K., & Escudé, M. (2016). Making through the lens of culture and power: Toward transformative visions for educational equity. *Harvard Educational Review, 86*(2), 206–232

Wilkinson, K., Anzivino, L., & Petrich, M. (2016). The big idea is their idea. In K. Peppler, E. Halverson, & Y. Kafai (Eds.), *Makeology: Makers as learners* (pp. 161–180). New York: Routledge.

Wohlwend, K., Peppler, K., Keune, A., & Thompson, N. (2017). Making sense and nonsense: Comparing mediated discourse and agential realist approaches to materiality in a preschool makerspace. *Journal of Early Childhood Literacy. 17*(3), 444–462.

29

Knowledge Building
Theory, Design, and Analysis

Carol K. K. Chan and Jan van Aalst

Knowledge building is an educational model that can be understood as a community's effort to advance the state of knowledge in that community (Bereiter, 2002; Scardamalia & Bereiter, 1994, 2014). In the 1970s and 1980s, cognitive science research had developed to the point that it could study complex learning in real-world contexts and inform the design of learning environments. Along with other approaches such as the Adventures of Jasper Woodbury series and Fostering Communities of Learners, knowledge building in the 1980s and 1990s was an attempt to bring about deep learning in complex domains and real classrooms (Lamon et al., 1996; McGilly, 1994). Today, knowledge building can be considered a model that has a family resemblance with many other approaches in the learning sciences, with its emphasis on building on prior knowledge, metacognition, regulation processes, collaboration, scaffolding, authentic learning contexts, the use of technology to extend students' cognitive systems, and transfer, as exemplified by many chapters in this handbook.

Scardamalia and Bereiter created a seminal instantiation of knowledge building in a computer-supported collaborative learning environment, originally called Computer-Supported Intentional Learning Environment (CSILE) and later Knowledge Forum® (described below). Since the 1990s, the model has influenced others and the term "knowledge building" has been adopted in various social-constructivist approaches that emphasize inquiry, problem solving, collaboration, and joint construction of knowledge, usually within designed tasks, curricula and projects. However, for Bereiter and Scardamalia (2014), knowledge building has a distinctive focus—knowledge building and knowledge creation are synonymous—and characterizes the kind of productive knowledge work found in scientific and research communities, including the practices community participants engage in to advance the knowledge of that community. The creation of knowledge is a collective product. They argued for an education agenda that helps children see their work as part of a civilization-wide effort to advance the knowledge frontier of the community. In this chapter, we review Scardamalia and Bereiter's knowledge-building model, including the theory and technology, design of knowledge building in classrooms, and analytic approaches used to examine and foster knowledge building. We also discuss important questions that the approach has raised for the learning sciences regarding theories of collective cognition, pedagogical tensions between well-structured and emergent approaches, and technological issues for assessing collective knowledge. The chapter concludes with considerations of future research directions.

Carol K. K. Chan and Jan van Aalst

Theory of Knowledge Building and Knowledge Forum

Theoretical Foundation

Knowledge building/creation originated in writing research in the 1980s, which distinguished "knowledge telling," in which students *retell* what they already knew, and "knowledge transformation" (Bereiter & Scardamalia, 1987a), in which students *restructure* their knowledge during writing. In the 1990s, research on intentional learning examined differences between task completion versus learning as an explicit goal; expert learners employ constructive learning efforts *over and above* task completion, and expertise involves reinvesting cognitive efforts to understand problems at progressively deeper levels and working at the edge of competence (Bereiter & Scardamalia, 1993). With the intent of making classroom communities places where knowledge production would be possible for school-aged children, Scardamalia and Bereiter (1994, 2006) developed a prototype computer-supported intentional learning environment (CSILE) in 1986, followed by Knowledge Forum, launched in 1997.

Two major epistemic dimensions are key to the knowledge-building/creation model.

1. **Learning and knowledge building.** Scardamalia and Bereiter (2006) distinguished between learning and knowledge building, with the former focusing on individual mental states and the latter on public ideas and theories. They argued that the common goal of learning, even among approaches that involve active and constructive processes, is for students to acquire knowledge of their intellectual heritage (e.g., understanding the law of supply and demand). In knowledge building, the key goal is to advance the state of community knowledge, while participants also learn. Knowledge building in schools is the educational variant of knowledge creation: the process by which new knowledge is created in science, engineering, medicine, and other fields of human endeavor. Knowledge creation as an educational goal, as Scardamalia and Bereiter (2006) put it, is "a civilization-wide human effort to extend the frontiers of knowledge." These authors argued that knowledge creation is not just for experts; students also need to learn the processes by which knowledge is created and that this is the kind of knowledge work that experts do. Of course, students are unlikely to make major scientific breakthroughs given that some discoveries have taken hundreds of years for scientists to uncover. Nevertheless, students, working as a community, can significantly advance the public ideas and knowledge of that community, and tackle some problems that have been historically significant. For example, Scardamalia and Bereiter (2006) referred to one student's comment on Knowledge Forum that the 19th-century scientist Mendel "worked on Karen's problem," to suggest that Karen, similar to Mendel, was investigating the same cutting-edge problem; her work was at the frontiers and on a continuing line with that of Mendel. In another often-cited example, elementary-school students, in studying the topic of light in science, wondered how refraction by tiny raindrops can give rise to a rainbow that spans the sky. Although the students did not make a new scientific discovery, they created novel and coherent explanations that extended the frontiers of knowledge, adding value to the class community (Bereiter & Scardamalia, 2010).
2. **Design mode and belief mode.** Bereiter (2002) argued that ideas should be regarded as "real" things, similar to bicycles and mobile telephones, and about which we can ask for what purposes a real thing can be used, how it can be tested, and how it can be modified. Likewise for ideas, leading to a key knowledge building principle: *all ideas are improvable*. In this respect, knowledge building resembles designing: knowledge builders attempt to create knowledge that is "more useful" than the knowledge with which they started with. "Usefulness" can be evaluated in terms of how and which ideas explain phenomena and bring about more testable predictions.

"Design-mode thinking" emphasizes that knowledge building, like design, is an open-ended journey (Scardamalia & Bereiter, 2006). For example, a manufacturer of cell phones produces the prototype of a certain design; however, the design work necessary for the next version begins almost immediately. Similarly, idea improvement is a continual process of inquiry; knowledge begets knowledge. In contrast, belief-mode thinking that involves reasoning, evidence, and evaluation of claims is more prevalent in schools, and is generally less open-ended than design-mode thinking. The discourse often stops when some arguments prevail; debate is a format that perpetuates belief-mode thinking. Knowledge building requires a discourse that develops new ideas, and therefore needs to focus on how these can be developed, tested, and improved; design-mode thinking involves an ever-deepening process of explanation and theory-building. This distinction between design- and belief-mode thinking is important for understanding the distinction between Bereiter and Scardamalia's conceptualization of knowledge building/creation as compared to other approaches to building knowledge together.

Knowledge Forum

Central to knowledge building is a communal knowledge space where students share their ideas and theories and work to improve them (Scardamalia & Bereiter, 1994, 2006). Knowledge Forum provides a digital space for this collective work: students can contribute, build on, reflect, synthesize, and "rise above," making further efforts towards knowledge advancement beyond their initial postings in the community. Students' sustained contributions and improvements to these digital and conceptual artifacts are essential to knowledge building. Technology is integral to knowledge-building theory and pedagogy, and Knowledge Forum is specifically designed to support knowledge creation processes (see Scardamalia & Bereiter, 2006, 2014).

The basic unit in a Knowledge Forum database is a *view*, essentially a canvas upon which notes are placed and where meta-information can be written (Figure 29.1). Within a specific view, students can write and create networks of notes to express their questions and ideas and build upon them to develop theories. The notes form networks similar to discussion threads in other online discourse environments. When writing a *note*, which is placed on a view, students can use modifiable *scaffolds* to help them focus on theory building (e.g., "my theory," "I need to understand," "new information"). For example, students can use the "my theory" scaffold to indicate which of their ideas need further testing or explanation. They can use "*keywords*" to denote key domain words. The canvas or view can include graphics and annotations—for example, to highlight the relationships between the networks of ideas, draw tentative conclusions, or point out the remaining questions.

Knowledge Forum includes features for supporting emerging ideas and synthesizing lines of inquiry for higher-level conceptualization. Different views can be linked and their relationships established. For example, one view may hold ideas early in the community's inquiry, and a later view the major advances on the same problem, or a collection of views could explore different aspects of light (e.g., the reflection and refraction of light by mirrors and lenses), with a super-view linked to these views synthesizing the community's knowledge advance. Knowledge Forum includes "reference notes" that are hyperlinks to other notes and "rise above" notes that support the participants' synthesizing different ideas. Rise-above notes/views and references help to create a "meta-discourse" for collective knowledge and to reformulate problems at successively higher levels. Knowledge Forum includes assessment tools, discussed later in the chapter, that help students assess how the community's ideas develop over time (for more details on Knowledge Forum, see Scardamalia & Bereiter, 2006, 2014).

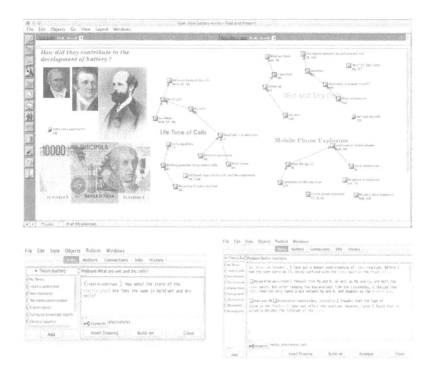

Figure 29.1 Features of Knowledge Forum

Note: A view is a collaborative inquiry space with notes and links and other information (top); A Knowledge Forum note with scaffolds (bottom left) and references notes (bottom right).

Classroom Design and Pedagogy for Knowledge Building

Design-Based Approach and Principle-Based Pedagogy

Knowledge-building pedagogy is designed in line with the theory of how knowledge is created and advanced in innovative communities. To examine and instantiate the theoretical framework, the knowledge-building model emphasizes the use of design-based research (DBR) methods focusing on principle-based pedagogy. As with other learning sciences approaches using DBR, knowledge building develops design principles in empirical classroom studies and tests the principles and practices to refine new designs for iterative improvement (see Puntambekar, this volume). In a review of 30 years of DBR in knowledge building, Chen and Hong (2016) traced the development of principle-based pedagogical principles for knowledge building; they are not just design parameters but integral to the development of the model.

Contrary to other inquiry-based pedagogy in which students work on pre-defined project tasks or problems, Scardamalia and Bereiter (2006) advocated what they called "principle-based pedagogy" focusing on developing workable principles that create the conditions that make the emergence of ideas more likely. Teachers and students co-construct the flow of inquiry as it unfolds and emerges, guided by a set of principles. Table 29.1 lists the principles based on the key ideas discussed earlier and evolving practices in classroom (see Scardamalia, 2002). These principles provide teachers a way of talking about their understanding of knowledge building; they work as a system rather than separately as isolated principles. Knowledge Forum provides a knowledge-creation space for realizing these principles. Students work in opportunistic and flexible ways rather than following scripted activities. For example, the principle "epistemic agency" states students negotiate the fit between their personal and others' ideas. Focusing on adaptive expertise, teachers can encourage epistemic agency in different ways, such as having students initiate experiments to test their ideas against scientific ones, or having students engage in "knowledge-building classroom talk" to contrast diverse and divergent ideas. The uses of these principles and related activities vary with the emergent questions and goals in the classroom community.

Design Considerations for Principle-Based Knowledge-Building Pedagogy

Several design considerations, synthesized based on classroom research, are important to creating principle-based knowledge-building classrooms (for more details, see Scardamalia, 2002; see also Chen & Hong, 2016). A key feature is to turn over high-level epistemic agency to students for collective idea improvement.

Emergent versus fixed curriculum. A progressive and emergent curriculum is needed to support idea development and to maximize the opportunity for knowledge creation (Caswell & Bielaczyc, 2001; Zhang, Scardamalia, Lamon, Messina, & Reeve, 2007). Unlike pedagogy using well-developed

Table 29.1 Knowledge Building Principles

Real ideas, authentic problems	Knowledge problems arise from students' efforts to understand their world; they work on real ideas and problems they care about.
Improvable ideas	All ideas are treated as improvable. Students work continually to improve the quality, coherence, and utility of ideas.
Epistemic agency	Students negotiate fit between their personal and others' ideas using contrast to spark knowledge advancement; take charge of motivation, planning, and evaluation.
Idea diversity	To understand an idea is to understand the ideas that surround it, including those that stand in contrast to it.
Rise above	Working towards more inclusive principles and higher-level formulation of problems for synthesis and emergent goals.
Community knowledge, collective responsibility	Contributions to shared, top-level community goals are as important as personal achievements; share responsibility for knowledge advance of the community.
Constructive use of authoritative sources	Creative work requires familiarity with current and cutting-edge knowledge of the field; students use authoritative information combined with a critical stance.
Knowledge building discourse	The discourse of knowledge building results in more than knowledge sharing; the knowledge itself is transformed by the discourse process.
Embedded and transformative assessment	Assessment is key to collective progress and embedded in day-to-day work; the community engages in its own assessment for transformative purposes.
Democratizing knowledge	All students are valued contributors to the community; the differences between have/have not are minimized; all are empowered to engage in innovative work.
Symmetric knowledge advancement	Knowledge advancements are symmetrical, with different parties across teams and communities advancing together.
Pervasive knowledge building	Knowledge building is not confined to particular occasions, but pervades mental life in and out of schools.

Source: Adapted from M. Scardamalia (2002), Collective cognitive responsibility. In B. Smith (Ed.), *Liberal education in a knowledge society* (Table 4.1, pp. 78–82). Chicago, IL: Open Court.

curriculum and inquiry-based materials, knowledge building does not use pre-designed materials. Each knowledge-building inquiry initiative, spanning a few months, is situated within some curriculum area; however, it is the students who are taking cognitive responsibility for the curriculum working on driving questions and inquiries into the core concepts, similar to scientists engaged in inquiry. Caswell and Bielaczyc (2001) examined idea improvement and how children collectively pursued inquiries into the study of evolution using an emergent curriculum, and discussed that the children's inquiries somewhat resembled the scientific progress to what Darwin did in his exploration. A knowledge-building curriculum relies a great deal on the emerging interests of the participants, and community knowledge is important. One year, a class studying light may become interested in how rainbows arise, but in the next year the question may be very different. In all cases, however, students work on the important concepts of the domain (e.g., refraction and reflection of light in different circumstances).

Idea-centered versus task-centered focus. Knowledge building focuses on idea-centered pedagogy; idea improvement is a key principle. Students' ideas, rather than tasks, are viewed as the center of classroom life (Scardamalia, 2002). There are no prescribed routines; the goal is to improve the community's ideas. Knowledge-building pedagogy encourages students to enact high-level epistemic agency—to refine their knowledge goals progressively as their inquiries unfold and to contribute, advance, and refine their collective ideas. Zhang et al. (2007) discussed the practices for idea improvement in a principle-based classroom. The students started with face-to-face discussions; different ideas were elicited and made public for improvement. Through both online and offline discourse, students pursued idea improvement: they formulated problems of understanding, set forth theories to be improved, identified constructive information, and compared different ideas and models. In line with design-mode thinking, the students continually refined and revised their ideas, hypotheses, and theories, thereby deepening their explanations. Their offline work varied and could include conducting experiments to test their ideas, reading to understand difficult information, and classroom discourse. These activities are not linear or prescribed, but conducted opportunistically, framed by the principles.

Meta-discourse and assessment. Central to innovative communities is reflection, rise-above, and assessment. Knowledge-building design involves a meta-level of discourse beyond problem solving; rise-above is a key principle. Knowledge-building pedagogy involves classroom talk as a *meta-discourse*, with students "talking about their talk," discussing their conversation on Knowledge Forum. Students can collectively monitor the community's progress and identify new lines of inquiry. Van Aalst and Chan (2007) designed an e-portfolio assessment tool using a set of four knowledge-building principles as the criteria to assess their collective work and advance on Knowledge Forum. The students constructed an e-portfolio note using "reference notes," with links to other Knowledge Forum notes, and explained why these were the high point of their collective work. Reflection and assessment helped the students engage in meta-discourse, synthesizing the best work of the community. More recently, Resendes, Scardamalia, Bereiter, Chen, and Halewood (2015) employed word-cloud visualization in formative assessment to help students conduct discourse about their Knowledge Forum work. Additional examples of assessment supported by technology are discussed in the next section.

Fixed groups versus opportunistic groups. Small collaborative group design is common in learning sciences and CSCL pedagogy, but knowledge building emphasizes designing for *distributed* and collective advance in community knowledge. Zhang, Scardamalia, Reeve, and Messina (2009) reported on a three-year DBR study of how a teacher changed his group collaboration structure within a knowledge-building classroom. Children worked in fixed assigned groups in year 1, interactive fixed groups in year 2, and opportunistic groups as a whole class in year 3; the opportunistic grouping was most effective for both diffusion of ideas and scientific accuracy of ideas. Siqin, van Aalst, and Chu (2015) compared this fixed versus opportunistic collaboration for tertiary students and found similar results, including more sophisticated knowledge-building inquiries within opportunistic community-based groups. Knowledge Forum allowed for interconnected views for opportunistic groups to work with ideas creatively. The students could work on different problems in different views; new views could be created as other goals emerged, and the students could synthesize knowledge in rise-above views. Cross-groups and community-based groupings supported by technology increased the emergence of ideas. Such designs also reflect the kinds of knowledge-creation dynamics in scientific communities with emergent interaction of ideas.

Role of teachers. Comparison of three idealized models helps to depict the roles of teachers in knowledge-building classrooms (Bereiter & Scardamalia, 1987b, cited in Chen & Hong, 2016). Teacher A is a "workbook" model common in schools, with teachers focusing on the routine of completing preset tasks and activities. Teacher B undertakes many good principles from the learning sciences (e.g., designs rich authentic problems, harnesses prior knowledge, and provides formative feedback) in the classroom. The Teacher C model, a knowledge-building approach, includes these good features, but makes it possible for students to do what Teacher B would do, but by carrying it out themselves. For example, rather than the teacher establishing authentic problems, the students may identify meaningful and cutting-edge problems of their community for investigation. Students are to carry out the executive functions for their progress not relying on teacher direction; epistemic agency and community knowledge are key to progress. The teacher's role is to highlight the epistemic needs of the students, helping them to "notice" what is significant in the community through modeling, co-reflection, and working as a fellow knowledge builder. With an emphasis on adaptive expertise, knowledge-building teachers do not rely on prescribed routines, but develop their flow by working on problems and ideas with their students.

Analyzing Knowledge Building and Technology-Supported Assessment

Theory-Driven Approaches and Analysis

Quantitative analyses. Central to knowledge-building theory is community knowledge. Analytic methods and technology-based tools have been developed to examine the nature of the network of notes and discourse in Knowledge Forum data. Accompanying Knowledge Forum is a suite of assessment applets that provide statistics on forum activities (e.g., notes read, key terms). Among them, the social network analysis (SNA) applet provides information on "network density," the proportion of the actual links of students' Knowledge Forum note-reading (or build-on) in relation to the maximum possible links that depict community interaction and connectedness (Teplovs, Donoahue, Scardamalia, & Philip, 2007). Zhang et al. (2009) used SNA techniques to analyze community process that included three dimensions: *awareness* of community contributions, complementary *contributions* to the community, and *distributed engagement* in the community (e.g., decentralized network), highlighting the importance of collective cognitive responsibility.

Another approach to examining community knowledge is by analyzing lexical measures—that is, students' use of shared key terms derived from an applet. Hong and Scardamalia (2014) found that higher forum engagement was associated with the use of specific domain words characterizing the development of expertise within the community. With the development of learning analytics, different tools such as the Knowledge-Building Discourse Explorer (KBDeX; Oshima, Oshima, & Matsuzawa, 2012) have been developed; and SNA techniques have been advanced to examine the conceptual relationship and network coherence among ideas. Such analysis explores how networks of ideas emerge in addition to individual contributions to network growth.

Qualitative analyses. Coding and analytic schemes attempt to capture indicators of what is important in knowledge-building theory and its pedagogical goals. One of the early code-and-count schemes examined epistemological inquiry, contrasting fact-seeking with explanation-seeking questions (Hakkarainen, 2004). Focus on questions and explanations is common in CSCL analysis; however, here the emphasis is epistemological, highlighting efforts towards *theory building* in which a major goal is for students to engage in *explanatory discourse* in knowledge building. The analysis of explanatory discourse is developed further through the "Ways of Contribution" scheme that includes the discourse categories of theorizing, working with information, and synthesizing (Chuy et al., 2011). These schemes are used for analysis; however, they also depict the kinds of productive knowledge-building discourse processes that students are encouraged to develop.

The qualitative analysis of discourse similarly emphasizes the assessment of community knowledge, and one approach works at distinguishing knowledge creation from other online discourse modes, including knowledge construction and knowledge sharing (van Aalst, 2009). Specifically, knowledge-sharing discourse refers to participants sharing opinions and information; knowledge-construction discourse refers to joint construction of understanding and shared meanings; and knowledge-creation discourse depicts the meta-discourse on how participants become cognizant of community issues and how they contribute to extending community knowledge. Recently, Bereiter and Scardamalia (2016) proposed a set of constructive dialogic moves towards knowledge creation, including problem definition, new ideas, "promisingness" evaluation, comparisons, critical discourse, high-level ideas, and meta-dialogue. These emphasize the notions of *theory building* and *meta-discourse* and point to new and possible ways of analyzing knowledge creation.

Principle-Based Approaches and Transformative Assessment

Knowledge-building assessment is premised on principles. A distinctive theme in knowledge building has been to develop approaches and tools that can be used by students for collective idea improvement, and that also aligns with the principle of epistemic agency. Scardamalia (2002) discussed the principle of assessment—it needs to be *embedded* in community practice with *concurrent* feedback to *transform* knowledge building. The Knowledge Forum applets that researchers used (Hong & Scardamalia, 2014) are also used by teachers and students for these purposes. One major approach to using assessment to promote knowledge building is the e-portfolio design (van Aalst & Chan, 2007). When students assess their own Knowledge Forum discourse and identify the best ideas from the community, they can better recognize what knowledge building entails that *transforms* their knowledge-building process. To help students move from an individual to a community focus, the Knowledge Connection Analyzer (KCA) tool was designed to help them reflect on and assess their discourse. Students can run KCA on databases using a set of four questions (e.g., "Are we putting our knowledge together?" and "How do our ideas develop over time?"). Students use the analytics information as evidence to help them reflect and set new goals for knowledge building (Yang, van Aalst, Chan, & Tian, 2016).

Two major assessment tools have recently been developed in conjunction with Knowledge Forum to support idea development for student reflection, assessment, and evaluation. The Idea Thread Mapper (ITM) originated from research on the "inquiry thread," analyzing a conceptual thread of Knowledge Forum notes focused on a principal problem (Zhang et al., 2007). With ITM, students can identify related notes on a specific problem and create idea threads that track idea development. They can document their journey of thinking and identify key themes and knowledge gaps (Zhang et al., in press). Knowledge creation and scientific discovery are based on evaluating promising ideas. A "promising idea" tool has been developed that supports young students in assessing and evaluating promising ideas in their community. With the use of learning analytics, these ideas are then aggregated and visualized to support students moving toward deeper inquiry (Chen, Scardamalia, & Bereiter, 2015). These different analytical approaches share the common theme of student agency in assessing the progress of their community work in the Knowledge Forum.

The Knowledge-Building Model: Comparisons, Synergies, and Tensions

Knowledge building bears resemblance with other inquiry-based models that highlight core learning sciences principles in pursuit of deep understanding, such as problem-based learning and project-based science. Nevertheless, knowledge building can be distinguished from other approaches by its epistemological focus on knowledge creation, principle-based pedagogy, and analytic aims.

Epistemic Focus

Various inquiry-based learning sciences models have focused on developing conceptual/domain knowledge, and including meta-knowledge and strategies within the domain. Examples are units on ecology that use inquiry in a virtual world intended to bring about key canonical understanding (see de Jong, Lazonder, Pedaste, & Zacharia, this volume) and designing inquiry tasks and problems to help students learn through problem solving (see Hmelo-Silver, Kapur, & Hamstra, this volume; Linn, McElhaney, Gerard, & Matuk, this volume). Knowledge building places less emphasis on pre-defined goals, but instead starts from the question of *how far* a community can advance from where it starts. An important goal of knowledge building is *epistemological*, that is, understanding how knowledge is created and the social nature of this process. Although in knowledge building students learn domain knowledge and problem solving, they also engage in the discourse necessary for establishing the community's next learning goals, synthesizing across multiple problems, evaluating the state of knowledge in the community, and reflecting on how the community has made knowledge advances.

Knowledge building as sustained inquiry for knowledge production has been compared to tacit knowledge transformation in organizations (Nonaka & Takeuchi, 1995) and activity system for expansive learning in workplaces (Engeström, Miettinen, & Punamäki, 1999) as different approaches to knowledge creation. Knowledge building is primarily about developing "community knowledge" that bears some resemblance to group cognition in CSCL (Stahl, 2006) but differs in emphasizing creation of ideas. Although CSCL and the learning sciences have often examined knowledge construction in small groups, recent research has been extended to explore collective knowledge construction within large groups and communities (see Cress and Kimmerle, this volume); the nature of collective knowledge and artifacts are fruitful areas for investigation. Recent developments in knowledge building have included the examination of self-organization at the interpersonal and idea levels (Scardamalia & Bereiter, 2014), which are also important research issues in complex systems (see Yoon, this volume); knowledge building involves the complex process of idea emergence and interaction. Examining knowledge building and its related approaches raises new questions for theories of collective cognition and knowledge creation in learning sciences.

Pedagogy and Design

Principle-based pedagogy in knowledge building/creation, rather than tasks and activities, raises provocative questions about pedagogical designs in the learning sciences. Similar controversies exist in the field regarding well-structured instruction versus inquiry-based designs (Hmelo-Silver, Duncan, & Chinn, 2007; Kirschner, Sweller, & Clark, 2006) and scripting versus non-scripting. There are other variants, such as the productive-failure approach using open-ended design followed by guided instruction (see Hmelo-Silver et al., this volume). Related models such as knowledge communities (see Slotta, Quintana, & Moher, this volume) also emphasize more teacher guidance. Knowledge building is one of the few pedagogical designs that advocates emergent and unstructured approaches to maximize students' creative work and collective agency.

One manifestation of the contrast has been focused on scripting approaches. Rather than considering scripting and non-scripting as polarized dimensions, it may be fruitful to consider a *continuum* of structured versus open-ended pedagogical approaches in ways that examine differences and relatedness. It may also be helpful to examine possibilities of how scripted pedagogy can incorporate high-order principles, and principle-based pedagogy in knowledge building developing some emergent-intermediate structures embedding activities with principles. Collins (2002) advocated balance between task-centered and idea- centered focus in CSCL classroom designs. Bereiter (2014) postulated *principled practical knowledge* that involves both practical know-how and coherence of scientific theory that might also address issues of general principles and practical activity.

Another way to address the controversy around levels and types of guidance is to consider the alignment of theory and design. For many learning sciences approaches, the goals are conceptual and metacognitive. For example, to learn difficult science content, guided inquiry may be relevant, and scripting (e.g., role-taking) may help with task engagement for problem solving. To develop creative expertise in knowledge building, students need an open-ended environment, and teachers need to develop adaptive expertise using principles to adapt to novel situations. Scripted and structured pedagogy may ensure higher task success but constrain emergence. Similarly, emergent pedagogy may not be the most effective for domain knowledge. Pedagogical and technological designs need to be informed by the underpinning epistemology. Reciprocally, using different designs and models in the learning sciences may illuminate and elicit different kinds of collaborative and creative processes, thereby enriching the theory and design of the learning sciences.

Analytic Aims and Approaches

Analyzing collaborations and online discourse are major research traditions within the learning sciences and CSCL. These knowledge-building approaches reflect the traditions and shifts from code-and-count to connected discourses and developments in learning analytics. The focus of analyzing knowledge building emphasizing community process, explanatory theory building, and meta-discourse highlights the distinctive characteristics of the model; theory-driven analysis also helps to illuminate processes for designs. Although these schemes are used for analysis, they can also provide useful pointers for developing productive discourse in knowledge-building and learning sciences classrooms.

A key theme in the analysis of knowledge building has been the focus on students assessing the growth of their own knowledge, supported by technology tools. This approach is aligned with the theoretical emphasis on epistemic agency and emergent design. Students and teachers can use such tools in opportunistic ways as the community needs arise. Increased attention has also been given in learning analytics on how teachers can use the information, and its use by students is just beginning. These knowledge-building examples may be relevant to informing these developments. Developing the analysis and tools to be used by participants in both *assessing* and *scaffolding* collaboration are important areas for further investigation.

Conclusions and Future Directions

Scardamalia and Bereiter's knowledge-building/-creation model aims to bring into education the goals and processes of knowledge-creation communities. Knowledge-building theory, evolving with Knowledge Forum, is instantiated with the principle-based pedagogy designed to mirror and support the process of creative expertise. Theory-driven and principle-based analysis of knowledge building, together with the development of analytical tools in support of collective agency, are integral to theory, principles, and pedagogy. The knowledge-building model provides an example of the integral and synergistic relationships between theory, design, and analysis, and this may be a fruitful area of inquiry in examining different and emerging learning sciences approaches.

The knowledge-building model, with its epistemic focus on knowledge creation, may help enrich theory development related to collective knowledge construction, an area of potential growth in the learning sciences. In terms of design, the principle-based pedagogy of knowledge building and the related controversy over structured and emergent design raises many important questions. These polarizing tensions need to be reexamined and the dialectics explored in light of different epistemological underpinnings and learning goals. How principle–structure–practice dialectics work together to facilitate adaptive expertise is an important question for the learning sciences. Methodologically, further analytical work is needed to examine what constitutes knowledge-creation discourse, to link analysis with design, and broaden the analyses through integrating different methodological approaches. Assessment and learning analytics and technology are rapidly growing areas in the learning sciences (see Pellegrino, this volume; Rosé, this volume), and how knowledge building can incorporate and contribute to these areas, and in particular integrate learning analytics and transformative assessments, requires further investigation.

The creation of new ideas and innovations is key to education; however, enacting collective agency for self-organization and emergent processes remains challenging. Designing knowledge-building and other learning sciences approaches in diverse sociocultural settings with different scales, from classrooms to schools and communities, may enrich the theory and design used to examine how the learning sciences work in the real world. The contributions of knowledge building also need to be tested beyond classrooms, in technology-supported knowledge communities and international networks.

Further Readings

Bereiter, C., & Scardamalia, M. (2014). Knowledge building and knowledge creation: One concept, two hills to climb. In S. C. Tan, H. J., So, & J. Yeo (Eds.), *Knowledge creation in education* (pp. 35–52). Singapore: Springer.
This chapter discusses how knowledge building postulated by Scardamalia and Bereiter is similar to knowledge creation in innovative organizations/communities, and examines the implications for education. The chapter helps to clarify the epistemological nature and goals of the model, and provides new perspectives on future directions in examining knowledge creation in the learning sciences.

Chen, B., & Hong, H. Y. (2016). Schools as knowledge building organizations: Thirty years of design research. *Educational Psychologist, 51*, 266–288.
This paper provides an overview of the knowledge-building model drawing from its 30 years of design-based research. The paper discusses the key goals of knowledge building for reframing education as a knowledge-creation enterprise, explicates the principle-based pedagogy, as well as reviews the educational benefits and impacts of knowledge building.

van Aalst, J. (2009). Distinguishing knowledge-sharing, knowledge-construction, and knowledge-creation discourses. *International Journal of Computer-Supported Collaborative Learning, 4*(3), 259–287.
This paper postulates a framework distinguishing knowledge-sharing, knowledge-construction, and knowledge-creation and analyzes Knowledge Forum discourse, identifying different discourse patterns. The framework is important for illuminating the theoretical nature of knowledge building and provides a new way for conceptualizing and analyzing CSCL discourse.

van Aalst, J., & Chan, C. K. K. (2007). Student-directed assessment of knowledge building using electronic portfolios. *Journal of the Learning Sciences, 16*(2), 175–220.
This paper examines the problems of how to design assessment to examine and scaffold knowledge building. In three related studies, the design involves students assessing their own knowledge advance identifying exemplary episodes guided by knowledge-building principles in their online discourse. This study helps to advance theory and design highlighting the dual role of assessment in characterizing and scaffolding collective learning.

Zhang, J., Scardamalia, M., Lamon, M., Messina, R., & Reeve, R. (2007). Socio-cognitive dynamics of knowledge building in the work of 9- and 10-year-olds. *Educational Technology Research & Development, 55*(2), 117–145.
This study examining elementary students' online discourse on Knowledge Forum illustrates what knowledge building entails and how it is manifested. A new analytic approach called inquiry thread, a series of notes addressing a conceptual problem, is developed, and analysis of discourse using knowledge-building principles provides evidence and demonstrates how young students can advance their collective knowledge.

NAPLeS Resources

Chan, C., van Aalst, J., *15 minutes about knowledge building* [Video file]. In *NAPLeS video series*. Retrieved from www.psy.lmu.de/isls-naples//video-resources/guided-tour/15-minute-chan_vanaalst/index.html
Scardamalia, M. & Bereiter, C. *Knowledge building: Communities working with ideas in design mode* [Webinar]. In *NAPLeS video series*. Retrieved from http://isls-naples.psy.lmu.de/intro/all-webinars/scardamalia-bereiter/index.html

References

Bereiter, C. (2002). *Education and mind in the knowledge age*. Mahwah, NJ: Lawrence Erlbaum.
Bereiter, C. (2014). Principled practical knowledge: Not a bridge but a ladder. *Journal of the Learning Sciences*, *23*(1), 4–17.
Bereiter, C., & Scardamalia, M. (1987a). *The psychology of written composition*. Hillsdale, NJ: Lawrence Erlbaum.
Bereiter, C., & Scardamalia, M. (1987b). An attainable version of high literacy: Approaches to teaching higher-order skills in reading and writing. *Curriculum Inquiry*, *17*(1), 9–30.
Bereiter, C., & Scardamalia, M. (1993). *Surpassing ourselves: An inquiry into the nature and implications of expertise*. Chicago, IL: Open Court.
Bereiter, C., & Scardamalia, M. (2010). Can children really create knowledge? *Canadian Journal of Learning and Technology*, *36*(1).
Bereiter, C., & Scardamalia, M. (2014). Knowledge building and knowledge creation: One concept, two hills to climb. In S. C. Tan, H. J. So, & J. Yeo (Eds.), *Knowledge creation in education* (pp. 35–52). Singapore: Springer.
Bereiter, C., & Scardamalia, M. (2016). "Good moves" in knowledge-creating dialogue. *QWERTY: Journal of Technology and Culture 11*(2), 12–26.
Caswell, B., & Bielaczyc, K. (2001). Knowledge forum: Altering the relationship between students and scientific knowledge. *Education, Communication and Information*, *1*(3), 281–305.
Chen, B., & Hong, H. Y. (2016). Schools as knowledge building organizations: Thirty years of design research. *Educational Psychologist*, *51*(2), 266–288.
Chen, B., Scardamalia, M., & Bereiter, C. (2015). Advancing knowledge: Building discourse through judgments of promising ideas. *International Journal of Computer-Supported Collaborative Learning*, *10*(4), 345–366.
Chuy, M., Resendes, M., Tarchi, C., Chen, B., Scardamalia, M., & Bereiter, C. (2011). Ways of contributing to an explanation-seeking dialogue in science and history. *QWERTY: Journal of Technology and Culture*, *6*(2), 242–260.
Collins, A. (2002). The balance between task focus and understanding focus: Education as apprenticeship versus education as research. In T. Koschmann, R. P. Hall, & N. Miyake (Eds.), *CSCL 2: Carrying Forward the Conversation* (pp. 43–47). Mahwah, NJ: Lawrence Erlbaum.
Cress, U., & Kimmerle, J. (2018). Collective knowledge construction. In F. Fischer, C. E. Hmelo-Silver, S. R. Goldman, & P. Reimann (Eds.), *International handbook of the learning sciences* (pp. 137–146). New York: Routledge.
de Jong, T., Lazonder, A., Margus Pedaste, M., & Zacharia, Z. (2018). Simulation, games, and modeling tools for learning. In F. Fischer, C. E. Hmelo-Silver, S. R. Goldman, & P. Reimann (Eds.), *International handbook of the learning sciences* (pp. 256–266). New York: Routledge.
Engeström, Y., Miettinen, R., & Punamäki, R. L. (Eds.). (1999). *Perspectives on activity theory*. Cambridge, UK: Cambridge University Press.
Hakkarainen, K. (2004). Pursuit of explanation within a computer-supported classroom. *International Journal of Science Education*, *26*(8), 979.
Hmelo-Silver, C. E., Duncan, R. G., & Chinn, C. A. (2007). Scaffolding and achievement in problem-based and inquiry learning: A response to Kirschner, Sweller, and Clark. *Educational Psychologist*, *42*(2), 99–107.
Hmelo-Silver, C. E., Kapur, M., & Hamstra, M. (2018). Learning through problem solving. In F. Fischer, C. E. Hmelo-Silver, S. R. Goldman, & P. Reimann (Eds.), *International handbook of the learning sciences* (pp. 210–220). New York: Routledge.
Hong, H. Y., & Scardamalia, M. (2014). Community knowledge assessment in a knowledge buiding environment. *Computers & Education*, *71*, 279–288.
Kirschner, P. A., Sweller, J., & Clark, R. E. (2006). Why minimal guidance during instruction does not work: An analysis of the failure of constructivist, discovery, problem-based, experiential, and inquiry-based teaching. *Educational Psychologist*, *41*(2), 75–86.
Lamon, M., Secules, T., Petrosino, A., Hakett, R., Bransford, J., & Goldman, S. (1996). Schools for thought: Overview of the international project and lessons learned from one of the sites. In L. Schauble & R. Glaser (Eds.), *Innovations in learning: New environments for education* (pp. 243–288). Mahwah, NJ: Lawrence Erlbaum.
Linn, Marcia C., McElhaney, K. W., Gerard, L., & Matuk, C. (2018). Inquiry learning and opportunities for technology. In F. Fischer, C. E. Hmelo-Silver, S. R. Goldman, & P. Reimann (Eds.), *International handbook of the learning sciences*. New York: Routledge.
McGilly, K. (1994). *Classroom lessons: Integrating cognitive theory and classroom practice*. Cambridge, MA: MIT Press.
Nonaka, I., & Takeuchi, H. (1995). *The knowledge creating company: How Japanese companies create the dynamics of innovation*. New York: Oxford University Press.
Oshima, J., Oshima, R., & Matsuzawa, Y. (2012). Knowledge building discourse explorer: A social network analysis application for knowledge building discourse. *Educational Technology Research and Development*, *60*(5), 903–921.

Pellegrino, J. W. (2018). Assessment of and for learning. In F. Fischer, C. E. Hmelo-Silver, S. R. Goldman, & P. Reimann (Eds.), *International handbook of the learning sciences* (pp. 410–421). New York: Routledge.

Puntambekar, S. (2018). Design-based research (DBR). In F. Fischer, C. E. Hmelo-Silver, S. R. Goldman, & P. Reimann (Eds.), *International handbook of the learning sciences* (pp. 383–392). New York: Routledge.

Resendes, M., Scardamalia, M., Bereiter, C., Chen, B., & Halewood, C. (2015). Group-level formative feedback and meta-discourse. *International Journal of Computer-Supported Collaborative Learning, 10*(3), 309–336.

Rosé, C. P. (2018). Learning analytics in the Learning Sciences. In F. Fischer, C. E. Hmelo-Silver, S. R. Goldman, & P. Reimann (Eds.), *International handbook of the learning sciences* (pp. 511–519). New York: Routledge.

Scardamalia, M. (2002). Collective cognitive responsibility for the advancement of knowledge. In B. Smith (Ed.), *Liberal education in a knowledge society* (pp. 67–98). Chicago, IL: Open Court.

Scardamalia, M., & Bereiter, C. (1994). Computer support for knowledge-building communities. *Journal of the Learning Sciences, 3*(3), 265–283.

Scardamalia, M., & Bereiter, C. (2006). Knowledge building: Theory, pedagogy, and technology. In R. K. Sawyer (Ed.), *The Cambridge handbook of the learning sciences* (pp. 97–115). New York: Cambridge University Press.

Scardamalia, M., & Bereiter, C. (2014). Knowledge building and knowledge creation: Theory, pedagogy and technology. In K. Sawyer (Ed.), *Cambridge Handbook of the learning sciences* (2nd ed., pp. 397–417). New York: Cambridge University Press.

Siqin, T., van Aalst, J., & Chu, S. K. W. (2015). Fixed group and opportunistic collaboration in a CSCL environment. *International Journal of Computer-Supported Collaborative Learning, 10*(2), 161–181.

Slotta, J. D., Quintana, R., & Moher, T. (2018). Collective inquiry in communities of learners In F. Fischer, C. E. Hmelo-Silver, S. R. Goldman, & P. Reimann (Eds.), *International handbook of the learning sciences* (pp. 308–317). New York: Routledge.

Stahl, G. (2006). *Group cognition: Computer support for building collaborative knowledge*. Cambridge, MA: MIT Press.

Teplovs, C., Donoahue, Z., Scardamalia, M., & Philip, D. (2007). Tools for concurrent, embedded and transformative assessment for knowledge building process and progress. In *Proceedings of the 8th International Conference on Computer-Supported Collaborative Learning*, 721–723.

van Aalst, J. (2009). Distinguishing knowledge-sharing, knowledge-construction, and knowledge-creation discourses. *International Journal of Computer-Supported Collaborative Learning, 4*(2), 259–287.

van Aalst, J., & Chan, C. K. K. (2007). Student-directed assessment of knowledge building using electronic portfolios. *Journal of the Learning Sciences, 16*(2), 175–220.

Yang, Y., van Aalst, J., Chan, C. K. K., & Tian, W. (2016). Reflective assessment in knowledge building by students with low academic achievement. *International Journal of Computer-Supported Collaborative Learning, 11*(3), 281–311.

Yoon, S. A. (2018). Complex systems and the learning sciences: Implications for learning, theory, and methodologies. In F. Fischer, C. E. Hmelo-Silver, S. R. Goldman, & P. Reimann (Eds.), *International handbook of the learning sciences* (pp. 157–166). New York: Routledge.

Zhang, J., Scardamalia, M., Lamon, M., Messina, R., & Reeve, R. (2007). Socio-cognitive dynamics of knowledge building in the work of 9- and 10-year-olds. *Educational Technology Research & Development, 55*(2), 117–145.

Zhang, J., Scardamalia, M., Reeve, R., & Messina, R. (2009). Designs for collective cognitive responsibility in knowledge-building communities. *Journal of the Learning Sciences, 18*(1), 7–44.

Zhang, J., Tao, D., Chen, M.H., Sun, Y., Judson, D., & Naqvi, S. (in press). *Journal of the Learning Sciences*.

30
Collective Inquiry in Communities of Learners

James D. Slotta, Rebecca M. Quintana, and Tom Moher

Introduction

Collective Inquiry is a pedagogical approach in which an entire classroom community (or potentially multiple classrooms) is engaged in a coherent curricular enterprise, with well-defined learning goals for both content and practice. Participants work individually or in small groups, holding a common understanding of their purpose as a learning community (Bielaczyc & Collins, 1999). As individual members add observations, ideas, or artifacts, the products of their efforts are integrated as a pooled community knowledge base. Students typically share a sense that the "whole is greater than the sum of its parts," as they build on their peers' contributions, organize content, synthesize ideas, identify gaps, and gain inspiration. This approach is related to the broader category of Inquiry Learning, as addressed by Linn, Gerard, McElhaney, and Mattuk (this volume).

There is an established international community of scholars investigating the learning community pedagogy, with contributions from Scandinavia (e.g., Lipponen & Hakkarainen, 2007), Europe (e.g., Cress & Kimmerle, 2007), Israel (Kali et al, 2015), Japan (Oshima, Oshima, & Matsuzawa, 2012), Hong Kong (e.g., van Aalst & Chan, 2007), the US (Chen & Zhang, 2016), and Canada (e.g., Scardamalia & Bereiter, 2006; Slotta, Tissenbaum, & Lui, 2013), amongst others. In the learning community approach, students are engaged as a scientific community, reminiscent of real-world science, and encouraged to develop their own inquiry progressions, building on one another's findings, collaborating with peers, and developing shared observational data.

This chapter will report on our own recent collaboration in which we developed a learning community curriculum to help elementary science students collectively investigate a simulated ecosystem embedded within their physical classroom space (e.g., in the walls or floor; Moher, 2006). No student working alone could understand these phenomena sufficiently, thereby establishing the pretext or need for cooperation and collaboration. Such an approach is well suited for collective inquiry, but does not in itself offer any solutions for *how* students can progress as a community, building on one another's ideas and gaining strength through their numbers. What should the community's objectives be when faced with such an object of inquiry, and how should inquiry progress? How should we represent community knowledge and scaffold inquiry practices and discourse? Our research investigates a model for the design of materials and activities that engaged students individually, in small groups, and as a whole class. We examine how knowledge was contributed and reused within the community, as well as what technology scaffolds could support these processes and reinforce collective inquiry. Our chapter begins with a review of learning communities, including a set of key challenges for collective inquiry, and describes how our own research has responded to those challenges, including an important role for scripting and orchestration.

Learning Communities for Collective Inquiry

The learning community approach positions learners as active constructors of knowledge within "a culture of learning in which everyone is involved in a collective effort of understanding" (Bielaczyc & Collins, 1999, p. 271). Learners are given high levels of agency and are responsible for developing their own questions and approaches to addressing those questions, for critiquing the ideas of peers, and for evaluating the progress within the community. Expertise does not reside solely with the teacher, but is rather distributed amongst all members (Brown & Campione, 1994). The teacher is a member of the community, and participates as a knowledgeable mentor. Artefacts, observations, and other products of student inquiry are often contributed to a community knowledge base—usually situated within a technology-mediated environment—where they become available for critique, improvement, and reuse. Slotta and Najafi (2013) articulated three common characteristics of learning communities: (1) an epistemic commitment to collective advancement, (2) a shared community knowledge base, and (3) common modes of discourse.

The learning community approach is well suited for designs in which students engage in in practices that mirror those of scientific communities, such as investigation and argumentation. Within such a community, students bring their diverse interests and expertise, with a shared understanding that their learning activities will align to advance the community's cause while at the same time helping individuals learn, and allowing everyone to benefit from the community's resources. With appropriate scaffolding, students can design their own experiments, interpret evidence to inform arguments, and synthesize knowledge from their peers. They are challenged to make the products of their work accessible and relevant within a community of peer investigators (Brown & Campione, 1994). Hence, this approach is well suited for 21st-century science education—engaging students directly in relevant STEM practices (e.g., working with data, collaborating with peers, interpreting evidence). Students' efforts ultimately feed back into the community, advancing the understandings of all members, leading to a sense of "collective cognitive responsibility" (Scardamalia, 2002).

Perhaps the most prominent example of collective inquiry is that of knowledge building communities (Scardamalia & Bereiter, 2006) which focuses on intentional learning and idea improvement. Knowledge Building (KB) is distinguished amongst learning community approaches by its "idea-centered" pedagogy, and reliance on students to determine the specific learning activities. This emphasis runs counter to the notion of scripting (Dillenbourg & Jermann, 2007), and instead includes parallel strands of student-driven inquiry. The teacher plays an extremely important role in KB, and student "knowledge work" is scaffolded by a technology environment called the Knowledge Forum® that is specifically designed to support such "knowledge work" (Scardamalia & Bereiter, 2006; Cress and Kimmerle, this volume, also discuss this research tradition).

Another well-recognized project is Fostering a Community of Learners (FCL), in which students are engaged as a scientific community of practice, with specific content and epistemic learning goals (Brown & Campione, 1994). FCL curricula are scripted around an iterative research cycle that consists of three interdependent stages: *research, share, and perform*. The cycle is launched by an anchoring event, in which the class shares in a common experience (e.g., watching a video or play, reading a work of fiction, or learning about an experiment) that is tied to the "big idea" of the unit (e.g., animal/habitat interdependence). Students conduct research and share knowledge through a variety of research activities, including reciprocal teaching, guided writing and composition, cross-age tutoring, and consultation with subject matter experts outside of the classroom (Bielaczyc & Collins, 1999). The "perform" stage of the cycle is motivated by a consequential task (e.g., designing a biopark), which requires that all students have learned the entire targeted conceptual domain, not just portions.

James D. Slotta, Rebecca M. Quintana, and Tom Moher

Several scholars have observed that it is challenging for teachers or researchers to enact a learning community approach (Slotta & Najafi, 2013; van Aalst & Chan, 2007). As observed by Kling and Courtright (2003, p. 221) "developing a group into a community is a major accomplishment that requires special processes and practices, and the experience is often both frustrating and satisfying for the participants." Slotta and Najafi (2013) argue that the pragmatic and epistemic challenges of shifting from a traditional mode of "knowledge transmission" into a mode of collective inquiry have contributed to a relatively low uptake of this approach amongst researchers and practitioners, and that there is a need for structural models that guide the design of individual, small group, and whole class activities through which students work as a community in collective inquiry.

Knowledge Communities and Inquiry (KCI)

We articulate four key challenges to a learning community pedagogy: (1) to establish an epistemological context where all members share an understanding of the collective nature of their learning, an awareness of how their individual efforts contribute, and how they can benefit personally; (2) to ensure that community knowledge is accessible as a resource for student inquiry (i.e., with effective, accessible, and timely representations); (3) to ensure that scaffolded inquiry activities advance the community's progress as well as all individual learners; (4) to foster productive teacher- and student-led discourse that helps individual students and the community as a whole make progress. We have developed the KCI model in response to these challenges, to guide the design of "collective inquiry" curricula that integrate whole class, small group and individual activities (Slotta & Najafi, 2013; Slotta & Peters, 2008). KCI curricula entail: (1) a knowledge base that is indexed to the targeted science domain, (2) an activity "script" that includes collective, collaborative, and individual inquiry activities in which students construct the knowledge base and then use it as a resource for inquiry, and (3) student-generated products that allow assessment of progress on targeted learning goals.

The notions of *scripting and orchestration* (Kollar, Fischer, & Slotta, 2007) help respond to the challenges of learning communities. In general, a pedagogical script serves to specify the media (e.g., worksheets, student-contributed content, or social media), activities (e.g., inquiry projects, class brainstorms, problem solving, modeling, argumentation, or reflection), grouping conditions (e.g., jigsaw) and activity sequences (e.g., brainstorm, followed by reflection, followed by a jigsaw group design, followed by a culminating project). The script is *"orchestrated"* by the instructor, and scaffolded by a technology environment, which helps track student progress, distribute instructions, materials and prompts, pause students for planned or spontaneous discussions, and collect and organize student work (Dillenbourg & Jermann, 2010). The orchestration of the script further depends upon in-the-moment decisions by the instructor, whose role is one of collaborator and mentor, responding to student ideas as they emerge, and orchestrating the flow of activities. Teachers are not just a "guide on the side" but rather have an explicitly scripted role at all times, in addition to responsibility for overall coordination of the curriculum. Large projected displays help teachers identify pedagogically meaningful signals from amidst the noise of student contributions, and help the community stay on target for learning goals (Slotta, Tissenbaum, & Lui, 2013).

KCI curricula typically span multiple weeks or months, and are developed through a sustained process of co-design (Roschelle, Penuel, & Shechtman, 2006) that includes researchers, teachers, and designers. Technology environments, such as wikis, are employed to give structure to the community's knowledge base and to scaffold collective knowledge building. Slotta and Peters (2008) engaged five sections of a 10th-grade biology course (n=108) in co-authoring wiki pages about human disease systems, ultimately producing a substantive "disease wiki" that served as a resource for their subsequent development and solution of peer-created medical cases. In this way, individual students are able to perceive their contributions within a broader collective effort, recognizing that they will benefit from the collective product and understanding the value of their individual contributions. The KCI script typically includes a major inquiry project, sometimes happening in the final phase of the curriculum, other times revisited throughout the curriculum that is carefully designed such that student products reflect their understanding and application of the targeted content and process learning goals.

Embedded Phenomena for Inquiry Communities

We recently began a collaboration where we applied KCI to support students in collectively investigate scientific phenomena, in the form of digital simulations, that are embedded within the physical space of their own classrooms (Moher, 2006). These simulations provide a location-based experience for scientific discovery learning and seek to "provide the opportunity for students to engage in spontaneous, harmless, and sustained investigation" (Malcolm, Moher, Bhatt, Uphoff, & López-Silva, 2008, p. 238). Students work collectively to monitor and manipulate the simulation in an effort to address their own inquiry questions. Known as Embedded Phenomena (EP), these unique objects of collective inquiry have been developed to situate investigations within the domains of seismology (*RoomQuake*), life sciences (*WallCology, Hunger Games*), astronomy (*HelioRoom*), and hydrology (*AquaRoom*).

Typically, EP persist over several weeks, with simulations running constantly, 24 hours a day, which provides students with opportunities for extended observation and systematic data collection. The design rationale behind such a temporal distribution is that it reinforces the concept that in nature, "things happen when they happen," and do not conform to the schedules of scientists, or even school cycles (Moher, 2006). Students could return from recess to find that the EP they are studying has undergone a major shift (e.g., catastrophic habitat destruction). As the simulation exhibits several changes (e.g., a series of earthquakes occur), a narrative unfolds, giving students opportunities to draw conclusions from their investigations, make comparisons with previously collected data, and engage in collaborative decision-making processes concerning how they might respond to the changes in the phenomena.

Our research collaboration, titled Embedded Phenomena for Inquiry Communities (EPIC) and began in 2010, includes learning scientists from several different research labs. Using EP as a source of inquiry, we have investigated collective inquiry scripts as well as technology-based orchestration supports for learners and teachers. We were particularly interested in the role of emergent visualizations of the community's aggregated knowledge (Cober, McCann, Moher, & Slotta, 2013), and the nature of teacher-led discourse that referred to those visualizations and served to advance community inquiry (Fong, Pascual-Leone, & Slotta, 2012). KCI served as a theoretical foundation, guiding our design of student inquiry, knowledge representations, and orchestration supports (Slotta & Najafi, 2013).

The next section describes our KCI script for the *WallCology* EP—a simulated ecosystem in which computer monitors are placed on each wall in the classroom, providing a form of X-ray "wallscope" that reveals hot and cold water pipes, as well as several different species of insects crawling around on those various surfaces, and vegetation (e.g., "mold" and "scum") that some insects are eating. Other insects are predators, and these food-web interactions are directly observable. Insects vary in terms of their preferred habitat (i.e., brick or pipes) and temperature tolerance (low, medium, high). Statistical information about each Wallscope habitat is available onscreen, in the form of population and temperature graphs as a function of time (see Figure 30.1).

One of the key technical and conceptual features of *WallCology* is that the simulations can be perturbed or changed over time, allowing an emulation of climate change, where the temperatures gradually or suddenly increases, or an "invasive species" that causes interesting or alarming readjustments in species population levels. An underlying biological model drives the simulation, developed in close collaboration with an expert biologist using the mathematics of a complex biological system of predators, prey, habitat conditions, and other factors. These dependencies make the inquiry environment sufficiently challenging to support a wide range of student investigations. Students can

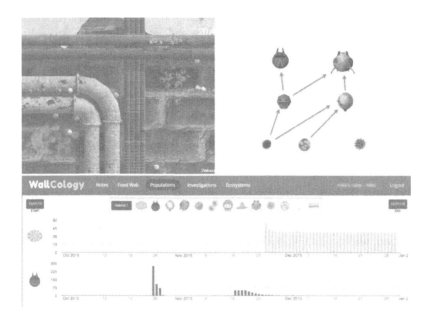

Figure 30.1 *WallCology* habitat viewed through a wallscope (top left); subset of *WallCology* species (top right); *WallCology* population graph (bottom)

identify and classify species, observe their habitat preferences, uncover food webs, and predict and evaluate species' responses to environmental changes. Finally, *WallCology* allows students to alter the state of the ecosystem, by adding or removing species, in order to respond to perturbations.

Working in pairs, students were scaffolded by a software environment called *Common Knowledge* (Fong et al., 2013), implemented on a tablet computer, which guided their food-web and predation observations, brainstorm discussions, access of the community knowledge base, and development of plans for responding to perturbations. An interactive whiteboard was located at the front of the room, providing summative views and interactive tools for sorting and presenting electronic contributions (Slotta et al., 2013).

We were interested in how these *WallCology* features could engage upper elementary students and teachers as a learning community, allowing students to investigate and report their findings, build knowledge with their peers, and develop a deep understanding of the relevant science content and practices. Our goal was to engage students in scientific investigations, evidence-based argumentation, collective knowledge building, and applications of their community knowledge within their own inquiries. One important feature of KCI is the use of dynamically assembled "aggregate representations" of student observations to provide an emergent, summative representation that allows a sense of progress and supports teacher-led discourse (Cober et al., 2013).

A KCI Script for WallCology: An Example of Collective Inquiry

Our team developed a KCI script, and corresponding orchestration supports, which included complex grouping conditions and activity sequences, and emphasized community progress and individual learning. We wanted to situate students' learning within the context of a scientific community, in which they work amongst peers to actively investigate the *WallCology* phenomena. The primary science learning goals included understanding habitats, species, and populations, as well as food webs, biodiversity, and ecosystems. Another important set of goals was concerned with engaging students in investigation and argumentation practices, including: interpreting graphs, reasoning from evidence, planning experiments, communicating findings, and collaborating with peers. Our overall design included three phases, each of which took between two and four weeks: (1) taking inventory of phenomena and constructing models of the food webs, including distinct trophic levels; (2) understanding implications of perturbation, such as temperature rise or habitat loss, in each of the ecosystems; and (3) investigating impact of changes to the ecosystems, such as adding new species or trapping and removing some existing species.

The first phase was conducted as a whole-class activity, with students familiarizing themselves with all four habitats to inform a community-wide knowledge base of the habitats, the species, and their interdependencies. The four ecosystems varied in terms of habitat conditions (how much pipe and brick, and what temperature) and also varied in terms of the population levels of different species (flora and fauna). Students were divided into four teams, with each taking stewardship of one wall (i.e., its habitat and denizen species), and tasked with understanding their habitat and species, then reasoning and problem-solving around any observed perturbations, making use of community-level knowledge, and sharing their own findings with the wider community. In the first phase, the student pooled their various observations into a collective knowledge base about all the various species and habitats—knowledge that would be crucial to their success in the latter two phases, where each team had to first understand a crisis that struck their habitat, and then intervene, creating a more balanced and healthy ecosystem.

Phase 1: Students walked into their classroom and discover the *WallCology* EP that was installed in their classrooms. The EP simulation ran continuously (i.e., all day) on the four monitors that were positioned on four different walls around the room, each displaying a distinct ecosystem. Students used the *Common Knowledge* tools to record observations, including details about the species' behavior, physical traits, habitat preferences, and food preferences. Wherever they witnessed a predation event, they recorded pairwise consumer–producer relationships, which were added to an aggregate food-web grid that appeared at the front of the room (i.e., tallying all observations in real time). Using that grid, each team then constructed their own food-web diagrams, consisting of the subset of species that were spotted within their ecosystem. Teachers then facilitated the construction of a whole-class food web, consisting of all species, using printed species and large paper that was affixed to the classroom wall for the remainder of the unit.

Phase 2: Students entered their classrooms and discovered that a major perturbation had taken place within their group's ecosystem. This was immediately apparent in terms of drastic changes in populations of some of the species in the ecosystems. Students interpreted the species and temperature graphs to deliberate what had happened to their ecosystem, which was drawn from one of four scenarios: habitat destruction, invasive vegetation, invasive predator, and climate change (i.e., temperature increase or decrease). After the students had come to some determination about their specific perturbation, the teacher led a discussion about real-world examples of ecosystem disruption (i.e., do to invasive species, climate change, etc.)—including some examples where ecologists had taken remediating measures. Using the *Common Knowledge* tool, the class brainstormed what constitutes a "healthy ecosystem" and the teacher helped students develop a community consensus, which was to be applied as a rubric to their own habitats (i.e., to measure their remediation).

Phase 3: In the final phase, teams could make changes to their ecosystems by either introducing a new species, or increasing or decreasing a species that was already present. The goal was to improve the overall health of the ecosystem, either by trying to return the ecosystem to its original state or by creating a more diverse ecosystem (i.e., with a robust combination of predators, herbivores, and resources). Findings from any team were shared within the community using *Common Knowledge*, which scaffolded each team in making one cycle of remediations, where each effort was then added to the community knowledge base, indexed to species and habitats. Each cycle began by designing a plan, included proposed steps and predictions about the outcome of their intervention (i.e., which species populations would increase or decrease, and why). At the end of that day's class, the teacher would implement the plans in each ecosystem, so that in the next class period—often with great excitement—students would discover the accuracy of their predictions and record their outcomes. Again, the reports were scaffolded by a new section of the *Common Knowledge* environment, which ensured that they would reflect on the failures and successes, compare against their predictions, and motivate the next intervention. In recording their results, students were asked to include relevant populations graphs and *WallCology* screen captures as evidence. Each team then shared their plan, predictions, and outcomes with their peers in a class presentation, using the *Common Knowledge* tool projected on the classroom's interactive whiteboard. This cycle was repeated several times, until each team was satisfied that they had improved the health of their ecosystem and achieved a desirable balance of populations.

Finding: A Role for Aggregate Representations of Community Knowledge

The *Common Knowledge* scaffolds were designed in close concert with our script, to provide orchestrational supports for students and teachers. This also allowed us to process the contents of the community knowledge base in order to create emergent, community-level views or representations that provided a sense of progress, allowing students and teachers to identify patterns, gaps, or conflicts in their collective products. For example, students worked in pairs to collect food-web observations, each of which took the form of a pairwise relationship (e.g., species X is eaten by species Y). As more and more of those observations were added (i.e., students wandering the room, observing predation events, and entering them using the *Common Knowledge* observation forms), we synthesized them into a table-like grid of all their aggregated contributions, which was displayed on the interactive whiteboard. As a result of students' distributed, independent observations, a collective product thus emerged, which became "greater than the sum of its parts"—revealing statistical patterns that could help resolve conflicts (e.g., if two student thought that an insect was a vegetarian, but there were eight observations of that species eating another insect), or suggest places where more effort was needed (e.g., if there were insufficient observations for certain species, the teacher could refer to the table to encourage students to fill in the gaps).

These aggregate representations made patterns within the data readily apparent to teachers and students, providing an important resource for whole class discussions. Teachers used them as a shared reference, highlighting areas of convergence and divergence, or gaps that required some attention. When the aggregate representations showed agreement in the data, teachers used them to facilitate discussions that allowed the class to reach consensus. Conversely, when the aggregate representations displayed disagreement, they provided direction for students on how to adjust their ongoing investigation. Divergence in the aggregate representation also provided a basis for discussion regarding best practices for inquiry, such negotiating acceptable levels of disagreement or planning how to resolve disagreements. In addition to providing a useful shared referent to guide discussions, the aggregate representations were used by students as an evidentiary database. For example, students referred to the aggregate representations of the producer–consumer relationships to construct their table group's food web.

Finding: Supporting Evidence-Based Arguments in a Scientific Community

An important goal of our research was to engage students in scientific arguments and explanations, using evidence from their *WallCology* investigations (e.g., the species population graphs showing changes in populations that resulted from their interventions). This occurred most prominently within phase 3, where students were scaffolded by the *Common Knowledge* environment, which included three distinct sections for (1) making a plan, (2) making and explaining predictions about species population changes, and (3) providing a report on the results of the investigation (i.e., how did the populations really change, and why did the changes vary from those predicted?). These reports were published in the community knowledge base and provided the basis for group presentations.

During these presentations, student groups reviewed their experiments in front of the community, with two primary goals: (1) to inform the other teams' planning (e.g., if other groups were planning manipulations involving the same species, and could learn from outcomes); and (2) to receive feedback and ideas about what they might try next (e.g., if students from other groups had done something similar or relevant, or had insights to offer about why a manipulation hadn't produced desired results). In analyzing students' presentations, we looked for three components: a claim (i.e., some conclusion or answer to their original question of how to make their ecosystem healthier), evidence (scientific data that are appropriate and sufficient to support the claim), and reasoning (a justification that connects evidence to claim). We used a customized rubric to evaluate students' scientific explanations, following the Claim, Evidence, Reasoning model outlined by McNeill and Krajcik (2011). In each of the two classrooms that we studied, teams showed consistent progress over four intervention cycles, learning from their own investigations and from the reports of their peers, and repeated this cycle four times. With each iteration, student groups were more strategic in their investigations as they became more knowledgeable about the species within their ecosystem, and about how to plan an effective manipulation. We found that students used an average of two claims in each of their presentations, with reasoning supported by evidence, including the results from other teams' investigations (Slotta, Quintana, Acosta, & Moher, 2016).

Conclusions: Classrooms as Learning Communities

KCI has been described here as a formal model for scripting and orchestration of collective inquiry, with the aim of transforming classrooms into learning communities. This model is under development through research such as the *WallCology* study reviewed above, and fits within a broader literature within the learning sciences including the FCL and KB models, which continue to receive attention from a widening circle of scholars (e.g., Kali et al, 2015). The challenges of establishing an epistemological "climate" of collective inquiry remain a major obstacle to both research and practice, reflected in Bereiter and Scardamalia's (2010) observation that it can take up to two years for a teacher to shift toward a collective epistemology. There are also real pedagogical challenges, which bring opportunities for research. How can teachers encourage autonomous inquiry while also ensuring progress on the well-defined learning goals? How can they make time for substantive inquiry given the content coverage demands? How can these learning community methods offer a means of reaching all students in the classroom, and enabling everyone to contribute and learn deeply? KCI research has investigated how community knowledge can be made visible and accessible to inform teacher-led discourse and guide inquiry progressions. We also explore the role of scripting and orchestration, to scaffold specific inquiry processes within the community, and ensure progress on the targeted learning goals. In a learning community approach, technology environments become more than just tools or scaffolds for specific learning processes, but rather serve as holistic frameworks for scaffolding student inquiry, capturing and processing the products of that inquiry, and making them available as consequential resources in subsequent activities.

Further Readings

Bielaczyc, K., & Collins, A. (1999). Learning communities in classrooms: A reconceptualization of educational practice. In C. M. Reigeluth (Ed.), *Instructional design theories and models: A new paradigm of instructional theory* (Vol. 2, pp. 269–292). London: Lawrence Erlbaum.

This seminal paper provides an early review of the key aspects of learning communities, introducing the notion, reviewing FCL, and identifying some core characteristics of the broad approach.

Brown, A. L., & Campione, J. C. (1994). Guided discovery in a community of learners. In K. McGilly (Ed.), *Classroom lessons: Integrating cognitive theory and classroom practice* (pp. 229–270). Cambridge, MA: MIT Press/Bradford Books.

James D. Slotta, Rebecca M. Quintana, and Tom Moher

This book introduces the FCL model, connects it to the psychological literature, and reviews early classroom research.

Cober, R., McCann, C., Moher, T., & Slotta, J. D. (2013). Aggregating students' observations in support of community knowledge and discourse. *Proceedings of the 10th International Conference on Computer-supported Collaborative Learning (CSCL)* (pp. 121–128). Madison, WI: ISLS.
This published proceedings paper reviews the authors' prior research in related topics.

Scardamalia, M. (2002). Collective cognitive responsibility for the advancement of knowledge. In B. Smith (Ed.), *Liberal Education in a Knowledge Society* (pp. 67–98). Chicago: Open Court.
This book chapter reviews the central tenets of knowledge building, clarifies the notion of collective cognitive responsibility, and articulates the teacher's role in a knowledge-building classroom.

Slotta, J. D., & Najafi, H. (2013). Supporting collaborative knowledge construction with Web 2.0 technologies. In C. Mouza & N. Lavigne (Eds.), *Emerging Technologies for the Classroom* (pp. 93–112). New York: Springer.
This book chapter reviews the KCI model and details two classroom implementations: (1) a semester-length climate change curriculum where 5 sections of a ninth-grade class worked in collective inquiry, and (2) a graduate level seminar in media design, where students build on an existing knowledge base, handed down from prior enactments of the course, and develop inquiry-oriented pedagogy for their own investigations of emerging media.

NAPLeS Resources

Chan, C., van Aalst, J., *15 minutes about knowledge building* [Video file]. In *NAPLeS video series*. Retrieved October 19, 2017, from www.psy.lmu.de/isls-naples//video-resources/guided-tour/15-minute-chan_vanaalst/index.html

Dillenbourg, P., *15 minutes about orchestrating CSCL* [Video file]. In *NAPLeS video series*. Retrieved October 19, 2017, from http://isls-naples.psy.lmu.de/video-resources/guided-tour/15-minutes-dillenbourg/index.html

Scardamalia, M., & Bereiter, C. *Knowledge building: Communities working with ideas in design mode* [Webinar]. In *NAPLeS video series*. Retrieved October 19, 2017, from http://isls-naples.psy.lmu.de/intro/all-webinars/scardamalia-bereiter/

Slotta, J. D., *Knowledge building and communities of learners* [Webinar]. In *NAPLeS video series*. Retrieved October 19, 2017, from http://isls-naples.psy.lmu.de/intro/all-webinars/slotta_video/index.html

References

Bereiter, C., & Scardamalia, M. (2010). Can children really create knowledge?. *Canadian Journal of Learning and Technology/La Revue canadienne de l'apprentissage et de la technologie*, *36*(1).

Bielaczyc, K., & Collins, A. (1999). Learning communities in classrooms: A reconceptualization of educational practice. In C. M. Reigeluth (Ed.), *Instructional design theories and models: A new paradigm of instructional theory* (Vol. 2, pp. 269–292). London: Lawrence Erlbaum.

Brown, A. L., & Campione, J. C. (1994). Guided discovery in a community of learners. In K. McGilly (Ed.), *Classroom lessons: Integrating cognitive theory and classroom practice* (pp. 229–270). Cambridge, MA: MIT Press/Bradford Books.

Chen, B., & Zhang, J. (2016). Analytics for knowledge creation: Towards epistemic agency and design-mode thinking. *Journal of Learning Analytics*, *3*(2), 139–163. http://dx.doi.org/10.18608/jla.2016.32.7

Cober, R., McCann, C., Moher, T., & Slotta, J. D. (2013). Aggregating students' observations in support of community knowledge and discourse. *Proceedings of the 10th International Conference on Computer-supported Collaborative Learning (CSCL)* (pp. 121–128). Madison, WI: ISLS.

Cress, U., & Kimmerle, J. (2007, July). A theoretical framework of collaborative knowledge building with wikis: a systemic and cognitive perspective. *Proceedings of the 8th International Conference on Computer Supported Collaborative Learning* (pp. 156–164). New Brunswick, NJ: ISLS.

Cress, U., & Kimmerle, J. (2018) Collective knowledge construction. In F. Fischer, C. E. Hmelo-Silver, S. R. Goldman, & P. Reimann (Eds.), *International handbook of the learning sciences* (pp. 137–146). New York: Routledge.

Dillenbourg, P., & Jermann, P. (2007). Designing integrative scripts. In F. Fischer, I. Kollar, H. Mandl & J. M. Haake (Eds.), *Scripting computer-supported collaborative learning: Cognitive, Computational and Educational Perspectives* (pp. 275–301). New York: Springer.

Dillenbourg, P., & Jermann, P. (2010). Technology for classroom orchestration. In M. S. Khine & I. M. Saleh (Eds.), *New Science of Learning* (pp. 525–552). New York: Springer.

Fong, C., Pascual-Leone, R., & Slotta, J. D. (2012). The Role of Discussion in Orchestrating Inquiry. *Proceedings of the Tenth International Conference of the Learning Sciences.* Sydney, ISLS. *2*, 64–71.

Kali, Y., Tabak, I., Ben-Zvi, D., Kidron, A., Amzalag, M., Baram-Tsabari, A., et al. (2015). Technology-enhanced learning communities on a continuum between ambient to designed: What can we learn by synthesizing multiple research perspectives? In O. Lindwall, P. Koschman, T. Tchounikine, & S. Ludvigsen (Eds.), *Exploring the Material Conditions of Learning: The Computer Supported Collaborative Learning Conference (CSCL)* (Vol. 2, pp. 615–622). Gothenburg, Sweden: ISCL.

Kling, R., & Courtright, C. (2003). Group behavior and learning in electronic forums: A sociotechnical approach. *The Information Society, 19*(3), 221–235.

Kollar, I., Fischer, F., & Slotta, J. D. (2007). Internal and external scripts in computer-supported collaborative learning. *Learning & Instruction, 17*(6), 708–721.

Linn, M. C., McElhaney, K. W., Gerard, L., & Matuk, C. (2018). Inquiry learning and opportunities for technology. In F. Fischer, C. E. Hmelo-Silver, S. R. Goldman, & P. Reimann (Eds.), *International handbook of the learning sciences* (pp. 221–233). New York: Routledge.

Lipponen, L., & Hakkarainen, K. (1997, December). Developing culture of inquiry in computer-supported collaborative learning. *Proceedings of the 2nd International Conference on Computer Support for Collaborative Learning* (pp. 171–175). Toronto, Ontario: ISCL.

Malcolm, P., Moher, T., Bhatt, D., Uphoff, B., & López-Silva, B. (2008, June). Embodying scientific concepts in the physical space of the classroom. *Proceedings of the (pp. International Conference on Interaction Design and Children (IDC)* (pp. 234–241), Chicago, IL.

Moher, T. (2006). Embedded phenomena: Supporting science learning with classroom-sized distributed simulations. *Proceedings of the SIGCHI Conference on Human Factors in Computing Systems (CHI)* (pp. 691–700). Montreal, Canada.

Oshima, J., Oshima, R., & Matsuzawa, Y. (2012). Knowledge Building Discourse Explorer: A social network analysis application for knowledge building discourse. *Educational Technology Research and Development, 60*(5), 903–921.

Roschelle, J., Penuel, W. R., & Shechtman, N. (2006). Co-design of innovations with teachers: Definition and dynamics. *Proceedings of the Seventh International Conference on Learning Sciences (ICLS)* (pp. 606–612), Bloomington, IN.

Scardamalia, M. (2002). Collective cognitive responsibility for the advancement of knowledge. In B. Smith (Ed.), *Liberal Education in a Knowledge Society* (pp. 67–98). Chicago: Open Court.

Scardamalia, M., & Bereiter, C. (2006). Knowledge building: Theory, pedagogy, and technology. In R. K. Sawyer (Ed.), *The Cambridge handbook of the learning sciences* (pp. 97–115). New York: Cambridge University Press.

Slotta, J. D., & Najafi, H. (2013). Supporting collaborative knowledge construction with Web 2.0 technologies. In C. Mouza & N. Lavigne (Eds.), *Emerging Technologies for the Classroom* (pp. 93–112). New York: Springer.

Slotta, J. D., & Peters, V. L. (2008). A blended model for knowledge communities: Embedding scaffolded inquiry. International Perspectives in the Learning Sciences: Cre8ing a learning world. Proceedings of the Eighth International Conference for the Learning Sciences (pp. 343–350), Utrecht, Netherlands: ISLS.

Slotta, J. D., Quintana, R. C., Acosta, A., & Moher, T. (2016). Knowledge construction in the instrumented classroom: Supporting student investigations of their physical learning environment. In C. K. Looi, J. L. Polman, U. Cress, & P. Reimann (Eds.), *Transforming Learning, Empowering Learners: The International Conference of the Learning Sciences (ICLS) 2016* (Vol. 2, pp. 1063–1070). Singapore: ISLS.

Slotta, J. D., Tissenbaum, M., & Lui, M. (2013, April). Orchestrating of complex inquiry: Three roles for learning analytics in a smart classroom infrastructure. *Proceedings of the Third International Conference on Learning Analytics and Knowledge* (pp. 270–274). Leuven, Belgium: ACM,.

van Aalst, J., & Chan, C. K. (2007). Student-directed assessment of knowledge building using electronic portfolios. *Journal of the Learning Sciences, 16*(2), 175–220.

31

Computer-Supported Argumentation and Learning

Baruch B. Schwarz

Argumentation and the Origins of Computer-Supported Collaborative Learning (CSCL)

Research on argumentation and learning developed approximately at the time CSCL became a field of inquiry. At first glance, these two domains emerged quite independently. Studies of the authentic practices of scientists and historians revealed the fundamentally argumentative and collaborative character of these domains, leading educators to argue for bringing argumentation into the classroom (e.g., Driver, Newton, & Osborne, 2000). Two dominant forms of argumentation tended to be incorporated: Toulmin's structural model (1958) or van Eemeren and colleagues' discursive, pragma-dialectical model (van Eemeren, Grootendorst, Henkenmans, Blair, Johnson, et al. 1996). According to the Toulmin structural model, arguments should be based on data, warranted, and should rebut counterarguments. This contrasts with the van Eemeren model that emphasizes argumentation as critical discussion governed by rules of talk: "a verbal and social activity of reason aimed at increasing (or decreasing) the acceptability of a controversial standpoint for the listener or reader, by putting forward a constellation of propositions intended to justify (or to refute) the standpoint before a rational judge" (van Eemeren et al., 1996, p. 5). The van Eemeren definition of argumentation points to a highly constrained process at the end of which a common conclusion may be endorsed (see Schwarz & Baker, 2016).

The merits of each model have been the subject of debate. Positive aspects of the structural model include that it is relatively easy to convey to teachers and students. Its graphical representation makes the components salient, facilitating the development of argumentative skills. However, the structural model does not capture the verbal and social aspects of argument. These are salient in the pragma-dialectical model. Over time, and despite the complexity of implementation, the discursive model gained a stronger foothold in the educational world. For example, Mercer's (1995) elaboration of *exploratory talk* and its ground rules has many similarities with the pragma-dialectical model. Schwarz and Baker (2016) reviewed the evolution of other similar talk practices in the educational system.

Early on in its history, the CSCL community hardly referred to argumentation. Collaboration was emphasized as a practice that would promote deep educational change. Ann Brown brought forward the idea of *co-construction of knowledge* and Marlene Scardamalia and Carl Bereiter, the metaphor of *knowledge building*. CSILE (Computer Supported Intentional Learning Environments) relied on the idea of knowledge building, on the nature of expertise, and on the socio-cultural dynamics of innovation (Chan & van Aalst, this volume; Scardamalia & Bereiter, 1991). CSILE, and its evolutionary successor Knowledge Forum©, intended to (a) make advanced knowledge processes accessible to all participants, (b) foster the creation and continual improvement of public artifacts or community knowledge, and (c) provide a community space for carrying out knowledge building work collaboratively. Users contributed individual notes that they could link to one another and "maps" of linked notes provided visual representations of knowledge in a public space. Individual notes were tagged with the initials of the contributor.

Although Knowledge Forum did not refer to argumentation, its appearance was followed by two interesting phenomena. First, the domain of argumentation for learning began to thrive. For example, the implementation of exploratory talk in long-term interventions led to impressive learning outcomes (Wegerif, Mercer & Dawes, 1999). Secondly, various CSCL tools were developed to support argumentation. Dan Suthers developed the system Belvedere (Suthers & Weiner, 1995), a set of graphical tools for collaborative inquiry and scientific argumentation. Suthers (2003) introduced the notion of *representational guidance* – the idea that certain external representations afford the enactment of collaborative practices. Such guidance was embodied in the icons in the Belvedere graphical interface that were labeled with the elements of Toulmin's (1958) structural model (e.g., claims, data) and differentiated by their unique shapes (Figure 31.1, left column). Students read a text similar to the once excerpted in the right column of Figure 31.1. They typed content into an icon they selected and organized the icons to reflect the structure of their initial argument (Figure 31.1, center panel: HIV virus causes AIDS). Belvedere operated asynchronously and thus reflected a cumulative structure across the community rather than a dialogic process. Although the map and the final hypothesis were co-constructed, Belvedere did not display the identity of the contributors.

Belvedere paved the way for various systems that afforded social argumentation in real time with visual representations of argumentation moves through ontologies representing the Toulmin components of argumentation (e.g., Schwarz & De Groot, 2007; van Amelsvoort, Andriessen, & Kanselaar, 2007). The book *Arguing to Learn: Confronting Cognitions in Computer-Supported Collaborative Learning Environments* (Andriessen, Baker & Suthers, 2003) is a collection of contributions that show that arguing to learn and CSCL enriched each other considerably. The parallel emergence of an emphasis

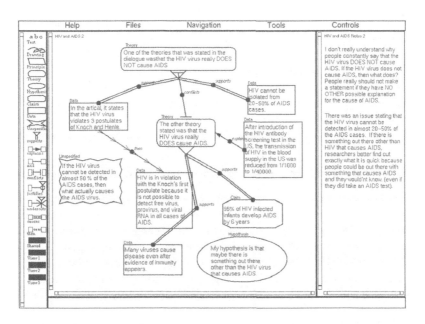

Figure 31.1 A map showing the co-construction of a scientific argument using Belvedere

on co-construction of knowledge building and the emergence of tools to represent argumentation were synergistic: Brown, Scardamalia and Bereiter's revolutionary ideas that equated learning with co-construction (or co-building) of knowledge infiltrated the scientific community. On the one hand, the observation of this co-construction in the discourse led educational psychologists to recognize the importance of argumentation in learning. On the other hand, computer software designers found in argumentative components natural building blocks with which to represent co-construction of knowledge.

The use of new tools for argumentation led to the emergence of new practices: discussions that occurred online generated artifacts in the form of argumentative maps that enabled the summarizing of discussions, collective reflection on the quality of discussions, or peer evaluation of arguments and of moves in discussions. In spite of these possibilities, unguided small groups rarely achieved high-level argumentation (e.g., de Vries, Lund & Baker, 2002). A recent meta-analysis (Wecker & Fischer, 2014) revealed that there was no overall effect on domain-specific learning. One explanation for the lack of effects is that software design often focused primarily on the structural components of argumentation, and tended to neglect the social, discursive aspect of argumentation (Fischer, Kollar, Stegmann, & Wecker 2013; Noroozi, Weinberger, Biemans, Mulder, & Chizari, 2013). The important research domain of collaboration scripts (Kollar, Wecker, & Fischer, this volume) nicely shows the strengths and the weaknesses of research on computer-supported argumentation. The advantage of external support tools for real-time argumentation is that they are less obtrusive and require less cognitive effort from the discussants themselves, as they are embedded in the discussion environment (Noroozi et al., 2012; McLaren et al., 2010). A typical support includes three text boxes, (claim, warrant, qualifier) that are arranged into one message or argumentation script. The scripts ask learners to post arguments, counterarguments and syntheses (see also Kollar, Wecker, & Fischer, this volume). Research has produced two important findings regarding collaborative scripts. First, they are mildly effective in terms of individual domain-specific learning gains. Second, they are effective in helping students learn how to collaborate or learn *about* argumentation (Weinberger, Stegmann, & Fischer, 2007).

In summary, this brief historical overview suggests that the results regarding computer-support for argumentation and learning are a bit disappointing: Learning tasks in CSCL were primarily directed to the building of content knowledge, yet the findings suggest that this goal was not attained beyond the level that was already achieved by other means. The design of the environments developed in this first period unearthed important ideas (e.g., representational guidance) but implementations mostly occurred in non-authentic contexts, and without regard to whether or to what extent learners needed the supports provided by the scripts or ontologies with which they were provided. In this early period of work, none of the experiments focused on the constitution of *communities of learners* – one of the underpinnings of both Brown's and Scardamalia and Bereiter's theories. Borge, Ong, and Rosé (2015) pointed out that research on collaborative scripts has contributed to the insight that what should also be at stake is the students' ability to modify their own dysfunctional discourse processes over time.

New Research Paths in Computer-Supported Argumentation for Learning

We present here some new research paths in computer-supported argumentation for learning. These research paths take into consideration lessons learned from the work summarized in the historical overview. The first path concerns the scaffolding of CSCL argumentation and the second concerns a ramification of the first path – the scaffolding of argumentation in multiple groups. The third path is about blended settings in classroom contexts, and the fourth path, about e-argumentation, dialogism and deliberative democracy. The second path is presented separately from the first because it brings small-group argumentation in the classroom context – a path that may lead to profound educational changes. All paths break assumptions or traditions.

To begin with, the first path breaks with the assumption that, with appropriate tools, collaboration rather than guidance, would promote collaborative knowledge construction.

The Present and Future of Scaffolding CSCL Argumentation

This direction reflects a move away from scripting and toward more investigations of human-provided interaction guidance. Face-to-face teacher scaffolding had traditionally been studied in one-to-one and small group settings. The passage to e-scaffold in the CSCL community focused on small groups in which students were invited to collaborate with each other. E-scaffold refers to on-line synchronous, adaptive guidance of the teacher. This contrasts with a-synchronous guidance provided by scripts. E-scaffolding has been shown to be effective in promoting collaborative argumentation, as evidenced by increases in group presentation of clear, sound arguments and counter-arguments, as well as interactions among students, especially in terms of stating personal standpoints and expressing agreement and/or disagreement with other students (e.g., Asterhan, Schwarz, & Gil, 2012). However, improvements tend to be restricted to precisely what was scaffolded. That is, if argumentation was targeted, interaction was not impacted and vice versa. In contrast, collaboration scripts have been shown to impact both interactions and argument quality (Weinberger, Ertl, Fischer, & Mandl 2005; Weinberger et al., 2007).

Asterhan and colleagues conjectured that differences in impact between human and computer provided guidance might be explained by the communication format: In the CSCL scripting approach, implicit guidance is embedded in the computer software, whereas human explicit instruction may be considered as too intrusive in an ongoing discussion. Figure 31.2 shows exemplary interventions of a teacher (in light-grey shapes) in a discussion from Asterhan, Schwarz and Gil's 2012 study. The students are constantly aware of the presence of the teacher as the teacher notes appear throughout the discourse. In contrast, the mode of group communication in the studies undertaken by Weinberger and his colleagues was a-synchronous. In the high pace of an *asynchronous* group discussion, guidance efforts that focus on regulating the interaction may go unnoticed or be disregarded easily. Interestingly, Schwarz, Schur, Pensso and Tayer (2011) observed a teacher who adopted a

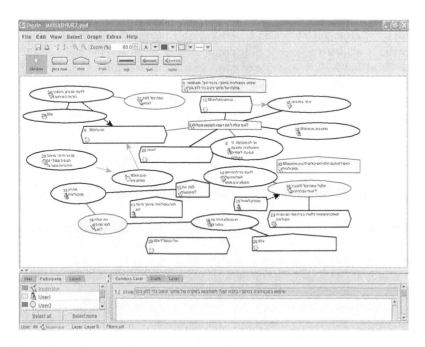

Figure 31.2 Persistent (light-grey) posts of interventions of a teacher scaffolding argumentation with Digalo

Source: Retrieved from Asterhan and Schwarz (2010).

combination of epistemic and interactional approaches in prompting his students to interact. The students learned a unit on the day-night cycle. The teaching was blended and included activities in various settings (individual, small group, whole class), some of them scaffolded by the teacher through the environment *Digalo* (Figure 31.2). The results were impressive, as conceptual change on the concept of day-night cycle were evident in individual tasks following the blended activities (Schwarz, Schur, Pensso, & Tayer, 2011). This study stresses that experienced teachers were less efficient than research students who functioned as teachers in their scaffolding of argumentation.

Another fundamental finding is that e-scaffolding of argumentation is different from face-to-face scaffolding. Teachers are more involved and more direct in their guidance (Asterhan & Schwarz, 2010). An example of this augmented involvement is the teachers' adoption of Devil's advocate strategies in e-scaffolding of argumentation, which appears less in face-to-face scaffolding. In their short-term study of CSCL-based scaffolded argumentation mentioned before, Asterhan et al. (2012) uncovered gender differences in behaviors in the context of CSCL-based human scaffolds of small-group argumentation. Gender differences in favor of girls were found on both the argumentative and the collaborative dimension of the discussions. To our knowledge, similar differences in favor of girls have not been detected in the context of face-to-face scaffolding.

E-Scaffolding Argumentation in Multiple Groups

E-scaffolding of argumentation does not fit well into classroom contexts. Unless argumentation is asynchronous, teachers cannot attend multiple groups arguing together. In the study just mentioned Asterhan and colleagues undertook, two groups of 5 and 6 students entered into discussion, and two teachers scaffolded it, one for each. E-scaffolding argumentation in a classroom context is a particular case of *teacher orchestration*. The domain of teacher orchestration is new (Dillenbourg, Prieto, & Olsen, this volume). It relies on the design of dedicated technologies and on the development of Learning Analytics techniques. We exemplify here e-scaffolding of argumentation in the ARGUNAUT environment, which was developed to help teachers support multiple group argumentation (Schwarz & Asterhan, 2011). This support was called *moderation* rather than scaffolding, to stress the dilemma of the teacher who is supposed to care for the progression of all groups and attempts to identify the needs of the discussants on-line in order to comply with the standards of critical discussions (e.g., justifying claims when asked to, or raising a challenging counter-argument when disagreeing with an argument brought forward) and at the same time, being minimally intrusive. ARGUNAUT provides awareness tools to teachers that help them inspecting participation, to what extent participants refer to each other, (dis-)agree as well as contribute to the content. Figure 31.3 displays different examples of awareness tools: Link Use displays the distribution of supporting, opposing, and neutral links and Group Relations displays the frequency of links between discussants. ARGUNAUT additionally includes an awareness Display Tab (not shown in Figure 31.3), a representation of contributions vertically organized per discussant according to chronological order, including deletions or modifications. With these awareness tools, the teacher can surf among different groups e-arguing and decide whether and where to intervene. Figure 31.3 shows also the moderator interface with which the moderator can choose a group, or a sub-group for sending messages.

Schwarz and Asterhan (2011) showed that the moderator could operate diverse strategies of moderation using ARGUNAUT: including a recalcitrant participant in participation to the discussion (by using a personal channel), broadening the discussion space, or deepening it, all this in a class in which up to four groups of four students discussed issues in a synchronous channel. These findings were encouraging. However, even experienced teachers had difficulties using the ARGUNAUT system because browsing of many groups, writing and sending messages, and tracing the evolution of discussions following the dispatching of the messages, is extremely demanding, even with the help of awareness tools (see Bodemer, Janssen, & Schnaubert, this volume, for an overview on awareness tools).

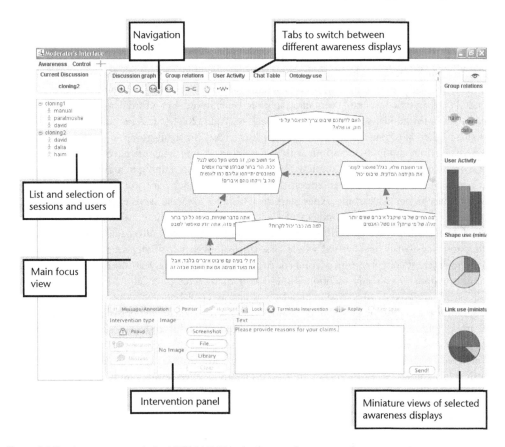

Figure 31.3 Awareness tools in ARGUNAUT help the moderator send messages to groups or individuals

Source: Retrieved from Schwarz and Asterhan (2011).

To help teachers in this demanding situation, researchers have begun providing tools for alerting them about events they might consider in their intervention through Learning Analytics techniques. According to a realistic scope suggested by McLaren et al., (2010), teachers will be able to work with up to 8 groups of 3–4 students. E-moderation will be partly triggered by alerts such as "X has not been active for 5 min," "Discussants seem not to challenge each other," or "The discussion is off topic". The use of Learning Analytics techniques in the domain of moderation of multiple discussions, and more generally in teachers' orchestration of collaborative activities is a promising research direction (see a discussion of the issue in Van Leeuwen, Van Wermeskerken, Erkens & Rummel, 2017).

Blended Settings in Classroom Contexts

The adaptive long-term implementation of argumentative activities in classrooms was missing in the beginning of research in computer-supported argumentation for learning. A new path has recently been initiated although, as we just saw, the on-line moderation of synchronous discussions in parallel is still hardly practical in classrooms. For this reason and for other pedagogical reasons, educationalists have suggested that CSCL-based instruction should not be exclusive: it should blend various settings, such as guided and unguided, online as well as offline discussions among small groups, individual argumentative writing, or teacher-led whole-class discussions. In such a context, e-discussions typically turn to be artifacts (maps, texts) to be used as resources for further activities. Studies that follow long-term blended settings in which computer-supported argumentation is central, are still rare. We report here on two such studies. Kuhn, Goh, Iordanou, & Shaenfield (2008) who had shown that individual argumentative skills develop in long-term programs based on electronic communication, designed an extended (3-year, twice weekly) curriculum in philosophy in order to afford dense practice in dialogic argumentation for middle-school students (Crowell & Kuhn, 2014). Students collaborated with peers who shared their positions on a series of social issues, both in small-group argument-building and reflective activities and in e-discussions with a succession of opposing-side peers. Annual assessments of individual argumentative skills on new topics showed students gaining in argumentative discourse skills across all 3 years of the intervention. In a recent study, Schwarz and Shahar (2017) helped a history teacher design a yearlong experiment in which his students were enculturated to historical reasoning through the extensive implementation of various argumentative activities. Instruction was blended as unguided e-discussions (generally around historical texts) alternated with argumentative writing and face-to-face teacher-led whole-class discussions in which all groups compared and discussed the different conclusions they reached. At the beginning of the year, the teacher often scripted the collaboration. We already mentioned the coercive character of collaborative scripts. However, collaborative scripts were integrated in blended settings, and collaborative scripts were presented *before* collaboration as games to be played in the framework of a *didactical contract* between the instructor and the learners, and reflected on *after* collaboration. The teacher led his students to impressive outcomes: high-level text-based critical discussions or evaluation of sources through a CSCL argumentation tool, and high-quality argumentative essays. These two studies function as existence proofs that instruction intensively based on CSCL argumentation can be effective, when settings are blended.

CSCL Tools, E-Argumentation, Dialogism, and Deliberative Democracy

One of the most salient weaknesses of research on computer-supported argumentation for learning is that it did not consider the changes in learners as a social entity. This is surprising since the constitution of a community of learners was one of the underpinnings of Brown's as well as Scardamalia and Bereiter's theories of knowledge construction and knowledge building (1991). The constitution of a community of learners necessitates time, and as seen before, for different reasons, the long-term implementation of CSCL-based argumentative activities in classrooms was not on the agenda of CSCL researchers. Effectiveness of e-argumentation with technological tools was exclusively understood from a rather short-term perspective on knowledge construction. However, several researchers have begun departing from this perspective. Wegerif (2007) elaborated a theoretical position according to which CSCL tools are aids for promoting dialogic teaching. By 'dialogic teaching' he meant a set of desirable practices identified by Alexander (2008), according to which when engaged in dialogue, students are attuned to each other; they listen with care and encourage the participation and the sharing of ideas of the others. They build on each other ideas, and while respecting the perspective of the minority, they strive for mutual understanding and conclusions. Like Wegerif, also other scientists saw in CSCL tools instruments with the potential to foster such types of behaviors (e.g. Mercer, Warwick, Kershner, & Staarman 2010; Pifarré & Staarman 2011).

There is hardly any research on the appropriation of such talk behaviors. An exception is a study undertaken by Slakmon and Schwarz (2014, 2017) on a yearlong philosophy course at the junior-high school level in which CSCL-based argumentation was blended with other dialectical activities. They adopted qualitative methods to show that CSCL affordances play a crucial role in establishing desired norms of talk, among them the suspension of immediate response, or a time to think things through. Major classroom identities such as 'weak', 'non-achiever', or 'disengaged' lost their imposing hold on the conversation unfolded in the virtual space. For example, as the teacher faded out from the virtual space, prolonged silence was no longer interpreted as a sign of disengagement or of weakness (like for silence in the classroom, where the teacher owns the floor). Students' proprietorship over the space was achieved through the reactions to their silence and non-participation, and later resulted in acts of high engagement. As the presence of the teacher faded from the virtual space, the appropriated discursive norms remained; the discussions continued to develop, without any reference to the students' status. Peer pressure encouraged students to join the collaborative effort without teacher intervention. Slakmon and Schwarz (2017) recently suggested a model on the role of CSCL in political education – on learning to live together, that is on how one realizes oneself as private and moves from this conception toward the togetherness of a public. This political becoming could be observed by tracing the development of spatial practices – how students exploited the public (and material) space that the CSCL tool provided. Slakmon and Schwarz showed that tracing spatial changes enables to scrutinize the freedom to act and exercise power over others, something crucial for the development of political agency. The spatial changes originated from the problem of density: as e-discussions develop, their representation became difficult to follow. Slakmon and Schwarz showed that this constraint led discussants to distinguish between private and public spheres. Figure 31.4 shows how spatial practices developed from random posting in which discussants located their posts randomly and shouted by putting magnified messages at the middle of the screen to a partition in which participants created a personal and a public space for their discussions.

Slakmon and Schwarz showed that the changes in spatial practices were accompanied by more utterances in discussions. In the partitioned discussions, the practice of referring to past contributions quadrupled. Gradually, as spatial practices developed, the students turned the mound of past contributions into a reservoir of collective memory. Past contributions became *an archive* from which selected contributions were chosen and woven back into the present moment of conversation. According to Slakmon and Schwarz, this phenomenon is crucial to citizenship as participants engage past actions in the ongoing issues of the present moment. They show that CSCL tools can boost the emergence of public and private spheres, and by such have a societal political dimension.

The philosopher Jürgen Habermas (1985) has pointed out the importance of argumentation for the rationalization of society towards the constitution of a *deliberative democracy*. New studies should investigate empirically whether CSCL tools could have a substantial democratic function. We admit here that this scope is audacious, but a research program targeted at the communal and societal dimensions of the iterated use of CSCL tools in argumentative activities in schools is an exciting and relatively untapped field.

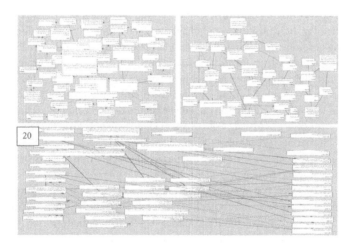

Figure 31.4 Evolution of spatial practices from random posting (accompanied with squashing and shouting) to partition, in which private and public spheres are delineated

Source: Slakmon and Schwarz (2016).

Baruch B. Schwarz

Issues That Need to Be Addressed in the Implementation of CSCL-Based E-Argumentation in Educational Contexts

We have described four current research paths in computer-supported argumentation for learning. We believe that in each of them, research will thrive. In this last section, we focus on general issues that urgently need to be addressed. A first general issue belongs to the psychological realm. Research in computer-supported argumentation for learning is often design-driven and characterized by cycles of system development and short-term implementations to test its effectiveness. There is little follow-up research that examines student argumentation in novel settings without the software support (but see Wecker & Fischer, 2011, and Iordanou, 2013, for exceptions). Moreover, as pointed out by Fischer et al. (2013), real-time process guidance should be carefully calibrated with existing *internal scripts* and existing skill development. Externally imposed collaboration support tools can be counter-effective when they inhibit the learner's autonomous application of skills that are already available and productive (Dillenbourg, 2002; Fischer et al., 2013). We are still far from the adaptive long-term implementation of CSCL-based argumentative activities in a classroom context. We should mention, however, a recent theoretical advancement in which Tchounikine (2016) opens an alternative perspective to Fischer's Script Theory of Guidance. He adopts a socio-cultural approach to consider scripts as artifacts learners interact with when engaging in a collaborative situation structured by a CSCL script. What learners consider is not the script, but their appropriation of the script – a cognitive process playing a role in both the recognition and the conceptualization of the task to be achieved. This appropriation does not depend on the script only. Different external aspects, like, e.g., institutional, domain, motivational, may influence the appropriation. This practical encounter between the psychological and the pedagogical is, in our view, crucial.

An additional issue is the fact that research on computer-supported argumentation for learning is disconnected from research in face-to-face argumentation for learning to a large extent. This was understandable, in the first steps of research on computer-supported argumentation of learning, since researchers were interested primarily in designing suitable tools. However, now design achievements in that domain are quite impressive, and it is possible to blend face-to-face and CSCL-based argumentation in classrooms. An example of substantial advancement in face-to-face argumentation and learning that might impinge on blended instruction is the research program that Richard Anderson has conducted for around 20 years around a setting he called *collective reasoning*, an open-format, peer-led approach to discussion intended to improve the quality of classroom talk, as small groups read stories and discuss controversial issues. Children take individual positions on the issue, present reasons and evidence for their positions, and challenge each other when they disagree. The teacher sits outside, offering coaching only when necessary. Face-to-face consecutive small group discussions around texts led to multiple effects (quality of discussions, reasoning skills, engagement, motivation, etc.). We do not cite here the numerous publications on these effects, except for the celebrated *snowballing* of collective reasoning – the propagation of collective reasoning from small groups to the whole class (Anderson et al., 2001), or the effects of iterated scaffolding (including fading out) of unguided collective reasoning (Jadallah, Anderson, et al. 2011). This kind of setting has not yet been investigated with CSCL tools. We alluded to important characteristics of CSCL tools as compared to face-to-face settings: they keep the history of interactions, to which discussants or teachers can refer; they provide a material space in which students locate their interventions. Such characteristics may result in very different effects. It is then urgent to implement CSCL-based argumentative activities in prolonged programs in schools, and to explore various effects of their iterated implementation.

We have suggested the idea that intensive CSCL-based argumentation may prepare citizens to contribute to the constitution of a deliberative democracy. Researchers in face-to-face classroom talk already claimed that high-quality talk practices like Accountable Talk (Michaels, O'Connor & Resnick, 2007) or deliberative argumentation (Schwarz & Baker, 2016) could contribute to prepare citizens to the constitution of a deliberative democracy. Here also, as suggested before, the characteristics of CSCL tools may uncover much richer dimensions to argumentation if their introduction in schools will be more systematic and as components of long-term programs. A research program in this direction would considerably enlarge the scope of the CSCL community.

Further Readings

Asterhan, C. S. C., & Schwarz, B. B. (2016). Argumentation for learning: Well-trodden paths and unexplored territories. *Educational Psychologist, 51*(2), 164–187.
This paper provides a state-of-the-art overview of research on argumentation for learning as well as new directions for investigation in this domain.

Mercier, H., & Sperber, D. (2017). *The enigma of reason.* Cambridge, MA: Harvard University Press.
In this provocative book, the authors wonder why reason has not evolved in other animals, and why humans frequently produce reasoned nonsense. They show that reason is not geared to solitary use, to arriving at better beliefs and decisions. Rather, reason helps justify beliefs and actions to others, convince them through argumentation, and evaluate the justifications and arguments that others address to us.

Osborne, J. (2010). Arguing to learn in science: The role of collaborative, critical discourse. *Science, 328*, 463–468.
This foundational paper shows that argumentation is at the heart of scientific activity and that it turns out to be central in science classrooms.

Schwarz, B. B., & Baker, M. J. (2016). *Dialogue, argumentation and education: History, theory and practice.* New York: Cambridge University Press.
This book includes a presentation of theories of argumentation adapted to education and a historical overview of talk practices in education. It depicts the ubiquity of argumentation in progressive pedagogies, exemplifies its productivity, reviews studies investigating conditions ensuring this productivity, and suggests that argumentative practices contribute to the constitution of a deliberative democracy.

NAPLeS Resources

Schwarz, B. B., *Argumentation and learning in CSCL* [Webinar]. In *NAPLeS video series*. Retrieved from http://isls-naples.psy.lmu.de/intro/all-webinars/schwarz_video/index.html

Schwarz, B. B., *15 minutes about argumentation and learning in CSCL* [Video file]. In *NAPLeS video series*. Retrieved from http://isls-naples.psy.lmu.de/video-resources/guided-tour/15-minutes-schwarz/index.html

Schwarz, B. B., *Interview about argumentation and learning in CSCL* [Video file]. In *NAPLeS video series*. Retrieved from http://isls-naples.psy.lmu.de/video-resources/interviews-ls/schwarz/index.html

References

Alexander, R. (2008). *Towards dialogic teaching: Rethinking classroom talk* (4th ed.). York, UK: Dialogos.

Anderson, R. C., Nguyen-Jahiel, K., McNurlen, B., Archodidou, A., Kim, S.-Y., Reznitskaya, A., & Gilbert, L. (2001). The snowball phenomenon: Spread of ways of talking and ways of thinking across groups of children. *Cognition and Instruction, 19*, 1–46.

Andriessen, J., Baker, M., & Suthers, D. (Eds.) (2003). Arguing to learn: Confronting cognitions in computer-supported collaborative learning environments. Dordrecht, Netherlands: Kluwer.

Asterhan, C. S. C. (2012). Facilitating classroom argumentation with computer technology. In R. Gillies (Ed.), *Pedagogies: New developments in the learning sciences* (pp. 105–129). New York: Nova Science.

Asterhan, C. S. C., & Schwarz, B. B. (2010). Online moderation of synchronous e-argumentation. *International Journal of Computer-Supported Collaborative Learning, 5*(3), 259–282.

Asterhan, C. S. C., Schwarz, B. B., & Gil, J. (2012). Guiding computer-mediated discussions in the classroom: Epistemic and interactional human guidance for small-group argumentation. *British Journal of Educational Psychology, 82*, 375–397.

Bodemer, D., Janssen, J., & Schnaubert, L. (2018) Group awareness tools for computer-supported collaborative learning. In F. Fischer, C. E. Hmelo-Silver, S. R. Goldman, & P. Reimann (Eds.), *International handbook of the learning sciences* (pp. 351–358). New York: Routledge.

Borge, M., Ong, Y. S. & Rosé, C. (2015). Activity design models to support the development of high quality collaborative processes in online settings. *The Computer-Supported Collaborative Learning Conference Proceedings* (pp. 427–434).

Brown, A. L. (1997). Transforming schools into communities of thinking and learning about serious matters. *American Psychologist, 52*(4), 399–413.

Chan, C., & van Aalst, J. (2018). Knowledge building: Theory, design, and analysis. In F. Fischer, C. E. Hmelo-Silver, S. R. Goldman, & P. Reimann (Eds.), *International handbook of the learning sciences* (pp. 295–307). New York: Routledge.

Crowell, A., & Kuhn, D. (2014). Developing dialogic argumentation skills: A three-year intervention study. *Journal of Cognition and Development, 15*, 363–381.

de Vries, E., Lund, K., & Baker, M. (2002). Computer-mediated epistemic dialogue: Explanation and argumentation as vehicles for understanding scientific notions. *Journal of the Learning Sciences, 11*, 63–103.

Dillenbourg, P. (2002). Over-scripting CSCL: The risks of blending collaborative learning with instructional design. In P. A. Kirschner (ed.), *Three worlds of CSCL: Can we support CSCL?* (pp. 61–91). Heerlen, Netherlands: Open Universiteit Nederland.

Dillenbourg, P., Prieto, L. P., & Olsen, J. K. (2018) Classroom orchestration. In F. Fischer, C. E. Hmelo-Silver, S. R. Goldman, & P. Reimann (Eds.), *International handbook of the learning sciences* (pp. 180–190). New York: Routledge.

Driver, R., Newton, P. & Osborne, J. (2000). Establishing the norms of scientific argumentation in classrooms. *Science Education, 84*, 287–312.

Fischer, F., Kollar, I., Stegmann, K., & Wecker, C. (2013). Toward a script theory of guidance in computer-supported collaborative learning. *Educational Psychologist, 48*(1), 56–66.

Habermas, J. (1985). *The theory of communicative action* (Vol. 1), Reason and the rationalization of society. Boston: Beacon Press.

Iordanou, K. (2013). Developing face-to-face argumentation skills: Does arguing on the computer help? *Journal of Cognition and Development, 14*(2), 292–320.

Jadallah, M., Anderson, R.C., Nguyen-Jahiel, K., Miller, M., Kim, I.-H., Kuo, L.-J., et al. (2011). Influence of a teacher's scaffolding moves during child-led small-group discussions. *American Educational Research Journal, 48*, 194–230.

Kollar, I., Wecker, C., & Fischer, F. (2018). Scaffolding and scripting (computer-supported) collaborative learning. In F. Fischer, C. E. Hmelo-Silver, S. R. Goldman, & P. Reimann (Eds.), *International handbook of the learning sciences* (pp. 340–350). New York: Routledge.

Kuhn, D., Goh, W., Iordanou, K., & Shaenfield, D. (2008). Arguing on the computer: A microgenetic study of developing argument skills in a computer-supported environment. *Child Development, 79*, 1310–1328. doi:10.1111/j.1467-8624.2008.01190.x

McLaren, B. M., Scheuer, O., & Mikšátko, J. (2010). Supporting collaborative learning and e-discussions using artificial intelligence techniques. *International Journal of Artificial Intelligence in Education, 20*, 1–46.

Mercer, N. (1995). *The Guided Construction of Knowledge: talk amongst teachers and Learners*. Clevedon: Multilingual Matters.

Mercer, N., Warwick, P., Kershner, R., & Staarman, J. K. (2010). Can the interactive whiteboard help to provide "dialogic space" for children's collaborative activity? *Language and Education, 24*(5), 367–384.

Michaels, S., O'Connor, C., & Resnick, L. (2007). Deliberative discourse idealized and realized: Accountable talk in the classroom and in civic life. *Studies in Philosophy of Education, 27*, 283–297.

Noroozi, O., Weinberger, A., Biemans, H. J. A., Mulder, M. & Chizari, M. (2012). Argumentation-based computer supported collaborative learning (ABCSCL): A synthesis of 15 years of research. *Educational Research Review, 7*(2), 79–106.

Noroozi, O., Weinberger, A., Biemans, H. J. A., Mulder, M., & Chizari, M. (2013). Facilitating argumentative knowledge construction through a transactive discussion script in CSCL. *Computers & Education, 61*, 59–76.

Pifarré, M., & Staarman, J. K. (2011). Wiki-supported collaborative learning in primary education: How a dialogic space is created for thinking together. *International Journal of Computer-Supported Collaborative Learning, 6*(2), 187–205.

Scardamalia, M., & Bereiter, C. (1991). Higher levels of agency for children in knowledge building: A challenge for the design of new knowledge media. *Journal of the Learning Sciences, 1*, 37–68.

Scheuer, O., Loll, F., Pinkwart, N., & McLaren, B. M. (2010). Computer-supported argumentation: a review of the state-of-the-art. *International Journal of Computer-Supported Collaborative Learning, 5*(1), 43–102.

Schwarz, B. B., & Asterhan, C. S. C. (2011). E-moderation of synchronous discussions in educational settings: A nascent practice. *Journal of the Learning Sciences, 20*(3), 395–442.

Schwarz, B. B., & Baker, M. J. (2016). *Dialogue, argumentation and education: History, theory and practice*. New York: Cambridge University Press.

Schwarz, B. B., & De Groot, R. (2007). Argumentation in a changing world. *International Journal of Computer-Supported Collaborative Learning, 2*(2–3), 297–313.

Schwarz, B. B., Schur, Y., Pensso, H., & Tayer, N. (2011). Perspective taking and argumentation for learning the day/night cycle. *International Journal of Computer Supported Collaborative Learning, 6*(1), 113–138.

Schwarz, B. B., & Shahar, N. (2017). Combining the dialogic and the dialectic: Putting argumentation into practice for classroom talk. *Learning, Culture and Social Interaction, 12*, 113–132.

Slakmon, B., & Schwarz, B. B. (2014). Disengaged students and dialogic learning: The role of CSCL affordances. *International Journal of Computer-Supported Collaborative Learning, 9*(2), 157–183.

Slakmon, B., & Schwarz, B. B. (2017). "You will be a polis": Political (democratic?) education, public space and CSCL discussions. *Journal of the Learning Sciences, 26*(2), 184–225.

Suthers, D. D. (2003). Representational guidance for collaborative inquiry. In J. Andriessen, M. Baker, & D. Suthers (Eds.), *Arguing to learn: Confronting cognitions in computer-supported collaborative learning environments* (pp. 27–46). Dordrecht, Netherlands: Kluwer.

Suthers, D. D., & Weiner, A. (1995). Groupware for developing critical discussion skills. In J. L. Schnase & E. L. Cunnius (Eds.), *Proceedings of CSCL '95, the First International Conference on Computer Support for Collaborative Learning* (pp. 341–348). Mahwah, NJ: Lawrence Erlbaum/University of Minnesota Press.

Tchounikine, P. (2016). Designing for appropriation: A theoretical account. *Human Computer Interaction.* doi: 10.1080/07370024.2016.1203263

Toulmin, S. (1958). *The Uses of Argument.* New York: Cambridge University Press.

Van Amelsvoort, M., Andriessen, J., & Kanselaar, G. (2007). Representational tools in computer-supported collaborative argumentation-based learning: How dyads work with constructed and inspected argumentative diagrams. *Journal of the Learning Sciences, 16*, 485–521.

van Eemeren, F. H., Grootendorst, R., Henkenmans, F. S., Blair, J. A., Johnson, R. H., et al. (1996). *Fundamentals of argumentation theory: A handbook of historical background and contemporary developments.* Hillsdale, NJ: Lawrence Erlbaum.

Van Leeuwen, A., Van Wermeskerken, M., Erkens, G., & Rummel, N. (2017). Measuring teacher sense making strategies of learning analytics: a case study. *Learning: Research and Practice, 3*(1), 42–58. doi:10.1080/23735082.2017.1284252

Wecker, C., & Fischer, F. (2011). From guided to self-regulated performance of domain-general skills: The role of peer monitoring during the fading of instructional scripts. *Learning and Instruction, 21*(6), 746–756.

Wecker, C., & Fischer, F. (2014). Where is the evidence? A meta-analysis on the role of argumentation for the acquisition of domain-specific knowledge in computer-supported collaborative learning. *Computers & Education, 75*, 218–228.

Wegerif, R. (2007). *Dialogic, educational and technology: Expanding the space of learning.* New York: Springer-Verlag.

Wegerif, R. B., Mercer, N., & Dawes, L. (1999). From social interaction to individual reasoning: an empirical investigation of a possible socio-cultural model of cognitive development. *Learning and Instruction, 9*(6), 493–516.

Weinberger, A., Ertl, B., Fischer, F., & Mandl, H. (2005). Epistemic and social scripts in computer-supported collaborative learning. *Instructional Science, 33*(1), 1–30.

Weinberger, A., Stegmann, K., & Fischer, F. (2007). Knowledge convergence in collaborative learning: Concepts and assessment. *Learning and Instruction, 17*(4), 416–426.

32
Theoretical and Methodological Frameworks for Computer-Supported Collaborative Learning

Heisawn Jeong and Kylie Hartley

Introduction

The pursuit of understanding and promoting computer-supported collaborative learning (CSCL) has attracted researchers from a variety of fields with varying interests, theoretical perspectives, and preferred methodological practices. Some seek to understand cognitive mechanisms of collaborative learning, while others focus on motivational dynamics of small group work and/or socio-cultural influences on collaboration. Others are interested in designing pedagogical and technological supports to facilitate collaborative learning. The goal of this chapter is to make sense of the complexity of CSCL research with an emphasis on its theoretical and methodological practices. It is informed by a series of content analyses of CSCL empirical research as well as recent conceptual reviews of CSCL (Arnseth & Ludvigsen, 2006; Jeong & Hmelo-Silver, 2012, 2016; Jeong, Hmelo-Silver, & Yu, 2014; Kirschner & Erkens, 2013; Puntambekar, Erkens, & Hmelo-Silver, 2011). The chapter reflects on how the diversity of CSCL research has contributed to the advancement of the field. We examine theory–method alignment and argue for increased efforts for theoretical integration and continued theory building.

Key Challenges of CSCL Research

What does it mean to understand CSCL? First, it involves understanding principles of collaborative learning. Successful collaboration is more than working in groups or assigning a group task to complete. Ideally, learners need to establish shared goals, engage in collaborative knowledge construction or problem solving, and monitor each other's progress during collaborative learning (Jeong & Hmelo-Silver, 2016). They also need to cultivate effective social relationships among themselves while maintaining productive engagement with the task(s). In this process, learners need to pay attention not only to their own learning processes but also to their partners'. This involves allocating attention and resources to their partners and their shared task that may have otherwise been spent trying to meet their individual learning goal(s). This can distract learners and lead to impasses and conflicts that they might not have experienced if they worked alone, but the very process of trying to understand each other and establish shared understanding can drive learners toward cognitive advancement and intersubjective meaning-making (Stahl, 2006; Suthers, 2006). Effective supports for collaboration require a deep understanding of how these processes may work. In our pursuit of understanding collaborative learning, we need to note that collaboration occurs in diverse contexts and can take a number of different forms in CSCL. It may involve direct face-to-face interaction among small groups of learners, but also distributed interactions among a large group of learners in online environments. Proliferation of social networking sites (SNS) and massive open online courses (MOOCs) suggests learners are increasingly interacting in online environments outside traditional school contexts. These environments provide important spaces for learning in which learners participate in personally interesting projects and build social networks.

The second component to understanding CSCL is to be aware of the premise that collaborative knowledge construction and problem solving can be effectively assisted by technology. The array of technology recruited for this purpose is quite diverse, ranging from software applications or systems that support collaboration (e.g., email, discussion forums, information sharing) to hardware (e.g. mobile devices, shared displays, tabletops; Jeong & Hmelo-Silver, 2012). As researchers take advantage of advancements in technologies, it became clear that CSCL is more than developing or using technological tools for collaboration. CSCL presents a complex socio-technical problem. Effective CSCL requires technologies that support core processes of collaborative learning (Jeong & Hmelo-Silver, 2016) as well as pedagogical principles that organize and structure learners' activities for desired learning goals. Collaborative norms and cultures influence the formation of groups and communities, rules of engagement and accountability, and/or ways to resolve conflicts and disagreements further. In sum, the main research challenge in CSCL involves the design of appropriate technological and pedagogical interventions to support effective collaborative learning based on understanding of collaborative learning. Addressing this requires multifaceted approaches; no single theoretical or methodological framework can address all aspects of CSCL. We examine next how different theoretical and methodological approaches have been involved in deepening our understanding of CSCL.

Theoretical Practices of CSCL

Learning sciences research has been guided by major learning theories. These theories explain what it means to learn, establish foundations for the core mechanisms of knowledge construction and problem solving (e.g., information acquisition, meaning-making activities), and attempt to better understand the role and impact of tools and technology in learning. Theoretical frameworks have a great significance on CSCL research. They guide how to assess and measure learning outcomes and provide rationales and justifications for certain pedagogical strategies and design of technologies. A diverse range of learning theories has guided CSCL research practices. Details of the major CSCL theoretical frameworks can be found throughout this handbook. The most influential learning theories for CSCL are constructivism, socio-cultural theories, and social psychology theories (Jeong et al., 2014; Danish & Gresalfi, this volume). Also present to a varying degree are information processing or cognitive theories, socio-cognitive communication, and motivation theories. In addition to these theories of learning and collaboration, there are pedagogical theories of CSCL that propose instructional approaches to facilitate collaborative learning such as scaffolding (Tabak & Kyza, this volume), scripts (Kollar, Wecker, & Fischer, this volume), or knowledge building (Chan & van Aalst, this volume; Slotta, Quintana, & Moher, this volume). These pedagogical design theories are guided by theories of learning, but further elaborate how to orchestrate computer-supported collaborative activities in classrooms (e.g., Dillenbourg, Prieto, & Olsen, this volume), overcome issues that arise in small group work, and/or increase the potentials and ideals of collaboration with the help of technology such as online-mentoring, participation in citizen science projects. Many of these theories are also inherently intermingled with design theories and addresses how tools can mediate collaborative learning processes and associated pedagogical interventions (Jeong & Hmelo-Silver, 2016).

In empirical research, the process of validating and testing theories encourages theory modification, elaboration, and sometimes the development of a new, more comprehensive theory (Shavelson & Towne, 2002). As a key methodology for the learning sciences, design-based research has emphasizes theory-based design (Puntambekar, this volume), and yet how theories are used in CSCL research vary a great deal as a recent synthesis effort has demonstrated (Jeong et al., 2014). CSCL researchers do not always emphasize theory-building and testing. Theories are often cited to align with particular research traditions and/or to justify the choice of research problems or design decisions. Many learning scientists consider themselves working in the valuable pursuit of "context of discovery" which is about formulating hypotheses in the first place rather than the "context of justification" that follows hypothetico-deductive model of justification (Hoyningen-Huene, 1987). Research may be theoretically motivated, but may not necessarily includes hypothesis testing or model building. Instead they focus on insights that emerge from an in-depth understanding of the specific learning contexts.

It is not always easy to figure out what it means to build and test theories. Different theoretical approaches often aim to explain different phenomena or problems and may complement rather than contradict each other. Constructivist theories may seek to explain cognitive processes and outcomes involved in collaborative learning, while motivation theories may seek to explain the role of collaboration on student motivation (Jones & Issroff, 2005). Theoretical approaches may also differ in their scopes or levels of explanations. Some may be called meso- or micro-level theories as they seek to explain a specific phenomenon such as why people perform better or worse in groups or to explore different factors that might affect group functioning such as gender or group composition. Some theories focus on tacit assumptions about the very nature of learning and tend to make general statements about learning and act more like a macro-level theory or paradigm. These general theories need to be formulated into testable claims about specific phenomena or variables, but figuring out how different theories are aligned with and related to each other is not always clear-cut, either. These factors make it challenging to make sense of how CSCL research has contributed to the theoretical advancement of the field.

Different theoretical approaches to CSCL allow us to examine CSCL from a wider perspective, and yet they need to be integrated to provide a coherent body of explanations for CSCL. They may reveal phenomena that may have been previously hidden and can serve as opportunities for reconciling differences that exist across different theoretical frameworks. Tensions and conflicts that may exists among different theories need to be identified and resolved (Arnseth & Ludvigsen, 2006). More active efforts to integrate different theoretical explanations and reconcile differences will contribute to the development of more comprehensive and sophisticated understanding of CSCL.

Methodological Practices of CSCL

CSCL is an interdisciplinary field in which a diverse range of research methods is used to study technology-mediated collaborative learning. Research methods are often distinguished in terms of their overall approach to research such as quantitative methods or qualitative methods. In order to understand how methodological practices are related to theoretical approaches, we need to examine methodological practices at a finer level. We focus here on differentiating methodological practices on research designs, settings, data sources, and analyses. These subcomponents deal with the major differences between qualitative and quantitative methods as well as major challenges of CSCL research. The goal of this section is to provide an overview of the methodological practices of CSCL based on a content meta-analysis of 400 CSCL empirical publications published between 2005 and 2009 (Jeong et al., 2014). Readers interested in the details of specific methodological traditions and techniques in CSCL are encouraged to refer to Section 3 of this handbook.

Research Design

Research design focuses on the objective and plan for data collection. Analysis of recent CSCL methodological practices indicates that the majority of CSCL research has descriptive goals (Jeong et al., 2014). Considering CSCL is a relatively new area of research, this is not surprising. Technologies are continually being developed. It makes sense to explore their affordances for collaborative learning as well as how learners may appropriate the technology. Descriptive studies may help to describe the phenomena qualitatively. For example, they may describe, among other things, the challenges involved in transitioning the innovations led by researchers to the hand of the teachers in the classroom (Smith, Underwood, Fitzpatrick, & Luckin, 2009) or students' personal experiences of the interaction on distance learning courses by interviewing them (Ferguson, 2010). Alternatively, descriptive studies may aim to describe a phenomenon quantitatively. They may seek to capture a pattern of relationships among variables such as the relationship between student academic self-concept and social presence and achievements in both face-to-face and online classrooms (Zhan & Mei, 2013) or the relationship between social loafing and group cohesion in online communities (Shiue, Chiu, & Chang, 2010).

Experimental designs aim to determine causal relationships among variables. An experimental design is usually adopted when there is a clear set of research questions and hypotheses. Roughly one-third of the studies used some form of experimental or quasi-experimental design (Jeong et al., 2014). CSCL experimental studies, for example, examine the effectiveness of the CSCL interventions such as anchored discussion system that link text annotations to related discussions (Eryilmaz et al., 2014) or specific functionality of the environment such as self-explanation question prompts embedded in an educational game (Adams & Clark, 2014). Studies also examine factors related to collaborative learning such as monitoring of others over self (Okita, 2014) or cultural backgrounds of dyads (Shi, Frederiksen, & Muis, 2013). Individual experiments may examine any of these variables, but only a narrow subset of the variables is tested in a given experiment. It can be difficult to understand how they all fit together and are related to the rest of the research. In experimentation, random assignments as well as careful control of extraneous variables in laboratory settings are means of achieving tight experimental control. The need for control, however, often leads to the creation of artificial or unrealistic learning situations and tasks, reducing the ecological validity of the study.

Design-based research (DBR) emerged from a need to engage in more principled and theoretically driven design works in classrooms and other real-world situations (Puntambekar, this volume). Shifting from traditional classrooms to knowledge building classrooms, for example, requires overcoming many challenges such as redesigning activity structures (Oshima et al., 2006) and conceptualizing Knowledge Forum as a space for epistemic games (Bielaczyc & Ow, 2014). Often this effort requires multiple rounds of redesign and adaptation. Data collection is made at each round with the goal of examining increasingly more sophisticated models of interventions and theories. DBR studies in CSCL often adopted quasi-experiments, pre-post comparisons, and case study approaches (Jeong et al., 2014), but assessing whether a particular design intervention may or may not improve students' learning outcomes at a particular point is just one piece of the overall research design. What is more important is to use the outcomes to continually build and refine the instructional practices and theories. It takes time for a new research paradigm to become a mainstream practice, especially when it demands lots of resources and efforts, as in the case of DBR (Jeong et al., 2014). Continued efforts are needed to fine-tune design-based research strategies and execution so that they can contribute to theory building as well as the creation of contextualized knowledge that can inform practices.

Settings

Much of CSCL research was conducted in classrooms rather than laboratories or other settings (e.g., online community, professional training; Jeong et al., 2014). We consider classrooms or other real-world practice settings as being the ultimate testing ground of CSCL effectiveness, as it is in those settings that the eventual success of CSCL would be evaluated. CSCL should continue to focus on them, but classrooms pose a number of challenges for researchers. Receiving permission from stakeholders, orchestrating data collection, and creating a tightly controlled environment are only some of the challenges. Researchers often do not have design options other than resorting to quasi-experimentation in classrooms. Lately, a rapid proliferation of online environments pose new methodological opportunities and challenges. For example, students are now interacting and collaborating across the boundaries of brick and mortar schools, cultures (Chia & Pritchard, 2014), or through various online communities (Shiue, Chiu, & Chang, 2010). These settings provide opportunities to study learner behaviors more comprehensively outside formal classrooms. Data collection often requires less planning and effort from the researchers, as the data is generated by the learners in the process of interacting with peers and engaging in the creation of artifacts such as Scratch programs or Wikipedia pages. But, cleaning and making sense of the data pose a bigger challenge in these settings. Diverse data sources need to be combined to achieve an integrated understanding of students' activities. Learning analytics is specifically devoted to addressing challenges and opportunities that arise from researching these settings (Rosé, this volume).

Data Sources and Analysis Methods

CSCL has considerably expanded the kinds of data sources used to assess the outcomes and processes of collaborative learning. Examples of data sources include video recordings of the interaction, synchronous and asynchronous messages, log data, and various digital artifacts created during CSCL, in addition to measures of students' outcomes such as multiple-choice tests and course grades. These diverse data sources can provide rich information about CSCL, but we need to be mindful of the challenges they present. Text messages and video recordings of the interaction, for example, do not easily lend itself to quantification except for surface characteristics such as word count. Researchers often resort to coding as a form of quantifying data (Vogel & Weinberger, this volume). For example, email and discussion board messages are coded for coherence, grounding moves, or self-regulation related dialogic actions, the results of which are often further subject to statistical testing (Shi et al., 2013; van der Meij, de Vries, Boersma, Pieters, & Wegerif, 2005). Many of them rely on applications of sophisticated statistical techniques such as structural equation modeling or multi-level analysis (de Wever & van Keer, this volume). Groups of techniques such as social network analysis and learning analytics examine log data and other forms of online traces of learner activities (Cho, Gay, Davidson, & Ingraffea, 2007; Fields, Kafai, & Giang, 2016). Automated coding and data extraction is making progress. Approximately three-fourths of CSCL studies analyzed data in a quantitative manner either in the form of coding or by relying on quantifiable measures of learning (Jeong et al., 2014). Quantification and use of statistical testing help researchers discern the complex relationships among variables and evaluate whether a given finding is generalizable beyond the study sample. Researchers need to be mindful of the fact, however, that quantified data often failed to reveal the full complexity of the phenomena and statistical significance does not necessarily translate into theoretical or practical significance. Efforts are needed to obtain evidence that can inform the theories and practices in meaningful ways.

Quantitative traditions are strong in CSCL research, but about half of the CSCL research also relied on some form of qualitative analysis (Jeong et al., 2014). For example, Smith et al. (2009) examined challenges involved in transitioning the innovation led by researchers to the hand of teachers in the classroom. The analysis involved looking for incidents that signal troubles getting the technology to work in the data, which included, but was not limited to, email exchanges, instant messages, logs, meeting notes, and video recordings. These incidents were then categorized and clustered to form higher-level categories, which in the end resulted in six major categories needed in setting up and running these projects in classrooms. There is a variety of qualitative traditions to choose from depending on the objectives and focus of the analyses (Green & Bridges, this volume), although many are done in a loosely defined manner without reference to specific analytic traditions (Jeong et al., 2014).

Quantitative and qualitative methods are born out of different research traditions, but they are often combined in a mixed method study (Dingyloudi & Strijbos, this volume). In So and Brush (2008), for example, regression analysis using questionnaire data showed that there is a positive relationship between students' perceived level of collaborative learning and satisfaction in blended environment. This analysis was supplemented by a qualitative analysis of student interviews, which showed that course structures, emotional supports, and communication medium additionally influenced students' experiences with the course. CSCL embraces mixed method in about one-third of the studies (Jeong et al., 2014). In its current practice, most of the mixed method approaches appear to be carried out to explore the complexity of the problem and/or seek converging evidence from multiple analysis techniques. In a more specific form of mixed method approach called the productive multivocality approach (Suthers, Lund, Rose, Teplovs, & Law, 2013, Lund & Suthers, this volume), researchers from different methodological traditions come together and analyze the same data corpus. It is done with the intention of making hidden assumptions and tensions between different approaches explicit. Such tensions can serve as an opportunity to deepen each other's understandings of different traditions as well as to resolve the differences in a productive manner. Continued efforts are needed to explore tensions in CSCL research and find ways for different methodological traditions to co-exist in a productive manner.

Theory–Method Alignments

Although it may not be obvious in individual studies, no research method is independent from the research traditions or theoretical frameworks in which the research is embedded. An overview of recent empirical CSCL investigations showed two theoretical frameworks—constructivism and socio-cultural theories—are often aligned with a specific set of methodological features (Jeong et al., 2014). These two theoretical frameworks, although they share a common interest in social interaction, have slightly different assumptions of what learning is, how it can be assessed, and the role of collaboration in learning. In the socio-cultural framework, thinking and learning is not possible without critical interaction. Meaning is constituted in the dialogue as participants respond to each other and to various contextual features of the interaction. The resulting knowledge is inseparable from the social and cultural contexts. Manipulating and assessing learning out of the contexts for experimental manipulation is meaningless in this perspective. Research studies categorized to be within the "socio-cultural classroom" cluster tended to support this theoretical framework and used descriptive designs that relied on data and analyses that could reveal the complexity of the problem beyond quantified summaries. In trying to understand the experiences of university distance education students, Ferguson (2010), for example, used open-ended surveys followed by extended in-depth asynchronous interviews with a subset of students. These data were analyzed inductively using grounded theory to identify themes important to students' experiences. These analyses revealed that students' reasons for online interactions often diverged from the intentions of the course designers.

In the constructivist framework, the focus is on learners' active construction of knowledge and social interaction is considered to be a way to facilitate active construction. The knowledge constructed during collaborative learning is assumed to become part of the learners' cognitive system in the form of more elaborate knowledge representations or problem solving skills, which can be assessed out of the learning context by asking learner to answer questions or solve problems. In Eryilmaz et al. (2014), for example, an anchored discussion system was developed to reduce the cost of interaction coordination, thereby increasing the resources for active construction. The effect of the system was tested by assessing individual students' domain-specific knowledge, and comparing their performance across two sections of the blended course that worked with the anchored or basic version of the discussion system. In general, research studies in the "constructivist classroom" cluster tend to use quasi-experimental design, questionnaire data, and/or use inferential statistics (Jeong et al., 2014).

The remaining two clusters showed distinctive methodological profiles, but were associated with a multitude of theoretical perspectives (e.g., social psychology, information processing theories) rather than one dominant perspective. Studies in the "eclectic descriptive" cluster adopted a number of different theoretical perspectives, thus called eclectic, ranging from social psychology theories emphasizing individual's self-perceptions and social presence (So & Brush, 2008; Zhan & Mei, 2013) to linguistic/communication theory that aimed to better understand the organizing structure of dialogues (van der Meij et al., 2005). These studies tended to collect questionnaire data in classrooms. The analysis techniques ranged from the analysis of survey responses about student academic self-concept and social presence (Zhan & Mei, 2013), coding of email messages along its rhetorical and semantic dimensions (van der Meij et al., 2005), to mixed methods analysis of survey and interview data (So & Brush, 2008).

The "eclectic experimental studies" cluster consisted of randomized experimental studies in laboratory settings. Studies in this cluster were again associated with a multitude of theoretical approaches such as a cognitive approach trying to develop and test an intelligent tutoring system promoting conceptual change (Dzikovska, Steinhauser, Farrow, Moore, & Campbell, 2014) or self-regulated learning theories being applied to understand the effects of the cultural composition of dyads (Shi et al., 2013). Studies in this cluster tended to be in controlled laboratory settings with participants randomly assigned to different treatment conditions. Analyses may have involved comparing experimental conditions on a number of measures, such as rate and frequency of information sharing (Cress, Barquero, Schwan, & Hesse, 2007), time spent on the task, test scores (Rummel & Spada, 2005), self-report questionnaire about tutoring quality (Dzikovska et al., 2014), or coding of the dialogue for self-regulated actions (Shi et al., 2013). Studies in this cluster relied heavily on inferential statistics to determine the significance and thus generality of the results beyond the samples of students participated in the study.

In sum, theory has wide implications for research practice. Theories can be about particular research topics and mechanisms, but they are also about basic assumptions of what learning is and how to study it. They direct attention to where and how to look when trying to better understand specific components of CSCL. Theoretical differences do not necessarily mean methodological differences. Some approaches diverge on their theoretical explanations, but may subscribe to a common methodology. Cognitive and social psychology theories of collaborative learning may provide different explanations about the key mechanisms of collaborative learning (e.g., motivation versus more active construction), but may use the same methods such as randomized experiments and statistical testing. On the other hand, socio-cultural and cognitive theories diverge not only on how to conceptualize learning and collaboration, but also how to study them. When theories differ both in their methods and explanations, as is often observed in interdisciplinary fields such as CSCL, cross-perspective communication becomes extremely challenging and yet important. Research methods come with certain assumptions and rationale for what counts as valid evidence. Theories with different methodological approaches are likely to approach research and data collection from different perspectives and use different criteria for data collection and evaluation. The pattern of alignments in theory–method clusters suggests there might be such a divide in CSCL. Although many studies attempt to bridge theoretical and methodological differences, we need to be vigilant about the existence of such divide and work toward reconciling the differences between theoretical and methodological approaches.

Taking Things Together

CSCL research is richly guided by theoretical frameworks, but this does not necessarily mean theoretical advancement. The nature of theories that guide much of CSCL research tends to be at the macro level. They are often used to justify selections of the problems and/or design rather than to generate hypothesis for exact empirical validations. Much of CSCL research seeks to discover rather than build and test competing theories. It is also unclear whether CSCL theories have sufficiently addressed areas of research important to CSCL. In spite of the fact that much of CSCL research has been technologically driven (Roschelle, 2013), for example, not much theoretical work has been done with respect to the role of technology mediation in supporting core mechanisms of collaborative learning, such as intersubjective meaning-making and share/co-regulation. Many researchers agree that students need to learn "with" technology, not "from" technology, but what it means to learn "with" technology is still poorly articulated and understood. In addition, although much effort has been directed at understanding cognitive mechanisms of collaborative learning, but we also need to better understand the affective dimensions of CSCL. Empirical as well as theoretical work is needed to understand how cognitive and affective dimensions of CSCL interact (Kirschner & Erkens, 2013). Theory building efforts are especially important in the current age of rapid knowledge production and big data. It is important for both individual researchers and the field as a whole to understand the composite shape of the knowledge generated by CSCL research and reflect on the connections between individual research and the theory to which the research will contribute.

In our efforts to build more sophisticated CSCL theories, we need to be mindful of the theoretical and methodological diversity in CSCL research. This diversity has enriched our understanding of collaborative learning and tool mediation, but the tensions between different approaches have not been fully understood and/or reconciled. Tensions are natural in interdisciplinary fields such as CSCL and can serve as productive impetus for the development of new ideas. We need to better understand the sources of these tensions and take advantage of them and work toward theoretical integration. CSCL theories differ in terms of how they conceptualize learning (e.g., acquisition versus participation), goals/units of collaborative learning (e.g., gains in individual student knowledge versus shared knowledge), and the role of the tools (e.g., external tools or partner in cognition). These differences need to be translated into testable and predictive theories and/or hypotheses with supportive evidence. These efforts are likely to have a limit, as some of them arise from irreconcilable epistemological differences. We may need to settle on productive coexistence rather than seamless integration. Ideally, the composite picture that emerges from integrating different perspectives can complement and strengthen the individual perspectives. This in turn can lead to the construction of a new, more comprehensive view on learning, collaboration, tool-mediation, and CSCL pedagogies, with the right balance between diversity and coherence. CSCL is uniquely positioned to address these challenges and can serve as a testbed for productive theoretical integration.

Further Readings

Arnseth, H. C., & Ludvigsen, S. (2006). Approaching institutional contexts: systemic versus dialogic research in CSCL. *International Journal of Computer-Supported Collaborative Learning, 1*(2), 167–185.
This paper contrasts two analytical approaches, systemic and dialogic approach in CSCL research. Systemic approach attempts to generate models of how specific features of the system works, whereas dialogic approach focuses on how meaning is constituted in social practice. Authors argue that differences in these analytical practices have consequences for the generation and assessment of findings.

Hmelo-Silver, C. E., Chinn, C., Chan, C. K. K., & O'Donnell, A. (Eds.). (2013). *International handbook of collaborative learning*. London: Taylor & Francis.
This handbook provides an overview of collaborative learning including technology supports for collaborative learning. It also provides an overview on study methods of collaborative learning and pedagogical approaches to collaborative learning.

Jeong, H., Hmelo-Silver, C. E., & Yu, Y. (2014). An examination of CSCL methodological practices and the influence of theoretical frameworks 2005–2009. *International Journal of Computer-Supported Collaborative Learning, 9*(3), 305–334.
This article reports on a content meta-analysis of CSCL research methods and theoretical frameworks. Four hundred CSCL empirical research papers published between 2005 and 2009 have been coded in terms of research design, settings, data sources, analysis methods, as well as theoretical frameworks. The analysis provide a bird-eye view of CSCL research practices. Cluster analysis also revealed four distinct theory-method clusters.

Kirschner, P. A., & Erkens, G. (2013). Toward a framework for CSCL research. *Educational Psychologist, 48*(1), 1–8.
This paper present a framework for CSCL research that consists of three dimensions—that is, levels of learning (i.e., cognitive, social, and motivational), unit of learning (i.e., individual, group/team, and community), and

pedagogical measures of learning (i.e., interactive, representational, and guiding). The framework is used to assess areas of CSCL research that require further theoretical research.

Suthers, D. D., Lund, K., Rosé, C. P., Teplovs, C., & Law, N. (Eds.). (2013). *Productive multivocality in the analysis of group interaction.* New York: Springer.

This book describes a multi-year project in which researchers from multiple methodological traditions make an effort to compare and contrast their analysis of the same data and to engage in dialogue with each other and consider how their different understandings can either complement or mutually elaborate on each other.

References

Adams, D. M., & Clark, D. B. (2014). Integrating self-explanation functionality into a complex game environment: Keeping gaming in motion. *Computers and Education, 73*, 149–159.

Arnseth, H. C., & Ludvigsen, S. (2006). Approaching institutional contexts: Systemic versus dialogic research in CSCL. *International Journal of Computer-Supported Collaborative Learning, 1*(2), 167–185.

Bielaczyc, K., & Ow, J. (2014). Multi-player epistemic games: Guiding the enactment of classroom knowledge-building communities. *International Journal of Computer-Supported Collaborative Learning, 9*(1), 33–62.

Chan, C., & van Aalst, J. (2018). Knowledge building: Theory, design, and analysis. In F. Fischer, C. E. Hmelo-Silver, S. R. Goldman, & P. Reimann (Eds.), *International handbook of the learning sciences* (pp. 295–307). New York: Routledge.

Chi, M. T. H., Siler, S. A., Jeong, H., Yamauchi, T., & Hausmann, R. G. (2001). Learning from human tutoring. *Cognitive Science, 25*, 471–533.

Chia, H. P., & Pritchard, A. (2014). Using a virtual learning community (VLC) to facilitate a cross-national science research collaboration between secondary school students. *Computers & Education, 79*, 1–15.

Cho, H., Gay, G., Davidson, B., & Ingraffea, A. (2007). Social networks, communication styles, and learning performance in a CSCL community. *Computers & Education, 49*(2), 309–329.

Cress, U., Barquero, B., Schwan, S., & Hesse, F. W. W. (2007). Improving quality and quantity of contributions: Two models for promoting knowledge exchange with shared databases. *Computers & Education, 49*(2), 423–440.

Danish, J., & Gresalfi, M. (2018). Cognitive and sociocultural perspective on learning: Tensions and synergy in the Learning Sciences. In F. Fischer, C. E. Hmelo-Silver, S. R. Goldman, & P. Reimann (Eds.), *International handbook of the learning sciences* (pp. 34–43). New York: Routledge.

de Wever, B., & van Keer, H. (2018). Selecting statistical methods for the Learning Sciences and reporting their results. In F. Fischer, C. E. Hmelo-Silver, S. R. Goldman, & P. Reimann (Eds.), *International handbook of the learning sciences* (pp. 532–541). New York: Routledge.

Dillenbourg, P., Prieto, L. P., Olsen, J. K. (2018) Classroom orchestration. In F. Fischer, C. E. Hmelo-Silver, S. R. Goldman, & P. Reimann (Eds.), *International handbook of the learning sciences* (pp. 180–190). New York: Routledge.

Dzikovska, M., Steinhauser, N., Farrow, E., Moore, J., & Campbell, G. (2014). BEETLE II: Deep natural language understanding and automatic feedback generation for intelligent tutoring in basic electricity and electronics. *International Journal of Artificial Intelligence in Education, 24*(3), 284–332.

Eryilmaz, E., Chiu, M. M., Thoms, B., Mary, J., & Kim, R. (2014). Design and evaluation of instructor-based and peer-oriented attention guidance functionalities in an open source anchored discussion system. *Computers and Education, 71*, 303–321.

Ferguson, R. (2010). Peer interaction: The experience of distance students at university level. *Journal of Computer Assisted Learning, 26*(6), 574–584.

Fields, D. A., Kafai, Y. B., & Giang, M. T. (2016). Coding by choice: A transitional analysis of social participation patterns and programming contributions in the online Scratch community. In U. Cress, J. Moskaliuk, & H. Jeong (Eds.), *Mass collaboration and education* (pp. 209–240). Cham, Switzerland: Springer International.

Green, J. L., & Bridges, S. M. (2018) Interactional ethnography. In F. Fischer, C. E. Hmelo-Silver, S. R. Goldman, & P. Reimann (Eds.), *International handbook of the learning sciences* (pp. 475–488). New York: Routledge.

Hoyningen-Huene, P. (1987). Context of discovery and context of justification. *Studies in History and Philosophy of Science Part A. 18*(4), 505–515.

Jeong, H., & Hmelo-Silver, C. E. (2012). Technology supports in CSCL. In J. van Aalst, K. Thompson, M. J. Jacobson, & P. Reinmann (Eds.), *The Future of Learning: Proceedings of the 10th International Conference of the Learning Sciences* (pp. 339–346). Sydney, NSW, Australia: International Society of the Learning Sciences.

Jeong, H., & Hmelo-Silver, C. E. (2016). Seven affordances of computer-supported collaborative learning: How to support collaborative learning? How can technologies help? *Educational Psychologist, 51*(2), 247–265.

Jeong, H., Hmelo-Silver, C. E., & Yu, Y. (2014). An examination of CSCL methodological practices and the influence of theoretical frameworks 2005–2009. *International Journal of Computer-Supported Collaborative Learning, 9*(3), 305–334.

Jones, A., & Issroff, K. (2005). Learning technologies: Affective and social issues in computer-supported collaborative learning. *Computers & Education, 44*(4), 395–408.

Kirschner, P. A., & Erkens, G. (2013). Toward a framework for CSCL research. *Educational Psychologist, 48*(1), 1–8.

Kollar, I., Wecker, C., & Fischer, F. (2018). Scaffolding and scripting (computer-supported) collaborative learning. In F. Fischer, C. E. Hmelo-Silver, S. R. Goldman, & P. Reimann (Eds.), *International handbook of the learning sciences* (pp. 340–350). New York: Routledge.

Lund, K., & Suthers, D. (2018). Multivocal analysis: Multiple perspectives in analyzing interaction. In F. Fischer, C. E. Hmelo-Silver, S. R. Goldman, & P. Reimann (Eds.), *International handbook of the learning sciences* (pp. 455–464). New York: Routledge.

Okita, S. Y. (2014). Learning from the folly of others: Learning to self-correct by monitoring the reasoning of virtual characters in a computer-supported mathematics learning environment. *Computers and Education, 71,* 257–278.

Oshima, J., Oshima, R., Murayama, I., Inagaki, S., Takenaka, M., Yamamoto, T., et al. (2006). Knowledge-building activity structures in Japanese elementary science pedagogy. *International Journal of Computer-Supported Collaborative Learning, 1*(2), 229–246.

Puntambekar, S. (2018). Design-based research (DBR). In F. Fischer, C. E. Hmelo-Silver, S. R. Goldman, & P. Reimann (Eds.), *International handbook of the learning sciences* (pp. 383–392). New York: Routledge.

Puntambekar, S., Erkens, G., & Hmelo-Silver, C. E. (2011). *Analyzing interactions in CSCL: Methods, approaches, and issues.* New York: Springer.

Roschelle, J. (2013). Special issue on CSCL: Discussion. *Educational Psychologist, 48*(1), 67–70.

Rosé, C.P. (2018). Learning analytics in the Learning Sciences. In F. Fischer, C. E. Hmelo-Silver, S. R. Goldman, & P. Reimann (Eds.), *International handbook of the learning sciences* (pp. 511–519). New York: Routledge.

Rummel, N., & Spada, H. (2005). Learning to collaborate: An instructional approach to promoting collaborative problem solving in computer-mediated settings. *Journal of the Learning Sciences, 14*(2), 201–241.

Shavelson, R. J., & Towne, L. (2002). *Scientific research in education.* Washington, DC: National Research Council.

Shi, Y., Frederiksen, C. H., & Muis, K. R. (2013). A cross-cultural study of self-regulated learning in a computer-supported collaborative learning environment. *Learning and Instruction, 23,* 52–59.

Shiue, Y. C., Chiu, C. M., & Chang, C. C. (2010). Exploring and mitigating social loafing in online communities. *Computers in Human Behavior, 26*(4), 768–777.

Slotta, J. D., Quintana, R., & Moher, T. (2018). Collective inquiry in communities of learners. In F. Fischer, C. E. Hmelo-Silver, S. R. Goldman, & P. Reimann (Eds.), *International handbook of the learning sciences* (pp. 308–317). New York: Routledge.

Smith, H., Underwood, J., Fitzpatrick, G., & Luckin, R. (2009). Classroom e-science: Exposing the work to make it work. *Educational Technology and Society, 12*(3), 289–308.

So, H.-J. J., & Brush, T. A. (2008). Student perceptions of collaborative learning, social presence and satisfaction in a blended learning environment: Relationships and critical factors. *Computers & Education, 51*(1), 318–336.

Stahl, G. (2006). *Group cognition: Computer support for building collaborative knowledge.* Cambridge, MA: MIT Press.

Strijbos, J. W., & Dingyloudi, F. (2018). Mixed methods research as a pragmatic toolkit: Understanding versus fixing complexity in the Learning Sciences. In F. Fischer, C. E. Hmelo-Silver, S. R. Goldman, & P. Reimann (Eds.), *International handbook of the learning sciences* (pp. 444–454). New York: Routledge.

Suthers, D. D. (2006). Technology affordances for intersubjective meaning making: A research agenda for CSCL. *International Journal of Computer-Supported Collaborative Learning, 1,* 315–337.

Suthers, D. D., Lund, K., Rosé, C. P., Teplovs, C., & Law, N. (Eds.). (2013). *Productive multivocality in the analysis of group interaction.* New York: Springer.

Tabak, I., & Kyza, E. (2018). Research on scaffolding in the learning sciences: A methodological perspective. In F. Fischer, C. E. Hmelo-Silver, S. R. Goldman, & P. Reimann (Eds.), *International handbook of the learning sciences* (pp. 191–200). New York: Routledge.

van der Meij, H., de Vries, B., Boersma, K., Pieters, J., & Wegerif, R. (2005). An examination of interactional coherence in email use in elementary school. *Computers in Human Behavior, 21*(3), 417–439.

Vogel, F., & Weinberger, A. (2018). Quantifying qualities of collaborative learning processes. In F. Fischer, C. E. Hmelo-Silver, S. R. Goldman, & P. Reimann (Eds.), *International handbook of the learning sciences* (pp. 500–510). New York: Routledge.

Zhan, Z., & Mei, H. (2013). Academic self-concept and social presence in face-to-face and online learning: Perceptions and effects on students' learning achievement and satisfaction across environments. *Computers and Education, 69,* 131–138.

33

Scaffolding and Scripting (Computer-Supported) Collaborative Learning

Ingo Kollar, Christof Wecker, and Frank Fischer

Why Does (Computer-Supported) Collaborative Learning Need to Be Scaffolded?

Having learners work on tasks in small groups yields a number of potential advantages over having them work individually. Group learning provides a natural context for learners to engage in activities such as explaining or building on other's contributions, through which they may further develop their knowledge and skills (Chi & Wylie, 2014). Meta-analytical evidence shows that collaborative learning has positive effects on variables such as academic achievement, attitudes towards learning, and transfer, as compared to individual learning (e.g., Pai, Sears, & Maeda, 2015).

Yet, many learners have difficulties collaborating productively: They often do not participate equally due to free-rider effects (e.g., Strijbos & De Laat, 2010) and tend to discuss superficially (e.g., Laru, Järvelä, & Clariana, 2012), leading to considerable variability in individual learning outcomes among group members (Weinberger, Stegmann, & Fischer, 2007). Research on collaborative learning is concerned with identifying ways to support and guide collaboration to help learners make full use of the potentials of collaborative learning, using both quantitative (e.g., Schellens, De Wever, Van Keer, & Valcke, 2007) and qualitative research methods (e.g., Hämäläinen & Arvaja, 2009). In the Learning Sciences, and particularly in research on computer-supported collaborative learning (CSCL), this support has been labelled as "scaffolding" or "scripting."

Defining Terms: Scaffolds and Scripts

The term "scaffolding" (see Tabak, this volume), first introduced by Wood, Bruner, and Ross (1976), is used to describe a "process that enables a child or novice to solve a task or achieve a goal that would be beyond his unassisted efforts" (p. 90). This definition is closely related to Vygotsky's Zone of Proximal Development, which is defined as the "distance between the actual developmental level as determined by independent problem solving and the level of potential development as determined through problem solving under adult guidance, or in collaboration with more capable peers" (Vygotsky, 1978, p. 86). While Vygotsky (1978) pointed to collaboration as a source of support *for individuals*, meanwhile the term "scaffolding" is used in a broad fashion to refer to many different kinds of support designed to help either individuals or groups learn productively in problem-solving contexts (e.g., Molenaar, Seegers, & Van Boxtel, 2014; Quintana et al., 2004).

Scaffolds may on the one hand provide support at the content level. A prototype for such scaffolds are worked-out examples—they present a specific problem, display all necessary steps to solve the problem, and present the final solution to the learners (Renkl & Atkinson, 2003). Other examples for content-related scaffolds are content schemes (Kopp & Mandl, 2011) or concept maps (Gijlers & de Jong, 2013). Nevertheless, scaffolds do not always provide support at the content level; they may also refer directly to the kinds of learning processes students are supposed to engage in during learning. For example, a learner may receive metacognitive prompts that ask her to reflect upon her learning process after she has worked on a certain task (e.g., Bannert, 2006). "Scripts" belong to such process-oriented scaffolds (O'Donnell & Dansereau, 1992), yet they are specific to *collaboration* processes. For example, during a discussion on whether nuclear power plants should be shut down, a collaboration script may ask one learner to produce an argument, and the other learner to find a counterargument, before both learning partners may be prompted to find a synthesis (see Table 33.1). Such "collaboration scripts" provide support that specifies and sequences collaboration phases, roles, and learning activities and distributes them among the members of learning groups (Fischer, Kollar, Stegmann, & Wecker, 2013).

Developing a Framework of Research on Scaffolding and Scripting (Computer-Supported) Collaborative Learning

Existing approaches to systematize scaffolding, such as the Scaffolding Design Framework (Quintana et al., 2004), target scaffolding of learning and problem-solving in general. Yet, a framework focusing specifically on scaffolding for *collaborative* learning appears to be missing. We propose a framework that is based on five guiding questions: (1) Who provides scaffolding in collaborative learning? (2) Who is the recipient of such scaffolding? (3) What learning activities are targeted by the scaffolding? (4) What are the intended effects of scaffolding? (5) What types of scaffolds for collaborative learning can be differentiated? In answering these questions, we draw on research on computer-supported collaborative learning (CSCL). Yet, we claim that the framework may just as well be applied to non-computer-supported learning contexts.

Who Provides Scaffolding in Collaborative Learning?

A first possible source of scaffolding in collaborative learning are *persons from inside the group*. This is the case if one learner provides support for his or her peer(s), e.g. by giving explanations (e.g., Ross, 2011) or by asking thought-provoking questions (King, 2007). Yet, learners typically need further guidance to scaffold their learning partners; for example, a CSCL script may assign a learner the task of monitoring her learning partner's learning activities (e.g., Wecker & Fischer, 2011).

A second source of scaffolding may be *persons outside of the group*, such as teachers or tutors (e.g., De Smet, Van Keer, & Valcke, 2009) or other groups (or single students) currently working on the same task. For example, a collaboration script may guide learners to first work in small groups on a given task, but in later phases have the different groups share and discuss their products with each other (e.g. Dillenbourg & Hong, 2008).

A third source of scaffolding are *technological artifacts*. The Learning Sciences have always put high hopes into technology, given its promise to design adaptable and adaptive learning environments that scaffold the learning activities of individuals, but also of larger groups (e.g., classrooms). Technological artifacts may include computer technology (e.g., role prompts that are built into a web-based learning environment; Schellens et al., 2007), but also non-digital technology such as pre-structured work sheets that may support the group in their learning activities.

Who Is the Recipient of the Scaffold?

Some scaffolds are presented *to groups as a whole*, having the effect that the support the group receives can immediately become a topic for discussion within the group. This typically is the case when collaborative learning is supported by a human tutor as, for example, in problem-based learning (Hmelo-Silver, 2004; see also Hmelo-Silver, Kapur, & Hamstra, this volume).

In other cases, scaffolds are presented not to the group as a whole, but rather to its *individual members*. On the one hand, there are scaffolds that present the support to each individual group member but, still, the support presented is the same for all members. For example, in a script used by Weinberger, Ertl, Fischer, and Mandl (2005), all students received the same script prompts during the course of the learning session, albeit at different points in time, as the roles students were to engage in rotated during collaboration. On the other hand, scaffolding may be presented to each individual group member, but be different for each of them. In this case, the support typically is complementary, as in a study by Wecker and Fischer (2011), in which students were asked to apply a certain argumentation strategy, while a specific member of each group was asked to monitor his or her partners' strategy execution and to give feedback.

What Learning Activities Are Targeted by the Scaffold?

Scaffolds are only effective when they trigger learning activities that are known to relate to knowledge or skill acquisition or other intended learning outcomes. Examples are argumentation, peer feedback, and joint regulation.

Argumentation. Producing and exchanging arguments and counterarguments and backing them up with evidence has been shown to be conducive for learning (e.g., Asterhan, Schwarz, & Gil, 2012). It is thus not surprising that scaffolds and scripts have been designed to promote argumentation. For example, Tsovaltzi, Puhl, Judele, and Weinberger (2014) developed a Facebook app that allowed students to label their contributions to a small-group discussion as claims, counterclaims, evidence, or rebuttals. Students who received the labeling option produced more arguments of high quality than students who did not.

Peer feedback. Feedback is among the teacher behaviors that have the largest effect on student achievement (Hattie, 2009). During collaborative learning, such feedback can also be given by peers (Strijbos, Narciss, & Dünnebier, 2010). Without guidance, though, peers often do not produce high-quality feedback (Patchan & Schunn, 2015). Research has shown better learning when peer feedback is scaffolded, e.g., by providing students with assessment rubrics (Hovardas, Tsivitanidou, & Zacharia, 2014) or feedback templates (Gielen & De Wever, 2015).

Joint regulation. Research on self-regulated learning has pointed to the importance of planning, monitoring and reflecting upon one's own learning (Zimmerman, 2008). During CSCL, such regulatory activities may occur at three levels (see Järvelä & Hadwin, 2013): first, at the *self-regulation* level, individuals regulate their own learning during collaboration; second, one collaborator may regulate the learning of a peer (*co-regulation*; Järvelä & Hadwin, 2013), e.g., by monitoring how he or she applies a certain strategy; and third, groups may engage in a joint, deliberate negotiation on how to regulate their learning, which is called *shared regulation*. Shared regulation seems to play a critical role for the effectiveness of CSCL (Järvelä, Malmberg, & Koivuniemi, 2016).

What Are the Learning Outcomes Targeted by a Scaffold?

In CSCL there are two basic types of learning outcomes: group level and individual level outcomes. Studies following the Knowledge Building paradigm mainly focus on *group-level outcomes* of CSCL. Zhang, Hong, Scardamalia, Teo, and Morley (2011), for example, described how elementary school classes built up group level knowledge by continuously sharing and extending their knowledge by aid of digital technology (see also Chan & van Aalst, this volume).

Studies concerned with *individual-level outcomes* often either refer to domain-specific knowledge (i.e., knowledge about the topics that are discussed within the group), or to rather general, "cross-domain" skills such as collaboration or argumentation skills. A meta-analysis by Vogel, Wecker, Kollar, and Fischer (2016) showed that CSCL scripts are effective with regard to both of these outcomes. Yet, effect sizes were higher for cross-domain skills.

What Types of Scaffolds Can Be Distinguished?

Not all interaction-related scaffolds are scripts. For example, group awareness tools (see Bodemer, Janssen, & Schnaubert, this volume) represent a less directive form of support. These tools capture certain information about single group members (e.g., their performance in a prior knowledge test) or their collaboration behavior (e.g., the number of contributions per group members) and mirror this information back to the group in order to indirectly influence collaboration, e.g., so that learners with higher scores in a pre-test on the topics to be discussed receive more questions by other learners when their superior knowledge test scores are presented to the rest of the group (e.g., Dehler, Bodemer, Buder, & Hesse, 2011).

When Scaffolds Work and When They Do Not: A Script Theory of Guidance Point of View

In a recent meta-analysis, Vogel et al. (2016) showed that the overall effect size of scripted vs. unscripted CSCL on the acquisition of domain-specific learning was significant and positive ($d = .20$). Yet, some studies also found null or even negative effects. This leads to the question as to what makes scaffolds and scripts effective and what might harm their effectiveness. One crucial aspect seems to be the extent to which the scaffold or script "fits" the learner's cognitive prerequisites and characteristics. In Vygotskian terms, the question is how scripts can be designed to create Zones of Proximal Development, i.e., to provide learners with the support they need, depending on their current proficiency levels.

According to the script theory of guidance (SToG), the main individual learning prerequisite that needs to be taken into account in that respect is the learners' internal collaboration scripts (Fischer et al., 2013), which describe their prior knowledge and expectations regarding how collaboration will evolve in the current situation. Building on Schank (1999), the SToG assumes such internal scripts to consist of four partially hierarchical components. At the top level, upon entering a new learning situation, learners select a (1) "play" from their dynamic memory to make immediate sense of the situation. For example, a learner who is asked to engage in a discussion about whether nuclear power plants should be shut down will likely select the play "debate" from his or her dynamic memory. Selecting that play then yields expectations about how the situation will evolve. On a still rather general level, selecting the "debate" play will generate expectations concerning the phases that are likely to be happening in the current situation. For example, the student may expect that there will first be a phase in which one party will express their arguments, followed by a phase in which the other party does the same, which is followed by a phase during which both parties try to find a compromise. Within the SToG, knowledge about such phases is stored in (2) "scenes." Once a certain scene is selected, expectations on a more fine-grained level are triggered, i.e., expectations about what activities are likely to be enacted in each phase. For example, during the "find compromise" scene, our hypothetical learner may expect that the strongest arguments will be reiterated and weighed until a joint position is reached. Knowledge about such activities (and how they are conducted), according to the SToG, is stored in (3) "scriptlets." Finally, our hypothetical learner may also have acquired knowledge in the past about what actor is likely to engage in what kinds of activities during the debate. For example, this learner may expect his or her learning partner to first offer her strongest argument, then she would respond with her own argument or rebuttal. Knowledge about the typical actors in a current situation is assumed to be stored in (4) "roles" components.

Internal scripts should not be regarded as stable cognitive structures. At any time, learners may reconfigure the different components of the internal script they have selected, depending on personal goals and perceived situational constraints. For example, at some point during the debate, the learner may lose interest in the topic and activate the goal to quickly bring the discussion to an end. In this case, the learner may skip the "find compromise" scene and replace it with a "comply with learning partner" scene and engage in activities that are different than ones originally planned. Such on-the-fly reconfigurations are possible at any time and at all four levels of an internal script.

This view on how internal scripts are configured and develop over time has important implications for the design of CSCL scripts. First, such externally provided scripts may trigger two processes. On the one hand, it may be that learners already have internal collaboration script components available that would enable them to engage in high-level activities, but just fail to apply them, perhaps because they do not notice that these internal script components could also be applied in the current situation. Here, an external collaboration script may provide scaffolding that will help the learner select these already present internal script components. On the other hand, it may also be that learners' dynamic memories do not include adequate internal script components at all. An external script will then have to present the targeted skills, even at the level of scriptlets, and provide the learner with (guided) practice opportunities to gradually build up these new internal script components and/or configurations.

Second, to increase the effectiveness of externally provided collaboration scripts, it is necessary to continuously assess the learners' internal scripts and to provide external scripts that "fit" the learners' current internal scripts in the sense of providing a Zone of Proximal Development (Vygotsky, 1978), to enable the learners to engage in a task somewhat beyond what they would be able to do on their own, without support. To systematize the different kinds of external scripts, the SToG uses the same four-level hierarchy again (see Table 33.1 for examples):

(1) *Play scaffolds* structure learning by aid of presenting learners a certain sequence of phases to go through during collaboration.
(2) *Scene scaffolds* provide support regarding the different activities that make up a certain scene.
(3) *Scriptlet scaffolds* help learners in their enactment of cognitive operations that make up the scriptlets.
(4) *Role scaffolds* guide learners in their engagement in certain activities that belong to their role.

Play and role scaffolds may thus be seen as similar to "macro scripts," while scene and scriptlet scaffolds may be considered as "micro scripts," as they provide very specific and fine-grained support on how to engage in certain phases during collaboration (see Dillenbourg & Hong, 2008).

According to the SToG, externally provided collaboration scaffolds are effective when they target internal script components for which subordinate internal script components exist in the learner's repertoire. For example, when learners already know how to enact certain activities (i.e., they already possess the necessary scriptlets in their repertoire), it will be enough to present a scaffold that just prompts the kinds and sequence of phases that are part of the play. If the learner does not know the cognitive operations necessary to perform certain activities, he or she will need a scriptlet scaffold that provides guidance in enacting these activities. Empirical evidence supports the claim that the effectiveness of externally provided collaboration scripts is moderated

Table 33.1 Examples for "Play Scaffolds," "Scene Scaffolds," "Scriptlet Scaffolds," and "Role Scaffolds" When Supporting Collaborators in the Process of Argumentation

Type of scaffold	Example	Explanation
Play scaffold	"Discuss whether nuclear power plants should be shut down. During your discussion, please first produce arguments in favor of shutting nuclear power plants down, then produce counter-arguments, and then try to find a synthesis."	The scaffold provides guidance on how to enact the play by suggesting a sequence of different phases of the collaboration. This is supposed to help learners build up or revise their knowledge about the play, which may guide their engagement in future debates.
Scene scaffold	For example, during the "*produce argument*" phase: "Please first describe the claim you want to make; then, provide evidence for the claim."	The scaffold provides guidance on how to enact a certain phase within the play by prompting the activities that are necessary to master that phase. This is supposed to help learners build up or revise their knowledge about that scene.
Scriptlet scaffold	For example, during "*provide evidence*" activity: "When providing evidence, make sure that the evidence comes from a reliable source and check carefully whether it really supports your claim."	The scaffold provides guidance on how exactly to conduct a certain activity within the scene by explicating the cognitive operations that are necessary during that activity. This is supposed to help learners build up or revise cognitive operations they can use in future situations in which they are supposed to produce arguments.
Role scaffolds	For *learner A*: "You will be taking the 'pro-nuclear-power-plants position." For *learner B*: "You will be taking the "con-nuclear-power-plants position."	The scaffold introduces different actors that are involved in the play and implicitly attaches certain activities to them. This is supposed to help learners build or revise knowledge about different actors in the play and the activities that they may be expected to engage in.

by the internal script level they target (Vogel et al., 2016). Yet, only few studies have attempted to directly measure internal collaboration scripts (Noroozi, Kirschner, Biemans, & Mulder, 2017, p. 24). Having such instruments available is a necessary precondition for further research on the interplay of internal and external scripts.

Making Scaffolds and Scripts Flexible to Support Self-Regulation of Collaboration

That scaffolds should create a Zone of Proximal Development by taking the learners' current and possible next proficiency levels into account is a long-held tenet of related research that has led to the introduction of the concept of "fading." "Fading" refers to the gradual reduction of support as the learner gains competence in the strategies that are targeted by the scaffold (e.g., Collins, Brown, & Newman, 1989, which helps him or her to become increasingly proficient at applying them in new situations without guidance (Pea, 2004).

Research, however, has shown that fixed fading regimes (i.e., fading mechanisms that are based on the number of enactments of certain skills or strategies) at best yield mixed results on learning (see Wecker & Fischer, 2011)—for many learners, fixed fading may simply come either too early or too late. The challenge is to design collaboration scripts with a fading mechanism that is contingent upon the learners' increasing proficiency in the targeted skills.

Adaptive Scripting

The idea of adaptive scripting is that the external collaboration script is automatically adjusted to the learners' current internal collaboration scripts. With the rapid progress in computer linguistics and learning analytics, research has explored how technology can be used to automatically assess the quality of the collaboration and use that information to continuously adjust the support that seems necessary for the group to succeed (see Rosé, this volume). For example, Rummel, Mullins, and Spada (2012) used the Cognitive Tutor Algebra in a dyadic scenario and extended it with an adaptive collaboration script. The adaptivity of the collaboration script was based on an ongoing, automated analysis of the dyads' problem solving. Once the group faced an impasse in their problem solving, the intelligent tutoring system noticed this and provided the group with reflection prompts. While the adaptive script had a positive effect compared to an unscripted condition on student collaboration and problem solving in a subsequent unscripted collaboration task, it did not yield further effects on individual knowledge. Karakostas and Demetriadis (2011), however, found positive effects on individual learning outcomes as well.

Adaptive scripting has also been implemented using natural language processing (NLP) techniques to continuously assess the quality of the group discourse and continuously adjust the support (see Rosé, this volume). To that end, NLP tools are fed with raw discourse data and codes from human coders (e.g., with respect to the quality of argumentation). The NLP tool then extracts language-based indicators that increase or decrease the likelihood for a certain code to be assigned to a certain discourse segment. As an example, Mu, Stegmann, Mayfield, Rosé, and Fischer (2012) showed how an NLP tool can be applied to assess several dimensions of collaboration and argumentation quality in student discourses on a certain topic with sufficient levels of reliability. The idea is that, once such NLP tools are sufficiently trained in the described way, they might also be used as collaboration is happening, and the resulting continuous diagnosis could be used to adjust the script to match the students' current need for support.

Adaptable Scripting

Even though adaptive tools may produce large effects on learning, their development is time- and cost-intensive. Also, by receiving adaptive support, students might not become aware of the quality of their collaboration. From a self-regulated learning perspective, this is unfortunate: If it is always an external agent that adjusts the support, the group will not be very likely to discuss their learning processes on a metacognitive level. Another approach towards flexible scripting is to look for ways that would help learners become better self-regulators, i.e., to gradually increase their proficiency in planning, monitoring, and reflecting upon their learning processes (see Järvelä, Hadwin, Malmberg, & Miller, this volume). This idea is taken up by *adaptable* scripting, which means that the decision on whether parts of the script are faded in or out is left to the learners themselves. In one study, Wang, Kollar, and Stegmann (2017) found that students in an adaptable script condition outperformed students who learned with a fixed script with respect to the acquisition of regulation skills. Further, this effect was mediated by an increased engagement in reflection that could be observed in discourse analyses and that was seemingly stimulated by scaffolds of the adaptable collaboration script. Nevertheless, it is likely that not all learners will be able to make efficient use of the opportunity to adapt a collaboration script. Further research is needed on how to support especially learners with lower self-regulation skills, when adaptable scripting is implemented.

Conclusions

We suggested to systematize research on scaffolding and scripting for (computer-supported) collaborative learning along five questions: (1) Who provides scaffolding? (2) Who is the recipient of scaffolding? (3) What learning activities are targeted by the scaffolding? (4) What are the intended outcomes of learning with scaffolding? (5) What types of scaffolds can be differentiated? By giving examples for how these different questions are answered by different approaches, we covered a range of scaffolding approaches from the literature.

Based on the SToG, we provided a promising account for understanding the (sometimes lacking) effects of scaffolds and scripts for collaborative learning on learning outcomes. The SToG can provide an articulate framework for the design of effective support for collaborative learning, by pointing out how to create Zones of Proximal Development with scaffolds targeted at the right internal collaboration script level.

However, future research has a lot left to achieve. For example, we urgently need sophisticated tools to reliably, objectively, and validly assess learners' internal scripts. It is theoretically questionable whether the learners' actual enactment of certain social-discursive activities is the most valid indicator for their internal scripts. For example, it may well be that students dispose of internal script components that would enable them to show a certain desired activity during collaboration; yet, due to changes in learners' actualized goals or perceived situational constraints, they may not engage in this activity. Such performance-based measures might need to be complemented with further techniques to measure internal collaboration scripts, such as tests that tap into the breadth of students' declarative knowledge about possible actions they might be able to undertake during collaboration.

Another underresearched question is how students "appropriate" the scaffolds and external scripts they are provided with (Tchounikine, 2016). Once a scaffold is offered to a group, students interpret that support in certain ways and consciously or unconsciously decide how to use the scaffold. Knowing more about the nature of such appropriation processes would support a more concise and unequivocal design of scaffolds and scripts for CSCL.

Practically, research on scaffolding and scripting CSCL has a lot to offer for the design of online learning environments such as MOOCs or online discussion forums. Highly valued collaborative processes like shared regulation (Järvelä et al., this volume) and knowledge co-creation (Slotta, Quintana, & Moher, this volume) may be stimulated and shaped by providing learners with guidance for their collaboration. Implementing such support in an adaptive or adaptable way is likely to add to the power of such environments.

Overall, research on scaffolding and scripting CSCL is a success story. Many insights have accumulated over the years, and still the field is vibrant and will produce further knowledge on effective designs for collaborative learning. Research on scripting is also a success story for the learning sciences with respect to its interdisciplinary nature (Hoadley, this volume). Psychologists, educational scientists, content experts, and computer scientists work together in theorizing how knowledge on collaboration might be represented in individuals and how effective, adaptive external guidance could be designed and implemented.

Further Readings

Dillenbourg, P., & Hong, F. (2008). The mechanics of CSCL macro scripts. *International Journal of Computer-Supported Collaborative Learning, 3*(1), 5–23.

A seminal article that provides an overview over three macro scripts for computer-supported collaborative learning along with a pedagogical design model that describes how macro scripts for CSCL can be computationally developed.

Fischer, F., Kollar, I., Stegmann, K., & Wecker, C. (2013). Toward a script theory of guidance in computer-supported collaborative learning. *Educational Psychologist, 48*(1), 56–66.

This article outlines a theory of collaboration scripts distinguishing between internal and external scripts and suggesting a set of hierarchically organized components to characterize internal scripts. It develops a set of principles as to how internal script components are developed and employed, and how external script components can guide the selection of internal script components.

Mu, J., Stegmann, K., Mayfield, E., Rosé, C., & Fischer, F. (2012). The ACODEA framework: Developing segmentation and classification schemes for fully automatic analysis of online discussions. *International Journal of Computer-Supported Collaborative Learning, 7*(2), 285–305.

This article describes an exemplary approach of an interdisciplinary collaboration on adaptive collaboration scripts using recent computer science and language technologies to automate the analysis of peer discussions.

O'Donnell, A. M., & Dansereau, D. F. (1992). Scripted cooperation in student dyads: A method for analyzing and enhancing academic learning and performance. In R. Hertz-Lazarowitz & N. Miller (Eds.), *Interaction in cooperative groups: The theoretical anatomy of group learning* (pp. 120–141). New York: Cambridge University Press.

O'Donnell and Dansereau provide the groundwork for introducing the script term to the scaffolding literature. By introducing the famous "MURDER" script, they show how providing learners with (external) scripts can help them engage in socio-cognitive processes that stand in a close relationship to academic learning outcomes.

Schellens, T., Van Keer, H., De Wever, B., & Valcke, M. (2007). Scripting by assigning roles: Does it improve knowledge construction in asynchronous discussion groups? *International Journal of Computer-Supported Collaborative Learning, 2*(2–3), 225–246.

This exemplary article describes a design experiment with online discussion groups in higher education involving an external collaboration script with role scaffolding. Results show a medium-sized effect in favor of small groups who were supported by the collaboration script.

NAPLeS Resources

Fischer, F., Wecker, C., & Kollar, I., *Collaboration scripts for computer-supported collaborative learning*. [Webinar]. In *NAPLeS video series*. Retrieved October 19, 2017, from http://isls-naples.psy.lmu.de/intro/all-webinars/fischer-kollar-wecker/index.html

Tabak, I., & Reiser, B., *15 minutes about scaffolding*. [Video file]. In *NAPLeS video series*. Retrieved October 19, 2017, from http://isls-naples.psy.lmu.de/video-resources/guided-tour/15-minutes-tabak-reiser/index.html

References

Asterhan, C. S. C., Schwarz, B. B., & Gil, J. (2012). Small-group, computer-mediated argumentation in middle-school classrooms: The effects of gender and different types of online teacher guidance. *British Journal of Educational Psychology, 82*(3), 375–397.

Bannert, M. (2006). Effects of reflection prompts when learning with hypermedia. *Journal of Educational Computing Research, 4*, 359–375.

Bodemer, D., Janssen, J., & Schnaubert, L. (2018) Group awareness tools for computer-supported collaborative learning. In F. Fischer, C. E. Hmelo-Silver, S. R. Goldman, & P. Reimann (Eds.), *International handbook of the learning sciences* (pp. 351–358). New York: Routledge.

Chan, C., & van Aalst, J. (2018). Knowledge building: Theory, design, and analysis. In F. Fischer, C. E. Hmelo-Silver, S. R. Goldman, & P. Reimann (Eds.), *International handbook of the learning sciences* (pp. 295–307). New York: Routledge.

Chi, M. T. H., & Wylie, R. (2014). The ICAP framework: Linking cognitive engagement to active learning outcomes. *Educational Psychologist, 49*(4), 219–243.

Collins, A., Brown, J. S., & Newman, S. E. (1989). Cognitive apprenticeship: Teaching the crafts of reading, writing, and mathematics. In L. B. Resnick (Ed.), *Knowing, learning, and instruction: Essays in honor of Robert Glaser* (pp. 453–494). Hillsdale, NJ: Erlbaum.

De Smet, M., Van Keer, H., & Valcke, M. (2009). Cross-age peer tutors in asynchronous discussion groups: a study of the evolution in tutor support. *Instructional Science, 37*(1), 87–105.

Dehler, J., Bodemer, D., Buder, J., & Hesse, F. W. (2011). Guiding knowledge communication in CSCL via group knowledge awareness. *Computers in Human Behavior, 27*(3), 1068–1078.

Dillenbourg, P., & Hong, F. (2008), The mechanics of macro scripts. *International Journal of Computer-Supported Collaborative Learning, 3*(1), 5–23.

Dillenbourg, P., & Tchounikine, P. (2008). Flexibility in macro-scripts for computer-supported collaborative learning. *Journal of Computer-Assisted Learning, 23*(1), 1–13.

Fischer, F., Kollar, I., Stegmann, K., & Wecker, C. (2013). Toward a script theory of guidance in computer-supported collaborative learning. *Educational Psychologist, 48*(1), 56–66.

Gielen, M., & De Wever, B. (2015). Scripting the role of assessor and assessee in peer assessment in a wiki environment: impact on peer feedback quality and product improvement. *Computers & Education, 88*, 585–594.

Gijlers, H., & de Jong, T. (2013). Using concept maps to facilitate collaborative simulation-based inquiry learning. *The Journal of the Learning Sciences, 22*(3), 340–374.

Hattie, J. (2009). *Visible learning. A synthesis of over 800 meta-analyses relating to achievement.* London: Routledge.

Hämäläinen, R., & Arvaja. M. (2009). Scripted collaboration and group-based variations in a higher education CSCL context. *Scandinavian Journal of Educational Research, 53*(1), 1–16.

Hmelo-Silver, C. E. (2004). Problem-based learning: What and how do students learn? *Educational Psychology Review, 16*, 235–266.

Hmelo-Silver, C. E., Kapur, M., & Hamstra, M. (2018) Learning through problem solving. In F. Fischer, C. E. Hmelo-Silver, S. R. Goldman, & P. Reimann (Eds.), *International handbook of the learning sciences* (pp. 210–220). New York: Routledge.

Hoadley, C. (2018). Short history of the learning sciences. In F. Fischer, C. E. Hmelo-Silver, S. R. Goldman, & P. Reimann (Eds.), *International handbook of the learning sciences*. New York: Routledge.

Hovardas, T., Tsivitanidou, O. E., & Zacharia, Z. C. (2014). Peer versus expert feedback: An investigation of the quality of peer feedback among secondary school students. *Computers and Education, 71*, 133–152.

Järvelä, S. & Hadwin, A. (2013). New frontiers: Regulating learning in CSCL. *Educational Psychologist, 48*(1), 25–39.

Järvelä, S., Hadwin, A., Malmberg, J., & Miller, M. (2018). Contemporary perspectives of regulated learning in collaboration. In F. Fischer, C. E. Hmelo-Silver, S. R. Goldman, & P. Reimann (Eds.), *International handbook of the learning sciences* (pp. 127–136). New York: Routledge.

Järvelä, S., Malmberg, J., & Koivuniemi, M. (2016). Recognizing socially shared regulation by using the temporal sequences of online chat and logs in CSCL. *Learning and Instruction, 42*, 1–11.

Karakostas, A., & Demetriadis, S. (2011). Enhancing collaborative learning through dynamic forms of support: the impact of an adaptive domain-specific support strategy. *Journal of Computer-Assisted Learning, 27*, 243–258.

King, A. (2007). Scripting collaborative learning processes: A cognitive perspective. In F. Fischer, I. Kollar, H. Mandl, & J. Haake (Eds.), *Scripting computer-supported collaborative learning: Cognitive, computational, and educational perspectives* (pp. 13–37). New York: Springer.

Kopp, B., & Mandl, H. (2011). Fostering argument justification using collaboration scripts and content schemes. *Learning and Instruction, 21*(5), 636–649.

Laru, J., Järvelä, S., & Clariana, R. B. (2012). Supporting collaborative inquiry during a biology field trip with mobile peer-to-peer tools for learning: a case study with K–12 learners. *Interactive Learning Environments, 20*(2), 103–117.

Molenaar, I., Sleegers, P. J. C., & Van Boxtel, C. A. M. (2014). Metacognitive scaffolding during collaborative learning: A promising combination. *Metacognition and Learning, 9*, 309–332.

Mu, J., Stegmann, K., Mayfield, E., Rosé, C., & Fischer, F. (2012). The ACODEA framework: Developing segmentation and classification schemes for fully automatic analysis of online discussions. *International Journal of Computer-Supported Collaborative Learning, 7*(2), 285–305.

Noroozi, O., Kirschner, P. A., Biemans, H. J. A., & Mulder, M. (2017). Promoting argumentation competence: extending from first- to second-order scaffolding through adaptive fading. *Educational Psychology Review, 30*(1), 153–176. doi:10.1007/s10648-017-9400-z

O'Donnell, A. M., & Dansereau, D. F. (1992). Scripted cooperation in student dyads: A method for analyzing and enhancing academic learning and performance. In R. Hertz-Lazarowitz & N. Miller (Eds.), *Interaction in cooperative groups: The theoretical anatomy of group learning* (pp. 120–141). New York: Cambridge University Press.

Pai, H., Sears, D., & Maeda, Y. (2015). Effects of small-group learning on transfer: A meta-analysis. *Educational Psychology Review, 27*(1), 79–102.

Patchan, M. M., & Schunn, C. D. (2015). Understanding the benefits of providing peer feedback: How students respond to peers' texts of varying quality. *Instructional Science, 43*(5), 591–614.

Pea, R. (2004). The social and technological dimensions of scaffolding and related theoretical concepts for learning, education, and human activity. *The Journal of the Learning Sciences, 13*(3), 423–451.

Quintana, C., Reiser, B. J., Davis, E. A., Krajcik, J., Fretz, E., Duncan, R. G., et al. (2004). A scaffolding design framework for software to support science inquiry. *The Journal of the Learning Sciences, 13*(3), 337–386.

Renkl, A., & Atkinson, R. K. (2003). Structuring the transition from example study to problem solving in cognitive skill acquisition: A cognitive load perspective. *Educational Psychologist, 38*(1), 15–22.

Rosé, C. P. (2018). Learning analytics in the Learning Sciences. In F. Fischer, C. E. Hmelo-Silver, S. R. Goldman, & P. Reimann (Eds.), *International handbook of the learning sciences* (pp. 511–519). New York: Routledge.

Ross, J. A. (2011). Explanation giving and receiving in cooperative learning groups. In R. M. Gillies, A. Ashman, & J. Terwel (Eds.), *The teacher's role in implementing cooperative learning in the classroom* (pp. 222–237). New York: Springer.

Rummel, N., Mullins, D., & Spada, H. (2012). Scripted collaborative learning with the cognitive tutor algebra. *International Journal of Computer-Supported Collaborative Learning, 7*(2), 307–339.

Schank, R. E. (1999). *Dynamic memory revisited*. New York: Cambridge University Press.

Schellens, T., Van Keer, H., De Wever, B., & Valcke, M. (2007). Scripting by assigning roles: Does it improve knowledge construction in asynchronous discussion groups? *International Journal of Computer-Supported Collaborative Learning, 2*(2–3), 225–246.

Slotta, J. D., Quintana, R., & Moher, T. (2018). Collective inquiry in communities of learners. In F. Fischer, C. E. Hmelo-Silver, S. R. Goldman, & P. Reimann (Eds.), *International handbook of the learning sciences* (pp. 308–317). New York: Routledge.

Strijbos, J.-W., & De Laat, M. F. (2010). Developing the role concept for computer-supported collaborative learning: An explorative synthesis. *Computers in Human Behavior, 26*(4), 495–505.

Strijbos, J.-W., Narciss, S., & Dünnebier, K. (2010). Peer feedback content and sender's competence level in academic writing revision tasks: Are they critical for feedback perceptions and efficiency? *Learning and Instruction, 20*(4), 291–303.

Tchounikine, P. (2016). Contribution to a theory of CSCL scripts: Taking into account the appropriation of scripts by learners. *International Journal of Computer-Supported Collaborative Learning, 11*(3), 349–369.

Tsovaltzi, D., Puhl, T., Judele, R., & Weinberger, A. (2014). Group awareness support and argumentation scripts for individual preparation of arguments in Facebook. *Computers & Education, 76*, 108–118.

Vogel, F., Wecker, C., Kollar, I., & Fischer, F. (2016). Socio-cognitive scaffolding with computer-supported collaboration scripts: A meta-analysis. *Educational Psychology Review, 29*(3), 477–511.

Vygotsky, L. S. (1978). *Mind and society*. Cambridge, MA: Harvard University Press.

Wang, X., Kollar, I., & Stegmann, K. (2017). Adaptable scripting to foster regulation processes and skills in computer-supported collaborative learning. *International Journal for Computer-Supported Collaborative Learning, 12*(2), 153–172.

Wecker, C., & Fischer, F. (2011). From guided to self-regulated performance of domain-general skills: The role of peer monitoring during the fading of instructional scripts. *Learning and Instruction, 21*(6), 746–756.

Weinberger, A., Ertl, B., Fischer, F., & Mandl, H. (2005). Epistemic and social scripts in computer-supported collaborative learning. *Instructional Science, 33*(1), 1–30.

Weinberger, A., Stegmann, K., & Fischer, F. (2007). Knowledge convergence in collaborative learning: Concepts and assessment. *Learning and Instruction, 17*(4), 416–426.

Wood, D. J., Bruner, J. S., & Ross, G. (1976). The role of tutoring in problem solving. *Journal of Child Psychiatry and Psychology, 17*(2), 89–100.

Zhang, J., Hong, H.-Y., Scardamalia, M., Teo, C. L., & Morley, E. A. (2011). Sustaining knowledge building as a principle-based innovation at an elementary school. *The Journal of the Learning Sciences, 20*(2), 262–307.

Zimmerman, B. J. (2008). Investigating self-regulation and motivation: historical background, methodological developments, and future prospects. *American Educational Research Journal, 45*(1), 166–183.

34

Group Awareness Tools for Computer-Supported Collaborative Learning

Daniel Bodemer, Jeroen Janssen, and Lenka Schnaubert

Group awareness (GA) refers to being informed about aspects of group members or the group—for example, the group members' current locations, activities, knowledge, interests, or feelings (Bodemer & Dehler, 2011; Gross, Stary, & Totter, 2005). Thus, it can cover various aspects of valid group- or person-related information that provide a context for primary activities. Establishing GA is recognized as an important prerequisite for successful collaboration in different fields. Thus, awareness-related learning sciences research aims at supporting GA by developing so-called group awareness tools (GATs) that focus specifically on information relevant to collaborative learning. By presenting GA information, such tools can tacitly guide learning processes.

The Concept of Group Awareness

The concept of GA has developed in the 1990s mainly in the research field of computer-supported cooperative work (CSCW) focusing on the perception and understanding of behavioral aspects such as activities of group members that were seen as crucial context for successful collaboration (e.g., Dourish & Bellotti, 1992). It complemented other awareness approaches developed in human factors research that focused on the perception and understanding of the material environment even in social settings (e.g., team situational awareness; Endsley, 1995). Computer-supported collaborative learning (CSCL) research seized the socio-behavioral perspective of the CSCW awareness approach and applied it to knowledge-related activities of group members (e.g., Ogata & Yano, 2000). While in early awareness-related learning sciences approaches, the term "knowledge" referred to artifacts in the external environment, research increasingly considers internal cognitive, emotional, or motivational states as target concepts for GA (see below; cf. Soller, Martínez, Jermann, & Muehlenbrock, 2005; Stahl, 2016).

GA information in the learning sciences covers a wide scope of behavioral, social, and cognitive data that learners can use to structure their learning processes successfully. It may specifically refer to stable characteristics of learning partners (e.g., interest, expertise, prior knowledge or beliefs) or more situational aspects (e.g. current location, availability, performance, engagement). Even if the information refers to group members, the concept of GA is focusing on individual cognitions, which sets it apart from group-level constructs like team mental models or common ground.

When a learner captures GA information during social interaction, it can be used immediately or stored in long-term memory for later usage. However, the available information is not always sufficient or salient enough for establishing GA that can be beneficially used for effective and efficient learning processes. This is why most research approaches concerned with GA and learning aim at developing GATs that help learners to perceive and understand essential information about their learning partners.

Types and Functions of Group Awareness Tools

A main goal of GA research in the learning sciences is to support learners' awareness of their learning partners for triggering beneficial learning processes. Therefore, various tools have been suggested and developed that provide learners with information on their learning partners during collaboration. GATs are distinguished by the type of information they focus on: cognitive GATs provide information directly related to the learning topic (e.g., a learning partner's knowledge, opinion, hypotheses) but also metacognitive information (e.g., task understanding, confidence judgments), while social GATs comprise socio-behavioral (e.g., participation, learning activities), socio-emotional (e.g., well-being, perceived friendliness) or socio-motivational (e.g., motivation, engagement) information. A typical example of a cognitive GAT is the Knowledge Awareness tool of Sangin and colleagues (Sangin, Molinari, Nüssli, & Dillenbourg, 2011). It collects information on a learner's knowledge by means of test items, calculates test scores regarding specific sub-topics and displays this information to another learner in form of a bar graph. Such information enables learners to structure the learning content, to model their learning partners' knowledge, and to adapt questions and explanations to the knowledge level of their learning partners. Another example with a social GA focus is the Radar tool of Phielix and colleagues (Phielix, Prins, Kirschner, Erkens, & Jaspers, 2011). Learners are provided with a radar diagram visualizing how each group member is rated by the group in terms of reliability, friendliness, cooperation, productivity, influence, and quality of contributions. Such information can directly influence the learning context, such as group climate when learners work towards equal distribution of contributions (preventing social loafing and free-riding).

Independent of the type of awareness that they intend to support, GATs usually process information in three consecutive steps (cf. Buder & Bodemer, 2008): (1) collecting GA information, (2) transforming it in a way that augments the original information, and (3) presenting the transformed data in a way that allows learners to take advantage of it. Each step involves specific research challenges.

Regarding data collection, GA information can result from different kinds of learner behaviors. Information can be harvested as a byproduct of the learners' activities such as searching, navigating, selecting, writing, or replying without requiring learners to provide additional information intentionally. For example, written text may be used to extract information on the writer (e.g., Erkens, Bodemer, & Hoppe, 2016). Many recommendation systems use this kind of non-obtrusive data collection. On the other hand, users can provide data intentionally, such as judging an article, their prior knowledge on a subject, or a learning partner's helpfulness (cf. the Radar tool of Phielix et al., 2011). Intentionally providing information gives learners the opportunity to portray themselves in an intended manner and to purposefully communicate individual opinions or needs. While this may account for authenticity and may additionally focus attention on the assessed concepts (e.g., own knowledge gaps), it may also corrupt the objectivity of the data, if it targets objectifiable information such as knowledge. Explicitly testing for learner characteristics is another approach requiring intentional activities, but aims at assessing objective data (e.g., the aforementioned Knowledge Awareness tool of Sangin et al., 2011).

When data is gathered, GATs frequently transform the information in a way that reduces complexity, enables social comparison, and permits the identification of specific patterns because learners may not be able to use the unmodified presentation of information for better learning. This is particularly important, when large datasets are collected (e.g., using discussion contributions and artifact modifications for identifying controversial perspectives). Transformation methods of GA tools range from rather simple methods (e.g. means, variances, correlations) to sophisticated approaches (e.g. text mining, network analysis).

After collecting and transforming GA information, it can be fed back to the learner, the learning group, or the learning community. The way of presenting this information largely depends on the target concept portrayed, e.g., its complexity, stability, or its relation to the learning task. For example, information on the group members' current participation requires constant updates, while information on their general readiness to help others may not. Furthermore, presenting GA information depends on the processes the tool designers want to foster, such as guiding attention or communication by rearranging and highlighting the data. For example, a GAT can present information on the learner in close proximity to group information, encouraging processes of social comparison between learners and their group. However, it can also rearrange the data in a way that suggests comparison processes between other group members or within learners (cf. Buder, 2011).

The effectiveness of GATs for collaborative learning is usually evaluated on a global tool level, disregarding the different functions and processes involved (see below). However, with a differentiated view, various functions of GATs can be identified and distinguished. For example, cognitive GATs can potentially support learning processes in various ways: (1) As a core function, providing knowledge-related information on learning partners might facilitate grounding and partner modelling processes during collaborative learning. However, (2) such information also refers to specific and often preselected content (e.g., a learning partner's hypothesis regarding a single element of the learning material), thereby cueing essential information about the learning material and constraining content-related communication. (3) When cognitive GATs provide information in a way that allows for comparing learning partners, they can guide learners to discuss particularly beneficial issues, such as diverging perspectives. (4) In addition to supporting collaboration processes, collecting and providing knowledge-related information may prompt learners to (re-)evaluate or refocus their individual learning processes.

Empirical Group Awareness Research: Methods and Findings

There is a variety of different approaches to studying the impact of GATs on CSCL. The majority of studies are highly controlled experiments conducted in the laboratory to systematically and validly explicate tool effects within one study session, with case studies mainly used to supplement experimental research. However, attempts have been made to foster external validity by bringing GA research into real-life contexts like schools (e.g., Phielix et al., 2011), or university courses (e.g., Lin, Mai, & Lai, 2015), often conducted over multiple sessions or longer periods of time. Additionally, some studies have been integrated into social media settings (e.g., Heimbuch & Bodemer, 2017). Studies researching effects of GATs mainly use computer-mediated communication scenarios (e.g., Sangin et al., 2011). Few use face-to-face settings, (e.g., Alavi & Dillenbourg, 2012) or blended learning settings (e.g., Puhl, Tsovaltzi, & Weinberger, 2015). Group sizes vary vastly, including dyads (e.g., Dehler, Bodemer, Buder, & Hesse, 2011), small groups of three to six students (e.g., Engelmann & Hesse, 2011), and whole seminars or classes (e.g., Puhl et al., 2015). Pseudo-collaborative studies supplement this research by eliminating the dynamics of actual learner interaction to systematically analyze tool effects in detail (e.g., Cress, 2005; Schnaubert & Bodemer, 2016).

The majority of studies researching GA focus on the implementation of GATs to foster specific kinds of awareness. Thus, these studies compare groups working with a GAT to groups without such a tool. Only few studies go beyond these very basic studies of GA support and systematically investigate specific tool features or functions. Additionally, individual characteristics moderating tool effects are rarely considered in GA research. Conceptualizations of GATs are as broad as the concept itself in defining, collecting, transforming, and presenting relevant data. The focus of most studies concerned with GATs is on the presented information rather than on the actual awareness, even if there are some studies that consider GA as a dependent variable (Janssen, Erkens, & Kirschner, 2011), as a mediator between tool provision and the adjustment of learning processes and learning outcomes (e.g., Sangin et al., 2011), or as an independent variable in terms of a treatment check (Engelmann & Hesse, 2011).

Mostly, it is assumed that providing GA information inevitably fosters awareness, but attempts to specifically assess awareness are rare. This methodological shortcut might be somewhat valid in some cases—for example, if it is observed that learners structure their learning processes according to the provided GA information. However, the validity of this inference highly depends on the close link between awareness information and task-related activities and is thus questionable without this link, e.g., when specific learner behaviors are ascribed to the presentation of very general GA information. Thus, some researchers related GATs to behavioral changes by asking learners about the perceived impact of GATs on the learning processes (e.g., Jongsawat & Premchaiswadi, 2014) or about tool usage (e.g., Jermann & Dillenbourg, 2008). Although this approach may demonstrate causal relations perceived by the learners, it still does not clearly infer GA. Some attempts have been made to account for this gap within the chain of effect by trying to assess GA directly, mostly by assessing whether learners access the provided GA information provided by the tool (e.g., Engelmann & Hesse, 2011) or how closely individual partner models portrayed partner characteristics after collaboration (e.g., Sangin et al., 2011). However, while objectively assessing the availability of GA information, this approach neglects the situational aspect of GA. Other approaches use self-assessment questionnaires asking for perceived GA (e.g., Janssen et al., 2011; Shah & Marchionini, 2010) and thus try to account for the awareness aspect but rely solely on self-report. Altogether, we can conclude that there have been multiple attempts to assess GA, but that further efforts are desirable to directly assess it during the process of collaboration and to consider GA as a mediating variable.

Apart from GA itself, a great variety of dependent variables is assessed within GA research in the learning sciences, with both learning processes and outcomes usually reported within one study. Hereby, assessed learning processes differ vastly between studies and may include communication processes (e.g., Gijlers & de Jong, 2009), navigation or selection processes of topics or items to discuss (e.g., Bodemer, 2011; Schnaubert & Bodemer, 2016), or participation quantity (e.g., Janssen et al., 2011). Learning outcomes mainly consist of individual knowledge tests or problem-solving performance after collaboration (e.g., Sangin et al., 2011). Alternatively, artifacts of the collaboration process may be evaluated, e.g., the quality of textual artifacts or a concept map constructed during collaboration may be scored (e.g., Engelmann & Hesse, 2011). Due to the interactivity of collaborative processes, issues of interdependence between individual learners need to be considered when deciding on appropriate statistical analyses (cf. De Wever & Van Keer, this volume).

Due to the variety of methods used in GA research, positive effects of GATs have repeatedly been found for different learning processes and outcomes in a number of educational settings, over various group sizes and in very different scenarios (e.g., face to face vs. online). Thus, GATs seem to have generalizable positive effects. However, the variety and variability of studies and target concepts make it hard to pinpoint the mechanisms and tool features responsible. Thus, it is not surprising—although there are some rather stable effects—that results differ widely between studies. Fine-grained cognitive awareness tools lead to guidance effects rather consistently insofar that learners change selection strategies of topics or items to attend to (e.g., Bodemer, 2011; Gijlers & de Jong, 2009), and effects on how learners shape their communication to discuss the material can be found frequently (e.g., Dehler et al., 2011; Sangin et al., 2011). However, replication studies are rare. Effects on learning outcomes like individual or group performance are common for cognitive GATs (e.g., Gijlers & de Jong, 2009; Sangin et al., 2011), but not consistent throughout or even within studies (e.g., Buder & Bodemer, 2008; Engelmann & Hesse, 2011). Cognitive GATs also seem to foster building an accurate partner model (e.g., Sangin et al., 2011; Schreiber & Engelmann, 2010). As for social and behavioral awareness tools, providing information on participation seems to change how much or equally learners within a group participate (e.g., Jermann & Dillenbourg, 2008; Michinov & Primois, 2005), although this may depend on the presentation format (e.g., Kimmerle & Cress, 2008, 2009). Again, positive effects on learning outcomes or the quality of group work results are found (e.g., Jongsawat & Premchaiswadi, 2014), but not consistently (e.g., Jermann & Dillenbourg, 2008; Phielix et al., 2011). However, effects of social GATs on learning outcomes are not always reported. Furthermore, it seems that social GATs have positive effects on the perception of the group and group cohesion (e.g., Leshed et al., 2009).

As discussed above, GATs may serve distinct functions within the learning process, but few studies have tried to pinpoint the effects or investigated specific aspects of the tool design. Positive examples are the studies of Kimmerle and Cress (2008, 2009), who compared different ways to edit and visualize the data presented (presentation format gradient and composition) and found differential effects on contribution behavior. Leshed and colleagues (2009) compared different visualization types, but found few differences. Aiming at extracting specific functions or mechanisms of GATs, Bodemer and Scholvien (2014) conducted a series of studies where they could find separate effects for information cueing and providing partner information. Although there are some studies reporting differential GAT effects, much more effort is needed to make generalizable statements as to who profits (more or less) from certain types of GA support.

Trends and Developments

Although the last decade of research on GATs within the learning sciences has shown the potential of GATs for various kinds of learning contexts, much remains unclear. As shown, attempts to systematically investigate features and functions of GATs are in their infancy. Similarly, differential effects are largely unknown, as well as effects on long-persisting groups. Thus, future research faces the task of shifting attention from inventing more and more tools and showing their effectiveness to comparing tool features and extracting precise mechanisms of GATs to ultimately provide efficient and precise support tailored to specific settings, tasks, and individuals, thereby generating highly effective tools that may be distributed throughout educational settings. To reach this goal, empirical and theoretical work is needed. The field lacks theories integrating the various kinds of GA into models of learning and thus precisely matching them with specific learning processes. Moreover, further empirical and theoretical research is needed to account for as yet isolated empirical findings, considering the chain of effects from tool provision to learning outcomes.

Related to this aspect, the combination with approaches and developments of other areas within the learning sciences can enrich GA research with regard to theoretical assumptions, empirical investigations and the development of GATs. Although these opportunities have not been taken to its full potential, there are several recent efforts that have been trying to integrate findings and incorporate them into instructional designs.

For example, research on collaboration scripts (Kollar, Wecker, & Fischer, this volume) appears to be in stark contrast with GA research, because scripts provide explicit guidance on how to coordinate or structure the learning process rather than implicitly guiding by providing information. However, current views integrate both approaches, acknowledging their different and complementary support mechanisms. Another example concerns research on (metacognitive) self-regulation that has recently been extended to CSCL, accounting for the need to support processes beyond coordination of knowledge construction (Järvelä & Hadwin, 2013; Lin, Lai, Lai, & Chang, 2016). Learners have to regulate collaboration, e.g., by constructing a shared understanding of the task, negotiating goals, or monitoring their progress (Miller & Hadwin, 2015). Approaches to foster socially shared self-regulated learning include enhancing GA by mirroring group processes and by encouraging learners to externalize their learning processes (Järvelä et al., 2016). In a recent attempt to combine the fields, Järvelä and colleagues (Järvelä et al., 2015) used the Radar tool (Phielix et al., 2011) and explicitly adapted it to support self- and shared regulation processes. Although experimental investigations of this support are still due, it promises a lot of potential to integrate techniques established in GAT research with information about individual and shared regulation processes to support CSCL. As a final example, learning materials have a major impact on learning processes not only in individual but also in collaborative settings. Thus, research on the instructional design of multiple, dynamic, and interactive learning material that is predominantly concerned with individuals needs to be connected to CSCL research. As GATs are particularly suited to be combined with content-related support measures due to their unobtrusive appearance and their focus on the social context, GA research already made some attempts to combine both fields (Bodemer, 2011). However, both fields of research still have a lot to learn from each other.

Of course, also considering and distinguishing approaches from other disciplines than the learning sciences can complement, specify, and enrich learning-related GA research. This has been particularly done for knowledge-related aspects of GA that have been linked to transactive memory, common ground, team mental models, and other approaches that concern knowledge about others' knowledge (e.g., Schmidt, 2002; Schmidt & Randall, 2016).

Advancements in the research fields of computer science and learning analytics offer new opportunities to gather user data non-intrusively by logging, preparing, and connecting large amounts of data. Such data may provide input for GATs (e.g., text mining methods, Erkens et al., 2016; social network analyses, Lin & Lai, 2013), improving not only the validity of the information portrayed, but also reducing the effort for learners and instructors (such as teachers), allowing broad usage. Furthermore, such methods may also be used to analyze the effects of GATs by analyzing and comparing user behavior and/or artifacts.

Conclusions

In this chapter, we defined GA as valid information on a learner's group or its members that is mentally present (aware) within the individual learner. The information may be group- or partner-related, may contain situational and/or stable characteristics that may be classified within a wide range of psychological concepts (e.g., social, motivational, emotional, cognitive, behavioral). Moreover, we portrayed GATs as (mostly) computer-based measures that collect, transform, and present such information to learners. Although positive effects have been found throughout a great variety of studies and settings, the future of GATs will require interacting efforts of multiple disciplines related to the learning sciences. Such interdisciplinary research activities can advance our knowledge about the mechanisms of GATs within learning (e.g., psychology), to use the potential of technologies to enhance the tools' effectiveness and efficiency (e.g., computer science) and to bring these tools into specific educational settings (e.g., educational sciences).

Further Readings

Bodemer, D. & Dehler, J. (Guest Eds.). (2011). Group awareness in CSCL environments [Special Section]. *Computers in Human Behavior, 27*(3), 1043–1117.
This special section is an integrative compilation of empirical CSCL research on group awareness, giving insight into complementary tools and methods.

Janssen, J., & Bodemer, D. (2013). Coordinated computer-supported collaborative learning: Awareness and awareness tools. *Educational Psychologist, 48*(1), 40–55. doi:10.1080/00461520.2012.749153
A systematic integration and discussion of various examples and variations of cognitive and social awareness tools found in literature.

Järvelä, S., Kirschner, P. A., Hadwin, A., Järvenoja, H., Malmberg, J., Miller, M., & Laru, J. (2016). Socially shared regulation of learning in CSCL: Understanding and prompting individual- and group-level shared regulatory activities. *International Journal of Computer-Supported Collaborative Learning, 11*(3), 263–280. doi:10.1007/s11412-016-9238-2
An article connecting socially shared regulation with various strands of CSCL support while discussing the relevance of self and group awareness.

Lin, J.-W., Mai, L.-J., & Lai, Y.-C. (2015). Peer interaction and social network analysis of online communities with the support of awareness of different contexts. *International Journal of Computer-Supported Collaborative Learning, 10*(2), 139–159. doi:10.1007/s11412-015-9212-4
A current empirical article investigating the differential impact of social and cognitive awareness information using social network analysis.

Schmidt, K. (2002). The problem with awareness. *Computer Supported Cooperative Work, 11*(3), 285–298. doi:10.1023/A:1021272909573
An earlier but significant conceptual contribution from the research field of computer-supported cooperative work.

References

Alavi, H. S., & Dillenbourg, P. (2012). An ambient awareness tool for supporting supervised collaborative problem solving. *IEEE Transactions on Learning Technologies, 5*(3), 264–274. doi:10.1109/TLT.2012.7

Bodemer, D. (2011). Tacit guidance for collaborative multimedia learning. *Computers in Human Behavior, 27*(3), 1079–1086. doi:10.1016/j.chb.2010.05.016

Bodemer, D., & Dehler, J. (2011). Group awareness in CSCL environments. *Computers in Human Behavior, 27*(3), 1043–1045. doi:10.1016/j.chb.2010.07.014

Bodemer, D., & Scholvien, A. (2014). Providing knowledge-related partner information in collaborative multimedia learning: Isolating the core of cognitive group awareness tools. In C.-C. Liu, H. Ogata, S. C. Kong, & A. Kashihara (Eds.), *Proceedings of the 22nd International Conference on Computers in Education ICCE 2014* (pp. 171–179). Nara, Japan.

Buder, J. (2011). Group awareness tools for learning: Current and future directions. *Computers in Human Behavior, 27*(3), 1114–1117. doi:10.1016/j.chb.2010.07.012

Buder, J., & Bodemer, D. (2008). Supporting controversial CSCL discussions with augmented group awareness tools. *International Journal of Computer-Supported Collaborative Learning, 3*(2), 123–139. doi:10.1007/s11412-008-9037-5

Cress, U. (2005). Ambivalent effect of member portraits in virtual groups. *Journal of Computer Assisted Learning, 21*(4), 281–291. doi:10.1111/j.1365-2729.2005.00136.x

De Wever, B., & van Keer, H. (2018). Selecting statistical methods for the Learning Sciences and reporting their results. In F. Fischer, C. E. Hmelo-Silver, S. R. Goldman, & P. Reimann (Eds.), *International handbook of the learning sciences* (pp. 532–541). New York: Routledge.

Dehler, J., Bodemer, D., Buder, J., & Hesse, F. W. (2011). Guiding knowledge communication in CSCL via group knowledge awareness. *Computers in Human Behavior, 27*(3), 1068–1078. doi:10.1016/j.chb.2010.05.018

Dourish, P., & Bellotti, V. (1992). Awareness and coordination in shared workspaces. In M. Mantel & R. Baecker (Eds.), *Proceedings of the 1992 ACM conference on Computer-supported cooperative work* (pp. 107–114). Toronto, Canada: ACM Press. doi:10.1145/143457.143468

Endsley, M. R. (1995). Toward a theory of situation awareness in dynamic systems. *Human Factors: The Journal of the Human Factors and Ergonomics Society, 37*(1), 32–64. doi:10.1518/001872095779049543

Engelmann, T., & Hesse, F. W. (2011). Fostering sharing of unshared knowledge by having access to the collaborators' meta-knowledge structures. *Computers in Human Behavior, 27*, 2078–2087. doi:10.1016/j.chb.2011.06.002

Erkens, M., Bodemer, D., & Hoppe, H. U. (2016). Improving collaborative learning in the classroom: Design and evaluation of a text mining based grouping and representing. *International Journal of Computer-Supported Collaborative Learning, 11*(4), 387–415. doi:10.1007/s11412-016-9243-5

Gijlers, H., & de Jong, T. (2009). Sharing and confronting propositions in collaborative inquiry learning. *Cognition and Instruction, 27*(3), 239–268. doi:10.1080/07370000903014352

Gross, T., Stary, C., & Totter, A. (2005). User-centered awareness in computer-supported cooperative work-systems: Structured embedding of findings from social sciences. *International Journal of Human-Computer Interaction, 18*(3), 323–360. doi:10.1207/s15327590ijhc1803_5

Heimbuch, S., & Bodemer, D. (2017). Controversy awareness on evidence-led discussions as guidance for students in wiki-based learning. *The Internet and Higher Education, 33*, 1–14. doi:10.1016/j.iheduc.2016.12.001

Janssen, J., Erkens, G., & Kirschner, P. A. (2011). Group awareness tools: It's what you do with it that matters. *Computers in Human Behavior, 27*(3), 1046–1058. doi:10.1016/j.chb.2010.06.002

Järvelä, S., & Hadwin, A. F. (2013). New frontiers: Regulating learning in CSCL. *Educational Psychologist, 48*(1), 25–39. doi:10.1080/00461520.2012.748006

Järvelä, S., Kirschner, P. A., Hadwin, A., Järvenoja, H., Malmberg, J., Miller, M., & Laru, J. (2016). Socially shared regulation of learning in CSCL: Understanding and prompting individual- and group-level shared regulatory activities. *International Journal of Computer-Supported Collaborative Learning, 11*(3), 263–280. doi:10.1007/s11412-016-9238-2

Järvelä, S., Kirschner, P. A., Panadero, E., Malmberg, J., Phielix, C., Jaspers, J., et al. (2015). Enhancing socially shared regulation in collaborative learning groups: Designing for CSCL regulation tools. *Educational Technology Research and Development, 63*(1), 125–142. doi:10.1007/s11423-014-9358-1

Jermann, P., & Dillenbourg, P. (2008). Group mirrors to support interaction regulation in collaborative problem solving. *Computers & Education, 51*(1), 279–296. doi:10.1016/j.compedu.2007.05.012

Jongsawat, N., & Premchaiswadi, W. (2014). A study towards improving web-based collaboration through availability of group awareness information. *Group Decision and Negotiation, 23*(4), 819–845. doi:10.1007/s10726-013-9349-3

Kimmerle, J., & Cress, U. (2008). Group awareness and self-presentation in computer-supported information exchange. *International Journal of Computer-Supported Collaborative Learning, 3*(1), 85–97. doi:10.1007/s11412-007-9027-z

Kimmerle, J., & Cress, U. (2009). Visualization of group members' participation how information-presentation formats support information exchange. *Social Science Computer Review, 27*(2), 243–261. doi:10.1177/0894439309332312

Kollar, I., Wecker, C., & Fischer, F. (2018). Scaffolding and scripting (computer-supported) collaborative learning. In F. Fischer, C. E. Hmelo-Silver, S. R. Goldman, & P. Reimann (Eds.), *International handbook of the learning sciences*. New York: Routledge.

Leshed, G., Perez, D., Hancock, J. T., Cosley, D., Birnholtz, J., Lee, S., et al. (2009). Visualizing real-time language-based feedback on teamwork behavior in computer-mediated groups. In *Proceedings of the SIGCHI Conference on Human Factors in Computing Systems* (pp. 537–546). New York: ACM Press. doi:10.1145/1518701.1518784

Lin, J.-W., & Lai, Y.-C. (2013). Online formative assessments with social network awareness. *Computers & Education, 66*, 40–53. doi:10.1016/j.compedu.2013.02.008

Lin, J.-W., Lai, Y.-C., Lai, Y.-C., & Chang, L.-C. (2016). Fostering self-regulated learning in a blended environment using group awareness and peer assistance as external scaffolds. *Journal of Computer Assisted Learning, 32*(1), 77–93. doi:10.1111/jcal.12120

Lin, J.-W., Mai, L.-J., & Lai, Y.-C. (2015). Peer interaction and social network analysis of online communities with the support of awareness of different contexts. *International Journal of Computer-Supported Collaborative Learning, 10*(2), 139–159. doi:10.1007/s11412-015-9212-4

Michinov, N., & Primois, C. (2005). Improving productivity and creativity in online groups through social comparison process: New evidence for asynchronous electronic brainstorming. *Computers in Human Behavior, 21*(1), 11–28. doi:10.1016/j.chb.2004.02.004

Miller, M., & Hadwin, A. (2015). Scripting and awareness tools for regulating collaborative learning: Changing the landscape of support in CSCL. *Computers in Human Behavior, 52*, 573–588. doi:10.1016/j.chb.2015.01.050

Ogata, H., & Yano, Y. (2000). Combining knowledge awareness and information filtering in an open-ended collaborative learning environment. *International Journal of Artificial Intelligence in Education, 11*(1), 33–46.

Phielix, C., Prins, F. J., Kirschner, P. A., Erkens, G., & Jaspers, J. (2011). Group awareness of social and cognitive performance in a CSCL environment: Effects of a peer feedback and reflection tool. *Computers in Human Behavior, 27*(3), 1087–1102. doi:10.1016/j.chb.2010.06.024

Puhl, T., Tsovaltzi, D., & Weinberger, A. (2015). Blending Facebook discussions into seminars for practicing argumentation. *Computers in Human Behavior, 53*, 605–616. doi:10.1016/j.chb.2015.04.006

Sangin, M., Molinari, G., Nüssli, M.-A., & Dillenbourg, P. (2011). Facilitating peer knowledge modeling: Effects of a knowledge awareness tool on collaborative learning outcomes and processes. *Computers in Human Behavior, 27*(3), 1059–1067. doi:10.1016/j.chb.2010.05.032

Schmidt, K. (2002). The problem with awareness. *Computer Supported Cooperative Work, 11*(3), 285–298. doi:10.1023/A:1021272909573

Schmidt, K., & Randall, D. (Eds.). (2016). Reconsidering "awareness" in CSCW [Special Issue]. *Computer Supported Cooperative Work (CSCW), 25*(4–5), 229–423.

Schnaubert, L., & Bodemer, D. (2016). How socio-cognitive information affects individual study decisions. In C.-K. Looi, J. Polman, U. Cress, & P. Reimann (Eds.), *Transforming Learning, Empowering Learners: The International Conference of the Learning Sciences (ICLS) 2016* (pp. 274–281). Singapore: International Society of the Learning Sciences.

Schreiber, M., & Engelmann, T. (2010). Knowledge and information awareness for initiating transactive memory system processes of computer-supported collaborating ad hoc groups. *Computers in Human Behavior, 26*(6), 1701–1709. doi:10.1016/j.chb.2010.06.019

Shah, C., & Marchionini, G. (2010). Awareness in collaborative information seeking. *Journal of the Association for Information Science and Technology, 61*(10), 1970–1986. doi:10.1002/asi.21379View

Soller, A., Martínez, A., Jermann, P., & Muehlenbrock, M. (2005). From mirroring to guiding: A review of state of the art technology for supporting collaborative learning. *International Journal of Artificial Intelligence in Education, 15*(4), 261–290.

Stahl, G. (2016). From intersubjectivity to group cognition. *Computer Supported Cooperative Work (CSCW), 25*(4–5), 355–384. doi:10.1007/s10606-016-9243-z

35
Mobile Computer-Supported Collaborative Learning

Chee-Kit Looi and Lung-Hsiang Wong

Introduction

Research in computer-supported collaborative learning (CSCL) has sought to understand the types of interaction in collaboration that leads to and explains positive learning outcomes. Such interactions include the cognitive efforts in establishing a joint understanding, explanations, argumentations, conflict resolution, knowledge construction, and artifact co-construction. Collaborative activities can be integrated into learning activity workflows or pedagogical scenarios that include individual, small-group, class-wide, or community-wide activities occurring in a variety of settings (e.g., classroom, home, workplace, field trips, off-campus community), modes (e.g., face-to-face or remote; synchronous or asynchronous), and devices (one device for each learner; one device for a group of learners; a few devices for each learner or for many learners).

Mobile learning provides the additional premises that learners are mobile, and mobile technologies are ubiquitous and ready-at-hand, enabling even more opportunities for learners to share and co-construct knowledge readily in different settings and modes. Such mobile technologies refer to handheld mobile devices such as smartphones, personal digital assistants (PDAs), and other portable computers. Unlike mainstream CSCL, which has developed to become an established area of research in the learning sciences, the field of CSCL using mobile technologies, known in the literature as mCSCL (mobile CSCL), is a peripheral area under the mainstream mobile learning field. In this field, the current foci of individual projects are not necessarily on collaboration-focused and CSCL-style analysis but more on exploiting the communication affordances of mobile devices to fulfill mobile learning goals. In addition, due to the mobile and fluidic nature of the learners, learning tasks, and learning contexts, mobile learners are typically engaged in multiple parallel, rapid, and ad hoc interactions which may be triggered by incidental encounters "in the wild," rather than attending to relatively well-structured decision processes within pre-designed and more controlled learning contexts (cf. Zurita, Autunes, Baloian, & Baytelman, 2007). Thus, besides situating mCSCL in mainstream CSCL frameworks, it is important for mCSCL researchers to make sense of the uniqueness of mobile learning in order to appropriately and effectively tackle the challenges of "collaboration on the move."

This chapter provides a summary of research and development in the field of mCSCL. We explore the synergies between CSCL approaches and mobile learning approaches. We will discuss the characteristics and affordances of mobile technologies, and contemporary mCSCL learning designs and practices, and postulate future trends and developments in mCSCL research.

Theorizing Mobile Learning

Sharples, Taylor, and Vavoula (2007) paved the way in developing a theory of mobile learning that re-conceptualizes learning by encompassing both learning supported by the mobile technology and learning characterized by the mobility of people and knowledge. Such a theory should be distinguished from other theories of learning by encompassing: (a) that learners learn across space as they take ideas and learning resources obtained in one location and apply or develop them in another; learners learn across time by revisiting knowledge acquired earlier in a different context, which provides a framework for lifelong learning; and learners move in and out of engagement with technology; (b) how impromptu sites of learning are created out of offices, classrooms and lecture halls; (c) a socio-constructivist approach that views learning as an active process of building knowledge and skills through practice in a supportive community; and (d) the ubiquitous use of personal and shared technologies. Based on these criteria, mobile learning is defined as "the processes of coming to know through conversations across multiple contexts amongst people and personal interactive technologies" (p. 225).

Thus, mobile learning is conceptualized as a process of coming to know through conversation across continually reconstructed contexts. Akin to Activity Theory (Engeström, 1987; Danish & Gresalfi, this volume), learning is conceptualized as interactions and negotiations between individuals, humans or non-humans, which occur in the form of evolving states of knowing as they are shaped by continuously negotiated goals in the changing contexts. The mobile technology provides a *shared conversational learning space* on the move, which can be used not only for single learners but also for learning groups and communities. The technology can also be utilized as mediating tools to demonstrate, talk over ideas, or proffer advice, or negotiate agreements.

The conversational framework of Laurillard (2007) holds that learners will be motivated to improve their collaborative practices if they can share their products with peers, and to enhance their conceptual understanding if they can reflect on their experiences by discussing their products with peers. Students' contextualized artifacts created in situ while on the move have the potential to go beyond facilitating just-in-time knowledge sharing to mediating future knowledge co-constructions (So, Seow, & Looi, 2009; Wong, Chen, & Jan, 2012). A typical collaborative mobile learning activity can provide more opportunities for digitally facilitated site-specific collaboration, and for ownership and control over what the learners do jointly, because the mobile devices digitally facilitate the link between the students and the data/products on the spot. However, while the conversational framework describes conversations for learning situated in one physical location, it does not address adequately the challenging issues of the constantly negotiated communication and interaction in the continually changing context in mCSCL (Sharples et al., 2007).

A key characteristic of mobile learning is the notion of seamless learning across contexts. Seamless learning includes collaborative learning with different partners across the different grains of space, time, and devices. The artifacts and postings created and built upon by learners are always re-contextualized depending on who the participants are, what learning spaces they are in, or whether they on the move, leading to and supporting a rich plethora of CSCL scenarios. For example, in the mobile-assisted Chinese idiom learning trajectory of Move, Idioms!, fifth-grade students worked in groups to co-create social media (student artifacts) that captured incidents within the school campus (a physical space) and utilized the idioms learned (Wong, Chin, Tan, & Liu, 2010). They then posted their social media online (cyberspace; re-contextualization), which will then open to the entire class to discuss and critique on the incidents, or review its linguistic accuracy (Wong et al., 2012). However, according to the meta-analysis of the mCSCL literature conducted by Song (2014) there are not many studies of "collaboration in the wild" that look at CSCL interactions with the same kind of rigor used for CSCL studies for more well-defined domains and tasks.

Learning and Collaboration Spaces Afforded by Mobile Devices

Core affordances of mainstream technology for collaborative learning based on theories of collaborative learning and CSCL practices have been proposed, such as affording learner opportunities to engage in a joint task, communicate, share resources, engage in productive collaborative learning processes, engage in co-construction, monitor and regulate collaborative learning, and find and build groups and communities (Jeong & Hmelo-Silver, 2016). The relevant question for us here is what the unique affordances of mobile technologies are, which extend as well as constrain the possibilities for collaboration beyond mainstream CSCL. One such affordance is that interacting on small mobile devices has the potential to increase shared attention and to build transactivity. When a mobile learning group or community is on the move and the learners are not always physically sticking together, without the mobile devices they could not otherwise communicate and collaborate, face-to-face or through desk-bound computers (e.g., Ogata et al., 2011; Spikol, 2008).

Despite this, some early mobile learning literature cautions about the potential challenge of using small mobile devices in mCSCL activities, particularly before the advent and proliferation of tablets and in the one-device-per-multiple-learners setting. Roschelle and Pea (2002) and Cole and Stanton (2003) comment that limited screen sizes would make the maintenance of shared attention problematic, as they do not easily allow multiple learners to view the same display or to share information. Such a doubt was valid, indeed noteworthy, in the early history of mobile learning, before the proliferation of tablet computers and when most of mobile learning designs adopted one-device-to-multiple-learners settings, perhaps due to the cost constraint.

Over the years, this limitation has been overcome by the advancement of techno-pedagogical designs. Cole and Stanton (2003), for example, contend that the techno-pedagogy can be designed around the ways that learners work and collaborate *around*, rather than *on*, a device. The device may best support more cooperative styles of working, where individual learners are responsible for their respective tasks, including greater interactions with the physical surroundings, but come together on occasion to share digital information. Devices can be used for short bursts of time (e.g., entering and comparing data, looking up and reviewing information, brief communication and sharing of artifacts with peers and remote people) to support physical learning activities in situ (Rogers, Connelly, Hazlewood, & Tedesco, 2010). Within uses of the devices themselves, mediation can be seen as broadly representational or broadly coordinative (Roschelle, Rosas, & Nussbaum, 2005). Representational mediation provides ways of organizing content that facilitate social and cognitive processing. Coordinative mediation allows ways to organize the flow of information among mCSCL devices to support the objectives of an activity.

Types of mCSCL Learning Designs

Wirelessly intercommunicated devices support constructivist educational activities through collaborative groups (Dede & Sprague, 1999), increasing motivation (Lai and Wu, 2006; Liu, Chung, Chen, & Liu, 2009), promoting interactive learning (Zurita and Nussbaum, 2004), developing cognitive skills (ordering, evaluating, synthesizing), and facilitating the control of the learning process and its relationship with the real world (Valdez et al., 2000). Research studies of mCSCL span various disciplines and populations of learners, such as geography (e.g., Reychav & Wu, 2015—involving middle school students), nursing (e.g., Lai & Wu, 2006—undergraduates), mathematics (e.g., Roschelle, Rafanan, Estrella, Nussbaum, & Claro, 2010—fourth graders), computer programming (e.g., Liu et al., 2009—graduate students), natural science (e.g., Sharples et al., 2014—fifth to eighth graders), humanities (e.g., So, Tan, & Tay, 2012—seventh graders), language (e.g., Ogata et al., 2011—undergraduates), graphic design (e.g., Hsu & Ching, 2012—undergraduates), business (e.g., Baloian & Zurita, 2012—undergraduates), and teacher education (e.g., Ke & Hsu, 2015—undergraduates). The research methods employed by the studies encompass design-based research (Roschelle et al.,

2010; Sharples et al., 2014; So et al., 2012), content analysis of learner interactions (Hsu & Ching, 2012; Liu et al., 2009), quasi-experimental (Baloian & Zurita, 2012; Lai & Wu, 2006; Ogata et al., 2011), mixed methods (Ke & Hsu, 2015), and quantitative analysis of learner perceptions and performances (Reychav & Wu, 2015).

Current mCSCL pedagogical design and practices can be categorized them into three main types: in-class mCSCL, out-of-class mCSCL, and mCSCL that bridges both in-class and out-of-class activities. In-class mobile CSCL typically augments the conventional face-to-face collaboration of physical classrooms with networked communication through devices (Boticki, Looi, & Wong, 2011; Nussbaum et al., 2009). In-class mobile CSCL democratizes participation in a classroom by enabling every student to participate in generating ideas and solutions through their personal devices. Out-of-class mobile CSCL supports situated, experiential, social, and inquiry views of learning in situ, typically in field trips and outdoor activities such as Ambient Wood (Rogers & Price, 2009) and LET'S GO! (Maldonado & Pea, 2010).

mCSCL can support seamless learning in bridging both in-class and out-of-class activities (Looi et al., 2010). Seamless learning refers to the synergistic integration of the learning experiences across various dimensions such as across formal and informal learning contexts, individual and social learning, and physical world and cyberspace. The basic premise of seamless learning is that it is not feasible nor productive to equip learners with all the knowledge and skills they need to have based on specific snapshots of episodic timeframe, location, or scenario.

Building on Zurita and Nussbaum (2004) and Hsu and Ching (2013), from a functional perspective as to how they support collaboration, we characterize the functions of the design that utilize mobile devices and technologies to support collaborative learning as depicted in Table 35.1. The broad range of function suggests that mCSCL activities can be situated in a broader curricular or learning workflow system in which mCSCL supports in-context interaction and context delivery and creation, as well as time and space for personalized and social learning.

Current Research into mCSCL

Research into mCSCL has been flourishing in the past decade. Yet the findings and implications presented in the literature are sporadic and largely meant for informing the research in, and practice of, mobile learning rather than trying to answer the big research questions posed by the mainstream CSCL field. The limited research may be due to a wide range of mobile learning approaches that foreground diversified forms and levels of learner collaborations, warranting a variety of research inquiries. In addition, most of the identified mCSCL studies were carried out from the perspective of learning technology with the focus on developing innovative mobile technologies to facilitate mCSCL as well as the (quasi-)experimental evaluations of learning effectiveness. Such studies typically ignored the analysis of collaborative process data, or performed simple analysis only offered limited insight regarding the dynamics of mCSCL. This section focuses on summarizing some salient findings and implications of prior mCSCL studies that contribute to CSCL research more broadly.

One aspect of research into mCSCL pertains to the size and the formation of groups, in view of the flexibility of catering for different group sizes and formations with the use of mobile technologies. In CSCL research, group size can have an effect on both students' influence on other members of the group and on students' performance (Cress, Kimmerle, & Hesse, 2009; Schwabe, Göth, & Frohberg, 2005). Several studies aimed to investigate optimal group sizes for mCSCL and the recurrent findings were that smaller groups tend to collaborate and co-learn more effectively. An early relevant attempt was reported in Schwabe et al. (2005), with the results showing that there is little significant evidence to the superiority of groups of two to groups of three in a location-based mobile learning game, in terms of learners' levels of engagement and immersion. However, groups of four are suboptimal in both accounts. Subsequent studies (e.g., Melero, Hernández-Leo, & Manatunga, 2015; Zurita & Nussbaum, 2007) yielded similar results. A common rationale postulated by these

Table 35.1 Functions, Design Strategies, and Examples of mCSCL

Functions of design	Design strategies	Examples
Enabling every learner to participate in collaborative activities	Support democratized participation of learners by enabling each to generate and post their contributions or solutions to a shared workspace	Clicker systems, and software systems or apps running on mobile devices, enable every learner to spontaneously make and post their contributions wherever they are, anonymously or otherwise, and thus enabling collaboration.
Organizing and presenting the collaborative task and materials	• Organize the individual portions of an assigned collaborative learning task, by supporting cooperative learning activities and other forms of collaboration • Collaboration takes the form of learners talking about the tasks, roles and representations on their mobile devices	In Roschelle et al. (2010), fraction problems in the multiple-choice format were displayed on individual devices held by individual learners during the discussion. They were able to show their individual portions of the tasks using their mobile devices during the discussion activity. They then determined if the fractions shown in multiple representation formats (e.g., numeral representation and pie representation) are equal.
Presenting the individual portions of an assigned learning task and learners interacting to create in-situ groups	Present information on devices to support learners interacting to form groups	In Boticki et al. (2011) and Wong, Boticki, Sun, and Looi (2011), learners are each assigned fractions or parts of Chinese characters, and they need to coordinate with others to form desired fractions or characters. Collaboration takes the form of forming emergent groups or patterns imposed by the task requirements.
Coordinating activity states or information flow that reinforces rules and flows for the learning and collaborative activities	• Enable and constrain the learning activities • Scaffold inquiry-based learning, enabling learners to work across settings and times, individually or in collaborative groups	In the mobile computer-based inquiry toolkit nQuire (Sharples et al., 2014), learners investigate an interest topic using a computer-based toolkit that guides the entire inquiry process and connects structured learning in the classroom with discovery and data collection at home or outdoors.
Facilitating communication and interaction in a shared workspace	Enable learners to send or post messages to other learners and the instructor, including commenting on each other's postings and artifacts	In Wong et al. (2010), with their personal smartphones, language learners share individually or co-created social media pertaining to their real-life encounters by utilizing recently learned vocabulary. They then comment on their peers' postings, and review (or correct) the vocabulary usage on the social network space.
Providing feedback for group learning	Provide feedback automated or otherwise to the appropriateness of students' answers	In Zurita and Nussbaum (2004), tasks are presented to groups of three learners to complete on their handheld devices. Each group works cooperatively. Once all group members have done their parts, feedback is provided to the group on the correctness of their work. This group-level feedback leads to discussions in the group to seek a consensus on revising each member's group to get them right.
Encouraging learner's mobility	Provide the flexibility and contextual affordances of learning and collaborating anywhere, anytime, be it in-class or out-of-class	In Looi et al. (2010), individual students bring their personal smartphones everywhere to perform a variety of individual and collaborative science learning activities, either in situ (usually self-directed) or in class (typically facilitated by the teachers), through or around their smartphones.
Harnessing the location-based features of mobile devices	Create a situated and experiential environment in which information, cues and contexts depend the locations and are shown on the mobile devices	One category of mCSCL design is location-based mobile-assisted game-based learning for outdoors and indoors. Squire and Jan (2007) present a location-based augmented reality game for student groups: to resolve a murder mystery around a lake. Through the learning trail, students develop investigation and inquiry skills with collaboration.

studies is that the larger the group size, the more social or socialization overheads took place (e.g., socializing discourses or longer time needed to reach consensus), diluting group members' attention to the learning tasks, such as their interactions with the authentic environments and the digital information provided by the device.

Groups do not have to be fixed before collaboration. mCSCL affords the emergent formation of groups. Another line of mCSCL research places its interest in grouping learners "on-the-fly." That is, depending on the techno-pedagogical design of the m-learning activities, mobile learner grouping could be more dynamic, ad hoc (e.g., groups may be reshuffled over time), and even context-bound. This might be different from many mainstream CSCL research in which learner grouping tends to be fixed for reasons such as that members from fixed groups have better bonding and mutual understanding, or the instructor can enjoy easier classroom/learning management or execute collaboration scripts more robustly. Nevertheless, most of the studies in this category seemed to be relatively technology-centric, with the common aim of developing innovative schemes, algorithms or technological tools for automated dynamic grouping. Despite offering sound engineering contributions, these development efforts are typically solution- rather than problem-driven, and often lack proper learning theory grounding. Specifically, mobile learning systems were developed to form homogeneous or heterogeneous learner groups based on their psychological traits (Wang, Lin, & Sun, 2007), learning behaviors (Huang & Wu, 2011), and social interaction behaviors (Hsieh, Chen, & Lin, 2010), among others. Some other systems group learners on certain combinations of the above-stated criteria plus the context or the location where the learners are to carry out their collaborative learning tasks (e.g., Amara, Macedo, Bendella, & Santos, 2015; Giemza, Manske, & Hoppe, 2013; Tan, Jeng, & Huang, 2010).

In contrast, Zurita, Nussbaum, and Salinas (2005) reported relatively well-grounded mCSCL work with a focus on dynamic grouping. Informed by the frameworks of Johnson and Johnson (1999) and Race (2000), the authors derived four separate grouping criteria for different objectives—namely, randomness (to achieve social and academic heterogeneity), preference (students are homogeneously grouped according to affinity with their classmates; to reduce choice heterogeneity), academic achievement (heterogeneous grouping; to foster intra-group peer learning), and sociability (students are grouped according to a teacher-defined affinity scale; to reduce social heterogeneity). During the classroom-based collaborative learning sessions, the teacher may change the grouping criteria any time, based on students' learning progress; and the system will reshuffle the student groups accordingly. The application of the work lies in the simple and fast way dynamic grouping can be applied with wirelessly interconnected device support, reshuffling participants in groups of different sizes chosen from a given set of students.

Future Research on mCSCL

Based on this review of existing work, we see several challenges for future research on mCSCL. We need synergy between contextualization and personalization of learning with mobile and ubiquitous technologies, and the affordances for collaboration. As mobile technologies afford individual personalized learning, from a design consideration point of view we need to consider how these elements pave the way or contribute to CSCL. A design framework for mCSCL should acknowledge that many social aspects of learning proceed better without technology and thus identify particular mediating roles for technology (Roschelle, Rosas, & Nussbaum, 2005). Technology should mediate group work to promote collaborative and not just personal uses of mobile technology.

In CSCL, there is a call for studying the interactions of collaborative learning to be understood, supported, and analyzed at multiple levels of time, space, and scale. Understanding the interactions in mCSCL would require understanding the interactions that happen in the different contexts of space, time, and devices. The attempt to bridge across levels of analysis—in CSCL theory, analysis, and practice—stands at the forefront of CSCL research today (Stahl, 2013). A microcosm for studying many of these issues is provided by mCSCL. However, there is a current dearth of in-depth mCSCL studies, but we hope to see more mCSCL research that can illuminate processes of collaboration for which interaction over mobile devices contributes an important part to the overall learning experiences.

Mobile CSCL studies that bridge formal and informal contexts face new methodological challenges with the need for data collection methods that can capture mobile CSCL processes and outcomes in continually moving and reconstructed contexts. Learners may carry and use their mobile devices as their personalized devices to do a range of activities, not all of which may be relevant for the analysis of CSCL interactions. The distributed and sparse nature of interactions through and over mobile devices poses a challenge for tracing the uptake of ideas and idea development processes. Advances in methodological collection and analyses of data coming from multiple learning spaces and devices are needed. Research methods arising from the development of learning analytics and big data analysis may provide some possible pathways to enable analyses of collaboration data. In summary, the learning sciences perspective will have much to offer to advancing mCSCL research.

Further Readings

Looi, C.-K., Seow, P., Zhang, B. H., So, H.-J., Chen, W., & Wong, L.-H. (2010). Leveraging mobile technology for sustainable seamless learning: A research agenda. *British Journal of Educational Technology, 42*(1), 154–169.
This article presents a vision of seamless learning, with mobile technology serving as a learning hub to bridge learning and collaboration in in-class and out-of-class settings, as well across other dichotomies of learning contexts. Elements of a global research agenda on seamless learning are articulated.

Roschelle, J., Rosas, R., & Nussbaum, M. (2005, May). Towards a design framework for mobile computer-supported collaborative learning. *Proceedings of the Conference on Computer Support for Collaborative Learning 2005: The Next 10 Years!* (pp. 520–524). Taipei, Taiwan: ISLS.
This article presents a design framework for mCSCL. The design framework proposes that mediation can occur in two complementary layers: social (e.g., rules and roles) and technological. The technological layer has two components, representational mediation and networked mediation. A particular design challenge is achieving an effective allocation of supporting structure to each layer as well as simple, transparent flow between them. The authors argue that the framework is sufficiently general to describe the most important mCSCL activities described in the literature.

Sharples, M., Scanlon, E., Ainsworth, S., Anastopoulou, S., Collins, T., Crook, et al. (2014). Personal inquiry: Orchestrating science investigations within and beyond the classroom. *Journal of the Learning Sciences, 24*(2). 308–341.
Mobile technology can be used to scaffold inquiry-based learning, enabling learners to work individually or collaboratively across settings and times. It can expand learners' opportunities to understand the nature of inquiry whilst they engage with the scientific content of a specific inquiry. This paper reports on the use of the mobile computer-based inquiry toolkit nQuire to manage the organization of the learning flow or scenario.

Song, Y. (2014). Methodological issues in mobile computer-supported collaborative learning (mCSCL): What methods, what to measure and when to measure?. *Educational Technology & Society, 17*(4), 33–48.
This article presents a review of mobile computer-supported collaborative learning (mCSCL) literature with the foci on investigating: (1) methods utilized in mCSCL research which focuses on studying, learning, and collaboration mediated by mobile devices; (2) whether these methods have examined mCSCL effectively; (3) when the methods are administered; and (4) what methodological issues exist in mCSCL studies. It attempts to bring to light methods that are more conducive to examining the effectiveness of mCSCL and thus in sustaining the practices.

Zurita, G., & Nussbaum, M. (2004). Computer supported collaborative learning using wirelessly interconnected handheld computers. *Computers & Education, 42*(3), 289–314.
This article presents early work on the unique affordances of mCSCL designs to support communication, negotiation, coordination, and interactivity in CSCL. It observes collaboration of young students in the classroom and identifies weaknesses in the aforementioned aspects, provides an mCSCL design, and investigate how mCSCL can reduce these weaknesses.

NAPLeS Resources

Dillenbourg, P. *15 minutes about orchestration* [Video file]. In *NAPLeS video series*. Retrieved October 19, 2017, from http://isls-naples.psy.lmu.de/video-resources/guided-tour/15-minutes-dillenbourg/index.html

Looi, C.-K. *15 minutes about seamless learning* [Video file]. In *NAPLeS video series*. Retrieved October 19, 2017, from http://isls-naples.psy.lmu.de/video-resources/guided-tour/15-minute-cee_kit_looi/index.html

Looi, C.-K. *Interview about seamless learning* [Video file]. In *NAPLeS video series*. Retrieved October 19, 2017, from http://isls-naples.psy.lmu.de/video-resources/interviews-ls/cee_kit_looi/index.html

References

Amara, S., Macedo, J., Bendella, F., & Santos, A. (2015). Dynamic group formation in mobile computer supported collaborative learning environment. *Proceedings of the 7th International Conference on Computer Supported Education* (pp. 530–539), Lisbon, Portugal.

Baloian, N., & Zurita, G. (2012). Ubiquitous mobile knowledge construction in collaborative learning environments. *Sensors, 12*, 6995–7014.

Boticki, I., Looi, C.-K., & Wong, L.-H. (2011). Supporting mobile collaborative activities through scaffolded flexibile grouping. *Educational Technology & Society, 14*(3), 190–202.

Cole, H., & Stanton, D. (2003). Designing mobile technologies to support co-present collaboration. *Personal and Ubiquitous Computing, 7*(6), 365–371.

Cress, U., Kimmerle, J., & Hesse, F. W. (2009). Impact of temporal extension, synchronicity, and group size on computer-supported information exchange. *Computers in Human Behavior, 25*(3), 731–737.

Danish, J. A., & Gresalfi, M. (2017). Cognitive and sociocultural perspective on learning: Tensions and synergy in the Learning Sciences. In F. Fischer, C. E. Hmelo-Silver, S. R. Goldman, & P. Reimann (Eds.), *International handbook of the learning sciences* (pp. 34–43). New York: Routledge.

Dede, C., & Sprague, D. (1999). If I teach this way am I doing my job? Constructivism in the classroom. *International Society for Technology in Education, 27*(1), 6–17.

Engeström, Y. (1987). *Learning by expanding: An activity-theoretical approach to developmental research*. Helsinki, Finland: Orienta-Konsultit.

Giemza, A., Manske, S., & Hoppe, H. U. (2013). Supporting the formation of informal learning groups in a heterogeneous information environment. *Proceedings of the 21st International Conference on Computers in Education* (pp. 367–375), Denpasar, Indonesia.

Hsieh, J.-C., Chen, C.-M., & Lin, H.-F. (2010). Social interaction mining based on wireless sensor networks for promoting cooperative learning performance in classroom learning environment. *Proceedings of the 6th IEEE International Conference on Wireless, Mobile, and Ubiquitous Technologies in Education* (pp. 219–221), Kaohsiung, Taiwan.

Hsu, Y.-C., & Ching, Y.-H. (2012). Mobile microblogging: Using Twitter and mobile devices in an online course to promote learning in authentic contexts. *International Review of Research in Open and Distributed Learning, 13*(4), 211–227.

Hsu, Y.-C., & Ching, Y.-H. (2013). Mobile computer-supported collaborative learning: A review of experimental research. *British Journal of Educational Technology, 44*(5), E111–E114.

Huang, Y.-M., & Wu, T.-T. (2011). A systematic approach for learner group composition utilizing U-Learning portfolio. *Educational Technology & Society, 14*(3), 102–117.

Jeong, H., & Hmelo-Silver, C. (2016) Seven affordances of computer-supported collaborative learning: How to support collaborative learning? How can technologies help? *Educational Psychologist, 51*(2), 247–265.

Johnson, D. W., & Johnson, R. T. (1999). *Learning together and along: Cooperative, competitive, and individualistic learning* (5th ed.). Boston, MA: Allyn & Bacon.

Ke, F., & Hsu, Y.-C. (2015). Mobile augmented-reality artifact creation as a component of mobile computer-supported collaborative learning. *The Internet and Higher Education, 26*, 33–41.

Lai, C.-Y., & Wu, C.-C. (2006). Using handhelds in a Jigsaw cooperative learning environment. *Journal of Computer Assisted Learning, 22*(4), 284–297.

Laurillard, D. (2007). Pedagogical forms of mobile learning: Framing research questions. In N. Pachler (Ed.), *Mobile learning: Towards a research agenda* (pp. 153–175). London: WLE Centre.

Liu, C.-C., Chung, C.-W., Chen, N.-S., & Liu, B.-J. (2009). Analysis of peer interaction in learning activities with personal handhelds and shared displays. *Educational Technology & Society, 12*(3), 127–142.

Looi, C.-K., Seow, P., Zhang, B. H., So, H.-J., Chen, W., & Wong, L.-H. (2010). Leveraging mobile technology for sustainable seamless learning: A research agenda. *British Journal of Educational Technology, 42*(1), 154–169.

Maldonado, H., & Pea, R. (2010). LET's GO! To the creek: Co-design of water quality inquiry using mobile science collaboratories. *Proceedings of the IEEE International Conference on Wireless, Mobile, and Ubiquitous Technologies in Education 2010* (pp. 81–87), Kaohsiung, Taiwan.

Melero, J., Hernández-Leo, D., & Manatunga, K. (2015). Group-based mobile learning: Do group size and sharing mobile devices matter? *Computers in Human Behavior, 4*, 377–385.

Nussbaum, M., Alvarez, C., McFarlane, A., Gomez, F., Claro, S., & Radovic, D. (2009). Technology as small group face-to-face Collaborative Scaffolding. *Computers & Education, 52*(1), 147–153.

Ogata, H., Li, M., Hou, B., Uosaki, N., El-Bishouty, M., & Yano, Y. (2011). SCROLL: Supporting to share and reuse ubiquitous learning log in the context of language learning. *Research and Practice in Technology Enhanced Learning, 6*(2), 69–82.

Race, P. (2000). *500 tips on group learning*. London: Kogan Page.

Reychav, I., & Wu, D. (2015). Mobile collaborative learning: The role of individual learning in groups through text and video content delivery in tablets. *Computers in Human Behavior, 50*, 520–534.

Rogers, Y., Connelly, K., Hazlewood, W., & Tedesco, L. (2010). Enhance learning: A study of how mobile devices can facilitate sensemaking. *Personal and Ubiquitous Computing, 14*(2), 111–124.

Rogers, Y., & Price, S. (2009). How mobile technologies are changing the way children learn. In A. Druin (Ed.), *Mobile technology for children* (pp. 3–22). Boston, MA: Morgan Kaufmann.

Roschelle, J., & Pea, R. (2002). A walk on the WILD side: How wireless handhelds may change computer-supported collaborative learning. *International Journal of Cognition and Technology, 1*(1), 145–168.

Roschelle, J., Rafanan, K., Estrella, G., Nussbaum, M., & Claro, S. (2010). From handheld collaborative tool to effective classroom module: Embedding CSCL in a broader design framework. *Computers & Education, 55*(3), 1018–1026.

Roschelle, J., Rosas, R., & Nussbaum, M. (2005). Towards a design framework for mobile computer-supported collaborative learning. *Proceedings of the International Conference on Computer-Support for Collaborative Learning 2005* (pp. 520–524), Taipei, Taiwan.

Schwabe, G., Göth, C., & Frohberg, D. (2005). Does team size matter in mobile learning? *Proceedings of the International Conference on Mobile Business* (pp. 227–234), Sydney, Australia.

Sharples, M., Scanlon, E., Ainsworth, S., Anastopoulou, S., Collins, T., Crook, C., et al. (2014). Personal inquiry: Orchestrating science investigations within and beyond the classroom. *Journal of the Learning Sciences. 24*(2), 308–341.

Sharples, M., Taylor, J., & Vavoula, G. (2007). A theory of learning for the mobile age. In R. Andrews & C. Haythornthwaite (Eds.), *The Sage handbook of e-learning research* (pp. 221–247). London: Sage.

So, H.-J., Seow, P., & Looi, C.-K. (2009). Location matters: Leveraging knowledge building with mobile devices and Web 2.0 technology. *Interactive Learning Environments, 17*(4), 367–382.

So, H.-J., Tan, E., & Tay, J. (2012). Collaborative mobile learning in situ from knowledge building perspectives. *Asia-Pacific Education Researcher, 21*(1), 51–62.

Song, Y. (2014). Methodological issues in mobile computer-supported collaborative learning (mCSCL): What methods, what to measure and when to measure? *Educational Technology & Society, 17*(4), 33–48.

Spikol, D. (2008). Playing and learning across locations: Identifying factors for the design of collaborative mobile learning. Licentiate thesis, Växjö University, Sweden.

Squire, K. D., & Jan, M. (2007). Mad city mystery: Developing scientific argumentation skills with a place-based augmented reality game on handheld computers. *Journal of Science Education and Technology, 16*(1), 5–29.

Stahl, G. (2013). Learning across levels. *International Journal of Computer-Supported Collaborative Learning, 8(1)*, 1–12.

Tan, Q., Jeng, Y.-L., & Huang, Y.-M. (2010). A collaborative mobile virtual campus system based on location-based dynamic grouping. *Proceedings of the 10th IEEE International Conference on Advanced Learning Technologies* (pp. 16–18), Sousse, Tunisia.

Valdez, G., McNabb, M., Foertsch, M., Anderson, M., Hawkes, M., & Raack, L. (2000). Computer-based technology and learning: Evolving uses and expectations. Retrieved from www.ncrel.org/tplan/cbtl/toc.htm

Wang, D. Y., Lin, S. J., & Sun, C. T. (2007). DIANA: A computer-supported heterogeneous grouping system for teachers to conduct successful small learning groups. *Computers in Human Behavior, 23*(4), 1997–2010.

Wong, L.-H., Boticki, I., Sun, J., & Looi, C.-K. (2011). Improving the scaffolds of a mobile-assisted Chinese character forming game via a design-based research cycle. *Computers in Human Behavior, 27*(5), 1783–1793.

Wong, L.-H., Chen, W., & Jan, M. (2012). How artefacts mediate small group co-creation activities in a mobile-assisted language learning environment? *Journal of Computer Assisted Learning, 28*(5), 411–424.

Wong, L.-H., Chin, C.-K., Tan, C.-L., & Liu, M. (2010). Students' personal and social meaning making in a Chinese idiom mobile learning environment. *Educational Technology & Society, 13*(4), 15–26.

Zurita, G., Autunes, P., Baloian, N., & Baytelman, F. (2007). Mobile sensemaking: Exploring proximity and mobile *Journal of Universal Computer Science, 13*(10), 1434–1448.

Zurita, G., & Nussbaum, M. (2004). Computer supported collaborative learning using wirelessly interconnected handheld computers. *Computers & Education, 42*(3), 289–314.

Zurita, G., & Nussbaum, M. (2007). A conceptual framework based on Activity Theory for mobile CSCL. *British Journal of Educational Technology, 38*(2), 211–235.

Zurita, G., Nussbaum, M., & Salinas, R. (2005). Dynamic grouping in collaborative learning supported by wireless handhelds. *Educational Technology & Society, 8*(3), 149–161.

36

Massive Open Online Courses (MOOCs) and Rich Landscapes of Learning

A Learning Sciences Perspective

Gerhard Fischer

Massive Open Online Courses (MOOCs)

MOOCs have generated a worldwide interest in learning and education. This interest has transcended narrow academic circles (e.g., the *New York Times* declared 2012 the "Year of the MOOC"; Pappano, 2012). As the costs of a residential university education have been growing dramatically, the promise of "free" MOOCs represented an exciting development. The different attributes used in the name provide a characterization of the objectives of MOOCS:

- *"massive,"* because they are designed to enroll very large number of students (i.e., thousands, often tens of thousands, and in some instances more than one hundred thousand);
- *"open,"* because anyone with an internet connection can sign up;
- *"online,"* being available on the internet and referring not just to the delivery mode but to the style of communication;
- *"courses,"* referring not only to content delivery (as it was the case with MIT's Open Courseware) but including other aspects (lectures, forums, peer-to-peer interaction, quizzes, exams, and credentials) associated with courses.

The name MOOC was created in 2008 by Dave Cormier, and the first examples were cMOOCs, followed by xMOOCs in 2011. The two approaches are grounded in two different design models (Daniel, 2012): cMOOCs are based on *connectivism* (the material being open, remixable, and evolvable, thereby giving learners an active role) (Siemens, 2005) and *networking* (connecting learners to each other to answer questions and collaborate on joint projects); whereas xMOOCs are based on an *instructionist, transmission-based approach* augmented with additional components (a detailed comparison between the two models can be found in Bates, 2014). At this point of time, xMOOCs are the focus of interest and attention and the arguments and examples discussed in this chapter are focused on them.

Some of the initial objectives articulated for MOOCS were (Fischer, 2014):

- represent first-class courses from the best professors coming from elite institutions;
- bring the best education in the world to the most remote corners of the planet;

- help professors to improve their classroom teaching by providing them with more data on what and how students in a course are doing;
- support communities among the students, thereby expanding their intellectual and personal networks; and
- provide students with insightful feedback in case they went wrong or got stuck in a problem-solving attempt.

Ancestors of MOOCs

The opinions about how innovative MOOCs are vary greatly. Radio and television were forms of distance learning that predated e-learning, with correspondence courses that were used for educational purposes to overcome distances and reach larger audiences. Many universities starting in the 1980s created special classrooms with video access for providing convenient and flexible education for working professionals by offering graduate degree programs and certificates in an accessible, online format. The following two specific developments played an important role:

- The Open University (OU) in the UK (founded in 1969; www.open.ac.uk) has been the pioneer of distance learning. It was "founded on the belief that communications technology could bring high quality degree-level learning to people who had not had the opportunity to attend traditional campus universities."
- The OpenCourseWare (OCW) initiative of MIT (started in 2002; http://ocw.mit.edu) was based on a commitment to put all the educational materials from MIT's undergraduate- and graduate-level courses online, partly free, and openly available to anyone and anywhere.

MOOCs Platform Providers

Over the last few years, numerous MOOCs platform providers have emerged as companies and non-profit organizations that partner with different universities and organizations worldwide to offer courses for anyone. Some of the most prominent providers are:

- Coursera (www.coursera.org/), a for-profit company offering over 1,500 courses from 140 partners across 28 countries in 2016;
- MIT's and Harvard's edX project (www.edxonline.org/), a non-profit company offering over 1,100 courses in 2016;
- Udacity (www.udacity.com/), focusing recently on nanodegree programs in which a certification can be earned in less than 12 months;
- FutureLearn (http://futurelearn.com/), a private company in the UK owned by the Open University, including non-university partners.

There are many similarities between these different platforms but there are also important differences from a learning science perspective. Over time, the companies by pursuing different strategies have contributed to a *diversification* of MOOCs (transcending the original distinction between xMOOCs and cMOOCs). Some providers focus on academic subjects and others provide vocational skills (with closer linkages to the job market), focus on everyday people or companies, and all of them experimenting with different business models and timing models.

MOOCs in the Context of Open, Online Learning Environments

Figure 36.1 provides an overview of open, online learning environments. MOOCs represent *one specific approach* in the "open, online courses" domain by having at least some of the attributes

Gerhard Fischer

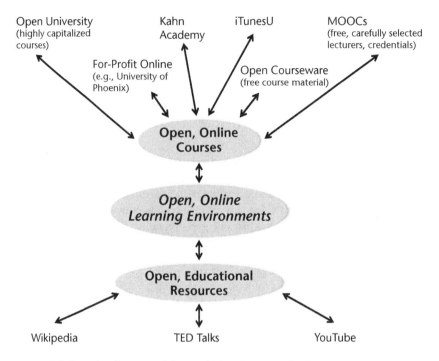

Figure 36.1 MOOCs in the Context of Open, Online Learning Environments

defining a course (e.g., lectures, forums, peer-to-peer interaction, quizzes, exams, and credentials). In contrast, open, educational resources serve different purposes; they offer information about specific, independent topics and questions requiring little cohesion between individual components.

Most of the discussions and analyses about MOOCS are based on *economic perspectives* (scalability, productivity, being "free") and *technology perspectives* (platforms supporting large number of students in online environments, enrichment components such as forums, peer-to-peer learning support, and automatic grading).

Few contributions have analyzed MOOCs from a learning science perspective and put them into a larger context with other approaches to learning and education. Some of the major expectations associated with MOOCs have been to enrich the landscape of learning opportunities and to reduce the digital divide by providing education for everyone by "making the knowledge of some of the world's leading experts available to anyone free of charge."

In their short time of existence, MOOCs deserve credit because they have woken up not only academia but also the media at large to bring online learning and teaching to the attention of the public. A special impact of MOOCs is their challenge to "force" residential, research-based universities to reflect, define, and emphasize their core competencies (Fischer & Wolf, 2015).

The special emphasis of this chapter is to assess MOOCs from a learning science perspective by locating them as one component in a rich landscape of learning. The expectation associated with this approach is that a symbiotic relationship can provide dividends and progress to two challenges: (1) that the future development of MOOCs can be grounded in insights from the learning sciences, and (2) that the research of the learning science can be enriched by exploring MOOCs as a specific and unique context for learning and teaching (Eisenberg & Fischer, 2014).

Rich Landscapes of Learning

One of the shortcomings of research in the learning science is that many approaches are too timid and not thinking radically enough by focusing too much on schooling and not paying enough attention to the multidimensional aspects of learning (Collins & Halverson, 2009; Resnick, 1987). Figure 36.2 provides an overview of the multidimensional aspects of learning leading to explore rich landscapes of learning—and the following paragraphs briefly describe the essential issues related to the different aspects.

Who Learns: People at Different Stages

The learner may be a student in different grades and institutions (ranging from K–12 to university education), a person working in industry, or curious citizens attempting to understand more about the world surrounding them. Some of the learners may be beginners (and general and uniform introductory courses will serve them well), whereas others may have a rich knowledge background and very specific objectives requiring more individualized instruction.

Why Learn: Different Objectives

Some people learn because they need to pass a test, fulfill the requirements of a course in school or university, and others learn because they are passionate about some activity (e.g., Collins & Halverson, 2009). *Individual MOOCs*, by their primarily instructionist nature, offer learners few opportunities for interest-driven learning. The evolving *space of all MOOCS* (approximately 7,000 courses were available at the end of 2016; www.class-central.com/report/mooc-stats-2016/) represent more courses than an individual university can offer, thereby covering niche topics in which a small number of learners will be interested.

Figure 36.2 Multidimensional Aspects of Learning

Gerhard Fischer

What to Learn: Exploring Personally Meaningful Problems and Acquiring Basic Skills and Core Competencies

In formal learning environments, students' learning is determined to a large extent by a curriculum (Resnick, 1987). Learners encounter few opportunities to gain experiences by exploring personally meaningful problems that need to be identified and framed. The engagement with personal meaningful problems should be complemented with learning opportunities to acquire basic skills and core competencies for the 21st century. These competencies do not primarily consist of learning and memorizing facts, but should be focused on: (1) acquiring and using information; (2) identifying, organizing, planning, and allocating resources; (3) collaborating with others; and (4) working with a variety of technologies.

How to Learn: Learning in Different Ways

Learning in today's world must conceptualize learning as an inclusive, social, informal, participatory, and creative lifelong activity. Many problems (specifically design problems; Simon, 1996) are unique and ill-defined and the knowledge to address them is not "out there," requiring contributions and ideas from all involved stakeholders. Learners in such settings must be *active contributors* rather than passive consumers, and the learning environments and organizations must foster and support mindsets, tools, and skills that help learners become empowered and willing to actively contribute (Jenkins, 2009; von Hippel, 2005).

Where to Learn: At the Right Places

Historically, schools provided the setting where individuals engaged in learning. The seeds of a new education system can be seen in the explosive growth of home schooling, workplace learning, distance education, adult education, and a variety of design spaces (e.g., museums, zoos, environmental centers). Research on everyday cognition demonstrates that the formal learning in schools and the informal learning in practical settings have important differences (National Research Council, 2009). What we discover about learning in schools is insufficient for a theory of human learning: schools are often focused on individual cognition, on memorization, and on learning general facts, whereas learning in the world at large needs to rely on shared cognition, use of powerful tools and external information sources, and situation-specific competencies.

When to Learn: At the Right Time

Information overload and the rapid change of our world have created new problems and new challenges for learning and education. People will have to keep learning new knowledge and skills throughout their lifetimes as their lives and jobs keep changing. New approaches are needed to circumvent the unsolvable problems of coverage and obsolescence (Goggins, Jahnke, & Wulf, 2013). Learning on demand is a promising approach for addressing these problems, because: (1) it contextualizes learning by allowing it to be integrated into work rather than relegating it to a separate phase, (2) it lets learners see for themselves the usefulness of new knowledge for actual problem situations, thereby increasing the motivation for learning new things, and (3) it makes new information relevant to the task at hand, thereby leading to more informed decision making, better products, and improved performance.

With Whom: Transcending to Individual Human Mind

In the past, most computational environments have focused on the needs of individual users. Systemic problems require more knowledge than any single person possesses, because the knowledge relevant to either frame or resolve these problems is usually distributed among stakeholders coming from different disciplines. The "Renaissance Scholar" (meaning the person who is knowledgeable in all relevant fields) no longer exists (Csikszentmihalyi, 1996). To deal with complex multidisciplinary problems, people need to use the powerful tools technology provides for finding, analyzing, manipulating, and communicating knowledge. This requires bringing different and often controversial points of view together to create a shared understanding among stakeholders, and it can lead to new insights, ideas, and artifacts.

MOOCs have the potential (some of it realized today, many aspects serving as design challenges for future MOOCs) to contribute to these different dimensions of multifaceted aspects of learning.

State of the Art

Conceptualizing MOOCs as components of rich landscapes of learning provides the foundation to differentiate an *internal* and an *external* view of MOOCs (Fischer, 2016). The internal view of MOOCs addresses numerous challenges directly associated with their strengths and weaknesses, whereas the external view is focused on the promise that research into MOOCs will inform *learning in all environments* and not just MOOCs.

Internal Versus External Views of MOOCS

An internal view of MOOCs. The internal view analyzes topics that are focused on MOOCs as a *specific* teaching and learning activity, rather than seeing them as a component of rich landscapes of learning. The internal view focuses on the following topics:

- distinguishing cMOOCs (fostering connections and collaborations among learners) and xMOOCs (efficiently delivering content to large audiences) (Bates, 2014);
- differentiating *basic services* provided for free (e.g., access to courseware) from *premium services* that require payment (e.g., access to projects, code-review and feedback, personal coaches, and verified certificates);
- identifying number of participants and calculating the *completion rates* for specific courses;
- analyzing the *educational background of participants* (empirical research uncovered the surprising finding that the largest group of participants in xMOOCS already have a master's degree);
- finding ways (by automating the process or by supporting peer grading) to *assess the achievements* of a large number of participants;
- taking advantage of capturing large amounts of data for *learning analytics* research (Siemens, 2012);
- supporting *local meet-up groups* (allowing participants in the same location to meet in person); and
- establishing nanodegree programs in which people (mostly from industry) can acquire specific knowledge and targeted skills without extended time requirements.

An external view of MOOCs. A learning science perspective puts the main emphasis on an external view of MOOCs. It provides frames of references for identifying the following themes:

- *different forms of learning* (lifelong, blended, collaborative) need to be supported and practiced (Bransford, Brown, & Cocking, 2001);
- *formal* learning in schools needs to be complemented by *informal* learning (National Research Council, 2009).
- *supply-oriented* ("push/delivery") models, in which learners are presented with knowledge that later may become relevant for them, need to be complemented by *"pull/learning on demand"* approaches (Collins & Halverson, 2009);

- *consumer-oriented cultures* need to be complemented by *participatory cultures* (Jenkins, 2009);
- *"learning about"* needs to be complemented by *"learning to be"* (Brown, 2005); and
- *"learning when the answer is known"* needs to be complemented by *"learning when the answer is not known"* (and exploring problems that have no answers) (Engeström & Sannino, 2010).

The different objectives summarized in this list represent *antinomies* (or design trade-offs) (Bruner, 1996)—pairs of truth, each worthwhile to pursue in different contexts, but also contradicting each other at a certain level, depending on the material to be learned, the students, the setting, and many other factors. The essential goal of the learning sciences in the face of new technologies such as MOOCs is to identify the various sides of the antinomies latent in the technology; once identified, we can use the technology in an informed way, research its role in learning, and design alternative or complementary technologies that mitigate the problems of one-sidedness.

Motivation for Participation

Why are MOOCs such a hot topic? One way to analyze this question is to explore the motivations of all the different stakeholders who are affected by the development of MOOCs: providers, teachers, students, parents, politicians, university administrators, and researchers in the learning sciences (these claims are supported by initial findings in several articles contained in DeCorte, Engwall, & Teichler, 2016).

Providers articulate a multitude of different reasons for being involved, including: (1) altruistic motivations (such as "education for everyone"); (2) addressing an exciting problem; (3) bringing fame to their institutions; and (4) exploring unique business opportunities.

Professors are interested in teaching MOOCS (http://chronicle.com/article/The-Professors-Behind-the-MOOC/137905/#id=overviews) based on some of the following motivations: (1) the reach and impact which they can achieve by reaching a very large number of students; (2) to face a new challenge and learn from it; (3) to avoid being left behind; (4) to increase their visibility and fame (maybe successful MOOC professors of the future will be treated like movie and sport stars?); and (5) to reap new rewards and benefits (e.g., getting tenure for the reputation and social capital that they earned by teaching a highly successful MOOC).

Learners (being students of all ages or working professionals) are motivated to participate in MOOCs: (1) based on intellectual curiosity; (2) to engage in lifelong learning; (3) to gain an understanding of specific knowledge relevant to problems which they face; (4) to exploit them as their only educational opportunities; and (5) to become members of interesting intellectual communities (maybe comparable to why people join book clubs?).

Parents (in most cases paying substantial amounts of money for the children's education) are interested to find out whether their children can get the same quality education for a fraction of the money that they have to pay for a conventional university education.

Politicians for public universities (or fundraisers for private ones) will welcome any change that will reduce the financial commitments needed by universities. University administrators will similarly welcome cost savings, but many at this moment are very concerned not to be left behind, rather than to deeply understand the impact of these developments on their own institutions.

Researchers in the learning sciences are often sharply divided in their opinions about MOOCs, but are provided with the opportunity to use MOOCs as relevant developments to rethink learning, teaching, and education (an attempt made with this chapter).

Big Data and Learning Analytics

The data revolution ("Big Data") provides insight to analyze and document human behavior to an extent considered impossible a few decades ago (but feared by some visionaries; Orwell, 1950). Google, Facebook, Amazon, Netflix, banks, and supermarkets (leave alone the National Security

Agency) know a lot about all people, their behavior, the information they have looked at, the stuff they have bought, and the places that they have visited.

MOOCs provide rich datasets about interactions, collaborations, and engagement that computational processes can exploit. *Learning analytics* (for more information, see Rosé, this volume) focuses on measuring, collecting, analyzing, and reporting data about learners and their contexts. It attempts to understand the background knowledge of learners and it adds an important data-gathering resource to online education as a dissemination method.

Opinions: Hypes and Underestimations

Will MOOCs end up being elixir or snake oil? The learning, teaching, and education domain has been populated by claims: (1) from info-enthusiasts promising that technology would revolutionize "education" and computers will replace teachers, and (2) from opposite claims by info-pessimists that computers in classrooms foster isolation, lack of creativity, rigid and sloppy thinking, and an overemphasis on abstract thinking (and consequent undervaluing of real-world experience).

The *hype* (Fischer, 2014) and *myths* (Daniel, 2012) around MOOCs is articulated in statements like:

- "There's a tsunami coming"—President John Hennessy of Stanford;
- "2012: the year of the MOOC" —*New York Times* (Pappano, 2012);
- "Technology is remaking every aspect of education, bringing top-notch courses to the world's poorest citizens and reshaping the way all students learn"—www.scientificamerican.com/editorial/digital-education.

The *underestimation* of MOOCs is expressed, for example, in the following opinion: "In fact, the absence of serious pedagogy in MOOCs is rather striking, their essential feature being short, unsophisticated video chunks, interleaved with online quizzes, and accompanied by social networking . . . If I had my wish, I would wave a wand and make MOOCs disappear, but I am afraid that we have let the genie out of the bottle" (Vardi, 2012, p. 5).

Empirical Research About MOOCs

Complementing the initial assumptions and opinions in some of the most widespread public media, empirical research is emerging that analyzes different aspects of MOOCs relevant from a learning sciences perspective (for a detailed recent review, see Veletsianos & Shepherdson, 2016). Most of the empirical studies so far have been focused on: (1) themes centered on participants, including learner behaviors, performances, participation and interaction, learner perceptions and preferences, learner experiences, motivation, educational background, and demographics; and (2) themes centered on the design of courses, including how the instructionist nature of the courses can be enriched with automatic and personal feedback, forums, peer-to-peer learning and grading, and the large-scale facilitation and support of learning communities.

A widespread argument broadly discussed as the most troubling aspects of MOOCS is their *low completion rates* (in many courses below 10%) (Breslow et al., 2013; Eriksson, Adawi, & Stöhr, 2016). The overemphasis and fallacy of this argument is the comparison with rates of courses taught in residential universities, because participation and performance in these two environments are fundamentally different. MOOCs allow free and easy registration, do not require formal withdrawals, and include a large number of students who may not have any interest in completing assignments and assessments. If we conceptualize MOOCs as the textbooks of the 21st century, the troubling aspects may be questioned, because nobody assumes that textbooks need to be read from beginning to end but they serve as a resource under the control of the learner.

Gerhard Fischer

Future Challenges, Trends, and Developments

Co-Evolution: Beyond Getting Stuck in "Gift-Wrapping"

New information and communication technologies have been heralded as the major driving forces behind innovation in learning and education. While the internet, smartphones, apps, 3-D printers, etc.) have caused an explosion of opportunities to improve learning and education by making established practices better and enabled new approaches and created new frameworks that were not possible or even conceivable before, many approaches have had only a minor impact for learning and education, based on the reduction to:

- *technology-centered developments* ignoring that technology alone does not determine social structures but only creates feasibility spaces for new social and cultural practice (Benkler, 2006)—changes in complex learning environments are not only dictated by technology; rather, they are the result of an incremental shift in human behavior and social organizations and as such require the *co-design* of social and technical systems;
- *gift-wrapping*, in which new media are used as add-ons to existing practices rather than as catalysts for fundamentally rethinking what education should and can be in the next century—the *"moocifying" of existing courses* represents the prime example of "gift-wrapping" (ignoring the fundamental assumption that distant learning is not classroom learning at a distance);
- a focus on *existing learning organizations* (such as schools and universities), thereby not exploring new possibilities such as e-learning environments (including MOOCs) in support of peer-support communities, and niche communities forming around special, idiosyncratic interests.

Finding the Needle in the Haystack: Personalization and Task Relevancy

The rapidly increasing number of open, online learning environments (see Figure 36.1), specifically of MOOCs, has created a unique and growing opportunity for learners to engage in self-study with individually tailored curricula. At the same time, this large and constantly evolving space has created the challenge as to how learners will find the best-matched learning resources (artifacts and humans) to their personal interest, and how they can be supported with guidance and advice by mentors and peers. While directory-style environments for courses provided by individual MOOCs platform providers and global directories of MOOCs (e.g., MOOC List; www.mooc-list.com) and Class Central (www.class-central.com) are important steps in the right directions, more support is need to assist learners in finding and assessing courses that are relevant to their tasks and compatible with their background knowledge.

Core Competencies of Residential, Research-Based Universities

Early visions about MOOCs predicted that they would eliminate a large percentage of residential universities. There is little evidence so far that this will happen and most of the more recent research activities are focused on *complementing* residential with online learning by identifying the core competencies of the two approaches. The appearance of MOOCs has created opportunities and necessities to reflect on the true value of residential university experiences provided by teacher–student and student–student interactions (Fischer & Wolf, 2015). In future emerging hybrid models, MOOCs could serve as the textbook of the 21st century and could support "flipped classroom" models. They could help residential universities move away from large lectures with learners listening to teachers, towards active learning environments characterized by personal attention from teachers and opportunities for participation. They could make a contribution to improve education outcomes in measurable ways at lower cost.

Conclusion

The most important contribution of MOOCs during their short lifespan is that they generated a broad and (so far) lasting discourse about learning, teaching, and education in which not only narrow, specialized academic circles participate, but the global media, universities administrators, and politicians got involved.

Rather than ignoring MOOCs and only grounding and evolving them in economic and technological perspectives, the research community in the learning sciences should get seriously involved with MOOCs and influence their evolution. Even the loudest critics of MOOCs do not expect them to fade away. More likely, they will morph into many different shapes (e.g., the "basic services" provided by MOOC 1.0 will be complemented by the "premium services" developed and envisioned in MOOC 2.0).

Researchers from the learning sciences should not only collect data about existing practices, but they should develop visions, explore important open issues, and investigate the pro and cons of different design choices. For example, what are the trade-offs between (1) an inexpensive educational infrastructure (in which students can easily afford at least a minimal education, and in which the resources associated with residential universities are scaled back), or (2) an expanded infrastructure (in which online education is complemented not only by residential universities, but by all the other components contributing to a rich landscape of learning, as indicated in Figure 36.1).

Major challenges for the learning sciences in the years to come that are grounded in the advent of open, online learning environments (and MOOCs specifically) are: (1) to create frames of reference to understand the role of MOOCs from a learning science perspective (in addition to economic and technological perspectives); (2) to identify the unique contributions of MOOCs to a rich landscape of learning; (3) to move beyond the exaggerated hype and total underestimation surrounding MOOCs; and (4) to analyze MOOCs as a forcing function in identifying the core competencies of residential, research-based universities. Experimentation will be needed to successfully integrate online education with residential education. In doing so, the learning sciences will make a contribution not only to understand the MOOC phenomena better, but contribute to fundamental challenges, such as isolating: (1) what it means to be educated in the digital age and (2) how interests, motivations, and collaborations can be stimulated to create rich learning environments in which people *want* to learn rather than *have* to learn.

Acknowledgments

The author would like to thank Christian Stöhr, Pierre Dillenburg, and the editors of the handbook for suggestions and critical feedback that contributed substantially to improving earlier versions of this chapter.

Further Readings

Collins, A. and R. Halverson (2009). *Rethinking education in the age of technology: The digital revolution and the school*. New York: Teachers College Press.
This book provides a vision for the future of learning. By transcending the narrow view of learning focused on school learning, the book illustrates which rich landscapes of learning can and should be pursued.

Cress, U., Jeong, H., & Moskaliuk, J. (Eds.). (2016) *Mass collaboration and education*. Heidelberg: Springer.
While MOOCs reach the masses, they are less successful in promoting and supporting mass collaboration. This book offers a comprehensive overview of mass collaboration by analyzing different theoretical approaches by describing a variety of case studies and investigating different methods to analyze processes.

DeCorte, E., Engwall, L., & Teichler, U. (Eds.). (2016). *From books to MOOCs? Emerging models of learning and teaching in higher education* (Wenner-Gren International Series, 88). London: Portland Press.

Gerhard Fischer

In this book, researchers from the field of the learning sciences, MOOCs developers, and MOOCs users critically analyze and discuss the state of the art of MOOCs from its beginning to 2015. Most of the contributions come from different European , providing evidence that MOOCs development represents an international phenomenon.

Hollands, F. M., & Tirthali, D. (2014). *MOOCs: Expectations and reality: Full report*. Center for Benefit- Cost Studies of Education, Teachers College, Columbia University, New York. Retrieved from http://cbcse.org/wordpress/wp-content/uploads/2014/05/MOOCs_Expectations_and_Reality.pdf
This report investigates the actual goals of institutions creating MOOCs or integrating them into their programs, and reviews the current evidence regarding whether and how these goals are being achieved, and at what cost.

Shah, D. (2015) MOOCs in 2015: Breaking down the numbers. *EdSurge*. Retrieved from www.edsurge.com/news/2015-12-28-moocs-in-2015-breaking-down-the-numbers
This article provides quantitative empirical data about the number of students who signed up for MOOC courses, the number of MOOC courses offered, distribution of subjects of MOOCs courses, providers of MOOCs courses, and ratings of courses offered.

NAPLeS Resources

Fischer, G., *Massively open online courses (MOOCs) as components of rich landscapes of learning* [Webinar]. In *NAPLeS video series*. Retrieved October 19, 2017, from http://isls-naples.psy.lmu.de/intro/all-webinars/fischer/index.html

References

Bates, T. (2014). *Comparing xmoocs and cmoocs: Philosophy and practice*, Retrieved from www.tonybates.ca/2014/10/13/comparing-xmoocs-and-cmoocs-philosophy-and-practice/
Benkler, Y. (2006). *The wealth of networks: How social production transforms markets and freedom*. New Haven, CN: Yale University Press.
Bransford, J. D., Brown, A. L., & Cocking, R. R. (Eds.). (2001). *How people learn—Brain, mind, experience, and school*. Washington, DC: National Academy Press.
Breslow, L., Pritchard, D. E., DeBoer, J., Stump, G. S., Ho, A. D., & Seaton, D. T. (2013). Studying learning in the worldwide classroom: Research into edX's first MOOC. *Research & Practice in Assessment, 8*.
Brown, J. S. (2005). *New learning environments for the 21st century*, Retrieved from www.johnseelybrown.com/newlearning.pdf
Bruner, J. (1996). *The culture of education*. Cambridge, MA: Harvard University Press.
Collins, A. & Halverson, R. (2009). *Rethinking education in the age of technology: The digital revolution and the school*, New York: Teachers College Press.
Csikszentmihalyi, M. (1996). *Creativity — Flow and the psychology of discovery and invention*. New York: HarperCollins.
Daniel, J. (2012). *Making sense of MOOCs: Musings in a maze of myth, paradox and possibility*. Retrieved from http://jime.open.ac.uk/articles/10.5334/2012-18/
DeCorte, E., Engwall, L., & Teichler, U. (Eds.). (2016). *From books to MOOCs? Emerging models of learning and leaching in higher Education* (Wenner-Gren International Series 88). London: Portland Press.
Eisenberg, M., & Fischer, G. (2014). MOOCs: A perspective from the learning sciences. In J. L. Polman, E. A. Kyza, D. K. O'Neill, I. Tabak, William R. Penuel, A. S. Jurow, et al. (Eds.), *Learning and Becoming in Practice: 11th International Conference of the Learning Sciences (ICLS)* (pp. 190–197), Boulder, CO.
Engeström, Y. & Sannino, A. (2010). Studies of expansive learning: Foundations, findings and future challenges. *Educational Research Review, 5*(1), 1–24.
Eriksson, T., Adawi, T., & Stöhr, C. (2016). "Time is the bottleneck": A qualitative study exploring why learners drop out of MOOCs. *Journal of Computing in Higher Education, 29*(1), 133–146. doi:10.1007/s12528-016-9127-8
Fischer, G. (2014). Beyond hype and underestimation: Identifying research challenges for the future of MOOCs. *Distance Education Journal (Commentary for a Special Issue, "MOOCS: Emerging Research"), 35*(2), 149–158.
Fischer, G. (2016). MOOCs as components of rich landscapes of learning. In E. DeCorte, L. Engwall, & U. Teichler (Eds.), *From books to MOOCs? Emerging models of learning and teaching in higher education* (pp. 43–54). London: Portland Press.
Fischer, G., & Wolf, K. D. (2015). What can residential, research-based universities learn about their core competencies from MOOCs (Massive Open Online Course). In H. Schelhowe, M. Schaumburg, &

J. Jasper (Eds.), *Teaching is touching the future—Academic teaching within and across disciplines* (pp. 65–75). Bielefeld, Germany: Universitätsverlag Webler.

Goggins, S., Jahnke, I., & Wulf, V. (Eds.). (2013). *Computer-supported collaborative learning at the workplace (CSCL@Work)*. Heidelberg, Germany: Springer.

Jenkins, H. (2009). *Confronting the challenges of participatory cultures: Media education for the 21st Century*. Cambridge, MA: MIT Press.

National Research Council (2009). *Learning science in informal environments—People, places, and pursuits*. Washington, DC: National Academy Press.

Orwell, G. (1950). *1984*. New York: Signet.

Pappano, L. (2012). The year of the MOOC. *The New York Times*. Retrieved from www.nytimes.com/2012/11/04/education/edlife/massive-open-online-courses-are-multiplying-at-a-rapid-pace.html

Resnick, L. B. (1987). Learning in school and out. *Educational Researcher, 16*(9), 13–20.

Rosé, C.P. (2018). Learning analytics in the Learning Sciences. In F. Fischer, C. E. Hmelo-Silver, S. R. Goldman, & P. Reimann (Eds.), *International handbook of the learning sciences* (pp. 511–519). New York: Routledge.

Siemens, G. (2005). Connectivism: A learning theory for the digital age. *Elearnspace*. Retrieved from www.elearnspace.org/Articles/connectivism.htm

Siemens, G. (2012). Learning analytics: Envisioning a research discipline and a domain of practice. *Proceedings of the Second International Conference on Learning Analytics and Knowledge (LAK '12)* (pp. 4–8). New York: ACM Press.

Simon, H. A. (1996). *The sciences of the artificial* (3rd ed.) Cambridge, MA: MIT Press.

Vardi, M. Y. (2012). Will MOOCs destroy academia? *Communications of the ACM (CACM), 55*(11), 5.

Veletsianos, G., & Shepherdson, P. (2016). A systematic analysis and synthesis of the empirical MOOC literature published in 2013–2015. *International Review of Research in Open and Distributed Learning, 17*(2).

von Hippel, E. (2005). *Democratizing innovation*. Cambridge, MA: MIT Press.

Section 3
Research, Assessment, and Analytic Methods

37
Design-Based Research (DBR)

Sadhana Puntambekar

Introduction

Since Ann Brown's seminal article on "design experiments" (Brown, 1992), several articles have been written to explain the main characteristics of what is now known in the learning sciences as design-based research (DBR). We have also seen special issues of well-regarded journals focused on DBR, as well as the publication of edited volumes. Much of the early literature on DBR, however, focused on the "what"—mainly discussing the unique characteristics of DBR, especially the dual role of refining an innovation, and the underlying theories of teaching and learning (DBRC, 2003). We are now seeing articles about the "how" of DBR (e.g., Sandoval, 2014), which is where I would like to focus this chapter. Specifically, I will discuss the notion of trajectories of studies in DBR. I will provide concrete examples and discuss how trajectories can be planned and executed, and what we can learn from the iterative development of an innovation.

Iterative development of an innovation in which the design is put to test in an authentic context is at the core of DBR. In each iteration, researchers are often interested in examining issues related to design, learning outcomes, and implementation, along with refining the theoretical underpinnings of their designs. But these intertwined goals also pose challenges (Phillips & Dolle, 2006). For example, when multiple goals are intertwined, researchers might collect large quantities of data (Dede, 2004), with the aim of answering multiple *types of* questions. Therefore, rigor has been questioned, since the analytic methods may not align well with the varied goals (Kelly, 2004). Another issue of concern when studying an innovation as it evolves is that it is often not possible to control variables, thereby limiting the robustness and the applicability of any findings (Dede 2004; Shavelson, Phillips, Towne, & Feuer, 2003).

I propose that design research is best described in terms of trajectories of studies, rather than a single study aimed at addressing multiple, intertwined goals. Carefully planned trajectories can address some of the methodological concerns about DBR. Rather than simultaneously studying features of a design and testing the underlying theoretical principles, studies along a trajectory can vary in size and scope. Specific studies along a trajectory can be designed to focus on design features, theoretical principles, or issues of implementation. Some studies along a trajectory might focus on the design, while other studies might focus on the underlying theory. Each study informs the next study, and helps to cumulatively build knowledge about the many aspects of understanding an innovation in context. I call these *informing cycles*, because each cycle informs the next set of studies. Taken together, the studies along a trajectory can add to the knowledge base about design, implementation, and teaching and learning in real-world contexts.

Accounts of DBR seldom focus on how studies are designed along a trajectory, and how each cycle informs the next. This is where I would like to focus this chapter. I will discuss the trajectory of one of my projects, CoMPASS (Puntambekar, Stylianou, & Goldstein, 2007), to illustrate the cycles of studies that we conducted. Each of our studies focused on specific research questions within the larger scope of the project, and its overarching theoretical framework. Each iteration illustrates how theory-driven design of an innovation might be combined with empirically driven research. It is not my intention to discuss details of our studies in this chapter. Details of the studies have been published elsewhere, and I have cited these studies throughout this chapter. I will discuss the studies only to help the reader understand how we planned our trajectories.

Trajectory of Studies in the CoMPASS Project

Iteration 1: Testing Our Key Conjectures

One of the distinguishing features of DBR is that principles and hypotheses about learning are embedded in the design of innovations. Designed innovations are often based on theoretical premises with the main theoretical constructs embodied in the designed artifacts. In the CoMPASS project, our motivating principle was based on research on knowledge representation, especially on research describing that experts' knowledge is represented in ways that shows richer organization, often around the central principles of the domain that can be generalized (e.g., Kozma, 2000; Newton & Newton, 2000). But science curricula fail to foster knowledge integration (Jacobson & Wilensky, 2006; Linn & Eylon, 2011), often resulting in students' ideas being fragmented across science topics. The aim of the CoMPASS project has therefore been to enable an in-depth, cohesive understanding of science content, rather than multiple disconnected facts. Our main conjecture was that when visual representations in software, and the accompanying curricula, emphasize key conceptual connections, students will develop a deeper understanding of science ideas and principles. This key idea was embedded in the designs of all our materials. First, in the electronic textbook—CoMPASS—we used visual representations to show students the connections between science ideas and principles. Visual representations (Bederson & Hollan, 1995; Furnas, 1995) in the form of concept maps were used to help make the connections visible to students, as well as to enable students to navigate through the text. Second, we designed instructional units broadly based on the principles of Learning by Design (Kolodner et al., 2003), with an overarching design challenge.

Our studies in the first two iterations constituted what Bielaczyc (2013) has termed existence proofs, i.e., we were implementing our units in a context in which it was developed. The school in which we were working was closely associated with research in the university. We developed a partnership with the teachers in this school during the development of our materials, getting their feedback on our software and instructional materials.

Studies 1 and 2: Principled Variation

In two studies of our first iteration, we wanted to test the key underlying principle at the core of our innovation: that the concept map visualizations will enable students to better understand the connections between science ideas, and help them easily navigate to concepts and principles related to their goals. We introduced a *principled variation*, in which the particular design element of concept map visualizations was varied. Introducing a principled variation is a way to focus on important elements of a design and study its effects on learning. In Studies 1 and 2, we compared two versions of the CoMPASS system—one with the map visualizations and the other with a traditional table of contents but no map visualizations. By having the same teacher teach classes in which students used different versions of the system, we were able to keep other aspects of our implementations, similar, so that there were fewer confounding factors.

In Study 1, an eighth-grade teacher used two versions of CoMPASS (Puntambekar, Stylianou, & Hübscher, 2003), one in each of his classes. Each version used a different visual representation: the maps version (N=22) used concept maps, and the no-maps version (N=21) used a list. Students used CoMPASS for the topic "Forces and Motion" for a total of three sessions over three days. In Study 2, we wanted to examine the effect of visualizations on learning: (1) at a different grade level, and (2) when students used CoMPASS for an extended length of time. A total of 156 sixth-grade students from eight classes, taught by two teachers, participated in this study. Students used CoMPASS over a seven-week period during a unit on simple machines. As students engaged in investigations designed for each simple machine (e.g., wedge, pulley, lever), they used CoMPASS to find the information that would help them in their investigations. For each teacher, two of the classes used a maps version of the system and the other two used a no-maps version. During the 10-week unit, there were a total of 12 sessions in which all the students used CoMPASS, each lasting approximately 20–25 minutes.

In both studies, we used pre- and post-tests of physics concepts along with a concept-mapping test to measure students' understanding of science. We also collected log files that kept track of the nodes that students visited and the time spent on each node. In both Study 1 and Study 2, an ANCOVA with pre-test scores as the covariate showed that the version of CoMPASS (maps or no-maps) did not affect students' scores on the multiple-choice portion of the pre- and post-tests. However, significant differences between the two groups were found in the open-ended questions in the post-test, with the maps group performing significantly better than the no-maps group (ES = 0.76 for sixth grade and 0.82 for eighth grade). Similarly, analysis of students' performance on a concept-mapping test showed that students in the maps group had a significantly deeper and richer understanding of the science content (ES = 0.09 for sixth grade and 0.16 for eighth grade). Students in the maps groups seemed to have understood the interconnected nature of the concepts and principles that they were learning (Puntambekar et al., 2007). Further analysis showed that students who used the maps version of the system spent a significantly greater amount of time on goal-related concepts, as opposed to students in the no-maps groups, who spent more time on general topic descriptions, leading to a shallower understanding of the science content (Puntambekar et al., 2003). A regression analysis showed that there was a significant positive relationship between students' understanding of the *function* of the maps (making relations between concepts visible), their use of the maps, and their scores on tests (Puntambekar et al., 2003). This meant that when students understood the affordances of the visual representations, they could navigate and learn successfully.

A point to note here is that, while we were putting the main premise underlying our innovation to test by introducing a principled variation, we conducted the studies at two grade levels to test the key conjectures underlying our innovation. Even though the studies were in the same school, limiting the demographic diversity, they helped us to better understand how the visualizations were used by students of different grade and age levels. Our results were consistent across grade levels, providing us evidence for the usefulness of visualizations for learning science content. But we also uncovered interesting relationships between students' understanding of the knowledge representation, and the effect of that understanding on their learning. We found that students needed more scaffolding to understand the affordances of the visual representations, in order for them to use the knowledge representation for learning and navigation. These new understandings led us to our third study in this iteration.

Study 3: Focused Study Within a Larger Implementation

We designed Study 3 based on what we learned about the need for additional scaffolding to help students understand the affordances of the concept-map visualizations in CoMPASS. Drawing from theories of metacognition in text and self-regulated learning (Azevedo & Cromley, 2004; Brown, Armbruster & Baker, 1986; Brown & Palincsar, 1987), in Study 3 we provided students with metacognitive support for navigation, to help them understand and use the affordances of the visualizations and to find science ideas related to their goals. We termed this support *metanavigation support* (Stylianou & Puntambekar, 2003, 2004).

We conducted Study 3 as a *focused study* with random assignment of conditions within the context of the larger implementation. The main reason for this was that we wanted to test the idea of providing metacognitive support based on students' navigation, which required real-time analysis of students' log files, and providing scaffolding based on this analysis. Since this was a time-intensive endeavor, we wanted to first test our conjecture that providing metanavigation support would help students understand the affordances of the visualizations and help them learn better. We therefore implemented a smaller study that lasted three days, within the larger 10-week implementation. In this study, classes taught by the same teachers were randomly assigned to one of the two conditions—metanavigation support or no support. Students' log files were collected and analyzed on the same day, and they were presented with prompts on paper for their next session on CoMPASS. Students took a test of comprehension and a test of science knowledge after this session. A total of 121 students participated in the study.

We found that metanavigation prompts enabled students to make coherent transitions using the concept-map visualizations, and gain a better understanding of content knowledge. Findings also revealed that reading comprehension, presence of metanavigation support, and prior domain knowledge significantly predicted students' understanding of science principles and the relationships among them.

Iteration 2: Understanding Teacher Facilitation

During all three studies in our first cycle, two members of the research team collected classroom observations to understand how the curriculum was working, and how teachers facilitated whole-class and small-group discussions. Our classroom observations showed that we needed to better support teachers to help them understand the underlying learning principles of our materials, as well as to help them with strategies for whole-class and small group facilitation. In the summer workshop following Study 3 (in Iteration 1), we worked with the teachers on issues such as allowing students to raise their own questions, helping students to explore and find information, and emphasizing the importance of open-ended questions so that students would not be looking for the "right" answers. The teachers spent two days on CoMPASS, familiarizing themselves with the software, reading the text, and giving feedback to the project team. We also created teacher guides that centered around the major principles and big ideas embedded in the curriculum, and helped teachers understand how they could reiterate the big ideas in whole-class discussions. Following the professional development, our focus moved to examining classroom implementations in greater detail. In the next iteration, we examined the strategies that teachers used in whole-class and small-group discussions.

As mentioned earlier, our main premise was helping students to understand connections among science ideas. Our instructional materials and software were designed to reflect this premise, and provided students with opportunities to understand the networks of ideas in science, rather than a set of discrete facts. But a main aspect of enabling students to make the connections between science principles is the facilitation provided by teachers during whole-class and small-group discussions (Hoffman, Wu, Krajcik, & Soloway, 2003; Lumpe & Butler, 2002; Palincsar & Magnusson, 2001). Teachers play a crucial role in reiterating and complementing the pedagogical principles embedded in the instructional materials (Sandoval & Daniszewski, 2004; Tabak, 2004; Tabak & Reiser, 1997). In Iteration 2, therefore, we specifically focused on understanding how teacher-led discussions could help (1) connect the activities within a curriculum unit and (2) enable deeper conceptual understanding by helping students make connections between science concepts and principles.

The participants were 132 sixth-grade students in seven science classes taught by two different teachers. Students used the simple machines curriculum over 10 weeks. We collected videos of all whole-class and small-group discussions during the unit. Analysis of pre- and post-tests showed that students in one teacher's classes performed significantly better on a science post-test that contained multiple-choice and essay items, as well as on the concept-mapping test (Puntambekar et al., 2007). Qualitative analyses of the enactments showed that there were important variations in the ways in which this teacher facilitated whole-class and small-group discussions—she helped students make connections among science ideas, and she also helped students understand how the different activities, such as conducting experiments, doing research, and making claims based on evidence, were related, so that students did not engage in each activity as an end in itself. Analyzing classroom interactions provided us with a context to help interpret students' scores on pre- and post-tests.

Iteration 3: Characterizing Implementations

In the first two iterations, we had the opportunity to field-test our design in a single school, albeit with different teachers and at different grade levels, to validate the key principles underlying our work. Following this, we moved from this limited context to the phase of practical implementations (Bielaczyc, 2013) by expanding our studies to schools in urban, rural, and suburban areas. In this phase, we implemented our units in two states within several school districts, including rural, urban, and suburban areas. Seven teachers teaching 27 classes participated in this study. Our aim was to understand if our innovation worked in other contexts in which the resources, teacher preparation, student population, and the degree of administrator support were quite different. We videotaped one target class per teacher, and collected observations from that class. We compared enactments across the multiple contexts, paying close attention to the ways in which the teachers in different contexts adjusted our designed curriculum (Bielaczyc, 2013), and the kinds of challenges that students as well as teachers faced. Following this cycle, we made several changes to our curriculum based on the analysis of videos and feedback from teachers, so that it could be used in different contexts. We refined strategies that teachers could use to facilitate whole-class discussions so that they were more concrete. This phase was important in our work, because, as described by Kolodner (2006), moving an innovation into classrooms outside of the designed context is challenging. An in-depth analysis of implementations in varied contexts helped us get a better understanding of the challenges teachers and students faced along with the changes we needed to make to our innovation. Particularly, we found that leading whole-class discussions was challenging for teachers, both in terms of focusing on the content ideas as well as developing strategies of helping students participate. Based on this cycle of studies, we made extensive revisions to our teacher materials, adding content-related information. We also found successful strategies that teachers used, and example videos from classrooms. We also conducted retrospective analyses of group discourse to understand how students collaborated and worked in small groups. This microgenetic analysis of a few targeted groups of students resulted in a better understanding of how we could support learning in small groups, both in terms of support from the teacher, and from the instructional materials (Bopardikar, Gnesdilow, & Puntambekar, 2011; Gnesdilow, Bopardikar, Sullivan, & Puntambekar, 2010). We have since been using these materials in several projects that followed the CoMPASS project.

Planning Trajectories in DBR

Size and Scope of Studies Within a Trajectory

The CoMPASS project is but one example of a trajectory along which studies were planned and conducted iteratively. Studies within a trajectory can be of different sizes and scopes. It is not always the case that studies involve a scaling-up; that is, the innovation is used in multiple contexts as the project matures. There could be other trajectories; each project could also have a different trajectory, determined by the underlying theoretical model, the relative significance of the design elements, and practical considerations. The following are examples of two research trajectories.

Example 1: Lab Studies to Classroom Studies

The trajectory in Ann Brown's work on reciprocal teaching and Fostering Communities of Learners (FCL) is perhaps one of the best examples of how theoretically driven lab studies could translate into classroom studies. The reciprocal teaching studies were based on a wealth of Brown's prior research on reading comprehension, the importance of effective strategies, and metacognition spanning several years of theory building and refinement. The reciprocal studies themselves followed a very interesting trajectory. Initial studies with reciprocal teaching took place with the researchers working with individual children. Later studies in resource rooms outside of classrooms were followed by studies in intact classrooms where teachers worked with small groups of children. It is important to note that in the reciprocal teaching studies spanning over a decade, the researchers conducted studies not only comparing reciprocal teaching with traditional forms of instruction but also studies in which components of the intervention were studied in detail, such as the value of presenting all strategies simultaneously or using reciprocal teaching for prevention of reading-related difficulties. Together, the studies led to theoretical advances in the study of comprehension, metacognition, and the use of strategies.

In classroom studies of reciprocal teaching, the researchers had to consider several interacting variables such as the nature and size of groups, instructional variations, nature of the text, learners, and their interactions. These studies were also different in scope because the instruction was carried out over several days, in contrast to earlier studies that included only a short intervention (1–2 days). This enabled the researchers to understand how the scaffolding provided by the teacher and the nature of peer groups influenced learning (Palincsar, 2003). All of these studies resulted in theory building and refinement, laying the groundwork for the next phase of Brown's work, FCL. Reciprocal teaching and the lessons learned about peers engaging in sharing ideas and building knowledge collaboratively formed the core of FCL. Reciprocal Teaching and FCL not only added to refinement of theory, but also added new forms of assessments. In addition, they helped focus the research community's attention on partnerships with teachers, administrators, and stakeholders. Most importantly, the FCL studies helped the researchers understand the complexity of classroom situations and the numerous factors therein, giving rise to the notion of design experiments!

Design-based research is about understanding how an innovation works in real-world contexts. But sometimes, issues arise during implementations that need to be tested in more controlled studies. Even when an innovation is being scaled up and implemented in multiple contexts, lab studies can be conducted to help researchers understand specific issues that arise. As discussed by Brown (1992), as key issues arose in FCL classrooms, the researchers conducted lab studies to systematically test hypotheses that were generated based on classroom studies. These could then inform the classroom studies, while at the same time refining the theoretical assumptions.

Example 2: Refining Specific Aspects of an Innovation

Another way trajectories of studies can be used is in refining understanding of both an innovation and the social structures of a classroom. Zhang, Scardamalia, Reeve, and Messina (2009) discuss a trajectory of three studies in which the design refinement happened in the participatory structures, or "social organization" examining different models of collaboration using Knowledge Forum®.

The aim of the studies that Zhang and his colleagues discuss was to understand the emergence of collective responsibility and the advancement of collective knowledge (cf. Chan & van Aalst, this volume). This required students' sustained participation in group discourse, sharing and refining ideas, examining multiple perspectives, and synthesis of ideas. As the object of study in this trajectory was the emergence of collective understanding, participant structures were varied to foster collective responsibility and knowledge construction. Zhang et al. tested three variants of participant structures: fixed groups, interacting groups, and opportunistic collaboration. After conducting a study with fixed groups in their first iteration, they found that there was a lack of cross-group interaction. Students mostly interacted with members of their own group, since the main goal was to advance collective knowledge. In the next iteration, therefore, opportunities for cross-group interaction were provided, so that students could contribute ideas to other groups and share information. However, it was found that the teacher still needed to coordinate these interactions. In the third iteration, groups were emergent in that all students were collectively responsible for advancing their understanding of the topic. Student groups convened based on their emerging goals and needs. Zhang et al. conducted detailed analyses of the discourse along dimensions indicative of collective responsibility.

In this case, all the studies along this trajectory happened in the same classroom taught by the same teacher. But, changes were made each year based on the previous study. The researchers were specifically interested in refining the participatory structures that supported collaborative work, and this goal formed the basis of the studies. By studying the participatory structures in classes taught by the same teacher, Zhang et al. were able to compare their iterative studies, and examine which frameworks of collaboration were most conducive to knowledge advancement.

Balancing Intertwined Goals

As mentioned earlier, DBR aims to refine elements of the design, as well as test the theoretical constructs underlying the design. But not all questions are, or should be, taken up in a single study. My discussion of examples of trajectories show how key conjectures about teaching and learning form the basis of studies in a trajectory, and how each study informs the next. In DBR, conjectures about learning and teaching could be embodied in materials and tools and/or participant structures. The trajectory of studies that Zhang et al. describe is an example of participant structures and how these could be refined through iterative studies.

In the CoMPASS project, our initial studies could be described as validation studies (McKenney, Nieveen, & van der Akker, 2006), or existence proofs (Bielaczyc, 2013). The key premise that was embodied in our innovation, that of helping students understand connections among science ideas, informed the design of our studies. We then moved to understanding classroom implementations in a limited context before expanding our studies to several school districts. In each iteration, methods were aligned with the questions that we were interested in for each study, combining the strengths of quantitative and qualitative methods. Each study was focused on a few research questions, so that data collection and analysis were aligned with the types of questions.

The important issue is to design the scope and scale of the studies within a trajectory so that we are *cumulatively* adding to learning theory, as well as understanding the main elements of the context in which our studies are being conducted. That might take the form of multiple iterations in the same classroom with the same teacher, or interspersing lab studies with classroom studies, or testing an innovation in different contexts. These decisions and the reasoning behind them is the key to DBR.

Conclusion

As I mentioned at the outset, design-based research is best characterized as a trajectory of studies, rather than a specific study or a type of study. The example from my own research that I discussed showed how research trajectories could be planned and implemented. The other examples that I discussed represent trajectories comprising studies of different scopes and sizes. I agree with Ormel, Roblin, McKenney, Voogt, and Pieters (2012) that there are few accounts that fully discuss design processes and the decisions that researchers make. To understand better how researchers plan trajectories, the choices they make, and how they address the constraints they face, we need more examples of research trajectories.

Further Readings

Brown, A. L. (1992). Design experiments: Theoretical and methodological challenges in creating complex interventions in classroom settings. *Journal of the Learning Sciences, 2*(2), 141–178.
This is still a must-read for anyone interested in DBR. It lays out the rationale for why research in complex settings such as classrooms is different, and the challenges of using experimental methods. The article also has excellent discussions on methodological issues such as grain size, and collecting qualitative and quantitative data, explained through examples from Brown's work.

Bielaczyc, K. (2013). Informing design research: Learning from teachers' designs of social infrastructure. *Journal of the Learning Sciences, 22*(2), 258–311.
This article explains expanding from studies that are conducted in smaller contexts, existence proofs, to practical implementations in DBR.

Dede, C. (2004). If design-based research is the answer, what is the question? *Journal of the Learning Sciences, 13*(1), 105–114.
This critique is important for us to think about the questions and the analytical methods we wish to use.

Puntambekar, S., Stylianou, A., & Goldstein, J. (2007). Comparing classroom enactments of an inquiry curriculum: Lessons learned from two teachers. *Journal of the Learning Sciences, 16*(1), 81–130.
I am including this as an example of one of our studies, in which we used both qualitative and quantitative analysis to understand students' learning and classroom enactments.

Zhang, J., Scardamalia, M., Reeve, R., & Messina, R. (2009). Designs for collective cognitive responsibility in knowledge-building communities. *Journal of the Learning Sciences, 18*(1), 7–44.
This is a great example of how trajectories could consist of studies in the same classroom, in which variations in the collaboration structures were introduced.

NAPLeS Resources

Puntambekar, S., *Design and design-based research* [Webinar]. In *NAPLeS video series*. Retrieved October 19, 2017, from http://isls-naples.psy.lmu.de/intro/all-webinars/puntambekar/index.html

References

Azevedo, R., & Cromley, J. G. (2004). Does training on self-regulated learning facilitate students' learning with hypermedia. *Journal of Educational Psychology, 96*, 523–535.

Bedersen, B. B., & Hollan, J. (1995). *Pad++: A zooming graphical interface for exploring alternate interface physics*. Paper presented at the Proceedings of the Seventh Annual ACM Symposium on User Interface Software and Technology, New York.

Bielaczyc, K. (2013). Informing design research: Learning from teachers' designs of social infrastructure. *Journal of the Learning Sciences, 22*(2), 258–311.

Bopardikar, A., Gnesdilow, D., & Puntambekar, S. (2011). Effects of using multiple forms of support to enhance students' collaboration during concept mapping. In H. Spada, G. Stahl, N. Miyake, & N. Law (Eds.) *Proceedings of the Computer Supported Collaborative Learning Conference* (pp. 104–111), Hong Kong: ISLS.

Brown, A. L. (1992). Design experiments: Theoretical and methodological challenges in creating complex interventions in classroom settings. *Journal of the Learning Sciences, 2*(2), 141–178.

Brown, A. L., Armbruster, B. B., & Baker, L. (1986). The role of metacognition in reading and studying. In J. Orasanu (Ed.), *Reading comprehension: From research to practice* (pp. 49–75). Hillsdale, NJ: Erlbaum.

Brown, A. L., & Palincsar, A. S. (1987). Reciprocal teaching of comprehension strategies: A natural history of one program for enhancing learning. In J. D. Day & J. G. Borkowski (Eds.), *Intelligence and exceptionality: New directions for theory, assessment, and instructional practice* (pp. 81–132). Westport, CN: Ablex.

Chan, C., & van Aalst, J. (2018). Knowledge building: Theory, design, and analysis. In F. Fischer, C. E. Hmelo-Silver, S. R. Goldman, & P. Reimann (Eds.), *International handbook of the learning sciences* (pp. 295–307). New York: Routledge.

Dede, C. (2004). If design-based research is the answer, what is the question? *Journal of the Learning Sciences*, *13*(1), 105–114.

Design-Based Research Collective (DBRC). (2003). Design-based research: An emerging paradigm for educational inquiry. *Educational Researcher*, *32*(1), 4–8.

Furnas, G. W. (1986). *Generalized fisheye views*. Paper presented at the Proceedings of the SIGCHI Conference on Human Factors in Computing Systems, New York.

Gnesdilow, D., Bopardikar, A., Sullivan, S., Puntambekar, S. (2010). Exploring Convergence of Science Ideas through Collaborative Concept Mapping. In K. Gomez, L. Lyons, & J. Radinsky (Eds.), *Proceedings of the Ninth International Conference of the Learning Sciences* (pp. 698–705), Chicago, IL.

Hoffman, J. L., Wu, H.-K., Krajcik, J. S., & Soloway, E. (2003). The nature of middle school learners' science content understandings with the use of on-line resources. *Journal of Research in Science Teaching*, *40*(3), 323–346.

Jacobson, M. J., & Wilensky, U. (2006). Complex systems in education: Scientific and educational importance and implications for the learning sciences. *Journal of the Learning Sciences*, *15*(1), 11–34.

Kelly, A. (2004). Design research in education: Yes, but is it methodological? *Journal of the Learning Sciences*, *13*(1), 115–128.

Kolodner, J. L. (2006). *The learning sciences and the future of education: What we know and what we need to be doing better*. Paper presented at the annual conference of the American Educational Research Association, San Francisco, CA.

Kolodner, J. L., Camp, P. J., Crismond, D., Fasse, B., Gray, J., Holbrook, J., (2003). Problem-based learning meets case-based reasoning in the middle-school science classroom: Putting learning by design(tm) into practice. *Journal of the Learning Sciences, 12*(4), 495–547.

Kozma, R. (2000). The use of multiple representations and the social construction of understanding in chemistry. In M. Jacobson & R. Kozma (Eds.), *Innovations in science and mathematics education: Advanced designs for technologies of learning* (pp. 11–46). Mahwah, NJ: Erlbaum.

Linn, M. C., & Eylon, B. S. (2011). *Science learning and instruction: Taking advantage of technology to promote knowledge integration*. New York: Routledge.

Lumpe, A. T., & Butler, K. (2002). The information seeking strategies of high school science students. *Research in Science Education*, *32*(4), 549–566.

McKenney, S., Nieveen, N., & van den Akker, J. (2006). Design research from a curriculum perspective. In J. van den Akker, K. Gravemeijer, S. McKenney, & N. Nieveen (Eds.), *Educational design research*, (pp. 67–90). New York: Routledge.

Newton, D. P., & Newton, L. D. (2000). Do teachers support causal understanding through their discourse when teaching primary science? *British Educational Research Journal*, *26*(5), 599–613.

Ormel, B. J., Roblin, N. N. P., McKenney, S. E., Voogt, J. M., & Pieters, J. M. (2012). Research–practice interactions as reported in recent design studies: Still promising, still hazy. *Educational Technology Research and Development*, *60*(6), 967–986.

McKenney, S., Nieveen, N., & van den Akker, J. (2006). Design research from a curriculum perspective. In J. van den Akker, K. Gravemeijer, S. McKenney, & N. Nieveen (Eds.), *Educational design research*, (pp. 67–90). New York: Routledge.

Palincsar, A. S. (2003). Ann L. Brown: Advancing a theoretical model of learning and instruction. In D. Zimmerman & D. H. Schunk, (Eds.), *Educational psychology: A century of contributions* (pp. 459–475). Mahwah, NJ: Erlbaum.

Palincsar, A. S. & Magnusson, S. J. (2001). The interplay of first-hand and second-hand investigations to model and support the development of scientific knowledge and reasoning. In S. M. Carver & D. Klahr (Eds.), *Cognition and instruction: Twenty-five years of progress* (pp. 151–194). Mahwah, NJ: Erlbaum.

Phillips, D. C., & Dolle, J. R. (2006). From Plato to Brown and beyond: Theory, practice, and the promise of design experiments. In L. Verschaffel, F. Dochy, M. Bockaerts, & S. Vosniadou (Eds.), *Instructional psychology: Past, present and future trends. Sixteen essays in honour of Erik De Corte* (pp. 277–292). Oxford, UK: Elsevier.

Puntambekar, S., Stylianou, A., & Goldstein, J. (2007). Comparing classroom enactments of an inquiry curriculum: Lessons learned from two teachers. *Journal of the Learning Sciences*, *16*(1), 81–130.

Puntambekar, S., Stylianou, A., & Hübscher, R. (2003). Improving navigation and learning in hypertext environments with navigable concept maps. *Human-Computer Interaction*, *18*(4), 395–426.

Sandoval, W. (2014). Conjecture mapping: An approach to systematic educational design research. *Journal of the Learning Sciences*, *23*(1), 18–36.

Sandoval, W. A., & Daniszewski, K. (2004). Mapping trade-offs in teachers' integration of technology-supported inquiry in high school science classes. *Journal of Science Education and Technology*, *13*(2), 161–178.

Shavelson, R. J., Phillips, D. C., Towne, L., & Feuer, M. J. (2003). On the science of education design studies. *Educational Researcher*, *32*(1), 25–28.

Stylianou, A., & Puntambekar, S. (2003). Does metacognitive awareness of reading strategies relate to the way middle school students navigate and learn from hypertext. *Annual Conference of the Northeast Educational Research Association*. Copenhagen, Denmark.

Stylianou, A., & Puntambekar, S. (2004). Understanding the role of metacognition while reading from nonlinear resources. *Sixth International Conference of the Learning Sciences (ICLS)*. Los Angeles, CA.

Tabak, I. (2004). Synergy: A complement to emerging patterns of distributed scaffolding. *Journal of the Learning Sciences*, *13*(3), 305–335.

Tabak, I., & Reiser, B. J. (1997). Complementary roles of software-based scaffolding and teacher–student interactions in inquiry learning. In R. Hall, N. Miyake, & N. Enyedy (Eds.) *Proceedings of Computer Support for Collaborative Learning '97* (pp. 289–298). Toronto, Canada.

Zhang, J., Scardamalia, M., Reeve, R., & Messina, R. (2009). Designs for collective cognitive responsibility in knowledge-building communities. *Journal of the Learning Sciences*, *18*(1), 7–44.

38
Design-Based Implementation Research

Barry Fishman and William Penuel

Design-based implementation research (DBIR) lies at the intersection of research, policy, and practice. DBIR emerges from a concern that many well-researched interventions, even those that are found to be effective in carefully designed randomized field trials, subsequently fail to produce the desired effects when employed in real-world settings. This is a gap between "what works" and "what works where, when, and for whom" (Means & Penuel, 2005, p. 181). In response, DBIR seeks to reconfigure the roles of researchers and practitioners to better support partnerships aimed at producing innovations that are effective, scalable, and sustainable and that be adapted successfully to meet the needs of diverse learners across diverse settings, in both formal and informal education.

Principles of DBIR

DBIR is built around four core principles, first introduced in Penuel, Fishman, Cheng, and Sabelli (2011). We argue that all four principles must be represented in order for research to be considered DBIR:

1. A focus on persistent problems of practice from multiple stakeholders' perspectives;
2. A commitment to iterative, collaborative design;
3. A concern with developing theory and knowledge related to both classroom learning and implementation through systematic inquiry; and
4. A concern with developing capacity for sustaining change in systems.

A focus on persistent problems of practice from multiple stakeholders' perspectives. One way that DBIR work inherently emphasizes equity is in its emphasis on identifying and negotiating problems of practice that respects the interest and experience of practitioners, researchers, and other stakeholders. Often, the manner in which research contributes to scholarly knowledge growth frames problems in ways that align with constructs or frames that make sense in an academic context, but are not easily recognizable in the complex, messy world of practice. For instance, researchers may be interested in better understanding how reflection works to deepen learner understanding. Teachers may think that this problem is interesting, but the real challenge they face is student performance on state tests. Joint identification and negotiation can identify ways in which these two problem areas overlap, enabling research to proceed in a way that serves the practical interest of all partners.

A commitment to iterative, collaborative design. As in design-based research (DBR), DBIR depends on the development of interventions that can be tested in real-world conditions and quickly iterated upon. Techniques similar to those in DBR, such as conjecture mapping (Sandoval, 2014) can be employed to assess whether designs are meeting their intended goals. What is critical is to have measurable goals and to employ data collection and analysis methods that allow for rapid and principled iteration.

A concern with developing theory and knowledge related to both classroom learning and implementation through systematic inquiry. Again, akin to DBR, DBIR stipulates that design work should contribute to the growing base of knowledge and theory, and design research has been framed as the iterative testing of "humble" theory (Cobb, Confrey, diSessa, Lehrer, & Schauble, 2003). Where DBIR and DBR diverge is in the nature of the theory under scrutiny. Where DBR most commonly focuses on learning and teaching within classrooms (or other kinds of learning environments), DBIR work also emphasizes theories related to how the innovation can be made usable and sustainable by teachers or others responsible for enacting or supporting it in the future, such as "infrastructuring," a socio-cultural theory that describes the process of creating and sustaining new cultural practices and supportive technological mechanisms. Star and Ruhleder (1996) describe infrastructure as: embedded, transparent, learned as part of membership in a community, linked with conventions of practice, and embodying standards and conventions.

A concern with developing capacity for sustaining change in systems. The fourth principle reiterates the importance of orienting the entire collaborative problem identification and design process towards what it means for the innovation to be useful past the time of initial development. While sustainability is an implicit goal of (most) education design work, it is not often an explicit element of research design or partnership formation.

Antecedents of DBIR

Though the term "DBIR" was first introduced in in 2011 *Educational Researcher* article (Penuel et al., 2011), we do not claim that DBIR is entirely new, or that it has no precedent among research approaches, especially in the Learning Sciences. We have already mentioned the close relationship between DBIR and DBR (cf. Puntambekar, this volume). DBIR also draws from research methodologies that emphasize partnership and participation, such as community-based participatory research, with its goal of fostering partnerships between the academy and the community to advance social change goals (Stewart & Shamdasani, 2006), or the participatory design tradition originating in Scandinavia (Simonson & Robertson, 2013). Evaluation research, especially utilization-focused evaluation, directs researchers to focus intended uses of interventions or innovations by desired end-users (Patton, 1997).

Two research traditions that are particularly important in the conceptualization of DBIR are improvement science and implementation research. Improvement science is a field that draws on insights and practices from management studies and healthcare. It focuses on learning from variation in outcomes that results from efforts to improve standard work practices in a field. Examples include practices intended to reduce infections during surgery and for reducing errors in manufacturing processes. In education, the Carnegie Foundation for the Advancement of Teaching has adapted improvement science methods for education, and they are engaged in developing networks organized around improvement of educational practices such as new teacher induction (Bryk, Gomez, Grunow, & LeMahieu, 2015). Implementation research has a long history in education and in related disciplines in the human sciences. It focuses on documenting and explaining variation in implementation of policies, practices, and programs, and draws on theories from political science, sociology, economics, and anthropology, to name several fields (Spillane, Reiser, & Reimer, 2002). A more thorough treatment of each of these research traditions and their relationship to DBIR may be found in the introductory chapter of the 2013 *National Society for the Study of Education Yearbook*, on DBIR (Fishman, Penuel, Allen, & Cheng, 2013).

Examples of DBIR

DBIR is currently employed to address a range of different types of educational challenges, including building teacher capacity, improving student outcomes at the scale of a local educational agency, facilitating community collaborations to enhance youths' out-of-school learning opportunities, and networks of organizations working on common problems with shared aims and strategies. In this section we introduce and discuss a range of examples of DBIR work.

Building teacher capacity to design curriculum in the Netherlands. In 2006, the Netherlands introduced new goals for secondary education. The goals were broad, and local education agencies were expected to make them concrete and to design curriculum to meet them. Teacher teams in particular were to play a significant role in curriculum design. But teacher teams would need support, since few had specialized expertise in designing curriculum that was aligned to external goals and internally coherent. In one small-scale DBIR project, researchers helped both to facilitate and study teachers' efforts to design a lesson series for interdisciplinary secondary courses (Huizinga, Handelzaltz, Nieveen, & Voogt, 2014). Their aim was to study the conditions and supports needed for effective teacher design teams. This type of investigation—one that examines key conditions needed for building teacher capacity—is a good example of a question that can be addressed through DBIR.

The research focused on interviews conducted with both facilitators and teachers at the conclusion of the design process and after teachers had enacted their lessons. Interviews focused on teachers' application of curriculum design expertise, using frameworks that drew from Shulman's (1986) pioneering work on teacher knowledge and from Remillard's (1999) analysis of teachers' approaches to curriculum. The study found that one of the most valuable scaffolds for supporting teachers in developing coherent curricula were a set of lesson templates. In addition, teachers valued having the external facilitator of their team provide feedback on the quality of the series, especially the degree to which they aligned with and met the new goals for student learning. Though small-scale, this study was part of a larger program of research on curricular co-design with teachers in the Netherlands that built a strong knowledge base about the value of co-design for teacher learning and conditions needed to support teacher teams (Voogt et al., 2015).

Connecting youth to out-of-school learning opportunities in the United States. In cities around the world, there are many different organized activities offered after school and in summer, but there are few places or organizations that provide ways for youth to learn about them. In the United States, concern about summer learning loss is growing, leading policymakers to strengthen and make summer learning opportunities more accessible for low-income youth. In Chicago, the mayor's office, funding agencies, and informal learning organizations created the Chicago City of Learning initiative to address these problems. They partnered with researchers at the Digital Youth Network (DYN; Barron, Gomez, Pinkard, & Martin, 2014) to build a centralized place for youth and families to search for learning opportunities and no way to document participation that could be aggregated together to create a citywide understanding of learning opportunities and participation. In this project, computer scientists and their research team work closely with local organizations to facilitate listing of program offerings, and they use data on youth search and improve the website where youth find activities. In addition, they have co-designed mobile programs—programs that can be offered in the short term in different parts of the city to increase access to specialized science, technology, engineering, and mathematics (STEM) programming.

The research on program locations and participation is focused on mapping and modeling the location of programs and youth participation. A team at DePaul University and the University of Colorado has developed easy-to-use metrics and a GIS interface for representing the diversity and accessibility of different types of programs by neighborhood (Pinkard et al., 2016). The partnership is using these metrics to inform iterations on the website design and to identify places where new programs for youth are needed to increase equity of opportunities.

Enhancing new teachers' experience in a network of districts in the United States. In the United States, new teacher retention is a significant problem, and research suggests it results in part from the poor feedback new teachers receive in their first few years. The Building Teaching Effectiveness Network (BTEN), led by the Carnegie Foundation for the Advancement of Teaching, was a partnership organized to address this problem. The network comprised representatives from the American Federation of Teachers, the Institute for Healthcare Improvement, two large, public, urban school districts, and a charter management organization. An early activity of the network illustrates the considerable effort of the team to define the problem of retention. They conducted analyses of the number of different sources of feedback teachers received, as well as the nature of that feedback. They documented ways that feedback was incoherent, infrequent, and not actionable for teachers.

The collaborative design process in BTEN followed principles adapted from improvement research in healthcare. Specifically, network participants worked together to design a protocol for a brief feedback encounter between a principal and a new teacher. They defined metrics for success that included a simple, "practical measure" to gauge teachers' response to the feedback, along with longer-term metrics of success, such as the intention to return to teaching the next year. Over multiple iterative Plan–Do–Study–Act (PDSA) cycles in a single principal–teacher pair, then five pairs, and ultimately 17 different pairs, the team revised the protocol. The research focused on documenting the conditions under which the use of PDSA cycles led to reliable improvements in new teachers' experiences and intentions to continue teaching (Hannan, Russell, Takahashi, & Park, 2015). It illustrates a key type of knowledge DBIR develops: conditions that support effective implementation of innovations.

Improving students' academic language in middle school in the United States. Many students struggle to "read to learn," that is, to engage in literacy practices in order to master subject-matter content. The Strategic Research Education Partnership Institute and one of its district partners, the Boston Public Schools, set out to tackle the challenge of developing students' specialized vocabulary in English after the district identified "academic language" as a cause of low test scores among middle-grades students. The group set up a design team comprising teachers, district leaders, and experts in literacy from Harvard University. The co-design process led to an intervention that was feasible for teachers to implement on a regular basis with students in small time increments and that engaged students in debate about personally relevant topics (Snow, Lawrence, & White, 2009).

The research that was conducted on the intervention, called Word Generation, included both an experimental study of its impact and studies focused on identifying sources of variability in implementation. A random assignment study—in which some teachers' classrooms were assigned by lottery to receive Word Generation, while others were not—found that the co-designed intervention had a significant impact on students' vocabulary (Snow et al., 2009). Another research team, led by a scholar of educational change, contributed to understanding the conditions under which its implementation was effective in improving student outcomes. This research has been used not only to inform the scaling within Boston, but also its spread to other districts, illustrating the ways that lessons learned about effective implementation can apply to new contexts.

Argumentative Grammars of DBIR

A central task for a community of scholars developing a new research approach is to collectively develop common understandings about the kinds of questions that the approach can answer, the methods and evidence that are needed to answer those questions, and the limits of the approach. These understandings constitute for a community of scholars something Kelly (2004) has called an "argumentative grammar," and refers to the logic for developing and warranting claims for that approach. In DBIR, there are likely to be multiple grammars, because there is no single type of study that is a DBIR study. As an approach, it encompasses different methods, each with its own logic.

The argumentative grammars of DBIR are still emerging, but the questions and methods appropriate for DBIR follow directly from the four key principles outlined at the start of this chapter. For example, one type of question DBIR can ask related to the first principle is "What do different stakeholders perceive as the major challenges to improving literacy instruction in the district?" A researcher might rely on any number of different sources of evidence to answer this question, from interviews to surveys, or use a technique such as the Delphi method, a structured approach to soliciting ideas from a diverse group of stakeholders (e.g., Penuel, Tatar, & Roschelle, 2004). But a convincing answer to the question posed demands more than just a large volume of evidence from one of these sources. Because the first principle focuses on "multiple stakeholders," it is important that the sample be representative of the different stakeholders—that is, it include representatives of groups of people who might benefit from or be harmed by literacy instruction or efforts to improve it. Also key to judging the adequacy of the conclusions is the degree to which the study analyzes the level of agreement or disagreement that exists among different stakeholders as to what are the most important pressing problems to address. That information is more likely to be of help in guiding a team to decide on a focus of joint work than a summary that only indicates the percentage of people who agree that something is an important problem.

Studies might also focus on the process of co-design (see also Gomez, Kyza, & Mancevice, this volume). Of particular concern for DBIR is the degree to which stakeholders have had a say in the design of innovations, and if so, how their perspectives were taken up in the design process. Documentation of participation in design through ethnographic observation, along with interviews of participants' perceptions of the process, can provide evidence to answer questions about participant voice and uptake (e.g., Severance, Penuel, Sumner, & Leary, 2016). Studies can also focus on design decisions, and a relevant question here for DBIR is the extent to which decisions about how to iterate on designs reflect evidence from implementation research (e.g., Vahey, Roy, & Fueyo, 2013). That evidence can be generated in different ways, including from systematic collection of the insights of practitioners. What is key, to be convincing for DBIR, is that the nature of evidence gathered and process for using it leads to decisions that have the potential to improve implementation of an innovation under development.

Studies that address the third principle—developing knowledge, tools, and theory related to both implementation and learning—are likely to adhere to standards for implementation and outcome research that are in wide circulation in education today. Implementation research is already a richly theorized tradition within education research that draws on perspectives from political science, sociology, economics, and anthropology (Honig, 2006; Hubbard, Mehan, & Stein, 2006; McLaughlin, 1987; Spillane et al., 2002). Research to evaluate interventions' impacts similarly is well developed in education. In the past decade, education research has refined strategies for warranting claims about impact, such as for determining the sample size needed (Schochet, 2008), how to implement random assignment effectively in schools (Roschelle, Tatar, Shechtman, & Knudsen, 2008), and how to analyze data appropriately, given different sources of variance that can explain student outcomes (Raudenbush & Bryk, 2002). Community understandings about good research design and analysis from implementation research and from impact studies apply to judgments about DBIR studies that answer questions related to enactment and impact.

Capacity-building studies within DBIR are still relatively rare. But they answer questions like "How can co-design build capacity of teachers to develop coherent units of instruction?" and "How can DBIR build networks of teacher leaders to improve schools?" Answers to such questions are likely to require data on the individual skill or human capital built through research activity, patterns of collaboration and help, or social capital produced. DBIR may also build capacity by creating material capital; that is, material tools and practices that are sustained over time.

Challenges in DBIR Work

DBIR is aspirational, in that its pursuit requires researchers to work beyond—or in spite of—the current infrastructures that exist within research and academia (O'Neill, 2016). How can longer cycles of collaboration in DBIR be supported in the context of shorter (and less reliable) funding support from federal agencies and private foundations? How can and should junior scholars be engaged in complexities of DBIR work? How does persistent instability (e.g., turnover, changes to priorities), especially in the most challenged urban educational systems, affect practitioners' ability to be fully engaged participants in DBIR?

As the field prepares new generations of researchers, we must provide opportunities for them to learn about key elements of partnership work, such as problem negotiation or co-design. These topics currently lie beyond standard "methods" training, and are most often learned through apprenticeship in existing research projects. Similarly, pre-service and in-service teachers might be prepared to think of themselves as co-designers of and collaborators in innovation.

A large challenge to the legitimization of DBIR work lies in the stances that education policymakers and funders often take towards implementation and innovation. Currently, a strong orientation to "fidelity" undervalues the importance of concepts such as local adaptation and mutual appropriation that are core to the view of innovation taken within DBIR. It is incumbent upon DBIR researchers to provide clear guidance on contextual and other factors that enable successful local adaptation.

Challenges similar to these were once (and in some cases still are) true for design-based research. We are confident that if DBIR is valued by the field, the challenges we identify will recede as a "new normal" for research partnerships in education emerges.

Future Trends in DBIR

As the use of DBIR continues to grow, it is important that scholarship focused on DBIR (not simply employing DBIR) continues to grow as well. As new exemplars from a range of contexts and content areas are developed, our understanding of what is common and what must vary will come into better focus. There is also a need for research *on* the various components of DBIR. Areas such as problem negotiation, capacity building, and infrastructuring require focused study and theorizing in order to develop guidance for how these components of DBIR might be pursued.

The field must continue to develop tools that generate fast, usable data, such as "practical measurements." We anticipate that emerging areas of "big data" scholarship such as learning analytics (Siemens, 2013; cf. Rosé, this volume) will make contributions in this area, helping to turn learner interactions with technological systems into inspectable and actionable information.

Finally, we look forward to the emergence of new institutional arrangements that facilitate partnership work, such as the Strategic Education Research Partnership (Donovan, Snow, & Daro, 2013) or the Stanford–San Francisco Unified School District Partnership (Wentworth, Carranza, & Stipek, 2016). These organizations will help guide and grow our focus on the changing institutional contexts of schooling and for learning out of school.

The Learning Sciences takes a multidisciplinary approach to the study of learning in context, and as such the field has a long tradition of embedding the study of cognition and performance within real-world learning environments. DBIR is an invitation to the field to refocus our work on the most challenging of those contexts to foster the design and development of interventions that have a lasting effect on learning and teaching.

Further Readings

Coburn, C. E., & Penuel, W. R. (2016). Research–practice partnerships in education: Outcomes, dynamics, and open questions. *Educational Researcher, 45*(1), 48–54. doi:10.3102/0013189X16631750

This paper presents evidence related to the outcomes and dynamics of research–practice partnerships in education and related fields. It also describes gaps in research on partnerships. The paper includes evidence that co-designed innovations using DBIR can impact student outcomes.

Gomez, L. M., Bryk, A. S., Grunow, A., & LeMahieu, P. G. (2015). *Learning to improve: How America's schools can get better at getting better.* Cambridge, MA: Harvard Education Press.

This book presents an overview of the Carnegie Foundation for the Advancement of Teaching's approach to Networked Improvement Communities, drawing on examples from K–12 and postsecondary education.

Russell, J. L., Jackson, K., Krumm, A. E., & Frank, K. A. (2013). Theories and research methodologies for design-based implementation research: Examples from four cases. In B. J. Fishman, W. R. Penuel, A.-R. Allen, & B. H. Cheng (Eds.), *Design-based implementation research: Theories, methods, and exemplars (Yearbook of the National Society for the Study of Education)* (pp. 157–191). New York: Teachers College Record.

This chapter provides an overview of theory and methods from implementation research and discusses how those theories and methods inform DBIR projects. The chapter appears within an edited volume that provides a broad overview of DBIR, with examples from a range of projects. Sections focus on designing across settings, designing across levels, forms of evidence, and infrastructures in support of DBIR.

Snow, C. E., Lawrence, J., & White, C. (2009). Generating knowledge of academic language among urban middle school students. *Journal of Research on Educational Effectiveness, 2*(4), 325–344. doi:10.1080/19345740903167042

A research collaboration organized by the Strategic Education Research Partnership Institute (SERP) that brought together school personnel facing a challenge in students' performance on a state standardized test with researchers in literacy and educational policy.

Vahey, P., Roy, G., & Fueyo, V. (2013). Sustainable use of dynamic representational environments: Toward a district-wide adoption of SimCalc-based materials. In S. Hegedus & J. Roschelle (Eds.), *Democratizing access to important mathematics through dynamic representations: Contributions and visions from the SimCalc research program* (pp. 183–202). New York: Springer.

This paper reviews a cycle of DBIR focused on increasing the sustainability of curricular materials in a research–practice partnership. It illustrates how a research team used sustainability evidence to refine its approach to supporting implementation of materials at scale.

NAPLeS Resources

Puntambekar, S., *Design and design-based research* [Webinar]. In *NAPLeS video series*. Retrieved October 19, 2017, from http://isls-naples.psy.lmu.de/intro/all-webinars/puntambekar/index.html

Penuel, W., *Design-based implementation research* (DBIR) [Video file]. *Introduction*. In *NAPLeS video series*. Retrieved October 19, 2017, from http://isls-naples.psy.lmu.de/video-resources/guided-tour/15-minutes-penuel/index.html

References

Barron, B., Gomez, K., Pinkard, N., & Martin, C. K. (2014). *The Digital Youth Network: Cultivating digital media citizenship in urban communities*. Boston, MA: MIT Press.

Bryk, A. S., Gomez, L. M., Grunow, A., & LeMahieu, P. (2015). *Learning to improve: How America's schools can get better at getting better*. Cambridge, MA: Harvard University Press.

Cobb, P., Confrey, J., diSessa, A., Lehrer, R., & Schauble, L. (2003). Design experiments in educational research. *Educational Researcher, 32*(1), 9–13.

Donovan, M. S., Snow, C. E., & Daro, P. (2013). The SERP approach to problem-solving research, development, and implementation. In B. J. Fishman, W. R. Penuel, A.-R. Allen, & B. H. Cheng (Eds.), *Design-based implementation research (Yearbook of the National Society for the Study of Education)* (pp. 400–425). New York: Teachers College Record.

Fishman, B., Penuel, W. R., Allen, A., & Cheng, B. H. (2013). Design-based implementation research: Theories, methods, and exemplars. *Yearbook of the National Society for the Study of Education* (Vol. 2). New York: Teachers College Record.

Gomez, K., Kyza, E. A., & Mancevice, N. (2018), Participatory design and the learning sciences. In F. Fischer, C.E. Hmelo-Silver, S. R. Goldman, & P. Reimann (Eds.), *International handbook of the learning sciences* (pp. 401–409). New York: Routledge.

Hannan, M., Russell, J. L., Takahashi, S., & Park, S. (2015). Using improvement science to better support beginning teachers: The case of the Building a Teaching Effectiveness Network. *Journal of Teacher Education, 66*(5), 494–508.

Honig, M. I. (2006). Complexity and policy implementation: Challenges and opportunities for the field. In M. I. Honig (Ed.), *New directions in education policy implementation: Confronting complexity* (pp. 1–23). Albany, NY: SUNY Press.

Hubbard, L., Mehan, H., & Stein, M. K. (2006). *Reform as learning: When school reform collided with organizational culture and community politics in San Diego.* New York: Routledge.

Huizinga, T., Handelzaltz, A., Nieveen, N., & Voogt, J. M. (2014). Teacher involvement in curriculum design: need for support to enhance teachers' design expertise. *Journal of Curriculum Studies, 46*(1), 33–57.

Kelly, A. E. (2004). Design research in education: Yes, but is it methodological? *Journal of the Learning Sciences, 13*(1), 113–128. doi:10.1207/s15327809jls1301_6

McLaughlin, M. W. (1987). Learning from experience: Lessons from policy implementation. *Educational Evaluation and Policy Analysis, 9,* 171–178.

Means, B., & Penuel, W. R. (2005). Research to support scaling up technology-based innovations. In C. Dede, J. Honan, & L. Peters (Eds.), *Scaling up success: Lessons from technology-based educational improvement* (pp. 176–197). New York: Jossey-Bass.

O'Neill, D. K. (2016). When form follows fantasy: Lessons for learning scientists from modernist architecture and urban planning. *Journal of the Learning Sciences, 25*(1), 133–152. doi:10.1080/10508406.2015.1094736

Patton, M. (1997). *Utilization-focused evaluation: The new century text.* Thousand Oaks, CA: Sage Publications Inc.

Penuel, W. R., Fishman, B., Cheng, B. H., & Sabelli, N. (2011). Organizing research and development at the intersection of learning, implementation, and design. *Educational Researcher, 40*(7), 331–337. doi:10.3102/0013189X11421826

Penuel, W. R., Tatar, D., & Roschelle, J. (2004). The role of research on contexts of teaching practice in informing the design of handheld learning technologies. *Journal of Educational Computing Research, 30*(4), 331–348.

Pinkard, N., Penuel, W. R., Dibi, O., Sultan, M. A., Quigley, D., Sumner, T., & Van Horne, K. (2016). *Mapping and modeling the abundance, diversity, and accessibility of summer learning opportunities at the scale of a city.* Paper presented at the Annual Meeting of the American Educational Research Association, Washington, DC.

Puntambekar, S. (2018). Design-based research (DBR). In F. Fischer, C. E. Hmelo-Silver, S. R. Goldman, & P. Reimann (Eds.), *International handbook of the learning sciences* (pp. 383–392). New York: Routledge.

Raudenbush, S. W., & Bryk, A. S. (2002). *Hierarchical linear models: Applications and data analysis methods* (2nd ed.). Thousand Oaks, CA: Sage.

Remillard, J. (1999). Curriculum materials in mathematics education reform: A framework for examining teachers' curriculum development. *Curriculum Inquiry, 29*(3), 315–342.

Roschelle, J., Tatar, D., Shechtman, N., & Knudsen, J. (2008). The role of scaling up research in designing for and evaluating robustness. *Educational Studies of Mathematics, 68*(2), 149–170.

Rosé, C.P. (2018). Learning analytics in the Learning Sciences. In F. Fischer, C. E. Hmelo-Silver, S. R. Goldman, & P. Reimann (Eds.), *International handbook of the learning sciences* (pp. 511–519). New York: Routledge.

Sandoval, W. (2014). Conjecture mapping: An approach to systematic educational design research. *Journal of the Learning Sciences, 23*(1), 18–36. doi:10.1080/10508406.2013.778204

Schochet, P. Z. (2008). Statistical power for random assignment evaluations of educational programs. *Journal of Educational and Behavioral Statistics, 33*(1), 62–87.

Severance, S., Penuel, W. R., Sumner, T., & Leary, H. (2016). Organizing for teacher agency in curriculum design. *Journal of the Learning Sciences, 25* (4), 531–564. doi:10.1080/10508406.2016.1207541

Shulman, L. S. (1986). Those who understand: Knowledge growth in teaching. *Educational Researcher, 15*(2), 4–14.

Siemens, G. (2013). Learning analytics: The emergence of a discipline. *American Behavioral Scientist, 57*(10), 1380–1400. doi:10.1177/0002764213498851

Simonsen, J., & Robertson, T. (Eds.). (2013). *Routledge International Handbook of Participatory Design* (1st ed.). New York: Routledge.

Snow, C., Lawrence, J., & White, C. (2009). Generating knowledge of academic language among urban middle school students. *Journal of Research on Educational Effectiveness, 2*(4), 325–344.

Spillane, J. P., Reiser, B. J., & Reimer, T. (2002). Policy implementation and cognition: Reframing and refocusing implementation research. *Review of Educational Research, 72,* 387–431.

Star, S. L., & Ruhleder, K. (1996). Steps Toward an ecology of infrastructure: Design and access for large information spaces. *Information Systems Research, 7*(1), 111–134. doi:10.1287/isre.7.1.111

Stewart, D. W., & Shamdasani, P. N. (2006). *Applied social research methods series.* Newbury Park, CA: Sage.

Vahey, P., Roy, G., & Fueyo, V. (2013). Sustainable use of dynamic representational environments: Toward a district-wide adoption of SimCalc-based materials. In S. Hegedus & J. Roschelle (Eds.), *Democratizing access to important mathematics through dynamic representations: Contributions and visions from the SimCalc research program* (pp. 183–202). New York: Springer.

Voogt, J. M., Laferrière, T., Breuleux, A., Itow, R. C., Hickey, D. T., & McKenney, S. E. (2015). Collaborative design as a form of professional development. *Instructional Science, 43*(2), 259–282.

Wentworth, L., Carranza, R., & Stipek, D. (2016). A university and district partnership closes the research-to-classroom gap. *Phi Delta Kappan, 97*(8), 66–69.

39
Participatory Design and the Learning Sciences

Kimberley Gomez, Eleni A. Kyza, and Nicole Mancevice

Introduction

Learning scientists seek to study learning in context, often by immersing themselves in research within schools and classrooms, in collaboration with practitioners and other stakeholders. As they inquire into how learning happens, and how it can be improved, learning scientists frequently engage in the design and study of learning environments—an approach which affords opportunities for deep learning. Researcher–practitioner collaborations are essential vehicles to facilitate this process, as teachers and researchers each bring diverse perspectives to the joint effort. In this chapter, we use the terms "participatory", "collaboration" and "co-design" to refer to educational curricular, software, programmatic, or other design efforts involving researchers and practitioners (e.g., teachers, administrators) working together to address an identified problem of practice. These collaborations frequently involve designing an instructional tool that not only considers the needs of the students, but also addresses the needs of the teachers who are ultimately responsible for using these tools in the classroom (Edelson, Gordin, & Pea, 1999). Initially, the contributions of teachers and researchers to the design process may be distinct: researchers pay particular attention to theory-driven decisions, and teachers bring their pragmatic views on how learning is realized in practice. Over time, however, these roles may mesh and broaden, and all contributors develop deeper knowledge and expertise (Herrenkohl, Kawasaki, & Dewater, 2010).

The design and research stance involved in researcher–practitioner collaborations can be unfamiliar for both researchers and practitioners (Penuel, Coburn, & Gallagher, 2013). We begin by briefly discussing the foundations of current co-design approaches and guiding principles undergirding collaboration in the learning sciences literature, with a special focus on researcher–practitioner collaborative activities. We then highlight recurrent themes in the teacher–researcher collaboration literature, and describe the challenges and tensions that can emerge. We conclude the chapter with a discussion of the design principles that shape and guide these efforts, and offer insights into implications for practice and policy.

Foundations

Emergence of Participatory Design in Education

Research into participatory design originated in the work of Nygaard and Bergo, who created a handbook for the trade union movement based on their work with the Iron and Metal Workers' Union in Scandinavia (Nygaard & Bergo, 1973). Their effort described workers' disadvantage as they struggled to participate in shaping production (Beck, 2001). To support effective negotiation with management (Nygaard & Bergo, 1973), participatory design efforts aimed to build trade unionists' deeper understanding of technical language and concepts about technology. As understandings about the importance of including end users earlier in the design process progressed, the notion of participation in design became more established in many applied fields, including architecture, civil planning, human-computer interaction, and the learning sciences.

In the early 1990s, researchers around the United States, in school-based efforts like Learning Through Collaborative Visualization (CoVis), the Center for Learning Technologies in Urban Schools (LeTUS), The Voyage of the Mimi, Schools for Thought, and the Supportive Scientific Visualization Environments for Education Project, were involved in participatory types of research, including software co-design, and professional development activities. For example, LeTUS (Marx et al., 2004), a collaborative project among Northwestern University and University of Michigan researchers and two large school systems (Chicago and Detroit), created contexts for curricular participatory design. These contexts, called work circles, represented an effort to democratize researcher–practitioner teams by explicitly recognizing, and drawing on, the distinct expertise and value of practitioners and researchers in joint curricular design efforts (Lewin, 1947). Undergirding these efforts was a commitment to creating curricular materials that were grounded in authentic contexts and pedagogy (Cognition & Technology Group, 1997) and in the interests of students and teachers (Rivet, Krajick, Marx, & Reiser, 2003). These early activities laid the groundwork for current co-design approaches, hereafter called participatory co-design (PCD), that focus on authentic problems of practice, and draw on the expertise of the group members (Kristiansen & Bloch-Poulsen, 2013).

Goals and Commitments of Co-Design

Typically, the object of activity in educational PCD is the development of an instructional tool or curriculum for specific contexts (e.g., D'Amico, 2010; Edelson et al., 1999; Kyza & Georgiou, 2014; Penuel & Yarnall, 2005). Researchers and teachers, as well as other key stakeholders (e.g., software designers, students, district administrators), collaborate to design or redesign an instructional innovation. As Recker (2013; Recker & Sumner, this volume) has noted, a central aim in involving teachers in design is to make instructional resources more useful and usable. Researcher–practitioner collaborations involve stakeholders as co-designers to address the needs of both students and teachers (Edelson et al., 1999; Penuel, Roschelle, & Shechtman, 2007), and to develop the practitioner capacity in, and ownership of, the instructional tools or curriculum (Kyza & Georgiou, 2014; Lui & Slotta, 2014).

PCD projects have historically been researcher-initiated. In this sense, the problem to be addressed may be identified at the launch of the co-design process (Penuel et al., 2007). Early conversations involve participants in defining the problem in greater detail, discussing pedagogical beliefs, and determining co-design goals (Cober, Tan, Slotta, So, & Könings, 2015). In co-design activities, researchers typically seek to co-create, or redesign, a locally useful instructional tool or curriculum. In each effort, it is important to consider whether all key stakeholders are involved in the process (Penuel et al., 2013).

Researcher–practitioner collaborations serve multiple goals. For example, pragmatically, teams need to attend to co-design tasks like the development of educational artifacts and tools. Research suggests that teachers indicate that this is of high priority to them (Gomez, Gomez, Cooper, Lozano, & Mancevice, 2016). At the same time, in co-design, researchers also aim to develop and validate theories. The collaborations pose opportunities and challenges, and require multiple levels of negotiation. The kinds of questions that can be explored are also highly dependent on the context of collaboration.

PCD educational artifacts are typically locally useful tools (Penuel & Yarnall, 2005). In many PCD contexts, researchers explicitly take a "distributed expertise" stance (e.g., Kyza & Nicolaidou, 2016), recognizing participants' expertise at different points of the design process (e.g., Lau & Stille, 2014). Researchers typically seek to maintain the team's focus on project goals (Penuel et al., 2007) as they structure, facilitate, and document the design process (Kyza & Georgiou, 2014; Kyza & Nicolaidou, 2016; Penuel et al., 2007). Practitioners' roles and responsibilities often relate to pedagogical and curriculum decisions and planning, as well as implementing and reflecting on designs (Cober et al., 2015). However, we acknowledge the uniqueness of design, as roles evolve to fit design stages and needs (e.g., Herrenkohl et al., 2010; Penuel et al., 2007).

To be successful in researcher–practitioner co-design efforts, all participants need to feel that their contributions are valued (Cober et al., 2015; Herrenkohl et al., 2010). Achieving this requires mutual trust, scaffolding of participation, and situated reflection. The co-design process involves an ongoing negotiation of participants' expectations and goals, design goals, and constraints of the local context (Edelson, 2002; Johnson, Severance, Penuel, & Leary, 2016). However, the authority role attributed to researchers, due to funding and research priorities, cannot be ignored (Gomez, Shrader, Williams, Finn, & Whitcomb, 1999; Penuel et al., 2007). Recognition of issues of ownership, control, and power are important for team dynamics; however, in cases where this authority was set aside in favor of parity, researchers observed a negative impact on co-design processes (Lau & Stille, 2014).

Theoretical commitments. Generally agreed upon theoretical commitments to PCD efforts in the learning sciences include a commitment to the value of all co-design participants' expertise and contributions, and a sociocultural theoretical commitment to learning from, and designing for, local contexts. Co-design has evolved, over the past two decades, with respect to membership, frequency and length of meetings, designed tools, targeted audience, aims for transformation and local reform with attention to classrooms (Gomez et al., 1999), schools (Coburn & Stein, 2010), and district practices (Penuel, Fishman, Cheng, & Sabelli, 2011). Current efforts retain an emphasis on democratic participation and broader participation among relevant stakeholders, attending directly to practice in authentic contexts, and have a focus on the learners who will experience the designed artifacts. Closely connected to the co-design commitment is an aim towards researcher and practitioner capacity building, particularly understanding which designs are most effective, when, how, and for whom. Together, through co-design, researchers and practitioners contribute to and/or refine theory and knowledge about practice (Gomez et al., 1999).

Methodological commitments. Methodologically, co-design efforts are documented through, and refined within, a design-based approach. Co-design typically involves several sequential but often overlapping phases of design, documentation, and iterative refinement. Inquiry into, and documentation of, this process is often described as design-based research (Barab & Squire, 2004). Within these phases, participants collectively engage in problem identification, and identify the tool(s) and/or processes that will be designed/redesigned. Often one or more of the co-design participants iteratively "try out" processes, tools, curricula, etc. in local rather than laboratory settings, to ensure that the designed artifact meets local needs (Gomez et al., 1999). The design rationale serves as a North Star, as the team juxtaposes the design and implementation with project aims, checking to ensure that the initial stimulus problem is being addressed. As teachers "try it out," efforts are reported to the group, analyzed, critiqued, and often iteratively refined until the group converges on mutually satisfactory results. Local setting testing, and iterative refinement, help to ensure that the tool meets the needs of everyday users—teachers, students, and/or administrators. Throughout the co-design effort, collective reflection, multi-voiced discussion and decision-making, and public critique are emphasized.

Research Themes

To date, with few exceptions, most educational PCD efforts have originated in Europe and North America. Research typically centers on teachers and teacher learning, rather than on researchers or other participants, and predominantly describes K-12 classroom efforts, designing within the STEM and language arts (e.g., literacy) domains. Among others, research has examined the design and use of new software tools, online virtual lab activities, project-based classroom learning, and assessments.

Studies have been primarily qualitative in nature, employing ethnographic and case study approaches, as researchers seek to understand how to facilitate the co-design process while investigating its tensions, challenges, opportunities, and outcomes. Several principal research themes and findings are evident:

- Forms of teacher participation and the conditions which might have supported their involvement (Cober et al., 2015);
- Teachers' perceptions about co-design and ownership (Cviko, McKenney, & Voogt, 2015; Gomez et al., 2015; Kyza & Georgiou, 2014);
- Student learning as the outcome of teachers' collaboration in co-design (Cviko, McKenney, & Voogt, 2014; Gomez et al., 2016; Kyza & Nicolaidou, 2016; Shanahan et al., 2016);
- Tensions, strengths, and challenges of teacher–researcher co-design sessions (Hundal, Levin, & Keselman, 2014);
- Impact of co-design on teachers' professional development (Johnson et al., 2016; Jung & Brady, 2016; Kyza & Nicolaidou, 2016);
- Power structures, equity, and parity in participatory research (Lau & Stille, 2014; Samuelson Wardrip, Gomez, & Gomez, 2015; Vines, Clarke, Wright, McCarthy, & Olivier, 2013).

Opportunities for Future Exploration

Although there is a growing literature on educational co-design research, further research on co-design processes, challenges, and benefits seems warranted. While PCD teams, which are often interdisciplinary in nature, frequently require boundary-crossing negotiations to co-exist and thrive, we know very little about *how* participants, and other stakeholders, negotiate, cross, and experience these boundary crossings. Recent studies (Bronkhorst, Meijer, Koster, Akkerman, & Vermunt, 2013; Jung & Brady, 2016; Lau & Stille, 2014; Penuel et al., 2007) suggest that inquiry into boundary crossings is an important area to investigate.

PCD work is resource-intensive, may span months, and even years, and often yields large datasets, raising issues of what should be investigated, and how. Researchers increasingly examine the development, efficacy, and impact of designed products, and also the *role* of design as professional development for practitioners (Gomez et al., 2015; Greenleaf, Brown, Goldman, & Ko, 2013; Kyza & Nicolaidou, 2016). Although some research suggests that teachers who participate in PCD report increased ownership and agency (Bronkhorst et al., 2013; Kyza & Georgiou, 2014), it is not clear that the PCD is an effective approach for all teachers. Teachers' perceptions of pedagogy seem to influence co-design commitment (Cviko et al., 2015). Other such variables should be investigated in future research.

Few studies consider PCD quality and failures, as well as successes (Kwon, Wardrip, & Gomez, 2014); we need to understand more about what works in PCD, and under what conditions. For example, we need to know more about the impact of co-design versus other transformational practice approaches on teacher perceptions of their PCD roles, and contributions. To date, PCD quality refers to the relationship between the PCD process, the product (e.g., tools or curriculum), and alignment with the group's initial goals (e.g., student outcomes). Future research should aim to fine-tune hypotheses regarding PCD's impact on participants, and on targeted reform initiatives.

Theory building about PCD should increasingly explore its sustainability and the likelihood of scaling up beyond initially targeted impact. Most co-design teams are small, with only a handful of researchers and teachers involved. We need to know more about the different support needs of classrooms, schools and districts that seek to engage in productive co-design activities (Fogleman, Fishman, & Krajcik, 2006; Kwon et al., 2014; Kyza & Georgiou, 2014). Recently, studies employing design-based implementation research explore these concerns (see Fishman & Penuel, this volume). However, more empirical evidence, using varied methods, is needed in order to build learning sciences theory on scale and sustainability in co-design.

Challenges and Tensions in Collaborations

PCD participants work towards creating a new, shared activity system (Greeno & Engeström, 2014) in which expertise is distributed and supports mutual learning. Several considerations are essential to supporting the new system.

Time, Scheduling, and Pacing

Designing curricula and instructional materials is often an intensive and lengthy time commitment. School leaders must provide space and time for PCD within teachers' in-school schedule (Stieff & Ryan, 2016) and out-of-school constraints (e.g., Kyza & Georgiou, 2014). Researchers must negotiate funder-created project timelines. In PCD, group goals for teacher enactment and collaborative refinement of curricular and instructional materials compete within these constraints, and may lead to teachers implementing materials or assessments that are still in the early stages of development (Ko et al., 2016). Related to this issue of pacing, indicators of co-design impact may be still emerging (Penuel et al., 2007).

Shared Language and Understanding Goals

As in most other group contexts, each PCD participant will have his or her own interpretation, and way of describing, the group's goals and expectations. The language, that researchers and teachers use to discuss design problems and their solutions, relate to their understanding of the goal and outcomes. Even among teachers at one school, there are often different communities of practice according to discipline and grade (Samuelson Wardrip et al., 2015). Ko et al. (2016) provide illustrative examples of how teams in the READI project grappled with different understandings of how to support students to make claims within a discipline.

To ensure that all participants can fully participate in design, researchers need to be aware of how issues of power (Kristiansen & Bloch-Poulsen, 2013) and equity may affect the co-design process (e.g., Penuel et al., 2007; Stieff & Ryan, 2016). For example, a participant may hesitate to disagree or share an experience with someone who is an authority figure in their school or district (Stieff & Ryan, 2016). Relatedly, teachers may defer to researchers as the experts on a particular topic (Penuel et al., 2007). Researchers have tried to foster equitable co-design processes by highlighting practitioner expertise in the co-design process (Gomez et al., 1999) by making transparent how practitioner recommendations were important to the co-design progress, and were incorporated in newly designed materials (Stieff & Ryan, 2016), and with respect to the co-design effort itself, by emphasizing the frequent absence of predetermined design paths and content (Penuel et al., 2007).

Implications for Practice and Policy

In this chapter, we have discussed co-design as a vehicle for sharing and building expertise, the development of usable, pragmatic designs that meet the needs of the target audience, and as a capacity-building professional development tool. The research we described focused on researcher–practitioner co-design partnerships. While the learning sciences, as a discipline, remains in the early stages of theory building about co-design, research on this important and exciting approach to researcher–practitioner collaborations has begun to yield useful design principles that speak to structuring co-design efforts and attend to social and organizational considerations. These principles reflect the challenges, tensions, and opportunities that can inform both policymakers and practitioners about important analytic and design considerations in future co-design efforts.

The design principles for effective collaborations that we present in this section reflect accumulated understandings extracted from the co-design studies reviewed in this chapter. These principles are associated with more effective co-design processes and researcher–practitioner relationships. We also offer insights regarding how these principles inform policy considerations on co-design and collaboration.

The Co-Design Process

- Co-design teams should establish and sustain honest relationships, trust, and mutual respect for the diversity of expertise contributed by each participant (Herrenkohl et al., 2010).
- Co-design teams should serve as contexts of situated learning, most often resembling learning communities and communities of practice (Fogleman et al., 2006).
- Co-design teams should focus on changing practice, and on student thinking, which are foci that help the different stakeholders unite on a shared vision for the co-design (Herrenkohl et al., 2010).
- Co-design team members, including researchers, must be explicit about what they hope to accomplish in the collaboration (Gomez et al., 2016).
- Co-design teams should establish mutually agreed-upon criteria for "what counts" as success (Blomberg & Henderson, 1990). These criteria are often context-based and can also reflect participants' positionality. A relevant concern is whether success criteria are iterative or summative in nature, or both.
- Co-design participation should be coupled with enactment to facilitate reflection about the co-design product and instructional practice, and to promote the enacting teachers' in-situ professional development (Kyza & Nicolaidou, 2016).
- Researchers should scaffold the co-design process (Cober et al., 2015); this scaffolding should include social and emotional support (Herrenkohl et al., 2010).

Organizational Considerations

- Systemic constraints, such as the co-design effort's fit with school priorities, should be taken into account; otherwise, teacher commitment to the team may diminish (Hundal et al., 2014; Jung & Brady, 2016; Penuel et al., 2007).
- Schools hosting co-design teams should allocate sufficient time for co-design work, the value of which is innovations that can be meaningfully used in teachers' classrooms. Traditionally, schools operate within a 6.5 - to 7-hour day structure. When teachers, or teachers and researchers seek to collaborate, they depend on planning time during the school day, allocated professional development day(s), or after-school time. District and school-level administrators should prioritize planning and scheduling to support co-design.
- Co-design team members should be aware that the co-design process is an iterative, collaborative, and practice-oriented endeavor, which requires substantial time and effort to be achieved, often occurring over several months or longer (Penuel et al., 2007).
- Co-design teams require sufficient resources. More than 20 years ago, Darling-Hammond and McLaughlin (1995) called for a redistribution of resources to "provide time for collegial work and professional learning" (p. 4). Too few state and local district policymakers prioritize the allocation of resources to support collaborative design at the local school level. State policymakers and university administrators share in this responsibility.
- Co-design efforts need dissemination mechanisms to share what has been learned, so that it can be of use to future researcher–practitioner collaborations. How co-design unfolds, and the benefits and challenges of these efforts, remains opaque to many educational researchers, school leaders, and practitioners. Dissemination that goes beyond conference presentations and publications is needed. Policymakers at the local district, and state levels, can signal the import of dissemination as a public good through linking expectations for sharing findings and, ideally, offering examples of white paper guidelines.

Conclusions

Learning scientists seek to understand learning in context, to build knowledge about learning, and to support learning. A particular approach to addressing issues of learning, in context, is the participatory co-design activity. In this brief chapter, our aim has been to characterize the foundations, theoretical commitments, and lessons learned from co-design research.

Researchers and practitioners bring diverse perspectives to collaborations as they analyze problems of practice, and co-design materials, tools, and processes to address complex local classroom teaching and learning concerns. By contrast to traditional approaches of non-context informed curricular design, in PCD the end-users are both the local context designers (practitioners) and those towards whom the design is directed (students). To arrive at a working design requires a great amount of investment of time and effort. While there is, yet, much to be learned about the long-term impact of co-design participation on practitioner professional learning, and its impact on the practice it seeks to transform (Cviko et al., 2015; Kyza & Nicolaidou, 2016), research suggests that the investment may be well worth the effort (Gomez et al., 2016; Kyza & Georgiou, 2014).

Further Readings

Cviko, A., McKenney, S., & Voogt, J. (2015). Teachers as co-designers of technology-rich learning activities for early literacy. *Technology, Pedagogy and Education, 24*(4), 443–459.

Cviko, McKenney, and Voogt conducted a case study to explore teachers' co-design experiences and students' learning outcomes. The authors conclude that the teachers' pedagogical approaches affected their co-design involvement and that co-designed activities had positive effects on student learning.

D'Amico, L. (2010). The Center for Learning Technologies in Urban Schools: Evolving relationships in design-based research. In C. E. Coburn & M. K. Stein (Eds.), *Research and practice in education: Building alliances, bridging the divide* (pp. 37–53). Lanham, MD: Rowman & Littlefield.

D'Amico presents a study of the Center for Learning Technologies in Urban Schools (LeTUS) project. The author highlights how the differing district contexts in Chicago and Detroit, and researchers' prior experiences with the districts, related to the co-design work in the two cities.

Herrenkohl, L. R., Kawasaki, K., & Dewater, L. S. (2010). Inside and outside: Teacher–researcher collaboration. *New Educator, 6*(1), 74–92.

Herrenkohl, Kawasaki, and Dewater present three moments from a teacher–researcher effort to characterize the nature of their collaboration. The authors suggest these efforts shifted teachers' and researchers' identities, and argue that teacher-researcher collaboration supports teacher and researcher professional learning.

Penuel, W. R., Roschelle, J., & Shechtman, N. (2007). Designing formative assessment software with teachers: An analysis of the co-design process. *Research and Practice in Technology Enhanced Learning, 2*(1), 51–74.

Penuel, Roschelle, and Shechtman provide a definition of educational co-design research, and describe characteristics of a co-design process.

Shanahan, C., Bolz, M. J., Cribb, G., Goldman, S. R., Heppeler, J., & Manderino, M., (2016). Deepening what it means to read (and write) like a historian: Progressions of instruction across a school year in an eleventh grade U.S. history class. *History Teacher, 49*(2), 241–270.

Shanahan and colleagues describe efforts to design history instruction that incorporated disciplinary reading and writing practices as part of Project READI. The authors provide an overview of learning goals that the history design team developed, and present an example of how one teacher on the design team integrated the learning goals in her regular history units.

References

Barab, S., & Squire, K. (2004). Design-based research: Putting a stake in the ground. *Journal of the Learning Sciences, 13*(1), 1–14.

Beck, E. E. (2001). *On participatory design in Scandinavian computing research* (Research Report No. 294). Oslo, Norway: University of Oslo, Department of Informatics.

Blomberg, J. L., & Henderson, A. (1990). Reflections on participatory design: Lessons from the trillium experience. In J. C. Chew & J. Whiteside (Eds.), *Proceedings of the SIGCHI Conference on Human Factors in Computing Systems* (pp. 353–359). New York: ACM Press.

Bronkhorst, L. H., Meijer, P. C., Koster, B., Akkerman, S. F., & Vermunt, J. D. (2013). Consequential research designs in research on teacher education. *Teaching and Teacher Education, 33*, 90–99.

Cober, R., Tan, E., Slotta, J., So, H. J., & Könings, K. D. (2015). Teachers as participatory designers: Two case studies with technology-enhanced learning environments. *Instructional Science, 43*(2), 203–228.

Coburn, C. E., & Stein, M. K. (Eds.). (2010). *Research and practice in education: Building alliances, bridging the divide.* Lanham, MD: Rowman & Littlefield.

Cognition and Technology Group at Vanderbilt. (1997). *The Jasper Project: Lessons in curriculum, instruction, assessment, and professional development.* Mahwah, NJ: Erlbaum.

Cviko, A., McKenney, S., & Voogt, J. (2014). Teacher roles in designing technology-rich learning activities for early literacy: A cross-case analysis. *Computers & Education, 72*, 68–79.

Cviko, A., McKenney, S., & Voogt, J. (2015). Teachers as co-designers of technology-rich learning activities for early literacy. *Technology, Pedagogy and Education, 24*(4), 443–459.

D'Amico, L. (2010). The Center for Learning Technologies in Urban Schools: Evolving relationships in design-based research. In C. E. Coburn & M. K. Stein (Eds.), *Research and practice in education: Building alliances, bridging the divide* (pp. 37–53). Lanham, MD: Rowman & Littlefield.

Darling-Hammond, L., & McLaughlin, M. W. (1995). Policies that support professional development in an era of reform. *The Phi Delta Kappan, 76*(8), 597–604.

Edelson, D. C. (2002). Design research: What we learn when we engage in design. *Journal of the Learning Sciences, 11*(1), 105–121.

Edelson, D. C., Gordin, D. N., & Pea, R. D. (1999). Addressing the challenges of inquiry-based learning through technology and curriculum design. *Journal of the Learning Sciences, 8*(3–4), 391–450.

Fishman, B., & Penuel, W. (2018). Design-based implementation research. In F. Fischer, C. E. Hmelo-Silver, S. R. Goldman, & P. Reimann (Eds.), *International handbook of the learning sciences* (pp. 393–400). New York: Routledge.

Fogleman, J., Fishman, B., & Krajcik, J. (2006). Sustaining innovations through lead teacher learning: A learning sciences perspective on supporting professional development. *Teaching Education, 17*(2), 181–194.

Gomez, K., Gomez, L. M., Cooper, B., Lozano, M., & Mancevice, N. (2016). Redressing science learning through supporting language: The biology credit recovery course. *Urban Education*. doi:10.1177/0042085916677345

Gomez, K., Gomez, L. M., Rodela, K. C., Horton, E. S., Cunningham, J., & Ambrocio, R. (2015). Embedding language support in developmental mathematics lessons: Exploring the value of design as professional development for community college mathematics instructors. *Journal of Teacher Education, 66*(5) 450–465.

Gomez, L., Shrader, G., Williams, K., Finn, L., & Whitcomb, J. (1999, March). Research for practice: Collaborative research design in urban schools. *Paper presented at the Spencer Foundation Training Conference*, New Orleans, LA.

Greenleaf, C., Brown, W, Goldman, S. R., & Ko, M. (2013, December). READI for science: Promoting scientific literacy practices through text-based investigations for middle and high school science teachers and students. *Paper presented at the NRC Workshop on Literacy for Science*, Washington, DC.

Greeno, J. G., & Engeström, Y. (2014). Learning in activity. In R. K. Sawyer (Ed.), *The Cambridge handbook of the learning sciences*, (2nd ed., pp. 128–147). Cambridge, UK: Cambridge University Press.

Herrenkohl, L. R., Kawasaki, K., & Dewater, L. S. (2010). Inside and outside: Teacher–researcher collaboration. *New Educator, 6*(1), 74–92.

Hundal, S., Levin, D. M., & Keselman, A. (2014). Lessons of researcher–teacher co-design of an environmental health afterschool club curriculum. *International Journal of Science Education, 36*(9), 1510–1530.

Johnson, R., Severance, S., Penuel, W. R., & Leary, H. (2016). Teachers, tasks, and tensions: Lessons from a research–practice partnership. *Journal of Mathematics Teacher Education, 19*(2), 169–185.

Jung, H., & Brady, C. (2016). Roles of a teacher and researcher during in situ professional development around the implementation of mathematical modeling tasks. *Journal of Mathematics Teacher Education, 19*(2), 277–295.

Ko, M., Goldman, S. R., Radinsky, J., James, K., Hall, A., Popp, J., et al. (2016). Looking under the hood: Productive messiness in design for argumentation in science, literature, and history. In V. Svihla & R. Reeve (Eds.), *Design as scholarship: Case studies in the learning sciences* (pp. 71–85). New York: Routledge.

Kristiansen, M., & Bloch-Poulsen, J. (2013). Participation in research-action: Between methodology and worldview, participation and co-determination. *Work & Education, 22*(1), 37–53. Retrieved from www.fae.ufmg.br/trabalhoeeducacao

Kwon, S. M., Wardrip, P. S., & Gomez, L. M. (2014). Co-design of interdisciplinary projects as a mechanism for school capacity growth. *Improving Schools, 17*(1), 54–71.

Kyza, E. A., & Georgiou, Y. (2014). Developing in-service science teachers' ownership of the PROFILES pedagogical framework through a technology-supported participatory design approach to professional development. *Science Education International, 25*(2), 55–77.

Kyza, E. A., & Nicolaidou, I. (2016). Co-designing reform-based online inquiry learning environments as a situated approach to teachers' professional development. *CoDesign*. doi:10.1080/15710882.2016.1209528

Lau, S. M. C., & Stille, S. (2014). Participatory research with teachers: Toward a pragmatic and dynamic view of equity and parity in research relationships. *European Journal of Teacher Education, 37*(2), 156–170.

Lewin, K. (1947). Group decisions and social change. In T. M. Newcomb & E. L. Hartley (Eds.), *Readings in social psychology* (pp. 330–344). New York: Henry Holt.

Lui, M., & Slotta, J. D. (2014). Immersive simulations for smart classrooms: Exploring evolutionary concepts in secondary science. *Technology, Pedagogy and Education, 23*(1), 57–80.

Marx, R. W., Blumenfeld, P. C., Krajcik, J. S., Fishman, B., Soloway, E., Geier, R., & Tal, R. T. (2004). Inquiry-based science in the middle grades: Assessment of learning in urban systemic reform. *Journal of Research in Science Teaching, 41*(10), 1063–1080.

Nygaard, K., & Bergo, O. T. (1973). Planlegging, styring og databehandling. In *Grunnbok for fagbevegelsen* [Planning, management and data processing. In *Basic reader for trade unions*], Vol. *1*. Oslo: Tiden norsk forlag.

Ormel, B. J. B., Roblin, N. N. P., McKenney, S. E., Voogt, J. M., & Pieters, J. M. (2012). Research–practice interactions as reported in recent design studies: Still promising, still hazy. *Educational Technology Research and Development, 60*(6), 967–986.

Penuel, W. R., Coburn, C. E., & Gallagher, D. J. (2013). Negotiating problems of practice in research–practice design partnerships. *Yearbook of the National Society for the Study of Education, 112*(2), 237–255.

Penuel, W. R., Fishman, B. J., Cheng, B. H., & Sabelli, N. (2011). Organizing research and development at the intersection of learning, implementation, and design. *Educational Researcher, 40*(7), 331–337.

Penuel, W. R., Roschelle, J., & Shechtman, N. (2007). Designing formative assessment software with teachers: An analysis of the co-design process. *Research and Practice in Technology Enhanced Learning, 2*(1), 51–74.

Penuel, W. R., & Yarnall, L. (2005). Designing handheld software to support classroom assessment: An analysis of conditions for teacher adoption. *Journal of Technology, Learning, and Assessment, 3*(5), 4–44.

Recker, M. (2013). Interview about teacher learning and technology [Video file]. In *NAPLeS video series*. Retrieved October 19, 2017, from http://isls-naples.psy.lmu.de/video-resources/interviews-ls/recker/index.html

Recker, M., & Sumner, T. (2018). Supporting teacher learning through design, technology, and open educational resources. In F. Fischer, C. E. Hmelo-Silver, S. R. Goldman, & P. Reimann (Eds.), *International handbook of the learning sciences* (pp. 267–275). New York: Routledge.

Rivet, A., Krajcik, J., Marx, R., & Reiser, B. (2003, April). Design principles for developing inquiry materials with embedded technologies. *Paper presented at the annual meeting of American Educational Research Association*, Chicago, IL.

Samuelson Wardrip, P., Gomez, L. M., & Gomez, K. (2015). We modify each other's lessons: The role of literacy work circles in developing professional community. *Teacher Development, 19*(4), 445–460.

Shanahan, C., Bolz, M. J., Cribb, G., Goldman, S. R., Heppeler, J., & Manderino, M., (2016). Deepening what it means to read (and write) like a historian: Progressions of instruction across a school year in an eleventh grade U.S. history class. *History Teacher, 49*(2), 241–270.

Stieff, M., & Ryan, S. (2016). Designing the Connected Chemistry Curriculum. In V. Svihla & R. Reeve (Eds.), *Design as scholarship: Case studies in the learning sciences* (pp. 100–114). New York: Routledge.

Vines, J., Clarke, R., Wright, P., McCarthy, J., & Olivier, P. (2013). Configuring participation: On how we involve people in design. *Proceedings of the SIGCHI Conference on Human Factors in Computing Systems* (pp. 429–438). New York: ACM Press.

40
Assessment of and for Learning

James W. Pellegrino

Chapter Overview and Goals

This chapter is about what constitutes high-quality and valid assessment from a learning sciences perspective. Such assessments are based upon three critical components that work together: (a) they are derived from theories and data about content-based cognition that indicate the knowledge and skills that should be assessed; (b) they include tasks and observations that can provide evidence and information about whether students have mastered the knowledge and skills of interest; and (c) they make use of qualitative and quantitative techniques for interpreting student performance that capture differences in knowledge and skill among students being assessed. Assessment design and use should be seen as a major form of conceptual research *within* the learning sciences. It should not be left to others with limited conceptions of what it means to know and to learn, and it requires multidisciplinary collaborations that integrate across disciplines including researchers in academic disciplines and measurement experts.

The first section differentiates the varying contexts and purposes of educational assessment. The second section discusses the key principle that all assessment involves a process of reasoning from evidence, and the third section then considers how that reasoning process should be driven by models of knowing and learning, including models expressed as learning progressions. The fourth section describes construct-centered design processes that serve to guide systematic development and interpretation of assessments, and the fifth section considers issues of validity in the design, use and interpretation of an assessment. The six section turns to the implications of the preceding material for (a) classroom assessment and (b) large-scale assessment. The final section briefly considers the importance of a careful and thoughtful approach to assessment design, use, and interpretation for research in the learning sciences.

Educational Assessment in Context

Assessment Purposes and Contexts

From teachers' classroom quizzes, mid-term or final exams, to nationally and internationally administered standardized tests, assessments of students' knowledge and skills have become a ubiquitous part of the educational landscape. Assessments of school learning provide information to help educators, administrators, policy makers, students, parents, and researchers judge the state of student learning and make decisions about implications and actions. The specific purposes for which an assessment will be used are important considerations in all phases of its design. For example, assessments used by instructors in classrooms to assist or monitor learning typically need to provide more detailed information than assessments whose results will be used by policy makers or accrediting agencies.

Assessment for learning. In the classroom context, instructors use various forms of assessment to inform day-to-day and month-to-month decisions about next steps for instruction, to give students feedback about their progress, and to motivate them (e.g., Black & Wiliam, 1998; Wiliam, 2007). One familiar type of classroom assessment is a teacher-made quiz, but assessment also includes more informal methods for determining how students are progressing in their learning, such as classroom projects, feedback from computer-assisted instruction, classroom observation, written work, homework, and conversations with and among students—all interpreted by the teacher in light of additional information about the students, the schooling context, and the content being studied.

These situations are referred to as *assessments for learning*, or the *formative use of assessment*. These assessments provide specific information about students' strengths and difficulties with learning. For example, statistics teachers need to know more than the fact that a student does not understand probability; they need to know the details of this misunderstanding, such as the student's tendency to confuse conditional and compound probability. Teachers can use information from these types of assessments to adapt their instruction to meet students' needs, which may be difficult to anticipate and are likely to vary from one student to another. Students can use this information to determine which skills and knowledge they need to study further and what adjustments in their thinking they need to make.

Assessment of individual achievement. Many assessments are used to help determine whether a student has attained a certain level of competency after completing a particular phase of education, whether it be a two-week curricular unit, a semester-long course, or 12 years of schooling. This is referred to as *assessment of individual achievement*, or the *summative use of assessment*. Some of the most familiar forms of summative assessment are those used by classroom instructors, such as end-of-unit or end-of-course tests, which often are used to assign letter grades when a course is finished. Large-scale assessments—which are administered at the direction of people external to the classroom, such as school districts, state boards of education, or national agencies—also provide information about the attainment of individual students, as well as comparative information about how one student performs relative to others. Because large-scale assessments are typically given only once a year and involve a time lag between testing and availability of results, the results seldom provide information that can be used to help teachers or students make day-to-day or month-to-month decisions about teaching and learning.

Assessment to evaluate programs and institutions. Another common purpose of assessment is to help administrators, policy makers, or researchers judge the quality and effectiveness of educational programs and institutions. Evaluations can be formative or summative; for example, instructional evaluation is formative when it is used to improve the effectiveness of instruction. Summative evaluations are increasingly used by school leaders and by policy makers to make decisions about individuals, programs, and institutions. For instance, public reporting of assessment results by school, district, or state are designed to provide information to parents and taxpayers about the quality and efficacy of their schools; these evaluations sometimes influence decisions about resource allocations.

Further Considerations of Purposes, Levels, and Timescales

No single type of assessment can serve all of the purposes and contexts reviewed above. Unfortunately, policy makers often attempt to use a single assessment for multiple purposes—either in the desire to save money, or to administer the assessment in less time, or to provide information to teachers to guide instructional improvement. The problem is that when a single assessment is used to serve multiple purposes, it ends up being sub-optimal for each specific purpose. The drive to identify a single one-size-fits-all assessment often results in inappropriate choices of assessments for instructional or research purposes, and this can in turn lead to invalid conclusions regarding persons, programs, and/or institutions.

The ultimate purpose of all assessments should be to promote student learning (e.g., Wiggins, 1998). But in some cases assessments are developed for evaluation purposes that are somewhat distant from this ultimate goal of promoting student learning. Ruiz-Primo, Shavelson, Hamilton, and Klein (2002) proposed a five-point continuum that reflects the proximity of an assessment to classroom instruction and learning: *immediate* (e.g., observations or artifacts from the enactment of a specific instructional activity), *close* (e.g., embedded assessments and semiformal quizzes of learning from one or more activities), *proximal* (e.g., formal classroom exams of learning from a specific curriculum), *distal* (e.g., criterion-referenced achievement tests such as required by the U.S. No Child Left Behind legislation), and *remote* (broader outcomes measured over time, including norm-referenced achievement tests and some national and international achievement measures, like PISA). Different assessments should be understood as different points on this continuum if they are to be effectively aligned with each other and with curriculum and instruction. In essence, an assessment is a test of transfer and it can be near or far transfer depending on where the assessment falls along the continuum noted above (see Hickey & Pellegrino, 2005).

The proximity of the assessment to moments of teaching and learning has implications for how and how well it can fulfill the different purposes of assessment (formative, summative, or program evaluation: Hickey & Pellegrino, 2005; NRC, 2003). For example, an assessment designed to aid teachers in diagnosing the state of student learning to modify instruction has to be contextualized relative to the curriculum and instructional materials that have been in use and it needs to be at a relatively fine grain size regarding specific aspects of knowledge and skill to be instructionally informative. Thus, it cannot cover large amounts of content superficially and it needs to use language and problem contexts familiar to the students. The capacity of such an assessment to function as a good and "fair" summative assessment for all students learning the same content is therefore limited. In contrast, a large-scale state or national achievement test needs to cover large amounts of content at a relatively coarse grain size and it cannot be curriculum- or context-dependent to be "fair" to all students tested. Thus, it must use problem formats that are general and curriculum-neutral. The capacity of such an assessment to provide instructionally useful information is therefore highly limited. Furthermore, it is typically far removed in time from when instruction and learning have transpired and thus its feedback capacity is similarly limited.

Assessment as a Process of Evidentiary Reasoning: The *Assessment Triangle*

Although the assessments used in various contexts, for differing purposes, and at different timescales often look quite different, they share certain common principles. One such principle is that assessment is always a process of reasoning from evidence. By its very nature, moreover, assessment is imprecise to some degree. For example, assessing educational outcomes is not as straightforward as measuring physical properties such as height or weight; the attributes to be judged are mental representations and processes that are not outwardly visible. An assessment is a tool designed to observe students' behavior and produce data that can be used to draw reasonable inferences about what students know. It is helpful to portray the process of reasoning from evidence using the *assessment triangle* (Figure 40.1; also see Pellegrino, Chudowsky, & Glaser et al., 2001). The vertices represent the three key elements underlying any assessment: a model of student *cognition* and learning in the domain of the assessment; a set of assumptions and principles about the kinds of *observations* that will provide evidence of students' competencies; and an *interpretation* process for making sense of the

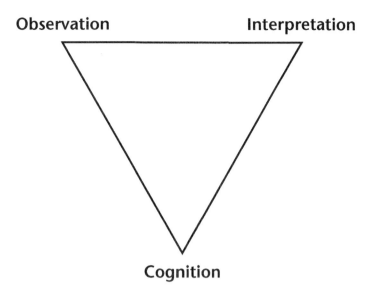

Figure 40.1 The Assessment Triangle

evidence in light of the assessment purpose and student understanding. An assessment cannot be designed and implemented, or evaluated, without incorporating all three (although, in many cases, one or more may remain implicit in the design rather than explicitly and consciously chosen).

The *cognition* corner of the triangle refers to theory, data, and a set of assumptions about how students represent knowledge and develop competence in a subject-matter domain (e.g., fractions, Newton's laws, thermodynamics). The use of the term *cognition* is not meant to define a particular model or view of the nature of knowing and learning, such as the information processing or rationalist model. Rather, it is meant to index whatever is the theoretical view of what it means to know and learn in some area of human endeavor. As such it serves to frame the reasoning from evidence process, since it guides design and selection of the observations, collection of the evidence, and rules for interpretation of the evidence and meaning making. In any particular assessment application, a theory of learning in the domain is needed to identify the set of knowledge and skills that is important to measure for the intended context of use, whether that be to characterize the competencies students have acquired at some point in time to make a summative judgment, or to make formative judgments to guide subsequent instruction so as to maximize learning. Such theories should be consistent with the latest scientific understanding of how learners represent knowledge and develop expertise in a domain (e.g., Reimann & Markauskaite, this volume).

The *observation* vertex of the assessment triangle represents a description or set of specifications for assessment *tasks* that will elicit illuminating responses from students. Every assessment is based on a set of assumptions and principles about the kinds of tasks or situations that will prompt students to say, do, or create something that demonstrates important knowledge and skills. Assessment tasks (whether answering a multiple-choice test, composing a one-paragraph essay, or responding to an oral question from the teacher) must be carefully designed to provide evidence that is linked to the model of learning and to support the kinds of inferences and decisions that will be made on the basis of the assessment results.

Every assessment is also based on assumptions and models for interpreting the evidence collected from the observations.

The *interpretation* vertex of the triangle encompasses all the methods and tools used to reason from observations. It expresses how the observations derived from a set of assessment tasks constitute evidence about the knowledge and skills being assessed. In the context of large-scale assessment, the interpretation method is usually a statistical model, which is a characterization or summarization of patterns one would expect to see in the data given varying levels of student competency. For example, on a state or national end-of-year achievement test, the performance of a student will be reported on a measurement scale that permits various comparisons, including comparisons across students who have taken different forms of the test in a given year, or the performance of students this year compared to those in prior years. In some cases, scores are also classified in terms of achievement levels such as basic, proficient, or advanced. Familiar examples of the use of a statistical model to derive a scale score are NAEP, the GRE, or SAT in US, and PISA internationally. In the context of classroom assessment, whether the activity is a conversation, a quiz, or a "formal" test, the interpretation is often made informally by the teacher, and is often based on an intuitive or qualitative model or simple quantitative model such as percent correct rather than a formal statistical one. Even so, teachers make coordinated judgments about what aspects of students' understanding and learning are relevant, how a student has performed one or more tasks, and what the performances mean about the student's knowledge and understanding. This occurs whether or not they may have used some quantitative metric like total points or percent correct.

The critical final point to note is that each of the three elements of the assessment triangle not only must make sense on its own, but also must align with the other two in a meaningful way to lead to an effective assessment and sound inferences.

Domain-Specific Learning: The Role of Learning Progressions

As argued above, the targets of inference for any given assessment should be largely determined by models of cognition and learning that describe how people relate to knowledge and develop competence in the domain of interest and what are the important elements of such competence. Starting with a model of learning is one of the main features that distinguish the approach to assessment design discussed here from typical approaches to test development. The model suggests the most important aspects of student achievement about which one would want to draw inferences, and provides clues about the types of assessment tasks that will elicit evidence to support those inferences (see also Pellegrino et al., 2001).

Consistent with these ideas, there has been a spurt of interest in the topic of "learning progressions." A variety of definitions of learning progressions (also called learning trajectories) now exist in the literature, with substantial differences in focus and intent (see Duncan & Rivet, this volume). Learning progressions are empirically grounded and testable hypotheses about how students' understanding of, and ability to use, core concepts and explanations and related disciplinary practices grow and become more sophisticated over time, with appropriate instruction. These hypotheses describe the pathways students are likely to follow as they master core concepts. The hypothesized learning trajectories are tested empirically to ensure their construct validity ("Does the hypothesized sequence describe a path most students actually experience given appropriate instruction?") and ultimately to evaluate their consequences ("Does instruction based on the learning progression produce better results for most students?").

Any hypothesized learning progression has implications for assessment, because effective assessments should be aligned with an empirically grounded model of knowing and learning. To be maximally useful, a learning progression should contain at least the following elements:

(1) *Target performances or learning goals*, which are the end points of a learning progression. They are defined by societal expectations, analysis of the discipline, and/or requirements for entry into the next level of education.
(2) *Progress variables*, which are the dimensions of understanding, application, and practice that are being developed and tracked over time. These may be core concepts in the discipline or practices central to literary, scientific, or mathematical work.

(3) *Levels of achievement*, which are intermediate steps in the developmental pathway(s) traced by a learning progression. These levels may reflect levels of integration or common stages that characterize the development of student thinking. They may be intermediate steps that are non-canonical but are stepping stones to canonical ideas.
(4) *Learning performances*, which are the kinds of tasks students at a particular level of achievement would be capable of performing. They provide specifications for the development of assessments by which students would demonstrate their knowledge and understanding.
(5) *Assessments*, which are the specific measures used to track student development along the hypothesized progression. Learning progressions should include an approach to assessment, as assessments are integral to their development, validation, and use.

As discussed throughout this handbook, research on cognition and learning has produced a rich set of descriptions of domain-specific learning and performance that can serve to guide both instruction and assessment design (see Duncan & Rivet, this volume; Herrenkohl & Polman, this volume). That said, there is much left to do in mapping out learning progressions for multiple areas of the curriculum in ways that can effectively guide the design of instruction and assessment (see Duncan & Rivet, this volume). Nevertheless, there is a fair amount known about student cognition and learning that we can make use of right now to guide how we design assessments, especially those that attempt to cover the progress of learning within and across grades. A recent paper by Bennett, Deane, and van Rijn (2016) on the CBAL assessment system (Cognitively Based Assessment of, for and as Learning) is an excellent example of such work. CBAL has concentrated on assessments of the English language arts and mathematics at the middle school level. The assessments have been designed with cognitive theory and data in mind, including the use of explicit learning progressions in these instructional domains. The Bennett et al. (2016) paper shows how cognitive and psychometric theory and methods can be used in complementary ways to design and validate an assessment program focused on critical aspects of student learning.

Assessment Development: Construct-Centered Design

The design of an actual assessment is a challenging endeavor that must be guided by theory and research about cognition in context, as well as practical prescriptions regarding the processes that lead to productive and potentially valid assessments for particular contexts of use. Design is always a complex process that applies theory and research to achieve near-optimal solutions under multiple constraints, some of which are outside the realm of science. Assessment design is influenced in important ways by variables such as its purpose (e.g., to assist learning, to measure individual attainment, or to evaluate a program); the context in which it will be used (e.g., classroom-, district-, or international-comparative); and practical constraints (e.g., resources and time).

The evidentiary reasoning logic embedded in the assessment triangle in Figure 40.1 is exemplified in the work of two groups of researchers that have generated frameworks for developing assessments: (a) the evidence-centered design (ECD) approach developed by Mislevy and colleagues (see, e.g., Mislevy & Haertel, 2006); and the construct-modeling approach developed by Wilson and his colleagues (see, e.g., Wilson, 2004). They both use a construct-centered approach to task development, and both closely follow the assessment triangle's logic of evidentiary reasoning.

Traditional approaches to assessment design tend to focus primarily on surface features of tasks, such as how they are presented to students, or the format in which students are asked to respond. In a construct-centered approach, the selection and development of assessment tasks, as well as the scoring rubrics and criteria, and the modes and style of reporting, are guided by the construct to be assessed and the best ways of eliciting evidence about a student's proficiency relative to that construct. In a construct-centered approach, the process of assessment design and development is characterized by the following developmental steps, which are common to both evidence-centered design and construct modeling:

- analyzing the cognitive domain that is the target of an assessment,
- specifying the constructs to be assessed in language detailed enough to guide task design,
- identifying the inferences that the assessment should support,
- laying out the type of evidence needed to support those inferences,
- designing tasks to collect that evidence, modeling how the evidence can be assembled and used to reach valid conclusions, and
- iterating through the above stages to refine the process, especially as new evidence becomes available.

Ultimately, the tasks designed in this way should allow students to "show what they know and can do" in a way that is as unambiguous as possible with respect to what the task performance implies about student knowledge and skill—i.e., the inferences about student cognition that are permissible and sustainable from a given set of assessment tasks or items.

Assessment Validity: Argumentation and Evidence

The ultimate goal of applying a theory-driven and evidence-based approach to the process of assessment design and use is to create tasks and situations that give us valid and reliable information about student learning. Thus, validity is central in all work on assessment. The joint AERA/APA/NCME Standards (1999 /2014) frame validity largely in terms of "the concept or characteristic that a test is designed to measure" (1999, p. 5). Contemporary educational measurement theorists have framed test validity as a reasoned argument backed by evidence (e.g., Kane, 2006, 2013). The particular forms of evidence are associated with the claims that one wishes to make about what a given assessment is and does and how its scores are to be interpreted. Some of those critical claims are related to the theoretical base underlying the design of a given assessment and those interpretive claims must be backed up by empirical evidence of various types that the observed performance does in fact reflect the underlying cognitive constructs. For assessments designed to support ongoing classroom teaching and learning, Pellegrino, DiBello, and Goldman (2016) have proposed a specific validity framework that identifies three related components—cognitive, instructional, and inferential —as follows:

> *Cognitive*—This component of validity addresses the extent to which an assessment taps important forms of domain knowledge and skill in ways that are not confounded with other aspects of cognition, such as working memory load. Cognitive validity should be based on what is known about the nature of student cognition and understanding in areas of the curriculum such as literacy, mathematics, and science, and how it develops over time with instruction to determine what knowledge and skills students are supposed to use and those that they actually do use when interacting with the assessment.
>
> *Instructional*—This component addresses the extent to which an assessment is aligned with curriculum and instruction, including students' opportunities to learn, as well as how it supports teaching practice by providing valuable and timely instruction-related information. Instructional validity should be based on evidence about alignment of the assessment with skills of interest as defined by standards and curricula, the practicality and usefulness for teachers, and the nature of the assessment as a guide to instruction.
>
> *Inferential*—This component is concerned with the extent to which an assessment reliably and accurately yields information about student performance, especially for diagnostic purposes.

Inferential validity should be based on evidence derived from various qualitative and quantitative analytic methods to determine whether task performance reliably aligns with an underlying conceptual measurement model that is appropriate to the intended interpretive use.

The compilation of evidence and construction of a *validity argument* for a given assessment should be an ongoing activity that begins during assessment design and continues through various iterations of use from pilot testing of the assessment materials and procedures to subsequent operational versions that might be used on various scales of implementation—classroom, school, district, state, and/or national. The three components noted above can serve as a guide to determining the validity of an assessment both prospectively, i.e., during its conceptualization and design, and retrospectively, i.e., for purposes of evaluation of the relative strengths and weaknesses of any given assessment that is being used by educators proximal to the processes of teaching and learning. For a more complete treatment of the various sources of evidence that can and should be compiled as part of the validation process, see Pellegrino et al. (2016).

Implications for Assessment Design and Use

The Design and Use of Classroom Assessment

Learning scientists generally argue that classroom assessment practices need to change to better support learning (also see Shepard, 2000). The content and character of assessments need to be significantly improved to reflect the latest empirical research on learning, together with societal innovations and expectations; and, given what we now know about learning progressions, the gathering and use of assessment information and insights should become a part of the ongoing teaching and learning process. This latter point further suggests that teacher education programs should provide teachers with a deep understanding of how to use assessment in their instruction. Many educational assessment experts believe that if assessment, curriculum, and instruction were more integrally connected, student learning would improve (e.g., Stiggins, 1997).

According to Sadler (1989), three elements are required if teachers are to successfully use assessment to promote learning:

1 a clear view of the learning goals (derived from the curriculum),
2 information about the present state of the learner (derived from assessment), and
3 action to close the gap (taken through instruction).

Each of these three elements informs the other. For instance, formulating assessment procedures for classroom use can spur a teacher to think more specifically about learning goals, thus leading to modification of curriculum and instruction. These modifications can, in turn, lead to refined assessment procedures, and so on. The mere existence of classroom assessment along the lines discussed here will not ensure effective learning. The clarity and appropriateness of the curriculum goals, the validity of the assessments in relationship to these goals, the interpretation of the assessment evidence, and the relevance and quality of the instruction that ensues are all critical determinants of the outcome.

Effective teaching must start with a model of knowledge and learning in the domain. For most teachers, the ultimate goals for learning are established by the curriculum, which is usually mandated externally (e.g., by state curriculum standards). But the externally mandated curriculum does not specify the empirically based cognition and learning outcomes that are necessary for assessment to be effective. As a result, teachers (and others responsible for designing curriculum, instruction, and assessment) must fashion intermediate goals that can serve as an effective route to achieving the externally mandated goals, and to do so effectively, they must have an understanding of how students relate to knowledge and develop competence in the domain (e.g., see Ufer & Neumann, this volume). Formative assessment should be based in cognitive theories about how people learn particular subject matter to ensure that instruction centers on what is most important for the next stage of learning, given a learner's current state of understanding.

The Design and Use of Large-Scale Assessment

Large-scale assessments are further removed from instruction but can still benefit learning if well designed and properly used. If the principles of design identified above were applied, substantially more valid, useful, and fair information would be gained from large-scale assessments. However, before schools and districts can fully capitalize on contemporary theory and research, they will need to substantially change how they approach large-scale assessment. Specifically, they must relax some of the constraints that currently drive large-scale assessment practices, as follows.

Large-scale summative assessments should focus on the most critical and central aspects of learning in a domain—as identified by curriculum standards and informed by cognitive research and theory. Large-scale assessments typically are based on models of learning that are less detailed than classroom assessments. For summative purposes, one might need to know whether a student has mastered the more complex aspects of multi-column subtraction, including borrowing from and across zero, whereas a teacher needs to know exactly which procedural errors lead to mistakes. Although policy makers and parents may not need all the diagnostic detail that would be useful to a teacher and student during the course of instruction, large-scale summative assessments should be based on a model of learning that is compatible with and derived from the same set of knowledge and assumptions about learning as classroom assessment.

Research on cognition and learning suggests a broad range of aspects of competency that should be assessed when measuring student achievement, many of which are essentially untapped by current assessments. Examples are knowledge organization, problem representation, strategy use, metacognition, and participatory activities (e.g., formulating questions, constructing and evaluating arguments, contributing to group problem-solving). These are important elements of contemporary theory and research on the acquisition of competence and expertise. Large-scale assessments should not ignore these aspects of competency and should provide information about these aspects of the nature of student understanding, rather than simply ranking students according to general proficiency estimates. If tests are based on a research-grounded theory of cognition and learning, those tests can provide positive direction for instruction, making "teaching to the test" productive for learning rather than destructive.

Unfortunately, given current constraints of standardized test administration, only limited improvements in large-scale assessments are possible. These constraints include: the need to provide reliable and comparable scores for individuals as well as groups; the need to sample a broad set of curriculum standards within a limited testing time per student; and the need to offer cost-efficiency in terms of development, scoring, and administration. To meet these kinds of demands, designers typically create assessments that are given at a specified time, with all students being given the same (or parallel) tests under strictly standardized conditions (often referred to as *on-demand* assessment). Tasks are generally of the kind that can be presented in paper-and-pencil format or via computer, that students can respond to quickly, and that can be scored reliably and efficiently. As a result, learning outcomes that lend themselves to being assessed in these ways are assessed, but aspects of learning that cannot be observed under such constrained conditions are not. Designing new assessments that capture the complexity of cognition and learning requires examining the assumptions and values that currently drive assessment design choices and breaking out of the current paradigm to explore alternative approaches to large-scale assessment, including innovative uses of technology (see, e.g., Quellmalz & Pellegrino, 2009).

The Role of Assessment in Learning Sciences Theory and Research

Learning sciences researchers need high-quality evidence that allows us to ask and answer critical questions about the outcomes of learning and instruction—what students know and are able to do. The first requirement for developing quality assessments is that the concepts and skills that signal progress toward mastery of a domain be understood and specified. These activities constitute fundamental components of a prospective learning sciences agenda as applied to multiple courses and content domains. Assessment of the overall outcomes of instruction is important to the learning sciences because it allows us to test program effectiveness. But it is more broadly important because the content of such assessments can drive instructional practice for better or for worse. Assessment of the impact of learning-sciences-based curricula is also important. Using assessment for program evaluation requires measures that are sensitive to learning and the impact of quality teaching. Researchers, educators, and administrators must therefore concern themselves with supporting research and development on effective and appropriate assessment procedures that can serve multiple purposes and functions as part of a coordinated system of assessments.

From a practical perspective regarding the use of assessment to guide instruction and learning, future research should explore: (a) how new forms of assessment can be made accessible for instructors and practical for use in classrooms, (b) how they can be made efficient for use in K–16+ teaching contexts, (c) how various new forms of assessment affect student learning, instructor practice, and educational decision making, (d) ways that instructors can be assisted in integrating new forms of assessment into their instructional practices and how they can best make use of information from such assessments, and (e) ways that structural features of instructional delivery in education (e.g., length of class time, class size and organization, and opportunities for students and/or instructors to work together) impact the feasibility of implementing new types of assessments and their effectiveness.

As a field, learning sciences continues to make progress on critical issues related to learning and instruction, with an increasing awareness of the importance of assessment in that enterprise. Hopefully this chapter's discussion of ways to think about the design and uses of assessment provides a useful set of ideas and approaches that can further advance the field of learning sciences research. There is a great deal at stake for the field of learning sciences by embracing the challenge of designing assessments that are aligned to our evolving conceptions of what it means to know and to learn. For one, assessment design forces us to be much more explicit about the nature of our constructs and how they are manifest in various aspects of student performance. This also provides the benefit of designing ways to gather evidence that can be used to test and demonstrate the efficacy of the learning environments and tools and technologies for learning that we design. Much of what we can do goes well beyond traditional ways in which student achievement has been assessed and thus we have the opportunity to also shape the future of educational assessment. It is important that learning scientists engage the educational assessment and policy communities when it comes to the design and use of tasks and situations that provide evidence of student accomplishments—whether that be for purposes of improving educational materials and tools in an era of new standards in mathematics, language arts, and science, or for designing and interpreting assessments that have national and international impact.

Further Readings

Bennett, R. E., Deane, P., & van Rijn, P. W. (2016). From cognitive-domain theory to assessment practice. *Educational Psychologist, 51*(1), 82–107.
This journal article is a discussion of the CBAL assessment development project and nicely illustrates how the use of theory and research on learning progressions, combined with principled processes of assessment design, can yield a set of instructionally useful assessments. It discusses design of those assessments and some of the empirical evidence in support of their validity with respect to the intended interpretive use.

National Research Council. (2003). *Assessment in support of learning and instruction: Bridging the gap between large-scale and classroom assessment*. Washington, DC: National Academies Press.

This is a short report from the U.S. National Research Council that discusses similarities and differences between classroom and large-scale assessments. It is a very nice discussion of the relevance and use of each type of assessment and some of the key issues in their respective design and use.

Pellegrino, J. W., Chudowsky, N., & Glaser, R. (Eds.). (2001). *Knowing what students know: The science and design of educational assessment*. Washington, DC: National Academies Press.

This is a major report from the U.S. National Research Council on how to conceptualize the nature of educational assessment and its design and use in education. It discusses the role of theory and research on learning in the conceptualization and design of educational assessments, the role of measurement theory, and the implications for assessment design. It also discusses the various purposes of assessment, including formative, summative, and program evaluation uses.

Pellegrino, J. W., DiBello, L. V., & Goldman, S. R. (2016). A framework for conceptualizing and evaluating the validity of instructionally relevant assessments. *Educational Psychologist, 51*(1), 59–81.

This journal article discusses what constitutes validity when one is focused on assessments that are designed to support processes of teaching and learning. It provides a conceptual framework that describes three aspects of validity—cognitive, instructional, and inferential—and the forms of evidence that could and should be obtained to establish the validity of assessments intended to support instruction. It also illustrates application of the framework to an example of assessments contained in a popular U.S. K–5 mathematics curriculum.

NAPLeS Resources

Pellegrino, J., *15 minutes about assessment* [Video file]. In *NAPLeS Video series*. Retrieved October 19, 2017, from http://isls-naples.psy.lmu.de/video-resources/guided-tour/15-minutes-pellegrino/index.html

Pellegrino, J., *Interview about assessment* [Video file]. In *NAPLeS Video series*. Retrieved October 19, 2017, from http://isls-naples.psy.lmu.de/video-resources/interviews-ls/pellegrino/index.html

References

American Educational Research Association, American Psychological Association, and National Council of Measurement in Education (AERA, APA, NCME). (1999/2014). *Standards for educational and psychological testing*. Washington, DC: American Educational Research Association.

Bennett, R. E., Deane, P., & van Rijn, P. W. (2016). From cognitive-domain theory to assessment practice. *Educational Psychologist, 51*(1), 82–107.

Black, P., & Wiliam, D. (1998). Assessment and classroom learning. *Assessment in Education, 5*(1), 7–73.

Duncan, R. G., & Rivet, A. E. (2018), Learning progressions. In F. Fischer, C. E. Hmelo-Silver, S. R. Goldman, & P. Reimann (Eds.), *International handbook of the learning sciences* (pp. 422–432). New York: Routledge.

Herrenkohl, L.R., & Polman J.L. (2018). Learning within and beyond the disciplines. In F. Fischer, C. E. Hmelo-Silver, S. R. Goldman, & P. Reimann (Eds.), *International handbook of the learning sciences* (pp. 106–115). New York: Routledge.

Hickey, D., & Pellegrino, J.W. (2005). Theory, level, and function: Three dimensions for understanding transfer and student assessment. In J. P. Mestre (Ed.). *Transfer of learning from a modern multidisciplinary perspective* (pp. 251–293). Greenwich, CO: Information Age Publishing.

Kane, M. T. (2006). Validation. In R. L. Brennan (Ed.), *Educational measurement* (4th ed., pp. 17–64). Westport, CT: Praeger.

Kane, M. T. (2013). Validating the interpretations and uses of test scores. *Journal of Educational Measurement, 50*(1), 1–73.

Mislevy, R. J., & Haertel, G. (2006). Implications of evidence-centered design for educational assessment. *Educational Measurement: Issues and Practice, 25*, 6–20.

National Research Council (NRC). (2003). *Assessment in support of learning and instruction: Bridging the gap between large-scale and classroom assessment*. Washington, DC: National Academies Press.

Pellegrino, J. W., Chudowsky, N., & Glaser, R. (Eds.). (2001). *Knowing what students know: The science and design of educational assessment*. Washington, DC: National Academies Press.

Pellegrino, J. W., DiBello, L. V., & Goldman, S. R. (2016). A framework for conceptualizing and evaluating the validity of instructionally relevant assessments. *Educational Psychologist, 51*(1), 59–81.

Quellmalz, E., & Pellegrino, J. W. (2009). Technology and testing. *Science, 323*, 75–79.
Reimann, P., & Markauskaite, L. (2018). Expertise. In F. Fischer, C. E. Hmelo-Silver, S. R. Goldman, & P. Reimann (Eds.), *International handbook of the learning sciences* (pp. 54–63). New York: Routledge.
Ruiz-Primo, M. A., Shavelson, R. J., Hamilton, L., & Klein, S. (2002). On the evaluation of systemic science education reform: Searching for instructional sensitivity. *Journal of Research in Science Teaching, 39*, 369–393.
Sadler, R. (1989). Formative assessment and the design of instructional systems. *Instructional Science, 18*, 119–144.
Shepard, L. A. (2000). The role of assessment in a learning culture. *Educational Researcher, 29*(7), 4–14.
Stiggins, R. J. (1997). *Student-centered classroom assessment*. Upper Saddle River, NJ: Prentice-Hall.
Ufer, S., & Neumann, K. (2018), Measuring competencies. In F. Fischer, C. E. Hmelo-Silver, S. R. Goldman, & P. Reimann (Eds.), *International handbook of the learning sciences* (pp. 433–443). New York: Routledge.
Wiggins, G. (1998). *Educative assessment: Designing assessments to inform and improve student performance*. San Francisco, CA: Jossey-Bass.
Wiliam, D. (2007). Keeping learning on track: Formative assessment and the regulation of learning. In F. K. Lester, Jr. (Ed.), *Second handbook of mathematics teaching and learning* (pp. 1053–1098). Greenwich, CT: Information Age Publishing.
Wilson, M. (2004). *Constructing measures: An item response modeling approach*. Mahwah, NJ: Erlbaum.

41
Learning Progressions

Ravit Golan Duncan and Ann E. Rivet

Learning Progressions and the Learning Sciences

Among features that distinguish the Learning Sciences are a focus on learning in authentic contexts and the role of design in shaping learning environments. Learning progressions (LPs) are hypothetical models of learning developed with the aim of informing the design of standards, curriculum, instruction, and assessment in K–16 settings. The promise of LPs to inform designs for learning across grades and grade bands is an attractive prospect to learning scientists, particularly in math and science, who seek to fundamentally change these learning settings.

The development, empirical testing, and refinement of these hypothetical models invariably involves the design of instructional materials, if not at the onset of LP development, then certainly at more extensive testing stages. As we discuss later, many existing progressions have been developed using design-based research methodology, a core methodology of the Learning Sciences (Design-Based Research Collective, 2003) through iterative and collaborative teaching experiments. Design-based research combines the simultaneous development of both ecologically valid theory and practical applications of the theory for real-world settings. The practical applications in this case are instructional interventions that embody the hypotheses about learning modeled in the progression. Iterative design cycles allow refinement of the progression, the development of a validity argument to support its assertions, and the production of instructional strategies and materials. Learning scientists, with their interdisciplinary focus, affinity for designs, and tenacity necessary to work in the complex and messy classroom contexts, are well suited for the prodigious task of researching learning trajectories and progressions.

The Historical Roots and Definition of Learning Progressions

Learning progressions have emerged over the past decade in response to researchers' and practitioners' need to more precisely represent the knowledge, skills, and practices that constitute current accounts of what we want students to know and be able to do, especially as expressed in the new mathematics (Common Core Standards Initiative; National Governors Association Center for Best Practices, & Council of Chief State School officers, 2010) and science standards in the US (NGSS Lead States, 2013). Learning progressions are central to implementing ongoing formative assessment, a process that, when well done, has been shown to have large effect sizes (Black & Wiliam, 1998).

Most existing progressions and trajectories deal with science or mathematics concepts (Heritage, 2008) and thus we shall focus on the discussion in these two domains. Note that "trajectories" has been the preferred term in the mathematics domain, whereas in science the preferred term is "progressions." We will use "learning progressions" (LPs) to refer to both.

There are several key features that characterize LPs in science and math (Corcoran, Mosher, & Rogat, 2009; Daro, Mosher, & Corcoran, 2011). First, LPs are organized around a few core disciplinary ideas and practices. Second, LPs describe the development of students' understandings as intermediate steps or levels between initial and final states. Progress along the levels is mediated by targeted instruction and curriculum and is not developmentally inevitable. Third, these levels are extensively grounded in research on student learning in the domain. It is important to note that LPs by their very nature are hypothetical; they are conjectural models of learning over time that need to be empirically validated.

In science, progressions have their roots in assessment and measurement where the need arose to have a well-defined theoretical model when developing assessments. Trajectories in math have a somewhat older history and have traditionally included both a conceptual model of progress and the instructional means by which to move students along the progression. The inclusion of instructional guides is less common in science progressions. The question of whether progressions should specify the instructional means needed to support learning remains an open issue, with some arguing that the progression itself can be used to inform the design of instruction, but does not need to include such specifications (Gunckel, Mohan, Covitt, & Anderson, 2012), while others arguie that instruction, and even professional development, are a necessary component of any progression (Lehrer & Schauble, 2012).

LPs in math and science build on older and established developmental constructs such as developmental corridors (Brown & Campione, 1994). The notion of deepening and broadening understanding over time in developmentally appropriate ways is at the heart of current LPs and their predecessors. Moreover, LPs differ from descriptions of scope and sequence (based on analyses of normative knowledge in the domain) in that they are grounded in research on how students actually develop understanding of core ideas in the domain. This is a critical distinction because the intermediate steps in a progression may include understandings that vary significantly from the canonical knowledge of the domain.

Similarities and Differences Across LPs

Examination of a variety of existing progressions shows that their core features can be operationalized in different ways. These features include: (a) the scope of the progression, (b) the type a of constructs (big ideas) included; (c) how progress along a progression is conceptualized; and (d) the methods used to develop and refine LPs. We discuss each of these features using examples drawn from science and math progressions.

Scope

Learning progressions describe students developing understandings for a particular slice of a domain and across a specific age range (span); we refer to this combination as the scope of a progression. The slice of the domain is often reflected in the number of ideas included in the progression (termed "constructs"). For example, an energy progression includes four interrelated constructs about energy: (a) forms, (b) transfer, (c) degradation, and (d) conservation (Neumann, Viering, Boone, & Fischer, 2013). Most progressions include multiple constructs, but a handful of progressions have only a single construct (e.g.. Clements & Sarama, 2009). Progressions typically span either a few years or several grade bands.

Along with scope, progressions can also vary in grain size; by this, we mean the size of the jump between each level. For example, a progression spanning eight years with three levels can be said to be at a coarser grain size than one that spans three years with four levels. One can also go to even a finer grain size by looking at sublevels. As such progressions can have a fractal-like quality that affords zooming into a particular level and revealing a progression within that level (that might span a year of instruction), and zooming out to reveal a broader progression that spans a longer time period (for an example, see Lehrer & Schauble, 2012).

These variations in the grain size of the constructs and levels of progressions have implications for designing instruction and assessment, as well as for the methods used to test and refine LPs. Such variation is natural given the different aims and intended audiences of existing progressions. A large grain-size progression is better suited for informing policy in terms of standards, curriculum sequences, and large-scale assessments (Alonzo, Neidorf, & Anderson, 2012), whereas a smaller grain size is more appropriate for the study of incremental learning processes and for use by teachers and curriculum developers (Furtak & Heredia, 2014).

Type of Levels

The current and growing set of progressions addresses many domains in science and math. Broadly we can categorize progressions into three types: (a) those dealing principally with content ideas (e.g., celestial motion progression (Plummer & Maynard, 2014); (b) those dealing principally with practices and discourse patterns (e.g., argumentation progressions by Berland and McNeill, 2010; and Osborne et al., 2016); and (c) those that attempt to combine content and practice. An example of the latter is the progression developed by Songer, Kelcey, and Gotwals (2009) that includes two distinct, but parallel, constructs: a content construct for biodiversity-related concepts such as food webs, habitats, and abundance of species; and an inquiry-reasoning construct that deals with the construction of evidence-based explanations. Taken together the two constructs describe the development of evidence-based explanations of biodiversity.

Why is there such a variety in the nature of constructs? It seems to us that this variance is, at least partly, due to the field's fledgling state and the discreet attempts of researchers to tackle "reasonable" slices of the domain that are within their area of expertise and that can be studied across the proposed age span. This typical state of affairs for a new field is also reflected in the existence of progressions that overlap in their focus or age span (e.g., the two argumentation progressions noted above). Another part of the answer relates to the ways in which researchers construe progress along the progression and conceptualize the role of the learning environment in promoting progress. In the following section we discuss these two issues.

Nature of Progress Along a Progression

There are several different ways in which levels and progress have been conceptualized for LPs. Here we discuss two conceptualizations that offer productive, yet distinct, ways of thinking about progress. However, both go beyond a simplistic view of progress as adding more ideas or increasing accuracy.

The first conceptualization is offered by Wiser, Smith, and Doubler (2012), who argue that progress along a progression entails major reconceptualizations of knowledge and beliefs because students' naïve understandings are often incommensurate with canonical scientific ideas. In this view, the difference between two successive levels in the progression is not that the second contains more elements of the canonical ideas, or is more similar to expert knowledge, but that each level is a network of understandings that is productive in that it sets the stage for the next level. Thus, intermediate levels of a progression, which Wiser et al. (2012) term "stepping stones," may contain ideas that are inaccurate, simplified, and incomplete but that still provide students with a productive means of explaining a variety of phenomena in the domain, and place students in a position to move towards the next level. For example, the first stepping stone in the progression for matter is characterized as the "compositional model," which entails understanding that objects are made up of materials and that materials have unique characteristics that are (mostly) maintained when they are broken into smaller pieces and, no matter how small, these material pieces have weight and take up space. This "compositional model" steppingstone understanding, while not the normative atomic molecular theory, still provides productive groundwork and conceptual leverage for understanding that matter is made of particles, which is the next stepping stone.

Most existing progressions define constructs and progress along them at the individual student level. However, an alternative conceptualization of progress is changes in discourse and practices at the community level; a few progressions (in mathematics mostly) describe progress at this group or community level. Lobato and Walters (2016) term this the "collective mathematical practices" approach, stating that "classroom mathematical practices are students' ways of operating, arguing, and using tools that function in the class as if they are taken-as-shared" (p. 84). The idea of taken-as-shared reflects the unit of analysis as the community and the trajectory does not deal with the learning of individuals. One example is the trajectory in statistical reasoning, specifically the use of bivariate data to generate data-based arguments, developed by Cobb and colleagues (Cobb & Gravemeijer, 2008). Their trajectory captures changes in classroom norms and mathematical practices associated with understanding considerations about how bivariate data are generated, how they can be represented (what is shown versus hidden in different representations), and how they can be interpreted (e.g., what counts as valid claims about the distribution of data given a specific data set and a representation). The trajectory includes both descriptions of the changing practices, as well as the tasks and tools needed to support shifts in the learning community's mathematical practices and norms. While we do not know of progressions in science (or other domains) that describe learning at the community level, we do not think there is any reason that such progressions could not be developed outside of mathematics.

Descriptions of what progresses in a progression (i.e., what is changing) are related to, yet distinct from, depictions of the movement itself. At first glance, many of the representations of LPs (lists of increasing levels) imply linear step-like movement in which prior simpler conceptions give way to more sophisticated ones. However, in most cases these representations are misleading in this regard and scholars have noted, early and often, that there are likely multiple possible paths, and that learning is multidimensional, context-dependent, and therefore not likely to be linear (e.g., Corcoran et al., 2009; Empson, 2011).

Salinas (2009) provides the metaphor of landscapes to emphasize the notion that there may be multiple paths between point A and point B of any learning progression. A metaphor of landscape suggests that, while there is space for multiple paths, there are also boundaries that constrain the space of possible paths (like the limitations on possible trails up a mountain; see Figure 41.1). Wiser et al. (2012) espouse a similar view and argue that early learning is heavily constrained by universal knowledge and learning experiences that shape further learning in childhood. While there is more than one path, there is not an infinite number of paths, given existing conceptual constraints, and these paths likely share some core milestones.

The landscape metaphor also allows for "backwards" movement. Battista (2011) captured this back and forth "zigzaggy" movement across levels, showing that progress is context dependent and, while a student may reason in more sophisticated ways on one task, a more complex task is still likely to elicit more "primitive" ways of reasoning. Thus, progress in this sense is akin to ecological succession in which older, less sophisticated ideas slowly give way to new more sophisticated ideas, but do not disappear completely, and students may revert to using less sophisticated ways of reasoning when presented with more difficult tasks or ones in new contexts. Sevian and Talanquer (2014) use another metaphor, that of a dynamic system with dynamic attractors that function as semi-stable states. They view the construction of progressions "as a process of identifying and characterizing both the

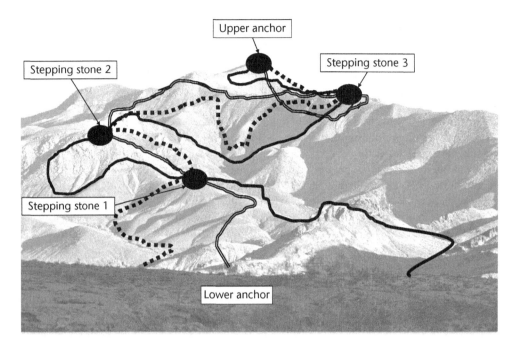

Figure 41.1 Learning Paths

Note: Analogy of mountain trails illustrates several key aspects of progress along a progression: (a) multiple possible paths, (b) common and stable stepping stone.

evolutionary path of such states from naive to sophisticated ways of thinking, as well as the internal constraints and external conditions (e.g., instruction) that support such evolution" (p. 14). This notion of dynamic attractors seems to us to be a particularly fruitful way of conceptualizing the levels of a progression, as it simultaneously conveys the complex, tentative, and messy nature of the learning process.

Methods for Developing and Refining LPs

While there is great variety in how learning progressions are defined, there is much less variability in how they are developed and studied. Overall, the development of almost all progressions starts with some review of existing research leading to an initial form of a hypothetical learning progression (Simon, 1995). In some cases, there is enough existing research to provide a fairly well-specified progression that requires empirical study and further refinement (e.g., Catley, Lehrer & Reiser, 2005; Smith, Wiser, Anderson, & Krijcik, 2006). In other cases, the research base is insufficient and further research is needed to formulate a coherent progression. This is where the methodological paths diverge into two distinct approaches: (a) assessment-driven cross-sectional studies of student thinking using interviews and written assessment to characterize levels of sophistication; and (b) short and long (i.e., longitudinal) teaching experiments using design-based research to characterize the development of students' ideas under specified instructional conditions.

An assessment–driven cross-sectional approach was used to develop both the carbon cycle progression (Mohan, Chen & Anderson, 2009) and the equipartitioning trajectory (Confrey & Maloney, 2015). Both were first developed based on clinical interviews with students across several grades and grade bands. It is important to note that these interviews were done under status quo curriculum and instruction and thus reflect what students can do under what might be considered less than ideal learning environments. Yet, these kinds of studies yield important and valuable information about patterns of reasoning and can substantially inform the development of a progression. Often interviews are supplemented by written assessment items that can readily be administered to larger samples of students and analyzed using advanced statistical models (e.g., Rasch models, latent class analyses, Bayesian networks) that provide information needed to develop a validity argument for the proposed levels of a progression (Wilson, 2005).

This approach highlights the important role of assessments and measurement in the study of LPs. The design of such items is not trivial and there are many different considerations that need to be taken into account. For example, it is difficult to design items that elicit responses that can demonstrate the full range of the progression. Often higher levels of a progression entail reasoning about invisible and unfamiliar entities (e.g., gases, molecules, genes) and may involve the use of some scientific terminology. However, using such concepts and terms in an item may render it confusing or inaccessible to learners at the lower levels of the progression. On the other hand, if an item does not explicitly cue students to draw on relevant concepts and terminology, even students that are able to provide sophisticated responses may opt to provide simpler ones that underrepresent what they are capable of doing. The solution is to develop items that are targeted at eliciting different levels of understanding (Jin & Anderson, 2012); however, that dramatically increases the number of items needed to "cover" the range of the progression and consequently the sample size needed to detect statistically significant patterns.

An alternative to assessment-driven cross-sectional studies is the teaching experiment approach. In this approach, researchers use the hypothetical progression to inform the development of carefully designed instructional interventions that are implemented in classroom settings and modified based on observed student performance. Such teaching experiments can occur over multiple years using a longitudinal study design (e.g., Wiser et al., 2012), or within one grade with units that span a few weeks (Shea & Duncan, 2013) to several months (Songer et al., 2009). Teaching experiments allow researchers to inform both theory (a refined progression) and practice (develop and refine an effective instructional intervention) through cycles of design-based research (Cobb, Confrey, diSessa, Lehrer, & Schauble, 2003). As researchers and teachers collaborate on the design of instructional interventions, they can develop and test, in tandem, specific conjectures of the progression. Teaching experiments take more of a data-driven or "bottom-up" approach to constructing and validating LPs. Obviously teaching experiments also involve assessments of student learning; however, these are often more qualitative (interview-based) and done on a smaller scale.

Teaching experiments by definition highlight the importance of instruction, and opportunities to learn, in promoting progress in learning. However, developing and studying progressions under specifically designed instructional conditions raises the question of whether learning paths would look different under dissimilar instructional conditions. On the one hand, if the conceptual constraints on learning underlying the progression are strong, then learning would simply be less efficient under instructional conditions that do not reflect the progression. On the other hand, there may be a plurality of progressions that reflect, and are specific to, instructional conditions and that are equally effective (Duncan & Gotwals, 2015). Further research is needed to settle this issue and answers may vary depending on the domain and focus of the progression.

Lastly, we wish to note that, while there may be differences in the methodologies used to develop the initial progression, it is often the case that the methodological paths converge. Those who initially used cross-sectional studies shift to using teaching experiments to further refine the progression (e.g., Confrey & Maloney, 2015; Gunckel et al., 2012), while those who initially relied solely on small-scale and qualitative analyses of teaching experiments turn to larger-scale assessment studies to mount support for the generalizability of their progression (e.g., Lehrer, Kim, Ayers, & Wilson, 2014). Ultimately for learning progressions to be useful as generalizable models of learning that can inform curriculum, instruction, assessment, and policy, they need to have the necessary evidentiary base. Such evidence can only be accrued through multiple cycles of research that employ different research designs to refine the progression and ultimately larger scale longitudinal studies.

Concerns and Criticisms

Learning progressions hold great promise as research-based models of learning paths that can inform the design of standards, curriculum, and assessment (Duncan & Rivet, 2013). However, for this potential to be realized the construct and consequential validity of LPs needs to be established. Construct validity refers to the extent to which the hypothetical LP accurately reflects how students actually conceptualize and learn the ideas and practices described by the progression. Consequential validity refers to the extent to which LPs are both useful, and appropriately used, tools for teaching, assessing, and policy making.

The issue with construct validity has to do with the extent to which the complexity of the learning process is reflected in the progression, and whether we can even describe student learning in terms of a few defined paths (Empson, 2011; Hammer & Sikorski, 2015). There is a concern here with a misrepresentation of learning as a predictable linear and straightforward movement towards more sophisticated ideas. Moreover, student performance is highly context dependent and can appear inconsistent; therefore "diagnosing" students' levels of reasoning is a tricky business (Steedle & Shavelson, 2009). Thus, there is a real risk of unintended data fitting (Shavelson & Kurpius, 2012). This suggests that LP researchers should employ multiple approaches to assessment and measurement and be aware of both the affordances and pitfalls of different measurement models.

The consequential validity concerns have to do with the ways in which LPs might be interpreted and used in different contexts and the consequences of such actions for students, teachers, and schools. One of the key critiques relates to the limited ways in which current LPs integrate cultural ways of knowing and doing, beyond noting them as the starting point or lower anchor (Anderson et al., 2012). This leads to the concern that current LPs are privileging certain ways of knowing and doing at the expense of others and thus exacerbating issues of inequity and access in science education. Research has shown that differences in prior knowledge, curriculum, and culture likely influence the paths students take. For example, Chinese and American students seem to take different paths though the carbon cycle progression (Gunckel et al., 2012), and differences between countries (Canada and China) were also noted in a learning progression for energy (Liu & Tang, 2004).

In the context of classrooms, there certainly is a risk of LPs being misinterpreted or misconstrued, and it is not trivial for teachers to use these models to inform their instruction and assessment practices (Furtak & Heredia, 2014). It seems that this concern can be, at least, partly addressed by providing supporting materials (tasks and assessments) and professional development with a pedagogical vision that includes multiple potential paths through the progression. The math trajectories tend to be more inclusive of such supports as part and parcel of the progression (Clements & Sarama, 2009; Daro et al., 2011); it seems progressions in other domains will likely benefit from developing and disseminating similar support systems.

Along with teachers, policy makers and assessment developers face similar challenges in making sense of using LPs. For example, while we have argued that incorrect but productive students' ideas can (and should) be a central part of intermediate levels of an LP, policy makers are reluctant to present "wrong" ideas in their standards (Foster & Wiser, 2012). In terms of assessment, there are concerns about the use of LP-based high-stakes assessments to "diagnose" students' levels of reasoning. Such uses are problematic for two reasons: (a) instructional contexts in the K–12 setting vary tremendously and not all learning environments are likely to provide adequate opportunities to learn; and (b) the ability to accurately diagnose students' levels of reasoning is questionable (e.g., Steedle & Shavelson, 2009). Therefore, making high-stakes decisions based on such assessment data is not recommended.

Conclusions

In closing, over the past decade LPs have become increasingly "popular." At the same time, there have been concerns that LPs are not ready for "prime time" (Shavelson & Kurpius, 2012). We agree with many of the criticisms raised and it is important to not lose sight of progress made in the Learning Sciences to conceptualize learning as being complex, contextual, and nuanced. However, we believe LP scholarship is still very young and invariably there are going to be differences in how they are conceptualized and presented. Consensus on at least some of these issues seems a bit premature. Furthermore, we worry that if we sit on our laurels and fret too much (we do need to fret some) about whether we have LPs just right, it will be too late and policy makers and public interest will have moved on to greener pastures and miss the opportunity to benefit from the insights gained from this budding field of inquiry. Current LPs appear to have provided some useful guidance for the development of the new standards in math and science and we anticipate that this young field will continue to evolve, improve, and contribute to learning, teaching, assessment, and policy.

Lastly, there are several aspects of LPs and LP research that could be improved. First, we need to be more explicit about mediating mechanisms of learning and change. How do later forms of reasoning and knowledge networks emerge from earlier ones? Second, we need to begin specifying the critical features of learning environments that can support productive movement along a progression and characterize the ways in which different instructional contexts (e.g., curriculum) might lead to learning along different paths. The extent to which progressions are dependent on and reflective of the specific learning environments used to develop them (teaching experiment contexts) is still an open question. Third, we need to begin connecting progression within and across domains. Collaborating across projects can help bring more coherence to LP research and development, and may provide the necessary infrastructure and capacity to engage in the still-much-needed, large-scale longitudinal studies of progressions.

Acknowledgment

This material is in part based upon Ann Rivet's work supported by (while serving at) the National Science Foundation. Any opinion, findings, and conclusions or recommendations expressed in this material are those of the author(s) and do not necessarily reflect the views of the National Science Foundation.

Further Readings

Confrey, J., & Maloney, A. P. (2015). A design research study of a curriculum and diagnostic assessment system for a learning trajectory on equipartitioning. *ZDM: The International Journal on Mathematics Education,* 47, 919–932.

This article highlights the important role of design-based research in developing and testing a learning trajectory. The article describes how the hypothetical trajectory informed the development of instructional materials and assessment tasks that could be subsequently used to test the various conjectures of the trajectory.

Corcoran, T., Mosher, F. A., & Rogat, A. (2009). *Learning progressions in science: An evidence-based approach to reform.* New York: Columbia University/Teachers College/Center on Continuous Instructional Improvement.

Daro, P., Mosher, F., & Corcoran, T. (2011). *Learning trajectories in mathematics: A foundation for standards, curriculum, assessment, and introduction.* Philadelphia, PA: Consortium for Policy Research in Education.

These reports, sponsored by the Consortium for Policy Research in Education, provide great introductions to progressions and trajectories. Building on the available research at the time, the reports sum up the state of the field and provide valuable insights about much needed future work. While not particularly recent, the definitions, critiques, and recommendations of these reports till hold true.

Empson, Susan B. (2011) On the idea of learning trajectories: Promises and pitfalls. *Mathematics Enthusiast, 8*(3), 571–598.

This article provides a brief historical overview and definition of trajectories and contrasts them with other developmental approaches to study of teaching and learning. It discusses some of the more heavily studied math learning trajectories and provides insights about the utility of the learning trajectory construct for math education.

Osborne, J., Henderson, J. B., MacPherson, A., Wild, A., Szu, E., & Yao, S.-Y. (2016). The development and validation of a learning progression for argumentation in science. *Journal of Research in Science Teaching, 53*(6), 821–846.

This article is a recent empirical study to develop and test a learning progression for argumentation. The progression includes parallel constructs for the construction and critique of argumentation with three major levels of sophistication regarding argument structure. The article presents findings from a large-scale, cross-sectional assessment study and provides a validity argument for the progression using Item Response Theory.

Smith, C., Wiser, M., Anderson, C. W., & Krajcik, J. (2006). Implications for children's learning for assessment: A proposed learning progression for matter and the atomic molecular theory. *Measurement, 14(1–2)*, 1–98.

The original version of this manuscript was a commissioned report for the National Research Council and is one of the first articles describing a learning progression. The article lays out a progression for matter and atomic molecular theory for K–8 that includes three big ideas: (a) matter and material kinds, (b) conservation and transformation of matter and material kinds, and (c) epistemology of atomic molecular theory-measurement, modeling, and argument.

NAPLeS Resources

Duncan, R. G., *Learning progressions* [Webinar]. In *NAPLeS video series*. Retrieved October 19, 2017, from http://isls-naples.psy.lmu.de/intro/all-webinars/duncan/index.html

References

Alonzo, A. C., Neidorf, T., & Anderson, C. W. (2012). Using learning progressions to inform large-scale assessment. In A. C. Alonzo & A. W. Gotwals (Eds.), *Learning progressions in science: Current challenges and future directions* (pp. 211–240). Rotterdam, Netherlands: Sense Publishers.

Anderson, C. W., Cobb, P., Calabrese Barton, A., Confrey, J., Penuel, W. R., & Schauble, L. (2012). *Learning Progressions Footprint Conference final report*. East Lansing, MI: Michigan State University.

Battista, M. T. (2011). Conceptualizations and issues related to learning progressions, learning trajectories, and levels of sophistication. *Mathematics Enthusiast, 8*(3), 507–570.

Berland, L. K., & McNeill, K. L. (2010). A learning progression for scientific argumentation: Understanding student work and designing supportive instructional contexts. *Science Education, 94*(5), 765–793.

Black, P., & Wiliam, D. 1998,*Inside the black box: Raising standards through classroom assessment*. London: School of Education, King's College.

Brown, A. L., & Campione, J. C. (1994). Guided discovery in a community of learners. In K. McGilly (Ed.), *Classroom lessons: Integrating cognitive theory and classroom practice* (pp. 229–270). Cambridge, MA: MIT Press/Bradford Books.

Catley, K. M., Lehrer, R., & Reiser, B. J. (2005). Tracing a prospective learning progression for developing understanding of evolution. Paper commissioned by the National Academies Committee on Test Design for K-12 Science Achievement, Washington, DC.

Clements, D., & Sarama, J. (2009). *Learning and teaching early math: The learning trajectories approach*. New York: Routledge.

Cobb, P., Confrey, J. diSessa, A., Lehrer, R., & Schauble, L. (2003). Design experiments in educational research. *Educational Researcher, 32*(1), 9–13.

Cobb, P., & Gravemeijer, K. (2008). Experimenting to support and understand learning processes. In A. E. Kelly, R. A. Lesh, & J. Y. Baek (Eds.), *Handbook of design research methods in education: Innovations in science, technology, engineering, and mathematics learning and teaching* (pp. 68–95). Mahwah, NJ: Erlbaum.

Confrey, J. & Maloney, A. (2015). A design research study of a curriculum and diagnostic assessment system for a learning trajectory on equipartitioning. *ZDM: The International Journal on Mathematics Education, 47*, 919–932.

Corcoran, T., Mosher, F. A., & Rogat, A. (2009). *Learning progressions in science: An evidence-based approach to reform*. New York: Columbia University/Teachers College/Center on Continuous Instructional Improvement.

Daro, P., Mosher, F., & Corcoran, T. (2011). *Learning trajectories in mathematics: A foundation for standards, curriculum, assessment, and introduction*. Philadelphia, PA: Consortium for Policy Research in Education.

Design-Based Research Collective. (2003). Design-based research: An emerging paradigm for educational inquiry. *Educational Researcher, 32*(1), 5–8, 35–37.

Duncan, R. G, & Gotwals, A. W. (2015). A tale of two progressions: On the benefits of careful comparisons. *Science Education, 99*(3), 410–416.

Duncan, R. G., & Rivet, A. E. (2013). Science learning progressions. *Science, 339*(6118), 396–397.

Empson, S. B. (2011) On the idea of learning trajectories: Promises and pitfalls. *Mathematics Enthusiast, 8*(3), 571–598.

Foster, J., & Wiser, M. (2012). The potential of learning progression research to inform the design of state science standards. In A. Alonzo & A. Gotwals (Eds.), *Learning progressions in science: Current challenges and future directions* (pp. 435–460). Rotterdam, Netherlands: Sense Publishers.

Furtak, E. M., & Heredia, S. (2014). Exploring the influence of learning progressions in two teacher communities. *Journal of Research in Science Teaching, 51*(8), 982–1020.

Gunckel, K. L., Mohan, L., Covitt, B. A., & Anderson, C. W. (2012). Addressing challenges in developing learning progressions for environmental science literacy. In A. Alonzo & A. W. Gotwals (Eds.), *Learning progressions in science: Current challenges and future directions* (pp. 39–75). Rotterdam, Netherlands: Sense Publishers.

Hammer, D., & Sikorski, T. (2015). Implications of complexity for research on learning progressions. *Science Education, 99*(3), 424–431.

Heritage, M. (2008). *Learning progressions: Supporting instruction and formative assessment*. Washington, DC: Council of Chief State School Officers.

Jin, H. & Anderson, C.W. (2012). A learning progression for energy in socio-ecological systems. *Journal of Research in Science Teaching, 49*, 1149–1180.

Lehrer, R., & Schauble, L., (2012). Seeding evolutionary thinking by engaging children in modeling its foundations. *Science Education, 96*(4), 701–724.

Lehrer, R., Kim, M.-J., Ayers, E., & Wilson, M. (2014). Toward establishing a learning progression to support the development of statistical reasoning. In J. Confrey & A. Maloney (Eds.), *Learning over time: Learning trajectories in mathematics education* (pp. 31–60). Charlotte, NC: Information Age Publishers.

Liu, X., & Tang, L. (2004). The progression of students' conceptions of energy: A cross grade, cross-cultural study. *Canadian Journal of Science, Mathematics, and Technology Education, 4*(1), 43–57.

Lobato, J., & Walters, C. D. (2016). A taxonomy of approaches to learning trajectories and progressions. In J. Cai (Ed.), *Compendium for research in mathematics education* (pp. 74–101). Reston, VA: National Council of Teachers of Mathematics.

Mohan, L., Chen, J., & Anderson, C. W. (2009). Developing a multi-year learning progression for carbon cycling in socio-ecological systems. *Journal of Research in Science Teaching, 46*(6), 675–698.

National Governors Association Center for Best Practices, & Council of Chief State School Officers. (2010). *Common Core State Standards for mathematics: Kindergarten introduction*. Retrieved from www.corestandards.org/Math/Content/K/introduction

Neumann, K., Viering, T., Boone, W. J., & Fischer, H. E. (2013). Towards a learning progression of energy. *Journal of Research in Science Teaching, 50*(2), 162–188.

NGSS Lead States (2013). *Next Generation Science Standards: For states, by states*. Washington, DC: National Academies Press.

Osborne, J., Henderson, J. B., MacPherson, A., Wild, A., Szu, E., & Yao, S.-Y. (2016). The development and validation of a learning progression for argumentation in science. *Journal of Research in Science Teaching, 53*(6), 821–846.

Plummer, J. D., & Maynard, L. (2014). Building a learning progression for celestial motion: An exploration of students' reasoning about the seasons. *Journal of Research in Science Teaching, 51*(7), 902–929.

Salinas, I. (2009). *Learning progressions in science education: Two approaches for development*. Paper presented at the Learning Progressions in Science (LeaPS) Conference, Iowa City, IA.

Sevian, H., & Talanquer, V. (2014). Rethinking chemistry: A learning progression on chemical thinking. *Chemistry Education Research and Practice, 15*(1), 10–23.

Shavelson, R. J., & Kurpius, A. (2012). Reflections on learning progressions. In A. C. Alonzo & A. W. Gotwals (Eds.), *Learning progressions in science: Current challenges and future directions* (pp. 13–26). Rotterdam, Netherlands: Sense Publishers.

Shea, N. A., & Duncan, R. G. (2013). From theory to data: The process of refining learning progressions. *Journal of the Learning Sciences, 22*, 7–32.

Simon, M. (1995). Reconstructing mathematics pedagogy from a constructivist perspective. *Journal for Research in Mathematics Education, 26*(2), 114–145.

Smith, C. L., Wiser, M., Anderson, C. W., & Krajcik, J. (2006). Implications of research on children's learning for standards and assessment: A proposed learning progression for matter and the atomic molecular theory. *Measurement: Interdisciplinary Research and Perspectives, 4*, 1–98.

Songer, N. B., Kelcey, B., & Gotwals, A. W. (2009). When and how does complex reasoning occur? Empirically driven development of a learning progression focused on complex reasoning about biodiversity. *Journal of Research in Science Teaching, 46*(6), 610–631.

Steedle, J., & Shavelson, R. (2009). Supporting valid interpretations of learning progression level diagnoses. *Journal of Research in Science Teaching, 46*(6), 699–715.

Wilson, M. (2005). *Constructing measures: An item response modeling approach.* Mahwah, NJ: Erlbaum.

Wiser, M., Smith, C. L., & Doubler, S. (2012). Learning progressions as tools for curriculum development: Lessons from the Inquiry Project. In A. C. Alonzo & A. W. Gotwals (Eds.), *Learning progressions in science: Current challenges and future directions* (pp. 359–403). Rotterdam, Netherlands: Sense Publishers.

42
Measuring Competencies

Stefan Ufer and Knut Neumann

Introduction

At the beginning of the 21st century the demands on the education of the individual have changed substantially (Bransford, Brown, & Cocking, 2000). The continued technological progress has led to a significant transformation of the workplace and an increasing need for a flexible workforce that can efficiently handle the complex technologies that are part of modern workplaces. Also, as social and environmental problems have become increasingly complex and interconnected, innovative strategies are needed to understand and manage them. Education cannot possibly provide individuals with all the individual abilities and skills needed to meet these challenges. Instead, education must help individuals develop the competence needed for continued learning and thus occupational and democratic participation (OECD, 1999). This shift towards competence as a desired outcome of education requires us also to rethink assessment. Innovative approaches, technologies, and formats are needed to measure competence constructs instead of measuring individual abilities or skills (NRC, 2014).

The extensive research delineating what constitutes competence in various domains and how it might be assessed was paralleled by substantial progress in the Learning Sciences focusing on learning and performance in specific situations. While competence research has developed methods to systematically address issues related to measuring complex, domain-specific learning outcomes in authentic situations in an unbiased and criterion-referenced way, these developments have not yet been taken up broadly in the Learning Sciences. The main goal of this chapter is to provide an overview of competence research and lay a foundation for stronger connections between the two research areas. We begin this chapter by elaborating on the ideas of competence and competence models. We then discuss issues around the measurement of competence, such as the identification of authentic contexts and the interpretation of observed behavior. Finally, we will elaborate on the importance of statistical models in addressing frequent issues in the measurement of competencies.

Delineating Competence: Competence Models

Competence. The idea of competence has been utilized in the context of education for more than 50 years, leading to a multitude of understandings. This renders it nearly impossible to provide a comprehensive definition (for an overview, see Winterton, Delamare-Le Deist, & Stringfellow, 2006).

White (1959), for example, understands competence as an individual's ability to interact effectively with the environment, Chomsky (1965) sees (language) competence as idealized capacity underlying language production in actual situations of performance, and McClelland (1973) defines competence in contrast to general cognitive abilities as the ability to perform concrete tasks such as driving a car. Obviously, one common feature of these understandings is the focus on (successful) performance in specific situations.

This domain-specific focus on competence made it popular in the fields of occupational education and training (for an overview, see Franke, 2008). General cognitive abilities such as intelligence had been found to be poor predictors of occupational success (McClelland, 1973). Competence in this sense incorporates the range of abilities and skills needed to perform successfully or obtain the abilities and skills needed to perform successfully in occupational situations (Nickolaus & Seeber, 2013); that is, competence also includes other underlying traits driving superior performance in occupational situations, such as motivation (Hartle, 1995, p. 107).

The idea of competence as an outcome of education gained massive popularity through the Programme for International Student Assessment (PISA). Prior to PISA, large-scale assessments such as the Trends in International Mathematics and Science Study (TIMSS) assessed students' knowledge and skills regarding the overlap of the curricula across the participating countries (Beaton, 1996). The PISA study, in contrast, aimed to assess to what extent students are prepared for future, lifelong learning and occupational and democratic participation (OECD, 1999). That is to say, PISA focuses on student competence, which (for science) was described as "the capacity to use scientific knowledge, to identify questions and to draw evidence-based conclusions in order to understand and help make decisions about the natural world and the changes made to it through human activity" (OECD, 1999, p. 60). The explicit focus on student competence, together with unexpectedly mediocre achievements, led to a discussion of the expected outcomes from general education in many countries (Neumann, Fischer, & Kauertz, 2010; Waddington, Nentwig, & Schanze, 2007).

In response to this discussion, the German-speaking countries introduced educational standards defining the expected outcomes of general education in terms of student competence (e.g., Klieme et al., 2003; KMK, 2003). For this purpose, Weinert's (2001) definition of competence was considered the most viable. According to this definition, competence reflects the (availability of the) cognitive abilities as well as the motivational and volitional attitudes needed to solve problems across a specified variety of different situations (p. 27ff.). The shift towards defining the aims of general education in terms of competence sparked a remarkable amount of research on students' competence in many different domains (for an overview, see Leutner, Grünkorn, Fleischer, & Klieme, 2017), sometimes modifying Weinert's definition: While Weinert (2001) included motivational aspects, Klieme and Leutner (2006), for example, restricted their definition exclusively to cognitive characteristics. Some features, however, are common all of the recent definitions of competence (Koeppen, Hartig, Klieme, & Leutner, 2008). First, competence is a latent construct that can only be inferred from an individual's actual performance in a range of different situations. Second, competence is specific to a domain in the sense that it explains an individual's performance only across situations from this domain. Third, competence incorporates abilities and skills that can be learned in one range of situations and transferred to a range of different situations from the same domain (Csapó, 1999). The latter requires to link multiple abilities and skills including metacognitive skills such as self-regulation (Erpenbeck, 1997) and non-cognitive dispositions such as beliefs (e.g., Baumert & Kunter, 2013). Most current conceptualizations of competence, however, focus solely on cognitive abilities and skill — mostly for the sake of assessment (Koeppen et al., 2008, p. 62).

Competence models. Defining what constitutes competence in a domain means to specify (a) the situations and demands that require the corresponding competence, and (b) criteria that specify what it means to cope with these demands successfully. These situations, however, may vary in terms of the abilities and skills required to master them successfully, making competence a multifaceted construct (Csapó, 2004). Thus, measuring competence requires differentiating the abilities and skills that underlie performance in different situations. The description of this "inner structure" of a competence construct (Mislevy, Steinberg, & Almond, 2002) is also referred to as a "competence model" (Koeppen et al., 2008). There are two fundamentally different approaches to specifying a competence model: (1) defining, from a normative perspective, how an individual should perform successfully across a set of situations from the domain in question (*normative models*, Schecker and Parchmann, 2006), or (2) empirically identifying patterns of abilities and skills underlying individuals' performance across a set of situations from the domain in question (*descriptive models*). Normative models are often constructed in a process of socio-political negotiation (Klieme et al., 2003), but can also emerge from theoretical considerations (e.g., Kremer et al., 2012; Nitsch et al., 2015) or a synthesis of existing research on students' competence in a domain (e.g., Hadenfeldt, Liu, & Neumann, 2014). A model underlying a set of standards for a domain is typically a normative model (e.g., KMK, 2003; NRC, 2012). Descriptive models are commonly constructed through the analysis of tasks typical for a domain, or from students' performance on such tasks; that is, based on empirical data, different levels or dimensions of competence are conceptualized in a data-driven way.

Structure models typically describe multiple areas or levels of competence in a domain, i.e., classes of situations that impose similar domain-specific demands. Mastering such sub-classes of situations requires parts of the overarching competence — these parts are sometimes also referred to as competencies. These may refer to situations that require varying subject-specific practices or the application of different concepts from a domain (CCSSI, 2011; KMK, 2003; Neumann et al., 2010). Other structure models specify different levels of competence that describe varying situational demands. These models are often also referred to as *level models* of competence. Examples are the (descriptive) proficiency level models derived from students' performance in TIMSS and PISA. Levels of student competence may be identified by grouping the tasks according to their empirical difficulty, and analyzing them for common features. Based on TIMSS data, Klieme, Baumert, Köller, and Bos (2000) identified five levels of student competence in science: applying everyday knowledge, explaining everyday phenomena in simple ways, basic scientific model conceptions, and applying scientific content knowledge, and using it in argumentation. Such level models make it possible to assign an individual to a specific level of competence based on observed performance. That is, they allow a criterion-referenced interpretation of an individual's performance.

Based on a structure model, *development models* describe how an individual's competence in a domain develops through instruction. This may simply mean that individuals are expected to successively develop competence in different areas (or different competencies respectively) or progress along a series of levels specified by the underlying structural model. In this sense, every learning progression represents a model of competence development (Duncan & Rivet, this volume). The force and motion learning progression described by Alonzo and Steedle (2009), for instance, may also be considered a model how individuals develop competence in the domain of force and motion. Like a learning progression, a model of competence development incorporates a clear developmental perspective that extends over a longer period of time and represents "increasingly sophisticated ways of reasoning within a content domain that follow one another as students learn" (Smith, Wiser, Anderson, & Krajcik, 2006, p. 1). Such models allow making inferences about an individuals' current level of competence development and the next potential level of competence development, thus informing formative assessment.

Measuring Competence: Conceptual Issues

Measuring competence means to infer an individual's ability to successfully perform in a domain from the individual's performance on a set of specific situations from the domain. Based on an existing model, these data may serve formative or summative purposes of assessment on the individual, class, district, or system level. Moreover, performance data may provide insights into the internal structure of a competence construct (i.e., levels, dimensionality, or development over time) to develop or refine a competence model.

The first and foremost criterion to measure an individual's competence in a domain is to observe this individual's performance in a sample of situations from this domain. This *holistic approach* requires assessment situations to be authentic and representative of the domain (McClelland, 1973; Shavelson, 2010); that is, the assessment situations must represent the domain sufficiently as defined by the underlying competence model, and each situation must allow the individual to demonstrate aspects of his or her competence. An *analytic approach* aims to provide information about the specific abilities, skills, or other individual resources that constitute competence in a domain. These resources

---

(a) Which of the following statements is true about organisms that are producers?
   ☐ They use energy from the sun to make food.
   ☐ They absorb energy from a host animal.
   ☐ They get energy from eating living plants.
   ☐ They get energy by breaking down dead plants and animals.

Source: https://nces.ed.gov/timss/pdf/timss2011_g8_science.pdf

(b) After going down a playground slide, you often land in sand.

Anna slid down the slide fast and left footprints in the sand. These footprints are shown in Picture A. Then, Anna smoothed out the sand and went down the slide slowly. These footprints are shown in Picture B.

Picture A—sliding fast   Picture B—sliding slow

Use your observations from Picture A and Picture B to construct an explanation to answer the question: How is Anna's energy when she was moving related to her speed when she hits the ground?

Source: Neumann, Kubsch, Fortus, Krajcik, and Nordine (2017)

Figure 42.1 Sample Tasks for Assessing (a) Content Knowledge and (b) Competence

are measured through specifically designed assessments (Blömeke, Gustafsson, & Shavelson, 2015). This approach can provide information about specific deficits in an individual's competence and thus give direction for further instructional support. However, assessing the different resources separately requires a large number of assessment tasks and may not provide insight into whether the individual is actually able to perform successfully in authentic situations. It is unclear, for example, if a student who has demonstrated modeling skills, knowledge of energy, and understanding of systems can actually model a system of energy transfers (see NRC, 2014). Blömeke et al. (2015) therefore propose to integrate both approaches and study the relation between the performance in authentic situations and the individual resources measured independently.

Assessments are needed to measure competence. Three foundational elements, comprising what is referred to as the "assessment triangle," are considered to underlie all assessments (NRC, 2001; Shavelson, 2010): *cognition, observations* and *interpretation* (see Pellegrino, this volume). *Cognition* refers to the theoretical assumptions about the construct to be assessed. This involves how an individual's abilities and skills are linked to successful performance in different situations from the domain, which may, for example, be described by a competence model. The tasks utilized to obtain evidence about an individual's competence, together with the individuals' performance on these tasks, represent the *observations* element. This may include practically any situation that allows individuals to demonstrate their competence. *Interpretation* refers to the methods and tools used to obtain conclusions about individuals' or groups of individuals' competence from the observations (NRC, 2001). Constructing reliable and valid assessments to measure competence in a domain requires the three elements to "be explicitly connected and designed as a coordinated whole" (p. 2).

From cognition to tasks. Assessing competence imposes different requirements on measurement instruments than assessing for example content knowledge. Declarative knowledge can be assessed with items like the one shown in Figure 42.1a, which simply asks students to choose the correct option. Assessment tasks need to require students to link multiple abilities and skills (such as to explain an observation using their knowledge about definitions of energy and the relationship between energy and matter; Figure 42.1b).

Obviously, even the task shown in Figure 42.1b does not comprehensively assess student competence in science. McClelland (1973) suggests selecting assessment tasks that adequately cover the universe of all potential situations from a domain. Typical examples are Objective Structured Clinical Examinations (OSCE) used in medical training. These consist of a set of short "assessment stations," each involving specific performance tasks like interviewing a standardized patient or performing a medical procedure. However, in most domains, it is a non-trivial task to ensure a representative sampling of situations (for OSCEs, see Cook, Zendejas, Hamstra, Hatala, & Brydges, 2014). In this case, a normative model of competence can help to ensure sufficient coverage of the different situations that constitute the domain.

From tasks to observations. The most valid information about the competence of an individual is yielded through performance tasks (NRC, 2014). The definition of a competence construct must determine criteria for assessing an individual's competence, based on the performance in a task. However, often not only the final product of a solution process bears information about an individual's competence, but also the way this solution is constructed. For example, there are more or less sophisticated and efficient strategies to arrive at the solution of a "rule of three" mathematics problem. Thus, apart from the final solution, log file data from computer-based assessments, video recordings of solution processes, participants' explanations of their solutions, or their actions in hands-on (e.g., experimentation) tasks may be considered. Empirical evidence shows, for example, that interviews, which allow to code students solution processes and explanations, lead to different interpretations of students' cognition than written tests (Opfer, Nehm, & Ha, 2012). However, since the main criterion for competence is task performance, drawing valid inferences based on processes observations requires empirically validated models that describe how different features of the solution processes are indeed related to the competence construct.

From observations to scores. Measuring competencies usually includes a quantification of an individual's performance or a quantitative rating of their solution processes. Depending on the type of observation data, a broad range of quantification techniques may be applied. For example, it is quite straightforward to code the solution of a mathematics calculation task, or a task requiring the extraction of clearly specified information from a text, which might have closed answer, multiple-choice and simple constructed answer formats. However, more complex answer formats such as longer texts, video data, observed behavior, or portfolios require deeper inferences by the coder. An OSCE, for example, usually provides rating scales for the examiner, and Schack et al. (2013) coded future teachers' diagnoses of students' numerical skills referring to the framework of professional vision. Similarly, coding other process data like log files or video data usually requires a sound understanding of process features that validly reflect an individual's competence. While complex answer formats and high-inference ratings may provide rich information about a person's competence, the required time to administer the items and to process the answers may limit the number of such tasks in an assessment. Moreover, since high-inference codings can vary substantially regarding their reliability (e.g., Cömert et al., 2016 for OSCEs; Shavelson & Dempsey-Atwood, 1976 for teacher behavior), measures have to be taken to control measurement errors. For example, multiple raters may code each single solution, or multiple indicators of performance or process quality can be applied in parallel.

From scores to interpretations. Single task scores usually carry substantial measurement error, since raters may misinterpret observed behavior, or indicators of competence may not be sufficiently well specified. The common solution is to aggregate the scores across several assessment tasks, which form a more or less representative sample of the universe of potential situations. Systematically varying the difficulty of tasks — for example, based on a level model of competence — can ensure that a wide spectrum of competence levels is covered, even if performance on a single task is coded only into few categories. However, aggregating scores across different tasks provides a useful measure of an individual's competences only if the variation between the scores can be attributed mainly to the individual who has taken the test. Generalizability Theory (Shavelson & Webb, 1991) helps determine what share of the variation in the scores is due to the differences between the individuals who solved the task, and to other factors, such as the task, the rater who rated the individual's performance, or combinations of these factors. Thus, it is possible to determine which factors actually influence the estimations of a person's competence and control potentially problematic factors.

Even though aggregated (or mean) scores over a set of tasks allow comparison of individuals' competence, these simple aggregate scores usually do not allow direct inferences regarding which kinds of situations an individual can actually master. If the assessment results should be used in a formative way to support further instructional choices, such criterial information is crucial. Competence level models provide such information systematically. Using appropriate statistical methods, it is possible to interpret the aggregate test score in terms of the demands posed by the most complex assessment tasks the person can master.

Analyzing Competence: Statistical Methods

Competence models are often derived from normative agreements as well as theoretical considerations about the learning, cognitive processing, and application of concepts in a domain. However, different models may be specified for a given competence construct. Mathematical competence, for example, can be described as one overarching competence, separate competencies that relate to different conceptual domains within mathematics (e.g., geometry, arithmetic, probability) or different mathematical practices (e.g., argumentation, modeling, or problem solving). These models will only yield valid assessment instruments and predictions about future performance if they reflect sufficiently well how people differ in their performance across different situations. Thus, the decision for a specified competence model will have to consider (a) the intended use of the model, (b) the complexity of the model, and (c) how well the model fits empirical data gathered from assessment studies.

Statistical models are required to derive estimates of competence from a group or individuals' performance on a set of assessment tasks. Such models also allow assessment of the extent a given cognitive competence model reflects observable differences in a person's performance. The main assumptions of most of these statistical models are (a) that an individual's competence can be described by one or more (numerical) estimates corresponding to latent variables in a statistical model, and (b) how these scores relate to the expected performance of an individual in a specific assessment situation. Two families of such statistical models are Structural Equation Modeling (SEM) or Item Response Theory (IRT).

Some of the most frequent challenges in the empirical study of competence models and statistical methods to deal with these challenges are the dimensionality of competence, criterial interpretation of scores, detecting and accounting for biased tasks, and linking different sets of assessment tasks.

Dimensionality of Competence

A model of competence may assume a single dimension of competence to describe persons' performance across different situations from a domain. This means that people who show higher performance in one relevant situation will necessarily also be expected to show at least slightly higher performance in other situations. For example, in a large-scale assessment, one overarching dimension of mathematical competence may be considered sufficient to compare educational systems. However, to inform teachers about what to focus on in further instruction, a multi-dimensional instrument differentiating separate scores for mathematical competence regarding numeric, geometric or probabilistic demands, for example, might be more helpful. Statistical indicators of model fit help to determine whether a one- or a multi-dimensional model explains differences in observed performance more efficiently (Hartig & Höhler, 2009). Multi-dimensional statistical models also provide a promising approach to disentangle interrelated skills and abilities that underlie performance in a domain—for example, general argumentation skills and domain-specific knowledge components.

Criterial Interpretation of Scores

As indicated above, it is important to link numerical competence scores to a criterial interpretation of the demands a person is able to master. IRT models help to determine difficulty parameters for items in an assessment, and ability parameters for each person taking the assessment, which can be interpreted on the same scale. A person with a given person parameter can be expected to solve a task that has a numerically equal difficulty parameter with a probability of 50%. If the person parameter is higher than the difficulty parameter, the person is more likely to solve the task correctly; if it is lower, the person is less likely to succeed (see Bond & Fox, 2001 for details). This interpretation makes it possible to infer which demands from the domain a person will be able to master with a given probability. If the tasks were constructed according to different levels of competence, this makes it possible to link a person's performance on the assessment directly to one of the competence levels, providing a criterial interpretation of the person's competence (e.g., Alonzo & Steedle, 2009; Kauertz, Fischer, Mayer, Sumfleth, & Walpuski, 2010). In research studies, for example, such an interpretation is promising to derive specific conclusions about which level of competence is necessary to benefit from one or the other intervention.

Detecting and Accounting for Biased Tasks

Constructing "fair" assessment tasks, which pose the same demands to all students, is a demanding challenge. Unusual terminology may disadvantage learners with low language skills, and tasks requiring specific strategies might be biased by the curriculum applied in a school. One challenge is to detect whether an assessment task is equally difficult for learners from different subgroups of the population. Such Differential Item Functioning (DIF) can be revealed, for example, by estimating separate difficulty parameters in subgroups of the population, e.g., learners with high and low language skills. This also allows the study of reasons for bias in assessment tasks. Haag, Heppt, Stanat, Kuhl, and Pant (2013), for example, investigated the influence of different linguistic characteristics of mathematics items on their bias with respect to the learners' language skills. If such a bias has been detected, it is also possible to control for these differences to a certain extent, as long as only a small number of tasks is affected by the bias.

Linking Different Sets of Assessment Tasks

For several reasons it is not always practicable that all participants work on all available assessment tasks when measuring a competence. For example, more tasks might be necessary to cover the whole competence construct than a participant can solve in the available time (e.g., in large-scale assessments). Moreover, researchers might want to use a second set of tasks in a longitudinal study, without having a participant solve a single item twice (e.g., Geller, Neumann, Boone, & Fischer, 2014). Finally, assessment tasks might be administered to the participants adaptively, to avoid administering tasks that are too hard or too easy. In all of these scenarios, different participants work on different sets of tasks, making it challenging to derive comparable competence estimates for all participants. If the difficulty parameters for the assessment tasks are available on a common scale, IRT models allow the generation of ability parameters (scores) for each person on the same scale (Kolen & Brennan, 2004). Thus, the resulting ability parameters may be compared meaningfully, even if the persons did not solve the same set of assessment tasks.

Conclusion

The notion of competence is different from other constructs that have been discussed in the past to explain student performance. Similar to the idea of three-dimensional learning (NRC, 2014), it explicitly emphasizes the capacity to integrate a multitude of different abilities and skills needed to cope with authentic situations. Assessing competence requires the specification of a competence model that guides the development of representative sets of authentic assessment tasks, allowing reliable and valid assessment of competence. Modern statistical methods allow researchers to approach some of the main challenges in the description of learning outcomes (e.g., dimensionality) or assessments (e.g., criterial interpretation, potential biases). These methods may, of course, be debated on the grounds of their underlying assumptions. However, what has to be kept in mind is that competence models, as well as statistical models, are *models*. This means they are tools that provide access to (surely more complex) phenomena that can be observed by researchers and practitioners. To what extent such a model is suitable to make a certain kind of inference about a person's competence is a matter of validity, which has to be argued based on empirical evidence as well as theoretical considerations.

The models provided by the extensive research on individuals' competence in many different domains, as well as the methods applied to interpret results from competence assessments, exemplify approaches to addressing some of the open methodological issues that have been identified in the Learning Sciences. Moreover, the Learning Sciences have developed tools to generate authentic and valid assessment tasks. Extending and intensifying the already existing overlap between the two disciplines may provide a fruitful stimulus for the Learning Sciences as well as for competence research.

Further Readings

Hadenfeldt, J. C., Neumann, K., Bernholt, S., Liu, X., & Parchmann, I. (2016). Students' progression in understanding the matter concept. *Journal of Research in Science Teaching, 53*(5), 683–708.
This paper presents a structural model of student understanding of matter and evidence about what understanding students have developed at different points in time throughout their K–12 education.

NRC. (2012). *A framework for K–12 science education: Practices, crosscutting concepts, and core ideas*. Washington, DC: National Academies Press.

This document contains a framework for national K–12 science standards in the US. It contains an interesting example of a normative, cognitive competence model that integrates competencies over a broad spectrum of scientific subjects such as biology, chemistry, and physics.

Obersteiner, A., Moll, G., Reiss, K., & Pant, H. A. (2015). *Whole number arithmetic—Competency models and individual development*. Paper presented at the Proceedings of the 23rd ICMI Study Conference: Primary Mathematics Study on Whole Numbers, Macao, China.

This contribution presents an example of a descriptive-level model of mathematical competence in the field of arithmetic, which has been generated from theoretical considerations and data from a large-scale assessment study.

Shavelson, R. J. (2010). On the measurement of competency. *Empirical Research in Vocational Education and Training, 2*(1), 41–63.

This contribution describes an Anglo-American perspective on competence measurement, based on holistic approaches and developed primarily in the field of vocational education. It provides a good overview of the practical issues in construction competence assessments.

Weinert, F. E. (2001). Concept of competence: A conceptual clarification. In D. S. Rychen & L. H. Salganik (Eds.), *Defining and selecting key competencies* (pp. 45–65). Göttingen: Hogrefe & Huber.

This contribution reviews different definitions of the competence construct. The contribution is a standard reference for the current notion of competence in educational research.

References

Alonzo, A. C., & Steedle, J. T. (2009). Developing and assessing a force and motion learning progression. *Science Education, 93*(3), 389–421.

Baumert, J., & Kunter, M. (2013). The COACTIV model of teachers' professional competence. In M. Kunter, J. Baumert, W. Blum, U. Klusmann, S. Krauss, & M. Neubrand (Eds.), *Cognitive activation in the mathematics classroom and professional competence of teachers: Results from the COACTIV project* (pp. 25–48). New York: Springer.

Beaton, A. E. (1996). *Mathematics achievement in the middle school years: IEA's Third International Mathematics and Science Study (TIMSS)*. Chestnut Hill, MA: ERIC.

Blömeke, S., Gustafsson, J.-E., & Shavelson, R. J. (2015). Beyond dichotomies. *Zeitschrift für Psychologie, 223*, 3–13.

Bond, T. G., & Fox, C. M. (2001). *Applying the Rasch model: Fundamental measurement in the human sciences*. Mahwah, NJ: Erlbaum.

Bransford, J. D., Brown, A., & Cocking, R. (2000). *How people learn: Mind, brain, experience and school expanded edition*. Washington, DC: National Academy Press.

Chomsky, N. (1965). *Aspects of the theory of syntax*. Cambridge, MA: Cambridge University Press.

Cömert, M., Zill, J. M., Christalle, E., Dirmaier, J., Härter, M., & Scholl, I. (2016). Assessing communication skills of medical students in objective structured clinical examinations (OSCE)—A systematic review of rating scales. *PloS one, 11*(3), 1–15.

Common Core State Standards Initiative (CCSSI) (2011). *Common core state standards for mathematics*. Retrieved January 24, 2018, from www.corestandards.org/wp-content/uploads/Math_Standards.pdf

Cook, D. A., Zendejas, B., Hamstra, S. J., Hatala, R., & Brydges, R. (2014). What counts as validity evidence? Examples and prevalence in a systematic review of simulation-based assessment. *Advances in Health Sciences Education, 19*(2), 233–250.

Csapó, B. (1999). Improving thinking through the content of teaching. In J. H. M. Hamers, J. E. H. van Luit, & B. Csapó (Eds.), *Teaching and learning thinking skills* (pp. 37–62). Lisse: Swets and Zeitlinger.

Csapó, B. (2004). Knowledge and competencies. In J. Letschert (ed.), *The integrated person: How curriculum development relates to new competencies* (pp. 35–49). Enschede: CIDREE.

Duncan, R. G., & Rivet, A. E. (2018). Learning progressions. In F. Fischer, C. E. Hmelo-Silver, S. R. Goldman, & P. Reimann (Eds.), *International handbook of the learning sciences* (pp. 422–432). New York: Routledge.

Erpenbeck, J. (1997). Selbstgesteuertes, selbstorganisiertes Lernen. *Kompetenzentwicklung, 97*, 309–316.

Franke, G. (2008). *Facetten der Kompetenzentwicklung*. Bielefeld, Germany: W. Bertelsmann.

Geller, C., Neumann, K., Boone, W. J., & Fischer, H. E. (2014). What makes the Finnish different in science? Assessing and comparing students' science learning in three countries. *International Journal of Science Education, 36*(18), 3042–3066. doi:10.1080/09500693.2014.950185

Haag, N., Heppt, B., Stanat, P., Kuhl, P., & Pant, H. A. (2013). Second language learners' performance in mathematics: Disentangling the effects of academic language features. *Learning and Instruction, 28*, 24–34. doi:10.1016/j.learninstruc.2013.04.001

Hadenfeldt, J. C., Liu, X., & Neumann, K. (2014). Framing students' progression in understanding matter: A review of previous research. *Studies in Science Education, 50*(2), 181–208.

Hartig, J., & Höhler, J. (2009). Multidimensional IRT models for the assessment of competences. *Studies in Educational Evaluation, 35*, 57–63.

Hartle, F. (1995). *How to re-engineer your performance management process*. London: Kogan Page.

Kauertz, A., Fischer, H. E., Mayer, J., Sumfleth, E., & Walpuski, M. (2010). Standardbezogene Kompetenzmodellierung in den Naturwissenschaften der Sekundarstufe I. *Zeitschrift für Didaktik der Naturwissenschaften* [Standards-based competence modeling in the natural sciences in lower secondary school. *Journal of Didactics of Science*], *16*, 143–152.

Klieme, E., Avenarius, H., Blum, W., Döbrich, P., Gruber, H., Prenzel, M., et al. (2003). *Zur Entwicklung nationaler Bildungsstandards [Regarding the development of national education standards]*. Berlin, Germany: Bundesministerium für Bildung und Forschung.

Klieme, E., Baumert, J., Köller, O., & Bos, W. (2000). Mathematische und naturwissenschaftliche Grundbildung: Konzeptuelle Grundlagen und die Erfassung und Skalierung von Kompetenzen. *TIMSS/III Dritte Internationale Mathematik-und Naturwissenschaftsstudie—Mathematische und naturwissenschaftliche Bildung am Ende der Schullaufbahn* (pp. 85–133). Berlin: Springer.

Klieme, E., & Leutner, D. (2006). Kompetenzmodelle zur Erfassung individueller Lernergebnisse und zur Bilanzierung von Bildungsprozessen. Beschreibung eines neu eingerichteten Schwerpunktprogramms der DFG. *Zeitschrift für Pädagogik, 52*(6), 876–903.

KMK. (2003). *Bildungsstandards im Fach Mathematik für den mittleren Schulabschluss [Educational Standards for middle secondary schools in mathematics]*. Munich: Luchterhand.

Koeppen, K., Hartig, J., Klieme, E., & Leutner, D. (2008). Current issues in competence modeling and assessment. *Zeitschrift für Psychologie [Journal of Psychology], 216*(2), 61–73. doi:10.1027/0044-3409.216.2.61

Kolen, M. J., & Brennan, R. L. (2004). *Test equating, scaling, and linking: Methods and practices* (2nd ed.). New York: Springer.

Kremer, K., Fischer, H. E., Kauertz, A., Mayer, J., Sumfleth, E., & Walpuski, M. (2012). Assessment of standard-based learning outcomes in science education: Perspectives from the German project ESNaS. In S. Bernholt, K. Neumann, & P. Nentwig (Eds.), *Making it tangible: Learning outcomes in science education* (pp. 201–218). Münster: Waxmann.

Leutner, D., Grünkorn, J., Fleischer, J., & Klieme, E. (2017). *Competence assessment in education*. Heidelberg: Springer International Publishing.

McClelland, D. C. (1973). Testing for competence rather than for "intelligence." *American Psychologist, 28*(1), 1–14.

Mislevy, R. J., Steinberg, L. S., & Almond, R. G. (2002). Design and analysis in task-based language assessment. *Language Testing, 19*(4), 477–496.

Neumann, K., Fischer, H. E., & Kauertz, A. (2010). From PISA to educational standards: The impact of large-scale assessments on science education in Germany. *International Journal of Science and Mathematics Education, 8*(3), 545–563.

Neumann, K., Kubsch, M., Fortus, D., Krajcik, J. S., & Nordine, J. (2017). *Assessing three-dimensional learning*. Paper presented at the Annual Conference of the National Association for Research in Science Teaching (NARST), San Antonio, TX.

Nickolaus, R., & Seeber, S. (2013). Berufliche Kompetenzen: Modellierungen und diagnostische Verfahren. In A. Frey, U. Lissmann, & B. Schwarz (Eds.), *Handbuch berufspädagogischer Diagnostik* (pp. 166–195). Weinheim: Beltz.

Nitsch, R., Fredebohm, A., Bruder, R., Kelava, A., Naccarella, D., Leuders, T., & Wirtz, M. (2015). Students' competencies in working with functions in secondary mathematics education—Empirical examination of a competence structure model. *International Journal of Science and Mathematics Education, 13*(3), 657–682. doi:10.1007/s10763-013-9496-7

NRC. (2001). *Knowing what students know: The science and design of educational assessment*. Washington, DC: National Academies Press.

NRC. (2012). *A framework for K–12 science education: Practices, crosscutting concepts, and core ideas*. Washington, DC: National Academies Press.

NRC. (2014). *Developing assessments for the next generation science standards*. Washington, DC: National Academies Press.

OECD. (1999). *Measuring student knowledge and skills: A new framework for assessment*. Paris, France: Organisation for Economic Co-operation and Development.

Opfer, J. E., Nehm, R. H., & Ha, M. (2012). Cognitive foundations for science assessment design: knowing what students know about evolution. *Journal of Research in Science Teaching, 49*(6), 744–777.

Pellegrino, J.W. (2018). Assessment of and for learning. In F. Fischer, C. E. Hmelo-Silver, S. R. Goldman, & P. Reimann (Eds.), *International handbook of the learning sciences* (pp. 410–421). New York: Routledge.

Schack, E. O., Fisher, M. H., Thomas, J. N., Eisenhardt, S., Tassell, J., & Yoder, M. (2013). Prospective elementary school teachers' professional noticing of children's early numeracy. *Journal of Mathematics Teacher Education, 16*(5), 379–397.

Schecker, H., & Parchmann, I. (2006). Modellierung naturwissenschaftlicher Kompetenz. *Zeitschrift für Didaktik der Naturwissenschaften, 12*(1), 45–66.

Shavelson, R. J. (2010). On the measurement of competency. *Empirical Research in Vocational Education and Training, 2(1)*, 41–63.

Shavelson, R. J., & Dempsey-Atwood, N. (1976). Generalizability of measures of teaching behavior. *Review of Educational Research, 46*(4), 553–611.

Shavelson, R. J., & Webb, N. M. (1991). *Generalizability theory: A primer* (Vol. 1). Thousand Oaks, CA: Sage.

Smith, C. L., Wiser, M., Anderson, C. W., & Krajcik, J. (2006). Implications of research on children's learning for standards and assessment: A proposed learning progression for matter and the atomic-molecular theory. *Measurement: Interdisciplinary Research & Perspective, 4*(1–2), 1–98.

Waddington, D. J., Nentwig, P., & Schanze, S. (2007). *Making it comparable*. Münster: Waxmann.

Weinert, F. E. (2001). Concept of competence: A conceptual clarification. In D. S. Rychen & L. H. Salganik (Eds.), *Defining and selecting key competencies* (pp. 45–65). Göttingen: Hogrefe & Huber.

White, R. W. (1959). Motivation reconsidered: The concept of competence. *Psychological Review, 66*(5), 297–333.

Winterton, J., Delamare-Le Deist, F., & Stringfellow, E. (2006). *Typology of knowledge, skills and competences: clarification of the concept and prototype*. Luxembourg: Office for Official Publications of the European Communities.

43

Mixed Methods Research as a Pragmatic Toolkit

Understanding Versus Fixing Complexity in the Learning Sciences

Filitsa Dingyloudi and Jan-Willem Strijbos

Educational research has been often criticized for being detached from practice, unable to address authentic educational issues raised in everyday learning and teaching settings, and not to contribute to educational practice and policy improvements (Lagemann, 2002; Lagemann & Shulman, 1999; Robinson, 1998). Educational researchers tend to be interested in rather narrow measures of learning, such as learning of pre-specified content and/or training of skills, implying that measuring learned content and/or skills is the only worthwhile measure of performance (Collins, Joseph, & Bielaczyc, 2004). As such, practice-detached educational research does not fully consider the complexity of the learning process due to contextual influences and complex outcomes (Robinson, 1998). This deficiency of educational research led to the establishment of the interdisciplinary field of the Learning Sciences (LS).

The LS studies learning phenomena (e.g., learning conditions, development, cognition) through various theoretical and research lenses by promoting, sustaining, and understanding innovations in messy real-life educational contexts (Lagemann, 2002; Sandoval & Bell, 2004). As such, LS bears a greater potential than traditional educational research to address real-practice issues: the development of knowledge with practical implications (Barab & Squire, 2004; Dix, 2007).

In the 1990s, LS researchers reflected on limitations of the existing "perspectives, paradigms, and methods . . . to provide adequate explanations" of interesting learning phenomena and expressed their frustration with the inadequacies of laboratory experimental research methods (Schoenfeld, 1992b, p. 137). Leading researchers such as Brown (1992) and Schoenfeld (1992a, 1992b) became interested in "real-world" learning phenomena along with their inherent complexity and messiness, calling for what Schoenfeld (1992b) describes as the four Cs; namely, "Conceptual Change amidst Creative Chaos" (p. 137). Schoenfeld (1992b) states that "the vast majority of professionals working in the learning sciences—frequently find ourselves in uncharted waters, looking at situations and behaviors we want to explain, but not having the necessary tools" (p. 138). To overcome the limitations of inadequate existing research perspectives, early LS researchers stressed the need to build evidence-based claims that result from *both* laboratory and naturalistic settings, thus calling for methodological expansion, pluralism, and inclusion (Barab & Squire, 2004).

What is evident in the early years of the LS is a need for mixing perspectives and research methodologies towards a more adequate understanding of real-world learning phenomena. More recently, Evans, Packer and Sawyer (2016) referred to LS as "an interdisciplinary approach to the study and facilitation of learning in authentic settings" (p. 1), accentuating the difference between authentic versus laboratory settings. The need to study learning in authentic settings led to the establishment of design-based research (DBR) (Barab, 2014; Barab & Squire, 2004; DBRC, 2003; Sandoval, 2004; cf. Puntambekar, this volume). Over the past decade DBR evolved into one of the main methodological paradigms of the LS and is often referred to as a "signature approach within the learning sciences" (Penuel, Cole, & O'Neill, 2016, p. 487).

DBR, as an interventionist paradigm, aims to conduct formative research in real-life teaching and learning settings "to test and refine educational designs and interventions based on theoretical principles derived from prior research" (Collins et al., 2004, p. 15). DBR is a pluralistic methodological innovation and includes a series of iterative, situated, and theory-based approaches to provide a greater understanding of the actual relationships among theory, artifacts, and practices (Barab & Squire, 2004; Bell, 2004; DBRC, 2003; diSessa & Cobb, 2004). In so doing, design-based researchers systematically refine various aspects of the designed context through experimentation (i.e., testing, observing, debugging of design) in naturalistic learning contexts up to the highest possible level of refinement (Barab & Squire, 2004; Cobb, Confrey, diSessa, Lehrer, & Schauble, 2003; Collins et al., 2004; DBRC, 2003; diSessa & Cobb, 2004). DBR aims to refine both theory and practice by pursuing the goals of developing effective learning settings while using them as "natural laboratories" to examine learning and teaching (Sandoval & Bell, 2004, p. 200). Irrespective of the value and contribution of DBR as an interventionist approach in the LS, this raises the question as to whether theory refinement in the LS is only accomplished through interventions. Do learning scientists merely intervene into learning contexts to refine the possibilities for learning? Is intervention all there is or should be?

As Lave (2014) suggested in her keynote speech at the 2014 International Conference of the Learning Sciences, prior to designing or intervening into any real-life learning situation, learning scientists first need to capture and understand the learning contexts per se in their inherent complexity. As Lave (2014) highlighted,

> When you formulate questions about learning that aren't about learning but are about the failure to learn, those aren't questions about how it is that learning takes place in the ongoing course of people's everyday lives, and so somehow, rather, addressing issues of failure before you ever get around to questions of what in fact is going on creates some real problems for your ability to address issues of learning. In the world in which we live I think that there are huge tensions and there is a contradiction between learning research in which what you as a scientist of learning would truly want to focus on, which is on questions of what is going on in the world, and questions which come from the impulse to fix it first and figure out what it is—if you ever have time—later.
>
> *(Lave, 2014)*

Understanding learning phenomena in complex learning environments—prior to adopting an interventionist approach (e.g., DBR) that aims to refine learning aspects, elements, or situations towards the best possible refinement for learning (as advocated by design-based researchers)—requires the inclusion of an exploratory/explanatory research approach to "understand first, refine later." Such a research approach is well represented by the mixed methods research (MMR) approach. MRR is a pragmatic research toolkit that offers learning scientists a wide range of research means to explore and understand learning phenomena that aim to inform practice *prior* to intervening in any learning situation. It should be highlighted that MMR is not the only means to explore and understand learning phenomena prior to an intervention. For example, qualitative approaches such as ethnography and ethnomethodology are amply used in the LS to explore and understand real-life learning (e.g., Lave, 2011), but they do so within mono-methodological confinements. Inclusion of MMR to the LS toolkit, in parallel with and/or combined with DBR or any other mono-methodological research approaches, responds to Barab and Squire's (2004) call for methodological pluralism and inclusion.

Filitsa Dingyloudi and Jan-Willem Strijbos

What is MMR?

MMR emerged in the 1990s and proposes the combination of qualitative and quantitative research approaches in one study through the use of multiple methodological approaches and ways of knowing towards a more complete understanding of multifaceted social or educational phenomena (Lund, 2012; Venkatesh, Brown, & Bala, 2013). Some MMR researchers treat MMR not only as a methodological approach or paradigm, but as a "mixed methods way of thinking [that is] generative and open, seeking richer, deeper, better understanding of important facets of our infinitely complex social world" (Greene, 2007, p. 20).

Over the years several researchers defined MMR (e.g., Creswell & Plano Clark, 2007, 2011; Greene, 2007; Johnson, Onwuegbuzie, & Turner, 2007; Morse, 2010; Morse & Niehaus, 2009; Tashakkori & Creswell, 2007; Tashakkori & Teddlie, 2010) but did not always fully agree on their definitions. Instead of a list of definitions (see Johnson et al., 2007), we support Creswell and Plano Clark's (2011, p. 5) position of defining MMR on the basis of six key features:

- collects and analyzes persuasively and rigorously both qualitative and quantitative data (based on research questions);
- mixes (or integrates or links) the two forms of data concurrently by combining them (or merging them), sequentially by having one build on the other, or embedding one within the other;
- gives priority to one or to both forms of data (in terms of what the research emphasizes);
- uses these procedures in a single study or in multiple phases of a program of study;
- frames these procedures within philosophical worldviews and theoretical lenses; and
- combines the procedures into specific research designs that direct the plan for conducting the study.

Likewise, Teddlie and Tashakkori (2010) identified two principles of MMR that move across the various definitions: (1) agreement upon the rejection of the "either–or" perspective (i.e., either quantitative/qualitative or qualitative/quantitative) across all levels of the research study; and (2) agreement on the iterative, cyclical research approach.

Why MMR?

The main reasons for employing MMR are to understand the (social) phenomenon of interest more comprehensively by aiming for a more complete picture, and more insightfully by aiming for a mixture of framing perspectives, ideas, and meanings (Greene, Kreider, & Mayer, 2005). MMR has been acknowledged as a valuable methodological approach for exploring complex social experiences and lived-in realities (Mason, 2006a), which is central to the LS.

Table 43.1 provides a comparison of MMR and DBR based on nine aspects: (1) the philosophical stance that is represented by each methodological approach, (2) the way knowledge is thought of, (3) the research approach it represents, (4) the overarching type of research design, (5) the implemented research methods, (6) the researcher(s)/research team typically involved, (7) the methodological aim, (8) the main research setting(s), and (9) the most representative research fields.

"Paradigmatic" Stance: A Pragmatic Stance to Methodology

Paradigms, as discussed in MMR (see Johnson, 2015), are not viewed as impositions of intertwined sets of philosophical assumptions that confine researchers' choices and decisions, as implied by Kuhn's work (1962), but rather as guiding tools of research activities that underlie a pluralist and compatibilist orientation to philosophical and theoretical elements (Johnson & Onwuegbuzie, 2004; Morgan, 2007; Teddlie & Tashakkori, 2010). In the MMR community several stances to methodology have

Table 43.1 Comparative Overview of MMR and DBR

	MMR	DBR
Philosophical stance	Pragmatism	Pragmatism
Knowledge	Pluralistic	Pluralistic
Research approach	Exploratory/explanatory	Interventionist
Research design	Formative	Iterative
Research methods	Mixed: qualitative and quantitative in a study	Either qualitative or quantitative or mixed in a study
Researcher(s)/research team	Researcher (and, if applicable, participants)	Researcher(s), practitioners, stakeholders (and, if applicable, participants)
Aim	Better understanding of phenomena	Refinement/improvement of learning contexts and of learning theory
Main setting(s)	Mixed (laboratory to authentic)	Mainly authentic (formal/informal)
Representative field(s)	Social/behavioral/human sciences	Learning Sciences

evolved: pragmatism, transformative-emancipation, dialectics, critical realism, dialectical pluralism 2.0, critical dialectical pluralism (see Johnson, 2015; Shannon-Baker, 2016). MMR researchers consider pragmatism to be the most representative stance in MMR (e.g., Tran, 2016).

In relation to LS, pragmatism is the most applicable due to its emphasis on practicality, inquiry (i.e., decisions leading to actions) and experience (i.e., interpretation of outcomes of actions), all situated within specific socio-historical contexts (Morgan, 2014). A pragmatic stance to MMR moves beyond the incompatibility thesis between qualitative and quantitative paradigms (Howe, 1988) and the implied confining purist stance to the qualitative and quantitative paradigms (Kuhn, 1962). Quantitative purists (e.g., Maxwell & Delaney, 2004; Schrag, 1992) are associated with positivism and argue for generalizability and objectivity of knowledge. Qualitative purists (e.g., Lincoln & Cuba, 2000; Schwandt, 2000) are associated with constructivism, interpretivism, and relativism, and argue for multiple constructed realities and subjectivity of knowledge. Pragmatism reflects a multiparadigmatic perspective that is represented by the freedom in moving across the quantitative and qualitative paradigms without being forced to follow a methodological dichotomy (Johnson, 2015; Morgan, 2007). The value of distinct elements in both qualitative and quantitative stances addressing different purposes is highly acknowledged, which in turn leads to their integration (Morgan, 2007; Onwuegbuzie & Leech, 2005).

The constituent elements of a pragmatic approach to methodology, as suggested by Morgan (2007), are (a) abduction, (b) intersubjectivity, and (c) transferability. Abduction refers to the reasoning that is in constant and multidirectional movement between data-driven induction and theory-driven deduction (Morgan, 2007, pp. 70–71). Intersubjectivity rejects the dichotomy between objectivity and subjectivity of knowledge and supports the compatibility of the existence of a common "real world" and a multitude of interpretations of this world by different individuals. Transferability moves beyond the dichotomy between context-specific and generalizable knowledge, to knowledge gained within a study that can be usable in a new set of circumstances while taking into consideration the factors that might affect its usability (Morgan, 2007, p. 72). All three elements of a pragmatic approach reflect an integrative nature in terms of theory and data, objectivity and subjectivity, context-specification and generalization; thus, rejecting any polarization between these else considered "dualities."

Finally, pragmatism emphasizes the importance of the research question in guiding the selection of appropriate research methods used to achieve the most informative answers to the question (Johnson & Onwuegbuzie, 2004, p. 18). Research questions, as representations of what researchers think they can ask and what they think they can observe, both influence and are influenced by the researcher's theoretical and methodological perspectives surrounding a study (Mason, 2006b). Therefore, an integration of the theoretical and epistemological perspectives—previously confined within the paradigms—also implies an integration of conceived realities represented in the research questions (Mason, 2006b).

Overall, a pragmatic stance to MMR integrates philosophical assumptions, epistemological beliefs and methods from various "paradigms" and serves as a logical, flexible, and productive approach that moves beyond paradigmatic confinements to more holistically address the research questions in a study. A pragmatic stance is appealing because it serves as a middle philosophical and methodological position between dogmatisms (Johnson & Onwuegbuzie, 2004).

MMR Designs: Typologies

Apart from the overall methodological stance—which is predominantly a pragmatic one—MMR researchers have also heavily debated how an MMR study could be designed and what kind of design elements can be distinguished. For example, Morse and Niehaus (2009) claim that a "mixed method design consists of a complete method (i.e., core component), plus one (or more) incomplete method(s) (i.e., supplementary component[s]) that cannot be published alone, within a single study" (p. 9). In contrast, Greene (2007) argues for a more flexible approach to designing a MMR study and highlights that there is no need of a prescriptive formula, but a "kind of mix that will best fulfil the intended purposes for mixing within the practical resources and contexts at hand" (Greene, 2007, p. 129).

Several MMR design typologies have been developed (e.g., Creswell, 2014; Creswell et al., 2003; Morse, 2010; Teddlie & Tashakkori, 2010) with main criteria being timing and purpose. Creswell (2014) identifies three basic designs (convergent parallel, explanatory sequential, exploratory

Table 43.2 Typology of Basic MMR Designs

Basic design type	Data collection	Data analysis	Integration purpose
Convergent parallel	Parallel/separate QUAN and QUAL	Separate	Confirmatory integration (e.g., confirming/ disconfirming results)
Sequential explanatory	Phase 1: QUAN Phase 2: follow-up QUAL	Separate	Explanatory integration (e.g., detailed explanation)
Sequential exploratory	Phase 1: QUAL Phase 2: follow-up QUAN	Separate	Exploratory integration (e.g., generalizability)

Note: QUAN = quantitative, QUAL = qualitative.

Table 43.3 Typology of Advanced MMR Designs Incorporating the Basic Designs

Advanced design type	Subtypes	Framework
Embedded	Convergent and/or sequential (nested in a larger study)	Larger study with either qualitative or quantitative orientation
Transformative	Convergent and/or sequential	Societal phenomena
Multiphase	Convergent, sequential, QUAN, and/or QUAL (in different project phases)	Longitudinal project/study

Note: QUAN = quantitative, QUAL = qualitative.

sequential) and three advanced designs that incorporate the basic forms (embedded, transformative, multiphase). Creswell's (2014) MMR design typology is relevant for MMR in LS due to its inclusiveness. An overview of basic and advanced designs is provided in Tables 43.2 and 43.3, respectively. The basic design types are described along three research aspects: data collection, data analysis, and integration purpose. The advanced design types are described on the basis of incorporated subtypes and framework.

These different MMR designs highlight the methodological flexibility and inclusion offered by MMR as a methodological approach and exemplify the pragmatic stance's principles. Apart from MMR's design flexibility, inclusion, and pragmatism, addressing MMR's typological criteria (see Creswell, 2014) fosters methodological systematicity as already called for by Schoenfeld (1992a, 1992b) in the 1990s.

Examples of MMR in the Learning Sciences

The role and contribution of MMR has been widely examined in the social and behavioral research, as well as in mainstream educational research (e.g., Creswell et al., 2003; Johnson & Onwuegbuzie, 2004). Yet, to date the role and contribution of MMR to LS has not been explicitly articulated. However, Schoenfeld (1992a)—in response to his reference to 'learning scientists exploring uncharted waters' (Schoenfeld, 1992b)—cautioned that this should not imply a methodological mess, and calls for systematicity. He suggests five (broadly described) standards that LS researchers should follow when describing any novel methods: (a) establishment of the context within which the phenomena have been studied, (b) description of the rationale for the employed methods, (c) description of the employed methods sufficiently enough for readers to be able to reapply them, (d) provision of sufficient amount of collected data for readers to be able to compare their understanding with that of the analysts', and (e) provision of a methodological discussion on the scope and methodological limitations, referring to issues of reliability and/or validity (Schoenfeld, 1992a).

Learning scientists, especially in the area of computer-supported collaborative learning (CSCL), contributed to the debate on the inadequacy of "one-method-only" by addressing MMR research questions related to the need for "understanding the how" before being able to "suggest the how." Jeong et al. (2014) conducted a content analysis on methodological practices in CSCL papers. They report that, although most papers discuss MMR, only 37% include mixed methods, and that predominantly on the analysis level. Since they rarely include it on the research design level and rarely draw any epistemological associations to MMR, their MMR practices do not necessarily imply a sophisticated methodological synthesis. According to the same analysis, DBR papers are also described as relying on mixed methods, but the how is left unspecified by the authors. As Jeong et al. (2004, p. 328) pinpoint, there is a need to develop "a more sophisticated way to combine different research traditions" and move from a coexistence level to a synthesis level, which could be accomplished through methodological integration on the epistemological and research design level. In order to exemplify MMR applications in LS, Table 43.4 gives an overview of five examples of MMR in LS in terms of design type, data collection/analysis, and integration purpose. Most examples reflect an explanatory orientation to the MMR design, which might be associated with the authors' strong quantitative background.

An example of a sequential explanatory MMR type is the study by Barron (2003), who analyzed the quality of group problem solving by quantifying qualitative discourse data for statistical analyses, followed by ideographic case studies of four groups (two successful and two unsuccessful groups) to further uncover cognitive and social mechanisms in the regulation of collaborative problem solving.

Another example that reflects a sequential explanatory is the study by Strijbos, Martens, Jochems, and Broers (2007). They inferred perceived group efficiency for groups with and without roles from statistical analyses of quantitative questionnaire data, inferred degree of coordination in those groups

Table 43.4 Examples of MMR in LS in Terms of Design Type, Data Collection/Analysis, and Integration Purpose

Study	Design type	Data collection/analysis	Integration purpose
Barron (2003)	Sequential explanatory	Phase 1: QUAN (statistical analysis of discourse data) Phase 2: QUAL (ideographic case studies)	Explanatory integration
Strijbos et al. (2007)	Sequential explanatory	Phase 1: QUAN (statistical analysis of questionnaire data and discourse data) Phase 2: QUAL (follow-up qualitative analysis of qualitative questionnaire data)	Explanatory integration
Zemel et al. (2007)	Sequential exploratory	Phase 1: QUAL (identification of long sequences with conversation analysis) Phase 2: QUAN (follow-up statistical analysis of discourse data in the identified long sequences)	Explanatory integration
Martinez et al. (2003)	Embedded/ Multiphase	Framed within laboratory studies Division into subprojects/evaluation phases Convergent parallel	Confirmatory integration
Suthers et al. (2013)	Convergent across analyses	Re-analysis of gathered data	Confirmatory integration

by quantifying qualitative discourse data, and inferred student experiences of group work using cross-case matrices to analyze qualitative questionnaire data.

Zemel, Xhafa, and Cakir (2007) exemplifies the sequential exploratory MMR type. They investigated chat-based problem solving, first applying qualitative conversation analysis to identify boundaries of long sequences and then quantifying qualitative discourse data in those long sequences for statistical analyses.

An example representing a systematic MMR approach in terms of its underlying philosophy, reflecting both an embedded (i.e., nested in laboratory studies) and a transformative multiphase (i.e., longitudinal project) MMR type, is the evaluation scheme developed by Martinez, Dimitriadis, Rubia, Gómez, and de la Fuente (2003). To investigate students' collaborative learning experiences, they incorporated quantitative, qualitative, and social network analyses within a single interpretative framework. Their MMR approach was on multiple levels (before, during, and after the study), triangulating data sources (i.e., questionnaires with open and closed questions, focus groups, classroom observations, log files, sociometrics), analytical methods (i.e., qualitative, partial qualitative, quantitative descriptive statistics, and social network analysis), and interpretation (partial conclusions and global conclusions).

Finally, a recent example that highlights the need for methodological plurality, along with the need for plurality in research traditions in general, is the "productive multivocality project" by Suthers, Lund, Penstein Rosé, Teplows, and Law (2013) (cf. Lund & Suthers, this volume). In the productive multivocality project, data gathered for a published study was made available for re-analysis by other researchers—who typically applied methods other than the original researchers—and the re-analyses were contrasted with the original analysis. Typically, the re-analyses not only partially converged and diverged from each other, but also from the original analyses, thus, partially confirming the initial conclusion but simultaneously generating new insights. The "productive multivocality project" is a case in point for the added value of the methodological pluralism in LS, but simultaneously also highlights the challenges faced by LS to act as an interdisciplinary community within which the different methods are used in an integrative way and not merely in an additive way.

It should be highlighted that these examples do not mean to be representative of all MMR designs/approaches in the LS community, but illustrative of how learning scientists applied different MMR designs. An analysis on MMR design representativeness in LS is worthwhile, but moves beyond the scope of this chapter.

Reflections and Future Considerations

LS and MMR seem to have some common foundations; that is, they both bring disciplines, paradigms, and methods within a pragmatic and practice-oriented framework. The examples show that learning scientists have made modest steps towards MMR approaches in their studies. However, LS researchers do so on a community level or on a single-study level, whereas MMR researchers do so (thus far) exclusively on a single-study level.

MMR offers the Learning Sciences a pragmatic toolkit for addressing MMR questions in order to explore and explain learning phenomena in a single study. If we consider LS a purely design or interventionist research field, MMR can methodologically contribute to LS with its integration in DBR: for example, prior-to-DBR, in-DBR, or post-DBR. If we consider LS as an exploratory and/or explanatory field as well, MMR can methodologically contribute to LS with its implementation either as a stand-alone approach (i.e., not in relation to DBR) or complementarily to other prominent methodological approaches in LS such as ethnography, ethnomethodology, or case studies, to name only a few. Learning scientists' call for expansion, plurality, and inclusion (see Barab & Squire, 2004; Schoenfeld, 1992b) can be addressed with the inclusion of MMR in LS, considering that all three aspects align with underlying principles of MMR. In fact, Lave's (2014) explicit call for understanding learning phenomena first and "fixing them" later (if at all) can be addressed via MMR approaches due to their explanatory and/or exploratory orientation as opposed to an interventionist or design approach.

We suggest that the systematic implementation of MMR as an exploratory/explanatory approach prior to any design or intervention decisions might let learning scientists be shaken by the complex before they shake the complex themselves. In addition, the degree and quality of "mixing methods" in a single study in LS can be enhanced by bringing together diverse expertise in a single study, as resembled by the multivocality project; both on an analytical and methodological level, moving from observations to computer simulations to experiments, and/or vice versa. Pragmatic research decisions that move beyond the incompatibility thesis can be an informative and flexible research guide for LS. More flexible, but still systematic MMR applications in LS, along with more room for MMR studies therein, might allow researchers to further develop methodological diversity (see NAPLeS Webinar Series Part C; NAPLeS, n.d.), not just in LS as a community but also in individual studies within LS.

Further Readings

Creswell, J. W. (2014). *Research design: Qualitative, quantitative, and mixed methods approaches* (4th ed.). Thousand Oaks, CA: Sage.
This book provides a comparative overview of qualitative, quantitative, and mixed methods research designs in the social sciences. Its target audience are students or researchers at the research planning stage.

Greene, J. C. (2007). *Mixing methods in social inquiry*. San Francisco, CA: Jossey-Bass.
This book provides an overview of mixed methods research by bringing together multiple theoretical frameworks and paradigms along with actual practices of mixed methods in the social sciences.

Jeong, H., Hmelo-Silver, C. E., & Yu, Y. (2014). An examination of CSCL methodological practices and the influence of theoretical frameworks 2005–2009. *International Journal of Computer-Supported Collaborative Learning, 9*(3), 305–334. doi:10.1007/s11412-014-9198-3
This paper provides an overview of methodological practices in CSCL during 2005–2009. The authors propose that the field can advance through meaningful synthesis of methodological practices (including MMR) and their theoretical frameworks.

Martinez, A., Dimitriadis, Y., Rubia, B., Gomez, E., & de la Fuente, P. (2003). Combining qualitative evaluation and social network analysis for the study of classroom social interactions. *Computers & Education, 41*(4), 353–368. doi:10.1016/j.compedu.2003.06.001
This study constitutes an advanced example of methodological integration through combining qualitative evaluation and social network analysis. The authors pay particular attention to the articulation of the mixed method component in their research design.

NAPLeS Resources

Strijbos, J. W., & Dingyloudi, F. *Mixed methods research* [Video file]. *Introduction and short discussion.* In *NAPLeS video series.* Retrieved from http://isls-naples.psy.lmu.de/video-resources/guided-tour/15-minutes-strijbos-dingyloudi/index.html

Strijbos, J. W., & Dingyloudi, F. *Mixed methods research* [Video file] *Interview.* In *NAPLeS video series.* Retrieved from http://isls-naples.psy.lmu.de/video-resources/interviews-ls/strijbosdingyloudi/index.html

References

Barab, S. (2014). Design-based research: A methodological toolkit for engineering change. In R. K. Sawyer (Ed.), *The Cambridge handbook of the learning sciences* (2nd ed., pp. 151–170). New York: Cambridge University Press.

Barab, S., & Squire, K. (2004). Design-based research: Putting a stake in the ground. *Journal of the Learning Sciences, 13*(1), 1–14. doi:10.1207/s15327809jls1301_1

Bell, P. (2004). On the theoretical breadth of design-based research in education. *Educational Psychologist, 39*(4), 243–253. doi:10.1207/s15326985ep3904_6

Barron, B. (2003). When smart groups fail. *Journal of the Learning Sciences, 12*(3), 307–359. doi:10.1207/S15327809JLS1203_1

Brown, A. L. (1992). Design experiments: Theoretical and methodological challenges in creating complex interventions in classroom settings. *Journal of the Learning Sciences, 2*(2), 141–178. doi:10.1207/s15327809jls0202_2

Cobb, P., Confrey, J., diSessa, A., Lehrer, R., & Schauble, L. (2003). Design experiments in educational research. *Educational Researcher, 32*(1), 9–13. doi:10.3102/0013189X032001009

Collins, A., Joseph, D., & Bielaczyc, K. (2004). Design research: Theoretical and methodological issues. *Journal of the Learning Sciences, 13*(1), 15–42. doi:10.1207/s15327809jls1301_2

Creswell, J. W., & Plano Clark, V. L. (2007). *Designing and conducting mixed methods research* (1st ed.). Thousand Oaks, CA: Sage.

Creswell, J. W., & Plano Clark, V. L. (2011). *Designing and conducting mixed methods research* (2nd ed.). Thousand Oaks, CA: Sage.

Creswell, J. W. (2014). *Research design: Qualitative, quantitative, and mixed methods approaches* (4th ed.). Thousand Oaks, CA: Sage.

Creswell, J. W., Plano Clark, V. L., Gutmann, M. L., & Hanson, W. E. (2003). Advanced mixed methods research designs. In A. Tashakkori & C. Teddlie (Eds.), *Handbook of mixed methods in social and behavioral research* (pp. 209–240). Thousand Oaks, CA: Sage.

Design-Based Research Collective (2003). Design-based research: An emerging paradigm for educational enquiry. *Educational Researcher, 32*(1), 5–8. doi:10.3102/0013189X032001005

diSessa, A. A., & Cobb, P. (2004). Ontological innovation and the role of theory in design experiments. *Journal of the Learning Sciences, 13*(1), 77–103. doi:10.1207/s15327809jls1301_4

Dix, K. L. (2007). DBRIEF: A research paradigm for ICT adoption. *International Education Journal, 8*(2), 113–124.

Evans, M. A., Packer, M. J., & Sawyer, R. K. (2016). Introduction. In M. A. Evans, M. J. Packer, & R. K. Sawyer (Eds.), *Reflections on the learning sciences: Current perspectives in social and behavioral sciences* (pp. 1–16). New York: Cambridge University Press.

Greene, J. C. (2007). *Mixing methods in social inquiry.* San Francisco, CA: Jossey-Bass.

Greene J. C., Kreider, H., & Mayer, E. (2005). Combining qualitative and quantitative methods in social inquiry. In B. Somekh & C. Lewin (Eds.), *Research methods in the social sciences* (pp. 274–281). London: Sage.

Howe, K. R. (1988). Against the quantitative–qualitative incompatibility thesis or dogmas die hard. *Educational Researcher, 17*(8), 10–16. doi:10.3102/0013189X017008010

Jeong, H., Hmelo-Silver, C. E., & Yu, Y. (2014). An examination of CSCL methodological practices and the influence of theoretical frameworks 2005–2009. *International Journal of Computer-Supported Collaborative Learning, 9*(3), 305–334. doi:10.1007/s11412-014-9198-3

Johnson, R. B. (2015). Conclusions: Toward an inclusive and defensible multi and mixed science. In S. Hesse-Biber & R. B. Johnson (Eds.), *The Oxford handbook of multimethod and mixed methods research inquiry* (pp. 688–704). New York: Oxford University Press.

Johnson, R. B., & Onweugbuzie, A. J. (2004). Mixed methods research: A research paradigm whose time has come. *Educational Researcher, 33*(7), 14–26. doi:10.3102/0013189X033007014

Johnson, R. B., Onwuegbuzie, A. J., & Turner, L. A. (2007). Toward a definition of mixed method research. *Journal of Mixed Methods Research, 1*(2), 112–133. doi:10.1177/1558689806298224

Kuhn, T. S. (1962). *The structure of scientific revolutions.* Chicago: University of Chicago Press.

Lagemann, E. C. (2002). *An elusive science: The troubling history of education research.* Chicago: University of Chicago Press.

Lagemann, E. C., & Shulman, L. S. (1999). *Issues in education research.* San Francisco: Jossey Bass.

Lave, J. (2011). *Apprenticeship in critical ethnographic practice.* Chicago: University of Chicago Press.

Lave, J. (2014). *Changing practice.* Keynote presented at the 11th International Conference of the Learning Sciences: Learning and Becoming in Practice, Boulder, CO.

Lincoln, Y. S., & Cuba, E. G. (2000). Paradigmatic controversies, contradictions, and emerging confluences. In N. K. Denzin & Y. S. Lincoln (Eds.), *Handbook of qualitative research* (pp. 163–188). Thousand Oaks, CA: Sage.

Lund, T. (2012). Combining qualitative and quantitative approaches: Some arguments for mixed methods research. *Scandinavian Journal of Educational Research, 56*(2), 155–165. doi:10.1080/00313831.2011.568674

Lund, K., & Suthers, D. (2018). Multivocal analysis: Multiple perspectives in analyzing interaction. In F. Fischer, C. E. Hmelo-Silver, S. R. Goldman, & P. Reimann (Eds.), *International handbook of the learning sciences* (pp. 455–464). New York: Routledge.

Martinez, A., Dimitriadis, Y., Rubia, B., Gomez, E., & de la Fuente, P. (2003). Combining qualitative evaluation and social network analysis for the study of classroom social interactions. *Computers & Education, 41*(4), 353–368. doi:10.1016/j.compedu.2003.06.001

Mason, J. (2006a). Mixing methods in a qualitatively driven way. *Qualitative Research, 6*(1), 9–25. doi:10.1177/1468794106058866

Mason, J. (2006b). *Six strategies for mixing methods and linking data in social science research* (NCRM Working Paper Series). Manchester: University of Manchester Press. Retrieved from http://eprints.ncrm.ac.uk/482/1/0406_six%2520strategies%2520for%2520mixing%2520methods.pdf

Maxwell, S. E., & Delaney, H. D. (2004). *Designing experiments and analyzing data: A model comparison perspective* (2nd ed.). Mahwah, NJ: Lawrence Erlbaum.

Morgan, D. L. (2007). Paradigms lost and pragmatism regained: Methodological implications of combining qualitative and quantitative methods. *Journal of Mixed Methods Research, 1*(1), 48–76. doi:10.1177/2345678906292462

Morgan, D. L. (2014). *Integrating qualitative and quantitative methods: A pragmatic approach.* Thousand Oaks, CA: Sage.

Morse, J. M. (2010). Simultaneous and sequential qualitative mixed method design. *Qualitative Inquiry, 16*(6), 483–491. doi:10.1177/1077800410364741

Morse, J. M., & Niehaus, L. (2009). *Principles and procedures of mixed methods design.* Walnut Creek, CA: Left Coast Press.

NAPLeS (n.d.). *Webinar series: Part C—Methodologies for the learning sciences.* Retrieved from http://isls-naples.psy.lmu.de/intro/4-methodologies/index.html

Onwuegbuzie, A. J., & Leech, N. L. (2005). On becoming a pragmatic researcher: The importance of combining quantitative and qualitative research methodologies. *International Journal of Social Research Methodology, 8*(5), 375–387. doi:10.1080/13645570500402447

Penuel, W. R., Cole, M., & O'Neill K. (2016). Introduction to the Special Issue. *Journal of the Learning Sciences, 25*(4), 487–496. doi:10.1080/10508406.2016.1215753

Puntambekar, S. (2018). Design-based research (DBR). In F. Fischer, C. E. Hmelo-Silver, S. R. Goldman, & P. Reimann (Eds.), *International handbook of the learning sciences* (pp. 383–392). New York: Routledge.

Robinson, V. (1998). Methodology and the research–practice gap. *Educational Researcher, 27*(1), 17–26. doi:10.3102/0013189X027001017

Sandoval, W. A. (2004). Developing learning theory by refining conjectures embodied in educational designs. *Educational Psychologist, 39*(4), 213–223. doi:10.1207/s15326985ep3904_3

Sandoval, W. A., & Bell, P. (2004). Design-based research methods for studying learning in context: Introduction. *Educational Psychologist, 39*(4), 199–201. doi:10.1207/s15326985ep3904_1

Schoenfeld, A. H. (1992a). On paradigms and methods: What do you do when the ones you know don't do what you want them to? Issues in the analysis of data in the form of videotapes. *Journal of the Learning Sciences, 2*(2), 179–214. doi:10.1207/s15327809jls0202_3

Schoenfeld, A. H. (1992b). Research methods in and for the Learning Sciences. *Journal of the Learning Sciences*, *2*(2), 137–139, doi:10.1207/s15327809jls0202_1

Schrag, F. (1992). In defense of positivist research paradigms, *Educational Researcher*, *21*(5), 5–8. doi:10.3102/0013189X021005005

Schwandt, T. A. (2000). Three epistemological stances for qualitative inquiry: Interpretivism, hermeneutics, and social constructionism. In N. K. Denzin & Y. S. Lincoln (Eds.), *Handbook of qualitative research* (pp. 189–213). Thousand Oaks, CA: Sage.

Shannon-Baker, P. (2016). Making paradigms meaningful in mixed methods research. *Journal of Mixed Methods Research*, *10*(4), 319–334. doi:10.1177/1558689815575861

Strijbos, J. W., Martens, R. L., Jochems, W. M. G., & Broers, N. J. (2007). The effect of functional roles on perceived group efficiency during computer-supported collaborative learning: A matter of triangulation. *Computers in Human Behavior*, *23*(1), 353–380. doi:10.1016/j.chb.2004.10.016

Suthers, D. D., Lund, K. Penstein Rosé, C., Teplows, C., & Law, N. (Eds.) (2013). *Productive multivocality in the analysis of group interactions*. New York: Springer.

Tashakkori, A., & Creswell, J. (2007). Exploring the nature of research questions in mixed methods research. *Journal of Mixed Methods Research*, *1*(3), 207–211. doi:10.1177/1558689807302814

Tashakkori A., & Teddlie, C. (2010). *Sage handbook of mixed methods in social and behavioral research* (2nd ed.). Thousand Oaks, CA: Sage.

Teddlie, C., & Tashakkori, A. (2010). Overview of contemporary issues in mixed methods research. In A. Tashakkori, & C. Teddlie (Eds.), *Sage handbook of mixed methods in social and behavioral research* (2nd ed.) (pp. 1–41). Thousand Oaks, CA: Sage.

Tran, B. (2016). The nature of research methodologies: Terms and usage within quantitative, qualitative, and mixed methods. In M. L. Baran & J. E. Jones (Eds.), *Mixed methods research for improved scientific study* (pp. 1–27). Hersey, PA: IGI Global.

Venkatesh, V., Brown, S. A., & Bala, H. (2013). Bridging the qualitative–quantitative divide: Guidelines for conducting mixed methods research in information systems. *MIS Quarterly*, *37*(1), 21–54.

Zemel, A., Xhafa, F., & Cakir, M. (2007). What's in the mix? Combining coding and conversation analysis to investigate chat-based problem solving. *Learning and Instruction*, *17*(4), 405–415. doi:10.1016/j.learninstruc.2007.03.006

44
Multivocal Analysis
Multiple Perspectives in Analyzing Interaction

Kristine Lund and Daniel Suthers

The role of interaction in learning has long been recognized, whether through deliberate design of social settings for learning (e.g., Slavin, 1990) or as intrinsic to human learning (e.g., Vygotsky, 1978). Human interaction is a complex process that lends itself to study from many different points of view. These points of view are also present within research on how learning occurs within groups, or how teachers, peers, or technology may facilitate collaborative learning. Researchers in a given tradition will choose to focus on particular aspects of human interaction that are emphasized by that tradition's theoretical and methodological framework. While there is nothing surprising about this per se, it's good science to want to understand how these particular aspects of human interaction fit together in a broader framework. However, a broader framework is not so easy to build, because each tradition teaches researchers different assumptions about what one should pay attention to and how one should do research. In this chapter, we present a collaborative approach to analysis that can help make these assumptions explicit and explore the level of integration that is possible between traditions. We begin by describing a selection of major methodological traditions that form claims about learning in groups, and then consider where and how integration can occur across traditions that are interested in collaborative learning.

Methodological Traditions in the Learning Sciences

In the classic hypothetico-deductive method, testing hypotheses about human interaction calls for an experimental set-up where independent variables are modified across a set of controlled situations in order to understand the effect this variation may have on dependent variables that are then measured. For example, such an approach is used for comparing what types of pedagogical or interactional support lead to stronger individual learning gains (Armbruster, Patel, Johnson, & Weiss, 2009), or for measuring how groups with particular characteristics can lead to higher-quality collaboration (cf. Vogel & Weinberger, this volume). Design-based research embodies hypotheses in socio-technical designs that are expected according to theory to influence an activity in certain ways (e.g., towards more effective learning), and observes how these designs play out in settings of interest (Design-Based Research Collective, 2003; Puntambekar, this volume). Discrepancies and surprises may lead to revisions of theory, and drive the next iteration of design and implementation, comparing to the previous iteration. Other analytic traditions dispense with the need for comparison. In classic ethnography, the researcher embeds him/herself into a context in order to understand it or a particular phenomenon in detail, primarily from participants' points of view, but sometimes through dialogue where participants and researchers perform interpretation on equal footing (cf. Green & Bridges' interactional ethnography, this volume). This latter approach is used, for instance, to understand the difficulties minorities face in informal contexts while learning (Valls & Kirikiades 2013). Ethnomethodological research, and conversation analysis that followed from it, is radically emic,[1] seeking to uncover how the analysis of participants' activity is embedded in that very activity (rather than generated by the researcher)—for example, to illustrate how expert and novice surgeons co-construct what it is they are seeing during a surgical act and where they should focus their attention (Koschmann & Zemel, 2011; Koschmann, this volume).

The above methods have often been opposed in light of the assumptions held by researchers in the respective traditions concerning the nature of human interaction and how they attempt to understand it (Lund, 2016). If traditions are based on different assumptions, the differences must be brought to light before any kind of integration can be considered. One way of distinguishing these traditions is by the relative importance they give to individual particularities of concrete cases. In the experimental approach, the assumption is to define constructs about human interaction based on the definition of an average behavior, and with a goal towards defining general laws (Lewin, 1931). These laws traditionally describe individual characteristics, but have recently extended their focus to group characteristics. In this approach, evidence is based on statistically unlikely differences between whole cohorts in distinct experimental conditions. In the ethnomethodological approach, the importance of a case study and its validity as proof cannot be evaluated by the frequency of its occurrence. Schegloff argues that "no number of other episodes that developed differently will undo the fact that, in these cases, it went the way it did with that exhibited understanding" (Schegloff, 1993, p. 101). These are two very different ways of making evidence-based claims about human interaction, and a fortiori, about learning. There are many other ways of distinguishing between the assumptions underlying methods. These differences in assumptions lead us to ask what a community such as the Learning Sciences can do in order to combine evidence-based claims of different natures so that the scientific knowledge we develop is stronger.

Mixed Methods and Multivocality

If we agree that we should combine evidence-based claims of different natures, we need to be aware of the dangers we may face in doing so. Researchers may attempt to combine theory, method, or data in incompatible ways. But if we approach the collective work done in the community of the Learning Sciences as a process of intersubjective meaning-making, it may become possible to productively interleave some of these research traditions that seek to understand human interaction, and seek to design technological artifacts to support learning (Lund, 2016; Suthers, 2006). Such intersubjective meaning-making can be achieved when researchers are able to turn their tradition-led activity into joint activity while co-constructing and sharing their interpretations across traditions. The value of doing this begins with, but goes beyond, mixed methods research (cf. Dingyloudi & Strijbos, this volume). A mixed methods approach has traditionally been defined as employing quantitative and qualitative analyses either in parallel or in succession and is in the first instance used to "triangulate," providing a broader view of the phenomena we are interested in and adding strength to results (Burke, Onwuegbuzie, & Turner, 2007). Yet, mixed methods may provide discrepant results (metaphorically, the triangulation fails to converge on a point), or incomparable results (metaphorically, each method operates on different planes that do not even intersect). These results are considered unproductive in the mixed methods approach, precisely because there is failure in constructing an integrated explanation of the observed phenomenon (Burke et al., 2007). Our perspective is different because we are working as a small community doing comparative analyses on the same corpus, where each person or author group represents a distinct analytic method. "Multivocal analysis"—analyses that embody different voices—can still give insight on the phenomena we are studying and the epistemologies underlying the methods we are using, even if triangulation fails to converge or even if results are incomparable because the methods operate on nonintersecting planes. Whereas mixed methods applied by a single agent can be forced to align without confronting tensions, the group analytic method that we call multivocal analysis *requires* that the methods be in dialogue with each other to explore alignments and misalignments and the different understandings and assumptions behind the latter (Suthers, Lund, Rosé, & Teplovs, 2013). In a mixed methods approach, multiple methods are elaborated together in service of the analytical goal. In a multivocal analysis, each method has its own spokesperson and way of approaching its goal (even if on some higher level there is a shared goal), and, although the corpus is shared, the underlying theories, methods, or results can only be articulated if assumptions are made explicit and if there is a way to focus in a shared manner on analyses. One of the essential differences, then, is what happens in case of failed triangulation. In mixed methods, there is an acknowledgment of the inability to construct an integrated explanation of the observed phenomenon. Paradoxes may indeed emerge from different data sources (i.e. quantitative and qualitative) and new modes of thinking can result, but in multivocal analyses this process is scaffolded with the goal of helping researchers identify how their assumptions have led to different research results, and this process takes place at the community level.

Interdisciplinarity in the Learning Sciences

In addition to the methodological traditions just described, the Learning Sciences encompass multiple disciplines. The fact that they are gathered together under the title "Learning Sciences" implies that dialogue and perhaps synthesis is desirable: achieving interdisciplinarity from multidisciplinarity. Multivocality is at its heart an approach to interdisciplinarity. To understand interdisciplinarity let us first consider what a discipline is. One way of looking at a discipline is to compare it to a community of practice (Lave & Wenger, 1991) where "groups of people share a concern, a set of problems or a passion about a topic and who deepen their knowledge and expertise in this area by interacting on an ongoing basis" (Wenger, McDermott, & Snyder, 2002, p. 4). Fiore (2008) reviews similar definitions of disciplines that refer to these components: core knowledge (e.g., "body of concepts, methods, and fundamental aims" [Toulmin, 1972, p. 139]) and practices of interacting together (e.g., "a communal tradition of procedures and techniques for dealing with theoretical or practical problems" [Toulmin, 1972, p. 142]). Van den Besselaar and Heimeriks define a disciplinary research field as "a group of researchers working on a specific set of research questions, using the same set of methods and a shared approach" (2001, p. 706).

An interdisciplinary community can also be understood in terms of domain, practice, and interaction. It must begin with the willingness of members of distinct traditions to interact with each other. But this interaction is concerned with the domain and associated practices at a meta-level. In addition to interacting to sustain a shared practice with respect to a domain as defined within a tradition, the interdisciplinary community also interacts to understand different ways of conceiving of the domain and distinct practices for approaching it (Derry, Schunn, & Gernsbacher, 2005). This meta-level interaction broadens participants' understanding of the domain and potentially creates a transcendent practice that could enable a new community to be formed and recognized (Klein, 1990).

Interdisciplinarity is needed to bring cohesion to multidisciplinary fields such as the Learning Sciences in a manner that respects and leverages their diversity. Individuals and communities studying human interaction must understand what assumptions we bring to the nature of knowledge and how it can be understood, and realize that other researchers have different assumptions. It's also crucial to recognize what motivates our analytical objectives and how and why we break down human interaction into units of action and units of *inter*action. Finally, the ways we choose to represent the interactions we study and how we may manipulate these representations during our analyses are essential for *orienting* how we can gain insight on human interaction.

The remainder of the chapter describes the origins of an approach to collaborative analysis called "multivocal analysis," along with examples that illustrate the ways in which an interdisciplinary outlook can be beneficial to research. We conclude the chapter with a summary of the challenges and the benefits of multivocal analysis and offer some perspectives on future work.

Multivocal Analysis

Multivocal analysis emerged from a five-year collaboration, the Productive Multivocality Project. This project involved 32 researchers who were engaged in analyzing group interaction in educational settings. Across the group of researchers multiple analytic traditions were in use. The multivocality project explored productive ways to engage in dialogue among these multiple traditions with the goal of enhancing the understanding of group interaction as well as that of surfacing strategies for multivocal analyses of human interaction.

Origins of the Productive Multivocality Project

Multivocal analysis has roots in earlier efforts, but was made explicit in the context of a long-term research collaboration around the analysis of group interaction. We called this collaboration the Productive Multivocality Project because it involved an effort to bring the various "voices" of multiple theoretical and methodological traditions into productive dialogue with each other. The work that emerged from the project is reported in detail in an edited volume (Suthers, Lund, Rosé, Teplovs, & Law, 2013).

The Productive Multivocality Project emerged over a series of workshops, the first of which was motivated by the observation that advances in shared representations, methods, and tools lead to progress in many scientific disciplines. The first workshop was convened to identify a common basis (conceptual model and representations) for shared tools for the analysis of learning through interaction. Failing to find sufficient commonality but finding each others' analyses to be of interest, in a second workshop we shifted our focus to productive understanding of the differences as well as commonalities between our approaches, by comparing analyses of shared datasets. Although this was a crucial strategy, we quickly found that sharing data was not enough: analysts "talk past" each other for different purposes. We realized that we were encountering a central problem in the coherence of the Learning Sciences, and committed as a group to continue our work together to address this problem. It was in a third workshop that the initial articulation of a core strategy for productive multivocality emerged. Analysts from diverse analytical traditions were assigned to shared corpora, as had been done in the second workshop. We added the requirement that the group address an analytic objective that was deliberately open to interpretation by the different traditions (e.g., "pivotal moments"). The dual focus on shared data and objectives was intended to motivate the analysts to compare and contrast their interpretations and perspectives. In this and two subsequent workshops we identified challenges to implementing this strategy and implemented and refined strategies for supporting productive cross-talk. For example, in order to compare analyses, analysts needed to eliminate nonessential differences and bring their analytic representations into alignment with each other, using software support if needed. Furthermore, we found that it was helpful to provide facilitators who helped both conceptually and practically with the work to be done, and ensured that the data providers' role was respected. Iteration of the process helped analysts revisit the comparison once gratuitous differences had been eliminated, and also allowed analysts to incorporate ideas from others when reconsidering their own analyses. In any given workshop, we had several groups of analysts, each focusing on their own shared dataset. After five workshops and ongoing online interaction and collaboration of project members spanning a total of five years and involving 32 researchers, we had developed a shared tenet and collection of strategies that we now refer to as "multivocal analysis." These are summarized below.

Tenets of Multivocal Analysis

The core idea is that scientific and practical advances in an area of study can be obtained if researchers working in multiple traditions—including traditions that have been assumed to be mutually incompatible—make a concerted effort to engage in dialogue with each other, comparing and contrasting their understandings of a given phenomenon and how these different understandings can either complement or mutually elaborate each other. Incompatibilities may remain, but are reduced to essential and possibly testable differences once nonessential differences have been identified. Multivocal analysis has benefits both for the individual analysts and the greater Learning Sciences community. These include confronting aspects of data not previously considered, challenging epistemological assumptions, fine-tuning analytic concepts, and a multidimensional understanding of the phenomenon being investigated and analytic constructs. The process enables greater dialogue and mutual understanding in the Learning Sciences.

Strategies of Multivocal Analysis

Strategies we developed for achieving *productive* multivocality in multivocal analysis are outlined here (cf. Suthers, Lund, Rosé, & Teplovs, 2013). These strategies are employed by and organize the work of *groups* of researchers as they engage in interdisciplinary dialogue about how they constitute and approach interaction data as an object of study, that is, they are strategies for the meta-level discourse needed for an interdisciplinary community of practice.

Analyze the same data. As other groups of researchers have recognized before us (e.g., Koschmann, 2011), *sharing data and comparing analyses* provides the possibility for dialogue regarding what we understand about human interaction. However, as others have also found, analyzing the same data is not enough.

Analyze from different traditions. Achieving the multivocality that makes explicit the epistemological foundations on which we build our science requires *assigning analysts from different traditions to the same data*. For example, a corpus might be analyzed from linguistic, ethnomethodological, and social network traditions. Our assumptions become explicit when we try to reach agreement on what we consider data, what data are worthy of analysis, what questions are appropriate, and how conclusions are drawn.

Push the boundaries of traditions without betraying them. A related strategy is to *push analysts outside their comfort zone, while maintaining the integrity of their traditions*. Analysts may be asked to take on settings, types of data, and/or research questions that differ somewhat from their normal practices. This involves risk for two reasons. The first is that some researchers do not appreciate working outside of their comfort zone and they may disengage from the group. The second reason is that those who see the value in it can sometimes modify their tradition in ways others in their tradition would not accept. For example, researchers who normally apply statistical analysis methods to "large N" data may be challenged by approaching small corpus data. Although the results may not be acceptable to their home discipline, their methods can yet inform dialogue with qualitative researchers if statistical results are treated as descriptive of the particular data rather than inferential.

Begin with a shared pre-theoretical analytic objective. Unless researchers can work toward a shared analytic objective, their analyses can be difficult to compare, because the questions being asked may be completely different. Researchers should *identify a shared but pre-theoretical concept as the analytic objective*. For example, in the third workshop discussed above the objective was to identify the *pivotal moments* in the collaboration dataset that were the subject of the joint analysis. We left what constituted a "pivotal moment" unspecified, other than specifying that such a moment (or event, episode, etc.) should be relevant to learning or collaboration. This resulted in researchers being able to examine whether or not they identified the same moments, where and why the moments and/or their criteria differed, and whether the moments identified by one analytic tradition might lead to refining another tradition.

Bring analytic representations into alignment with each other and the original data. Initially, each analyst performs analyses from his or her own perspective on the shared dataset, using their preferred representation, depending on their assumptions. We then asked researchers to *bring their analytic representations into alignment*. (Analytic representations may include transcripts, segmentation of those transcripts, codes applied to segments, graphs indicating relationships between segments, tables of statistics summarizing properties, etc.) Comparison of analyses is facilitated if researchers can identify where their own analytic representations of the original phenomenon address the same temporal, spatial, and semantic spans as other researchers' representations. For example, in order to focus on the pivotal moments we used the Tatiana analytic software (Dyke, Lund, & Giradot, 2009) to help us to visualize multiple categorical codings and uptake graphs and locate them on the same timeline. The common visual representation made clear where the identified "pivotal moments" aligned across analyses and where they did not. The differences triggered productive discussion about the reasons for lack of alignment.

Assign a facilitator/provocateur. A commitment to pay attention to the analyses of others is necessary for any meaningful collaboration of persons claiming to do multivocal analyses. We countered researchers' natural tendency to focus more on their own work by assigning a facilitator to each group of analysts who were sharing data. Facilitators manage the collaboration, a process that may include doing some of the work necessary to compare results such as aligning analytic representations and pointing out differences that the analysts should discuss.

Eliminate gratuitous differences. Nonessential differences between two analyses are those that do not reflect core commitments of the analytic traditions. Such differences include having chosen to analyze different temporal segments of human interaction (when the same segments could have been selected without betraying either tradition), or having given different names to the same entities (to the extent that the names do not carry theoretical implications). When analytic representations are aligned and correspondences for naming conventions are found, these differences can be eliminated, and researchers can focus on both what they share and why they don't share their differences. Other differences, such as including or excluding private communications or nonverbal actions, may be more "essential," as they reflect the epistemological assumptions of the researcher. Productive does not necessarily imply agreement: going through the process of separating nonessential and essential differences makes the latter more salient for all involved.

Iterate. The benefits of many of the foregoing strategies are better realized if researchers iterate their analyses, especially if iteration includes a phase of attempted representational alignment or identification of gratuitous differences. Researchers may also iterate after becoming clearer about the epistemological foundations of their views on human interaction and learning. This might come about in a number of ways, including adopting analytical concepts from other traditions or taking up different conceptions of key constructs such as "pivotal moment."

Attend to the needs of the data providers. Data providers are valuable actors within multivocal analysis, and they need support. Facilitators can help by reminding other analysts that data providers had their own objectives when they initially produced the data that is now being shared. Once data have been shared, it may be tempting for new analysts to criticize how the data were produced, because they may have different criteria for data production and different analytic needs. Data sharing is risky: data providers' traditions should be respected. Respectful sharing of results and iteration wherein analyses are revised can lead to new understandings of value to all participants.

Reflect on practice. Perhaps most importantly, while recognizing that methods have biases, our Productive Multivocality Project argued that researchers have (and should use) agency when they apply the methods. This means that researchers are not deterministically bound to the traditions that originally derived the methods. Although methods include practices associated with their use (e.g., how to select questions worth asking, how to map an interaction to analytic notations representing the interaction, how to transform these representations from one form to another, and how to interpret them), practices reflect theoretical and epistemological commitments (Kuhn, 1962/1970). We argue that multivocal analyses allow for an explicit examination of these commitments (Yanchar & Williams, 2006). Here, the strategy is to remove one's methodological eyeglasses and *view and dialogue about methods as object-constituting, evidence-producing, and argument-sustaining tools.*

Multivocal Analysis in Action

This section provides examples of multivocal analysis as it was applied in different contexts. The first is a brief example from the original collaborative project, while others are exterior to the project.

Peer-led team learning for chemistry. The comparison of analyses in this group from the Productive Multivocality Project (Rosé, 2013) incited researchers to challenge the assumptions each of them made about how they operationalized analytic constructs such as social positioning, idea development, and leadership. Approaches ranged from qualitative descriptive analyses (Sawyer, Frey, & Brown, 2013) to quantitative social network analyses (Oshima, Matsuzawa, Oshima, & Niihara, 2013), passing through coding and counting schemes (Howley, Mayfield, Rosé, & Strijbos, 2013) that combined both quantitative and qualitative aspects. The confrontation and comparison of these approaches allowed all the researchers to address from different angles the articulation of complex reasoning, social interconnectedness, and hierarchy, thus broadening their collective comprehension of these analytic constructs.

Epistemic agency in collaborative learning. One way a method of doing research spreads is if a member of the initial group introduces the method to new colleagues and applies it to new data. Oshima, an original member of the multivocality project, billed some new work of his involving the study of epistemic agency in collaborative learning as multivocal (Oshima, Oshima, & Fujita, 2015). They first used social network analysis to identify pivotal moments of discourse where it looked like students could engage in an epistemic action in order to search for missing knowledge. Then, they performed qualitative analyses of the discourse surrounding the pivotal moments in order to understand how knowledge was pursued. This, however, illustrates how multivocality differs from a mixed methods approach. They used different approaches, but these were applied successively toward the common goal of analyzing shared epistemic agency in collaborative learning and to test its effectiveness. All of the researchers involved worked towards this goal by putting together a set of methods that set out a path toward reaching it. In the chemistry example above, collective analysis benefited from a productive tension between multiple voices that had different orientations in dealing with social positioning, idea development, and leadership, even if they had a shared pre-theoretical analytic objective.

Visual analytics for teachers' dynamic diagnostic pedagogical decision-making. Whereas the productive multivocality project focused mainly on collaboration between researchers (see Law & Laferrière, 2013 for a critique), Vatrapu, Teplovs, Fujita, and Bull (2011) extended the stakeholders involved in multivocality to include teachers, design-based researchers, and visual analytics experts to explore the roles of these different stakeholders. Whereas above we wrote of different voices in a multivocal analysis involving different researcher perspectives, here the concept of "voices" is extended to include the perspective of those *designing* the affordances of visual analytic tools to support how teachers do diagnosis and make pedagogical decisions in the classroom.

Kristine Lund and Daniel Suthers

The Future of Multivocal Analysis

The Learning Sciences is a field ripe for supporting multivocal analyses of human interaction and learning. It is beneficial for the study of a particular phenomenon to examine the phenomenon through the theoretical and methodological lenses of different disciplines. However, achieving this takes experience, knowledge of neighboring disciplines that focus on different aspects of the same phenomenon, and a willingness to engage in the difficult conceptual work that involves comparing the foundations of different traditions and finding where it would be productive to use the differences to better understand the phenomenon under study.

We propose that this reflection be undertaken collectively in our community and supported as an explicit goal within the Learning Sciences. The leveraging of multiple perspectives in the analysis of human interaction will help us give our work broader coherence and make our field stronger. Readers wishing to embark on multivocal analysis will in the first instance need collaborators from different disciplines. They are then advised to consult the references indicated in the Further Readings.

Looking to the future, it would be beneficial to extend multivocal analysis to include other research traditions not represented in the productive multivocality project, and to settings, questions, and consequently data types beyond small group interaction. These might include, for example, analyses of workplace practices, learning analytics in online learning settings, and informal learning in social media settings. Vatrapu et al. (2011) have already shown how multivocality can be extended to stakeholders other than researchers; strategies for doing this can be borrowed from other communities (e.g., action research, participatory design—cf. Gomez, Kyza, & Mancevice, this volume), and developed further through multiple case studies. The greatest challenge for multivocal analysis (and indeed interdisciplinarity) may be incentive structures. Natural tendencies to focus on and promote one's own approach are reinforced by academic promotion and funding practices. Incentives for multi- and interdisciplinary work need to be promoted, through the policies of both professional societies and government agencies.

For the Learning Sciences to move toward greater coherence as a field, there needs to be empirically grounded dialogue between methodological and disciplinary traditions. These methods and disciplines may differ in how they construe and approach the object of study, and thus dialogue requires strategies for revealing both commonalities and differences. Although individuals may have productive careers without grappling with these problems, a collective program of productive multivocality is essential to the success of the Learning Sciences as a whole. We hope that the strategies outlined in this chapter provide a useful guide for moving toward greater interdisciplinarity through multivocal efforts.

Further Readings

Lund, K., Rosé, C. P., Suthers, D. D., & Baker, M. (2013). Epistemological encounters in multivocal settings. In D. D. Suthers, K. Lund, C. P. Rosé, C. Teplovs, & N. Law (Eds.), *Productive multivocality in the analysis of group interactions* (pp. 659–682). New York: Springer.
This chapter examines different ways in which epistemological engagement can be achieved. They do not all imply agreeing on how to do research.

Oshima, J., Oshima, R., & Fujita, W. (2015). A multivocality approach to epistemic agency in collaborative learning. In O. Lindwall, P. Häkkinen, T. Koschmann, P. Tchounikine, & S. Ludvigsen (Eds.), *Proceedings of the 11th International Conference on Computer Supported Collaborative Learning (CSCL '2015), Exploring the Material Conditions of Learning* [Vol. 1, pp. 62–69]. June 7–11, Gothenburg, Sweden: International Society of the Learning Sciences.
This article illustrates how a researcher who has experienced multivocal analysis becomes open to new methodologies and is able to incorporate them into a mixed methods approach.

Rosé, C. P., & Lund, K. (2013). Methodological pathways for avoiding pitfalls in multivocality. In D. D. Suthers, K. Lund, C. P. Rosé, C. Teplovs & N. Law (Eds.), *Productive multivocality in the analysis of group interactions* (pp. 613–637). New York: Springer.
This chapter gives advice on avoiding pitfalls during constitution of research teams, when teams give presentations, and when data gets transferred and shared.

Suthers, D. D., Lund, K., Rosé, C. P., & Teplovs, C. (2013). Achieving productive multivocality in the analysis of group interactions. In D. D. Suthers, K. Lund, C. P. Rosé, C. Teplovs, & N. Law (Eds.), *Productive multivocality in the analysis of group interactions* (pp. 577–612). New York: Springer.

This chapter contains a summary of the project and strategies, and is freely available on the web by agreement with the publisher. This chapter is an excellent starting point for further exploration of the multivocal approach and project.

Suthers, D. D., Lund, K., Rosé, C. P., Teplovs, C., & Law, N. (Eds.). (2013). *Productive multivocality in the analysis of group interactions*. New York: Springer.

This is the book that presents the Productive Multivocality Project. It contains three introductory chapters; five sections, each containing a detailed case study on a different dataset and involving different combinations of researchers and traditions; and a collection of summary and retrospective chapters.

NAPLeS Resources

Lund, K., Rosé, C. P., & Suthers, D. D., *Multivocality in analysing interaction* [Webinar]. In *NAPLeS video veries*. Retrieved October 19, 2017, from www.psy.lmu.de/isls-naples//intro/all-webinars/lund-rose-suthers/index.html

Note

1 "It proves convenient—though partially arbitrary—to describe behavior from two different standpoints, which lead to results which shade into one another. The etic viewpoint studies behavior as from outside of a particular system, and as an essential initial approach to an alien system. The emic viewpoint results from studying behavior as from inside the system" (Pike, 1967, p. 37).

References

Armbruster P., Patel, M., Johnson, E., & Weiss, M. (2009). Active learning and student-centered pedagogy improve student attitudes and performance in introductory biology. *CBE Life Sciences Education, 8*, 203–213.

Burke, J., Onwuegbuzie, A. J., & Turner, L. A. (2007). Toward a definition of mixed methods research. *Journal of Mixed Methods Research, 1*(2), 112–133.

Derry, S. J., Schunn, C. D., Gernsbacher, M. A. (2005). (Eds). *Interdisciplinary collaboration: An emerging cognitive science*. Mahwah, NJ: Lawrence Erlbaum.

Design-Based Research Collective. (2003). Design-based research: An emerging paradigm for educational inquiry. *Educational Researcher, 32*(1), 5–8.

Dingyloudi, F., & Strijbos, J. W. (2018). Mixed methods research as a pragmatic toolkit: Understanding versus fixing complexity in the Learning Sciences. In F. Fischer, C. E. Hmelo-Silver, S. R. Goldman, & P. Reimann (Eds.), *International handbook of the learning sciences* (pp. 444–454). New York: Routledge.

Dyke, G., Lund, K., & Girardot, J.-J. (2009). Tatiana: An environment to support the CSCL analysis process. In C. O'Malley, P. Reimann, D. Suthers, & A. Dimitracopoulou (Eds.), *Computer Supported Collaborative Learning Practices: CSCL 2009 Conference Proceedings* (pp. 58–67). Rhodes, Greece: International Society of the Learning Sciences.

Fiore, S. (2008). Interdisciplinarity as teamwork: How the science of teams can inform team science. *Small Group Research, 39*(3), 251–277.

Gomez, K., Kyza, E. A., & Mancevice, N. (2018), Participatory design and the learning sciences. In F. Fischer, C. E. Hmelo-Silver, S. R. Goldman, & P. Reimann (Eds.), *International handbook of the learning sciences* (pp. 401–409). New York: Routledge.

Green, J. L., & Bridges, S. M. (2018). Interactional ethnography. In F. Fischer, C. E. Hmelo-Silver, S. R. Goldman, & P. Reimann (Eds.), *International handbook of the learning sciences* (pp. 475–488). New York: Routledge.

Heritage, J. (1984). *Garfinkel and ethnomethodology*. Cambridge, UK: Polity Press.

Howley, I., Mayfield, E., Rosé, C. P., & Strijbos, J. W. (2013). A multivocal process analysis of social positioning in study group interactions, In D. D. Suthers, K. Lund, C. P. Rosé, C. Teplovs, & N. Law (Eds.), *Productive multivocality in the analysis of group interactions* (Vol. 15, pp. 205–223). New York: Springer.

Klein, J. T. (1990). *Interdisciplinarity: History, theory, and practice*. Detroit, MI: Wayne State University.

Koschmann, T. (2011). *Theories of learning and studies of instructional practice*. New York: Springer.

Koschmann, T. (2018). Ethnomethodology: Studying the practical achievement of intersubjectivity. In F. Fischer, C. E. Hmelo-Silver, S. R. Goldman, & P. Reimann (Eds.), *International handbook of the learning sciences* (pp. 465–474). New York: Routledge.

Koschmann, T., & Zemel, A. (2011). "So that's the ureter". The informal logic of discovering work. *Ethnographic Studies, 12*, 31–46.

Kuhn, T. S. (1962/1970). *The structure of scientific revolutions*. Chicago, IL: University of Chicago Press.

Lave, J. & Wenger, E. (1991). *Situated learning: Legitimate peripheral participation*. Cambridge, UK: Cambridge University Press.

Law, N., & Laferrière, T. (2013). Multivocality in interaction analysis: Implications for practice. In D. D. Suthers, K. Lund, C. P., Rosé, C. Teplovs, & N. Law (Eds.), *Productive multivocality in the analysis of group interactions*. (Vol. *15*, pp. 683–699). New York: Springer.

Lewin, K. (1931). The conflict between Aristotelian and Galileian modes of thought in contemporary psychology. *Journal of General Psychology, 5*, 141–177.

Lund, K. (2016). Modeling the individual within the group: An interdisciplinary approach to collaborative knowledge construction. *Habilitation à diriger des recherches*. Grenoble, France: Université Grenoble Alpes.

Oshima, J., Matsuzawa, Y., Oshima, R. & Niihara, Y. (2013). Application of social network analysis to collaborative problem solving discourse: An attempt to capture dynamics of collective knowledge advancement. In D. D. Suthers, K. Lund, C. P. Rosé, C. Teplovs, & N. Law (Eds.), *Productive multivocality in the analysis of group interactions* (Vol. *15*, pp. 225–242). New York: Springer.

Oshima, J., Oshima, R., & Fujita, W. (2015). A multivocality approach to epistemic agency in collaborative learning. In (Eds.). O. Lindwall, P. Häkkinen, T. Koschmann, P. Tchounikine, S. Ludvigsen, *Proceedings of the 11th International Conference on Computer Supported Collaborative Learning (CSCL '2015), Exploring the Material Conditions of Learning* (Vol. *1*, pp. 62–69). June 7–11, Gothenburg, Sweden: International Society of the Learning Sciences.

Pike, K. (1967). *Language in relation to a unified theory of the structure of human behavior*. The Hague, Netherlands: Mouton.

Puntambekar, S. (2018). Design-based research (DBR). In F. Fischer, C. E. Hmelo-Silver, S. R. Goldman, & P. Reimann (Eds.), *International handbook of the learning sciences* (pp. 383–392). New York: Routledge.

Rosé, C. (2013). A multivocal analysis of the emergence of leadership in chemistry study groups. In D. D. Suthers, K. Lund, C. P. Rosé, C. Teplovs, & N. Law (Eds.), *Productive multivocality in the analysis of group interactions* (Vol. *15*, pp. 243–256). New York: Springer.

Sawyer, K., Frey, R., & Brown, P. (2013) Knowledge building discourse in Peer-Led Team Learning (PLTL) groups in first-year General Chemistry. In D. D. Suthers, K. Lund, C. P. Rose, C. Teplovs, & N. Law (Eds.) *Productive multivocality in the analysis of group interactions* (Vol. *15*, pp. 191–204). New York: Springer.

Schegloff, E. A. (1993). Reflections on quantification in the study of conversation. *Research on Language and Social Interaction, 26*(1), 99–128.

Slavin, R. E. (1990). *Cooperative learning: Theory, research, and practice*. Englewood Cliffs, NJ: Prentice-Hall.

Suthers, D. D. (2006). Technology affordances for intersubjective meaning-making: A research agenda for CSCL. *International Journal of Computer Supported Collaborative Learning, 1*(3), 315–337.

Suthers, D. D., Lund, K., Rosé, C. P., & Teplovs, C. (2013). Achieving productive multivocality in the analysis of group interactions. In D. D. Suthers, K. Lund, C. P. Rosé, C. Teplovs, & N. Law (Eds.), *Productive multivocality in the analysis of group interactions* (pp. 577–612). New York: Springer.

Suthers, D. D., Lund, K., Rosé, C. P., Teplovs, C., & Law, N. (2013). *Productive multivocality in the analysis of group interactions*. New York: Springer.

Toulmin, S. (1972). *Human understanding*, Vol. *1*, *The collective use and development of concepts*. Princeton, NJ: Princeton University Press.

Valls, R., & Kirikiades, L. (2013). The power of Interactive Groups: how diversity of adults volunteering in classroom groups can promote inclusion and success for children of vulnerable minority ethnic populations. *Cambridge Journal of Education, 43*(1), 17–33.

van den Besselaar, P., & Heimeriks, G. (2001). Disciplinary, multidisciplinary, interdisciplinary: Concepts and indicators. In M. Davis & C. S. Wilson (Eds.), *ISSI 2001, 8th International Conference of the Society for Scientometrics and Informetrics* (pp. 705–716). Sydney: UNSW.

Vatrapu, R., Teplovs, C., Fujita, N., & Bull, S. (2011). Towards visual analytics for teachers' dynamic diagnostic pedagogical decision-making. *Proceedings of the First International Conference on Learning Analytics & Knowledge* (pp. 93–98). New York: ACM Press.

Vogel, F., & Weinberger, A. (2018). Quantifying qualities of collaborative learning processes. In F. Fischer, C. E. Hmelo-Silver, S. R. Goldman, & P. Reimann (Eds.), *International handbook of the learning sciences* (pp. 500–510). New York: Routledge.

Vygotsky, L. S. (1978). *Mind in society: The development of higher psychological processes*. Cambridge, MA: Harvard University Press.

Wenger, E. C., McDermott, R., & Snyder, W. C. (2002). *Cultivating communities of practice: A guide to managing knowledge*. Cambridge, MA: Harvard Business School Press.

Yanchar, S. C., & Williams, D. D. (2006). Reconsidering the compatibility thesis and eclecticism: Five proposed guidelines for method use. *Educational Researcher, 35*(9), 3–12.

45

Ethnomethodology
Studying the Practical Achievement of Intersubjectivity

Timothy Koschmann

We are, each and every one of us, engaged in a never-ending project of making sense of the world around us. And, because we share this world with others, we are obliged to coordinate our understandings with theirs. The current chapter explores how the practical methods through which this coordination is done might be subjected to study. Take, for example, the following fragment in which we find an educator, designated "RL," conversing with two students, Wally and Jewel. The focus of the discussion is how they might represent some data on a sheet of graph paper, and Wally offers the suggestion they make "a stem-and-leaf graph."[1]

```
RL:     You don't agree with us Wally? What's your idea?
Wally:  We draw a stem-and-leaf graph?
RL:     You should what?
Wally:  We should draw a ⌈stem-and-leaf
Jewel:                  ⌊Okay draw a stem-and-leaf
```

Let us see how this might be related to the notion of intersubjectivity. We take intersubjectivity to be the degree to which interlocutors are able to understand, in congruent ways, the matters about which they are interacting. It was once held uncontroversially that there is a one-to-one correspondence between words and meanings and, on this view, maintenance of intersubjectivity is not an issue. But the correspondence theory of meaning has fallen into disrepute in contemporary philosophy of language. Wittgenstein (1958), for instance, denied that there can be any such correspondence and argued: "§43. For a *large* class of cases—though not for all—in which we employ the word 'meaning' it can be defined thus: the meaning of a word is its use in the language." But if the meanings of words and expressions are established in use and can vary from situation to situation, how then is intersubjectivity ever possible?

Though philosophers of language continue to wrestle with such questions, the American sociologist Harold Garfinkel took as a given that we can and do achieve intersubjectivity, but considered the question of just how this is done to be a critical one and one that could be investigated empirically. As Heritage (1984) explained:

> Instead of beginning from the assumption that the terms of a language invoke a fixed domain of substantive content and that their intelligibility and meaning rest upon a shared agreement between speakers as to what this content consists of, Garfinkel proposed an alternative *procedural*

version of how description works. In this alternative version, he argues that the intelligibility of what is said rests upon the hearer's ability to make out what is meant from what is said according to *methods* which are tacitly relied on by both speaker and hearer.

(p. 144, original author's emphasis)

Garfinkel's procedural version of how description works can be applied more generally to characterize how meaningful action is produced. In the same way that speakers and hearers have methods that enable them to coordinate the mutual intelligibility of their talk, Garfinkel argued that "the activities whereby members produce and manage settings of organized everyday affairs are identical with members' procedures for making those settings 'account-able'" (Garfinkel, 1967, p. 1). It might seem that Garfinkel was merely substituting *account-able* for the word *meaningful*, but this would miss the point. He was, in fact, advancing a radically different model of meaning production.

The notion of 'accountability' is a foundational one for Ethnomethodology (EM), the school of sociological inquiry that Garfinkel founded (Garfinkel, 1967, Heritage, 1984; Livingston, 1987). In brief, we are accountable to each other to conduct ourselves in coherent ways. This conforms with the way the term is conventionally used. But it has another sense, as well; conduct is account-able to the extent that it stands as an account of what it is. EM's notion of accountability draws on both senses. The "settings of everyday affairs" of which Garfinkel wrote are concrete situations in which we are accountable (in the conventional sense) to conduct ourselves in ways that are sensible to others. At the same time, we collaboratively manage these settings by designing our conduct in such a way that it serves as an account of what it is that we are doing together. Garfinkel (1967) wrote: "The appropriate image of a common understanding is therefore an operation rather than a common intersection of overlapping sets" (p. 30). As he posited, the methods we draw upon to recognize what others are attempting to do are the same methods we employ in crafting our own actions to make them recognizable to others. EM devotes itself to the study of such methods.

Returning to the fragment at the beginning of the chapter, we observe the three participants working together to make sense of what they might be talking about. But what can we say about the methods they are using to do so? RL's "You don't agree with us Wally?" is delivered with questioning intonation. It would seem to assign next-speakership to Wally, but before Wally can respond, RL displaces the initial query by supplying another. But, the first query provides context for interpreting the second and the two together make relevant some sort of response on Wally's part. Not just any response, but one fitted to RL's compound query. Wally's response not only displays a certain understanding of the task in which they are engaged, but also his understanding of the action performed by RL's prior turn.

When intersubjectivity falters, we have procedures for accomplishing its "repair" (Schegloff, 1991; Schegloff, Jefferson, & Sacks, 1977) and we see RL drawing on one such procedure in his next turn when he seeks clarification of Wally's prior response. As before, note how Wally demonstrates his understanding of the repair initiated by RL. RL had targeted something and Wally displays a candidate understanding of what it might be by recycling RL's "You should. . ." construction in his response, in effect matching "draw a stem-and-leaf" with the interrogative pronoun "what." Though Wally has still not elaborated on what he is taking a "stem-and-leaf graph" to be, he has provided enough to get a go-ahead from Jewel. Presumably, whatever it is or could be will be worked out in good time. In Garfinkel's (1967) words, Wally's expression exhibits "retrospective and prospective possibilities" (p. 41).

The example illustrates some of the rudiments of how one might begin to conduct a study into how meaning-in-use is accomplished in a concrete case. Though not developed very far, the analysis is sufficient to show some features of the participants' exchange that might not have been apparent at first blush. The value of an analytic account, then, lies in what it brings us to see. Unlike a conventional educational research report, a report summarizing such a study does not require a separate methods section. It is not that such analytic accounts lack method, but rather that the methodology is built into the analytic account itself—when we read it and come to recognize what it is holding up to our scrutiny, we have effectively replicated the study.

The issues related to establishing local intersubjectivity have deep significance for all social interaction, including that carried out in a scholarly register. As an example, consider this proposal from a recently published National Academy of Science (NAS) report in the United States: "When the goal is to prepare students to be able to be successful in solving new problems and adapting to new situations, then *deeper learning* is called for" (Pellegrino & Hilton, 2012, p. 70, emphasis added). Other examples from the Learning Sciences literature might include "learning for understanding" (Perkins & Unger, 1999), "active learning" (Johnson, Johnson, & Smith, 1998), "collaborative learning" (Dillenbourg, 1999), and "constructivism" (Jonassen, 1999). Though all of these recommendations seem immanently reasonable, how would we begin to advise a teacher with respect to how to actually *do* deeper learning, etc. on a turn-by-turn basis in the classroom? It is not a simple matter of requiring greater specification, but one of finding meaning-in-use in a practical setting, and this is a different problem entirely. So, in our conversations with practitioners or, for that matter, even with our colleagues, we find ourselves grappling with the challenges of working out together just what we are talking about in a way that is reminiscent of the participants in the 'stem-and-leaf' fragment. There is no recess from the work of intersubjectivity.

Four Approaches to Studying the Practical Achievement of Intersubjectivity

EM supplies a theoretical foundation upon which an empirical program could be built, but it is not, as the name might suggest, a research methodology per se. It poses an important and research-able question, but we must turn elsewhere for the analytic tools to pursue it. *Conversation Analysis* (CA), for example, offers both methods and past findings, which can be employed in studying the practical achievement of intersubjectivity. Efforts to extend CA beyond talk-in-interaction to include embodied aspects of communication are sometimes taken up under the banner of *multimodal interaction* or *multimodal CA* (Mondada, 2011). A third approach, known as *Context Analysis*, has its disciplinary roots not in sociology, but in anthropology. Like multimodal CA, it examines embodied aspects of sense-making and local accountability. Much work in LS has been pursued under the title *Interaction Analysis* (Hall & Stevens, 2016). Interaction Analysis (IA) has connections to all of the analytic traditions mentioned so far—CA, multimodal CA, and Context Analysis—but differs in having a specific orientation to issues of learning and instruction. In this section, we explore the historical and theoretical underpinnings of these four analytic traditions and present some concrete examples of each. The practical details of how one might go about employing these approaches is beyond the scope of this chapter, but the interested reader can learn more by consulting the references provided.

Conversation Analysis

Pioneering work in CA began in the late 1960s. In the intervening half-century, a substantial body of findings has been developed.[2] CA was founded to elucidate the "machinery" (Sacks, 1992, p. 113) that makes everyday conversation possible. It utilizes transcription conventions (Jefferson, 2004) that capture features of the talk—timing, intonation, stress, tempo—that contribute in important ways to sense-making. CA-based studies in instructional settings have pursued a variety of questions. Here, three of the prominent concerns are considered: describing the social organization of the classroom, formulating questioning strategies in instruction, and the role of correction in classroom interaction.

The social organization of classrooms. Macbeth's (2000) chapter on classrooms as "installations" is as good a place as any to begin our exploration of CA-based inquiry into the social organization of classrooms. He argues that classrooms are "social technologies for the production of competence, fluency and knowing action" (p. 23). In an earlier article, Macbeth (1991) documented classroom teachers' practices for asserting authority. Turn-taking is foundational, not only to classroom control, but also to establishing the institutional character of the classroom. In a seminal CA report, Sacks, Schegloff, and Jefferson (1974) described an algorithm (the "simplest systematics") whereby a listener analyzes a turn-in-progress in order to determine the point at which a transition to a new speaker might be relevant. McHoul (1978) sought to give an account of how this algorithm is modified in the classroom and becomes a field for enacting teacher control. Heap (1993), building on McHoul's proposal, documented how the turn-taking system in the classroom can be ambiguous for students.

Formulating questioning strategies. Questioning strategies are crucially important to classroom interaction both because teacher questions are the primordial instrument of assessment and because it is through assessment that what counts as knowledge is established. Mehan (1979) offered this example:

```
Speaker A:    What time is it, Denise?
Speaker B:    2:30
Speaker A:    Very good, Denise
```

We have no trouble recognizing which speaker is the teacher in this exchange. The pedagogical character of the exchange becomes apparent in the third turn, where we discover that the leading query was not information seeking, but rather has an assessment motive. Mehan described such questions as "known information questions." He christened the three-part structure it launches the "initiation-reply-evaluation" (IRE) sequence. It is a staple of classroom recitation and there is a substantial body of CA-based work devoted to its study (e.g., Heap, 1988; Hellermann, 2005; Lee, 2007; Macbeth, 2004; Waring, 2009). One focus of interest has been in exploring alternative arrangements that can be enacted in the third turn (e.g., Zemel & Koschmann, 2011).

Correction and repair. We already briefly touched on repair mechanisms in the context of the 'stem-and-leaf' fragment. Repair and correction were taken up in a second seminal paper in the CA canon, that of Schegloff et al. (1977).[3] Schegloff (1991) was to later describe repair as "the last systematically provided opportunity to catch (among other problems) divergent understandings that embody breakdowns of intersubjectivity, that is, trouble in social shared cognition of the talk and conduct in the interaction" (p. 158). When teachers correct, it is an instance of what Schegloff et al. termed "other-initiated" (p. 365) repair. A considerable amount of attention has been directed to understanding how this gets done. McHoul (1990) sought to show how repair is coordinated with IRE sequences. Macbeth (2004) argued against this conceptualization, positing that the organizations of classroom correction and conversational repair are fundamentally distinct. Correction in the classroom need not always be overt and can sometimes be performed in an "embedded" (Jefferson, 1982) fashion. "Re-voicing" (O'Connor & Michaels, 1993) is one mechanism whereby this can be accomplished. It might also be noted in passing that correction sequences are not always initiated by teachers (Koschmann, 2016).

Before leaving the topic of CA-based approaches to study the practical achievement of intersubjectivity, it seems useful to mention a related area of work that also involves a different sort of interaction. Computer-mediated communication is an important part of computer-supported collaborative learning (CSCL) research and there has been some work examining online chat-based interaction in a CA-informed way (e.g., Zemel & Koschmann, 2103; Stahl, 2009). Though online chat-based interaction does not follow the same turn-taking rules as talk-in-interaction (cf. Garcia & Jacobs, 1999), it is orderly and it serves in its own right as a site for interactional sense-making.

Multimodal CA

The linguistic anthropologist, Charles Goodwin, published an early study that argued that speaker's and recipient's gaze play consequential roles in the production of a turn at talk (Goodwin, 1980). A later report, Goodwin and Goodwin (1986), documented the critical role of pointing gestures in the achievement of mutual orientation. These early efforts to extend the analytic interests of CA beyond talk into the complexities of bodily and contextual relevancies have been carried in a variety of directions by other researchers.

As Streeck (2009) has described, we utilize our hands in a number of ways in sense-making. First and most straightforwardly, we use our hands themselves in "making sense of the world at hand" (p. 8). Examples include Kreplak and Mondemé's (2014) account of a group of blind visitors participating in a guided tour of an art museum and Lindwall and Eckström's (2012) description of learning to knit. We can also employ our hands to point out or demonstrate features of the material environment for others; what Streeck describes as "disclosing the world in sight" (2009, p. 8). Descriptions of such demonstrations can be found in Goodwin's (2003) account of gesture use at an archeological dig and Goodwin and Goodwin's (1997) report of instruction provided to the jury in interpreting video presented at the Rodney King trial.

Koschmann and LeBaron (2002) studied how learners use their hands and bodies in putting together explanations. This had been expanded upon in later work (e.g., Arnold, 2012; Ivarsson, 2017). Such gestures, what McNeill (1992) described as "iconics" (p. 12), are more elaborate than the simple points described earlier. McNeill provided examples of iconic gestures being employed by mathematicians within their interaction. Much recent work has focused on gesture production within mathematics instruction and the potential for it to facilitate, foster, and signal emergent understanding (e.g., Abrahmson, 2009; Alibali & Nathan, 2012, this volume).

Inscription might be thought of as a persistent form of gesture, as visual signs available for future reference. Such representational practices have received considerable attention in the literature on practical sense-making. Goodwin's (2003) description of pointing at the archeological worksite mentioned earlier, for example, included an instance in which pointing with a trowel resulted in the production of traced lines in the soil. Greiffenhagen's (2014) description of the presentation of a mathematical proof at the blackboard offers another instructive example of the use of inscription in sense-making. Other authors (e.g., Koschmann & Mori, 2014; Lindwall & Lymer, 2008; Murphy, 2005; Roschelle, 1992) provide examples of learners making sense of *provided* representations. Yet others (e.g., Macbeth, 2011a; Moschkovich, 2008; Zemel & Koschmann, 2013) examine situations in which learners build sense through their own representational constructions.

To conclude the discussion of multimodal approaches to studying interaction, we return to the topic that Streeck (2009) described as "making sense of the world at hand" (2009, p. 8), but this time focusing not on the manual action, but rather on the things to be found there. Sense-making with objects is central to analyses focusing on the use of manipulatives in mathematics education (e.g., Koschmann & Derry, 2016; Roth & Gardner, 2012). Finally, Lynch and Macbeth (1998) provide an account of the production of physics demonstrations in an elementary classroom. These represent foundational inquiries into how we build a world held in common.

Context Analysis

Whereas multimodal CA seeks to incorporate bodily and environmental features into analyses of talk-in-interaction, Context Analysis treats the embodied aspects of interaction as primary.

McDermott (1976) succinctly observed, "people constitute environments for each other" (p. 283). He explained:

> What is going on in any given situation is available to the analyst in the participants' behavior as they formulate, orient to and hold each other accountable to the order of their behavior. In this way they establish a group state of readiness for acting upon whatever happens next. Every next event occurs and is responded to within the context of what the members of the group are doing to organize each other.
>
> *(p. 24)*

It was Scheflen (1973) who first theorized that this interactional work is systematically organized and studiable. His approach was heavily influenced by Birdwhistell's (1970) earlier research on "kinesics." Kendon (1990), a prominent contemporary practitioner, reported that, in its development, Context Analysis was strongly influenced by the work of the American sociologist, Erving Goffman. Goffman's (1983) proposal to initiate studies of the "interaction order" had a profound influence on the development of CA as well (Schegloff, 1992), and, as we will see, IA. McDermott (1976) examined the production of a reading lesson in an elementary school classroom. His report serves as an accessible tutorial on how this approach might be employed in an instructional setting.

Interaction Analysis

The label for this final approach comes from a frequently cited article by Jordan and Henderson (1995). The article served to lay out the foundational principles and methodology for a descriptive, video-analytic approach designed "to identify regularities in the ways in which participants utilize the resources of the complex social and material world of actors and objects within which they operate" (p. 41). Jordan and Henderson took a particular interest in "learning as a distributed, ongoing social process, in which evidence that learning is occurring or has occurred must be found in understanding the ways in which people collaboratively do learning and recognize learning as having occurred" (p. 42). They labelled this approach Interaction Analysis (IA).[4] IA is less an alternative to the three approaches described previously than an amalgamation of them. It employs the transcription conventions of CA and, like CA, it is an EM-informed approach. IA, however, focuses on phenomena, including learning, that go beyond the structural features of talk-in-interaction. As with the case of multimodal CA, IA treats material artifacts and embodied conduct as obligatory aspects of the analysis. Jordan and Henderson list "participant frameworks" (pp. 67–69) as an important focus of IA. These they describe as "fluid structures of mutual engagement and disengagement characterized by bodily alignment (usually face-to-face), patterned eye-contact, situation-appropriate tone of voice, and other resources the situation may afford" (p. 67). The notion of 'participation frameworks' came originally from Goffman (1981) and, in the way in which it is employed in IA, suggests an immediate affinity with the theoretical framing of Context Analysis.

Empirical studies into how participants "do learning and do recognize learning as having occurred" (p. 42) are both diverse and extensive. Melander and Sahlström (2009), in an approach quite consistent with the stated goals of IA, sought to theorize learning in terms of changes in participation structures. In an example drawn from jazz performance, Klemp et al. (2016) sought to document how a "mis-take" can be conceptualized as a kind of learning. Roschelle (1992), in a classic CSCL study, documented how two students collaboratively achieved a practical understanding of the notion of acceleration while conducting experiments in a computer-based simulation environment. Koschmann and Zemel (2009) and Zemel and Koschmann (2013) explored how students talk about some matter being discovered, while they are still in the process of discovering it. Koschmann and Derry (2016), addressing the issue of transfer of training, described how past learning can be made instructably relevant for present needs and purposes. Finally, Macbeth (2011b) discusses the forms of competence and understanding that are antecedent to and that undergird all claims of learning and instruction.

The four analytic traditions summarized here—CA, multimodal CA, Context Analysis, and IA—are designed to produce descriptive analytic accounts of the sort discussed earlier. Such accounts are inextricably tied to particular occasions. In some circles, descriptions of single cases are treated as, at best, weak evidence and, more often, unscientific. Such a view, however, ignores the fact that meaning-and-use and intersubjectivity are essentially situated matters that cannot be studied isolated from the settings within which they were produced. To try to examine how intersubjectivity is achieved in the abstract, then, is to risk "the loss of the phenomenon" (Garfinkel, 2002, p. 253). Given that there can be no instruction without intersubjectivity, this is a loss we can ill afford to accept.

Further Readings

Garfinkel, H. (1967). *Studies in ethnomethodology*. Englewood Cliffs, NJ: Prentice-Hall.
Garfinkel, H. (2002). *Ethnomethodology's program: Working out Durkheim's aphorism*. Lanham, MD: Rowman & Littlefield.
Turning to the primary literature, I would direct the interested reader to the two major works that bookended Garfinkel's career, *Studies in Ethnomethodology* and *Ethnomethodology's Program*. Both represent attempts on his part to clarify the nature of the program: one written to announce its launch; the other, published near the end of his career, to summarize its contributions.

Heritage, J. (1984). *Garfinkel and ethnomethodology*. Cambridge, UK: Polity Press.
Livingston, E. (1987). *Making sense of ethnomethodology*. London: Routledge & Kegan Paul.
For those seeking a more thorough introduction to EM, these two books are both excellent guides.

Macbeth, D. (2011a). A commentary on incommensurate programs. In T. Koschmann (Ed.), *Theories of learning and studies of instructional practice* (pp. 73–103). New York: Springer Science.
As an example of an EM-informed, empirical study, I recommend Macbeth. It draws on the dataset from which the "stem-and-leaf" fragment came and illustrates two important features of descriptive research. First, it is exemplary in the way in which it invites the reader to look a little more deeply into the exchanges that are described there. At the same time, in revisiting previously analyzed data, it demonstrates how the work of careful analysis never ends.

NAPLeS Resources

Koschmann, T., *Conversation and interaction analysis/ethnomethodological approaches* [Webinar]. In *NAPLeS video series*. Retrieved October 19, 2017, from http://isls-naples.psy.lmu.de/intro/all-webinars/koschmann_all/index.html

Koschmann, T., *15 minutes about conversation and interaction analysis/ethnomethodological approaches* [Video file]. In *NAPLeS video series*. Retrieved October 19, 2017, from http://isls-naples.psy.lmu.de/video-resources/guided-tour/15-minutes-koschmann/index.html

Koschmann, T., *Interview about conversation and interaction analysis/ethnomethodological approaches* [Video file]. In *NAPLeS video series*. Retrieved October 19, 2017, from http://isls-naples.psy.lmu.de/video-resources/interviews-ls/koschmann/index.html

Notes

1 The fragment comes from Excerpt 5 described in Lehrer and Schauble (2011).
2 See Pomerantz and Fehr (2011) or Lindwall, Lymer, and Greiffenhagen (2015) for accessible introductions.
3 Another topic of particular interest to the LS community that bears a relation to conversational repair is argumentation, in that informal arguments are frequently constructed using repair sequences. Though I will not take up argumentation in this chapter, interested readers are encouraged to seek out some of the carefully done studies on children's argumentation that can be found in the CA literature (e.g., M. Goodwin, 1990).
4 This is not to be confused with an earlier approach to studying classroom interaction of the same name (i.e., Flanders, 1970), which was based on coding procedures. Writing from a perspective informed by EM, Heap (1982) offers a critique of studying sense-making in this way.

References

Abrahmson, D. (2009). Embodied design: Constructing means for constructing meaning. *Educational Studies in Mathematics, 70*, 27–47. doi:10.1007/s10649-008-9137-1

Alibali, M., & Nathan, M. J. (2012). Embodiment in mathematics teaching and learning: Evidence from learners' and teachers' gestures. *Journal of the Learning Sciences, 21*, 247–286.

Alibali, M. W., & Nathan, M. (2018). Embodied cognition in learning and teaching: action, observation, and imagination. In F. Fischer, C. E. Hmelo-Silver, S. R. Goldman, & P. Reimann (Eds.), *International handbook of the learning sciences* (pp. 75–85). New York: Routledge.

Arnold, L. (2012). Dialogic embodied action: Using gesture to organize sequence and participation in instructional interaction. *Research on Language and Social Interaction, 45*, 269–296.

Birdwhistell, R. (1970). *Kinesics and context*. Philadelphia, PA: University of Pennsylvania Press.

Dillenbourg, P. (1999). What do you mean by "collaborative learning"? In P. Dillenbourg (Ed.), *Collaborative learning: Cognitive and computational approaches* (pp. 1–19). New York: Pergamon.

Flanders, N. A. (1970). *Analyzing teaching behavior*. Reading, MA: Addison-Wesley.

Garcia, A., & Jacobs, J. (1999). The eyes of the beholder: Understanding the turn-taking system in quasi-synchronous computer-mediated communication. *Research on Language and Social Interaction, 32*, 337–368.

Garfinkel, H. (1967). *Studies in ethnomethodology*. Englewood Cliffs, NJ: Prentice-Hall.

Garfinkel, H. (2002). *Ethnomethodology's program: Working out Durkheim's aphorism*. Lanham, MD: Rowman & Littlefield.

Goffman, E. (1981). *Forms of talk*. Philadelphia, PA: University of Pennsylvania Press.

Goffman, E. (1983). The interaction order. *American Sociological Review, 48*, 1–17.

Goodwin, C. (1980). Restarts, pauses, and the achievement of a state of mutual gaze at turn-beginning. *Sociological Inquiry, 50*, 272–302.

Goodwin, C. (2003). Pointing as situated practice. In S. Kita (Ed.), *Pointing: Where language, culture, and cognition meet* (pp. 217–242). Mahwah, NJ: Lawrence Erlbaum.

Goodwin, C., & Goodwin, M. H. (1997). Contested vision: The discursive constitution of Rodney King. In B.-L. Gunnarsson, P. Linell, & B. Nordberg (Eds.), *The construction of professional discourse* (pp. 292–316). New York: Longman.

Goodwin, M. H. (1990). *He-said-she-said: Talk as social organization among black children*. Bloomington, IN: Indiana University Press.

Goodwin, M. H., & Goodwin, C. (1986). Gesture and co-participation in the activity of searching for a word. *Semiotica, 62*, 51–75.

Greiffenhagen, C. (2014). The materiality of mathematics: Presenting mathematics at the blackboard. *British Journal of Sociology, 65*, 502–528.

Hall, R., & Stevens, R. (2016). Interaction analysis approaches to knowledge in use. In A. A. diSessa, M. Levin, & N. J. S. Brown (Eds.), *Knowledge and interaction: A synthetic agenda for the learning sciences* (pp. 72–108). New York: Routledge.

Heap, J. (1982). The social organization of reading assessment: Reasons for eclecticism. In G. Payne & E. C. Cuff (Eds.), *Doing teaching: The practical management of classrooms* (pp. 39–59). London: Batsford Academic and Educational.

Heap, J. (1988). On task in classroom discourse. *Linguistics and Education, 1*, 177–198.

Heap, J. (1993). Seeing snubs: An introduction to sequential analysis of classroom interaction. *Journal of Classroom Interaction, 27*, 23–28.

Hellermann, J. (2005). Syntactic and prosodic practices for cohesion in series of three-part sequences in classroom talk. *Research on Language and Social Interaction, 38*, 105–130.

Heritage, J. (1984). *Garfinkel and ethnomethodology*. Cambridge, UK: Polity Press.

Ivarsson, J. (2017). Visual practice as embodied practice. *Frontline Learning Research, 5*, 12–27.

Jefferson, G. (1982). On exposed and embedded correction in conversation. *Studium Linguistik, 14*, 58–68.

Jefferson, G. (2004). Glossary of transcript symbols with an introduction. In G. Lerner (Ed.), *Conversation analysis: Studies from the first generation* (pp. 13–31). Amsterdam, Netherlands: John Benjamins.

Johnson, D. W., Johnson, R. T., & Smith, K. A. (1998). *Active learning: Cooperation in the college classroom*. Edina, MN: Interaction Book Company.

Jonassen, D. (1999). Designing constructivist learning environments. In C. M. Reigeluth (Ed.), *Instructional-design theories and models* (Vol. 2, pp. 215–240). New York: Routledge.

Jordan, B., & Henderson, A. (1995). Interaction analysis: Foundations and practice. *Journal of the Learning Sciences, 4*, 39–104.

Kendon, A. (1990). *Conducting interaction: Patterns of behavior in focused encounters*. New York: Cambridge University Press.

Klemp, N., McDermott, R., Duque, J., Thibeault, M., Powell, K., & Levitin, D. J. (2016). Plans, takes and mis-takes. *Éducation & Didactique, 10*, 105–120.

Koschmann, T. (2016). "No! That's not what we were doing though." Student-initiated, other correction. *Éducation & Didactique, 10*, 39–48.

Koschmann, T., & Derry, S. (2016). "If green was A and blue was B": Isomorphism as an instructable matter. In R. Säljö, P. Linell, & Å. Mäkitalo (Eds.), *Memory practices and learning: Experiential, institutional, and sociocultural perspectives* (pp. 95–112). Charlotte, NC: Information Age Publishing.

Koschmann, T., & LeBaron, C. (2002). Learner articulation as interactional achievement: Studying the conversation of gesture. *Cognition & Instruction, 20*, 249–282.

Koschmann, T., & Mori, J. (2016). "It's understandable enough, right?" The natural accountability of a mathematics lesson. *Mind, Culture and Activity, 23*, 65–91. doi:10.1080/10749039.2015.1050734

Koschmann, T., & Zemel, A. (2009). Optical pulsars and black arrows: Discoveries as occasioned productions. *Journal of the Learning Sciences, 18*, 200–246. doi:10.1080/10508400902797966

Kreplak, Y., & Mondemé, C. (2014). Artworks as touchable objects: Guiding perception in a museum tour for blind people. In M. Nevile, P. Haddington, T. Heinemann, & M. Rauniomaa (Eds.), *Interacting with objects: Language, materiality, and social activity* (pp. 295–318). Amsterdam: John Benjamins.

Lee, Y.-A. (2007). Third turn position in teacher talk: Contingency and the work of teaching. *Journal of Pragmatics, 39*, 1204–1230.

Lehrer, R., & Schauble, L. (2011). Designing to support long-term growth and development. In T. Koschmann (Ed.), *Theories of learning and studies of instructional practice* (pp. 19–38). New York: Springer.

Lindwall, O., & Ekström, A. (2012). Instruction-in-interaction: The teaching and learning of a manual skill. *Human Studies, 35*, 27–49.

Lindwall, O., & Lymer, G. (2008). The dark matter of lab work: Illuminating the negotiation of disciplined perception in mechanics. *Journal of the Learning Sciences, 17*, 180–224.

Lindwall, O., Lymer, G., & Greiffenhagen, C. (2015). The sequential analysis of instruction. In N. Markee (Ed.), *The handbook of classroom discourse and interaction* (pp. 142–157). New York: John Wiley.

Livingston, E. (1987). *Making sense of ethnomethodology*. London: Routledge & Kegan Paul.

Lynch, M., & Macbeth, D. (1998). Demonstrating physics lessons. In J. G. Greeno & S. V. Goldman (Eds.), *Thinking practices in mathematics and science learning* (pp. 269–298). Mahwah, NJ: Lawrence Erlbaum.

Macbeth, D. (1991). Teacher authority as practical action. *Linguistics and Education, 3*, 281–313.

Macbeth, D. (2000). Classrooms as installations: Direct instruction in the early grades. In S. Hester & D. Francis (Eds.), *Local education order: Ethnomethodological studies of knowledge in action* (pp. 21–72). Philadelphia, PA: John Benjamins.

Macbeth, D. (2004). The relevance of repair for classroom correction. *Language and Society, 33*, 703–736.

Macbeth, D. (2011a). A commentary on incommensurate programs. In T. Koschmann (Ed.), *Theories of learning and studies of instructional practice* (pp. 73–103). New York: Springer Science.

Macbeth, D. (2011b). Understanding understanding as an instructional matter. *Journal of Pragmatics, 43*, 438–451.

McDermott, R. (1976). *Kids make sense: An ethnographic account of the interactional management of success and failure in one first-grade classroom* (Unpublished Ph.D. thesis). Stanford University, Palo Alto, CA.

McHoul, A. (1978). The organization of turns at formal talk in the classroom. *Language and Society, 7*, 183–213.

McHoul, A. (1990). The organization of repair in classroom talk. *Language and Society, 19*, 349–377.

McNeill, D. (1992). *Hand and mind: What gestures reveal about thought*. Chicago: University of Chicago Press.

Mehan, H. (1979). "What time is it, Denise?" Asking known information questions in classroom discourse. *Theory into Practice, 18*, 285–294.

Melander, H., & Sahlström, F. (2009). In tow of the blue whale: Learning as interactional changes in topical orientation. *Journal of Pragmatics, 41*, 1519–1537.

Mondada, L. (2011). Understanding as an embodied, situated and sequential achievement in interaction. *Journal of Pragmatics, 43*(2), 542–552. doi:10.1016/j.pragma.2010.08.019

Moschkovich, J. N. (2008). "I went by twos, he went by one": Multiple interpretations of inscriptions as resources for mathematical discussions. *Journal of the Learning Sciences, 17*, 551–587. doi:10.1080/10508400802395077

Murphy, K. M. (2005). Collaborative imagining: The interactive use of gestures, talk and graphic representation in architectural practice. *Semiotica, 156*, 113–145.

O'Connor, M. C., & Michaels, S. (1993). Aligning academic task and participation status through revoicing: Analysis of a classroom discourse strategy. *Anthropology and Education Quarterly, 24*, 318–335.

Pellegrino, J., & Hilton, M. L. (2012). *Education for life and work; Developing transferable knowledge and skills in the 21st century*. Washington, DC: National Academies Press.

Perkins, D. N., & Unger, C. (1999). Teaching and learning for understanding. In C. M. Reigeluth (Ed.), *Instructional-design theories and models* (Vol. 2, pp. 91–114). New York: Routledge.

Pomerantz, A., & Fehr, B. J. (2011). Conversation analysis: An approach to the analysis of social interaction. In T. A. van Dijk (Ed.), *Discourse Studies: A multidisciplinary introduction* (2nd ed., pp. 165–190). Thousand Oaks, CA: Sage.

Roschelle, J. (1992). Learning by collaboration: Convergent conceptual change. *Journal of the Learning Sciences, 2*, 235–276.

Roth, W.-M., & Gardner, R. (2012). "They're gonna explain to us what makes a cube a cube?" Geometrical properties as contingent achievement of sequentially ordered child-centered mathematics lessons. *Mathematics Education Research Journal, 24*, 323–346.

Sacks, H. (1992). *Lectures on conversation* (Vol. 1, G. Jefferson, Ed.). Oxford, UK: Blackwell.

Sacks, H., Schegloff, E., & Jefferson, G. (1974). The simplest systematics for the organization of turn-taking for conversation. *Language, 50*, 696–735.

Scheflen, A. E. (1973). *Communicational structure: Analysis of a psychotherapy session*. Bloomington, IN: Indiana University Press.

Schegloff, E. (1991). Conversation analysis and socially shared cognition. In L. Resnick, J. Levine, & S. Teasley (Eds.), *Perspectives on socially shared cognition* (pp. 150–171). Washington, DC: American Psychological Association.

Schegloff, E. (1992). Introduction. In G. Jefferson (Ed.), *Harvey Sacks: Lectures on conversation* (Vol. 1, pp. ix–lxii). Oxford, UK: Blackwell.

Schegloff, E., Jefferson, G., & Sacks, H. (1977). The preference for self-correction in the organization of repair in conversation. *Language, 53*, 361–382.

Stahl, G. (2009). Deictic referencing in VMT. In G. Stahl (Ed.), *Studying virtual math teams* (pp. 311–326). New York: Springer.

Streeck, J. (2009). *Gesturecraft: The manu-facture of meaning*. Amsterdam: John Benjamins.

Waring, H. Z. (2009). Moving out of IRF (Initiation–Response–Feedback): A single case analysis. *Language Learning, 59*, 796–824.

Wittgenstein, L. (1958). *Philosophical investigations* (2nd ed., G. E. M. Anscombe, Trans.). Malden, MA: Blackwell.

Zemel, A., & Koschmann, T. (2011). Pursuing a question: Reinitiating IRE sequences as a method of instruction. *Journal of Pragmatics, 43*, 475–488. doi:10.1016/j.pragma.2010.08.022

Zemel, A., & Koschmann, T. (2013). Recalibrating reference within a dual-space interaction environment. *Computer-Supported Collaborative Learning, 8*, 65–87. doi:10.1007/s11412-013-9164-5

46
Interactional Ethnography

Judith L. Green and Susan M. Bridges

Introduction

This chapter lays out the *governing principles of operation and conduct* (Heath, 1982) and the theoretical perspectives guiding the iterative, recursive, and abductive (IRA) logic and actions (Agar, 2006) that constitute interactional ethnography (IE) as a *logic-of-inquiry*. This logic-of-inquiry guides outsiders (ethnographers) as they seek to develop understandings of what insiders need to know, understand, produce, and predict as they learn with, and from, others in educational and social environments (Heath, 1982; Street, 1993). Specifically, an IE logic-of-inquiry supports researchers in exploring what is being constructed in and through micro-moments of *discourse-in-use*, *historical roots* of observed phenomena, and *macro-level actors and sources* that support and/or constrain opportunities for learning afforded to, constructed by, and taken up (or not) by participants in purposefully designed educational programs (e.g., Bridges, Botelho, Green, & Chau, 2012; Castanheira, Crawford, Dixon, & Green, 2000; Green, Skukauskaite, Dixon, & Cordóva, 2007). This goal, as we will demonstrate, complements the goals of learning scientists who seek to develop situated understandings of learning as a social and cognitive construction (Danish & Gresalfi, this volume) and a design-based research approach (Puntambekar, this volume).

To accomplish these goals, IE researchers seek to develop grounded understandings of learning as a socially constructed process, and how learning processes vary with actors (participants) and events being constructed. This approach also explores goals of participants (and institutional actors) as they propose and develop meanings, interactions, and activity across configurations of actors, times, and events. By drawing on Bridges et al.'s. (2012; Bridges, Green, Botelho, & Tsang, 2015) studies, in which we participated as internal (Bridges) and external (Green) ethnographers along with an inter-professional team of dental educators/researchers (Botelho, Chau, Tsang), we make visible how an IE epistemological approach guides research decisions within an ongoing study and program of research. By (re)constructing Bridges et al.'s (2012) IE *logic-in-use*, and situating the study in the ongoing program of research (see Figure 46.1), we provide a grounded illustration of the theoretical perspectives that guide *principles of conduct* undertaken by IE researchers as they engage in different levels of analysis to build theoretical inferences from particular studies. Through this process, we also demonstrate how IE meets the following *principles of operation* framed by Heath (1982) for ethnographic studies in education: *stepping back from ethnocentrism*, *bounding units of analysis*, and *making connections*.

By grounding the presentation of *principles of conduct* in the research program of Bridges and her colleagues, we construct a *telling case* (Mitchell, 1984) of IE as a logic-of-inquiry, and how observations, decisions, and actions of an IE research team lead to the construction of a developing *logic-in-use*. Through this telling case of IE as an epistemological approach, we make visible what constitutes an IE logic-of-inquiry, and how emerging questions within an IE program of research result from, and lead to, interdependent chains of *iterative, recursive, and abductive processes*. As we will demonstrate, these processes involve the ongoing construction of data sets from the *ethnographic archive* as an *ethnographic space* (Agar, 2006). This *telling case* of IE as a logic-of-inquiry, therefore, makes visible the ongoing and developing nature of IE studies and the purposefully developed role of archived records (e.g., video records, formal and ongoing interviews, documents brought to and constructed in developing events, artifacts and resources participants engage with, and physical/technological records from online learning environments) in an IE-grounded program of research.

Theoretical Roots of IE as a Logic of Inquiry

In this section, we focus on the *principles of conduct* that *orient* IE researchers as they seek to enter new and unknown contexts, or to build an IE base to their ongoing research program in developing and purposefully designing educational programs. These *orienting principles of conduct* were proposed by *ethnographers in education*, Shirley Brice Heath and Brian Street, individually (Heath, 1982; Street, 1993) and collectively (Heath & Street, 2008) and constitute a *chain of principles* and *implicated actions* that ethnographers *in* education (and other social settings) draw on as they seek to gain emic (insider) understandings of what constitutes *members' (insider) knowledge*:

- Suspending known categories to construct understandings of local and situated categories and referential meanings of actions being developed by participants;
- Acknowledging differences between what they as ethnographers know and what the actor(s) in the context know;
- Constructing new ways of knowing that are grounded in local and situated ways of knowing, being and doing the processes and practices of everyday life within a social group or configuration of actors;
- Developing ways of representing what is known by local actors and what the ethnographers learn from the analysis at different levels of analytic scale.

These *principles of conduct* constitute an *orienting set of goals* for ethnographic studies in education. They also serve to make visible a social constructionist as well as a sociocultural approach to studying social, cultural, and linguistic phenomena that shape, and are shaped by (Fairclough, 1992) what participants in particular learning environments *count as learning and knowledge* (cf. Heap, 1991; Kelly, 2016).

Central to the view of the social constructionist and sociocultural perspectives that guide an IE logic-of-inquiry are a series of arguments that invite researchers to (re)think how they view the concept of culture and the nature of discourse (Agar, 2006; Bloome & Clarke, 2006; Kelly, 2016). In this section, we present a set of theoretical arguments that are central to understanding the roles of discourse in the social construction of knowledge central to an IE logic-of-inquiry. The first conceptual argument draws on work of Michael Agar (1994), who framed *culture as a conceptual system*. He defines this system as a *languaculture* and argues that:

> The *langua* in languaculture is about discourse, not just about words and sentences. And the *culture* in a languaculture is above meanings that include, but go well beyond, what the dictionary and the grammar offer . . . Culture is a conceptual system whose surface appears in the words of people's language. That was the similarity against which differences around the world would be investigated.
> (p. 87)

To this argument, we add that of Brian Street (1993), who conceptualized *culture as a verb* (1993). Although not a comprehensive view of the perspectives on language and culture guiding IE researchers, these two conceptual arguments orient IE researchers to explore culture as an ongoing social construction.

Complementing this view of culture is work on the nature of speech genres (discourse) (Bakhtin, 1979/1986) in the construction of knowledge, and the processes through which knowledge is individually and collectively constructed in moment-by-moment and over-time communication among participants in sustaining social groups. Bakhtin (1979/1986) captures this process in the following argument:

> Sooner or later what is heard and actively understood will find its response in the subsequent speech or behavior of the listener. In most cases, genres of complex cultural communication are intended precisely for this kind of actively responsive understanding with delayed action. Everything that we have said here also pertains to written and read speech, with the appropriate adjustments and additions.
>
> (p. 60)

This conceptual argument frames ways of understanding discourse as a social construction and how, over time, particular texts (spoken, written, visual, and read) become recognized, and engaged with, as authoritative texts, that is, they frame what is socially and academically significant in particular areas of study as well as ways of communicating with others. This argument also points to the fact that what is visible to participants (and by extension to ethnographers) in any one moment is not evidence that the person has (or has not) heard what was proposed or what the person may (or may not) be thinking (Frederiksen & Donin, 2015). As Bakhtin argues, most complex communication is not meant for immediate response; therefore, IE researchers seek evidence across times and events that what was proposed at one point in time and was taken up (or not) in subsequent events individually and/or collectively.

Bakhtin's argument, therefore, orients IE researchers to the value in exploring the roots of discourse and how across times and events of interconnected opportunities for learning, consequential progressions are constructed (Putney et al., 2000). Bakhtin's conceptualization of the over-time nature of discourse also relates to directions across disciplines that focus on understanding what counts as knowledge of concepts, processes, and practices within and across disciplinary settings (Kelly, 2016) and how, through discourse, local knowledge (Mercer & Hodgkinson, 2008) is constructed.

The implications of these arguments for identifying and bounding units of analysis as well as for making connections among levels of analysis can be seen in the concept of *intertextuality* as defined by Bloome and Egan-Robertson (1993). Building on Bakhtin, they argue that participants in interaction propose, recognize, acknowledge, and interactionally accomplish what is socially, academically, and interpersonally/personally significant. Intertextuality, therefore, involves tracing the roots of particular references and actions both anchored in the present and signaled as relevant for understanding past as well as future actions and knowledge constructions.

This argument also points to the importance of how discourse is *textually inscribed* (Bucholtz, 2000; Ochs, 1979) in the process of analysis. In the following sections, we return to this issue as we unfold levels of analysis undertaken by Bridges and colleagues (2012) in the study that serves as an anchor for exploring IE as a logic-of-inquiry. As part of this process, we present ways of graphically mapping the flow of interactions and conduct within a developing event (Baker & Green, 2007) and for different levels of interactional scale (time, human, and historical) that are intertextually related to what is being proposed, undertaken, and interactionally accomplished. By (re)constructing analyses inscribed in Bridges et al. (2012), we demonstrate how discourse serves as an anchor for constructing *maps of the inter-relationships* between and among discourse and actions in and across time in order to build warranted accounts of how learning of particular concepts and processes as well as practices of a group are collectively and personally constructed (or not) (Bridges et al., 2012, 2015; Bridges, Jin, & Botelho, 2016; Green et al., 2012; Heap, 1995; Kelly, 2016).

Judith L. Green and Susan M. Bridges

Principles of Conduct for Bounding Studies Within an Ongoing Ethnography

In the remaining sections of this chapter, we present a series of *principles of conduct* guiding IE research today. We illustrate these principles by taking a grounded approach in (re)constructing the chain of actions that Bridges et al. (2012) undertook. As part of this (re)construction, we present a set of additional principles and actions that make visible the need to explore the contexts in which a particular study is grounded, given the ongoing nature of ethnographic programs of research.

Constructing Telling Cases: Principle of Conduct 1

Mitchell's (1984) definition of *telling* case is central to the decision of how to construct ethnographically the boundaries of a case study:

> Case studies are the detailed presentation of ethnographic data relating to some sequence of events from which the analyst seeks to make some theoretical inference. The events themselves may relate to any level of social organization: a whole society, some section of a community, a family or an individual. What distinguishes [telling] case studies from more general ethnographic reportage is the detail and particularity of the account. Each case study is a description of a specific configuration of events in which some distinctive set of actors have been involved in some defined situation at some particular point of time.
>
> *(p. 222)*

In this definition, Mitchell frames a telling case study as not defined by size but by a series of decisions and actions that lay a foundation for developing *theoretical inferences* from detailed ethnographic analyses (personal communication, Brian Street, November 2016). In the following sections, we unfold the layers of detailed ethnographic analyses and the principles of conduct that supported Bridges et al. (2012), as they identified a telling case and followed an individual through collective and personal learning opportunities in a problem-based learning (PBL) program in undergraduate dental education (see also Hmelo-Silver, Kapur, & Hamstra, this volume). As we unfold each level of analysis and dimensions of social organization being examined, we make visible how IE researchers construct *empirically grounded connections* between these levels to develop *theoretical inferences* from these analyses that form consequential progressions to knowledge construction (Putney, Green, Dixon, Duran, & Yeager, 2000).

(Re)presenting the Logic-in-Use: Principle of Conduct 2

To (re)construct the roots of the *logic-in-use* reported in Bridges et al. (2012), as authors who had different roles in this study (i.e., internal and external ethnographers), we engaged in a series of ongoing dialogues as Bridges (re-)entered the archive of the ongoing IE grounded research program on technology-enabled undergraduate dental education. The goal of (re-)entering the archive and to (re)construct a *historical map* of *the ongoing project* was twofold. First, it provided a historical grounding for selection and construction of the telling case reported in Bridges et al. (2012). Second, it made visible how the *principles of operation* guiding the overarching ethnographic project were themselves guided by *principles of conduct* that led to the constructions of graphic maps of the levels of social organization identified as intertextually tied to particular moments of discourse.

As indicated in Figure 46.1, Bridges engaged her colleagues in developing a new program of research upon her appointment as the "educationist" for the undergraduate dental education program. By reconstructing the timeline of actions prior to the telling case study reported in Bridges et al. (2012), Bridges makes visible in Figure 46.1 the ongoing nature of *entering a site* and the chains of interdependent actions undertaken to construct the *archive as an ethnographic space* with expanding boundaries of time, levels of decisions and social organization, role of different researchers, and

- **ROOTS OF THE INTERDISCIPLINARY RESEARCH AGENDA (2007–present)**

- 2007 Bridges' appointment to Faculty of Dentistry as "educationist" (challenge to learn about undergraduate dental education and problem-based curriculum designs)
- OBSERVING as an educational ethnographer entering the unknown (dental discipline) to identify the phenomenon of potential research interest
- ESTABLISHING a goal for the building research agenda: Learning from and with clinical educators (Botelho & Dyson) & IT Officer on recent developments (3-D digital learning objects; adapting a Learning Management System for PBL)
- IDENTIFYING a guiding question: How do students collectively and individually learn in PBL when using online resources?
- FORMALIZING the research agenda: 2007–08 New Faculty Seed Grant (Bridges, Dyson, Botelho)
- DEVELOPING the team's common language and theoretical base for an IE research project

Grant research question: How do undergraduate dental students use the features of video and QTVR materials in WebCT to support learning in PBL modules?

STEP 1. Developing the project archive (video recordings and artefact collection) (2008–09)

- COLLECTING: Video recordings, artefacts of student work, copies of curriculum documents (see Table 2)
- RICH POINT #1: Ethnographer selects anchor event(s) for team archive analysis based on observed phases of activity linked to visual PBL inquiry materials (videos, study casts)
- NOTE MAKING: Team meeting to gain insider understanding from team disciplinary experts/ cultural guides (learning with and from)
- SELECTING a telling case by identifying the anchor moment/rich point/frame clash in Tutorial 2 when Year 3 student (S4) is asked to draw an image on a whiteboard to explain a complex concept then returns to board to correct/refine drawing
- POSING new questions e.g., *What are the actions Student 4 took to move from the unknown to the known?* (starting point is the known and tracing back over the archive)

STEP 2. Constructing the data set for the telling case

- **BOUNDING units of analysis by locating the chains of intertextually tied events**
- COLLECTIVE dialogic video and transcript analysis (ethnographer; clinical faculty as disciplinary/cultural guide; research assistant) to confirm significance of anchor event;
- LOCATING the pathways leading to and/ or from the anchor event;
- EXAMINING the S4's discourse and actions across the three events.
- **TRACING the actor (S4) within and across time by engaging in an iterative recursive and abductive process:**
- IDENTIFYING from Tutorial 1 Student 4's related questioning sequence (see Figures 2–6);
- TRACING actions (from T1 to T2) based on Camtasia screen recording of online activity (WebCT plus search engines) during self-directed learning (SDL);
- MAPPING online searching web content and associated discourse to the T2 drawing and explanation;
- REVISITING Tutorial 1 and noting silence following questioning linked topically to online search activity

STEP 3. Event mapping and analysis and reporting

- MEMBER CHECKING: Inviting PBL case writer for checking and technical analysis
- CONSTRUCTING the event map graphic with associated texts (visual and discursive)
- THEORIZING by drawing on explanatory theories to understand the phenomenon identified
- IDENTIFYING theoretical and practical implications and directions for future research

Figure 46.1 The Ethnographic Process as Undertaken by a Developing, Interdisciplinary IE Team

Judith L. Green and Susan M. Bridges

Table 46.1 Locally Constructed Data Set for Telling Case in Bridges et al. (2012, p. 105)

Events	Location	Timing (problem cycle)	Data source	Student identifiers (Year 3)		Length
Tutorial 1 (T1)	Scheduled university tutorial room	Day 1 (AM)	Video + audio	$n = 8$	S1—S8	1:35:50
Self-directed learning (SDL) (1st of 3 sessions)	University student computer laboratory	Day 1 (PM)	Video (whole group) screen capture (Camtasia)	$n = 6$	S1 S4 S7 S8 S9 S10	0:29:57 0:29:37 0:30:52 0:30:57 0:29:52 0:29:20
Tutorial 2 (T2)	Scheduled university tutorial room	Day 9 (PM)		$n = 8$	S1-S8	2:08:01

relationships to institutional goals. This timeline also demonstrates the centrality of her role as *internal ethnographer*, who framed IE as a basis of a *reflexive approach* to institutional research to inform curriculum developments on PBL and educational technologies within the institution.

This process of constructing *graphic (re)presentations*, as the remaining sections will demonstrate, has become a defining *principle of conduct* for an IE logic-of-inquiry. As indicated in Figure 46.1, by graphically (re)presenting each analytic process at particular levels of social organization, Bridges made transparent how her IE team identified a particular level of phenomenon that created a question for them, what she called a "noticing," and how they engaged in a process of identifying *bounded units of analysis* that they were then able to trace across times and events. Table 46.1 (7.1 in the original) provides an illustration of a *graphic (re)presentation* of the data set constructed from the larger project archive that the team identified as serving the basis for the analyses undertaken in Bridges et al. (2012).

As indicated in this table, the structure of the ethnographic archive provided a basis for the IE team to provide contextual information about events observed and recorded, location of events, timing of the PBL cycle, students participating in tutorials, sources of data (videos and screen grabs), as well as information about each *intertextually tied sequence of activity* in which the *anchor student* (S4) was engaged in different ways. These two levels of mapping, mapping the archive itself and constructing a data set from the archive, as analyses in the next sections will show, demonstrate the need to situate an ethnographic study in the history of the larger context in which it is embedded.

Mapping Intertextual Relationships at Multiple Levels of Analytic Scale: Principle of Conduct 3

In this section, we present different levels of social organization identified and analytic processes that Bridges et al. (2012) undertook to explore the roots of S4's (the anchor student) shift in displaying her developing understanding of a key dental concept. Specifically, we make visible how the IE team undertook a series of *backward* and *forward mapping* processes to make connections across levels of social organization and to create an intertextual web for making theoretical inferences about the sources of actions and intertextual processes that led to observed changes in S4's response to the facilitator's questions (Green & Meyer, 1991; Putney et al., 2000) and group collaborations within one PBL group (8 students).

The anchor level of maps resulting from this process is presented in Figure 46.2. As indicated in this figure, the central unit of analysis was time within the program.

As indicated in Figure 46.2, the point in time that the PBL instructional cycle under study occurred was in the third year undergraduate Bachelor of Dental Surgery (BDS) program in Module III. By locating this within instructional sequences that preceded this event, this IE team created

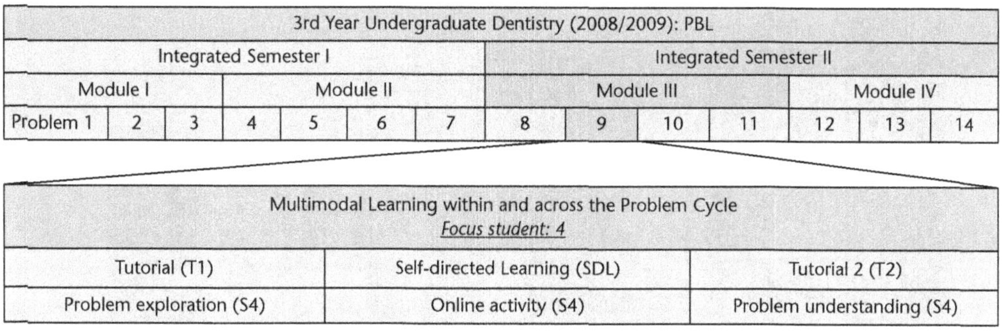

3rd Year Undergraduate Dentistry (2008/2009): PBL													
Integrated Semester I							Integrated Semester II						
Module I		Module II					Module III				Module IV		
Problem 1	2	3	4	5	6	7	8	9	10	11	12	13	14

Multimodal Learning within and across the Problem Cycle *Focus student: 4*		
Tutorial (T1)	Self-directed Learning (SDL)	Tutorial 2 (T2)
Problem exploration (S4)	Online activity (S4)	Problem understanding (S4)

Figure 46.2 A Graphic Inscribing Events Bounding the Telling Case

a potential for tracing the roots of what was proposed to students within and across each Module that preceded the "noticing"/"aha" moment. Figure 46.3 (Bridges et al., 2012, p. 107) provides a graphic (re)presentation of the next levels of analysis, the construction of timelines and event maps of the developing structure of each social organization space in which the anchor student participated: Tutorials 1 and 2 (T1 and T2) and the Self-Directed Learning (SDL) space.

This level of mapping provided an anchor for exploring the local intertextual web of activity that was then analyzed for the discourse-in-use. As indicated in Figure 46.3, the purple color establishes where in the flow of activity the intertextual chain of events was identified. This level of mapping made visible *visual (e.g., screen captures of object, physical configuration of actors, and digital resources oriented to by participants)* texts that anchored, and thus became an actor in the developing event that both the collective, and S4 in the self-directed learning space, oriented to, and engaged with, in the program. By including these visual (re)presentations, the IE team added levels of contextualization cues (Gumperz & Behrens, 1992) to the meanings of concepts in developing contexts in which they were being proposed and made available to the students.

This intertextual approach enabled development of new angles for analysis of developing events and discourse. Mapping these interconnected texts and events, therefore, provided an *ethnographic space* for exploring the roots of and developing texts of this PBL cycle of activity. By including visual texts, the authors also provided evidence of the range of visual texts that were afforded students, collectively and individually, and to which they oriented within the PBL, as well as the SDL online system and databases.

Making Connections Through Discourse Analyses in an Intertextual Web of Activity: Principle of Conduct 4

Missing from the previous levels of mapping is the actual discourse that was being proposed, recognized, acknowledged (or not), and interactionally accomplished (see also Bloome et al., 2005; Koschmann, this volume). In this section, we present two analyses, one of Tutorial 1 and one of Tutorial 2, to demonstrate what the discourse analysis made possible to explore that other levels of mapping and map construction did not. By adding a transcript and locating the *point of conversation* in the developing tutorial sessions, the IE team brought the analytic focus to the specific moments in which the discourse framed an anchor for examining *how* the students and their facilitator shaped the direction of the tutorials in this single PBL cycle. The two figures presented in this section include the discourse to make visible the developing understandings of the anchor student at particular points in the PBL cycle as well as observed processes of facilitation.

As indicated in the discourse excerpt, after S4 completed reading the case scenario, the chairperson (Student 6, S6) invited them to discuss visual texts (*radiograph, photo,* and *study cast*) presented as inquiry materials. In this discourse segment, S6, under the observation of the tutorial's facilitator, established a set of parameters for the *collective work*. To explore the proposed actions for engaging in this PBL learning

3rd Year Undergraduate Dentistry (2008/2009): PBL													
Integrated Semester I							Integrated Semester II						
Module I			Module II				Module III				Module IV		
Problem 1	2	3	4	5	6	7	8	9	10	11	12	13	14

Multimodal Learning within and across the Problem Cycle *Focus student: 4*								
Tutorial (T1)			Self-directed Learning (SDL)			Tutorial 2 (T2)		
Problem exploration (S4)			Online activity (S4)			Problem understanding (S4)		
Time	Modality	Fact/ Hypothesis	Time	Modality	Topic/ Keywords	Time	Modality	Synthesis
0:04:48	Problem Statement (print)	Referred to see a dentist	0:07:55	eLibrary > Dental Update	Factors affecting jaw growth	0:15:08	Visualizations: Whiteboard drawings	Lateral view of skull
0:13:32	Clinical Photograph	Reduced lower facial height	0:09:01	Discussion	Class III malocclusion	0:51:22		Rotational growth in mandible
0:26:05	Cephalometric Radiograph	Relationship of jaws	0:13:59	eLibrary > Orthodontic Cyber Journal	Factors affecting jaw growth	0:52:22		Forward rotation in males/ Decreased vertical height
0:37:13	Study Casts	Displaced, open-bite	0:14:42	eLibrary > Science Direct	Factors affecting jaw growth	0:58:09	Discussion	Growth measurement
0:51:39	Cephalometric Radiograph	Positions of ANS & PNS	0:15:49	eLibrary > Journal of Orthodontics	Factors affecting jaw growth	1:23:24		Class I, II & III malocclusion
1:00:28	Discussion	Oral condition/ Pronunciation & mastication difficulties	0:20:53	eLibrary > Science Direct	Growth of jaw bones	1:40:29	Cephalometric & Panoramic Radiograph	Functional appliance rejection/ Molar development
1:12:22		Orthognathic surgery	0:21:29	Discussion: Science Direct	Growth/ Orthodontic treatment	1:53:45	Discussion	Clinical implications of antidepressant drugs
1:14:42		Low compliance to complicated surgery	0:22:21	Discussion	Factors affecting jaw growth	1:55:58		Bipolar disorder/ Depression & anxiety
			0:26:08	eLibrary > Science Direct	Growth/ Orthodontic treatment			

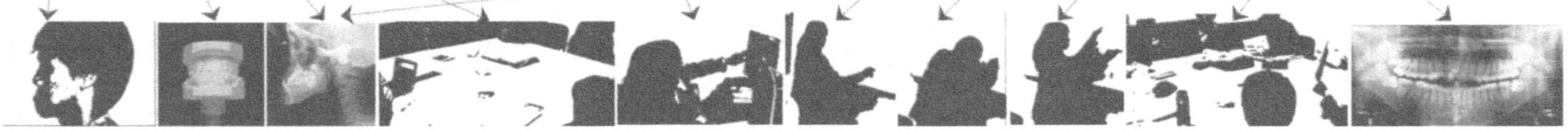

Figure 46.3 Connecting Event Maps to Visual Activity Anchors

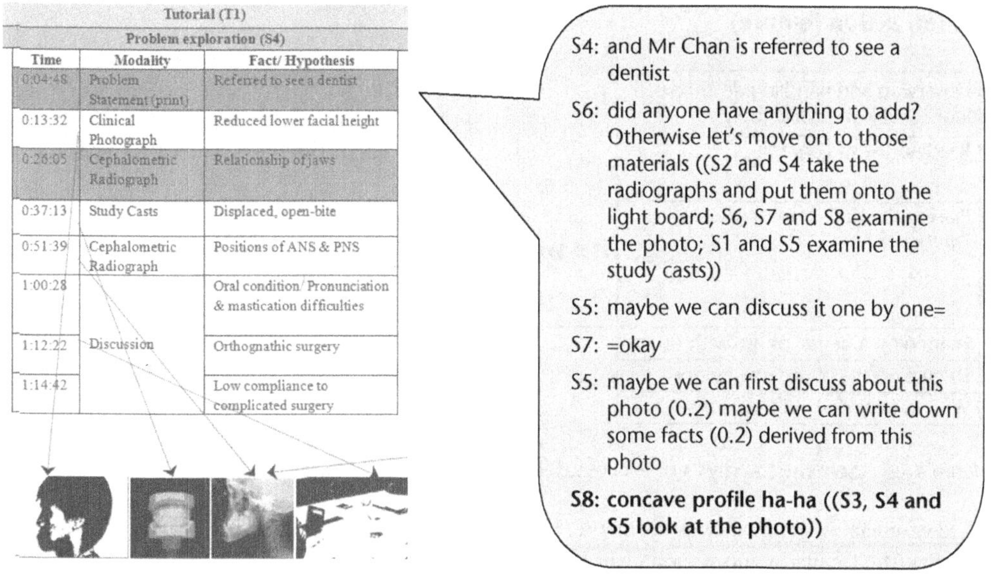

Figure 46.4 Making Connections Between Activity, Modalities, and Discourse-in-Use (Tutorial 1, Figure 46.3)

process, we undertook a new approach to analysis, one that extends what was reported in Bridges et al. (2012). We (re)constructed the following logic that Spradley (1980) proposed to explore relationships among the features of culture and to identify specific types of cultural patterns (see Table 46.2).

The analysis that follows is designed to demonstrate how cultural patterns-in-the-making can be identified as they are proposed, recognized, acknowledged, and interactionally accomplished by identifying actions, meanings, reasons proposed, and places for actions, among other cultural features. We focus on making visible how small chains of interaction provide a rich base for identifying cultural processes and practices as well as identifying disciplinary language among other cultural features by adopting this *logic as a principle of conduct*. To demonstrate how this logic of analysis supports

Table 46.2 Semantic Analysis as Principle of Conduct 6

1	Strict inclusion: X is a kind of Y
	kinds of actors, activities, events, objects, relationships, goals, time
2	Spatial: X is a part of Y
	parts of activities, places, events, objects/artifacts
3	Rationale: X is a reason for doing Y
	reasons for actions, carrying out activities, using objects, arranging space, seeking goals
4	Location for action: X is a place for doing Y
	places for activities, where people act, events are held
5	Function: X is used for Y
	uses for objects, events, acts, activities, places
6	Means–end: X is a way to do Y
	ways to organize space, to act, to become actors, to acquire information
7	Sequence: X is a step in Y
	steps for achieving goals in an act, an event, an activity, in becoming an actor
8	Attribution: X is an attribute of Y
	characteristics of objects, places, time, actors, activities, events
9	Cause–effect: X is a result of Y
	results of activities, acts, events, feelings

X (an action feature)

Observing and handling *dental casts* individually and collectively		
Reading the radiograph		**Cover term derived from the actions**
Placing the *radiograph* in the lightboard	*is a way of*	Engaging with multimodal inquiry texts during live PBL tutorials for exploring dental problems present in the PBL scenario
Examining a dental photo with others		
Discussing the photo one by one		
Writing facts derived from photo		

Figure 46.5 Semantic Analysis of Processes Requested by the Facilitator

IE researchers in stepping back from the known, we undertook a semantic analysis of *multimodal texts* that the facilitator and session chair proposed to students in the transcript excerpt of Tutorial 1.

This brief example demonstrates how, in small moment-by-moment interactions, what is being communicated is not mere words. As Fairclough (1992) argued, *utterances* are simultaneously *texts, discourse processes*, and *social practices* that lead to a process of text construction within a developing event. Although the student chairperson (S6) proposed the set of actions listed in Figure 46.5, how and in what ways students read, interpreted, and acted on (or not) was not always visible in that moment but was visible in subsequent events.

To understand the need for tracing configurations of actors and what is interactionally being proposed and accomplished across across times and events, we draw on Gumperz (1982), who, in framing Interactional Sociolinguistics, proposed ways of understanding developing interactions among participants. He argued that people bring *linguistic, cultural*, and *social presuppositions* to any interaction based on past experiences. In this brief segment from Tutorial 1, which was embedded within a series of PBL cycles across an educational program, what became visible was how the anchor student (S4) and student chairperson (S6) engaged in a process of negotiating with other students how, and in what ways, the task might be undertaken to ground their discussion in the professional objects/artifacts that dentists read and interpret to make decisions about what clinical decisions and actions to take. What was not visible is how the students took up and subsequently drew on these proposed actions in their future work. Therefore, in this brief analysis of the initiating interchange among the tutorial students, we made visible *conceptually guided* ways of deriving features of culture being constructed by participants, and for exploring developing professional education events as places for *individual–collective development*.

Using this analysis as an anchor, we now explore a *frame clash* that S4 faced in Tutorial 2. As stated previously, student (S4) became silent during the Tutorial 1, but in Tutorial 2 she was able to participate in academically expected ways. The following excerpt (see original chapter, Bridges et al., 2012, Excerpt 5) captures a point of *frame clash* or confrontation in expectations (Tannen, 1993) that S4 experienced in Tutorial 2, when she struggled to verbally (re)present the dental concept.

F: So:: males will tend to rotate which direction?
S4: Clockwise (.)
F: Vertical=vertical growth, (.) increase in vertical growth?
S4: Decrease (.)
F: Decrease but it's clockwise?
S4: You mean the mandible grows in anticlockwise but it=I don't know how to describe, it is (.) forward growth of mandible?=

Interactional Ethnography

In this discursive exchange between S4 and the dental educator acting as facilitator (F), S4 struggled to verbally explain the growth of the mandible. What occurred next provided evidence that the facilitator created a *rich point* for S4 (and, by extension, the ethnographer); that is, he shifted his request from a verbal statement (the frame clash for S4) to a request for a visual display by saying *just draw it*, so that S4 might display her understanding of this complex dental concept (see Figure 46.6). By (re)formulating the directive, the facilitator helped resolve the frame clash for S4. This was indicated in S4's drawing as directed and, significantly, her subsequent autonomy in self-correcting and modifying the initial drawing of the phenomenon.

This chain of talk and activity led the IE team to explore further the roots of the drawing and associated reasoning (backward mapping). Analysis of events between the tutorial sessions led the team to identify how S4 had engaged in online searches and exploration of different texts in the SDL spaces. This set of analyses also framed the need to explore literature on multimodal texts (Kress, 2000, 2010).

This exchange and the actions across the three social organizational spaces provided an empirical basis for constructing a series of warranted accounts (Heap, 1991) that laid a foundation for Bridges et al. (2012) to construct theoretical understandings of the agency of actors in and across developing events that point to the necessity of stepping back from traditional views of observable moments to undertake multi-faceted levels of analysis.

F: =Just draw it (.) arrow direction ((S4 moves to draw on the board while S8 & S1 discuss between themselves))

F: This is no::t rotation, it is AP, you are just shifting the whole thing AP.

S4: the bac=forward rotation is here↑ a::nd backward rotation ((finishes and walks back to her seat))

S4: I didn't draw it round enough, I'll draw it again ((returns and draws another one))

S4: This is (.) backward.

F: And this is female=

S4: =Yeah=

F: =Okay

S4: And forward is (.) you imagine it's (.) yeah another (one)

F: (But) then the:: vertical dimension is different (.) This is increased in facial height if this is backward. But you're talking about male is increased on vertical dimension right?=

S4: =Oh, decrease=

F: =Male is decreased. Okay?

S4: I dunno. It's written in the book that, the cranial facial growth in adult and when other dimensions cease, the vertical change still predominate ((looking up at the facilitator)) and there is a tendency for a male to have forward rotation (.) yeah

F: That means decrease in vertical dimension=

S4: =Mmm (0.10.0) ((S4 whispers to S6))

F: Okay

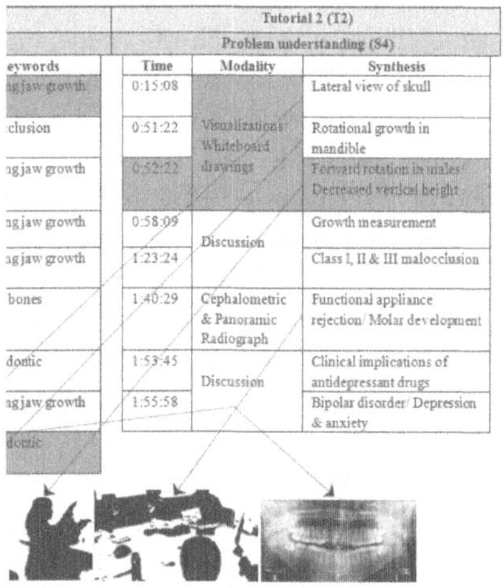

Tutorial 2 (T2)			
Problem understanding (S4)			
eywords	Time	Modality	Synthesis
ig jaw growth	0:15:08		Lateral view of skull
clusion	0:51:22	Visualization: Whiteboard drawings	Rotational growth in mandible
ig jaw growth	0:52:22		Forward rotation in males/ Decreased vertical height
ig jaw growth	0:58:09	Discussion	Growth measurement
ig jaw growth	1:23:24		Class I, II & III malocclusion
bones	1:40:29	Cephalometric & Panoramic Radiograph	Functional appliance rejection/ Molar development
dontic	1:53:45	Discussion	Clinical implications of antidepressant drugs
ig jaw growth	1:55:58		Bipolar disorder/ Depression & anxiety
dontic			

Figure 46.6 Tracing the Interactional Chain From a Frame Clash

Judith L. Green and Susan M. Bridges

Expanding the IE Logic of Inquiry: On Reflexivity as Principle of Conduct 5

In (re)constructing Bridges et al. and in extending that analysis through an analysis based on Spradley's (1980) features of culture, we illustrated how the *iterative, recursive,* and *abductive* logic-in-use was guided by the *principles of conduct* presented previously. By presenting this logic-in-use, we made transparent, not just visible, the principles of conduct that guided the IE team in constructing local and situated understandings of what counted as disciplinary and professional knowledge in this purposefully designed educational environment. We also made visible why Bridges' team drew on advances in theoretical perspectives on the multimodal nature of texts (spoken, written, oral, visual) (Jewitt, 2014; Kress, 2000, 2010) which framed a theory of semiosis (Kress, 2000, 2010) to support their interpretation (explanation) of particular patterns identified that earlier IE studies had not addressed—specifically, Kress's theoretical perspective created ways of understanding *intervisual links*. Additionally, the team drew on Wertsch's (1994) notion of *semiotic mediation* to develop an explanation of how meanings of texts (spoken and written) are socially negotiated. By including these theories, they developed a set of *explanatory theories* to support theoretical inferences from the patterns that the team identified both in the tutorials and in the previously unexamined spaces between the tutorials in the self-directed learning environments. By engaging in a reflexive process that identified the need for a theoretical perspective on the engagement and use of multimodal texts, Bridges et al. (2012) made visible an additional *principle of conduct—the need to return to literature to construct explanatory theories for the patterns identified through the IE logic-in-use, i.e., patterns not previously known but that levels of analysis undertaken indicated were necessary to develop theoretically informed understandings of such patterns*. Thus, this principle of conduct made visible that IE as a logic-of-inquiry is not a fixed theoretical perspective but one that continues to develop as new studies are undertaken that lead to the need for new or expanded theoretical understandings of the phenomenon under study (Heap, 1995).

Currently, Bridges and the team have expanded their IE logic-of-inquiry to address the particular context of their study—PBL and educational technologies (see Bridges et al., 2016). The expanded logic-of-inquiry includes conceptual perspectives that focus them on questions of the conceptual alignment between PBL as a social constructivist learning design (Hmelo-Silver et al., this volume), while continuing to be grounded in IE's conceptual foundation presented in this chapter. This goal has led to the identification of recent developments related to the socially constructed nature of learning in inquiry-based, small-group discussions, among other settings, processes that intersect with the goals of the Learning Sciences (Hoadley, this volume). This focus has also brought forward a set of discourse-based theoretical perspectives that support analysis of epistemic understandings (Kelly, 2016) of knowledge as being socially constructed through (inter)actional and (inter)textual activity among actors in purposefully designed educational learning spaces.

Acknowledgment

We would like to thank W. Douglas Baker (Eastern Michigan University) for his editorial comments throughout this process. His insights and suggestions were critical to the development of this chapter and the description of IE as an epistemology, not method.

Further Readings

Bloome, D., Carter, S. P., Christian, B. M., Otto, S., & Shuart-Faris, N. (2005). *Discourse analysis and the study of classroom language and literacy events: A microethnographic perspective*. Mahwah, NJ: Lawrence Erlbaum.
This volume provides frames a microethnographic approach to examining how literacy events, identities, and power relationships are socially constructed in the discourse in use in classrooms with diverse learners.

Gee, J. P., & Green, J. L. (1998). Discourse analysis, learning, and social practice: A methodological study. *Review of Research in Education, 23,* 119–169.
This article provides a synthesis of methodological issues involved in discourse studies of learning in educational context by proposing a framework for exploring the *material, activity, semiotic,* and *social* dimension of learning in social settings

Kaur, B. (Ed.). (2012). *Understanding teaching and learning: Classroom research revisited.* Rotterdam/Boston/Taipei: Sense Publishers.
This edited volume includes an interdisciplinary group of researchers, including LS researchers, who were invited to: critically (re)examine and elaborate theories guiding the work of Graham Nuthall on teaching–learning relationships in classrooms; propose future directions for classroom research and assessment; and provide models designed to attain a more complex representation of the individual and diverse learners in the social/cultural milieu of classrooms and schools.

Skukauskaite, A., Liu, Y., & Green, J. (2007). Logics of inquiry for the analysis of video artefacts: Researching the construction of disciplinary knowledge in classrooms. *Pedagogies: An International Journal, 2*(3), 131–137.
This special issue of *Pedagogies* provides an international set of video-enabled ethnographic studies that explore how video-based discourse analysis supports exploration of micro–macro relationships that support and/or constrain the opportunities for learning disciplinary knowledge in K–12 classrooms.

Walford, G., (2008) (Ed.), *How to do educational ethnography.* London: Tufnell Press.
This volume is a compilation of methodological issues and directions in ethnographic research processes involved in designing, collecting, analysing, and reporting ethnographic studies in education.

NAPLeS Resources

Green, J. L. & Bridges, S. M., *15 minutes about interactional ethnography* [Video file]. In *NAPLeS video series.* Retrieved October 19, 2017, from http://isls-naples.psy.lmu.de/video-resources/guided-tour/15-minutes-bridges_green/index.html

Green, J. L. & Bridges, S. M., *Interview about interactional ethnography* [Video file]. In *NAPLeS video series.* Retrieved October 19, 2017, from http://isls-naples.psy.lmu.de/video-resources/interviews-ls/bridges_green/index.html

References

Agar, M. (1994). *Language shock: Understanding the culture of conversation.* New York: Quill.
Agar, M. (2006) An ethnography by any other name. *Forum Qualitative Sozialforschung/Forum: Qualitative Social Research, 7*(4, September, Art. 36). Retrieved from http://nbn-resolving.de/urn:nbn:de:0114-fqs0604367
Baker, W. D. & Green, J. (2007). Limits to certainty in interpreting video data: Interactional ethnography and disciplinary knowledge, *Pedagogies: An International Journal, 2*(3), 191–204. doi:10.1080/15544800701366613
Bakhtin, M. M. (1979/1986). *Speech genres and other late essays* (V. W. McGee, Trans.). Austin, TX: University of Texas Press.
Bloome, D., Carter, S. P., Christian, B. M., Otto, S., & Shuart-Faris, N. (2005). *Discourse analysis and the study of classroom language and literacy events: A microethnographic perspective.* Mahwah, NJ: Lawrence Erlbaum.
Bloome, D., & Clarke, C. (2006). Discourse-in-use. In J. Green, G. Camilli, & P. Elmore (Eds.), *Handbook of complementary methods in education research* (pp. 227–242). Washington, DC/Mahwah, NJ: AERA/Lawrence Erlbaum.
Bloome, D., & Egan-Robertson, A. (1993). The social construction of intertextuality in classroom reading and writing lessons. *Reading Research Quarterly, 28*(4), 305–333.
Bridges, S., Botelho, M., Green, J. L., & Chau, A. C. M. (2012). Multimodality in problem-based learning (PBL): An interactional ethnography. In S. Bridges, C. McGrath, & T. L. Whitehill (Eds.), *Problem-based learning in clinical education: The next generation* (pp. 99–120). Dordrecht: Springer.
Bridges, S. M., Green, J. Botelho, M. G., & Tsang, P. C. S. (2015). Blended learning and PBL: An interactional ethnographic approach to understanding knowledge construction in-situ. In A. Walker, H. Leary, C. E. Hmelo-Silver, P. A. Ertmer (Eds.), *Essential readings in problem-based learning: Exploring and extending the legacy of Howard S. Barrows* (pp. 107–130). West Lafayette, IL: Purdue Press.
Bridges, S. M., Jin, J., & Botelho, M. G. (2016). Technology and group processes in PBL tutorials: An ethnographic study. In Bridges, S. M., Chan, L. K., Hmelo-Silver, C. (Eds.), *Educational technologies in medical and health sciences education.* (pp. 35–55). Dordrecht: Springer.
Bucholtz, M. (2000). The politics of transcription. *Journal of Pragmatics, 32*, 1439–1465.
Castanheira, M. L., Crawford, T., Dixon, C. N., & Green, J. L. (2000). Interactional ethnography: An approach to studying the social construction of literate practices. *Linguistics and Education, 11*(4), 353–400. doi: 10.1016/s0898-5898(00)00032-2
Danish, J., & Gresalfi, M. (2018). Cognitive and sociocultural perspective on learning: tensions and synergy in the Learning Sciences. In F. Fischer, C. E. Hmelo-Silver, S. R. Goldman, & P. Reimann (Eds.), *International handbook of the learning sciences* (pp. 34–43). New York: Routledge.

Fairclough, N. (1992). Intertextuality in critical discourse analysis. *Linguistics and Education*, 4(3–4), 269–293.

Frederiksen, C., & Donin, H., (2015). Discourse and learning in contexts of educational interaction, In N. Markee (Ed.), *Handbook of classroom discourse & interaction* (pp. 96–114). Oxford: Wiley-Blackwell.

Green, J., & Meyer, L. (1991). The embeddedness of reading in classroom life: Reading as a situated process. In C. Baker & A. Luke (Eds.), *Towards a critical sociology of reading pedagogy* (pp. 141–160). Amsterdam: John Benjamins.

Green, J. L., Skukauskaite, A., & Baker, W. D. (2012). Ethnography as epistemology: An introduction to educational ethnography. In J. Arthur, M. J. Waring, R. Coe & L. V. Hedges (Eds.), *Research methodologies and methods in education* (pp. 309–321). London: Sage.

Green, J., Skukauskaite, A., Dixon, C., & Córdova, R., (2007). Epistemological issues in the analysis of video records: Interactional ethnography as a logic of inquiry. In R. Goldman, R. Pea, B. Barron, & S. J. Derry (Eds.), *Video research in the learning sciences* (pp. 115–132), Mahwah, NJ: Lawrence Erlbaum.

Gumperz, J. J. (1982). *Discourse strategies* (Vol. 1). New York: Cambridge University Press.

Gumperz, J. J., & Behrens, R. (1992). Contextualization and understanding. In A. Duranti & C. Goodwin (Eds.), *Rethinking context: Language as an interactive phenomenon* (pp. 229–252). Cambridge, UK: Cambridge University Press.

Heap, J. (1991). A situated perspective on what counts as reading. In C. Baker & A. Luke (Eds.), *Towards a critical sociology of reading pedagogy* (pp. 103–139). Philadelphia, PA: John Benjamins.

Heap, J. L. (1995). The status of claims in "qualitative" educational research. *Curriculum Inquiry*, 25(3), 271–292.

Heath, S. B. (1982). Ethnography in education: Defining the essentials. In P. Gilmore & A. A. Glatthorn (Eds.), *Children in and out of school: Ethnography and education* (pp. 33–55). Washington, DC: Center for Applied Linguistics.

Heath, S. B., & Street, B. V. (2008). *On ethnography: Approaches to language and literacy research*. New York: Teachers College/NCRLL.

Hmelo-Silver, C. E., Kapur, M., & Hamstra, M. (2018) Learning through problem solving. In F. Fischer, C. E. Hmelo-Silver, S. R. Goldman, & P. Reimann (Eds.), *International handbook of the learning sciences* (pp. 210–220). New York: Routledge.

Hoadley, C. (2018). A short history of the learning sciences. In F. Fischer, C. E. Hmelo-Silver, S. R. Goldman, & P. Reimann (Eds.), *International handbook of the learning sciences* (pp. 11–23). New York: Routledge.

Jewitt, C. (2014). *The Routledge handbook of multimodal analysis* (2nd ed.). London/New York. Routledge.

Kelly, G. J. (2016). Methodological considerations for the study of epistemic cognition in practice. In J. A. Greene, W. A. Sandoval, & I. Braten (Eds.), *Handbook of epistemic cognition* (pp. 393–408). New York: Routledge.

Koschmann, T. (2018). Ethnomethodology: Studying the practical achievement of intersubjectivity. In F. Fischer, C. E. Hmelo-Silver, S. R. Goldman, & P. Reimann (Eds.), *International handbook of the learning sciences* (pp. 465–474). New York: Routledge.

Kress. G. (2000). Multimodality. In B. Cope & M. Kalantzis (Eds.), *Multiliteracies.* (pp. 182–202). London: Routledge.

Kress. G. (2010). *Multimodality: A social semiotic approach to contemporary communication*. London/New York. Routledge.

Mercer, N., & Hodgkinson, S. (2008). *Exploring talk in schools*. London: Sage.

Mitchell, C. J. (1984). Producing data. In R. F. Ellen (Ed.), *Ethnographic research: A guide to general conduct* (pp. 213–293). New York: Academic Press.

Ochs, E., (1979). Transcription as theory. In E. Ochs & B. Schieffelin (Eds.), *Developmental pragmatics* (pp. 43–72). New York: Academic.

Puntambekar, S. (2018). Design-based research (DBR). In F. Fischer, C. E. Hmelo-Silver, S. R. Goldman, & P. Reimann (Eds.), *International handbook of the learning sciences* (pp. 383–392). New York: Routledge.

Putney, L., Green, J. L., Dixon, C., Duran, R., & Yeager, B. (2000). Consequential progressions: Exploring collective individual development in a bilingual classroom, In C. Lee & P. Smagorinsky (Eds.), *Constructing meaning through collaborative inquiry: Vygotskian perspectives on literacy research* (pp. 86–126). New York: Cambridge University Press.

Spradley, J. (1980). *Participant observation*. Fort Worth, TX: Harcourt Brace Jovanovich.

Street, B. (1993) Culture is a verb. In D. Graddol (Ed.), *Language and Culture* (pp. 23–43). Clevedon, UK: Multilingual Matters/BAAL.

Tannen, D. (Ed.). (1993). *Framing in discourse*. New York: Oxford University Press.

Wertsch, J. (1994). Mediated action in sociocultural studies. *Mind, Culture and Activity*, 1, 202–208.

47

Video Research Methods for Learning Scientists
State of the Art and Future Directions

Sharon J. Derry, Lana M. Minshew, Kelly J. Barber-Lester, and Rebekah Duke

---

The easy availability of affordable, usable, portable, high-quality video technology is markedly influencing research in the learning sciences. Video technologies—which include recording, editing, archiving, and analysis tools—provide researchers with increasingly powerful ways of collecting, sharing, studying, and presenting detailed cases of teaching and learning, in both formal and informal educational settings, for both research and teaching purposes. Half the studies published in the *Journal of the Learning Sciences* since 2010 included collection and analysis of video data, and 366 video research papers have appeared in the last three proceedings of the International Conference of the Learning Sciences. Video research accounts for a substantial share of the scholarly activity within the learning sciences.

Video enables learning scientists to capture and study fine points of interactional processes, including talk, eye gaze, body posture, tone of voice, facial expressions, use of tools, production of artifacts, and maintenance of joint attention (Barron, Pea, & Engle, 2013). However, even a few hours of video data contain so much detail that it is easy for researchers to become overwhelmed unless they have practical tools and strategies for focusing their work.

Our chapter addresses challenges for educational design research (McKenney & Reeves, 2012) that collects video as a major data source. We offer research strategies based on previously published guidelines (e.g., Barron et al., 2013; Derry et al., 2010; Derry, Sherin, & Sherin, 2014) vetted through our own experience. We then look at emerging trends related to video as "big data," including instrumentation, learning analytics, data management, and ethics. Our goal is to orient designer-researchers who are embarking on the exciting adventure of video research within a rapidly changing technical landscape.

Video Research on Teaching and Learning

Planning

Video is used to study many types of learning environments, ranging from classrooms, to museums, to laboratories (e.g., Kisiel, Rowe, Wartabedian, & Kopczak, 2012; Minshew, Derry, Barber-Lester, & Anderson, 2016; van de Pol, Volman, Oort, & Beishuizen, 2014; Zahn, Pea, Hesse, & Rosen, 2010). Each type of environment presents unique problems that the researcher must address with thoughtful advance planning.

Sharon J. Derry, Lana M. Minshew, Kelly J. Barber-Lester, and Rebekah Duke

Theoretically-motivated research questions can help researchers maintain focus in what can otherwise become an overwhelming sea of detail. Research questions inform decisions related to equipment choice, timing and amount of recording, camera placement, and other data collection. Constraints on video data collection are created not only by the physical layout of the environment, but also by ethical and institutional review requirements and what recording equipment must be used. Advance scouting of the environment will provide an understanding of logistical problems that must be solved related to availability of electrical outlets, spaces for tripods, adequacy of lighting, audio challenges, natural patterns of activity that could interfere with video collection, and arrangement of the environment to avoid recording non-consenting participants.

Consider the planning of a single researcher wanting to study teachers' gestures in whole-class mathematics discussions in a crowded classroom using two available cameras with one wide-angle lens and some available microphones. Based on research questions and study of the environment the researcher chooses to manage one wireless follow camera focusing on the teacher wearing a Bluetooth lapel microphone. A backup battery will be kept at the ready. A single wide-angle lens camera will be positioned on a tripod taped to the floor next to an outlet. The wide-angle camera will focus on one side of the classroom arranged to seat only those students with informed consent. A student will help start both cameras simultaneously. The researcher will note critical events immediately following each recording session. Now, ready, set, record!

Equipment

The availability and quality of video recording devices has increased significantly over time as the price for these devices has decreased, making video research possible for researchers on almost any budget. High-quality, commercial cameras are available for those with generous budgets and special needs. However, basic home video recorders or even smartphone recorders are adequate for many types of research. We utilize home video recorders and sturdy tripods to collect whole-class and small-group video data in classrooms. We have used as many as four cameras running simultaneously to capture macro-level classroom events alongside micro-level small-group interactions.

Home recorders have limitations with regard to sound quality provided through built-in microphones. Audio enhancing devices such as table, lapel, and boom microphones improve sound quality in noisy settings such as classrooms. The environment and research goals determine what audio support is needed. For example, if there is need to focus on entire classrooms and feature the teacher, a camera with a wide-angle lens and a lapel microphone on the teacher are often used. If the target of research is students in small groups, one preferred method is to position cameras to focus on students arranged in a semicircle around a small table, with a table microphone to capture sound.

Increasingly common are studies that utilize the cameras embedded in students' laptop computers along with specialized screencast software to capture the faces and conversations of subjects as they interact around computer-based activities. These data may be analyzed alongside screen capture and log files of their computer manipulations (e.g., Lee, Pallant, Tinker, & Horwitz, 2014; Malkiewich, Lee, Slater, Xing, & Chase, 2016). Some researchers utilize first-person perspective cameras, like Go-Pros®, to capture video representing the subject's perspective (Choi, Land, & Zimmerman, 2016). An increasing number of alternative video recording approaches, including inexpensive 360-degree immersive cameras and various recording enhancements for smartphones, are flooding the market and being adopted by learning science researchers.

Selection: Aiming, Panning, and Zooming

Decisions about how to focus the camera during recording are, essentially, data sampling choices. The angles from which video is captured, and any panning and zooming during data collection, may significantly influence research outcomes. Too much panning and zooming can make video data difficult to watch, eliminate important contextual elements, and potentially miss significant elements that become outside of the frame. A guide on camera work by Rogers Hall is provided in Derry et al. (2010).

Field Notes

In a typical design study, large quantities of video data are amassed rapidly, which could quickly become difficult to navigate and use. Field notes, taken in timed increments (e.g., every five minutes) or with time stamps, are a good aid in managing video data. Field notes can guide researchers to moments of interest within a video collection. Theoretical perspective and research questions should guide the taking of field notes. For example, when researching student argumentation, the researcher may note instances of students utilizing evidence or counterargument. Later, field notes can point to specific targets for analysis within a compatibly time-stamped video.

Storing and Archiving

Although high-quality field notes provide the first step in indexing video, another important practical consideration is how to store and archive the video to support future research. Our team uses visual representations that capture the workflow of our implemented designs, along with an associated tagging system, to organize a large data corpus (Barber-Lester, Derry, Minshew, & Anderson, 2016). Visual representations and tags together allow for rapid systematic selection of data for research between iterations in design studies. Student artifacts, video files, lesson plans, and assessments are all represented and archived in our system. An important idea underlying this work is the value of intermediate representations in video research (Barron et al., 2013; Derry et al., 2010).

Analytic Frames

Much video research in the learning sciences is concerned with what is learned during interaction and what aspects of interaction promote or interfere with learning. How one chooses to approach analyses of interaction depends on one's research questions and theoretical commitments. An interesting perspective for thinking about these commitments was provided by Enyedy and Stevens (2014), who described three frameworks. From the cognitive tradition are studies in which interaction is of interest primarily as a window into participants' cognitions. In these studies, viewed as extensions of think-aloud methods, video data are examined to draw inferences about thought processes revealed in discourse. A second category of research pays more attention to the interactions and may sequentially code data to reveal discourse patterns and other interactional structures (e.g., Mehan, 1979). Studies in this category may seek relationships between repeatable discourse structures and distally measured learning outcomes. A third type of research, increasingly common in the learning sciences, focuses on the more ambitious goal of understanding the complex system of collaborative interaction, including mediating tools from the environment, and how these patterns give rise to practices and conceptual understanding. Researchers with this goal are often interested in rich interpretive description and may not be interested in correlating these insights with distal measures of learning (e.g., Alac' & Hutchins, 2004; see also Green & Bridges, this volume; Koschmann, this volume; Puntambekar, this volume).

Selection

Whatever one's framing assumptions, an important step in conducting video analysis requires selecting which video and supporting data from a larger corpus to examine closely. Sometimes the selection strategy must ensure that analytic results are representative of a larger group. For example, we are currently examining sampled interactions from small collaborative groups both prior to and after a school-based intervention, to assess whether and how the intervention influenced collaborative processes within an entire grade. In this case, we strived for systematic and unbiased selection from the video corpus.

This approach stands in contrast to less systematic sampling. For example, researchers have sometimes selected just a single video clip to study in depth, not because it is representative of a clearly defined population, but because it is a rich example of interaction that interests the researcher(s) (e.g., Koschmann & Derry, 2017). One might argue that, since we currently know so little about human interaction, all such examples are worthy of study.

Transcription

Another analytic decision the video researcher makes is whether or not to transcribe. Most video researchers do transcribe, although this step is by no means universal. Many transcription approaches are available and the choice of which to use depends on research questions and theoretical commitments. When video is transcribed prior to analysis, the transcription itself may become the main data source. Thus, the researcher must decide which parts of the video are important and should be represented in the transcript and which parts not to include. For example, Big D discourse analysis (Gee, 1999) involves study of language plus actions, interactions, gestures, and intonations indicative of meaning and positioning, all of which must be represented in a transcript that is a primary data source. A complete notation system, such as Jeffersonian transcription, must be used in such cases. Barron et al. (2013) provide a table of common transcription choices.

Transcribing small-group interaction is a common challenge for video researchers. Most transcripts of collaborative interaction are arranged on a timeline with the discourse contribution of each subject appearing in a separate row or column, so that segments of overlapping talk are evident. Other enhancements to the transcript are made as needed for the research.

Sometimes the preparation of a detailed transcript is itself the analytic process that the researcher carries out personally for the purpose of developing a detailed understanding of the video. Analytic transcription of video may be shared and debated among collaborators as they gradually perfect their transcript and their understanding of the data.

Even detailed approaches to video analysis may not include developing a transcript at all but rather viewing and reviewing the video directly. This is facilitated by technologies that enable researchers to find, play, and replay the video, sometimes in slow motion, sometimes without sound, to deeply examine interactions. In this type of analysis, often collaborative, a transcript might only support video study. Automated transcription is increasing in accuracy and may sometimes be used for this purpose, but the sound quality from research video is often not of sufficient quality to support automated transcription of collaborative interaction.

Coding

As with transcription, researchers must also decide whether or not to code their video data and, if so, what approach to adopt. There may be times when coding is absolutely necessary, and times when it is inappropriate for the study at hand (Saldana, 2013). For example, if video data are used to measure whether a treatment leads to improvements in students' use of scientific evidence in argument, then coding for quantity and quality of evidential argument would be appropriate. If the goal is to provide a rich and thick description of students' scientific reasoning, then coding may not help.

Video coding can be more or less deductive or inductive. If research is testing particular theoretical conjectures or hypotheses, a predetermined coding system can be applied to data. If the researcher is approaching the data without strong hypotheses, then systematic coding approaches based on grounded theory, which range from open to more focused coding, allow patterns to emerge from the data (Glaser & Strauss, 1967). The research literature contains many examples of video coding systems developed in classroom research (e.g., Wu, Krajcik, & Soloway, 2001). A classic example is from the famous Third International Mathematics and Science Study (TIMSS) described by Stigler, Gallimore, and Hiebert (2000), in which coding enabled comparisons of teaching across cultures.

Social Analysis Process

Video research is greatly enhanced by collaborative analysis. Collaborative analysis in coding studies is essential to establish reliability and validity of codes and the coding process. Most methods of social video analysis without coding represent some variation on interaction analysis (IA), as described by Jordan and Henderson (1995) and others. IA sessions involve multiple researchers in viewing and discussing video together. A lead researcher usually assembles a group that includes various experts, furnishes the video and a transcript, and informs the group of the research purpose. In a typical procedure, the group closely watches a short video clip together while taking notes, and with stops and re-viewings as requested. Afterward, group members have five minutes or more to write reflections. Then researchers share their reflections verbally followed by a whole-group discussion. Session recordings become data used by the group leader to write a video analysis, which is later shared with the group for comment.

Technological Tools

A major focus of learning science research has been development of innovative technologies for supporting video research, primarily technologies for editing, archiving and analyzing video. Some were inspired by Goldman's (Goldman, 2007; Goldman-Segall, 1998) ORION (originally called Learning Constellations), an early innovative system for working collaboratively with video clips to develop interpretive analyses. Years of research with video technologies have produced important and creative ideas (Pea, 2006). Today many video handling and analysis functions, such as searching, importing, syncing with transcripts, coding, and annotation, are available within proprietary commercial systems, such as NViVo or the web-based Dedoose, which support researchers internationally and meet institutional review board (IRB) requirements. Good video research can also be conducted with basic capabilities provided by Microsoft and Apple tools. Recently, data mining and analytics tools have spawned new possibilities in video analysis, presenting new technological challenges, as will be discussed.

Reporting

There have been notable experiments with video reporting formats, such as special issues of journals inviting multiple analyses of a video that was also supplied to readers (Koschmann, 1999). Some researchers have pioneered video websites to supplement and enhance their publications (Goldman-Segall, 1998). However, most video research leaves the video behind when analyses are reported. This is partly in response to ethical and regulatory protections for the privacy of human subjects who are recorded. Anonymous transcripts of video may be provided in appendices, and excerpts of transcripts may be interspersed within text. Screen shots from video are effective for illustrating such things as gesture, spatial orientation, and facial expression. In addition to screen shots, illustrations are sometimes used (Hall, Stevens, & Torralba, 2002).

Many studies report both quantitative and qualitative analyses of video data, which is advocated in guideline publications (Derry et al., 2010). While most journals specify standard formats for reporting quantitative data, standardized formats for reporting qualitative analyses of video are not common. One frequently used approach is play-by-play analysis, in which segmented transcripts of video are presented in narrative order, interspersed with interpretive discussion of each segment. Transcript segments of collaborative interaction may be presented on a timeline with the discourse contribution of each subject unfolding in a separate row or column. Because journal pages are limited, video researchers are challenged to find concise ways of clearly presenting convincing analyses of complex video phenomena.

Conciseness may not be the goal of video research that aims for thick description (Geertz, 1973). Still, layered description of complex phenomena requires well-organized and interesting narrative that uses video data as evidence, as illustrated in the classic piece on professional vision by Goodwin (1994).

Conducting Research on Learning with/Through Video

Most learning scientists collect video recordings as data sources. However, learning scientists also conduct educational design research on learning environments that employ video as key components of systems for learning and/or assessment (e.g., Seidel, Blomberg, & Renkl, 2013; Zahn et al., 2010). These environments typically aim to provide learners with a wider range of complex real-world experiences than would be available to them otherwise. In teacher education, for example, videos of K–12 classrooms provide pre-service learners with common objects to discuss in their classes, providing broader experiences than they would encounter through field placements alone. Here we provide some guidance for conducting design research with multimedia learning environments that employ video.

Theoretical Frameworks

We mention three theoretical frameworks that offer methodological guidance for studying video as a tool for learning. One framework, by Schwartz and Hartman (2007), suggests that specific types of "designed video" (p. 335) are needed to achieve specific learning goals. Their scheme identifies four types of video learning outcomes (saying, engaging, doing, and seeing), suggests assessment approaches for each type, and identifies genres of video suitable for achieving desired learning outcomes. For example, advertisements, trailers, and narratives represent video genres suitable for engaging learners' interest, which can be assessed with preference inventories.

A framework by Mayer and Moreno (2003) suggests principles of design as well as research questions for studies of multimedia environments. Their framework posits a dual-channel cognitive architecture in which: (1) humans process visual and verbal material in different channels; (2) the amount of material that can be processed by each channel at once is limited; and (3) meaningful learning involves actively building representations that conceptually integrate visual and verbal material. This cognitive architecture makes it likely that complex learning with video and text will create cognitive overload, so design principles for managing overload are needed. Mayer and Moreno validated design principles for efficiently engaging both channels in multimedia laboratory settings. Derry et al. (2014) applied and extended this theory to create design and assessment ideas for research on video-based teacher professional development.

Another theoretical perspective that continues to influence video-learning research is Cognitive Flexibility Theory (CFT; Spiro, Feltovich, Jacobson, & Coulson, 1991; Spiro, Collins, Thota, & Feltovich, 2003). CFT suggests strategies for advanced knowledge acquisition that develop the ability to flexibly see situations in multiple complex ways. CFT strategies also attempt to compact experience to accelerate development beyond what might be accomplished through traditional instruction and field-based experience. Two CFT strategies are *domain criss-crossings* and *small multiples instruction*. With domain criss-crossing, a technology presents learners with a large number of short video cases (mini-cases) that represent a target concept. With small multiples, students repeatedly bring learned concepts together, in different blends, to interpret complex cases. These strategies might be combined in an online environment or during a class discussion aided by technologies. CFT approaches have been tested in experimental and quasi-experimental studies, faring well on multiple measures in relation to comparison groups (e.g., Derry et al., 2006; Goeze, Zottmann, Fischer, & Schrader, 2010; Palincsar et al., 2007). CFT principles suggest important hypotheses and design conjectures for educational design studies.

Example Research

In the learning science field, research on learning from video has focused, not on creating high-quality video productions, but rather on developing design principles for learning from minimally edited video depicting real-world practice. Many approaches emphasize building learning communities and cultures of discourse that engage in collaborative analysis of video cases. For example, Sherin and colleagues (e.g., Sherin & van Es, 2009) conduct research on teacher learning in the context of video clubs, designed professional development sessions in which colleagues watch and discuss video excerpts from each other's classrooms. Research on video clubs illustrates how learning occurs and is measured at both individual and system level. Also significant is the proliferation of online environments that feature collaborative learning with video. For example, Derry and Hmelo-Silver (e.g., Hmelo-Silver, Derry, Bitterman, & Hatrak, 2009) used video cases online to develop pre-service teachers' abilities to employ learning science concepts to analyze and design instruction. Their studies documented the effectiveness of this online approach and illustrated use of complex video-based assessments in research. Learning to design effective collaborative environments that use authentic video sources to promote and measure learning is an important program of research in the learning sciences.

Analytics and Data Mining

Advances in computation that have accompanied the emergence of "big data" hold promise for improving video research. Of immediate relevance are Educational Data Mining and Learning Analytics, which involve measurement, collection, analysis, and reporting of data about learners and their contexts for purposes of understanding learners and optimizing learning environments (Siemens & Baker, 2012; Rosé, this volume). These fields have overlapping purposes but work with different types of algorithms.

Video repositories provide one context for application. With appropriate attention to ethical standards for sharing video (Derry et al., 2010), a video research database can be an important resource to be mined and studied repeatedly long after data collection is finished. The Video Mosaic repository for mathematics learning research and education illustrates this idea (Maher, 2009). Educational Data Mining or Learning Analytics might be used to search a large video data corpus and automatically select data relevant to specific research questions. When a video feature is mined, it undergoes analysis through a computer algorithm, and that information can be automatically annotated on the video, which can be selected for analysis (Calderara, Cucchiara, & Prati, 2008). For example, finding audio stream patterns in classroom videos can identify and select clips for study of classroom discussion (Li & Dorai, 2005).

Educational Data Mining and Learning Analytics have strong potential for supporting video analysis. Indexing, coding, and pattern finding are tedious and time-consuming, even using video analysis software. Big data methodologies offer the possibility of automatic selection and annotation, data transformation, and rapid analysis. Even when raw video data is not automatically selected and analyzed, manual video annotations can be mined for patterns.

Complex computer-vision algorithms show promise for educational video analytics, but these developments have occurred largely for the purpose of surveillance. Surveillance systems have been used in educational research to track and study people in museum environments (e.g., Beaumont, 2005). One barrier to using surveillance systems to study learning is that algorithms are proprietary and unavailable for academic use (Borggrewe, 2013). Challenges include ethical considerations related to video recording in public spaces and obtaining informed consent for research.

Multimodal Learning Analytics (Worsley et al., 2016) suggests the possibility of advanced computational approaches to analysis of multimodal data that includes video as a major data source. In addition to mining patterns from video, Multimodal Learning Analytics incorporates use of additional data generated from capture devices that are synchronized with video collection. Multimodal data can include indicators of attention, stress, emotions, or any number of other data streams.

Learning scientists have not fully explored the potential for Learning Analytics, Educational Data Mining, and Multimodal Learning Analytics in video research, possibly because many educational researchers lack expertise in these methods and the learning curve is steep. Many studies that use these approaches require researchers to build their own software, posing a challenge for those without software-creation skills or personnel. As graduate programs shift their goals and standard "big data" tools and algorithms become more commonplace, they will likely prove important for video research in the learning sciences.

Concluding Comments

Many, if not most, research projects in the learning sciences include a substantial video component and this trend is likely to continue. Our chapter shared some major challenges we have faced in conducting video-based educational design projects, as well as solutions ranging from standard practices to newly invented ones. An important avenue for future research is to examine these challenges in light of solutions offered by emerging computational technologies, which could greatly advance the power of video research in the foreseeable future.

Further Readings

Derry, S. J., Pea, R., Barron, B., Engle, R., Erickson, F., Goldman, R., et al. (2010). Conducting video research in the learning sciences: Guidance on selection, analysis, technology, and ethics. *Journal of the Learning Sciences, 19*, 1–51.
An interdisciplinary conference of scholars convened to provide guidelines for judging the video research proposed to and funded by the National Science Foundation. The conference generated a lengthy report http://drdc.uchicago.edu/what/video-research-guidelines.pdf. This condensed article by a subset of scholars summarized major findings.

Maher, C. (2009, November) *Video mosaic collaborative.* [Video file]. Retrieved December 13, 2016, from www.youtube.com/watch?v=FQqx8Gw610g
Dr. Carolyn Maher discusses Video Mosaic, a collaborative video repository used for teaching and research that houses the Robert B. Davis Institute for Learning Collection of videos depicting students' learning mathematical concepts captured through a longitudinal study from 1992 to the present.

Siemens, G., & Baker, R. S. J. (2012). Learning analytics and educational data mining: Towards communication and collaboration. *Proceedings of the Second International Conference on Learning Analytics and Knowledge* (pp. 252–254). New York: ACM Press.
Growing interest in big data has strong implications for the future of video research. Two lines of work, Educational Data Mining and Learning Analytics, have developed separately. This paper offers an accessible introduction to these fields.

Stigler, J. W., Gallimore, R., & Hiebert, J. (2000). Using video surveys to compare classrooms and teaching across cultures: Examples and lessons from the TIMSS video studies. *Educational Psychologist, 35*(2), 87–100.
This piece contains a description and discussion of the research methods, including code development and coding of video data, employed in the third international mathematics and science study (TIMSS), which used video to compare classroom practices in Germany, Japan, and the United States.

Zahn, C., Pea, R., Hesse, F., & Rosen, J. (2010) Comparing simple and advanced video tools as supports for complex collaborative design processes. *Journal of the Learning Sciences, 19*(3), 403–440.

An experiment analyzed video and other data from 24 collaborating dyads to compare history learning in two designed learning environments based on different video tools: the advanced video tool WebDIVER and a simple video playback tool. The advanced tool fostered better understanding and student products, and was more efficient.

References

Alac', M. & Hutchins, E. (2004). I see what you are saying: Action as cognition in fMRI brain mapping practice. *Journal of Cognition & Culture, 4*(3), 229–661.

Barber-Lester, K. J., Derry, S. J., Minshew, L. M. & Anderson, J. (2016). Exploring visualization and tagging to manage big datasets for DBR: A modest proposal with significant implications. In C.-K. Looi, J. Polman, U. Cress, & P. Reimann (Eds), *Transforming learning, empowering learners: Proceedings of The International Conference of the Learning Sciences (ICLS) 2016* (pp. 966–969). Singapore, June.

Barron, B., Pea, R., & Engle, R. (2013). Advancing understanding of collaborative learning using data derived from video records. In C. Hmelo-Silver, C. A. Chinn, A. M O'Donnell, & C. Chan (Eds.), *The international handbook of collaborative learning* (pp. 203–219). New York: Routledge.

Beaumont, E. (2005). Using CCTV to study visitors in the New Art Gallery, Walsall, UK. *Surveillance and Society, 3*, 251–269.

Borggrewe, S. (2013). *Movement analysis of visitors using location-aware guides in museums* (master's thesis). Media Computing Group, Computer Science Department, RWTH Aachen University.

Calderara, S., Cucchiara, R., & Prati, A. (2008). Bayesian-competitive consistent labeling for people surveillance. *IEEE Transactions on Pattern Analysis and Machine Intelligence, 30*(2), 354–360.

Choi, G. W., Land, S. M., & Zimmerman, H. T. (2016). Educational affordances of tablet-mediated collaboration to support distributed leadership in small group outdoor activities. In C.-K. Looi, J. Polman, U. Cress, & P. Reimann (Eds), *Transforming learning, empowering learners: Proceedings of the International Conference of the Learning Sciences (ICLS) 2016* (pp. 882–885), Singapore, June.

Derry, S. J., Hmelo-Silver, C. E., Nagarajan, A., Chernobilsky, E., & Beitzel, B. (2006). Cognitive transfer revisited: Can we exploit new media to solve old problems on a large scale? *Journal of Educational Computing Research, 35*, 145–162.

Derry, S. J., Pea, R., Barron, B., Engle, R., Erickson, F., Goldman, R., et al. (2010). Conducting video research in the learning sciences: Guidance on selection, analysis, technology, and ethics. *Journal of the Learning Sciences, 19*, 1–51.

Derry, S. J., Sherin, M. G., & Sherin, B. L. (2014). Multimedia learning with video. *Cambridge handbook of multimedia learning* (2nd ed., pp. 785–812). New York: Cambridge University Press.

Enyedy, N., & Stevens, R. (2014). Analyzing collaboration. In K. Sawyer (Ed.), *The Cambridge handbook of the learning sciences* (2nd ed., pp. 191–212). New York: Cambridge University Press.

Gee, J. P. (1999). *An introduction to discourse analysis: Theory and method.* New York: Routledge.

Geertz, C. (1973). Thick description: Toward an interpretive theory of culture. In C. Geertz (Ed.), *The interpretation of cultures: Selected essays* (pp. 3–30). New York: Basic Books.

Glaser, B. G., & Strauss, A. L. (1967). *The discovery of grounded theory: Strategies for qualitative research.* New Brunswick, NJ: Transaction Publishers.

Goeze, A., Zottmann, J. M., Vogel, F., Fischer, F., & Schrader, J. (2014). Getting immersed in teacher and student perspectives? Facilitating analytical competence using video cases in teacher education. *Instructional Science, 42*(1), 91–114.

Goldman, R. (2007). ORION™, an online digital video data analysis tool: Changing our perspectives as an interpretive community. In R. Goldman, R. Pea, B. Barron, & S. J. Derry (Eds.), *Video research in the learning sciences* (pp. 507–520). Mahwah, NJ: Erlbaum.

Goldman-Segall, R. (1998). *Points of viewing children's thinking: A digital ethnographer's journey.* Mahwah, NJ: LEA.

Goodwin, C. (1994). Professional vision. *American Anthropologist 96*(3), 606–633.

Green, J. L., & Bridges, S. M. (2018) Interactional ethnography. In F. Fischer, C. E. Hmelo-Silver, S. R. Goldman, & P. Reimann (Eds.), *International handbook of the learning sciences* (pp. 475–488). New York: Routledge.

Hall, R., Stevens, R., & Torralba, A. (2002). Disrupting representational infrastructure in conversations across disciplines. *Mind, Culture, and Activity, 9*(3), 179–210.

Hmelo-Silver, C., Derry, S., Bitterman, A., & Hatrak, N. (2009). Targeting transfer in a STELLAR PBL course for preservice teachers. *Interdisciplinary Journal of Problem-based Learning 3*(2), 24–42.

Jordan, B., & Henderson, A. (1995). Interaction analysis: Foundations and practice. *Journal of the Learning Sciences, 4*(1), 39–103.

Kisiel, J., Rowe, S., Wartabedian, M. A., & Kopczak, (2012). Evidence for family engagement in scientific reasoning at interactive animal exhibits. *Science Education, 96*(6), 1047–1070.

Koschmann, T. (Ed.). (1999). Meaning making [Special issue]. *Discourse Processes, 27*(2).
Koschmann, T. (2018). Ethnomethodology: Studying the practical achievement of intersubjectivity. In F. Fischer, C. E. Hmelo-Silver, S. R. Goldman, & P. Reimann (Eds.), *International handbook of the learning sciences* (pp. 465–474). New York: Routledge.
Koschman, T., & Derry, S. J. (2017). "If Green was A and Blue was B": Isomorphism as an instructable matter. In A. Makitalo, P. Linell, & R. Saljo (Eds.), *Memory practices and learning* (pp. 95–112). Charlotte, NC: Information Age Publishing.
Lee, H. S., Pallant, A., Tinker, R., & Horwitz, P. (2014). High school students' parameter space navigation and reasoning during simulation-based experimentation. In J. L. Polman, E. A. Kyza, D. K. O'Neill, I. Tabak, W. R. Penuel, A. S. Jurow, et al. (Eds.), *Learning and becoming in practice: The International Conference of the Learning Sciences (ICLS) 2014* (Vol. 1, pp. 681–688), Boulder, CO.
Li, Y., & Dorai, C. (2005). Video frame identification for learning media content understanding. *2005 IEEE International Conference on Multimedia and Expo* (pp. 1488–1491). Amsterdam, July. doi:10.1109/ICME.2005.1521714
Maher, C. (Nov, 2009). Video Mosaic Collaborative. Retrieved December 13, 2016, from www.youtube.com/watch?v=FQqx8Gw610g on
Malkiewich, L., J., Lee, A., Slater, S., Xing, C., & Chase, C. C. (2016). No lives left: How common game features could undermine persistence, challenge-seeking and learning to program. In C.-K. Looi, J. Polman, U. Cress, & P. Reimann (Eds.), *Transforming learning, empowering learners: Proceedings of the International Conference of the Learning Sciences (ICLS) 2016* (pp. 186–193), Singapore, June.
Mayer, R. E., & Moreno, R. (2003). Nine ways to reduce cognitive load in multimedia learning. *Educational Psychologist, 38(1)*, 43–52.
McKenney, S., & Reeves, R. C. (2012). *Conducting educational design research.* New York: Routledge.
Mehan, H. (1979). *Learning lessons: Social organization in the classroom.* Cambridge, MA: Harvard University Press.
Minshew, L., Derry, S., Barber-Lester, K. J., & Anderson, J. (2016). Designing for effective collaborative learning in high-needs rural schools. In C.-K. Looi, J. Polman, U. Cress, & P. Reimann (Eds.), *Transforming learning, empowering learners: Proceedings of the International Conference of the Learning Sciences (ICLS) 2016* (pp. 978–981), Singapore, June.
Palincsar, A. P., Spiro, R. J., Kucan, L., Magnusson, S. J., Collins, B. P., Hapgood, S., et al. (2007). Research to practice: Designing a hypermedia environment to support elementary teachers' learning of robust comprehension instruction. In D. McNamara (Ed.), *Reading comprehension strategies: Theory, interventions, and technologies.* Mahwah, NJ: Lawrence Erlbaum.
Pea, R. D. (2006). Video-as-data and digital video manipulation techniques for transforming learning sciences research, education, and other cultural practices. In J. Weiss, J. Nolan, J. Hunsinger, & P. Trifonas (Eds). *The international handbook of virtual learning environments* (pp. 1321–1393). Dordrecht: Springer.
Puntambekar, S. (2018). Design-based research (DBR). In F. Fischer, C. E. Hmelo-Silver, S. R. Goldman, & P. Reimann (Eds.), *International handbook of the learning sciences* (pp. 383–392). New York: Routledge.
Rosé, C. P. (2018). Learning analytics in the Learning Sciences. In F. Fischer, C. E. Hmelo-Silver, S. R. Goldman, & P. Reimann (Eds.), *International handbook of the learning sciences* (pp. 511–519). New York: Routledge.
Saldana, J. (2013). *The coding manual for qualitative researchers.* Thousand Oaks, CA: SAGE.
Schwartz, D. L., & Hartman, K. (2007). It's not television anymore: Designing digital video for learning and assessment. In R. Goldman, R. Pea, B. Barron, & S. Derry (Eds.), *Video research in the learning sciences* (pp. 335–348). Hillsdale, NJ: Erlbaum.
Seidel, T., Blomberg, G., & Renkl, A. (2013). Instructional strategies for using video in teacher education. *Teaching and Teacher Education, 34*, 56–65.
Sherin, M. G., & van Es., E. A. (2009). Effects of video club participation on teachers' professional vision. *Journal of Teacher Education 60*(1), 20–37.
Siemens, G., & Baker, R. S. J. d. (2012). Learning analytics and educational data mining: Towards communication and collaboration. *Proceedings of the Second International Conference on Learning Analytics and Knowledge* (pp. 252–254). New York: ACM Press.
Spiro, R. J., Collins, B. P., Thota, J. J., & Feltovich, P. J. (2003). Cognitive flexibility theory: Hypermedia for complex learning, adaptive knowledge application, and experience acceleration. *Educational Technology, 43*, 5–10.
Spiro, R. J., Feltovich, P. J., Jacobson, M. J., & Coulson, R. L. (1991). Cognitive flexibility, constructivism, and hypertext: Random access instruction for advanced knowledge acquisition in ill-structured domains. *Educational Technology, 31*(5), 24–33.
Stigler, J. W., Gallimore, R., & Hiebert, J. (2000). Using video surveys to compare classrooms and teaching across cultures: Examples and lessons from the TIMSS video studies. *Educational Psychologist, 35*(2), 87–100.

van de Pol, J., Volman, M., Oort, F., & Beishuizen, J. (2014). Teacher scaffolding in small-group work: An intervention study. *Journal of the Learning Sciences, 23*(4), 600–650. doi:10.1080/10508406.2013.805300

Worsley, M., Abrahamson, D., Bilkstein, P., Grover, S., Schneider, B. & Tissenbaum, M. (2016). Situating multimodal learning analytics. In C.-K. Looi, J. Polman, U. Cress, & P. Reimann (Eds.), *Transforming learning, empowering learners: Proceedings of the International Conference of the Learning Sciences (ICLS) 2016* (pp. 1346–1349), Singapore, June.

Wu, H. K., Krajcik, J. S., & Soloway, E. (2001). Promoting understanding of chemical representations: Students' use of a visualization tool in the classroom. *Journal of Research in Science Teaching, 38*(7), 821–842.

Zahn, C., Pea, R., Hesse, F., & Rosen, J. (2010) Comparing simple and advanced video tools as supports for complex collaborative design processes. *Journal of the Learning Sciences, 19*(3), 403–440.

48

Quantifying Qualities of Collaborative Learning Processes

Freydis Vogel and Armin Weinberger

Why Analyze Collaborative Learning Processes?

Collaborative learning, especially with the support of computers, is conducive to the development of higher-order thinking, of key competencies, and of learner agency (Cohen, 1994; King, 2007; Roschelle, 2013). Over the past 30 years, numerous studies showed how learners may help each other achieve higher learning gains than learning individually, provided the right goal structures are in place (Slavin, 2010) and provided the learning tasks require learners to work together (Cohen, 1994). In addition to collaborative problem-solving tasks (Rosen & Foltz, 2014), current research highlights the role of argument and discussion for deep elaboration of learning materials (e.g., Stegmann, Wecker, Weinberger, & Fischer, 2012). In argumentative knowledge construction, generating and sharing arguments are related to understanding multiple perspectives of an issue, using and linking concepts to analyze a problem, and learning how to argue (Kuhn & Crowell, 2011; Noroozi, Weinberger, Biemans, Mulder & Chizari, 2012).

Regardless of whether approaches to collaborative learning focus on motivation or cognition, all of these theoretical models entail, implicitly or explicitly, hypotheses about collaborative learning processes. Consideration of learners' interactions (Dillenbourg, Baker, Blaye, & O'Malley, 1995) is important for several reasons.

First, understanding and facilitating collaborative learning is difficult without such consideration. To understand the mechanisms in collaborative learning, it is important to link context variables – such as type of task, individual differences, and incentives – to the learners' interactions.

Second, some phenomena may emerge only in learners' interactions, such as co-constructed knowledge, or learners' social skills and internal scripts (Fischer, Kollar, Stegmann, & Wecker, 2013). Although the extent to which evaluation of learning can be based only on group processes is debatable, the analysis of collaborative learning processes in addition to the analysis of learning outcomes can certainly help in assessing learners' approaches and understanding.

Third, theoretical models have more or less explicit hypotheses about processes of collaborative learning, so these models cannot be put to the test without analyzing processes as mediators and moderators of learning. For instance, process analyses are required for understanding the interplay of mutual support versus mutually posing challenges to peers' understandings (Piaget, 1969; Vygotsky, 1978). However, in experimental research, hypotheses about the relation between the independent variable (or the influencing factor) and the dependent variable (usually operationalized as the learning outcomes) have often been investigated without specifying the learning processes.

Thus, various research syntheses report on the effects of specific types of collaborative learning on learning outcomes but rarely report on the relationships between learning processes and learning outcomes (Johnson & Johnson, 2009; Murphy, Wilkinson, Soter, Hennessey, & Alexander, 2009; Vogel, Wecker, Kollar, & Fischer, 2017). Analysis of collaborative learners' overt behavior, including verbal and non-verbal communication, may offer a view – however distorted and limited – on how learners interpret and co-regulate peer interaction and how they process information together.

The Procedure of Analyzing Behavior and Dialogue in the Learning Sciences

This section takes a step-by-step tour through the procedure of analyzing interactions and dialogue. Identifying relevant collaborative learning processes and how to operationalize these processes requires a firm theoretical basis. Qualitative research in the Learning Sciences and beyond has substantially contributed to identifying specific processes or activities as relevant for learning (see also Dingyloudi & Strijbos, this volume; Green & Bridges, this volume; Koschmann, this volume). However, attaining high external validity and making predictions is difficult with qualitative approaches. To do this, it is necessary to quantify the qualities of the collaborative learning processes for a large sample of learners or at least a rather large sub-sample of the initial sample of a study (Chi, 1997).

Selection and Operationalization of Collaborative Learning Processes to Be Observed

To observe and analyze collaborative learning process data, several decisions need to be made and justified: What is the theoretical basis of the analysis? What are the relevant process variables potentially influencing learning? How can these variables be operationalized? How can learning process data be collected and recorded? These decisions precede the development of a "coding scheme" that delineates whether and how the data should be segmented and how each segment can be assigned to one category for each variable. The following sections present the above questions and strategies that can be followed to come to a decision for each question.

What Is the Theoretical Basis of the Analysis?

Deciding which variables to analyze can build on approaches that specify relevant learning activities (e.g., Chi & Wylie, 2014; Fischer et al., 2013). For instance, the use of so-called interactive activities (Chi & Wylie, 2014), transactivity (King, 1998), or convergent conceptual change (Roschelle, 1992) during the learning process is ascribed to positively affect the learning outcome. These beneficial learning activities mainly presume that learners build upon each other's contributions. Based on a Vygotskyan theoretical approach to collaborative learning (Vygotsky, 1978), by building on each other's contributions, learners will mutually help each other to achieve higher levels of development. From a Piagetian point of view (Piaget, 1969), learners will rather encounter socio-cognitive conflicts when building on each other's contradictory contributions and will learn through their effort to find a solution for the socio-cognitive conflict that might also lead to convergent conceptual change (Roschelle, 1992).

The collaborative learning activities of questioning and explaining (Gelmini-Hornsby, Ainsworth, & O'Malley, 2011; Webb et al., 2009) as well as reciprocal peer-tutoring (King, 1998; Palincsar & Brown, 1984; Walker, Rummel, & Koedinger, 2011), can be seen as instances of Vygotsky's (1978) theoretical approach. The reciprocal peer-tutoring structures communication between learners in a way that uncovers the learners' difficulties with comprehension and enables learners to resolve the difficulties by reciprocal support.

An extended version of reciprocal peer-tutoring targets uncovering the different perspectives and perceptions and overcoming them by discourse (King, 1998). This builds on the Piagetian idea that resolving socio-cognitive conflicts can be beneficial for learning (Mugny & Doise, 1978). Such conflicts emerge, for instance, when students from different domains are brought together to solve interdisciplinary problems (Noroozi, Teasley, Biemans, Weinberger, & Mulder, 2013; Rummel, Spada, & Hauser, 2009). In addition, incidences of overcoming socio-cognitive conflicts occur when learners in small groups are asked to critically reflect on partner arguments, balance arguments, and negotiate a joint solution (Asterhan & Schwarz, 2009; Kollar et al., 2014; Noroozi et al., 2012; Vogel et al., 2017).

What Are the Relevant Process Variables Potentially Influencing Learning?

To find the relevant variables that should be coded, it is important to clarify which theoretical approach should be followed in the context of a given study such as the argumentative knowledge construction approach (Andriessen, Baker, & Suthers, 2003; Schwarz & Baker, 2016; Weinberger & Fischer, 2006).

To illustrate how the procedure of analyzing behavior and dialogue in the Learning Sciences may look in practice, we exemplify the procedure for a specific case. In this case, learners in groups of four are asked to discuss a complex problem of conflicting interests on genetically modified food to learn more about scientific, ecological, and social aspects of using genetic engineering. The topic comprises many different perspectives for which a compromise might not easily be reached. Building on the argumentative knowledge construction approach, learners elaborate these multiple perspectives and resolve socio-cognitive conflicts to construct knowledge (Andriessen et al., 2003; Piaget, 1969).

How Can These Variables Be Operationalized?

Also building on the argumentative knowledge construction approach, different argumentative moves can be considered as categories of analysis, for example, statements and arguments for the different positions followed by counter-arguments and syntheses (e.g., Leitão, 2000). Beyond this specific analysis of the formal-rational argument structure, the arguments can then be coded with respect to multiple dimensions, including intuitive and emotional facets of discourse, which may also be represented in non-verbal behavior (Lund, Rosé, Suthers, & Baker, 2013).

How Can Learning Process Data Be Collected and Recorded?

The form of data collection will differ, depending on the form of communication (co-present, online synchronous, online asynchronous, oral, written, etc.). In computer-supported asynchronous communication (e.g., a forum), a log file might automatically record each step the learner may take, whereas in a co-present, face-to-face situation it might be useful to video- and audio-record the learners. In the proposed example about groups of four learners having a face-to-face discussion about genetically modified food, audio recordings that are then transcribed into text would be a coherent choice. Video-recording would be even richer because it provides more information about the learners' non-verbal communication.

Segmentation and Coding

The next steps include determining how to segment the data. The segments will be the smallest units of analysis on which to apply the coding scheme. The coding scheme must be developed based on the operationalization of the learning process. Segmentation and coding should be performed by more than one researcher to improve the objectivity of the rating process (De Wever, Schellens, Valcke, & Van Keer, 2006). Further, inter-rater reliability should be reported (Strijbos, Martens, Prins, & Jochems, 2006; Weinberger & Fischer, 2006). The analysis can also be supported by machines designed to rate like human raters (Rosé et al., 2008).

The first part of the coding scheme comprises how to segment the data. In many cases, segmentation rules are based on the syntax of the process data or on surface-level features. An example for syntactical segmentation would be to set the borders of the segments at the end of sentences marked by punctuation. One example of surface-level segmentation would be to define the border of the segments by turn taking. Obviously, these kinds of segmentation rules won't have many ambiguous cases regarding how to realize the segmentation (Strijbos et al., 2006). However, the resulting segments must enable the coding.

The granularity of the segmentation must be decided next. For instance, focusing on questions and answers may afford a medium granularity of segmentation. Overly fine-grained segments (e.g., words) and coarse-grained segments (e.g., the whole session) would forbid coding the segment in question. To decide on a specific level of granularity, the collaborative learning activities may be governed on different levels of specificity. Activities could be assigned to sequences of questions and answers on a rather low level of specificity. On a higher level of specificity, the activities could be further categorized into questions and answers and/or their internal syntactic structure. The specificity of the collaborative learning activities would dictate the high or low level of granularity chosen for the segmentation.

The script theory of guidance (Fischer et al., 2013) can help to decide the level of specificity. This theory proposes that learners use internal scripts as flexible cognitive schemas in collaborative learning. The learners' internal scripts consist of script components on different hierarchical levels with increasing specificity—namely, the play, scene, and scriptlet level. Each of the three hierarchical levels includes information about the structure of the internal script components one level below. That means, for instance, that on the play level there is information about the different scenes and their structure and sequence (Fischer et al., 2013; Vogel et al., 2016). By applying the script levels to the different levels of operationalization of argumentation in the learning process, we can map each level of operationalization to a script level. Thus, script levels may correspond to segmentation on different levels of granularity (see Table 48.1).

In the example shown in Table 48.1, the learning process is operationalized by the dialectical use of the pro-argument, counter-argument, and integration scenes, which can be seen as components of the play level. This might lead to setting the borders of the segments on a rather broad level – for instance, at the points of turn taking.

A coding scheme usually contains the description of the raw data, an explanation of the segmentation process of the data, and the coding rules and how to apply them to the data. More specifically, for the coding rules, descriptions are needed for the different dimensions, categories, and codes that operationalize the learning process based on the theory in question. Dimensions need to be applied to each segment of the data in parallel, and each dimension usually has at least two categories that

Table 48.1 Argumentative Activities Within the Collaborative Learning Process on Different Script Levels

Script level	Script play Inquiry argumentation/argumentative knowledge construction					
Play components	**Pro-argumentation scene**			**Counter-argumentation scene**		...
Scene components	Claim- scriptlet	Ground- scriptlet	Warrant- scriptlet	Claim- scriptlet	Ground-scriptlet	...
Scriptlet components	*relevance,* ...	*relevance,* *reliability,* ...	*logic,* ...	*relevance,* ...	*relevance,* *reliability,* ...	...

describe different levels of the dimension. The categories of each dimension are usually mutually exclusive, meaning that each segment can be rated with only one category per dimension. For each category within each dimension, a code must be defined that will be used to assign the categories of each dimension to the segments of the data. The categories of one dimension can be either nominal (pro-argumentation, counter-argumentation, etc.); ordinal (first-order counter-argument, second-order counter-argument, etc.) or on an ordinal/interval/ratio level (quality of the argument estimated on a Likert scale). Furthermore, the coding scheme needs explanations and examples for each category of each dimension. Examples can be generated from theory in a top–down process or they can be selected from the coding training material in a bottom-up process. A template can easily be used to present the coding information (see Table 48.2).

Segments are often ambiguous, and coders may disagree on assigning the segment to one specific category. Further rules determine what to do in those cases. For this purpose, different categories of one dimension can be hierarchically ordered; a rule might state that in the case of the occurrence of more than one category in one segment, the highest-order category must be chosen. Other rules might also make sense, such as falling back on additional indicators in the discourse data. Importantly, these rules need to be documented in the coding scheme and should allow for a distinct decision about which category to use.

Table 48.2 Template for a Coding Scheme for One Dimension with Different Categories, Including Their Descriptions, Examples, and Codes

	Dimension: "Argumentation Scene"		
Code	Category	Description	Example
1	Pro-Argumentation Scene	Segments in which the learners pose their standpoints and arguments without relating to the contributions of the others.	"My opinion is that genetically transformed food should not be given to babies because we don't have any long-time experiences about the impact of the food on one's health."
2	Counter-Argumentation Scene	Segments in which the learners mainly pose a critique or counter-argument against the argumentation of the learning partner.	"No, you are wrong. Many experiments have been done with mice and primates and showed no health problems when consuming genetically transformed food."
3	Integration Scene	Segments in which different arguments are counterbalanced and/or a conclusion is drawn from the integration of different arguments.	"You might be right, saying that studies did not find any health problems for mice and primates. Nevertheless, we do not know to what extent that finding is also valid for human babies. Therefore, it would be a question of possible benefits and costs that come with feeding human babies genetically modified food."
...	...	...	...
99	No Argumentation	Segment in which no argumentation, counter-argumentation, or integration about the topic of genetically modified food can be found.	...

After the coding scheme has been developed, at least two independent coders should be trained on it. The coding training continues until sufficient inter-rater reliability is reached (Strijbos & Stahl, 2007). To measure the inter-rater reliability, different approaches can be taken depending on the level of data that is produced by the coding and the number of trained coders. For data on the interval/ratio or ordinal level, the intraclass correlation can be used no matter how many raters (e.g., Shrout & Fleiss, 1979). For data on the nominal level, Cohen's kappa (Cohen, 1960) and/or Fleiss's kappa (Fleiss, 1971) can be used.

The coding training follows the sequence of (1) joint discussion and changing the coding scheme; (2) coding 10% of the actual process data by each coder individually; (3) comparing the coding of all coders, detecting differences, and calculating the inter-rater reliability. If the inter-rater reliability does not reach sufficient values, start again with step (1). After reaching sufficient values, the data corpus can then be allocated to individual coders. To reduce the bias of the coded data due to differences in coding style (e.g., strictness of coding), data from different experimental conditions should then be distributed equally among the coders. To make the coding as objective and unbiased as possible, the coders must not know which condition they will be coding or whether they are coding pre- or post-test data. For more details about assessment and reporting of inter-rater reliability, see Lombard, Snyder-Duch, and Bracken (2002).

Tools to Support Coding

In their effort to bring together approaches from computer linguistics and the analysis of collaborative learning processes in the Learning Sciences, Rosé and colleagues (2008) proposed automatic coding. This can facilitate the otherwise very time-consuming manual coding procedure, thus increasing objectivity while speeding up the process. For a more detailed description, see the NAPLeS video recordings with Carolyn Rosé about learning analytics and educational data mining in learning discourses (http://isls-naples.psy.lmu.de/intro/all-webinars/rose_all/index.html). Various tools have been developed to support the coding process, as discussed next.

The nCoder tool (http://n-coder.org) is an internet platform that offers coder training and rule-based semi-automated coding. The tool is open for registered users, and it offers the development, implementation, and validation of automated coding schemes. It is especially designed for massive amounts of text data and promises that the amount of hand-coded data needed to establish validity and reliability of the automated coding is kept to a minimum. To use the nCoder tool for quantifying qualities of the learning process, the process data must be available in text format and already be segmented. The data can be uploaded to the platform, and a coding scheme with examples for each code that can be applied to the segments of the data can be entered into the system.

LightSIDE (http://ankara.lti.cs.cmu.edu/side/download.html) is a powerful software for text mining and machine learning. It can be freely downloaded and used after a registration process. LightSIDE requires prior segmentation of the text corpus. Also, LightSIDE needs manually coded data in order to learn the coding rules and calculate inter-rater reliability. Yet, beyond searching for predefined values within each segment, the software can learn the coding rules by feeding it manually coded data and algorithms that will form the basis of machine coding learning (Mayfield & Rosé, 2013).

Tatiana (Trace Analysis Tool for Interaction Analysts; http://tatiana.emse.fr/) is software that focuses on the analysis of multi-modal computer-mediated human interaction and can be freely downloaded from the webpage. This software was created by the European project LEAD (Dyke, Lund, & Girardot, 2009). Instead of substituting the human coder, the Tatiana software supports human coders in their effort to analyze collaborative learning interactions in a multi-modal way. The software can synchronize the different data and artifacts that emerge during collaborative learning, such as video data, audio data, log files, and much more. Then the synchronized data and the specific coding and interpretation of each type of data from different points of view can be represented within one screen. Furthermore, Tatiana helps researchers find traces of collaborative learning interaction by including all multi-modal data and views.

Outlook on Aggregation of Data and Statistical Models for Testing Hypotheses

After coding the process data comes the aggregation of data, such as identifying frequencies of categories. The theoretical approach, the research questions, and the hypotheses should be used as bases for decisions about the aggregation, such as which of the coded dimensions and categories should be used, which unit of analysis should be taken, and how data for each unit of analysis should be aggregated. The unit of analysis can be chosen differently; hence, it will also influence the interpretation of the outcome. For instance, the unit of analysis could be the learning group, the individual learner, the individual learner per time slot, or the single segments.

The most straightforward way to aggregate the data might be to count the occurrences of each code within the learning process of each learner or small group. However, simple counting leads to loss of information on the sequence of activities or the duration of segments. New methods, such as statistical discourse analysis (Chiu, 2008), allow integration of this kind of information. As an introduction to statistical discourse analysis, the recording of a webinar held by Ming Ming Chiu in the context of the ISLS NAPLeS webinar series is recommended (http://isls-naples.psy.lmu.de/intro/all-webinars/chiu/index.html).

To answer research questions about the influence of specific facets of the collaborative learning process on learning, standard statistical procedures such as ANOVAs and regressions are still commonly used (Cen, Ruta, & Powell, 2016). However, these procedures may be suboptimal because the assumption of independence of predictor variables is violated (Cress, 2008). Multi-level analysis is a method that can overcome the problem of variable dependence (Cress, 2008). Building on the regression model, multi-level analysis allows developing models with predictors on multiple levels, such as sentences in messages in discussions by individuals in groups in certain experimental conditions. For a deeper elaboration on the method of multi-level analysis, see De Wever and Van Keer (this volume) and the Further Readings section.

To not only analyze the properties of the single units of analysis but also identify and quantify the connections among units of analysis, the so-called method of social network analysis or epistemic network analysis is increasingly being used. With these analyses, the units and their relations can be represented in dynamic network models (de Laat, Lally, Lipponen, & Simons, 2007). For further insights about this method, see Shaffer (this volume) and the Further Readings section.

Also appropriate for the analysis of collaborative learning process data are further group level methods for analyzing knowledge convergence (Weinberger, Stegmann, & Fischer, 2007; Zottmann et al. 2013). These further methods are applicable to small groups, they take the interdependence of the data into account, and they also analyze the strength of the interdependence as an important part of the collaborative learning process.

Summary and Conclusion

The quantification of qualities for learning process data is an approach for testing hypotheses derived from process-oriented models of collaborative learning. The procedure of analyzing learning process data has been gradually developed and has attained high levels of objectivity. Aggregation of coded data can go beyond simple counting of the occurrences of relevant categories, and new methods for the analysis of the aggregated data can provide insights into various aspects of the collaborative learning process.

Nevertheless, there is room for continuous development of the procedure of analyzing learning process data. Advanced methods to describe and interpret collaborative learning processes may integrate multiple modalities of learner behavior as well as multiple subjective and objective data sources. Bringing together different approaches to address the multiple modalities of learner interactions, Suthers, Lund, Rosé, Teplovs, and Law (2013) have also highlighted how researchers have converged regarding how and what to analyze in collaborative learning processes. This consolidation of analysis methods promises to systematically advance our understanding of the implicit and explicit process assumptions on collaborative learning, including novel scenarios and arrangements.

Further Readings

Chiu, M. M., & Khoo, L. (2005). A new method for analyzing sequential processes: Dynamic multi-level analysis. *Small Group Research, 36*, 600–631. doi:10.1177/1046496405279309
Chiu and Khoo provide a broader perspective on multi-level analysis in this article. In particular, dynamic multi-level analysis is introduced as a method for analyzing sequential collaborative learning processes. This method not only helps to overcome the problem of dependent variables in collaborative learning, but also provides an opportunity to handle the sequential characteristics of process data.

Cress, U. (2008). The need for considering multi-level analysis in CSCL research—An appeal for the use of more advanced statistical methods. *International Journal of Computer-Supported Collaborative Learning, 3*(1), 69–84. doi:10.1007/s11412-007-9032-2
Cress focuses on methodological issues that arise when statistical models are used to test hypotheses about collaborative learning. This paper presents more insight about problems with traditional statistical models, such as ANOVA or regression analysis, and explains how multi-level analysis can help to overcome these problems.

Shaffer, D. W., Collier, W., & Ruis, A. R. (2016). A tutorial on epistemic network analysis: Analyzing the structure of connections in cognitive, social and interaction data. *Journal of Learning Analytics, 3*(3), 9–45. doi:10.18608/jla.2016.33.3
In this paper, Shaffer and colleagues offer a tutorial on epistemic network analysis as another method that can be used to analyze rich collaborative learning data, with which cognitive, social, and interaction levels can be connected.

Strijbos, J. W., & Stahl, G. (2007). Methodological issues in developing a multi-dimensional coding procedure for small group chat communication. *Learning & Instruction, 17*, 394–404. doi:10.1016/j.learninstruc.2007.03.005
Strijbos and Stahl focus in this paper on methodological issues that arise when developing coding schemes for analyzing dyadic learning conversations. It addresses readers who would like to deepen their understanding of the development of coding schemes, the unit of analysis, or the determination of inter-rater reliability.

Valcke, M. & Martens, R., (2006). Methodological issues in researching CSCL [Special Issue]. *Computers & Education, 46*(1). doi:10.1016/j.compedu.2005.04.004
In this paper, Valcke and Martens provide a detailed overview on the methodological issues in analyzing collaborative learning processes—recommended as an introductory reading.

NAPLeS Resources

Chiu, M. M., *Statistical discourse analysis (SDA)* [Webinar]. In *NAPLeS video series*. Retrieved October 19, 2017, from http://isls-naples.psy.lmu.de/intro/all-webinars/chiu/index.html

Rosé, C. P., *15 minutes about learning analytics and educational data mining in learning discourses* [Video file]. In *NAPLeS video series*. Retrieved October 19, 2017, from http://isls-naples.psy.lmu.de/video-resources/guided-tour/15-minutes-rose/index.html

Rosé, C. P., *Interview about learning analytics and educational data mining in learning discourses* [Video file]. In *NAPLeS Video series*. Retrieved October 19, 2017, from http://isls-naples.psy.lmu.de/video-resources/interviews-ls/rose/index.html

Rosé, C. P., *Learning analytics and educational data mining in learning discourses* [Webinar]. In *NAPLeS video series*. Retrieved October 19, 2017, from http://isls-naples.psy.lmu.de/intro/all-webinars/rose_all/index.html

Weinberger, A., & Stegmann, K., *Behavior and dialog analyses-quantifying qualities* [Webinar]. In *NAPLeS video series*. Retrieved October 19, 2017, from http://isls-naples.psy.lmu.de/intro/all-webinars/stegmann-weinberg2/index.html

References

Andriessen, J., Baker, M., & Suthers, D. (2003). *Arguing to learn: Confronting cognitions in computer-supported collaborative learning environments*. Dordrecht: Kluwer.

Asterhan, C. S. C., & Schwarz, B. B. (2009). Argumentation and explanation in conceptual change: Indications from protocol analyses of peer to peer dialog. *Cognitive Science, 33*(3), 374–400. doi:10.1111/j.1551-6709.2009.01017.x

Cen, L., Ruta, D., & Powell, L (2016). Quantitative approach to collaborative learning: Performance prediction, individual assessment, and group composition. *International Journal of Computer-Supported Collaborative Learning, 11*(2), 187–225. doi:10.1007/s11412-016-9234-6

Chi, M. (1997). Quantifying qualitative analyses of verbal data: A practical guide. *Journal of the Learning Sciences, 6*, 271–315. doi:10.1207/s15327809jls0603_1

Chi, M. T. H., & Wylie, R. (2014). The ICAP framework: Linking cognitive engagement to active learning outcomes. *Educational Psychologist, 49*(4), 219–243. doi:10.1080/00461520.2014.965823

Chiu, M. M. (2008). Flowing toward correct contributions during groups' mathematics problem solving: A statistical discourse analysis. *Journal of the Learning Sciences, 17*(3), 415–463. doi:10.1080/10508400802224830

Cohen, G. C. (1994). Restructuring the classroom: Conditions for productive small groups. *Review of Educational Research, 64*(1), 1–35. doi:10.3102/00346543064001001

Cohen, J. (1960). A coefficient of agreement for nominal scales. *Educational and Psychological Measurement, 20*, 37–46. doi:10.1177/001316446002000104

Cress, U. (2008). The need for considering multi-level analysis in CSCL research—An appeal for the use of more advanced statistical methods. *International Journal of Computer-Supported Collaborative Learning, 3*, 69–84. doi:10.1007/s11412-007-9032-2

de Laat, M., Lally, V., Lipponen, L., & Simons, R.-J. (2007). Investigating patterns of interaction in networked learning and computer-supported collaborative learning: A role for social network analysis. *International Journal of Computer-Supported Collaborative Learning, 2*(1), 87–103. doi:10.1007/s11412-007-9006-4

De Wever, B., Schellens, T., Valcke, M., & Van Keer, H. (2006). Content analysis schemes to analyze transcripts of online asynchronous discussion groups: A review. *Computers & Education, 46*(1), 6–28. doi:10.1016/j.compedu.2005.04.005

De Wever, B., & Van Keer, H. (2018). Selecting statistical methods for the learning sciences and reporting their results. In F. Fischer, C. E. Hmelo-Silver, S. R. Goldman, & P. Reimann (Eds.), *International handbook of the learning sciences* (pp. 532–541). New York: Routledge.

Dillenbourg, P., Baker, M., Blaye, A., & O'Malley, C. (1995). The evolution of research on collaborative learning. In P. Reimann & H. Spada (Eds.), *Learning in humans and machines: Towards an interdisciplinary learning science* (pp. 189–211). Oxford, UK: Elsevier.

Dingyloudi, F., & Strijbos, J. W. (2018). Mixed methods research as a pragmatic toolkit: Understanding versus fixing complexity in the Learning Sciences. In F. Fischer, C. E. Hmelo-Silver, S. R. Goldman, & P. Reimann (Eds.), *International handbook of the learning sciences* (pp. 444–454). New York: Routledge.

Dyke, G., Lund, K., & Girardot, J.-J. (2009). Tatiana: an environment to support the CSCL analysis process. In C. O'Malley, D. D. Suthers, P. Reimann, & A. Dimitracopoulou (Eds.), *Computer supported collaborative learning practices: CSCL 2009 Conference Proceedings* (pp. 58–67). Rhodes: International Society of the Learning Sciences.

Fischer, F., Kollar, I., Stegmann, K., & Wecker, C. (2013). Toward a script theory of guidance in computer-supported collaborative learning. *Educational Psychologist, 48*(1), 56–66. doi:10.1080/00461520.2012.748005

Fleiss, J. L. (1971), Measuring nominal scale agreement among many raters. *Psychological Bulletin, 76*, 378–382. doi:10.1037/h0031619

Gelmini-Hornsby, G., Ainsworth, S., & O'Malley, C. (2011). Guided reciprocal questioning to support children's collaborative storytelling. *International Journal of Computer-Supported Collaborative Learning, 6*(4), 577–600. doi:10.1007/s11412-011-9129-5

Green, J. L., & Bridges, S.M. (2018) Interactional ethnography. In F. Fischer, C. E. Hmelo-Silver, S. R. Goldman, & P. Reimann (Eds.), *International handbook of the learning sciences* (pp. 475–488). New York: Routledge.

Johnson, D. W., & Johnson, R. T. (2009). An educational psychology success story: Social interdependence theory and cooperative learning. *Educational Researcher, 38*(5), 365–379. doi:10.3102/0013189X09339057

King, A. (1998). Transactive peer tutoring: Distributing cognition and metacognition. *Educational Psychology Review, 10*(1), 57–74. doi:10.1023/A:1022858115001

King, A. (2007). Scripting collaborative learning processes: A cognitive perspective. In F. Fischer, I. Kollar, H. Mandl, & J. M. Haake (Eds.), *Scripting computer-supported collaborative learning—Cognitive, computational, and educational perspectives* (pp. 13–37). New York: Springer.

Kollar, I., Ufer, S., Reichersdorfer, E., Vogel, F., Fischer, F., & Reiss, K. (2014). Effects of collaboration scripts and heuristic worked examples on the acquisition of mathematical argumentation skills of teacher students with different levels of prior achievement. *Learning and Instruction, 32*(1), 22–36. doi:10.1016/j.learninstruc.2014.01.003

Koschmann, T. (2018). Ethnomethodology: studying the practical achievement of intersubjectivity. In F. Fischer, C. E. Hmelo-Silver, S. R. Goldman, & P. Reimann (Eds.), *International handbook of the learning sciences* (pp. 465–474). New York: Routledge.

Kuhn, D., & Crowell, A. (2011). Dialogic argumentation as a vehicle for developing young adolescents' thinking. *Psychological Science, 22*(4), 545–552. doi:10.1177/0956797611402512

Leitão, S. (2000). The potential of argument in knowledge building. *Human Development, 43*(6), 332–360. doi:10.1159/000022695

Lombard, M., Snyder-Duch, J., & Bracken, C. C. (2002). Content analysis in mass communication: Assessment and reporting of intercoder reliability. *Human Communication Research, 28*, 587–604. doi:10.1111/j.1468-2958.2002.tb00826.x

Lund, K., Rosé, C. P., Suthers, D. D., & Baker, M. (2013). Epistemological encounters in multivocal settings. In D. D. Suthers, K. Lund, C. P. Rosé, C. Teplovs, & N. Law (Eds.), *Productive multivocality in the analysis of group interactions* (pp. 659–682). New York: Springer.

Mayfield, E., & Rosé, C. P. (2013). LightSIDE: Open source machine learning for text accessible to non-experts. In M. D. Shermis & J. Burstein (Eds.), *Handbook of automated essay evaluation: Current applications and new directions* (pp. 124–135). New York: Routledge.

Mugny, G., & Doise, W. (1978). Socio-cognitive conflict and structure of individual and collective performances. *European Journal of Social Psychology, 8*, 181–192. doi:10.1002/ejsp.2420080204

Murphy, P. K., Wilkinson, I. A., Soter, A. O., Hennessey, M. N., & Alexander, J. F. (2009). Examining the effects of classroom discussion on students' comprehension of text: A meta-analysis. *Journal of Educational Psychology, 101*(3), 740–764. doi:10.1037/a0015576

Noroozi, O., Teasley, S. D., Biemans, H. J. A., Weinberger, A., & Mulder, M. (2013). Facilitating learning in multidisciplinary groups with transactive CSCL scripts. *International Journal of Computer-Supported Collaborative Learning, 8*(2), 189–223. doi:10.1007/s11412-012-9162-z

Noroozi, O., Weinberger, A., Biemans, H. J. A., Mulder, M., & Chizari, M. (2012). Argumentation-based computer-supported collaborative learning (ABCSCL). A systematic review and synthesis of fifteen years of research. *Educational Research Review, 7*(2), 79–106. doi:10.1016/j.edurev.2011.11.006

Palincsar, A. S., & Brown, A. L. (1984). Reciprocal teaching of comprehension-fostering and comprehension-monitoring activities. *Cognition and Instruction, 1*(2), 117–175. doi:10.1207/s1532690xci0102_1

Piaget, J. (1969). *The psychology of the child*. New York: Basic Books.

Roschelle, J. (1992). Learning by collaborating: Convergent conceptual change. *Journal of the Learning Sciences, 2*(3), 235–276. Retrieved from http://www.jstor.org/stable/1466609

Roschelle, J. (2013). Discussion [Special Issue on CSCL]. *Educational Psychologist, 48*(1), 67–70. doi:10.1080/00461520.2012.749445

Rosé, C. P., Wang, Y.-C., Cui, Y., Arguello, J., Stegmann, K., Weinberger, A., & Fischer, F. (2008). Analyzing collaborative learning processes automatically: Exploiting the advances of computational linguistics in computer-supported collaborative learning. *International Journal of Computer-Supported Collaborative Learning, 3*(3), 237–271. doi:10.1007/s11412-007-9034-0

Rosen, Y., & Foltz, P. W. (2014). Assessing collaborative problem solving through automated technologies. *Research and Practice in Technology Enhanced Learning, 9*(3), 389–410. doi:10.1007/978-3-319-33261-1_5

Rummel, N., Spada, H., & Hauser, S. (2009). Learning to collaborate while being scripted or by observing a model. *International Journal of Computer-Supported Collaborative Learning, 4*(1), 69–92. doi:10.1007/s11412-008-9054-4

Schwarz, B. B., & Baker, M. J. (2016). *Dialogue, argumentation and education: History, theory and practice*. Cambridge, UK: Cambridge University Press.

Shrout, P. E., & Fleiss, J. L. (1979) Intraclass correlations: Uses in assessing rater reliability. *Psychological Bulletin, 86*, 420–428.

Slavin, R. E. (2010). Co-operative learning: What makes groupwork work? In H. Dumont, D. Istance, & F. Benavides (Eds.), *The nature of learning: Using research to inspire practice* (pp. 161–178). Paris: OECD.

Stegmann, K., Wecker, C., Weinberger, A., & Fischer, F. (2012). Collaborative argumentation and cognitive elaboration in a computer-supported collaborative learning environment. *Instructional Science, 40*(2), 297–323. doi:10.1007/s11251-011-9174-5

Strijbos, J. W., Martens, R. L., Prins, F. J., & Jochems, W. M. G. (2006). Content analysis: What are they talking about? *Computers & Education, 46*(1), 29–48. doi:10.1016/j.compedu.2005.04.002

Strijbos, J. W., & Stahl, G. (2007). Methodological issues in developing a multi-dimensional coding procedure for small group chat communication. *Learning & Instruction, 17*, 394–404. doi:10.1016/j.learninstruc.2007.03.005

Suthers, D. D., Lund, K., Rosé, C. P., Teplovs, C., & Law, N. (2013). *Productive multivocality in the analysis of group interactions*. New York: Springer.

Vogel, F., Kollar, I., Ufer, S., Reichersdorfer, E., Reiss, K., & Fischer, F. (2016). Developing argumentation skills in mathematics through computer-supported collaborative learning: The role of transactivity. *Instructional Science, 44*(5), 477–500. doi:10.1007/s11251-016-9380-2

Vogel, F., Wecker, C., Kollar, I., & Fischer, F. (2017). Socio-cognitive scaffolding with collaboration scripts: A meta-analysis. *Educational Psychology Review 29*(3), 477–511. doi:10.1007/s10648-016-9361-7

Vygotsky, L. S. (1978). *Mind and society: The development of higher mental processes*. Cambridge, MA: Harvard University Press.

Walker, E., Rummel, N., & Koedinger, K. R. (2011). Designing automated adaptive support to improve student helping behaviors in a peer tutoring activity. *International Journal of Computer-Supported Collaborative Learning, 6*, 279–306. doi:10.1007/s11412-011-9111-2

Webb, N. M., Franke, M. L., De, T., Chan, A. G., Freund, D., Shein, P., & Melkonian, D. K. (2009). "Explain to your partner": Teachers' instructional practices and students' dialogue in small groups. *Cambridge Journal of Education, 39*(1), 49–70. doi:10.1080/03057640802701986

Weinberger, A., & Fischer, F. (2006). A framework to analyze argumentative knowledge construction in computer-supported collaborative learning. *Computers and Education, 46*(1), 71–95. doi:10.1016/j.compedu.2005.04.003

Weinberger, A., Stegmann, K., & Fischer, F. (2007). Knowledge convergence in collaborative learning: Concepts and assessment. *Learning and Instruction, 17*(4), 416–426. doi:10.1016/j.learninstruc.2007.03.007

Zottmann, J. M., Stegmann, K., Strijbos, J.-W., Vogel, F., Wecker, C., & Fischer, F. (2013). Computer-supported collaborative learning with digital video cases in teacher education: The impact of teaching experience on knowledge convergence. *Computers in Human Behaviour, 29*(5), 2100–2108. doi:10.1016/j.chb.2013.04.014

49

Learning Analytics in the Learning Sciences

Carolyn P. Rosé

Introduction: Scope and Positioning

Over the past two decades, the fields of Educational Data Mining (Baker & Yacef, 2009) and Learning Analytics (Siemens & Baker, 2012) have gained prominence in research, policy, and public literature. The fields evolved from the applied disciplines of Machine Learning, Intelligent Tutoring Systems and Data Mining, which in turn have their own roots in Applied Statistics, Education, Psychology, Cognitive Science, and Computational Linguistics.

This chapter explores the niche domain that Learning Analytics is establishing within the Learning Sciences specifically and offers a taste of the impact it is having within that sphere. The goal is to increase appreciation of the ways in which this emerging area can be a means for bridge building between the communities of Educational Data Mining and Learning Analytics on the one side and the Learning Sciences on the other.

This chapter will not attempt to offer a comprehensive review of research in the areas of Learning Analytics or Educational Data Mining. Such reviews have already been published in other venues. Similarly, though Discourse Analytics (Buckingham-Shum, 2013; Buckingham-Shum, de Laat, de Liddo, Ferguson, & Whitelock, 2014) and computational approaches to analysis of discourse for learning as a subfield of Learning Analytics will be referred to throughout as examples, reviews, and more intensive introductions to that area have also been published elsewhere, and this chapter will not attempt to recount all of that work. Instead, the purpose of this chapter is to approach the topic from a conceptual and methodological level, proposing a vision for where the field of Learning Sciences may have a unique role to play within the broad and intertwining landscape. Because of the emphasis on discourse, interested readers may also be interested in the chapter on behavior and dialogue analysis (Vogel & Weinberger, this volume).

Because of the technical nature of work in Learning Analytics, this is a golden opportunity for bridge building with other related research communities. Towards this goal, leaders from fields focusing on technology supported learning and applications of modeling technologies to problems in education have partnered in recent years with the International Society of the Learning Sciences. Fewer connections have so far been made with societies that do core research in analytic technologies such as Knowledge Discovery and Data Mining and the Association for Computational Linguistics. One goal of this chapter is to motivate the desire for such longer-distance bridge building in the future as a next step. We will return to this theme at the end of the chapter.

At its core, a fundamental tenet of Learning Analytics that resonates with the values of the Learning Sciences is the idea that decision making regarding the administration of learning at all levels should be guided by data—both data about learners in general, and data about specific learners. This idea can be appropriated through a multiplicity of methodological approaches within the Learning Sciences, but within Learning Analytics it is inherently quantitative in its conceptualization. In order to support design and decision making, causal models and appropriate generalizations are needed, although these claims may come with strong caveats (for an alternative view, see the chapters on design-based research and design-based implementation research in this volume; Puntambekar, and Fishman and Penuel, respectively). The strong emphasis on empiricism grounded in big data advocated by data mining researchers can sometimes be misunderstood as an advocacy of atheoretical approaches. As a learning scientist representing one strand within that broader community, I invite multivocal approaches where both quantitative and situated perspectives and approaches have a voice in the conversation (cf. Suthers, Lund, Rosé, Teplovs & Law, 2013; Lund & Suthers, this volume). I caution against a bottom–up, atheoretical empiricism. In contrast, I would stress the role of rich theoretical frameworks for motivating operationalizations of variables within models I build and use. And I strive for interdisciplinary perspectives on these operationalizations that afford intensive exchange between the Learning Sciences and neighboring fields of data mining, computational linguistics, and other areas of computational social sciences.

Not all learning scientists would agree with this view or even this usage of the term "model," which here refers to a mathematical formulation designed to represent behavioral trends and regularities. However, the aim of this chapter is not to challenge a reader's usage of the term "model" in their own work but rather to encourage an appreciation of the ways in which Learning Analytics may provide a synergistic approach within a multivocal landscape.

To that end, this chapter is meant to illuminate certain problems that occur when the partnership between communities is not proceeding in a healthy way, with the hope that with reflection some of these issues can be addressed. An example can be observed in a recent article where the voice of the authors takes on an identity as learning scientists that paint a picture of a dystopian future brought about by technologies of Learning Analytics, framed in that article as an outside voice (Rummel, Walker, & Aleven, 2016). In the vision of that article, the problem of machine learning has been solved, and the result is very negative because educational practitioners at different levels abdicate decision making to the resulting perfected, though entirely atheoretical, models.

This image of a dystopian future illustrates many ways in which the relationship between communities can become dysfunctional, specifically in a lack of understanding of the history, goals, and objectives of each. In this case, the very idea that the problem of machine learning will or ever could be solved with respect to modeling learners already neglects a very important inherent limitation in computational approaches to fully modeling human behavior, namely, the ability of humans to make choices that violate norms. Second, the idea that modeling is inherently atheoretical neglects the rich history in the behavioral sciences of quantitative approaches that are nevertheless grounded in theory. The same computational machinery from applied statistics that is used for hypothesis testing within the behavioral sciences underlies the field of machine learning as well. Thus, while it is always possible to find examples of machine learning applied to behavioral data in an atheoretical and naive way, it is by no means antithetical to applied machine learning to leverage theoretical insights in setting up their models, and some researchers within the field of Learning Sciences argue for just such a theory-motivated approach to machine learning (Gweon, Jain, McDonough, Raj, & Rosé, 2013). Finally, the idea of humans abdicating to machine learning models, besides being reminiscent of science fiction movies like *Terminator*, neglects the history of artificial intelligence since the early 1980s, where there are ample illustrations of why it is unwise to abdicate important decision making to machines. Best practices in computer science since that time have always advocated for manual override.

The hope of this chapter is to encourage a more functional relationship between communities, where the field of Learning Sciences accepts Learning Analytics as an area where researchers bring different skills and potential than was present in the Learning Sciences community before its emergence and where researchers drawing from different areas of expertise listen to one another and challenge one another as partners rather than adversaries.

The work of the field of Learning Sciences is to contribute to theories of how people learn, and then to work within those theoretical frameworks to effect positive change in the world. Computational tools, which include machine learning approaches, can serve as lenses through which researchers may make observations that contribute to theory, as machinery used to encode operationalizations of theoretical constructs, and as languages to build assessments that measure the world in terms of these operationalizations. They are just augmentations of human observation. They are more limited in the sense that, in applying them, a reduction of the richness of signal in the real world occurs as a necessary discretization takes place. However, they are also more powerful in the sense of the speed and ubiquity of the observation that is possible. Thus, though the Learning Sciences community may include those who can fundamentally extend the abilities of computational approaches, the work of the community gets done by using these approaches to advance separate goals. However, that does not mean it is advisable for Learning Sciences researchers to be disinterested consumers of this technology. It would be more advantageous to strive to understand how to most fully appropriate this technology as it grows and changes so that the work of the community can be accomplished more effectively. It is not the work of the machine learning community to understand or contribute to theories about how people learn. However, members of those communities may make effective partners in the work to effect change in the world. Both sides of the bridge must take an active role in the exchange.

Greater mutual understanding is needed to move towards more active exchange. One area where greater mutual understanding is needed is with respect to the contrast between maximizing interpretability (i.e., learning scientists using a model to contribute to theory through empirical work) versus maximizing predictive accuracy (i.e., machine learning researchers using a model to do a task with high accuracy regardless of whether the model is valid by behavioral research standards). In the first case, the models are a means to drawing a conclusion, whereas in the second case the model formulation is of interest as a technical accomplishment.

From the standpoint of a type of Learning Sciences research where statistical models are used for hypothesis testing, the value in a model is measured in terms of how much it has yielded in terms of facilitating the production of the primary research products. In the fields of Machine Learning and Data Mining, researchers in general are not aware of the theoretical models that suggest variables of interest, questions, hypotheses, and insights about confounds that must be avoided. Both traditions place a premium on what they call rigor in evaluation and achievement of generalization of results across data sets. A shared value is further the reproducibility of results. Nevertheless, because of differences in goals, the notion of rigor that has developed within the two types of communities has important differences. In order to maintain a high standard of rigor within the Machine Learning and Data Mining communities, research in the field has produced practices for standardization of metrics, sampling, and avoidance of over-fitting or over-estimation of performance through careful separation of training and testing data at all stages of model development and refinement. A corresponding striving for rigor that is closer to our own community has produced, in addition to the development of statistical machinery for analysis, an elaborate process for validation of such modeling approaches and collection of practices for careful application and interpretation of the results. For communication and collaboration between communities, it is important to consider these differences and how they should guide the multivocal conversation between communities.

History of Learning Analytics Within the Learning Sciences

Before 2005, there was scant representation of machine learning within the International Society of the Learning Sciences (ISLS) conferences. Nevertheless, since the beginning of the Society, leaders such as Pierre Dillenbourg have been advocating for the relevance of machine learning to this field. For example, at the Computer-Supported Collaborative Learning conference in 2005, attention was given to a vision for the next 10 years, and Pierre Dillenbourg was a spokesperson in that session, offering one of the visionary lectures on this topic. Automating analysis of collaborative processes for making support adaptive and dynamic was one of the topics discussed in plenary papers discussing the vision for the future. Nevertheless, in the same conference, none of the sessions were named in a way that acknowledged machine learning or other modeling technologies as constituting an area. Instead, papers offering modeling technologies were "hidden" within other sessions on topics like Argumentation or Interactivity. This was in keeping with practices from earlier conferences from the Society.

From 2005 onwards, however, a trend of increasing attention occurred. There was a plenary keynote and two workshop keynotes on dynamic support for collaboration at the first Kaleidoscope CSCL Rendez Vous, held in Villars, Switzerland. The trend of increasing attention continued at the 2007 conference, where papers in the area became more frequent. They appear in sessions entitled "Tools & Interfaces," "Methods of Scaffolding," "CSCL for Science Learning," and in the posters session. In particular, "Methods of Scaffolding" gives evidence of beginnings of a vision that a new form of scaffolding for collaboration was becoming possible. In this session, Frank Fischer, leader in the area of scripted collaboration, published a paper on the vision for fading of scripted support for collaboration. Interested readers may refer also to the chapter on scaffolding and scripting collaboration (Kollar, Wecker & Fischer, this volume). There was also a symposium on "Adaptive Support for Collaborative Learning," where a vision for developing technology for this purpose was explored. A major shift is apparent by the 2009 conference, where we see two workshops with related work, one on "Intelligent Support for CSCL" and another on "Interaction Analysis and Visualization," where automated and semi-automated analysis technologies were featured topics. In the main conference, a session was included on "Scripts & Adaptation" and another on "Data Mining and Process Analysis.

Beginning with the 2015 conference, multiple symposia have featured the specific question as to the unique flavor and representation of Learning Analytics in the Learning Sciences. At the 2015 conference, a pair of symposia explored trends within the field of Learning Analytics and invited reflection, discussion, and feedback from the ISLS community. In the 2016 conference, the specific focus of an invited symposium was regarding "Analytics of Social Processes," where papers discussed how analytics are applied to collaboration and other interaction data at multiple levels, including administration and policy, teaching, learning, and curriculum design. In all cases, the agenda was set by longstanding initiatives within the Learning Sciences, and analytics served to support and enhance these efforts. In many cases, the work was at an early stage, but with glimmers of hope for a future where the partnership between fields could yield fruit at all of these levels.

A Theory Grounded Methodology

The partnership between fields of Data Mining and Machine Learning on the one hand, and the interdisciplinary field of Learning Sciences on the other, begins with data of common interest. This chapter takes a very quantitative approach to the use of data within the Learning Sciences. With that in mind, best practices for the application of machine learning serve as rules of engagement for this interaction. Here we outline such a process, which begins with issues regarding operationalization of variables, then building models, then testing and validation, which normally transitions into troubleshooting and conducting error analyses, which then leads to iteration.

Before any modeling technology can be applied to processes occurring in a learning setting, behavior traces must first be recorded. This behavior trace data may come in the form of images or video, audio, physiological sensor data, text, or clickstream data. With the exception of clickstream data, these behavior traces are largely unstructured. Machine learning and data mining paradigms cannot be applied to unstructured data until it is first preprocessed and transformed into a structured representation, which could be thought of as a form of ontology. Most typically, that representation first requires segmenting the behavior stream into units, and then variables are extracted from those units. Thus, a data point is a vector representation constructed from a list of attribute–value pairs. This first stage of setting up the data trace for application of modeling paradigms is arguably the most critical. And, unfortunately, it is the step where the biggest errors of process occur, which may render the data useless for effective modeling. Typically, vectors constructed for machine learning are composed of very large numbers of variables, often thousands or even tens of thousands, in contrast to the small numbers of very carefully operationalized variables that are more typically used in behavioral research. These vectors become unwieldy, which makes the process of validation and interpretation challenging.

Once the data are represented as vectors of variables, these vectors can be compared with one another. Cluster analyses, which are well-known methodologically in behavioral research fields, can be applied to these representations. Additionally, supervised learning approaches can be used to identify patterns in these feature values that are associated with some dependent variable, which is referred to as a class value. Both for the purpose of obtaining clusters that are meaningful and for being able to achieve high predictive accuracy with a supervised model, it is important that the set of data points that should be grouped together are close to one another within the vector space and far away from data points that are not part of the same grouping. Problems occur when either data points that should be grouped together are not close within the vector space, or data points that should not be grouped together are close within the vector space. When this is the case, the problem can be addressed through introduction or removal of some of the variables from the representation of the data. Sometimes the information in its native form, as found in the raw data trace, do not enable this clean grouping of data points, but a reformulation of that information through replacement of some variables for ones that are more nuanced, context-sensitive, and meaningful could. Sometimes the problem is that low-level features that correlate with what would be meaningful features act as proxies. Those proxies function effectively for predictive accuracy in context, but do not generalize well, and may even reduce predictive accuracy substantially when models are applied in contexts that differ from the ones where the training data were obtained.

As an example, let us consider a study of collaborative problem solving for fraction addition where pairs of sixth graders over two days competed for a pair of movie passes, which was described in an earlier introduction to linguistic analysis of collaboration (Howley, Mayfield, & Rosé, 2013). This kind of problem-solving task has been a frequent context for collaborative learning studies (Harrer, McLaren, Walker, Bollen, & Sewall, 2006; Olsen et al., 2014). What is important to note is that there is a plethora of different approaches that could be taken to extract variables from the textual data traces of student behavior in the study. Many different approaches may lead to models that predict with high accuracy who learned more, but they vary in terms of how interpretable and actionable the findings are.

Let's now consider the details of this study from the perspective of the multiplicity of lenses that have been applied to the interpretation of its data. In this study, the students collaborated online in a problem-solving environment that allowed each student to participate from her own computer. She could interact with the partner student through chat. And each could contribute to the problem solving through a graphical user interface. In this study, on each of two lab days, the students worked through a series of story problems. In between problems, in the experimental condition, a conversational computer agent asked a question to each student along the lines of "Student1, which do you find more entertaining, books or movies?" or "Student2, would you prefer a long car ride or an airplane ride?" The answers to the questions would be used to tailor the story problem. For example, "Jan packed several books to amuse herself on a long car ride to visit her grandma. After 1/5 of the trip, she had already finished 6/8 of the books she brought. How many times more books should she have brought than what she packed?" In the control condition, there were no such questions about personal preferences directed at the students in between problems. Instead the story problem templates were filled in by randomly picking from among the answers to the questions that would have been given if the students were in the experimental condition. In this way, the math-relevant aspects of the treatment were the same across conditions.

The finding in this study was that students in the experimental condition learned marginally more (Kumar, Gweon, Joshi, Cui, & Rosé, 2007). An observation was made that there was a great deal of belligerent language exchanged between students in this study, typically coming from the more capable peer in many pairs. A correlational analysis showed that the amount of belligerent language a student uttered significantly correlated with how much that student learned (Prata et al., 2009). One naive interpretation of features extracted from the text data and used to predict learning from this data would suggest that behaving belligerently contributes to learning. Fortunately, we can confidently conclude that this does not make sense as an explanation. Another observation from this data was that students learned more when they practiced more on concepts that were indicated by the pretest to be difficult. This purely cognitive explanation is consistent with prior work and makes sense, but may not be actionable in this context. It begs the question as to why some students got less practice on the skills they needed practice on.

In digging deeper, it became clear that there was a social explanation for the learning result. When some students became belligerent, their corresponding partner student shifted to a less authoritative position in the interaction with respect to active contribution to the problem solving and discussion (Howley et al., 2013). While authoritativeness by itself as a measure of interaction dynamics did not predict learning directly in this study, the shift to a less authoritative stance was associated with a difference in problem-solving strategy. In pairs with a large authoritativeness differential, the less authoritative student, upon reaching an impasse, abdicated to the other student, whereas in other groups with less of a differential, the student got feedback and tried again. The belligerence and shift in authoritativeness was primarily observed in the control condition.

As a final analysis we noted that the students in the control condition who were the victims of belligerent behavior from their partner student learned significantly less than their partner students and all students in the experimental condition. Thus, we see that the experimental manipulation had an effect on the social climate, that then had an effect on problem-solving behavior, that then had an effect on learning. This explanation makes sense and is actionable—it gets to the source of the problem and offers a solution, namely creating a playful environment through some manipulation like the conversational computer agent in this study to support productive social dynamics.

In all forms of learning analytics and educational data mining, when the data is unstructured we must make choices about what variables we will extract from the data trace and use in the modeling to make predictions. In the example above, we see that some choices would lead to conclusions that are actionable and make sense, and others would not. A naive approach to extracting features from the text, such as using each word that occurs at least once as a variable, which is an approach that frequently serves as a baseline representation, would lead to suboptimal results. In order to achieve an interpretable and actionable model, more thought and care regarding the representation of the data before machine learning is required.

The case illustrates some important aspects of applied machine learning that should be kept in mind. For example, in this study, amount of belligerent language uttered by a student predicted learning. In this study, the most likely reason for this correlation is that the partner student's reaction to the belligerent language created more opportunities for the student who uttered the belligerent language to get practice. Thus, in some ways this low-level feature of the interaction served as a proxy for amount of practice, at least in the control condition. In another context, belligerent language uttered might not correlate with amount of practice, and thus in that other context we would not expect to see the same correlation. Thus, if a feature for belligerent language was included in a vector representation of a student's participation, a machine learning model would assign weight to that feature since it makes an accurate prediction in that data. But placing weight on that feature would lead the model to make wrong choices in a different context where the correlation between belligerent language and practice would not hold. This phenomenon of weights being attributed to proxies and then not generalizing to other contexts is called over-fitting, and avoiding over-fitting like this is one of the goals of much of the best practices in applied machine learning.

The dream of big data in education was that data mining would help to answer questions in the Learning Sciences faster, or with more data, potentially leading to more nuanced findings. Provided that appropriate methodology is used to create the models that would be used for this purpose, this vision could certainly be realized. It is important to note, however, the upfront cost associated with creating such models in thinking about the appropriate role they might play within the research landscape. In order to train a model to make accurate predictions, a substantial amount of data must be labeled, frequently by hand. If a researcher plans to run many studies using the same paradigm, this upfront investment might still be worth it. More frequently, the value in automated analysis tools is in enabling interventions triggered by real-time interaction data.

Into the Future: An Agenda for Learning Analytics in the Learning Sciences

Advances in the field of machine learning and related fields like language technologies have enabled some of the recent advances in Learning Analytics. However, these fields have their own history, which intersects with that of the Learning Sciences. As learning scientists look to the future with the hope of drawing more and more from these sister fields, we would do well to consider their own trajectory and how it might synergize with that of the Learning Sciences going forward.

Longtime members of the fields of machine learning and data mining, as outgrowths of the broader field of artificial intelligence, have observed the paradigm shift that took place after the mid-1990s. Initially, approaches that combined symbolic and statistical methods were still of interest. However, with the increasing focus on very large corpora and leveraging of new frameworks for large-scale statistical modeling, symbolic and knowledge-driven methods were largely left by the wayside. Along with older symbolic methods that required carefully crafted rules, the concept of knowledge source became strongly associated with the antithesis of empiricism. On the positive side, the shift towards big data came with the ability to build real-world systems relatively quickly. However, as knowledge-based methods were replaced with statistical models, a grounding in theory grew more and more devalued, and instead a desire to replace theory with an almost atheroetical empiricism became the zeitgeist.

About a decade ago, just as an atheoretical empiricism was becoming not only the accepted norm, but memories of a more theory-driven past were sinking into obscurity, an intensive interest in analysis of social data and integration of artificial intelligence with social systems began to grow. More recently, the tide has begun to turn back in ways that have the potential to benefit the Learning Sciences. For example, a growing appreciation of the connection between constructs from the social sciences and work on natural language processing applied to social data has surfaced in recent years, such as data from a variety of social media environments. As this work draws significantly from research literature that informs our understanding of how social positioning takes place within conversational interactions, we are in a better position to track how students work together to create a safe or unsafe environment for knowledge building, how students take an authoritative or non-authoritative stance within a collaborative interaction, or how students form arguments in ways that either build or erode their working relationships with other students. The time is ripe for bridge building and partnership between fields.

Acknowledgments

This work was funded in part by NSF grants IIS-1546393 and ACI-1443068.

Carolyn P. Rosé

Further Readings

Baker, R. S., & Yacef, K. (2009). The state of educational data mining in 2009: A review and future visions. *JEDM-Journal of Educational Data Mining, 1*(1), 3–17.
This paper presents one of the first surveys of the field of educational data mining, which has its roots in the Artificial Intelligence in Education community. Major thrusts of work in this field related to cognitive modeling and mining of clickstream data from Intelligent Tutoring Systems.

Erkens, M., Bodemer, D., & Hoppe, H. U. (2016). Improving collaborative learning in the classroom: Text mining based grouping and representing. *International Journal of Computer-Supported Collaborative Learning, 11*(4), 387–415.
This recent empirical article demonstrates how to use a text mining approach, namely Latent Dirichlet allocation, for grouping students for collaborative learning. It serves as a practical application of a learning-analytic technology to instruction.

Gweon, G., Jain, M., McDonough, J., Raj, B., & Rosé, C. P. (2013). Measuring prevalence of other-oriented transactive contributions using an automated measure of speech style accommodation, *International Journal of Computer Supported Collaborative Learning, 8*(2), 245–265.
This empirical study presents an example of how theories from cognitive and social psychology can motivate operationalizations of language constructs and specific modeling approaches. In particular, it presents an approach to rating prevalence of transactive contributions in collaborative discourse from speech data.

Siemens, G., & d Baker, R. S. J. d. (2012, April). Learning analytics and educational data mining: towards communication and collaboration. *Proceedings of the Second International Conference on Learning Analytics and Knowledge* (pp. 252–254), Vancouver, BC, Canada.
This paper presents a more updated account of work in the area of modeling technologies applied to learning data. It spans two fields, namely Educational Data Mining and Learning Analytics, and presents a vision for the synergies between these two communities.

Van Leeuwen, A., Janssen, J., Erkens, G., & Brekelmans, M. (2014). Supporting teachers in guiding collaborating students: Effects of learning analytics in CSCL. *Computers & Education, 79*, 28–39.
This article presents another practical application of Learning Analytics to instruction. This time the study investigates how teachers use visualizations produced through an analytic process in the classroom during instruction.

The LearnSphere project website is also a useful resource for data infrastructure and analytic tools for educational data mining and learning analytics (http://learnsphere.org/).

NAPLeS Resources

Rosé, C. P., *Learning analytics and educational data mining in learning discourses* [Webinar]. In *NAPLeS video series*. Retrieved October 19, 2017, from http://isls-naples.psy.lmu.de/intro/all-webinars/rose_all/index.html

Rosé, C. P., *15 minutes about learning analytics and educational data mining in learning discourses* [Video file]. In *NAPLeS video series*. Retrieved October 19, 2017, from http://isls-naples.psy.lmu.de/video-resources/guided-tour/15-minutes-rose/index.html

Rosé, C. P., *Interview about learning analytics and educational data mining in learning discourses* [Video file]. In *NAPLeS video series*. Retrieved October 19, 2017, from http://isls-naples.psy.lmu.de/video-resources/interviews-ls/rose/index.html

References

Baker, R. S., & Yacef, K. (2009). The state of educational data mining in 2009: A review and future visions. *JEDM-Journal of Educational Data Mining, 1*(1), 3–17.
Buckingham-Shum, S. (2013). *Proceedings of the 1st International Workshop on Discourse-Centric Analytics, workshop held in conjunction with Learning, Analytics and Knowledge 2013*, Leuven, Belgium.
Buckingham-Shum, S., de Laat, M., de Liddo, A., Ferguson, R., & Whitelock, D. (2014). *Proceedings of the 2nd International Workshop on Discourse-Centric Analytics, workshop held in conjunction with Learning, Analytics and Knowledge 2014*, Indianapolis, IN.
Fishman, B., & Penuel, W. (2018). Design-based implementation research. In F. Fischer, C. E. Hmelo-Silver, S. R. Goldman, & P. Reimann (Eds.), *International handbook of the learning sciences* (pp. 393–400). New York: Routledge.

Gweon, G., Jain, M., McDonough, J., Raj, B., & Rosé, C. P. (2013). Measuring prevalence of other-oriented transactive contributions using an automated measure of speech style accommodation, *International Journal of Computer Supported Collaborative Learning, 8*(2), 245–265.

Harrer, A., McLaren, B. M., Walker, E., Bollen, L., & Sewall, J. (2006). Creating cognitive tutors for collaborative learning: Steps toward realization. *User Modeling and User-Adapted Interaction, 16*(3–4), 175–209.

Howley, I., Mayfield, E., & Rosé, C. P. (2013). Linguistic analysis methods for studying small groups. In C. Hmelo-Silver, A. O'Donnell, C. Chan, & C. Chin (Eds.), *International handbook of collaborative learning*. New York: Taylor and Francis.

Kollar, I., Wecker, C., & Fischer, F. (2018). Scaffolding and scripting (computer-supported) collaborative learning. In F. Fischer, C. E. Hmelo-Silver, S. R. Goldman, & P. Reimann (Eds.), *International handbook of the learning sciences* (pp. 340–350). New York: Routledge.

Kumar, R., Gweon, G., Joshi, M., Cui, Y., & Rosé, C. P. (2007). Supporting students working together on math with social dialogue. *Proceedings of the ISCA Special Interest Group on Speech and Language Technology in Education Workshop (SLaTE)*, Farmington, PA.

Lund, K., & Suthers, D. (2018). Multivocal analysis: Multiple perspectives in analyzing interaction. In F. Fischer, C. E. Hmelo-Silver, S. R. Goldman, & P. Reimann (Eds.), *International handbook of the learning sciences* (pp. 455–464). New York: Routledge.

Olsen, J. K., Belenky, D. M., Aleven, V., & Rummel, N. (2014, June). Using an intelligent tutoring system to support collaborative as well as individual learning. *International Conference on Intelligent Tutoring Systems* (pp. 134–143). Honolulu, Hawaii: Springer International Publishing.

Prata, D., Baker, R., Costa, E., Rose, C., & Cui, Y. (2009). Detecting and understanding the impact of cognitive and interpersonal conflict in computer supported collaborative learning environments. In T. Barnes, M. Desmarais, C. Romero, & S. Ventura (Eds.), *Proceedings of the 2nd International Conference on Educational Data Mining* (pp. 131–140), Cordoba, Spain.

Puntambekar, S. (2018). Design-based research (DBR). In F. Fischer, C. E. Hmelo-Silver, S. R. Goldman, & P. Reimann (Eds.), *International handbook of the learning sciences* (pp. 383–392). New York: Routledge.

Rummel, N., Walker, E., & Aleven, V. (2016). Different futures of adaptive collaborative learning support. *International Journal of Artificial Intelligence in Education, 26*(2), 784–795.

Siemens, G., & Baker, R. S. J. d. (2012, April). Learning analytics and educational data mining: towards communication and collaboration. *Proceedings of the Second International Conference on Learning Analytics and Knowledge* (pp. 252–254), Vancouver, BC, Canada.

Suthers, D., Lund, K., Rosé, C. P., Teplovs, C., & Law, N. (Eds.) (2013). *Productive multivocality in the analysis of group interactions*. New York: Springer.

Vogel, F., & Weinberger, A. (2018). Quantifying qualities of collaborative learning processes. In F. Fischer, C. E. Hmelo-Silver, S. R. Goldman, & P. Reimann (Eds.), *International handbook of the learning sciences* (pp. 500–510). New York: Routledge.

50
Epistemic Network Analysis
Understanding Learning by Using Big Data for Thick Description

David Williamson Shaffer

Introduction

The advent of massive open online courses (MOOCs), educational games and simulations, computer-based tests, and other computational tools for learning and assessment means that the data available to learning scientists is growing exponentially (e.g., Fields & Kafai, this volume; G. Fischer, this volume). At the same time, the new fields of learning analytics, data mining, and data science more generally have proposed a range of statistical and computational approaches to analyzing such data, some of which are described elsewhere in this handbook (e.g., Rosé, this volume).

One critical feature of the learning sciences, however, is that researchers conduct analyses from a theoretical perspective. Moreover, the theories that learning scientists use are designed to investigate how learning happens not just in laboratory settings, but in the real world of classrooms and after-school centers, of parents and teachers and students, of children in Makerspaces and collaborating online. As Maxwell (2008) explains, qualitative methods are particularly appropriate for addressing questions that "focus on *how* and *why* things happen, rather than *whether* there is a particular difference or relationship or how much it is explained by other variables" (p. 232). Such analyses depend on what Geertz (1973b) popularized as *thick description*. Thick description is an attempt to understand how and why people acted in some specific time and place: an explanation of the experiences, assumptions, emotions, and meaning-making in lived experience that makes sense of how and why events unfolded. When applied to data about students, teachers, classrooms, informal educational settings, and individual learning environments, this approach leads to powerful insights about how learning works (e.g., Green & Bridges, this volume; Koschmann, this volume) Critically, though, such qualitative analyses are typically done by hand. Researchers identify concepts and categories of experience that are meaningful to the participants in some setting, and then look for patterns that explain how the people they are observing make sense of what is happening.

It is possible, of course, to create thick descriptions in this way from Big Data by analyzing only a very small subset of the information collected. But it is also possible to use statistical and computational techniques to bring qualitative insights to bear on large corpuses of data that learning scientists now have available. In what follows, I describe *epistemic network analysis* (ENA), a network analysis technique being used by a growing community of learning sciences researchers to support thick descriptions based on Big Data about learning.

Theoretical Foundations of ENA

ENA is based on four ideas about how human action is situated in the world that come together in the theory of *epistemic frames*. These four ideas are relatively uncontroversial at this point in the development of the learning sciences, and thus are set forth here as axioms or postulates (see Shaffer, 2017 for a comprehensive overview; Danish & Gresalfi, this volume; Hoadley, this volume).

First, learners are always embedded within a *culture* or cultures. Human beings traffic in symbols: in action, in talk, in writing, and in making things that mean something to ourselves and to others. A long tradition of work in the learning sciences looks at how learning is a form of enculturation, in which students become situated in *communities of practice* (e.g., Lave & Wenger, 1991). From this perspective, learning means becoming progressively more adept at using what Gee (1999) calls the Big-D discourse of some community: a way of "talking, listening, writing, reading, acting, interacting, believing, valuing, and feeling (and using various objects, symbols, images, tools, and technologies)" (p. 23) that is particular to some group of people who share a common culture.

Second, learners are always embedded in *discourse*. Activity in a cultural setting is always expressed in action: in talk, in the creation and manipulation of artifacts, in gestures, in movement, in anything that is perceptible in the world. Becoming enculturated—and likewise studying how enculturation happens—means looking at the record of such activity to understand how these events in the world are interpreted within the Big-D Discourse of a community (Gee, 1999). Goodwin (1994) argues that a critical part of this process is developing *professional vision*: "socially organized ways of seeing and understanding events that are answerable to the distinctive interests of a particular social group" (p. 606). He refers to these socially organized ways of seeing as *codes*, and the study of learning, from this perspective, requires understanding when and how people use the codes from some community of practice.

Third, learners are always embedded in *interaction*. Learning is fundamentally an interpersonal process, where learners are engaged with others. Typically such interactions are happening directly with teachers or mentors and with peers. But, as Pea (1993) suggests, artifacts that others have created carry intelligence within them, and even a student looking up information online or solving a problem with a calculator is in a conversation with others who "respond" to a student's actions in particular ways, albeit in mediated form (e.g., Cress & Kimmerle, this volume).

Finally, learners are always embedded in *time*. Culture is expressed (and thus constructed) temporally: it unfolds through sequences of action and response—one student speaks, another replies, a third summarizes, or a teacher interrogates, a student responds, and the teacher evaluates. But even when a single person is working in relative isolation—relative because no human activity takes place completely cut off from social convention—one step in solving a problem follows idea is informed by, and also transforms, the ideas that came before it.

The theory of epistemic frames (Shaffer, 2017) extends these ideas by suggesting that, while understanding a Big-D Discourse requires making sense of the codes of some community, it is not enough to simply *identify* the codes that the community uses. We understand culture—and thus create thick descriptions—by understanding how codes are systematically *related* to one another. Cultures—what Geertz (1973a) calls "organized systems of significant symbols" (p. 46)—are composed of symbols that interact to form a web of meanings. That is, we understand symbols in terms of other symbols (Deacon, 1998), and thus, in order to understand a Big-D Discourse, we need to understand how codes are systematically related to one another as people participate in a culture.

An *epistemic frame* is a formal description of how codes are related to one another, which are uncovered by examining how the codes from a Big-D Discourse are systematically related to one another in the discourse of some person or group of people. For example, Shaffer (2017) describes part of the epistemic frame of journalism by explaining how, when investigative journalists talk about their role as watchdogs, they refer to the values of giving voice to those without one, and being there for people who need a reporter, as well as the practice of warning sources about the dangers of talking to a reporter and the knowledge that journalists need solid evidence to take on the police. From the perspective of the learning sciences, then, we can assess the extent to which journalism students have (or have not) adopted the Big-D Discourse of investigative journalism by seeing whether they make the same connections in their discourses over time.

ENA provides a way of modeling epistemic frames by mapping the way that codes are connected to one another in discourse. Suthers and Desiato (2012) point out that activity takes place within what they call a *recent temporal context*: the previous events that provide a common ground within which actions are interpreted; that is, people act in response to things that happened only a little while earlier in a conversation or stream of activity. So if a student asks a question about how an investigative journalist can take on the police, and a veteran journalist responds by talking about solid evidence, the veteran journalist is making a connection between the two codes because they are both within the same recent temporal context. By mapping connections between codes in this way, ENA provides a model of the epistemic frame of a learner or learners—and thus a way to describe how learning is simultaneously embedded in culture, discourse, interaction, and time.

The Pragmatics of ENA

What follows summarizes a much more detailed discussion of the mechanics of ENA presented in Shaffer and Ruis (2017). ENA models the structure of connections among codes within some recorded discourse as a network. In an ENA network, *nodes* represent individual codes in some Big-D Discourse; *links* between nodes represent the strength of association of those codes within the specific discourse being modelled; that is, ENA shows how some person or group of people make connections that matter in the Big-D Discourse of some community. The strength of association, or *connection*, between two nodes is proportional to the relative frequency of their *co-occurrence* in the data being modeled, where the co-occurrence of two codes means that they both are interpretations of data in the same recent temporal context.

In what follows I will present an example based on a sample dataset available online as part of the ENA software. The data is described more fully in documentation that accompanies the dataset. Briefly, this is data from 44 first-year college students enrolled in an introductory engineering course. During the course, students participated in two simulations where they role-played as interns at an engineering firm. In the simulations, students worked in groups, communicating via email and chat. The dataset contains the logfile entries for all student chat messages from one of these simulation games, called *RescuShell*, in which students designed a robotic exoskeleton for rescue workers. Notably, some of the students in the sample used *RescuShell* in the first part of the semester; others in the sample used *RescuShell* in the second part of the semester, after they had already used another engineering simulation.

To create an ENA model:

1. Data is segmented into lines in a data table. Figure 50.1 shows data from part of the conversation among students in one group in *RescuShell*. This segmentation was already done, in the sense that the log file was already organized into lines where each line is a turn of talk. This is often the case with data in the learning sciences; however, it is important to be sure that the segmentation chosen is compatible with the coding scheme being used.[1]
2. Each line of data is coded. There are many ways to accomplish this, including hand coding and the use of automated classifiers. In the *RescuShell* data, the codes were Data, Technical Features, Performance Parameters, Client Requests, Design Reasoning, and Collaboration, which are shown in the columns of the same name in Figure 50.1. The quality of an ENA model depends on the quality of the coding, and any codes should be appropriately validated (for more on coding and related issues, see Vogel & Weinberger, this volume).

		Data	Technical Features	Performance Parameters	Client Requests	Design Reasoning	Collaboration
Christina	I think NiCd was best option overall	0	1	0	0	1	0
Derek	I liked the Nickel Cadmium battery as well. I think it was harder to choose a sensor	0	1	0	0	1	0
Nicholas	I think a low cost device would be the best option, even if we have to sacrifice some performance aspects	1	0	1	0	1	0
Christina	I think Piezoelectric could be a good choice because it has great safety and and good agility and recharge interval	0	1	1	0	1	0
Derek	I think safety is probably the most important, if the user isn't safe using device I don't think it's worth the risk of everything else	1	0	1	1	1	0
Derek	Look at which of the two options we want to combine together and see what the combinations of the designs create in order to get a better idea of which combinations create what results	1	0	0	0	1	0
Nicholas	I agree, we can try a few variations to figure out the best option with these exoskeletons	0	0	0	0	1	0

Figure 50.1 Excerpt from the Sample Dataset from *RescuShell*

3   The lines of data are segmented into *conversations*. A conversation represents a grouping of lines that could be related to one another in the dataset, such as the turns of talk of one group of students from one activity. This step accounts for the fact that learning sciences datasets often contain data about multiple groups or individuals who are working at the same *time*, but not necessarily *interacting* with one another. In the *RescuShell* data, conversations were defined as all of the lines of chat for a single group within a single activity in the simulation, although the variables in the dataset that indicate group and activity are not shown in the excerpt in Figure 50.1.

4   Each line of data is associated with its recent temporal context. ENA defines recent temporal context in terms of a *stanza*, where the stanza for a line of data, l, is the other lines in the data that are part of the recent temporal context for line l. The ENA software models stanzas by using a *moving window* of fixed size w; that is, each line of data l is associated with the w-1 lines of data in the conversation that precede line l (creating a total window size of w lines). The choice of window size, w, depends on the nature of the data—specifically, what size window best captures recent temporal context in the recorded discourse.

5   In the *RescuShell* data, the window size chosen was 7. This can be seen in the excerpt in Figure 50.1. Nicholas says: "I agree, we can try a few variations to figure out the best option with these exoskeletons." In saying that "we can try a few variations," he is referring to the previous lines in which the group discussed the importance of both Performance Parameters such as safety, agility, and recharge intervals, as well as particular Technical Features such as choice of battery or sensors used—going all the way back to Christina's comment that "NiCd [batteries] was best option overall" six turns of talk previously.

6   An adjacency matrix is constructed for each line of data, showing the connections between codes in the line of data and codes in its recent temporal context. Specifically, the ith row and jth column of the adjacency matrix for line l is a 1 if the ith code in the model appears in line l and the jth code in the model appears in the stanza for line l; otherwise, the ith row and jth column is a 0. This computation is done in the ENA web tool or rENA package.[2]

For Nicholas' final comment in Figure 50.1, for example, the adjacency matrix would show connections between Design Reasoning, which is the only code present in that line of data, and all of the other codes in the data except Collaboration, because the other codes all appear in the stanza while Collaboration does not.

Adjacency matrices are summed for lines of data that correspond to the people or groups of interest (the units of analysis) in the data. The entries in the summed adjacency matrix for each person or group thus represent the number of times that person or group made a connection between each pair of codes in the model. Again, this is done in the ENA software, resulting in a single network model for each unit of analysis in the data.[3] For example, in the *RescuShell* data, we might compare the networks of individual students working in groups, or compare the epistemic networks between groups.

An ENA network thus represents the frequency with which each pair of codes is temporally co-located in the data for each person or group; that is, an ENA network models the temporal correlation structure of codes, or the level of interaction between codes, for people or groups of interest in the dataset. This network approach has mathematical and conceptual advantages over more typical multivariate models, where interactions between variables are typically (and in some ways necessarily) secondary to the underlying variables themselves (see Collier, Ruis, & Shaffer, 2016; Jaccard & Turrisi, 2003).

Unlike network analysis techniques that are optimized to analyze very large networks with thousands or millions of nodes, an ENA model has a fixed and relatively small set of nodes. It is therefore possible to project the adjacency matrices as points in a high-dimensional space, and use a dimensional reduction to identify which dimensions of the network space reveal differences of interest between networks. For example, a dimensional reduction can be used to identify the ways in which the epistemic frames of novices and experts are different in some domain.

Analyzing an ENA model thus entails three further steps:

7   An appropriate dimensional reduction needs to be chosen. Typical choices are a *singular value decomposition*, which shows the maximum variance in the networks being modeled, or a *means rotation*, which shows the maximum difference between two groups of networks. Networks are typically normalized before doing a dimensional reduction, which removes differences in networks due only to variation in the volume of data for different networks—for example, because some people talk more than others. In the *RescuShell* example, we might choose a dimensional reduction that shows the difference between students who used *RescuShell* at the beginning of the semester and those who used it after using another engineering simulation.

    The result of the dimensional reduction is an *ENA space*, where the network for each unit of analysis is represented as a point, and the distances between units are a measure of the similarity or difference between the connections in the two networks.

8   Differences between networks are compared in the resulting ENA space. Networks can be compared and differences measured and modeled using a wide range of inferential statistics of the kind discussed elsewhere in this handbook (De Wever & Van Keer, this volume). For example, in the graph in Figure 50.2, each point represents the position within the ENA space of the epistemic network of a single student who used *RescuShell*. The graph shows that there are statistically significant differences in discourse between students who used *RescuShell* at the beginning of the semester (darker points, $mean_x = -0.09$) and those who used it after another role-playing simulation (lighter points, $mean_x = 0.11$, $p < 0.001$). However, we need one additional step to interpret the dimensions of the ENA space—and thus the meaning of this difference between groups of students.

9   The resulting network models are visualized in the ENA space as *network graphs*, which make it possible to see what properties of the networks—and thus of the epistemic frames they model—account for the differences between networks or groups of networks. For example, Figure 50.3

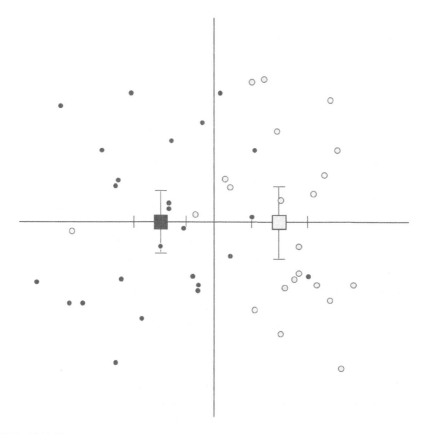

Figure 50.2 ENA Space

Note: The graph compares the discourse pattern of students using *RescuShell* at the beginning of the semester (darker points) and those who used it after using another role-playing simulation (lighter points). The means of the two groups (shown as squares with confidence intervals) show that there is a difference between the two groups; however, a second, coordinated representation is needed to interpret the difference between groups.

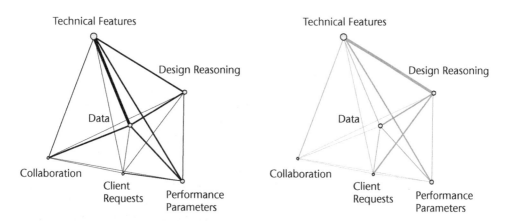

Figure 50.3 Network Graphs for Two Students

Note: One student used *RescuShell* at the beginning of the semester (darker-colored lines) and one used it after using another role-playing simulation (lighter-colored lines).

shows the network graphs for two students. The graph with darker lines, left, is from a student who was using *RescuShell* at the beginning of the semester. The graph with lighter lines, right, is from a student who was using *RescuShell* after another role-playing simulation. Note that connections to collaboration are more prominent in the darker graph, suggesting that a significant difference between these two groups is that students using *RescuShell* at the beginning of the semester spend more time talking about how to work together than those with more experience.

In this sense, the network graphs provide both an explanation for and interpretation of the ENA space, and a critical step in using ENA is returning to the original data to confirm the results of an analysis. While there is not space here to provide an extended qualitative example, note that the sample data in Figure 50.1 above was from a group of students who had already completed another engineering role-playing simulation. In the example, they made plans for testing design prototypes without discussing roles and responsibilities or otherwise communicating about their collaborative process.

Notice also that the weighting of the network graphs in Figure 50.3 corresponds to the location of points in the ENA space in Figure 50.2. The darker network graph makes strong connections to the node for Collaboration, which is located in the lower left portion of the space. Similarly, the networks with darker points in the ENA space are located in the left part of the space. The lighter network graph makes strong connections to the node for Design Reasoning, which is located in the upper right portion of the space. Similarly, the networks with lighter points in the ENA space are located in the right part of the space. Thus we can interpret the x-dimension of the space as showing the difference between students who focused on collaboration and those who focused on design reasoning when making sense of their design work.

In other words, ENA takes advantage of a set of network modeling and visualization techniques that are specifically designed to highlight salient features of the way codes are used in discourse—in this case, to show that students with more design experience spend less time organizing their collaboration and relatively more time solving the design problem using ideas about data and design to maximize the performance of a device by choosing appropriate technical features. Clearly there were many analytical choices that were made in constructing this model, which there is not room to report and justify here. Shaffer and Ruis (2017), from which this example was adapted, gives further details about this specific analysis, and Shaffer (2017) provides a more comprehensive overview of conceptual issues in using ENA.

ENA Analysis: A Case Study

To make this somewhat abstract presentation of ENA more concrete, this section of the chapter provides a summary of a previously published study using ENA.

Quardokus Fisher et al. (2016) studied Industrially-Situated Virtual Laboratory (ISVL) projects. In these projects, students work in teams with a simulation of a manufacturing process. In the ISVL, each team is guided by a *coach*: a more experienced engineer whose job is to help "enculturate students to the expectations of industrial project work." ISVL projects let students solve authentic engineering tasks, integrating their understanding of science topics through an iterative process of experimentation, analysis, and reflection. Quardokus Fisher and her colleagues focused on how coaches used particular mentoring techniques, such as questioning and directive dialogue, to guide students to think about important engineering content, such as the rates of kinetic reactions, experimental design, and the choice of input parameters to an experiment, which they operationalized in codes for guiding, kinetics, experimental design, experimental design, collaboration, input parameters, and data collection.

Quardokus Fisher and her colleagues used ENA to model the connections between codes in their data. The researchers analyzed 27 coaching sessions, 14 by one coach and 13 by another, and used

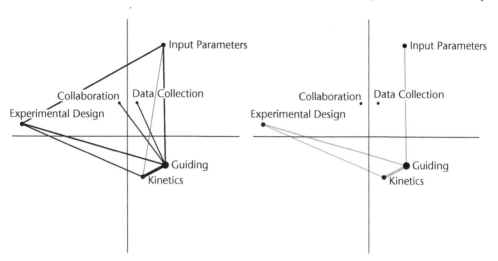

Figure 50.4 Models Showing the Discourse Patterns of Two Different Coaches in an Industrially Situated Virtual Laboratory Project

Source: Adapted from Quardokus Fisher et al. (2016).

that data to look for similarities and differences between the two coaches' approach to mentoring engineering students. The resulting network graphs are shown in Figure 50.4.

These network graphs show the mean strength of connection among codes for the coaching sessions of the two different coaches (the network with heavy, black lines, on the left, and the network with light, gray lines, on the right). The model confirmed and extended what the researchers saw in their data. Both the Heavy Coach and the Light Coach focused their guidance largely on helping students understand the rate of kinetic reactions, which was central to the experiments students were conducting. Both coaches also guided students to think about experimental design and input parameters in an experimental setting. But the Heavy Coach integrated these topics. The black network shows that averaged across coaching sessions for 14 different teams, the Heavy Coach facilitated connections between input parameters, experimental design, and kinetics, as indicated by the more robust network of connections in the red network. Facilitation from the Light Coach, in contrast, was less integrated, and specifically did not connect input variables to either experimental design or kinetics. The researchers explain that the Heavy Coach systematically "preferred to use 'Input Parameters' as an access point for discussion of the project," which the Light Coach did not.

These two network graphs both represent the *mean* discourse network across multiple coaching sessions. As shown in Figure 50.5, Quardokus Fisher and her colleagues were able to create an ENA space to compare all coaching sessions by the Heavy Coach (black points, $mean_y = 0.21$) and all coaching sessions by the Light Coach (gray points, $mean_y = -0.23$). The researchers used inferential statistics to warrant that the difference between the two samples—the difference between the discourse of these two coaches—was statistically significant ($p < 0.001$).

Lessons for the Learning Sciences

ENA is one type of statistical tool available to learning scientists to support qualitative analyses. It has been used to model a wide range of phenomena, including students' complex thinking and collaborative problem solving in simulations of urban planning (Bagley & Shaffer, 2015; Nash & Shaffer, 2011), and other STEM professional practices (Hatfield, 2015); surgery trainees' complex thinking and collaboration during a simulated procedure (D'Angelo, 2015); gaze coordination during

David Williamson Shaffer

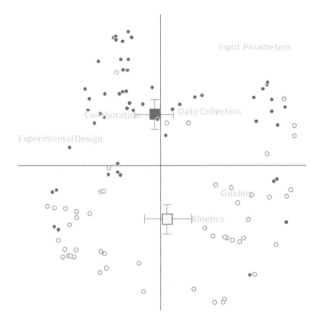

Figure 50.5 Model Showing Statistically Significant Differences in the Discourse Patterns of Two Coaches

Source: Adapted from Quardokus Fisher et al. (2016).

collaborative work (Andrist, Collier, Gleicher, Mutlu, & Shaffer, 2015); and the ways in which students work together to find information on the internet (Knight et al., 2014).

There are many conceptual and practical considerations that have been discussed far too briefly in the presentation here: identifying codes, organizing and segmenting data, and so on. Vogel and Weinberger (this volume) provide an overview of such considerations; Shaffer (2017) provides a detailed discussion. The purpose of the presentation in this chapter has been to suggest that a critical turn in the learning sciences is to use Big Data for Thick Description, and that a critical component of linking quantitative and qualitative analyses in this way is using a statistical tool that is based in a theory of the underlying processes of learning being examined. The point of contact in these examples was *epistemic frame theory*, which suggests that understanding the relationship between codes is a critical part of understanding a culture of learning. ENA is not the only way to model how meaning is constructed through connections in discourse; there are other tools and techniques that learning scientists can use for this purpose. But to fulfill the remit of learning sciences research, any such tool would need to operationalize some theory about how people make meaning in the world.

All formal analyses are simplifications of reality. Like any scientific instrument, ENA makes it possible to analyze one kind of phenomenon—in this case, Big Data about people's actions online, threaded discussions and forums, Twitter, Facebook, and other social media, and a host of data that are collected from games, simulations, and other immersive environments—by combining the techniques of ethnographic and statistical analysis. As a field we have much to gain from such an approach. What we potentially lose is the certainty that a person or group of people has read and interpreted all of the data—though reading and interpreting the volume of data produced by many modern learning and assessment contexts would be impossible regardless.

But, unlike purely statistical or computational techniques, when we analyze Big Data with an approach of the kind described here, the models are based on a theory of learning.[4] They are not just models of what people *do*, but how of how people make meaning. The models are not just patterns that a researcher happened to find in the data, but warrants that an examination of some discourse tells us something about a culture of learning.

Acknowledgments

This work was funded in part by the National Science Foundation (DRL-0918409, DRL-0946372, DRL-1247262, DRL-1418288, DUE-0919347, DUE-1225885, EEC-1232656, EEC-1340402, REC-0347000), the MacArthur Foundation, the Spencer Foundation, the Wisconsin Alumni Research Foundation, and the Office of the Vice Chancellor for Research and Graduate Education at the University of Wisconsin–Madison. The opinions, findings, and conclusions do not reflect the views of the funding agencies, cooperating institutions, or other individuals.

Further Readings

Andrist, S., Collier, W., Gleicher, M., Mutlu, B., & Shaffer, D.W. (2015). Look together: Analyzing gaze coordination with epistemic network analysis. *Frontiers in Psychology, 6*, 1016. doi: 10.3389/fpsyg.2015.01016

Csanadi, A., Eagan, B., Shaffer, D., Kollar, I., & Fischer, F. (2017). Collaborative and individual scientific reasoning of pre-service teachers: New insights through epistemic network analysis (ENA). In B. K. Smith, M. Borge, E. Mercier, and K. Y. Lim (Eds.). (2017). *Making a Difference: Prioritizing Equity and Access in CSCL, 12th International Conference on Computer Supported Collaborative Learning (CSCL) 2017* (Vol. 1). Philadelphia, PA: International Society of the Learning Sciences.

Hatfield, D. (2015). The right kind of telling: An analysis of feedback and learning in a journalism epistemic game. *International Journal of Computer-Mediated Simulations, 7*(2), 1–23.

Quardokus Fisher, K., Hirshfield, L., Siebert-Evenstone, A. L., Arastoopour, G., & Koretsky, M. (2016). Network analysis of interactions between students and an instructor during design meetings. *Proceedings of the American Society for Engineering Education* (p. 17035). ASEE.

These articles are all good examples of ENA in use, and span a range of domains of interest and research questions. Andrist et al. applies ENA to a multimodal dataset, including eye-tracking data. Csanadi et al. examines the effects of temporality on an ENA analysis. Hatfield looks at using ENA on text data. Quardokus Fisher et al. is described in the main text of this chapter.

Shaffer, D. W. (2017) *Quantitative Ethnography*. Madison, WI: Cathcart Press.

Quantitative Ethnography gives an overview of ENA, with details about coding, segmentation, and statistical issues in applying ENA to qualitative data.

http://www.epistemicnetwork.org/
This website about epistemic network analysis contains articles, user guides, tutorials, and sample data, including the *RescuShell* data described in this chapter.

https://cran.r-project.org/web/packages/rENA/index.html
This is a statistical package for epistemic network analysis.

NAPLeS Resources

Shaffer, D. W., *Tools for quantitative ethnography* [Webinar]. In *NAPLeS video series*. Retrieved October 19, 2017, from http://isls-naples.psy.lmu.de/intro/all-webinars/shaffer_video/index.html

Notes

1 Shaffer (2017) provides a comprehensive overview of data segmentation in the context of ENA.
2 It is also possible to construct weighted models where the value of the ith row and jth column is proportional to the number of times the ith code and jth code appear in the stanza. However, that is beyond the scope of the discussion here.
3 It is also possible to model network trajectories—that is, a longitudinal sequence of networks that show change over time. Again, that is beyond the scope of the discussion here.
4 For more on the importance of theory in learning analytics, see Wise and Shaffer (2015).

References

Andrist, S., Collier, W., Gleicher, M., Mutlu, B., & Shaffer, D. (2015). Look together: Analyzing gaze coordination with epistemic network analysis. *Frontiers in Psychology, 6*, 1016. doi: 10.3389/fpsyg.2015.01016

Bagley, E. A., & Shaffer, D. W. (2015). Stop talking and type: Comparing virtual and face-to-face mentoring in an epistemic game. *Journal of Computer Assisted Learning, 26*(4), 369–393.

Collier, W., Ruis, A., & Shaffer, D. W. (2016). Local versus global connection making in discourse. *12th International Conference on Learning Sciences (ICLS 2016)* (pp. 426–433). Singapore: ICLS.

Cress, U. & Kimmerle, J. (2018) Collective knowledge construction. In F. Fischer, C. E. Hmelo-Silver, S. R. Goldman, & P. Reimann (Eds.), *International handbook of the learning sciences* (pp. 137–146). New York: Routledge.

D'Angelo, A.-L. (2015). *Evaluating operative performance through the lens of epistemic frame theory* (Unpublished master's thesis). University of Wisconsin-Madison.

Danish, J., & Gresalfi, M. (2018). Cognitive and sociocultural perspective on learning: Tensions and synergy in the Learning Sciences. In F. Fischer, C. E. Hmelo-Silver, S. R. Goldman, & P. Reimann (Eds.), *International handbook of the learning sciences* (pp. 34–43). New York: Routledge.

De Wever, B., & Van Keer, H. (2018). Selecting statistical methods for the learning sciences and reporting their results. In F. Fischer, C. E. Hmelo-Silver, S. R. Goldman, & P. Reimann (Eds.), *International handbook of the learning sciences* (pp. 532–541). New York: Routledge.

Deacon, T. W. (1998). *The symbolic species: The co-evolution of language and the brain*. W. W. Norton & Company.

Fields, D. A., & Kafai, Y. B. (2018). Games in the learning sciences: Reviewing evidence from playing and making games for learning. In F. Fischer, C.E. Hmelo-Silver, S. R. Goldman, & P. Reimann (Eds.), *International handbook of the learning sciences* (pp. 276–284). New York: Routledge.

Fischer, G. (2018) Massive Open Online Courses (MOOCs) and rich landscapes of learning: A learning sciences perspective. In F. Fischer, C. E. Hmelo-Silver, S. R. Goldman, & P. Reimann (Eds.), *International handbook of the learning sciences* (pp. 363–380). New York: Routledge.

Gee, J. P. (1999). *An introduction to discourse analysis: Theory and method*. London: Routledge.

Geertz, C. (1973a). The impact of the concept of culture on the concept of man. In C. Geertz, *The interpretation of Cultures: Selected essays* (pp. 33–54). New York: Basic Books.

Geertz, C. (1973b). Thick description: Toward an interpretive theory of culture. In C. Geertz, *The interpretation of cultures: Selected essays* (pp. 3–30). New York: Basic Books.

Goodwin, C. (1994). Professional vision. *American Anthropologist, 96*(3), 606–633.

Green, J. L., & Bridges, S. M. (2018) Interactional ethnography. In F. Fischer, C. E. Hmelo-Silver, S. R. Goldman, & P. Reimann (Eds.), *International handbook of the learning sciences* (pp. 475–488). New York: Routledge.

Hatfield, D. (2015). The right kind of telling: An analysis of feedback and learning in a journalism epistemic game. *International Journal of Computer-Mediated Simulations, 7*(2), 1–23.

Hoadley, C. (2018). Short history of the learning sciences. In F. Fischer, C. E. Hmelo-Silver, S. R. Goldman, & P. Reimann (Eds.), *International handbook of the learning sciences*. New York: Routledge.

Jaccard, J., & Turrisi, R. (2003). *Interaction effects in multiple regression*. Thousand Oaks, CA: Sage.

Knight, S., Arastoopour, G., Shaffer, D. W., Shum, S. B., & Littleton, K. (2014). Epistemic networks for epistemic commitments. *Proceedings of the International Conference of the Learning Sciences*, Boulder, CO.

Koschmann, T. (2018). Ethnomethodology: Studying the practical achievement of intersubjectivity. In F. Fischer, C. E. Hmelo-Silver, S. R. Goldman, & P. Reimann (Eds.), *International handbook of the learning sciences* (pp. 465–474). New York: Routledge.

Lave, J., & Wenger, E. (1991). *Situated learning: Legitimate peripheral participation*. New York: Cambridge University Press.

Maxwell, J. A. (2008). Designing a qualitative study. In L. Bickman & D. J. Rog (Eds.), *The Sage handbook of applied social research methods* (2nd ed., pp. 214–253). Los Angeles: Sage.

Nash, P., & Shaffer, D. W. (2011). Mentor modeling: The internalization of modeled professional thinking in an epistemic game. *Journal of Computer Assisted Learning, 27*(2), 173–189.

Pea, R. D. (1993). Practices of distributed intelligence and designs for education. In G. Salomon (Ed.), *Distributed cognitions: Psychological and educational considerations* (pp. 47–87). Cambridge, UK: Cambridge University Press.

Quardokus Fisher, K., Hirshfield, L., Siebert-Evenstone, A. L., Arastoopour, G., & Koretsky, M. (2016). Network analysis of interactions between students and an instructor during design meetings. *Proceedings of the American Society for Engineering Education* (p. 17035). New Orleans, IN: ASEE.

Rosé, C.P. (2018). Learning analytics in the Learning Sciences. In F. Fischer, C. E. Hmelo-Silver, S. R. Goldman, & P. Reimann (Eds.), *International handbook of the learning sciences* (pp. 511–519). New York: Routledge.

Shaffer, D. W. (2017). *Quantitative ethnography*. Madison, WI: Cathcart Press.

Shaffer, D. W., & Ruis, A. R. (2017). Epistemic network analysis: A worked example of theory-based learning analytics. In C. Lang, G. Siemens, A. F. Wise, & D. Gašević (Eds.), *Handbook of learning analytics* (pp. 175–187). Society for Learning Analytics Research (SoLAR).

Suthers, D. D., & Desiato, C. (2012). Exposing chat features through analysis of uptake between contributions. *45th Hawaii International Conference on System Science* (pp. 3368–3377). New Brunswick, NJ: Institute of Electrical and Electronics Engineers (IEEE).

Vogel, F., & Weinberger, A. (2018). Quantifying qualities of collaborative learning processes. In F. Fischer, C. E. Hmelo-Silver, S. R. Goldman, & P. Reimann (Eds.), *International handbook of the learning sciences* (pp. 500–510). New York: Routledge.

Wise, A. F., & Shaffer, D. W. (2015). Why theory matters more than ever in the age of big data. *Journal of Learning Analytics, 2*(2), 5–13.

51
Selecting Statistical Methods for the Learning Sciences and Reporting Their Results

Bram De Wever and Hilde Van Keer

Introduction

This chapter presents four recommendations to consider when selecting, conducting, and reporting on statistical analyses in the field of the learning sciences. The aim of the chapter is realistic and consequently not to provide you with an exhaustive list of statistical methods that can be used in learning sciences research and all ways of reporting them. Since there are many methods and different ways of handling a variety of data within this broad research field, needless to say it is impossible to provide a full overview of all statistical methods used and thought relevant in the learning sciences or to provide a complete overview of recommendations for future work. Rather, we would like to put forward some elements that should be taken into consideration when selecting appropriate data analysis techniques and when reporting the results of studies conducted within this field. These elements—we call them recommendations, but they can also be read as matters to reflect on—will be illustrated with a selection of studies that we consider "good practices". In the following, we briefly elaborate on typical characteristics of learning sciences research, before presenting four recommendations related to these features. Then, we illustrate our recommendations by means of some selected exemplifying articles, and conclude with a short discussion section and five commented further references.

Characteristics of Research in the Learning Sciences

It is not a straightforward task to describe what research in the learning sciences generally or habitually looks like. We try to do so, anyway, and obviously acknowledge that these enumerated characteristics are subject to discussion. Although we are confident that the characteristics listed here are more or less agreed upon within the field, we are well aware of the fact that this is also a personal interpretation and that other scholars may come up with other or more characteristics and might emphasize different aspects.

First, quantitative research in the learning sciences is frequently characterized by studying the effects of different interventions in (quasi-)experimental studies. This implies that statistical techniques focusing on comparing groups are often used. Moreover, the focus is typically on human beings as participants and the studies are frequently conducted with students and/or teachers in (close to) authentic educational settings. In addition, in our field we often want to investigate development and changes over time (e.g., learning growth, increase in knowledge and competence). In what follows, we will link these specific characteristics of learning science research to guidelines for selecting and applying statistical techniques and for reporting on the results.

Recommendations for Selecting, Applying, and Reporting Statistical Methods and Results

Before we start discussing specific recommendations, we want to highlight that selecting, applying, and reporting statistical methods and results is only one part of conducting a good study. Having a solid theoretical framework and problem statement, well-formulated research questions, and a well-considered research design (just to name a few) are other, and equally important, characteristics.

The first three recommendations regarding statistical methods that we would like to put forward are related to considering the complexity of the data. More particularly, when performing research in the field of the learning sciences, your data may be complex in several ways. First, as human beings are generally the study subjects, there is a wide range of individual characteristics that might influence or confound the learning and instruction processes and outcomes under investigation. Second, many studies in the learning sciences are characterized by a strong focus on (semi-)authentic settings. This implies that the learning of the individuals under study is often not investigated in isolated settings or in the lab, but that learning and instruction are studied within specific authentic or authentic-like class or school contexts. Third, research in the learning sciences is often based on investigating the learning growth, development, and progress of people, and is accordingly interested in tracking changes over time. These three issues are discussed consecutively below, as they are all related in one way or another to considering the complexity of your data. Thereafter, we also present a fourth issue that is more related to reporting statistical methods and results.

The first recommendation we would like to put forward is to consider all potentially relevant variables in order to do justice to the complexity of your data as much as possible. There are of course many research design elements that can be important in view of tackling this issue, such as randomly assigning participants to research conditions or setting up matched designs, just to name a few examples. However, when it comes to the data analysis part, it is necessary to gather information on, and take into account, (theoretically) important influencing variables. Adding these variables to your statistical model is imperative in view of aiming to study phenomena within their complex context and in this respect increasing the validity of your research results. While a selection of variables is often necessary, given the limited resources and time, we still must be cautious not to oversimplify reality and aim to capture all relevant information, to take that information into account when analyzing. This is often realized by adding covariates, either as control variables or as specific variables under study. Carefully considering which variables might be meaningful and critically reflecting on their status is important, as they can explain or moderate specific intervention effects. They are, for example, essential in view of aptitude-by-treatment interactions (i.e., some instructional strategies may be more or less effective for specific students, as students' characteristics may interact with the instructional approaches and consequently individual learners may react differently to specific treatments; Jonassen & Grabowski, 1993; see also Bouwer, Koster, & Van den Bergh, 2018; Kollar et al., 2014). Taking these interactions into account and reporting on them is important in view of understanding complex relations that are often present within learning sciences research.

The second recommendation is to consider and explicitly model the complexity of your data when it comes to the grouping or nesting of your participants or measurements. This is often related to the (quasi-)experimental (authentic) situations in which your data is gathered. Students may, for instance, be collaborating in different groups, be part of different classes, or go to different schools. If this is the case, it is a good practice to take this nesting of individual students within groups (e.g., small collaborative groups, classes, schools, or any combination of those) into account by applying multilevel modeling. Multilevel modeling (you can also explore the search terms "hierarchical linear models" or "mixed-effects models") is a technique that takes into account this nestedness of the data (Cress, 2008; Hox, 1995). Obviously, the nestedness can also be present with other participants (e.g., teachers may be working in the same departments or at the same schools), but also within participants (e.g., different measurement occasions or different outcome variables may be nested).

The third recommendation is applicable for longitudinal research focusing on individuals' evolution over time, such as research on learning growth. If this is the aim of your research, make sure to take growth or progress explicitly into account throughout the study. This not only has implications for the design of your study (i.e., opt for a pretest-posttest design, consider the use of a retention test to study long-term effects, or check for transfer and generalizability), but has consequences for selecting your analysis techniques and reporting your findings as well. In this respect, careful consideration regarding analyzing differences in time is required. Simply calculating the difference between the pretest and posttest and performing an analysis on this difference score (often labeled "growth" or "increase") may not do the trick. Large increases may be due to a high posttest score, or a low pretest score, for example. Readers need to know how scores evolved during time. There are several techniques that deal with these issues, such as (multilevel) repeated measures or (piecewise) growth modeling (see, e.g., Gee, 2014; Hedeker & Gibbons, 2006), and the decision on which ones to use is of utmost importance. Modeling the complexity of your data is statistically challenging. However, your results will be much more valid, as your statistical model will be a lot closer to reality. In addition, reporting these results should be accessible for readers, which brings us to our next guideline.

Our fourth recommendation is to use everyday language—and, if possible, clarifying figures—when reporting and displaying your research results. As we argued in our previous recommendations, advanced and complicated statistical methods may be needed to model your data in an appropriate way. However, although you should detail the specific techniques for interested readers in your method section, we believe it is also important to report on your data in a way that it is logically, unambiguously, and clearly interpretable for all readers. In this respect, we deem that choosing an appropriate way to visualize results, such as graphs, figures, or tables, is often a useful addition to the use of everyday language (see Doumont, 2009). However, this does not imply that results should be oversimplified. The best scientific reports are not the ones saying simple things in a hard way, but those that explain complex things (e.g., statistical procedures) in an easy way.

Examples of Good Practices Illustrating the Recommendations

To illustrate the four recommendations we outlined above, we now present several studies as examples of good practice. The selection of the studies we will present is based on a convenience sample. We did not use a systematic approach for selecting them; on the contrary, we selected studies within the field of the learning sciences that we came across by reading literature in the field, or that we have knowledge of because we were co-authoring them. We do realize there is some bias towards intervention studies here (as this is the research area where we come from), although many of the examples are relevant for other types of studies as well. In all the examples, the participants under investigation are students. It is, however, important to mention that our recommendations are unquestionably also applicable for studies within the learning sciences focusing on teachers, teacher trainers, and other instructors.

A first study that exemplifies some of our recommendations is Kollar et al.'s (2014) study, "Effects of collaboration scripts and heuristic worked examples on the acquisition of mathematical argumentation skills of teacher students with different levels of prior achievement," whose aim was to investigate two different instructional approaches regarding their ability to support the development of mathematical argumentation skills. The latter were conceptualized as consisting of two different components, both of which are measured within this study—namely, an individual-mathematical and a social-discursive component. The sample consisted of 101 beginning student mathematics teachers. An experimental study was setup within a five-day period of an introductory course, consisting of three 45-minute treatment sessions, based on a 2 × 2 factorial design, with collaboration scripts versus without scripts, and heuristic worked examples versus problem solving.

In this study, learners' prior achievement (i.e., high school GPA) was explicitly considered. The authors refer explicitly to research on Aptitude-Treatment-Interactions (ATI) and to the "Matthew effect," a well-known phenomenon within ATI used to describe situations in which "students with higher prior achievement benefit more from a given kind of instruction than learners with lower prior achievement" (Kollar et al., 2014, p. 23). In this respect, this is an illustration of recommendation 1: to consider the characteristics of the participants, and specifically check for how these characteristics are influencing the results and interacting with the treatment.

Regarding our third recommendation, learning gain was modeled by entering pretest measures as covariates in the models. As discussed above, in addition, prior high school achievement was added as a covariate. The specific grouping of the students (cf. recommendation 2) was not statistically modeled. However, the authors did pay attention to this issue by establishing homogenous dyads to "reduce potential noise in the data that is produced if some students would collaborate with peers that were comparable to them while others would form dyads with students with considerably higher or lower GPA, which would later on be difficult to partial out" (Kollar et al., 2014, p. 27).

The authors spent considerable attention to explaining the rather complex interaction effects, showing that, for the argumentation components, the learning gain in a specific condition was dependent on the prior achievement, while in other conditions it was not. Regression slopes were

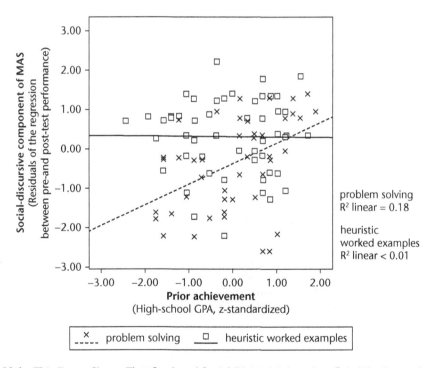

Figure 51.1 This Figure Shows That Students' Social-Discursive Learning Gain Was Dependent on Prior Achievement in the Problem-Solving Condition (Dashed Regression Line), but not in the Heuristic Worked-Examples Condition (Full and Close to Horizontal Line). This Particular Figure Helps, Together with the Detailed Explanations in the Text, to Present the Rather Complicated Interaction Effects in an Understandable Way. Reprinted with Permission From Kollar et al. (2014, p. 30).

calculated and compared, which is also illustrated with figures in the article. Without summarizing all results of this study, and without being able to provide the full context of the study, in Figure 51.1 we present an illustration of one of the article's figures. Figure 51.1 shows that social-discursive learning gain was dependent on prior achievement in one of the conditions (i.e., the problem-solving condition, indicated with the dashed regression line), while in the other condition this was not the case (i.e., the heuristic worked examples condition, indicated by the full and close-to-horizontal line). Together with the detailed explanations in the text, this figure helps to present to the readers, in an understandable way (cf. recommendation 4), the rather complicated interaction effects.

The second study that we would like to put forward as an example is Raes, Schellens, and De Wever's (2013) "Web-based collaborative inquiry to bridge gaps in secondary science education," whose main aim was to investigate the implementation of a web-based inquiry project in secondary school classes, with a specific focus on students' gender, achievement level, and the academic track they are in. The sample consisted of 370 students from 19 secondary school classes (grades 9 and 10). The study used a quasi-experimental pretest-posttest design and focused on the differential effects of web-based science inquiry for different existing groups of students (i.e., boys versus girls, low versus high achievers, science-track versus general-track students). The intervention consisted of four 50-minute lessons in which a web-based inquiry project was implemented.

This study pays specific attention to the multilevel nature of the data by using multilevel modeling (cf. recommendation 2). Students were working in dyads throughout the collaborative inquiry project, and these dyads were part of specific classes, so this was considered. At the same time, the repeated measures design (i.e., each student had results on a pretest and a posttest) was taken into account. This led to a four-level model: measurement occasions (level 1) "were clustered within students (level 2), which were nested within dyads (level 3), which in turn were nested within classrooms (level 4)" (p. 330).

The three independent variables—i.e., gender, achievement level, and tracking—are used as explanatory variables. More specifically, they were used to form eight different categories (see Figure 51.2) in order to take into account all important influencing variables at the same time (cf. recommendation 1), i.e., students are not only male or female, they are at the same time high or low achiever, and enrolled in a general or a science track. Given that the study aimed to investigate whether the intervention was able to close gaps between groups of students reported in earlier literature, it was important to identify

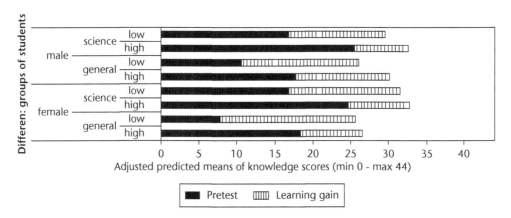

Figure 51.2 Example of a Graphical Representation. The Graph Shows the Adjusted Predicted Means of the Knowledge Scores at Pretest and Posttest for Different Groups of Students. At a Glance, the Reader Is Able to See the Pretest Scores, the Posttest Scores, and the Learning Gain for All Groups. Reprinted with Permission From Raes et al. (2013).

those gaps on the one hand, and how the intervention influenced them on the other. Therefore, a specific model was constructed in which students' achievement at pretest was modeled (i.e., allowing the reader to see how the different groups already scored differently at pretest), together with students' increase in knowledge between pretest and posttest (i.e., allowing the readers to see the increase, but also the scores at posttest; cf. recommendation 3).

What is especially interesting in this study is that there is a specific explanation on how the (rather complex) multilevel table needs to be interpreted on the one hand (there are even some examples explaining how the estimated means need to be calculated given the parameters in the multilevel model), and there is a clear and easy-to-understand graphical representation of these results on the other (cf. recommendation 4). In Figure 51.2, we represent this graphical representation, showing the adjusted predicted means of the knowledge scores at pretest and posttest for the different groups of students. In one single view, the reader can see how high the pretest scores were for the different groups, how high the posttest scores were, and how high the learning gain was for all groups. Without reading the article in detail, you can conclude in a visual way that the gaps between groups at pretest were in general larger than at posttest.

The third study that we present as an example is Merchie and Van Keer's (2016) "Stimulating graphical summarization in late elementary education: The relationship between two instructional mind-map approaches and student characteristics," which aimed to examine the (differential) effectiveness of two instructional mind-mapping approaches in view of fostering graphical summarization skills in elementary education. The sample consisted of 35 fifth- and sixth-grade Flemish (Belgium) teachers and their 644 students from 17 different elementary schools. A randomized quasi-experimental repeated-measures design (i.e., pretest, posttest, retention test) was set up, with two experimental conditions and one control condition. After teacher training provided by the researchers, students in the experimental conditions followed a 10-week teacher-delivered instructional treatment working with either researcher-provided or student-generated mind maps. Students in the control condition received a customary teaching repertoire (i.e., no systematic and explicit graphical summarization strategy instruction), providing an objective comparison baseline. Schools were randomly assigned to one of the three research conditions.

The data were analyzed by applying multilevel piecewise growth analysis. Multilevel analysis was applied, to consider the nested data due to the sampling of students within teachers and classes (cf. recommendation 2). Because in elementary education in Flanders complete classes are taught by one classroom teacher, the teacher and class level coincide in the present study. However, bearing in mind the repeated measures design of the study (cf. recommendation 3), three-level analyses were performed, with measurement occasions (i.e., pre-, post-, and retention test) (level 1) clustered within students (level 2), in their turn nested within classes/teachers (level 3). Since only a limited number of classes within a school participated in the study, no significant variance at school level was found and schools were not included as a fourth level in the analyses.

With regard to the third recommendation, this article explicitly studies students' growth and focuses on the differential impact of two experimental conditions as compared to a control condition. In this respect, the researchers specifically opted for multilevel piecewise growth analysis to model growth and evolution in each of the research conditions, and mutually compare this. In view of these analyses, the time span from pretest to retention test was split into a first phase covering students' growth from pretest to posttest, and a second phase reflecting students' growth from posttest to retention test. These phases were included in the model as repeated-measures dummy variables with correlated residuals at the student level.

With respect to the first recommendation, learner characteristics and aptitude-by-treatment interactions were explicitly considered. Student background characteristics (e.g., gender, home language, general achievement level) and the interaction of these with the growth in the different research conditions were especially included in the multilevel piecewise growth model. In this respect, the authors consider that various student-level characteristics might be related to students' graphical summarization skills (e.g., non-native speakers with a lower proficiency in the instructional language might experience more difficulties with graphical summarization, leading to lower summarization scores). At the same time, however, they also acknowledge—and therefore model—that the impact of the experimental instructional treatments might be dissimilar for students with different characteristics, since in educational studies and in view of formulating guidelines for educational practice it is important to consider whether different groups of learners benefit more or less from a particular intervention approach. In the present study, for example, learners with lower (verbal) abilities (e.g., low achievers or non-native speakers) might benefit more from working with worked examples in the researcher-provided mind-map condition, while high achievers might profit more by actively generating mind maps themselves.

As to our fourth recommendation, the researchers report on the results by providing the statistical parameter information in tables and additional appendices. However, the different statistical parameters and how to interpret these are discussed and clarified in the text as well. In this respect, readers with a more limited statistical background are also able to interpret the figures in the tables. Moreover, the differences between the research conditions are also clearly demonstrated by means of illustrations from completed student test materials and by means of graphs visualizing how students' graphical summarization skills are changing from pretest to posttest and retention test and how these trajectories vary for the different research conditions. Furthermore, additional appendices are included, which contribute to a better understanding of the differences between the experimental interventions on the one hand and of the applied measurement instruments on the other.

A fourth example of good practice in view of our recommendations describes a study, "Effects of a strategy-focused instructional program on the writing quality of upper elementary students," by Bouwer et al. (2018). This intervention study aimed to evaluate the effects of a comprehensive program for the teaching of writing of several different types of texts (i.e., descriptives, narratives, and persuasive letters) in an authentic general educational setting. The intervention was implemented by a sample of 76 teachers over an eight-week period in 60 fourth- to sixth-grade classrooms. More specifically, the main focus of the content of the program was on teaching students a strategy for writing, supplemented with the teaching of self-regulatory skills, and instruction in text structure. The mode of instruction was observational learning, complemented with explicit strategy instruction and guided practice with extensive scaffolding. A switching replication design with two groups and three measurement occasions was applied. Students ($N = 688$) and teachers ($N = 31$) in Group 1 worked with the intervention program in the first period of eight weeks (i.e., between the first and second measurement occasion). Group 2 served as a control group during this period in which teachers ($N = 45$) and students ($N = 732$) engaged in their regular writing activities and routines. During the second phase of eight weeks (i.e., between the second and third measurement occasion), the intervention switched between groups, such that the teachers and students in Group 2 started to work with the writing program, while those in Group 1 returned to their regular writing activities. Students' writing skills were assessed using three different types of texts (i.e., descriptions, narratives, and persuasive letters) at each measurement occasion.

When comparing the characteristics of this study with our abovementioned guidelines and recommendations, we observe that the study explicitly takes into consideration the hierarchical organization of the data (cf. recommendation 2). More particularly, writing scores are cross-classified with students and tasks, and students are nested within classes. Therefore, the data are analyzed by applying cross-classified multilevel models, allowing writing scores to vary within and between students, between tasks, and between classes.

Furthermore, when studying the impact of the intervention on the quality of students' texts, learner characteristics and aptitude by treatment interactions were also explicitly considered (cf. recommendation 1). More specifically, the authors studied whether the effectiveness of the program differed between grades and between male and female students. In addition, they also investigated whether the effect of the program depended on the proficiency of the writer.

In line with our third recommendation, the researchers of this study focus on growth in the quality of students' writing and on whether the effect of the intervention was maintained over time. With respect to the latter, the third measurement occasion served as a posttest for students in Group 2, but was considered as a delayed posttest for students in Group 1, with which the authors were able to measure retention and the maintenance effect of the intervention. Moreover, it especially needs mentioning that the differential growth for both Group 1 and 2, and the maintenance effect in Group 1—also broken down for the grades that students are in—are clearly visualized by including graphical representations of the growth trajectories, next to an overview of the parameters presented in a table (cf. recommendation 4).

Discussion and Conclusion

Writing a (rather short) chapter on statistical methods for the learning sciences required us to limit ourselves to a small selection of techniques. We based this selection on what we believe are important techniques deserving attention in future learning sciences research, as they can deal with specific core aspects of this research field (e.g., multilevel modeling for taking into account nesting and full complexity of the data, growth modeling for analyzing changes over time). However, in making this selection we, unfortunately, also had to disregard other statistical techniques that can be of equal importance for future studies, such as: exploratory and confirmatory factor analysis (EFA/CFA) in view of developing and validating instruments; structural equation modeling (SEM) aimed at establishing causal, mediating, or moderating relationships between variables; item response theory (IRT) in view of developing tests, cluster analyses, meta-analyses techniques, non-parametric analyses, amongst others. Some of these, such as specific techniques for data mining and learning analytics, are described in other chapters of this handbook (e.g., Rosé, this volume). Others are to be explored outside of this book.

Nevertheless, we believe the four presented recommendations are important in view of learning sciences research, especially since they touch upon the heart of some essential features distinguishing the learning sciences from other disciplines. In this respect, the present chapter is intended to create or increase awareness on the typicality of our field and the related implications for data analysis and reporting of the results.

Further Readings

Baguley, T. (2012). *Serious stats: a guide to advanced statistics for the behavioral sciences*. Basingstoke: Palgrave Macmillan.
There are many good handbooks on statistics and quantitative data analysis, so suggesting just one as a further reference is not straightforward. However, we believe that *Serious Stats* is a nice all-rounder for statistical procedures in our field. It deals, amongst other things, with (repeated measures) ANOVA, ANCOVA, and multilevel modeling. It further provides code for R and syntaxes for SPSS (statistical software, see below), as well. In addition, the handbooks of Kutner, Nachtsheim, Neter, and Li (2004) and Singer and Willett (2003) can also be consulted for more detailed information respectively on regression and repeated measures from a longitudinal perspective.

Cress, U. (2008). The need for considering multilevel analysis in CSCL research—An appeal for the use of more advanced statistical methods. *International Journal of Computer-Supported Collaborative Learning, 3*(1), 69–84. doi:10.1007/s11412-007-9032-2
This is an interesting introduction in multilevel modeling, contextualized for the learning sciences, that we recommend to read alongside more general statistical introductions to multilevel modeling (the article itself contains references to important publications on multilevel modeling in general). The article explains several reasons on why—when students are collaborating in groups—individual observations of students are not independent from each other, as groups may be composed differently, they may share a common fate (that is different from other groups), or group members may influence each other during the learning process. Given that the latter, the reciprocal influence, is actually intended in collaborative learning, the article also appeals to applying multilevel

modeling. A specific example (i.e., data set) is used to explain what multilevel modeling is, how it works, and why it is necessary. The article also presents the mathematical/statistical background, including the formulas, but does this in a clear way, illustrating the models with the specific example.

Dalgaard, P. (2008). *Introductory statistics with R* (2nd ed.). New York: Springer.
While we agree that SPSS is probably the best-known software package in our field, and packages such as MLwiN, HLM, and Mplus especially are excellent packages for the multilevel modeling and growth modeling we have put forward, we opted for the R software in this reference (www.r-project.org/). R is free software (GNU-GPL), runs on many operating systems, has a lot of possibilities and a growing community of users. Moreover, there are many environments built around R—we especially like Rstudio in this regard (www.rstudio.com/). There are various resources on the internet, and we further refer to *Introductory Statistiscs with R* as a reference work. Although the author states that there is some biostatistical bias in the choice of the examples, it is a thorough introduction to start working with R.

Doumont, J. (2009). *Trees, maps, and theorems. Effective communication for rational minds*. Kraainem: Principiae.
In line with our recommendation to use clear everyday language and accessible figures to report and illustrate research results, we refer the readers further to this book. One of the chapters particularly deals with effective graphical displays and pays attention to how graphs can be designed for comparing data, showing evolution, or comparing groups. Specific attention is also paid to drafting the caption. In this respect, Dumont argues that the caption should not be a descriptive title, but rather a clear sentence explaining what is to be seen in the figure or graph, answering the so-called "so what" question of the reader (i.e., "why are you showing this to me?", p. 149).

Gee, K. A. (2014). Multilevel growth modeling: An introductory approach to analyzing longitudinal data for evaluators. *American Journal of Evaluation, 35*(4), 543–561. doi:10.1177/1098214014523923
This article presents an applied introduction to multilevel growth modeling (also referred to as mixed-effects regression models or growth curve models), which is one method of analyzing longitudinal data collected over time on the same individuals. The introductory concepts are grounded, and the method is illustrated in the context of a longitudinal evaluation of an early childhood care program. The article provides a concise and accessible "how-to" approach. Gee more specifically informs the reader on data requirements, visualizing change, specifying multilevel growth models, and interpreting and displaying the results. It has to be noted that the article focuses on continuous outcomes only. Readers are referred to the work of Hedeker and Gibbons (2006) for details on multilevel growth models with non-continuous outcomes.

NAPLeS Resources

De Wever, B. *Selecting statistical methods for the learning sciences and reporting their results* [Video file]. *Introduction and short discussion*. In *NAPLeS video series*. Retrieved October 19, 2017, from http://isls-naples.psy.lmu.de/video-resources/guided-tour/15-minutes-dewever/index.html

De Wever, B. *Selecting statistical methods for the learning sciences and reporting their results* [Video file]. *Interview*. In *NAPLeS video series*. Retrieved October 19, 2017, from http://isls-naples.psy.lmu.de/video-resources/interviews-ls/dewever/index.html

References

Bouwer, R., Koster, M., & Van den Bergh, H. (2018). Effects of a strategy-focused instructional program on the writing quality of upper elementary students in the Netherlands. *Journal of Educational Psychology, 110*(1), 58–71. http://dx.doi.org/10.1037/edu0000206

Cress, U. (2008). The need for considering multilevel analysis in CSCL research—An appeal for the use of more advanced statistical methods. *International Journal of Computer-Supported Collaborative Learning, 3*(1), 69–84. doi:10.1007/s11412-007-9032-2

Doumont, J. (2009). *Trees, maps, and theorems. Effective communication for rational minds*. Kraainem: Principiae.

Gee, K. A. (2014). Multilevel growth modeling: An introductory approach to analyzing longitudinal data for evaluators. *American Journal of Evaluation, 35*(4), 543–561. doi:10.1177/1098214014523923

Hedeker, D., & Gibbons, R. D. (2006). *Longitudinal data analysis*. Hoboken, NJ: John Wiley & Sons.

Hox, J. J. (1995). *Applied multilevel analysis*. Amsterdam: TT-publikaties. http://joophox.net/publist/amaboek.pdf

Jonassen, D. H., & Grabowski, B. L. (1993). *Handbook of individual differences, learning and instruction*. Hillsdale, NJ: Erlbaum.

Kollar, I., Ufer, S., Reichersdorfer, E., Vogel, F., Fischer, F., & Reiss, K. (2014). Effects of collaboration scripts and heuristic worked examples on the acquisition of mathematical argumentation skills of teacher students with different levels of prior achievement. *Learning and Instruction, 32*, 22–36. doi:10.1016/j.learninstruc.2014.01.003

Kutner, M., Nachtsheim, C., Neter, J., & Li, W. (2004). *Applied linear statistical models* (5th ed.). Boston: McGraw-Hill.

Merchie, E., & Van Keer, H. (2016). Stimulating graphical summarization in late elementary education: The relationship between two Instructional mind-map approaches and student characteristics. *Elementary School Journal, 116*, 487–522.

Raes, A., Schellens, T., & De Wever, B. (2013). Web-based collaborative inquiry to bridge gaps in secondary science education. *Journal of the Learning Sciences, 23*(3), 316–347. doi:10.1080/10508406.2013.836656

Rosé, C. P. (2018). Learning analytics in the Learning Sciences. In F. Fischer, C. E. Hmelo-Silver, S. R. Goldman, & P. Reimann (Eds.), *International handbook of the learning sciences* (pp. 511–519). New York: Routledge.

Singer, J. D., & Willett, J. B. (2003). *Applied longitudinal data analysis: Modeling change and event occurrence.* New York: Oxford University Press.

Index

academic language 396
accountability 27, 466
Accountable Talk 326
acquisition metaphor 44
action-concept congruencies 81
actions: gesture as simulated 80–81; imagined or simulated 79–80; observing 78–79
activity system 57
Activity Theory 360
ACT-R (adaptive control of thought-rational) 202
adaptable scripting 346
adaptive expertise 59, 148
adaptive guidance 193
adaptive scripting 346
adaptivity 148, 247
Additive Factors Model 187
affective spectacle exhibits 239–240, 242
affinity spaces 278
Agar, M. 476
agent-based models 162
AIR model 28–29
Aleven, V. 187, 202
Alexander, R. 324
Allen, S. 119
Alonzo, A. C. 435
analytic approach 436
Anderson, J. 56
Anderson, R. 326
apprenticeship learning 44–50, 192
aquaria 239–240, 242
Archer, M. S. 140, 141
ARGUGNAUT 322–323
Arguing to Learn (Andriessen, Baker, & Suthers) 18, 319
argumentation 342
argumentative grammar 396–397
argument structuring tools 226–227
articulation 49

Assaraf, O. B. 159–160
assessment: 4C/ID model and 174; construct-centered design and 415–416; continuous 229; design and use of 417–418; domain-specific learning and 414–415; embodied perspective on 81–82; knowledge building and 300, 304; large-scale 411, 412, 418; LPs and 428; measuring competencies 433–440; multi-representational learning and 101; as process of evidentiary reasoning 412–414; purposes and contexts of 410–412; technology-supported 301–302; theory and research and 419; transformative 302; validity of 416–417
assessment-driven cross-sectional approach 426–427
assessment triangle 412–414, 437
Asterhan, C. S. C. 321, 322
augmented reality (AR) 238
automated guidance 227–228
automaticity 56
autonomous learning 221–223, 228
Azevedo, R. 194

Baker, M. J. 318
Bakhtin, M. 140, 477
Ball, N. 71
Bandura, A. 203
Bang, M. 109
Bannert, M. 130
Barab, S. 445
Barnes, J. 40–41
Barron, B. 449, 492
Barrows, H. S. 214
Battista, M. T. 425
Baumert, J. 435
BeeSign project 40
behaviorism 12
behavior schemas 58
belief mode 296–297

542

Bell, B. S. 193
Belvedere 319
Bennett, R. E. 415
Bereiter, C. 15, 59, 139, 295–298, 300, 302, 303–304, 315, 318, 320, 324
Bergo, O. T. 401–402
Berthold, K. 260
Bevan, B. 290
Bielaczyc, K. 300, 384
Big Data 374–375, 495–496, 520–529
Billett, S. 148
Birdwhistell, R. 470
Bivall, P. 257
Black, J. B. 257
Blair, K. P. 69
Blake, R. L. 213
blended learning environments 251
blended settings 323–324
Blikstein, P. 288
Blömeke, S. 437
Bloome, D. 477
Bodemer, D. 355
Bofferding, L. 69
Bonn, D. 216
Bonnard, Q. 183
Borge, M. 320
Bos, W. 435
Boshuizen, H. P. V. 92
boundary objects/boundary work 110–111
Bouwer, R. 538–539
Boxerman, J. Z. 119
Brand-Gruwel, S. 92
Bransford, J. D. 15, 59, 60, 181, 210
Bråten, I. 24, 91, 92
Bridges, S. M. 475, 477, 478–480, 483, 486
Britt, M. A. 88
Broers, N. J. 449–450
Bromme, R. 25
Brown, A. L. 15, 18, 318, 320, 324, 383, 388, 444
Brown, J. S. 14, 15, 57
Brown, M. W. 268
Bruner, J. S. 340
Brush, T. 214
Brush, T. A. 335
Building Teaching Effectiveness Network (BTEN) 396
Bull, S. 461
Burke, Q. 279

Caballero, D. 184
Cakir, M. 450
Cambridge Handbook of the Learning Sciences (Sawyer) 18
Campione, J. C. 15
Carey, S. 13
Caswell, B. 300
causality 141
CBAL (Cognitively Based Assessment of, for and as Learning) 415

Chan, C. K. K. 300
Chase, W. G. 55
Cheche Konnen project 27
Chen, B. 298, 300
Cheng, B. H. 393
Chi, M. T. H. 55, 158, 159
child development, apprenticeship as metaphor for 46–47
Ching, Y.-H. 362
Chinn, C. A. 25, 28
Chiu, M. M. 506
Chomsky, N. 13, 434
Chu, S. K. W. 300
"chunking" 55
CILT (Center for Innovative Learning Technology) 17
Claim, Evidence, Reasoning model 315
Clark, D. B. 277, 278
classroom logistics 184
classroom orchestration 180–188
classroom usability 183–184
clockwork-versus-complex systems understanding 160–161
coaching 49
Cobb, P. 425
co-construction of knowledge 318, 320
Coder movement 276
co-design 397, 401–407
coding scheme and tools 502–506
co-evolution model 142
cognition: assessment and 413, 437; embodied 75–82; neuroarchitecture of 65–66
cognitive apprenticeship framework 48–50
cognitive flexibility research 49
Cognitive Flexibility Theory (CFT) 494–495
cognitive load theory 96, 177, 184, 202
cognitive modeling 48, 50
cognitive neuroscience foundations 64–71
cognitive perspective: embodied cognition and 75; on expertise 54–56; knowledge construction and 138, 142–143; on learning 34–41
cognitive science, development of 12–13
cognitive strategies 172
Cognitive Studies in Education Program (CSEP) 15
cognitive task analysis (CTA) 174
cognitive theories 35–36
Cognitive Tutor Algebra 346
cognitive tutors 39, 49
cognitivism 12
Cohen, J. 505
Cole, H. 361
Cole, M. 15
collaboration *see* computer-supported collaborative learning (CSCL); participatory design
collaboration script 180, 195, 321, 340–347, 355, 364, 534–535
collaborative learning processes, quantifying qualities of 500–507
collective inquiry 308–315

Index

collective knowledge construction 137–144
collective reasoning 326
Collins, A. 15, 18, 47, 49, 57, 303
communities of practice 45–46
community/system level 27
CoMPASS project 384–387, 389
competence models 434–435
competencies, measuring 433–440
complementarity 161
complexity, levels of 172
complex learning, 10 steps to 174
complex systems 157–164
computational thinking 224–225
computer-based training (CBT) 246
computer-supported argumentation 318–327
computer-supported collaborative learning (CSCL): analytic aims and 304; argumentation and 318–327; Conversation Analysis and 468–469; development of 138; emergence of 14, 15–17; frameworks for 330–337; group awareness tools for 351–356; knowledge building and 303; MMR and 449; mobile 359–365; orchestration and 180; scaffolding and scripting for 340–347; technological tools for 132–133
computer-supported collaborative work (CSCW) 133
Computer-Supported Intentional Learning Environment (CSILE) 295, 296, 318–319
computing, impact of advances in 12–13
concept maps 162
constraints, multiple 181
construct-centered design 415–416
constructionism 261, 289
constructivism 138, 170, 335
construct-modeling approach 415–416
context, sociocultural approaches and 36
Context Analysis 467, 469–470
contextual inquiry 183
Conversation Analysis (CA) 467–469
COPES architecture (Conditions, Operations, Products, Evaluations, and Standards) 130
co-regulated learning (CoRL) 127, 128–129, 342
Cormier, D. 368
Cornelius, L. 30
correction and repair 468
Courtright, C. 310
CRAFT 176
Cress, U. 355
Creswell, J. W. 446, 448–449
critical realism 140, 141
Cromley, J. G. 194
Cross, N. 110
Cudendet, S. 183
cultural-historical activity theory (CHAT) 149, 150
cultures/cultural settings 107–108, 476–477, 521
Curriculum Customization Service (CCS) 269, 271–274, 340
customization 235

cyberlearning 267–268
cybernetics 12–13

Danish, J. A. 40
Darling-Hammond, L. 59, 60, 406
data collection and methods 37–38
data mining 269, 495–496, 511, 517
Day, J. 216
Deane, P. 415
decision-making 238
DeGroot, A. D. 55
de Jong, T. 259, 260
de la Fuente, P. 450
Delazer, M. 67, 70
deliberative democracy 324–325
deliberative practice 58–59
Demetriadis, S. 346
Denessen, E. 280
Denner, J. 280
Derry, S. J. 215, 470–471, 491, 494, 495
descriptive models 435
Desiato, C. 522
design, inquiry and 225–226
design-based implementation research (DBIR) 393–398
design-based research (DBR) methods: co-design and 403; frameworks for 330–337; hypotheses and 455; increased visibility of 18; inquiry learning and 223; MMR and 445, 446, 449, 451; multidisciplinarity and 106; orchestration and 184; principle-based pedagogy and 298–301; problem solving and 86; simulations and 261; trajectories of studies and 383–390
Design-Based Research Collective 18
Design Experiment Consortium 18
designing for learning 39–41
design mode 296–297
design principles: motivation, engagement, and interest and 123; for orchestration 183; for PBL 215; regulated learning and 132–133
developmental psychology 12
De Wever, B. 506, 536–537
Dewey, J. 12, 221, 236
dialogism 324–325
Diaz, A. 184
DiBello, L. V. 416
Differential Item Functioning (DIF) 439–440
differentiated scaffolding 193, 195
digital society, multiple sources in 86–92
Dillenbourg, P. 18, 183, 184, 513–514
dimensionality of competence 439
Dimitriadis, Y. 450
disciplinary thinking and learning 106–109
disciplines, learning within and beyond 106–111
discursive model of argumentation 318
dispositions 148
distance effect 67
distributed cognition 150
distributed scaffolding 193

Dixon, C. 290
DIY (do-it-yourself) culture 285–291
D'Mello, S. K. 249
Dodick, J. 159–160
Do-Lenh, S. 183
domain specificity 54–55, 56
domain-specific learning 414–415
dominant objectivist approach 170
double-content examples 204
Doubler, S. 424–425
Dougherty, D. 286
Dreyfus, H. L. 106
Dreyfus, S. E. 106
duality of structure 141
Duguid, P. 15, 57
Duncan, S. 278

e-argumentation 324–327
Eberhardt, J. 214
Eckström, A. 469
ecosystemic approach 234–242
educational data mining (EDM) 269, 495–496, 511, 517
educational technology, embodied perspective on 82
Egan-Robertson, A. 477
Eilam, B. 97
Eisenstadt, M. 55
Elby, A. 26
electroencephalogram (EEG) 65
Embedded Phenomena (EP) 311–312
Embedded Phenomena for Inquiry Communities (EPIC) 311–312
embodied cognition 75–82
emergence 140, 157, 159, 161
emergent curriculum 299–300
empowerment 181–182, 183
engagement 116–123, 132
Engeström, Y. 139, 149
Enyedy, N. 37, 491
epistemic agency 298–299
epistemic aims 29–30
epistemic cognition (EC) 24–31, 91
epistemic focus 303
epistemic frames 521–522, 528
epistemic ideals 29
epistemic network analysis (ENA) 520–529
epistemic practices and objects 151–152, 153
epistemological development 24–31
e-portfolio 300, 302
Ericsson, K. A. 59
Ertl, B. 342
Eryilmaz, E. 335
e-scaffolding 321–323
ethnicity 122–123, 163
ethnomethodology (EM) 456, 465–471
Evans, M. A. 444–445
evidence-centered design (ECD) approach 415–416
Ewenstein, B. 152
examples-based learning 201–206

exhibits 234–242
expertise/expertise development 48, 54–60, 97, 107, 147–148, 149–150, 296
exploration, fostering 49
exploratory talk 318–319
extrinsic constraints 183–184
Eysink, T. H. S. 260

fading 195, 345–346
Fairclough, N. 484
feedback: ITS and 247, 248–249; peer 342
Feinstein, N. 109
Feltovich, P. 55
Fenton, K. D. 71
Ferguson, R. 335
Ferrara, F. 79
Fields, D. A. 280, 280–281
Fiore, S. 457
Fischer, F. 326, 342, 343, 346, 514
Fishman, B. 393
five-step dialogue frame 248
fixed curriculum 299–300
fixed groups 300–301
Fleiss, J. L. 505
flow state 118
formative assessment 411, 418, 422
Fostering a Community of Learners (FCL) 309, 388
4C/ID (four-component instructional design) model 169–177
frames/framings 237, 241
Fraser, K. 176
Fredricks, J. A. 119
Freud, S. 11
Fujita, N. 461
functional MRI (fMRI) 65

Gallimore, R. 493
games 256–262, 276–282
gamification approaches 236
Garfinkel, H. 465–466
Gazzaniga, M. S. 64
Gee, E. R. 276, 281, 282
Gee, J. P. 521
Geertz, C. 520
gender 122, 163, 322
Generalizability Theory 438
Generalized Intelligent Framework for Tutoring (GIFT) 250–251
Gerjets, P. 90, 101
Gesture as Simulated Action framework 80–81
gestures 469
Giddens, A. 140, 141
Gil, J. 321
Glaser, P. 55
Glenberg, A. M. 78, 80
Goffman, E. 470
Goh, W. 324
Goldman, R. 493
Goldman, S. R. 28, 30, 108, 416

Index

Gómez, E. 450
Goodwin, C. 469, 494, 521
Goodwin, M. H. 469
Gottlieb, E. 26, 109
Gotwals, A. W. 159, 424
Grabner, R. H. 68
Graesser, A. C. 249
Gravel, B. 287–288
Greene, J. A. 24, 26, 194
Greene, J. C. 448
Greeno, J. 15, 150
Greiffenhagen, C. 469
Gresalfi, M. S. 40–41
Grimes, S. M. 280–281
Grotzer, T. A. 158
group awareness tools 343, 351–356
group cognition 19, 139–140
guidance 226, 227–228
Gumperz, J. J. 484
Gutwill, J. P. 119, 290

Haag, N. 440
Habermas, J. 325
"hacker culture" 288
Hadwin, A. F. 130, 131
Halewood, C. 300
Hall, R. 491
Hamilton, L. 412
Hammer, D. 26
Han, I. 257
Hand, V. 108
Harmsen, R. 261
Hartman, K. 494
Hatano, G. 59
Häussler, P. 122
Hawking, S. 157
Heap, J. 468
Heath, S. B. 475, 476
Heimeriks, G. 457
Henderson, A. 470, 493
Hendricks-Lee, M. 58
Heppt, B. 440
Heritage, J. 465–466
Herrenkohl, L. R. 30, 109–110
Hidi, S. 119
Hiebert, J. 493
history: contrasted to science 28–30; of learning sciences 11–21
Hmelo-Silver, C. E. 159, 160, 214, 495
Hoadley, C. 123
Höchstötter, N. 90
Hoffmann, L. 122
holistic approach 436
Holmes, N. 216
Hong, H. Y. 298, 301, 342
Hoogerheide, V. 205
Hope, J. M. 110
horizontal expertise 148, 149, 150
Howard-Jones, P. A. 71

How People Learn 401–107
Hsu, Y.-C. 362
human-computer interaction (HCI) 183–184, 237
Hung, W. 215

idea-centered focus 300, 303, 309
implementation research 394, 397
improvement science 394
Inagaki, K. 59
indie games 282
Indigenous epistemologies 109
individual knowledge construction 138–139, 142
individual level 28
Industrially-Situated Virtual Laboratory (ISVL) 526–527
inert knowledge problem 210
information-delivery exhibits 234–236, 240–241
information problem solving 86–92
information processing theory 212
information resources 88
infrastructure 394
"initiation-reply-evaluation" (IRE) sequence 468
inquiry learning 221–229, 308
Institute for Research on Learning (IRL) 15
Instructional Architect (IA) 269, 270–271, 273–274, 340
instructional design, embodied perspective on 81
instructional explanations 205
instructional support, software tools and 260
Integrated Theory of Text and Picture Learning 96
integrative goals, models for 170
integrative model 87–89
intelligent tutoring system (ITS) 186–187, 246–252
intentional learning 296
interactional ethnography 475–486
interactional level 27–28
Interaction Analysis (IA) 467, 470, 470–471, 493
interactivity 247
interdisciplinarity 457–458
interest 116–123, 132
internalizing external interactions 191–192
internal scripts 343–344
International Society of the Learning Sciences 17
interpretation, assessment and 413–414, 437, 438, 439
intersubjectivity 465–471
intertextuality 477
intuition 59, 106
Iordanou, K. 324
Isohätälä, J. 131
Item Response Theory (IRT) 439
Ivry, R. B. 64

Jacobson, M. 159, 161
Järvelä, S. 131, 132, 355
Järvenoja, H. 131
Jefferson, G. 468
Jelsma, O. 169
Jeong, H. 449

Jochems, W. M. G. 449–450
Johnson, D. W. 364
Johnson, R. T. 364
joint regulation 342–343
Jonassen, D. 215
Jordan, B. 470, 493
Judele, R. 342

Kafai, Y. B. 276, 279, 280
Kammerer, Y. 90
Kamp, E. 259
Kaptelinin, V. 14
Kapur, M. 211, 214
Karakostas, A. 346
Kareev, Y. 55
Kauffman, S. 158
Kelcey, B. 424
Kelly, A. E. 396
Kendon, A. 470
Kienhues, D. 25
Kimmerle, J. 355
King, A. 139
Kirschner, P. A. 174
Klein, S. 412
Klemp, N. 470
Klieme, E. 434, 435
Kling, R. 310
Klopfer, E. 163
Knorr Cetina, K. 151
knowing/knowledge: co-construction of 318, 320; cognitive approaches to 35; content of 58; expertise and 58; organization of 55–56; procedural 148; propositional 148; sociocultural approaches to 36; structures for 58; tacit 59; types of 48
Knowledge Awareness tool 352
knowledge building 295–305, 309, 318, 324, 342
knowledge communities and inquiry (KCI) model 310–314
knowledge components (KCs) 249
knowledge creation metaphor 140
Knowledge Forum 295, 296, 297, 309, 319, 388–389
knowledge integration framework 223
knowledge-intensive work 147, 148, 149, 151–152
Ko, M. 405
Koedinger, K. R. 202
Kollar, I. 195, 343, 346, 534–536
Köller, O. 435
Kolodner, J. 14, 16, 387
Konopasky, A. 288
Koschmann, T. 14, 15, 16, 469, 470–471
Kozlowski, S. W. J. 193
Kozma, R. 97–98
Krajcik, J. 315
Kreplak, Y. 469
Kress, G. 486
Kuhl, P. 440
Kuhn, D. 324

Kuhn, T. S. 446
Kyle, W. C., Jr. 109

Lab for Comparative Human Cognition 15
Lacy, J. E. 287
Lagemann, E. C. 12, 18
Lajoie, S. P. 131
landscape metaphor 425
language comprehension 76, 78
Latour, B. 107
Laurillard, D. 360
Lave, J. 15, 45, 149, 445, 451
Law, N. 450, 507
Lazonder, A. W. 259, 260, 261
learner attributes 88
learning analytics 375, 511–517
learning communities, collective inquiry in 308–315
learning progressions 414–415, 422–429
learning sciences: history of 11–21; methodologies of 26–27; research in 1–8
learning supports 161–163
learning tasks 171–172
LeBaron, C. 469
Lee, C. D. 39–40, 108
Lee, C. L. 259
Lee, V. R. 38, 82, 119
legitimacy 27
Lehrer, R. 27
Leontiev, A. N. 139
Leshed, G. 355
Leutner, D. 434
level models of competence 435
Levy, S. T. 161
Lewandowski, D. 90
Lindwall, O. 469
Linn, M. C. 107–108, 163, 262
Liu, L. 159
Lleras, A. 78
Lobato, J. 425
Loibl, K. 50
Loyens, S. M. M. 205
Luhmann, N. 140, 141, 142
Lund, K. 21, 450, 507
Lundeberg, M. 214
Lynch, M. 469

Macbeth, D. 468, 469, 471
machine learning 511–517
macro-scripts 180–181
magnetic resonance imaging (MRI) 65
Maker Movement 276, 285–291
makerspaces 287–288
Malmberg, J. 131, 132
Mandl, H. 342
Mangun, G. R. 64
manipulatives 78, 81
Manz, E. 27
Marathe, S. 159
Martens, R. L. 449–450

Index

Martin, L. 290
Martinez, A. 450
massive open online courses (MOOCs) 368–377
mathematical thinking and learning 67–69
Maxwell, J. A. 520
Mayer, R. E. 97, 261, 494
Mayfield, E. 346
McCarthy, L. 278
McClelland, D. C. 434, 437
McCrudden, M. T. 91
McDermott, R. 470
McHoul, A. 468
McKenney, S. E. 389–390
McLaren, B. M. 202, 323
McLaughlin, M. W. 406
McNeill, D. 469
McNeill, K. L. 195, 315
Mehan, H. 468
Melander, H. 470
memory systems 66
mental models 172
Mercer, N. 318
Merchie, E. 537–538
Mertl, V. 109–110
Messina, R. 300, 388–389
metacognition 25, 30, 48, 131, 133, 386
metacognitive awareness 130–131
meta-discourse 300, 302
metanavigation support 386
micro-scripts 180, 181
Mikšátko, J. 323
Miller, M. 131
Minecraft 281, 282
Mislevy, R. J. 415
Mitchell, C. J. 478
mixed methods research (MMR) 444–451, 456–457
mobile interpretation systems 235
modeling: examples of 201–202, 203–205; inquiry learning and 224–225; student 249; tools for 256–262
moderation 322–323
Molfese, D. L. 65
Mondemé, C. 469
Montessori, M. 12
Moog, R. 259
Moos, D. C. 194
Moreno, R. 494
Morgan, D. L. 447
Morley, E. A. 343
Morse, J. M. 448
motion capture technologies 82
motivation 35–37, 116–123, 132, 211, 374
Mu, J. 346
Mulder, Y. G. 259
Mullins, D. 346
multidisciplinary partnerships 222
multidisciplinary research 24
multiliteracies 289
multimedia learning 96–102

Multimedia Learning Theory 96
multimodal data 133–134
multimodal interaction/CA 467, 469
Multimodal Learning Analytics 496
multiple representations 96–102
multiple source comprehension 86–92
multivocal analysis 455–462

Najafi, H. 309, 310
Nasir, N. I. S. 108
Nathan, M. J. 78
natural history museums 234–236
natural language processing tools 227, 228, 346
natural user interfaces (NUIs) 98, 100
Nemirovsky, R. 79
Network of Academic Programs in the Learning Sciences (NAPLeS) webinar 123
network theory 65–66
neuroarchitecture 65–66
neuroethics, educational 71
neuromyths, educational 70
neuroscience methods 64–65
Niehaus, L. 448
Nonaka, I. 137–138, 141
nonlinearity 158–159, 247
non-routine problems 148
normative assumptions, challenging 24, 25
normative models 435
Nussbaum, M. 184, 362, 364
Nygaard, K. 401–402

Objective Structured Clinical Examinations (OSCE) 437, 438
observation, assessment and 413, 437–438
Olsen, J. 119
Olsen, J. K. 187
Ong, Y. S. 320
Onwu, G. O. M. 109
OpenCourseWare (OCW) 369
open educational resources (OERs) 267–274, 340, 370
Open University (OU) 369
Oppenheimer, F. 235
opportunistic groups 300–301
orchestration, classroom 180–188, 310, 322
Order out of Chaos (Prigogine and Stengers) 158
organizational approaches 161
Ormel, B. J. 389–390
Oshima, J. 461
out-of-school learning opportunities 395

Paas, F. 169
Packer, M. J. 444–445
pair programming 280
Pant, H. A. 440
Papaevripidou, M. 258–259
Papert, S. 13, 15
Parks, A. H. 216
participation metaphor 44

participatory design 401–407
part-task practice 173–174
Pavlov, I. 12
Pea, R. D. 361, 521
pedagogical design capacity 268
peer communication 248
peer feedback 342
peer production 268–269
Pellegrino, J. W. 416, 417
Pensso, H. 321–322
Penstein Rosé, C. 450
Penuel, W. R. 393
Peppler, K. 111, 290
perceptual patterns 55
Perry, N. E. 132
perspective-taking 240
Peters, V. L. 310
Petrich, M. 290
phenomenological exploration 237–238
phenomenological inquiry exhibits 236–238, 241–242
phenomenology 76
Phielix, C. 352
Piaget, J. 12, 76, 236, 501, 502
Pieters, J. M. 389–390
Plan–Do–Study–Act (PDSA) cycles 396
planes, multiple 181
plane switching 186–187
"planned discovery" exhibits 236
Plano Clark, V. L. 446
plasticity 66
play scaffolds 344
Polman, J. L. 110
polyphonic collaborative learning 140
Porsch, T. 25
Poyas, Y. 97
practices, focus on 24, 25
pragmatism 446–448
preparation for future learning 148
Prieto, L. P. 184
Prigogine, I. 158
principle-based approaches 302
principle-based pedagogy 298–301
problem-based learning (PBL) 211–217
problem representation 58
problem size effect 67, 68
problem solving 78, 89–92, 210–217
problem-solving principles 55–56
procedural information 173
proceduralization 56
procedural knowledge 148
productive failure (PF) 211, 212–213, 214–215, 216
productive interactions 180
Productive Multivocality Project 450, 458–461
productive success 211–212, 216–217
professional perception 58
Programme for International Student Assessment (PISA) 434
prompting 49, 193–195, 260

propositional knowledge 148
Puckett, C. 287–288
Puhl, T. 342

qualitative analyses 301–302
quantitative analyses 301
Quardokus Fisher, K. 526–527
questioning strategies 468
questions, broadening range of 24, 25

Race, P. 364
Radar tool 352, 355
Raes, A. 195, 536–537
reciprocal teaching 388
Recker, M. 402
redundancy, avoiding 204
redundant scaffolding 193
Reeve, R. 300, 388–389
reflection 49, 182, 215, 226
regulated learning 127–134
Reimann, P. 130
Reiser, B. J. 108
Remillard, J. 267, 395
Renkl, A. 202, 260
Renninger, K. A. 119, 122, 132
representational guidance 319
RescuShell 522–526
research in learning sciences, evolution of 1–8
Resendes, M. 300
Resnick, M. 159, 289
re-voicing 468
rich landscapes of learning 371–373
Rickety, D. 259
Rinehart, R. W. 25
Rizzolatti, G. 79
Roberts, D. A. 109
Roblin, N. N. P. 389–390
Rogoff, B. 46–47, 192
role-playing 238
role scaffolds 344
Roll, I. 216
Roschelle, J. 361, 470
Rosé, C. P. 320, 346, 505, 507
Rosenbaum, E. 289
Rosenberg, S. 26
Ross, G. 340
Rouet, J.-F. 88
Rousseau, J.-J. 221
routine expertise 59
Rubia, B. 450
Ruhleder, K. 394
Ruijters, M. C. P. 154
Ruis, A. R. 522
Ruiz-Primo, M. A. 412
Rummel, N. 187, 346
Rusk, N. 288
Russ, R. S. 38
Ryoo, K. 262

Index

Sabelli, N. 393
Sacks, H. 468
Sadler, R. 417
Sahlström, F. 470
Salden, R. J. C. M. 202
Salinas, I. 425
Salinas, R. 364
Sandoval, W. A. 24
Sangin, M. 352
Satwicz, T. 278
Sawyer, R. K. 444–445
Saye, J. 214
scaffolding: collaborative learning and 140; collective inquiry and 308, 309, 310, 312, 313–315; CoMPASS project and 385; CSCL and 321–323, 340–347; description of 49, 172; exhibits and 238–239; experiential and inquiry learning and 260; interest and 121; knowledge building and 297, 304; problem solving and 211–212, 213, 215, 216; research on 191–196
Scardamalia, M. 14–15, 59, 139, 295–298, 300–302, 304, 315, 318, 320, 324, 342, 388–389
scene scaffolds 344
Schack, E. O. 438
Schank, R. E. 14, 17, 343
Schauble, L. 27
Scheflen, A. E. 470
Schegloff, E. A. 456, 468
Scheiter, K. 101
Schellens, T. 536–537
Scheuer, O. 323
Schmidt, H. G. 214
Schmied, A. 71
Schoenfeld, A. H. 444, 449
Scholvien, A. 355
Schunk, D. H. 130
Schur, Y. 321–322
Schwabe, G. 362
Schwartz, D. L. 68, 69, 181, 210, 494
Schwarz, B. B. 318, 321–322, 324–325
science, history contrasted to 28–30
science centers 236–238
scripting 303, 309, 310, 312–314, 324, 326, 340–347, 355
script theory of guidance (SToG) 343–345, 347, 503
seamless learning 360, 362
search processes 88, 89–90
seductive allure of neuroscience explanations (SANE) 70
segmentation 502–505
self-explanations 205
self-regulated learning (SRL) 127–128, 342, 346, 355
sequencing learning activities 48–49
Serious Games movement 276
Sevian, H. 425–426
Shaenfield, D. 324
Shaffer, D. W. 521, 528
Shahar, N. 324
Shapiro, R. B. 111

Sharma, K. 184
Sharples, M. 360
Shavelson, R. J. 412
Sheridan, K. 287, 288
Sherin, B. L. 38
Sherin, M. G. 495
Shulman, L. S. 60, 395
signifying 39–40
Simon, H. A. 55
Simons, P. R.-J. 154
simulations 143, 162–163, 237–238, 241–242, 256–262, 311–312
Siqin, T. 300
situated cognition 44
situated-cognition perspective 75
situated learning 44
Situated Learning (Lave and Wenger) 45
situated perspective, on expertise 56–57
situativity 24, 26
Skinner, B. F. 12, 13, 47
Slakmon, B. 324–325
Slavin, R. E. 139
Slotta, J. D. 309, 310
Smith, B. K. 14
Smith, C. L. 424–425
Smith, H. 334
So, H.-J. J. 335
Sobocinski, M. 131
social context of learning 49
social frames 38
social interactions 238–239
social learning theory 78–79
socially shared regulation of learning (SSRL) 127, 129, 133
social nature of EC 24, 25–26
social network analysis (SNA) 143, 301
social organization of classrooms 468
social planes 181
social practice stance 148–149
social systems theory 140, 141–142
sociocultural approaches, knowledge construction and 139
sociocultural learning 44
sociocultural perspective: CSCL and 335; on expertise 56–57; on learning 34–41; learning at work and 149
sociocultural theories 36–37
sociological approaches, knowledge construction and 140–142
software prompts 192
solution generation effect 214
Song, Y. 360
Songer, N. B. 159, 424
Sonnenberg, C. 130
sourcing/sourcing processes 30, 89–92
Spada, H. 346
split attention effect 97, 204
Spradley, J. 483, 486
Squire, K. 277, 445

550

Stahl, G. 19, 139–140
Stalbovs, K. 101
Stanat, P. 440
Stanton, D. 361
Star, S. L. 110, 394
statistical methods, selecting 532–539
statistical methods for analyzing competence 438–440
Steedle, J. T. 435
Stegmann, K. 346
Steinkuehler, C. 277, 278
Stember, M. 110
STEM learning, supporting 234–242
Stengers, I. 158
Stenseth, T. 91
stepping stones 424–425
Stevens, R. 278, 491
Stigler, J. W. 493
Stokes, D. E. 1
Streeck, J. 469
Street, B. 476, 478
Strijbos, J. W. 449–450
Strømsø, H. I. 91
Structural Equation Modeling (SEM) 439
structural model of argumentation 318, 319
structuration theory 140, 141
structure, behavior, function (SBF) framework 160
structure models 435
student-driving inquiry 309
summative assessment 411, 418
supportive information 172–173
Suthers, D. D. 319, 450, 507, 522
Sweller, J. 203
symmetry 68–69
synergistic scaffolding 193, 195
systems thinking approach 159–160

Tabak, I. 108
Takeuchi, H. 137–138
Talanquer, V. 425–426
Tashakkori, A. 446
task-center focus 300, 303
task interpretation 88
task specificity 54–55
Tayer, N. 321–322
Taylor, J. 360
Taylor, K. H. 111
Tchounikine, P. 326
teacher expertise 57–60
teacher retention 396
teacher support 267–274, 395
teaching experiment approach 427
technological artifacts 341
technology: assessment and 301–302; for exhibits 240–242; impact of advances in 12–13, 15; in learning through problem solving 215–216; mCSCL 359–365; opportunities for 221–229
Teddlie, C. 446

Ten Steps to Complex Learning (van Merriënboer & Kirschner) 169
Teo, C. L. 342
Teplovs, C. 450, 461, 507
ter Vrugte, J. 259, 260
theory-based design 332
theory building 297, 301, 302
thick description 520–529
Thomas, L. E. 78
Thorndike, E. 47
tinkering 290
Tolman, E. C. 12
tools 56–57
Toulmin, S. 318, 319
Training Complex Cognitive Skills (van Merriënboer) 169
transfer 35, 36
transfer-appropriate processing 212
transformative assessment 302
Trausan-Matu, S. 140
trialogical approach 140
triggers 119, 121
Tripto, J. 159–160
Trout, L. 259
Tsang, J. M. 69
Tsovaltzi, D. 342
tutored problem-solving practice 202
Tutwiler, M. S. 158

use diffusion 269
"use oriented" perspective 1

validity 416–417, 428
van Aalst, J. 300
van den Besselaar, P. 457
van der Meijden, H. 280
van Eemeren, F. H. 318
van Gog, T. 205
Van Keer, H. 506, 537–538
VanLehn, K. 248
van Meriënboer, J. J. G. 169, 174
van Rijn, P. W. 415
Van Strien, J. 92
van Wermeskerken, M. 205
variability 172
Varma, S. 68, 163
Vasudevan, V. 280
Vatrapu, R. 461, 462
Vavoula, G. 360
Vernon, D. T. A. 213
vertical expertise 147–148, 149
video research methods 489–496
Vizner, M. 287–288
Vogel, F. 343, 528
Voogt, J. M. 389–390
Vos, N. 280
Vossoughi, S. 288
Vygotsky, L. 44, 46, 57, 107, 139, 181, 191, 192, 340, 501

Index

Walker, A. 214
WallCology 311–315
Walters, C. D. 425
Wang, X. 346
Watson 47
Web-based Inquiry Science Environment (WISE) 107–108, 162–163
Wecker, C. 342, 343
Wegerif, R. B. 324
Weinberger, A. 321, 342, 528
Weinert, F. E. 434
Wen, Y. 184
Wenger, E. 15, 45, 149
Werner, B. 90
Wertsch, J. 486
What We Know About CSCL and Implementing It in Higher Education (Strijbos, Kirschner, & Martens) 18
White, R. W. 434
Whitehead, A. N. 210
whole-task models 170
Whyte, J. 152
Wilensky, U. 161
Wilkinson, K. 290
Wilson, M. 415
Wineburg, S. 26, 107, 109
Winne, P. H. 130
Winters, F. I. 194
Wiser, M. 424–425

Wittgenstein, L. 465
Wolbring, G. 71
Wood, D. J. 192, 194, 340
Word Generation 396
work, learning at 147–154
worked examples 201–202, 203–204
World of Warcraft (WoW) 278
Worsley, M. 288
Wundt, W. 11

Xhafa, F. 450

Yew, E. H. 214
Yinger, R. 58
Yoon, S. 159, 161, 163
Young, M. F. 277
youth culture perspective 111

Zacharia, Z. C. 257, 259
Zeitz, C. M. 55
Zemel, A. 450, 470
Zhang, J. 300, 301, 342, 388–389
Zhang, M. 214
Zimmerman, B. J. 130
Zone of Proximal Development 46, 49, 81, 191–192, 213, 340, 343, 344, 345, 347
zoos 239–240, 242
Zurita, G. 362, 364